Lecture Notes in Artificial Intelligence 12461

Subseries of Lecture Notes in Computer Science

More information about this subseries at http://www.springer.com/series/1244

Yuxiao Dong · Georgiana Ifrim ·
Dunja Mladenić · Craig Saunders ·
Sofie Van Hoecke (Eds.)

Machine Learning and Knowledge Discovery in Databases

Applied Data Science and Demo Track

European Conference, ECML PKDD 2020
Ghent, Belgium, September 14–18, 2020
Proceedings, Part V

Editors
Yuxiao Dong
Microsoft Research
Redmond, WA, USA

Georgiana Ifrim
University College Dublin
Dublin, Ireland

Dunja Mladenić
Jožef Stefan Institute
Ljubljana, Slovenia

Craig Saunders
Amazon Alexa Knowledge
Cambridge, UK

Sofie Van Hoecke
Ghent University
Kotrijk, Belgium

ISSN 0302-9743 ISSN 1611-3349 (electronic)
Lecture Notes in Artificial Intelligence
ISBN 978-3-030-67669-8 ISBN 978-3-030-67670-4 (eBook)
https://doi.org/10.1007/978-3-030-67670-4

LNCS Sublibrary: SL7 – Artificial Intelligence

This Springer imprint is published by the registered company Springer Nature Switzerland AG
The registered company address is: Gewerbestrasse 11, 6330 Cham, Switzerland

Preface

This edition of the European Conference on Machine Learning and Principles and Practice of Knowledge Discovery in Databases (ECML PKDD 2020) is one that we will not easily forget. Due to the emergence of a global pandemic, our lives changed, including many aspects of the conference. Because of this, we are perhaps more proud and happy than ever to present these proceedings to you.

ECML PKDD is an annual conference that provides an international forum for the latest research in all areas related to machine learning and knowledge discovery in databases, including innovative applications. It is the leading European machine learning and data mining conference and builds upon a very successful series of ECML PKDD conferences.

Scheduled to take place in Ghent, Belgium, due to the SARS-CoV-2 pandemic, ECML PKDD 2020 was the first edition to be held fully virtually, from the 14th to the 18th of September 2020. The conference attracted over 1000 participants from all over the world. New this year was a joint event with local industry on Thursday afternoon, the AI4Growth industry track. More generally, the conference received substantial attention from industry through sponsorship, participation, and the revived industry track at the conference.

The main conference programme consisted of presentations of 220 accepted papers and five keynote talks (in order of appearance): Max Welling (University of Amsterdam), Been Kim (Google Brain), Gemma Galdon-Clavell (Eticas Research & Consulting), Stephan Günnemann (Technical University of Munich), and Doina Precup (McGill University & DeepMind Montreal).

In addition, there were 23 workshops, nine tutorials, two combined workshop-tutorials, the PhD Forum, and a discovery challenge.

Papers presented during the three main conference days were organized in four different tracks:

- Research Track: research or methodology papers from all areas in machine learning, knowledge discovery, and data mining;
- Applied Data Science Track: papers on novel applications of machine learning, data mining, and knowledge discovery to solve real-world use cases, thereby bridging the gap between practice and current theory;
- Journal Track: papers that were published in special issues of the journals *Machine Learning* and *Data Mining and Knowledge Discovery*;
- Demo Track: short papers that introduce a new system that goes beyond the state of the art, accompanied with a video of the demo.

We received a record number of 687 and 235 submissions for the Research and Applied Data Science Tracks respectively. We accepted 130 (19%) and 65 (28%) of these. In addition, there were 25 papers from the Journal Track, and 10 demo papers (out of 25 submissions). All in all, the high-quality submissions allowed us to put together an exceptionally rich and exciting program.

The Awards Committee selected research papers that were considered to be of exceptional quality and worthy of special recognition:

- Data Mining best paper award: "Revisiting Wedge Sampling for Budgeted Maximum Inner Product Search", by Stephan S. Lorenzen and Ninh Pham.
- Data Mining best student paper award: "SpecGreedy: Unified Dense Subgraph Detection", by Wenjie Feng, Shenghua Liu, Danai Koutra, Huawei Shen, and Xueqi Cheng.
- Machine Learning best (student) paper award: "Robust Domain Adaptation: Representations, Weights and Inductive Bias", by Victor Bouvier, Philippe Very, Clément Chastagnol, Myriam Tami, and Céline Hudelot.
- Machine Learning best (student) paper runner-up award: "A Principle of Least Action for the Training of Neural Networks", by Skander Karkar, Ibrahim Ayed, Emmanuel de Bézenac, and Patrick Gallinari.
- Best Applied Data Science Track paper: "Learning to Simulate on Sparse Trajectory Data", by Hua Wei, Chacha Chen, Chang Liu, Guanjie Zheng, and Zhenhui Li.
- Best Applied Data Science Track paper runner-up: "Learning a Contextual and Topological Representation of Areas-of-Interest for On-Demand Delivery Application", by Mingxuan Yue, Tianshu Sun, Fan Wu, Lixia Wu, Yinghui Xu, and Cyrus Shahabi.
- Test of Time Award for highest-impact paper from ECML PKDD 2010: "Three Naive Bayes Approaches for Discrimination-Free Classification", by Toon Calders and Sicco Verwer.

We would like to wholeheartedly thank all participants, authors, PC members, area chairs, session chairs, volunteers, co-organizers, and organizers of workshops and tutorials for their contributions that helped make ECML PKDD 2020 a great success. Special thanks go to Vicky, Inge, and Eneko, and the volunteer and virtual conference platform chairs from the UGent AIDA group, who did an amazing job to make the online event feasible. We would also like to thank the ECML PKDD Steering Committee and all sponsors.

October 2020

Tijl De Bie
Craig Saunders
Dunja Mladenić
Yuxiao Dong
Frank Hutter
Isabel Valera
Jefrey Lijffijt
Kristian Kersting
Georgiana Ifrim
Sofie Van Hoecke

Organization

General Chair

Tijl De Bie	Ghent University, Belgium

Research Track Program Chairs

Frank Hutter	University of Freiburg & Bosch Center for AI, Germany
Isabel Valera	Max Planck Institute for Intelligent Systems, Germany
Jefrey Lijffijt	Ghent University, Belgium
Kristian Kersting	TU Darmstadt, Germany

Applied Data Science Track Program Chairs

Craig Saunders	Amazon Alexa Knowledge, UK
Dunja Mladenić	Jožef Stefan Institute, Slovenia
Yuxiao Dong	Microsoft Research, USA

Journal Track Chairs

Aristides Gionis	KTH, Sweden
Carlotta Domeniconi	George Mason University, USA
Eyke Hüllermeier	Paderborn University, Germany
Ira Assent	Aarhus University, Denmark

Discovery Challenge Chair

Andreas Hotho	University of Würzburg, Germany

Workshop and Tutorial Chairs

Myra Spiliopoulou	Otto von Guericke University Magdeburg, Germany
Willem Waegeman	Ghent University, Belgium

Demonstration Chairs

Georgiana Ifrim	University College Dublin, Ireland
Sofie Van Hoecke	Ghent University, Belgium

Nectar Track Chairs

Jie Tang	Tsinghua University, China
Siegfried Nijssen	Université catholique de Louvain, Belgium
Yizhou Sun	University of California, Los Angeles, USA

Industry Track Chairs

Alexander Ypma	ASML, the Netherlands
Arindam Mallik	imec, Belgium
Luis Moreira-Matias	Kreditech, Germany

PhD Forum Chairs

Marinka Zitnik	Harvard University, USA
Robert West	EPFL, Switzerland

Publicity and Public Relations Chairs

Albrecht Zimmermann	Université de Caen Normandie, France
Samantha Monty	Universität Würzburg, Germany

Awards Chairs

Danai Koutra	University of Michigan, USA
José Hernández-Orallo	Universitat Politècnica de València, Spain

Inclusion and Diversity Chairs

Peter Steinbach	Helmholtz-Zentrum Dresden-Rossendorf, Germany
Heidi Seibold	Ludwig-Maximilians-Universität München, Germany
Oliver Guhr	Hochschule für Technik und Wirtschaft Dresden, Germany
Michele Berlingerio	Novartis, Ireland

Local Chairs

Eneko Illarramendi Lerchundi	Ghent University, Belgium
Inge Lason	Ghent University, Belgium
Vicky Wandels	Ghent University, Belgium

Proceedings Chair

Wouter Duivesteijn	Technische Universiteit Eindhoven, the Netherlands

Sponsorship Chairs

Luis Moreira-Matias	Kreditech, Germany
Vicky Wandels	Ghent University, Belgium

Volunteering Chairs

Junning Deng	Ghent University, Belgium
Len Vande Veire	Ghent University, Belgium
Maarten Buyl	Ghent University, Belgium
Raphaël Romero	Ghent University, Belgium
Robin Vandaele	Ghent University, Belgium
Xi Chen	Ghent University, Belgium

Virtual Conference Platform Chairs

Ahmad Mel	Ghent University, Belgium
Alexandru Cristian Mara	Ghent University, Belgium
Bo Kang	Ghent University, Belgium
Dieter De Witte	Ghent University, Belgium
Yoosof Mashayekhi	Ghent University, Belgium

Web Chair

Bo Kang	Ghent University, Belgium

ECML PKDD Steering Committee

Andrea Passerini	University of Trento, Italy
Francesco Bonchi	ISI Foundation, Italy
Albert Bifet	Télécom Paris, France
Sašo Džeroski	Jožef Stefan Institute, Slovenia
Katharina Morik	TU Dortmund, Germany
Arno Siebes	Utrecht University, the Netherlands
Siegfried Nijssen	Université catholique de Louvain, Belgium
Michelangelo Ceci	University of Bari Aldo Moro, Italy
Myra Spiliopoulou	Otto von Guericke University Magdeburg, Germany
Jaakko Hollmen	Aalto University, Finland
Georgiana Ifrim	University College Dublin, Ireland
Thomas Gärtner	University of Nottinghem, UK
Neil Hurley	University College Dublin, Ireland
Michele Berlingerio	IBM Research, Ireland
Elisa Fromont	Université de Rennes 1, France
Arno Knobbe	Universiteit Leiden, the Netherlands
Ulf Brefeld	Leuphana Universität Luneburg, Germany
Andreas Hotho	Julius-Maximilians-Universität Würzburg, Germany

Program Committees

Guest Editorial Board, Journal Track

Ana Paula Appel	IBM Research – Brazil
Annalisa Appice	University of Bari Aldo Moro
Martin Atzmüller	Tilburg University
Anthony Bagnall	University of East Anglia
James Bailey	The University of Melbourne
José Luis Balcázar	Universitat Politècnica de Catalunya
Mitra Baratchi	University of Twente
Srikanta Bedathur	IIT Delhi
Vaishak Belle	The University of Edinburgh
Viktor Bengs	Paderborn University
Batista Biggio	University of Cagliari
Hendrik Blockeel	KU Leuven
Francesco Bonchi	ISI Foundation
Ilaria Bordino	UniCredit R&D
Ulf Brefeld	Leuphana Universität Lüneburg
Klemens Böhm	Karlsruhe Institute of Technology
Remy Cazabet	Claude Bernard University Lyon 1
Michelangelo Ceci	University of Bari Aldo Moro
Loïc Cerf	Universidade Federal de Minas Gerais
Laetitia Chapel	IRISA
Marc Deisenroth	Aarhus University
Wouter Duivesteijn	Technische Universiteit Eindhoven
Tapio Elomaa	Tampere University
Stefano Ferilli	University of Bari Aldo Moro
Cesar Ferri	Universitat Politècnica de València
Maurizio Filippone	EURECOM
Germain Forestier	Université de Haute-Alsace
Marco Frasca	University of Milan
Ricardo José Gabrielli Barreto Campello	University of Newcastle
Esther Galbrun	University of Eastern Finland
Joao Gama	University of Porto
Josif Grabocka	University of Hildesheim
Derek Greene	University College Dublin
Francesco Gullo	UniCredit
Tias Guns	VUB Brussels
Stephan Günnemann	Technical University of Munich
Jose Hernandez-Orallo	Universitat Politècnica de València
Jaakko Hollmén	Aalto University
Georgiana Ifrim	University College Dublin
Mahdi Jalili	RMIT University
Szymon Jaroszewicz	Polish Academy of Sciences

Michael Kamp	Monash University
Mehdi Kaytoue	Infologic
Marius Kloft	TU Kaiserslautern
Dragi Kocev	Jožef Stefan Institute
Peer Kröger	Ludwig-Maximilians-Universität Munich
Meelis Kull	University of Tartu
Ondrej Kuzelka	KU Leuven
Mark Last	Ben-Gurion University of the Negev
Matthijs van Leeuwen	Leiden University
Marco Lippi	University of Modena and Reggio Emilia
Claudio Lucchese	Ca' Foscari University of Venice
Brian Mac Namee	University College Dublin
Gjorgji Madjarov	Ss. Cyril and Methodius University of Skopje
Fabrizio Maria Maggi	Free University of Bozen-Bolzano
Giuseppe Manco	ICAR-CNR
Ernestina Menasalvas	Universidad Politécnica de Madrid
Aditya Menon	Google Research
Katharina Morik	TU Dortmund
Davide Mottin	Aarhus University
Animesh Mukherjee	Indian Institute of Technology Kharagpur
Amedeo Napoli	LORIA
Siegfried Nijssen	Université catholique de Louvain
Eirini Ntoutsi	Leibniz University Hannover
Bruno Ordozgoiti	Aalto University
Pančе Panov	Jožef Stefan Institute
Panagiotis Papapetrou	Stockholm University
Srinivasan Parthasarathy	Ohio State University
Andrea Passerini	University of Trento
Mykola Pechenizkiy	Technische Universiteit Eindhoven
Charlotte Pelletier	Univ. Bretagne Sud/IRISA
Ruggero Pensa	University of Turin
Francois Petitjean	Monash University
Nico Piatkowski	TU Dortmund
Evaggelia Pitoura	Univ. of Ioannina
Marc Plantevit	Claude Bernard University Lyon 1
Kai Puolamäki	University of Helsinki
Chedy Raïssi	Inria
Matteo Riondato	Amherst College
Joerg Sander	University of Alberta
Pierre Schaus	UCLouvain
Lars Schmidt-Thieme	University of Hildesheim
Matthias Schubert	LMU Munich
Thomas Seidl	LMU Munich
Gerasimos Spanakis	Maastricht University
Myra Spiliopoulou	Otto von Guericke University Magdeburg
Jerzy Stefanowski	Poznań University of Technology

Giovanni Stilo	Università degli Studi dell'Aquila
Mahito Sugiyama	National Institute of Informatics
Andrea Tagarelli	University of Calabria
Chang Wei Tan	Monash University
Nikolaj Tatti	University of Helsinki
Alexandre Termier	Univ. Rennes 1
Marc Tommasi	University of Lille
Ivor Tsang	University of Technology Sydney
Panayiotis Tsaparas	University of Ioannina
Steffen Udluft	Siemens
Celine Vens	KU Leuven
Antonio Vergari	University of California, Los Angeles
Michalis Vlachos	University of Lausanne
Christel Vrain	LIFO, Université d'Orléans
Jilles Vreeken	Helmholtz Center for Information Security
Willem Waegeman	Ghent University
Marcel Wever	Paderborn University
Stefan Wrobel	Univ. Bonn and Fraunhofer IAIS
Yinchong Yang	Siemens AG
Guoxian Yu	Southwest University
Bianca Zadrozny	IBM
Ye Zhu	Monash University
Arthur Zimek	University of Southern Denmark
Albrecht Zimmermann	Université de Caen Normandie
Marinka Zitnik	Harvard University

Area Chairs, Research Track

Cuneyt Gurcan Akcora	The University of Texas at Dallas
Carlos M. Alaíz	Universidad Autónoma de Madrid
Fabrizio Angiulli	University of Calabria
Georgios Arvanitidis	Max Planck Institute for Intelligent Systems
Roberto Bayardo	Google
Michele Berlingerio	IBM
Michael Berthold	University of Konstanz
Albert Bifet	Télécom Paris
Hendrik Blockeel	Katholieke Universiteit Leuven
Mario Boley	MPI Informatics
Francesco Bonchi	Fondazione ISI
Ulf Brefeld	Leuphana Universität Lüneburg
Michelangelo Ceci	Università degli Studi di Bari Aldo Moro
Duen Horng Chau	Georgia Institute of Technology
Nicolas Courty	Université de Bretagne Sud/IRISA
Bruno Cremilleux	Université de Caen Normandie
Andre de Carvalho	University of São Paulo
Patrick De Causmaecker	Katholieke Universiteit Leuven

Nicola Di Mauro	Università degli Studi di Bari Aldo Moro
Tapio Elomaa	Tampere University
Amir-Massoud Farahmand	Vector Institute & University of Toronto
Ángela Fernández	Universidad Autónoma de Madrid
Germain Forestier	Université de Haute-Alsace
Elisa Fromont	Université de Rennes 1
Johannes Fürnkranz	Johannes Kepler University Linz
Patrick Gallinari	Sorbonne University
Joao Gama	University of Porto
Thomas Gärtner	TU Wien
Pierre Geurts	University of Liège
Manuel Gomez Rodriguez	MPI for Software Systems
Przemyslaw Grabowicz	University of Massachusetts Amherst
Stephan Günnemann	Technical University of Munich
Allan Hanbury	Vienna University of Technology
Daniel Hernández-Lobato	Universidad Autónoma de Madrid
Jose Hernandez-Orallo	Universitat Politècnica de València
Jaakko Hollmén	Aalto University
Andreas Hotho	University of Würzburg
Neil Hurley	University College Dublin
Georgiana Ifrim	University College Dublin
Alipio M. Jorge	University of Porto
Arno Knobbe	Universiteit Leiden
Dragi Kocev	Jožef Stefan Institute
Lars Kotthoff	University of Wyoming
Nick Koudas	University of Toronto
Stefan Kramer	Johannes Gutenberg University Mainz
Meelis Kull	University of Tartu
Niels Landwehr	University of Potsdam
Sébastien Lefèvre	Université de Bretagne Sud
Daniel Lemire	Université du Québec
Matthijs van Leeuwen	Leiden University
Marius Lindauer	Leibniz University Hannover
Jörg Lücke	University of Oldenburg
Donato Malerba	Università degli Studi di Bari "Aldo Moro"
Giuseppe Manco	ICAR-CNR
Pauli Miettinen	University of Eastern Finland
Anna Monreale	University of Pisa
Katharina Morik	TU Dortmund
Emmanuel Müller	University of Bonn
Sriraam Natarajan	Indiana University Bloomington
Alfredo Nazábal	The Alan Turing Institute
Siegfried Nijssen	Université catholique de Louvain
Barry O'Sullivan	University College Cork
Pablo Olmos	University Carlos III of Madrid
Panagiotis Papapetrou	Stockholm University

Andrea Passerini	University of Turin
Mykola Pechenizkiy	Technische Universiteit Eindhoven
Ruggero G. Pensa	University of Torino
Francois Petitjean	Monash University
Claudia Plant	University of Vienna
Marc Plantevit	Université Claude Bernard Lyon 1
Philippe Preux	Université de Lille
Rita Ribeiro	University of Porto
Celine Robardet	INSA Lyon
Elmar Rueckert	University of Lübeck
Marian Scuturici	LIRIS-INSA de Lyon
Michèle Sebag	Univ. Paris-Sud
Thomas Seidl	Ludwig-Maximilians-Universität Muenchen
Arno Siebes	Utrecht University
Alessandro Sperduti	University of Padua
Myra Spiliopoulou	Otto von Guericke University Magdeburg
Jerzy Stefanowski	Poznań University of Technology
Yizhou Sun	University of California, Los Angeles
Einoshin Suzuki	Kyushu University
Acar Tamersoy	Symantec Research Labs
Jie Tang	Tsinghua University
Grigorios Tsoumakas	Aristotle University of Thessaloniki
Celine Vens	KU Leuven
Antonio Vergari	University of California, Los Angeles
Herna Viktor	University of Ottawa
Christel Vrain	University of Orléans
Jilles Vreeken	Helmholtz Center for Information Security
Willem Waegeman	Ghent University
Wendy Hui Wang	Stevens Institute of Technology
Stefan Wrobel	Fraunhofer IAIS & Univ. of Bonn
Han-Jia Ye	Nanjing University
Guoxian Yu	Southwest University
Min-Ling Zhang	Southeast University
Albrecht Zimmermann	Université de Caen Normandie

Area Chairs, Applied Data Science Track

Michelangelo Ceci	Università degli Studi di Bari Aldo Moro
Tom Diethe	Amazon
Faisal Farooq	IBM
Johannes Fürnkranz	Johannes Kepler University Linz
Rayid Ghani	Carnegie Mellon University
Ahmed Hassan Awadallah	Microsoft
Xiangnan He	University of Science and Technology of China
Georgiana Ifrim	University College Dublin
Anne Kao	Boeing

Javier Latorre	Apple
Hao Ma	Facebook AI
Gabor Melli	Sony PlayStation
Luis Moreira-Matias	Kreditech
Alessandro Moschitti	Amazon
Kitsuchart Pasupa	King Mongkut's Institute of Technology Ladkrabang
Mykola Pechenizkiy	Technische Universiteit Eindhoven
Julien Perez	NAVER LABS Europe
Xing Xie	Microsoft
Chenyan Xiong	Microsoft Research
Yang Yang	Zhejiang University

Program Committee Members, Research Track

Moloud Abdar	Deakin University
Linara Adilova	Fraunhofer IAIS
Florian Adriaens	Ghent University
Zahra Ahmadi	Johannes Gutenberg University Mainz
M. Eren Akbiyik	IBM Germany Research and Development GmbH
Youhei Akimoto	University of Tsukuba
Ömer Deniz Akyildiz	University of Warwick and The Alan Turing Institute
Francesco Alesiani	NEC Laboratories Europe
Alexandre Alves	Universidade Federal de Uberlândia
Maryam Amir Haeri	Technische Universität Kaiserslautern
Alessandro Antonucci	IDSIA
Muhammad Umer Anwaar	Mercateo AG
Xiang Ao	Institute of Computing Technology, Chinese Academy of Sciences
Sunil Aryal	Deakin University
Thushari Atapattu	The University of Adelaide
Arthur Aubret	LIRIS
Julien Audiffren	Fribourg University
Murat Seckin Ayhan	Eberhard Karls Universität Tübingen
Dario Azzimonti	Istituto Dalle Molle di Studi sull'Intelligenza Artificiale
Behrouz Babaki	Polytechnique Montréal
Rohit Babbar	Aalto University
Housam Babiker	University of Alberta
Davide Bacciu	University of Pisa
Thomas Baeck	Leiden University
Abdelkader Baggag	Qatar Computing Research Institute
Zilong Bai	University of California, Davis
Jiyang Bai	Florida State University
Sambaran Bandyopadhyay	IBM
Mitra Baratchi	University of Twente
Christian Beecks	University of Münster
Anna Beer	Ludwig Maximilian University of Munich

Adnene Belfodil	Munic Car Data
Aimene Belfodil	INSA Lyon
Ines Ben Kraiem	UT2J-IRIT
Anes Bendimerad	LIRIS
Christoph Bergmeir	Monash University
Max Berrendorf	Ludwig Maximilian University of Munich
Louis Béthune	ENS de Lyon
Anton Björklund	University of Helsinki
Alexandre Blansché	Université de Lorraine
Laurens Bliek	Delft University of Technology
Isabelle Bloch	ENST - CNRS UMR 5141 LTCI
Gianluca Bontempi	Université Libre de Bruxelles
Felix Borutta	Ludwig-Maximilians-Universität München
Ahcène Boubekki	Leuphana Universität Lüneburg
Tanya Braun	University of Lübeck
Wieland Brendel	University of Tübingen
Klaus Brinker	Hamm-Lippstadt University of Applied Sciences
David Browne	Insight Centre for Data Analytics
Sebastian Bruckert	Otto Friedrich University Bamberg
Mirko Bunse	TU Dortmund University
Sophie Burkhardt	University of Mainz
Haipeng Cai	Washington State University
Lele Cao	Tsinghua University
Manliang Cao	Fudan University
Defu Cao	Peking University
Antonio Carta	University of Pisa
Remy Cazabet	Université Lyon 1
Abdulkadir Celikkanat	CentraleSupelec, Paris-Saclay University
Christophe Cerisara	LORIA
Carlos Cernuda	Mondragon University
Vitor Cerqueira	LIAAD-INESCTEC
Mattia Cerrato	Università di Torino
Ricardo Cerri	Federal University of São Carlos
Laetitia Chapel	IRISA
Vaggos Chatziafratis	Stanford University
El Vaigh Cheikh Brahim	Inria/IRISA Rennes
Yifei Chen	University of Groningen
Junyang Chen	University of Macau
Jiaoyan Chen	University of Oxford
Huiyuan Chen	Case Western Reserve University
Run-Qing Chen	Xiamen University
Tianyi Chen	Microsoft
Lingwei Chen	The Pennsylvania State University
Senpeng Chen	UESTC
Liheng Chen	Shanghai Jiao Tong University
Siming Chen	Frauenhofer IAIS

Saso Dzeroski	Jožef Stefan Institute
Mohamed Elati	Université d'Evry
Lukas Enderich	Robert Bosch GmbH
Dominik Endres	Philipps-Universität Marburg
Francisco Escolano	University of Alicante
Bjoern Eskofier	Friedrich-Alexander University Erlangen-Nürnberg
Roberto Esposito	Università di Torino
Georgios Exarchakis	Institut de la Vision
Melanie F. Pradier	Harvard University
Samuel G. Fadel	Universidade Estadual de Campinas
Evgeniy Faerman	Ludwig Maximilian University of Munich
Yujie Fan	Case Western Reserve University
Elaine Faria	Federal University of Uberlândia
Golnoosh Farnadi	Mila/University of Montreal
Fabio Fassetti	University of Calabria
Ad Feelders	Utrecht University
Yu Fei	Harbin Institute of Technology
Wenjie Feng	The Institute of Computing Technology, Chinese Academy of Sciences
Zunlei Feng	Zhejiang University
Cesar Ferri	Universitat Politècnica de València
Raul Fidalgo-Merino	European Commission Joint Research Centre
Murat Firat	Technische Universiteit Eindhoven
Francoise Fogelman-Soulié	Tianjin University
Vincent Fortuin	ETH Zurich
Iordanis Fostiropoulos	University of Southern California
Eibe Frank	University of Waikato
Benoît Frénay	Université de Namur
Nikolaos Freris	University of Science and Technology of China
Moshe Gabel	University of Toronto
Ricardo José Gabrielli Barreto Campello	University of Newcastle
Esther Galbrun	University of Eastern Finland
Claudio Gallicchio	University of Pisa
Yuanning Gao	Shanghai Jiao Tong University
Alberto Garcia-Duran	Ecole Polytechnique Fédérale de Lausanne
Eduardo Garrido	Universidad Autónoma de Madrid
Clément Gautrais	KU Leuven
Arne Gevaert	Ghent University
Giorgos Giannopoulos	IMSI, "Athena" Research Center
C. Lee Giles	The Pennsylvania State University
Ioana Giurgiu	IBM Research - Zurich
Thomas Goerttler	TU Berlin
Heitor Murilo Gomes	University of Waikato
Chen Gong	Shanghai Jiao Tong University
Zhiguo Gong	University of Macau

Hongyu Gong	University of Illinois at Urbana-Champaign
Pietro Gori	Télécom Paris
James Goulding	University of Nottingham
Kshitij Goyal	Katholieke Universiteit Leuven
Dmitry Grishchenko	Université Grenoble Alpes
Moritz Grosse-Wentrup	University of Vienna
Sebastian Gruber	Siemens AG
John Grundy	Monash University
Kang Gu	Dartmouth College
Jindong Gu	Siemens
Riccardo Guidotti	University of Pisa
Tias Guns	Vrije Universiteit Brussel
Ruocheng Guo	Arizona State University
Yiluan Guo	Singapore University of Technology and Design
Xiaobo Guo	University of Chinese Academy of Sciences
Thomas Guyet	IRISA
Jiawei Han	University of Illinois at Urbana-Champaign
Zhiwei Han	fortiss GmbH
Tom Hanika	University of Kassel
Shonosuke Harada	Kyoto University
Marwan Hassani	Technische Universiteit Eindhoven
Jianhao He	Sun Yat-sen University
Deniu He	Chongqing University of Posts and Telecommunications
Dongxiao He	Tianjin University
Stefan Heidekrueger	Technical University of Munich
Nandyala Hemachandra	Indian Institute of Technology Bombay
Till Hendrik Schulz	University of Bonn
Alexander Hepburn	University of Bristol
Sibylle Hess	Technische Universiteit Eindhoven
Javad Heydari	LG Electronics
Joyce Ho	Emory University
Shunsuke Horii	Waseda University
Tamas Horvath	University of Bonn and Fraunhofer IAIS
Mehran Hossein Zadeh Bazargani	University College Dublin
Robert Hu	University of Oxford
Weipeng Huang	Insight
Jun Huang	University of Tokyo
Haojie Huang	The University of New South Wales
Hong Huang	UGoe
Shenyang Huang	McGill University
Vân Anh Huynh-Thu	University of Liège
Dino Ienco	INRAE
Siohoi Ieng	Institut de la Vision
Angelo Impedovo	Università "Aldo Moro" degli studi di Bari

Muhammad Imran Razzak	Deakin University
Vasileios Iosifidis	Leibniz University Hannover
Joseph Isaac	Indian Institute of Technology Madras
Md Islam	Washington State University
Ziyu Jia	Beijing Jiaotong University
Lili Jiang	Umeå University
Yao Jiangchao	Alibaba
Tan Jianlong	Institute of Information Engineering, Chinese Academy of Sciences
Baihong Jin	University of California, Berkeley
Di Jin	Tianjin University
Wei Jing	Xi'an Jiaotong University
Jonathan Jouanne	ARIADNEXT
Ata Kaban	University of Birmingham
Tomasz Kajdanowicz	Wrocław University of Science and Technology
Sandesh Kamath	Chennai Mathematical Institute
Keegan Kang	Singapore University of Technology and Design
Bo Kang	Ghent University
Isak Karlsson	Stockholm University
Panagiotis Karras	Aarhus University
Nikos Katzouris	NCSR Demokritos
Uzay Kaymak	Technische Universiteit Eindhoven
Mehdi Kaytoue	Infologic
Pascal Kerschke	University of Münster
Jungtaek Kim	Pohang University of Science and Technology
Minyoung Kim	Samsung AI Center Cambridge
Masahiro Kimura	Ryukoku University
Uday Kiran	The University of Tokyo
Bogdan Kirillov	ITMO University
Péter Kiss	ELTE
Gerhard Klassen	Heinrich Heine University Düsseldorf
Dmitry Kobak	Eberhard Karls University of Tübingen
Masahiro Kohjima	NTT
Ziyi Kou	University of Rochester
Wouter Kouw	Technische Universiteit Eindhoven
Fumiya Kudo	Hitachi, Ltd.
Piotr Kulczycki	Systems Research Institute, Polish Academy of Sciences
Ilona Kulikovskikh	Samara State Aerospace University
Rajiv Kumar	IIT Bombay
Pawan Kumar	IIT Kanpur
Suhansanu Kumar	University of Illinois, Urbana-Champaign
Abhishek Kumar	University of Helsinki
Gautam Kunapuli	The University of Texas at Dallas
Takeshi Kurashima	NTT
Vladimir Kuzmanovski	Jožef Stefan Institute

Anisio Lacerda	Centro Federal de Educação Tecnológica de Minas Gerais
Patricia Ladret	GIPSA-lab
Fabrizio Lamberti	Politecnico di Torino
James Large	University of East Anglia
Duc-Trong Le	University of Engineering and Technology, VNU Hanoi
Trung Le	Monash University
Luce le Gorrec	University of Strathclyde
Antoine Ledent	TU Kaiserslautern
Kangwook Lee	University of Wisconsin-Madison
Felix Leibfried	PROWLER.io
Florian Lemmerich	RWTH Aachen University
Carson Leung	University of Manitoba
Edouard Leurent	Inria
Naiqi Li	Tsinghua-UC Berkeley Shenzhen Institute
Suyi Li	The Hong Kong University of Science and Technology
Jundong Li	University of Virginia
Yidong Li	Beijing Jiaotong University
Xiaoting Li	The Pennsylvania State University
Yaoman Li	CUHK
Rui Li	Inspur Group
Wenye Li	The Chinese University of Hong Kong (Shenzhen)
Mingming Li	Institute of Information Engineering, Chinese Academy of Sciences
Yexin Li	Hong Kong University of Science and Technology
Qinghua Li	Renmin University of China
Yaohang Li	Old Dominion University
Yuxuan Liang	National University of Singapore
Zhimin Liang	Institute of Computing Technology, Chinese Academy of Sciences
Hongwei Liang	Microsoft
Nengli Lim	Singapore University of Technology and Design
Suwen Lin	University of Notre Dame
Yangxin Lin	Peking University
Aldo Lipani	University College London
Marco Lippi	University of Modena and Reggio Emilia
Alexei Lisitsa	University of Liverpool
Lin Liu	Taiyuan University of Technology
Weiwen Liu	The Chinese University of Hong Kong
Yang Liu	JD
Huan Liu	Arizona State University
Tianbo Liu	Thomas Jefferson National Accelerator Facility
Tongliang Liu	The University of Sydney
Weidong Liu	Inner Mongolia University
Kai Liu	Colorado School of Mines

Shiwei Liu	Technische Universiteit Eindhoven
Shenghua Liu	Institute of Computing Technology, Chinese Academy of Sciences
Corrado Loglisci	University of Bari Aldo Moro
Andrey Lokhov	Los Alamos National Laboratory
Yijun Lu	Alibaba Cloud
Xuequan Lu	Deakin University
Szymon Lukasik	AGH University of Science and Technology
Phuc Luong	Deakin University
Jianming Lv	South China University of Technology
Gengyu Lyu	Beijing Jiaotong University
Vijaikumar M.	Indian Institute of Science
Jing Ma	Emory University
Nan Ma	Shanghai Jiao Tong University
Sebastian Mair	Leuphana University Lüneburg
Marjan Mansourvar	University of Southern Denmark
Vincent Margot	Advestis
Fernando Martínez-Plumed	Joint Research Centre - European Commission
Florent Masseglia	Inria
Romain Mathonat	Université de Lyon
Deepak Maurya	Indian Institute of Technology Madras
Christian Medeiros Adriano	Hasso-Plattner-Institut
Purvanshi Mehta	University of Rochester
Tobias Meisen	Bergische Universität Wuppertal
Luciano Melodia	Friedrich-Alexander Universität Erlangen-Nürnberg
Ernestina Menasalvas	Universidad Politécnica de Madrid
Vlado Menkovski	Technische Universiteit Eindhoven
Engelbert Mephu Nguifo	Université Clermont Auvergne
Alberto Maria Metelli	Politecnico di Milano
Donald Metzler	Google
Anke Meyer-Baese	Florida State University
Richard Meyes	University of Wuppertal
Haithem Mezni	University of Jendouba
Paolo Mignone	Università degli Studi di Bari Aldo Moro
Matej Mihelčić	University of Zagreb
Decebal Constantin Mocanu	University of Twente
Christoph Molnar	Ludwig Maximilian University of Munich
Lia Morra	Politecnico di Torino
Christopher Morris	TU Dortmund University
Tadeusz Morzy	Poznań University of Technology
Henry Moss	Lancaster University
Tetsuya Motokawa	University of Tsukuba
Mathilde Mougeot	Université Paris-Saclay
Tingting Mu	The University of Manchester
Andreas Mueller	NYU
Tanmoy Mukherjee	Queen Mary University of London

Ksenia Mukhina	ITMO University
Peter Müllner	Know-Center
Guido Muscioni	University of Illinois at Chicago
Waleed Mustafa	TU Kaiserslautern
Mohamed Nadif	University of Paris
Ankur Nahar	Indian Institute of Technology Jodhpur
Kei Nakagawa	Nomura Asset Management Co., Ltd.
Haïfa Nakouri	University of Tunis
Mirco Nanni	KDD-Lab ISTI-CNR Pisa
Nicolo' Navarin	University of Padova
Richi Nayak	Queensland University of Technology
Mojtaba Nayyeri	University of Bonn
Daniel Neider	MPI SWS
Nan Neng	Institute of Information Engineering, Chinese Academy of Sciences
Stefan Neumann	University of Vienna
Dang Nguyen	Deakin University
Kien Duy Nguyen	University of Southern California
Jingchao Ni	NEC Laboratories America
Vlad Niculae	Instituto de Telecomunicações
Sofia Maria Nikolakaki	Boston University
Kun Niu	Beijing University of Posts and Telecommunications
Ryo Nomura	Waseda University
Eirini Ntoutsi	Leibniz University Hannover
Andreas Nuernberger	Otto von Guericke University of Magdeburg
Tsuyoshi Okita	Kyushu Institute of Technology
Maria Oliver Parera	GIPSA-lab
Bruno Ordozgoiti	Aalto University
Sindhu Padakandla	Indian Institute of Science
Tapio Pahikkala	University of Turku
Joao Palotti	Qatar Computing Research Institute
Guansong Pang	The University of Adelaide
Pance Panov	Jožef Stefan Institute
Konstantinos Papangelou	The University of Manchester
Yulong Pei	Technische Universiteit Eindhoven
Nikos Pelekis	University of Piraeus
Thomas Pellegrini	Université Toulouse III - Paul Sabatier
Charlotte Pelletier	Univ. Bretagne Sud
Jaakko Peltonen	Aalto University and Tampere University
Shaowen Peng	Kyushu University
Siqi Peng	Kyoto University
Bo Peng	The Ohio State University
Lukas Pensel	Johannes Gutenberg University Mainz
Aritz Pérez Martínez	Basque Center for Applied Mathematics
Lorenzo Perini	KU Leuven
Matej Petković	Jožef Stefan Institute

Bernhard Pfahringer	University of Waikato
Weiguo Pian	Chongqing University
Francesco Piccialli	University of Naples Federico II
Sebastian Pineda Arango	University of Hildesheim
Gianvito Pio	University of Bari “Aldo Moro”
Giuseppe Pirrò	Sapienza University of Rome
Anastasia Podosinnikova	Massachusetts Institute of Technology
Sebastian Pölsterl	Ludwig Maximilian University of Munich
Vamsi Potluru	JP Morgan AI Research
Rafael Poyiadzi	University of Bristol
Surya Prakash	University of Canberra
Paul Prasse	University of Potsdam
Rameshwar Pratap	Indian Institute of Technology Mandi
Jonas Prellberg	University of Oldenburg
Hugo Proenca	Leiden Institute of Advanced Computer Science
Ricardo Prudencio	Federal University of Pernambuco
Petr Pulc	Institute of Computer Science of the Czech Academy of Sciences
Lei Qi	Iowa State University
Zhenyue Qin	The Australian National University
Rahul Ragesh	PES University
Tahrima Rahman	The University of Texas at Dallas
Zana Rashidi	York University
S. S. Ravi	University of Virginia and University at Albany – SUNY
Ambrish Rawat	IBM
Henry Reeve	University of Birmingham
Reza Refaei Afshar	Technische Universiteit Eindhoven
Navid Rekabsaz	Johannes Kepler University Linz
Yongjian Ren	Shandong University
Zhiyun Ren	The Ohio State University
Guohua Ren	LG Electronics
Yuxiang Ren	Florida State University
Xavier Renard	AXA
Martí Renedo Mirambell	Universitat Politècnica de Catalunya
Gavin Rens	Katholiek Universiteit Leuven
Matthias Renz	Christian-Albrechts-Universität zu Kiel
Guillaume Richard	EDF R&D
Matteo Riondato	Amherst College
Niklas Risse	Bielefeld University
Lars Rosenbaum	Robert Bosch GmbH
Celine Rouveirol	Université Sorbonne Paris Nord
Shoumik Roychoudhury	Temple University
Polina Rozenshtein	Aalto University
Peter Rubbens	Flanders Marine Institute (VLIZ)
David Ruegamer	LMU Munich

Matteo Ruffini	ToolsGroup
Ellen Rushe	Insight Centre for Data Analytics
Amal Saadallah	TU Dortmund
Yogish Sabharwal	IBM Research - India
Mandana Saebi	University of Notre Dame
Aadirupa Saha	IISc
Seyed Erfan Sajjadi	Brunel University
Durgesh Samariya	Federation University
Md Samiullah	Monash University
Mark Sandler	Google
Raul Santos-Rodriguez	University of Bristol
Yucel Saygin	Sabancı University
Pierre Schaus	UCLouvain
Fabian Scheipl	Ludwig Maximilian University of Munich
Katerina Schindlerova	University of Vienna
Ute Schmid	University of Bamberg
Daniel Schmidt	Monash University
Sebastian Schmoll	Ludwig Maximilian University of Munich
Johannes Schneider	University of Liechtenstein
Marc Schoenauer	Inria Saclay Île-de-France
Jonas Schouterden	Katholieke Universiteit Leuven
Leo Schwinn	Friedrich-Alexander-Universität Erlangen-Nürnberg
Florian Seiffarth	University of Bonn
Nan Serra	NEC Laboratories Europe GmbH
Rowland Seymour	Univeristy of Nottingham
Ammar Shaker	NEC Laboratories Europe
Ali Shakiba	Vali-e-Asr University of Rafsanjan
Junming Shao	University of Science and Technology of China
Zhou Shao	Tsinghua University
Manali Sharma	Samsung Semiconductor Inc.
Jiaming Shen	University of Illinois at Urbana-Champaign
Ying Shen	Sun Yat-sen University
Hao Shen	fortiss GmbH
Tao Shen	University of Technology Sydney
Ge Shi	Beijing Institute of Technology
Ziqiang Shi	Fujitsu Research & Development Center
Masumi Shirakawa	hapicom Inc./Osaka University
Kai Shu	Arizona State University
Amila Silva	The University of Melbourne
Edwin Simpson	University of Bristol
Dinesh Singh	RIKEN Center for Advanced Intelligence Project
Jaspreet Singh	L3S Research Centre
Spiros Skiadopoulos	University of the Peloponnese
Gavin Smith	University of Nottingham
Miguel A. Solinas	CEA
Dongjin Song	NEC Labs America

Arnaud Soulet	Université de Tours
Marvin Ssemambo	Makerere University
Michiel Stock	Ghent University
Filipo Studzinski Perotto	Institut de Recherche en Informatique de Toulouse
Adisak Sukul	Iowa State University
Lijuan Sun	Beijing Jiaotong University
Tao Sun	National University of Defense Technology
Ke Sun	Peking University
Yue Sun	Beijing Jiaotong University
Hari Sundaram	University of Illinois at Urbana-Champaign
Gero Szepannek	Stralsund University of Applied Sciences
Jacek Tabor	Jagiellonian University
Jianwei Tai	IIE, CAS
Naoya Takeishi	RIKEN Center for Advanced Intelligence Project
Chang Wei Tan	Monash University
Jinghua Tan	Southwestern University of Finance and Economics
Zeeshan Tariq	Ulster University
Bouadi Tassadit	IRISA-Université de Rennes 1
Maryam Tavakol	TU Dortmund
Romain Tavenard	Univ. Rennes 2/LETG-COSTEL/IRISA-OBELIX
Alexandre Termier	Université de Rennes 1
Janek Thomas	Fraunhofer Institute for Integrated Circuits IIS
Manoj Thulasidas	Singapore Management University
Hao Tian	Syracuse University
Hiroyuki Toda	NTT
Jussi Tohka	University of Eastern Finland
Ricardo Torres	Norwegian University of Science and Technology
Isaac Triguero Velázquez	University of Nottingham
Sandhya Tripathi	Indian Institute of Technology Bombay
Holger Trittenbach	Karlsruhe Institute of Technology
Peter van der Putten	Leiden University & Pegasystems
Elia Van Wolputte	KU Leuven
Fabio Vandin	University of Padova
Titouan Vayer	IRISA
Ashish Verma	IBM Research - US
Bouvier Victor	Sidetrade MICS
Julia Vogt	University of Basel
Tim Vor der Brück	Lucerne University of Applied Sciences and Arts
Yb W.	Chongqing University
Krishna Wadhwani	Indian Institute of Technology Bombay
Huaiyu Wan	Beijing Jiaotong University
Qunbo Wang	Beihang University
Beilun Wang	Southeast University
Yiwei Wang	National University of Singapore
Bin Wang	Xiaomi AI Lab

Jiong Wang	Institute of Information Engineering, Chinese Academy of Sciences
Xiaobao Wang	Tianjin University
Shuheng Wang	Nanjing University of Science and Technology
Jihu Wang	Shandong University
Haobo Wang	Zhejiang University
Xianzhi Wang	University of Technology Sydney
Chao Wang	Shanghai Jiao Tong University
Jun Wang	Southwest University
Jing Wang	Beijing Jiaotong University
Di Wang	Nanyang Technological University
Yashen Wang	China Academy of Electronics and Information Technology of CETC
Qinglong Wang	McGill University
Sen Wang	University of Queensland
Di Wang	State University of New York at Buffalo
Qing Wang	Information Science Research Centre
Guoyin Wang	Chongqing University of Posts and Telecommunications
Thomas Weber	Ludwig-Maximilians-Universität München
Lingwei Wei	University of Chinese Academy of Sciences; Institute of Information Engineering, CAS
Tong Wei	Nanjing University
Pascal Welke	University of Bonn
Yang Wen	University of Science and Technology of China
Yanlong Wen	Nankai University
Paul Weng	UM-SJTU Joint Institute
Matthias Werner	ETAS GmbH, Bosch Group
Joerg Wicker	The University of Auckland
Uffe Wiil	University of Southern Denmark
Paul Wimmer	University of Lübeck; Robert Bosch GmbH
Martin Wistuba	University of Hildesheim
Feijie Wu	The Hong Kong Polytechnic University
Xian Wu	University of Notre Dame
Hang Wu	Georgia Institute of Technology
Yubao Wu	Georgia State University
Yichao Wu	SenseTime Group Limited
Xi-Zhu Wu	Nanjing University
Jia Wu	Macquarie University
Yang Xiaofei	Harbin Institute of Technology, Shenzhen
Yuan Xin	University of Science and Technology of China
Liu Xinshun	VIVO
Taufik Xu	Tsinghua University
Jinhui Xu	State University of New York at Buffalo
Depeng Xu	University of Arkansas
Peipei Xu	University of Liverpool

Yichen Xu	Beijing University of Posts and Telecommunications
Bo Xu	Donghua University
Hansheng Xue	Harbin Institute of Technology, Shenzhen
Naganand Yadati	Indian Institute of Science
Akihiro Yamaguchi	Toshiba Corporation
Haitian Yang	Institute of Information Engineering, Chinese Academy of Sciences
Hongxia Yang	Alibaba Group
Longqi Yang	HPCL
Xiaochen Yang	University College London
Yuhan Yang	Shanghai Jiao Tong University
Ya Zhou Yang	National University of Defense Technology
Feidiao Yang	Institute of Computing Technology, Chinese Academy of Sciences
Liu Yang	Tianjin University
Chaoqi Yang	University of Illinois at Urbana-Champaign
Carl Yang	University of Illinois at Urbana-Champaign
Guanyu Yang	Xi'an Jiaotong - Liverpool University
Yang Yang	Nanjing University
Weicheng Ye	Carnegie Mellon University
Wei Ye	Peking University
Yanfang Ye	Case Western Reserve University
Kejiang Ye	SIAT, Chinese Academy of Sciences
Florian Yger	Université Paris-Dauphine
Yunfei Yin	Chongqing University
Lu Yin	Technische Universiteit Eindhoven
Wang Yingkui	Tianjin University
Kristina Yordanova	University of Rostock
Tao You	Northwestern Polytechnical University
Hong Qing Yu	University of Bedfordshire
Bowen Yu	Institute of Information Engineering, Chinese Academy of Sciences
Donghan Yu	Carnegie Mellon University
Yipeng Yu	Tencent
Shujian Yu	NEC Laboratories Europe
Jiadi Yu	Shanghai Jiao Tong University
Wenchao Yu	University of California, Los Angeles
Feng Yuan	The University of New South Wales
Chunyuan Yuan	Institute of Information Engineering, Chinese Academy of Sciences
Sha Yuan	Tsinghua University
Farzad Zafarani	Purdue University
Marco Zaffalon	IDSIA
Nayyar Zaidi	Monash University
Tianzi Zang	Shanghai Jiao Tong University
Gerson Zaverucha	Federal University of Rio de Janeiro

Javier Zazo	Harvard University
Albin Zehe	University of Würzburg
Yuri Zelenkov	National Research University Higher School of Economics
Amber Zelvelder	Umeå University
Mingyu Zhai	NARI Group Corporation
Donglin Zhan	Sichuan University
Yu Zhang	Southeast University
Wenbin Zhang	University of Maryland
Qiuchen Zhang	Emory University
Tong Zhang	PKU
Jianfei Zhang	Case Western Reserve University
Nailong Zhang	MassMutual
Yi Zhang	Nanjing University
Xiangliang Zhang	King Abdullah University of Science and Technology
Ya Zhang	Shanghai Jiao Tong University
Zongzhang Zhang	Nanjing University
Lei Zhang	Institute of Information Engineering, Chinese Academy of Sciences
Jing Zhang	Renmin University of China
Xianchao Zhang	Dalian University of Technology
Jiangwei Zhang	National University of Singapore
Fengpan Zhao	Georgia State University
Lin Zhao	Institute of Information Engineering, Chinese Academy of Sciences
Long Zheng	Huazhong University of Science and Technology
Zuowu Zheng	Shanghai Jiao Tong University
Tongya Zheng	Zhejiang University
Runkai Zheng	Jinan University
Cheng Zheng	University of California, Los Angeles
Wenbo Zheng	Xi'an Jiaotong University
Zhiqiang Zhong	University of Luxembourg
Caiming Zhong	Ningbo University
Ding Zhou	Columbia University
Yilun Zhou	MIT
Ming Zhou	Shanghai Jiao Tong University
Yanqiao Zhu	Institute of Automation, Chinese Academy of Sciences
Wenfei Zhu	King
Wanzheng Zhu	University of Illinois at Urbana-Champaign
Fuqing Zhu	Institute of Information Engineering, Chinese Academy of Sciences
Markus Zopf	TU Darmstadt
Weidong Zou	Beijing Institute of Technology
Jingwei Zuo	UVSQ

Program Committee Members, Applied Data Science Track

Deepak Ajwani	Nokia Bell Labs
Nawaf Alharbi	Kansas State University
Rares Ambrus	Toyota Research Institute
Maryam Amir Haeri	Technische Universität Kaiserslautern
Jean-Marc Andreoli	Naverlabs Europe
Cecilio Angulo	Universitat Politècnica de Catalunya
Stefanos Antaris	KTH Royal Institute of Technology
Nino Antulov-Fantulin	ETH Zurich
Francisco Antunes	University of Coimbra
Muhammad Umer Anwaar	Technical University of Munich
Cristian Axenie	Audi Konfuzius-Institut Ingolstadt/Technical University of Ingolstadt
Mehmet Cem Aytekin	Sabancı University
Anthony Bagnall	University of East Anglia
Marco Baldan	Leibniz University Hannover
Maria Bampa	Stockholm University
Karin Becker	UFRGS
Swarup Ranjan Behera	Indian Institute of Technology Guwahati
Michael Berthold	University of Konstanz
Antonio Bevilacqua	Insight Centre for Data Analytics
Ananth Reddy Bhimireddy	Indiana University Purdue University - Indianapolis
Haixia Bi	University of Bristol
Wu Bin	Zhengzhou University
Thibault Blanc Beyne	INP Toulouse
Andrzej Bobyk	Maria Curie-Skłodowska University
Antonio Bonafonte	Amazon
Ludovico Boratto	Eurecat
Massimiliano Botticelli	Robert Bosch GmbH
Maria Brbic	Stanford University
Sebastian Buschjäger	TU Dortmund
Rui Camacho	University of Porto
Doina Caragea	Kansas State University
Nicolas Carrara	University of Toronto
Michele Catasta	Stanford University
Oded Cats	Delft University of Technology
Tania Cerquitelli	Politecnico di Torino
Fabricio Ceschin	Federal University of Paraná
Jeremy Charlier	University of Luxembourg
Anveshi Charuvaka	GE Global Research
Liang Chen	Sun Yat-sen University
Zhiyong Cheng	Shandong Artificial Intelligence Institute

Silvia Chiusano	Politecnico di Torino
Cristian Consonni	Eurecat - Centre Tecnòlogic de Catalunya
Laure Crochepierre	RTE
Henggang Cui	Uber ATG
Tiago Cunha	University of Porto
Elena Daraio	Politecnico di Torino
Hugo De Oliveira	HEVA/Mines Saint-Étienne
Tom Decroos	Katholieke Universiteit Leuven
Himel Dev	University of Illinois at Urbana-Champaign
Eustache Diemert	Criteo AI Lab
Nat Dilokthanakul	Vidyasirimedhi Institute of Science and Technology
Daizong Ding	Fudan University
Kaize Ding	ASU
Ming Ding	Tsinghua University
Xiaowen Dong	University of Oxford
Sourav Dutta	Huawei Research
Madeleine Ellis	University of Nottingham
Benjamin Evans	Brunel University London
Francesco Fabbri	Universitat Pompeu Fabra
Benjamin Fauber	Dell Technologies
Fuli Feng	National University of Singapore
Oluwaseyi Feyisetan	Amazon
Ferdinando Fioretto	Georgia Institute of Technology
Caio Flexa	Federal University of Pará
Germain Forestier	Université de Haute-Alsace
Blaz Fortuna	Qlector
Enrique Frias-Martinez	Telefónica Research and Development
Zuohui Fu	Rutgers University
Takahiro Fukushige	Nissan Motor Co., Ltd.
Chen Gao	Tsinghua University
Johan Garcia	Karlstad University
Marco Gärtler	ABB Corporate Research Center
Kanishka Ghosh Dastidar	Universität Passau
Biraja Ghoshal	Brunel University London
Lovedeep Gondara	Simon Fraser University
Severin Gsponer	Science Foundation Ireland
Xinyu Guan	Xi'an Jiaotong University
Karthik Gurumoorthy	Amazon
Marina Haliem	Purdue University
Massinissa Hamidi	Laboratoire LIPN-UMR CNRS 7030, Sorbonne Paris Cité
Junheng Hao	University of California, Los Angeles
Khadidja Henni	Université TÉLUQ

João Mendes-Moreira	University of Porto
Ioannis Mitros	Insight Centre for Data Analytics
Elena Mocanu	University of Twente
Hebatallah Mohamed	Free University of Bozen-Bolzano
Roghayeh Mojarad	Université Paris-Est Créteil
Mirco Nanni	KDD-Lab ISTI-CNR Pisa
Juggapong Natwichai	Chiang Mai University
Sasho Nedelkoski	TU Berlin
Kei Nemoto	The Graduate Center, City University of New York
Ba-Hung Nguyen	Japan Advanced Institute of Science and Technology
Tobias Nickchen	Paderborn University
Aastha Nigam	LinkedIn Inc
Inna Novalija	Jožef Stefan Institute
Francisco Ocegueda-Hernandez	National Oilwell Varco
Tsuyoshi Okita	Kyushu Institute of Technology
Oghenejokpeme Orhobor	The University of Manchester
Aomar Osmani	Université Sorbonne Paris Nord
Latifa Oukhellou	IFSTTAR
Rodolfo Palma	Inria Chile
Pankaj Pandey	Indian Institute of Technology Gandhinagar
Luca Pappalardo	University of Pisa, ISTI-CNR
Paulo Paraíso	INESC TEC
Namyong Park	Carnegie Mellon University
Chanyoung Park	University of Illinois at Urbana-Champaign
Miquel Perelló-Nieto	University of Bristol
Nicola Pezzotti	Philips Research
Tiziano Piccardi	Ecole Polytechnique Fédérale de Lausanne
Thom Pijnenburg	Elsevier
Valentina Poggioni	Università degli Studi di Perugia
Chuan Qin	University of Science and Technology of China
Jiezhong Qiu	Tsinghua University
Maria Ramirez-Loaiza	Intel Corporation
Manjusha Ravindranath	ASU
Zhaochun Ren	Shandong University
Antoine Richard	Georgia Institute of Technology
Kit Rodolfa	Carnegie Mellon University
Mark Patrick Roeling	Technical University of Delft
Soumyadeep Roy	Indian Institute of Technology Kharagpur
Ellen Rushe	Insight Centre for Data Analytics
Amal Saadallah	TU Dortmund
Carlos Salort Sanchez	Huawei
Eduardo Hugo Sanchez	IRT Saint Exupéry
Markus Schmitz	University of Erlangen-Nuremberg/BMW Group
Ayan Sengupta	Optum Global Analytics (India) Pvt. Ltd.
Ammar Shaker	NEC Laboratories Europe

Manali Sharma	Samsung Semiconductor Inc.
Jiaming Shen	University of Illinois at Urbana-Champaign
Dash Shi	LinkedIn
Ashish Sinha	IIT Roorkee
Yorick Spenrath	Technische Universiteit Eindhoven
Simon Stieber	University of Augsburg
Hendra Suryanto	Rich Data Corporation
Raunak Swarnkar	IIT Gandhinagar
Imen Trabelsi	National Engineering School of Tunis
Alexander Treiss	Karlsruhe Institute of Technology
Rahul Tripathi	Amazon
Dries Van Daele	Katholieke Universiteit Leuven
Ranga Raju Vatsavai	North Carolina State University
Vishnu Venkataraman	Credit Karma
Sergio Viademonte	Vale Institute of Technology, Vale SA
Yue Wang	Microsoft Research
Changzhou Wang	The Boeing Company
Xiang Wang	National University of Singapore
Hongwei Wang	Shanghai Jiao Tong University
Wenjie Wang	Emory University
Zirui Wang	Carnegie Mellon University
Shen Wang	University of Illinois at Chicago
Dingxian Wang	East China Normal University
Yoshikazu Washizawa	The University of Electro-Communications
Chrys Watson Ross	University of New Mexico
Dilusha Weeraddana	CSIRO
Ying Wei	The Hong Kong University of Science and Technology
Laksri Wijerathna	Monash University
Le Wu	Hefei University of Technology
Yikun Xian	Rutgers University
Jian Xu	Citadel
Haiqin Yang	Ping An Life
Yang Yang	Northwestern University
Carl Yang	University of Illinois at Urbana-Champaign
Chin-Chia Michael Yeh	Visa Research
Shujian Yu	NEC Laboratories Europe
Chung-Hsien Yu	University of Massachusetts Boston
Jun Yuan	The Boeing Company
Stella Zevio	LIPN
Hanwen Zha	University of California, Santa Barbara
Chuxu Zhang	University of Notre Dame
Fanjin Zhang	Tsinghua University
Xiaohan Zhang	Sony Interactive Entertainment
Xinyang Zhang	University of Illinois at Urbana-Champaign
Mia Zhao	Airbnb
Qi Zhu	University of Illinois at Urbana-Champaign

Hengshu Zhu	Baidu Inc.
Tommaso Zoppi	University of Florence
Lan Zou	Carnegie Mellon University

Program Committee Members, Demo Track

Deepak Ajwani	Nokia Bell Labs
Rares Ambrus	Toyota Research Institute
Jean-Marc Andreoli	NAVER LABS Europe
Ludovico Boratto	Eurecat
Nicolas Carrara	University of Toronto
Michelangelo Ceci	Università degli Studi di Bari Aldo Moro
Tania Cerquitelli	Politecnico di Torino
Liang Chen	Sun Yat-sen University
Jiawei Chen	Zhejiang University
Zhiyong Cheng	Shandong Artificial Intelligence Institute
Silvia Chiusano	Politecnico di Torino
Henggang Cui	Uber ATG
Tiago Cunha	University of Porto
Chris Develder	Ghent University
Nat Dilokthanakul	Vidyasirimedhi Institute of Science and Technology
Daizong Ding	Fudan University
Kaize Ding	ASU
Xiaowen Dong	University of Oxford
Fuli Feng	National University of Singapore
Enrique Frias-Martinez	Telefónica Research and Development
Zuohui Fu	Rutgers University
Chen Gao	Tsinghua University
Thomas Gärtner	TU Wien
Derek Greene	University College Dublin
Severin Gsponer	University College Dublin
Xinyu Guan	Xi'an Jiaotong University
Junheng Hao	University of California, Los Angeles
Ziniu Hu	University of California, Los Angeles
Chao Huang	University of Notre Dame
Hong Huang	UGoe
Neil Hurley	University College Dublin
Guillaume Jacquet	Joint Research Centre - European Commission
Di Jiang	WeBank
Song Jiang	University of California, Los Angeles
Jihed Khiari	Johannes Kepler Universität Linz
Mark Last	Ben-Gurion University of the Negev
Thach Le Nguyen	The Insight Centre for Data Analytics
Vincent Lemaire	Orange Labs
Camelia Lemnaru	Universitatea Tehnică din Cluj-Napoca
Bowen Liu	Stanford University

Yin Lou	Ant Financial
Yao Ma	Michigan State University
Brian Mac Namee	University College Dublin
Susan Mckeever	Technological University Dublin
Edgar Meij	Bloomberg L.P.
Gabor Melli	Sony PlayStation
Decebal Constantin Mocanu	University of Twente
Elena Mocanu	University of Twente
Luis Moreira-Matias	Kreditech
Latifa Oukhellou	IFSTTAR
Chanyoung Park	University of Illinois at Urbana-Champaign
Julien Perez	NAVER LABS Europe
Chuan Qin	University of Science and Technology of China
Jiezhong Qiu	Tsinghua University
Zhaochun Ren	Shandong University
Kit Rodolfa	Carnegie Mellon University
Robert Ross	Technological University Dublin
Parinya Sanguansat	Panyapiwat Institute of Management
Niladri Sett	University College Dublin
Manali Sharma	Samsung Semiconductor Inc.
Jerzy Stefanowski	Poznań University of Technology
Luis Teixeira	Fraunhofer Portugal AICOS
Roberto Trasarti	ISTI-CNR, Pisa
Anton Tsitsulin	University of Bonn
Grigorios Tsoumakas	Aristotle University of Thessaloniki
Marco Turchi	Fondazione Bruno Kessler
Vishnu Venkataraman	Credit Karma
Yue Wang	Microsoft Research
Xiang Wang	National University of Singapore
Dingxian Wang	East China Normal University
Ying Wei	The Hong Kong University of Science and Technology
Le Wu	Hefei University of Technology
Jian Xu	Citadel
Carl Yang	University of Illinois at Urbana-Champaign
Yang Yang	Northwestern University
Haiqin Yang	Ping An Life
Chung-Hsien Yu	University of Massachusetts Boston
Chuxu Zhang	University of Notre Dame
Xiaohan Zhang	Sony Interactive Entertainment

Sponsors

amazon | science

\Orchestrating a brighter world

Qualcomm
AI research

visitgent

gentcongres

Contents – Part V

Applied Data Science: E-Commerce and Finance

Demo Track

Applied Data Science: Social Good

Confound Removal and Normalization in Practice: A Neuroimaging Based Sex Prediction Case Study

Shammi More[1,2], Simon B. Eickhoff[1,2], Julian Caspers[3], and Kaustubh R. Patil[1,2](✉)

[1] Institute of Neuroscience and Medicine (INM-7), Forschungszentrum Jülich, Jülich, Germany
{s.more,s.eickhoff,k.patil}@fz-juelich.de

[2] Institute of Systems Neuroscience, Medical Faculty, Heinrich Heine University Düsseldorf, Düsseldorf, Germany

[3] Department of Diagnostic and Interventional Radiology, University Hospital Düsseldorf, Düsseldorf, Germany
julian.caspers@med.uni-duesseldorf.de

Abstract. Machine learning (ML) methods are increasingly being used to predict pathologies and biological traits using neuroimaging data. Here controlling for confounds is essential to get unbiased estimates of generalization performance and to identify the features driving predictions. However, a systematic evaluation of the advantages and disadvantages of available alternatives is lacking. This makes it difficult to compare results across studies and to build deployment quality models. Here, we evaluated two commonly used confound removal schemes–whole data confound regression (WDCR) and cross-validated confound regression (CVCR)–to understand their effectiveness and biases induced in generalization performance estimation. Additionally, we study the interaction of the confound removal schemes with Z-score normalization, a common practice in ML modelling. We applied eight combinations of confound removal schemes and normalization (pipelines) to decode sex from resting-state functional MRI (rfMRI) data while controlling for two confounds, brain size and age. We show that both schemes effectively remove linear univariate and multivariate confounding effects resulting in reduced model performance with CVCR providing better generalization estimates, i.e., closer to out-of-sample performance than WDCR. We found no effect of normalizing before or after confound removal. In the presence of dataset and confound shift, four tested confound removal procedures yielded mixed results, raising new questions. We conclude that CVCR is a better method to control for confounding effects in neuroimaging studies. We believe that our in-depth analyses shed light on choices associated with confound removal and hope that it generates more interest in this problem instrumental to numerous applications.

Keywords: Confound removal · Generalization · Interpretability · Sex classification · Neuroimaging application

Y. Dong et al. (Eds.): ECML PKDD 2020, LNAI 12461, pp. 3–18, 2021.
https://doi.org/10.1007/978-3-030-67670-4_1

1 Introduction

A critical challenge in applied machine learning is controlling for confounding effects as not removing them can lead to biased predictions and interpretations. This is especially true for biological data as common underlying processes introduce shared variance between the measurements, giving rise to confounding effects and blurring the boundaries between signals and confounds. Nevertheless, when confounds can be identified, removing their effects can lead to unbiased models and better understanding of the underlying biological processes.

In the field of neuroimaging, predictive analysis using machine learning has gained popularity for decoding phenotypes with a clear application to understand brain organization and its relationship to behavior and disease [9,14,41] with a twofold aim, (1) to establish brain-phenotype relationship by estimating the generalization performance, and (2) to identify brain regions explaining the variance of the phenotype. Cross-validation (CV) is employed for the first goal while the second goal is usually achieved by identifying predictive features, e.g., features with a high weight assigned by a linear model. Specifically, in addition to information uniquely associated with the target (true signal) neuroimaging features may also contain information from nuisance sources, e.g., brain size, confounding the relationship between the neuroimaging signal and the target. In this case, both goals can yield biased results as a successful prediction might be driven by the confounding signal rather than the true signal (Fig. 1a). Thus, the confounding effects need to be removed to estimate generalizability and to gain interpretability in an unbiased way. Various alternatives exist for confound removal and are integrated within ML pipelines. However, the pros and cons of these possibilities are not well understood.

Confounding can be controlled in the experiment design phase before data collection by randomization, restriction and matching [27]. However, this is not always feasible, e.g. when all the confounds are not known. Confounds can be controlled for after data acquisition. One way is to add them as additional predictors to capture the corresponding variance. However, this approach is not suitable for predictive modelling because it is designed to control in-sample rather than out-of-sample (OOS) properties. Another method is post-hoc counterbalancing i.e., taking a subset in which there is no empirical relationship between the confound and the target [35]. Advanced techniques such as the anti-mutual information sampling [10] and stratification using pooling analysis by the Mantel-Haenszel formula [38] have been proposed. However, these methods lead to data loss and are not feasible with a small sample and a large number of confounds. Specifically, when matching sexes according to brain size, these methods will represent extremes of the population and not the whole population. Of note, confound removal can be seen as supplementary to debiasing and fair learning [2,16,18] but here we do not investigate this angle.

One of the most common confound control approaches while using all the data is "regressing out" their variance from the features before learning, referred to as confound regression [35] or image correction [28]. In this method, a linear regression model is fitted on each feature separately with the confounds as predictors, and the corresponding residuals are used as new "confound-removed" features. This approach can be implemented in two possible ways. The first scheme is whole data confound regression (WDCR), regresses out confounds from the entire dataset at once [28,35,37] followed by CV to estimate the generalization performance. WDCR has yielded inconsistent results, from a substantial drop in performance [17,37] to a similar or slightly lower performance compared to the models without confound control [28]. This discrepancy is possibly due to differences in the strength of the relationship between the confounds, the features, and the target and implementation differences. WDCR leads to "data-leakage" as the information from the whole data is used to create the confound-removed features before CV. However, the "aggressive" confound removal by WDCR has been proposed to be desirable [25].

To alleviate issues with WDCR, a CV-consistent scheme, cross-validated confound regression (CVCR) has been proposed in which the linear confound regression models are estimated within CV using only the training subset, and applied to both the training and the validation subsets. This avoids information leaking from training into validation sets. Although both WDCR and CVCR schemes have been used in neuroimaging studies [20,35,45], there is a lack of information regarding how they affect the generalization estimates and interpretability with one study recommending WDCR [25] while another recommending CVCR [35].

Moreover, whether to apply a feature normalization and standardization procedures, like Z-scoring (Zero mean and unit-variance features), before confound removal or after has not been investigated. It is known that in the specific case of normalization using rank-based inverse normal transformation (INT) after confound regression may reintroduce confounding effects [24]. Such reintroduction of confounding effects can be counterproductive for model generalizability and interpretability. Furthermore, the ability of an algorithm to learn from the data might differ depending upon when normalization is applied. This lack of understanding about the interaction between confound regression and normalization makes it difficult to design ML pipelines. Lastly, building models when one suspects a shift in the covariates and/or in the relationship between the confounds, the features and the target has not been studied. Several design possibilities can be imagined and need to be evaluated.

In this work we empirically investigate three facets of the confound removal issue, (1) evaluation of the two confound removal schemes, WDCR and CVCR, for their effectiveness in removing the confounding signal and estimation of generalization performance, (2) interaction of confound removal schemes with normalization, and (3) model deployment when covariate and confounding effect shift is suspected. We consider prediction of sex from resting-state functional magnetic resonance imaging (rfMRI) data while controlling for two confounds, brain size and age. We aim to answer an important biological question "are male and female brains functionally different after controlling for the apparent difference in brain size?". With systematic evaluation of a real-world problem reporting positive as well as negative results, we hope to attract the attention of the machine learning community to the critical problem of confound removal.

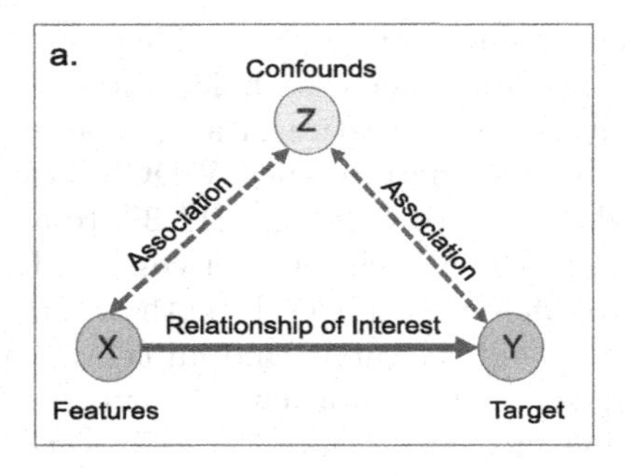

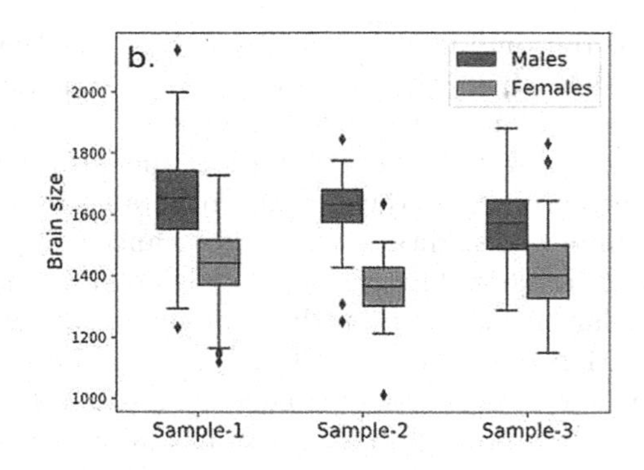

Fig. 1. (a) Confounding effect: Confound Z influences both the features X and the target Y. In the presence of Z, the actual relationship between X and Y is masked. For sex classification, brain size is a confound (Z) as it is associated with both rfMRI features (X) and sex (Y). (b) Significant sex difference in brain size in the three data samples used in this study.

2 Sex Classification and Brain Size

There are reports on differences in cognition and psychopathology between sexes [33], such as differences in spatial tasks [22], females being more vulnerable to depression [26] and autism being more prevalent in males [42]. These differences may influence diagnostic practices and help developing sex-specific treatments, making understanding neurobiology of sex differences essential. Accordingly there has been an increasing interest in finding sex differences in structural and functional properties of the brain [29,30,41].

Functional magnetic resonance imaging (fMRI) is a non-invasive technique used to study functional–i.e. time dependent–changes in brain activity by taking 3D MRI images in succession. Even unregulated processes in the resting brain, i.e., resting-state fMRI (rfMRI), show stable and individualized synchronies [12]. Such functional activities have been related to cognition and several phenotypes, especially using the functional connectivity (FC) (see Sect. 4.2). Based on whole-brain FC, the sex prediction accuracy of 75–80% was achieved with discriminative features mainly located in the Default mode network (DMN) [41,45]. Another study with a lower prediction accuracy of 62% found discriminative FC in motor, sensory, and association areas [6]. Smith and colleagues [34] reported a higher prediction accuracy of 87%. A recent study reported sex prediction accuracy of 98% using multi-label learning, i.e., sex in conjunction with nine other cognitive, behavioural and demographic variables [8].

Brain size is highly correlated with sex, with larger total brain volume in males compared to females [4,29]; and is encoded in MRI data. Figure 1b shows the difference in brain size between sexes for the data samples used in the current study. In such a scenario, even if a model decodes sex from MRI data significantly above chance, there is no clear understanding of the unique contribution of the functional features independent of brain size. It is likely that the prediction is driven partly by brain size in addition to the functional differences. Zhang and colleagues [45] have shown that the sex prediction accuracy drops from 80% to 70% after regressing out brain size from FC, indicating an apparent effect of brain size in sex prediction. Hence, there is clearly a need to study sex prediction using rfMRI while controlling for brain size.

3 Experimental Setup

3.1 Study Design

With a limited and contrasting literature, there is a lack of knowledge of how to perform confound removal. Here we aimed to evaluate two confound removal schemes (WDCR and CVCR) and their interaction with the commonly used Z-score feature normalization. We evaluated eight pipelines in total (Fig. 2a);

1. No confound removal, no Z-scoring (NCR-NZ)
2. No confound removal, with Z-scoring (NCR-Z)
3. WDCR, no Z-scoring (WDCR-NZ)
4. WDCR, Z-scoring after confound removal (WDCR-ZAC)
5. WDCR, Z-scoring before confound removal (WDCR-ZBC)
6. CVCR, no Z-scoring (CVCR-NZ)
7. CVCR, Z-scoring after confound removal (CVCR-ZAC)
8. CVCR, Z-scoring before confound removal (CVCR-ZBC)

We applied these pipelines for predicting an individual's sex using features derived from rfMRI data while controlling for two confounds brain size and age. We performed two evaluations; (1) CV to estimate the generalization performance and compared it with prediction on an OOS dataset, and (2) OOS prediction with covariate and confound shift as a model deployment scenario. The prediction performance was evaluated using AUC, F1-score and balanced accuracy.

For evaluation-1, we used a publicly available database (HCP, see Sect. 4.1) and carefully derived sample-1 (N = 377) and sample-2 (N = 54). After standard preprocessing two types of features were extracted from rfMRI data, Regional Homogeneity (ReHo) and FC (see Sect. 4.2). Each feature set was analyzed separately using Ridge Regression and Partial Least Square Regression with all eight pipelines. The generalization performance was estimated on sample-1 using 10 times repeated 10-fold CV. The OOS performance was evaluated on sample-2. By comparing the CV and OOS results, we can comment on whether the CV procedure can reliably estimate the generalization performance.

As the confounds were linearly removed from the features in a univariate way (see Sect. 3.2) multivariate confounding effects might still remain. We, therefore, assessed the effectiveness of confound removal pipelines in removing univariate and multivariate confounding effects. The Pearson correlation between each

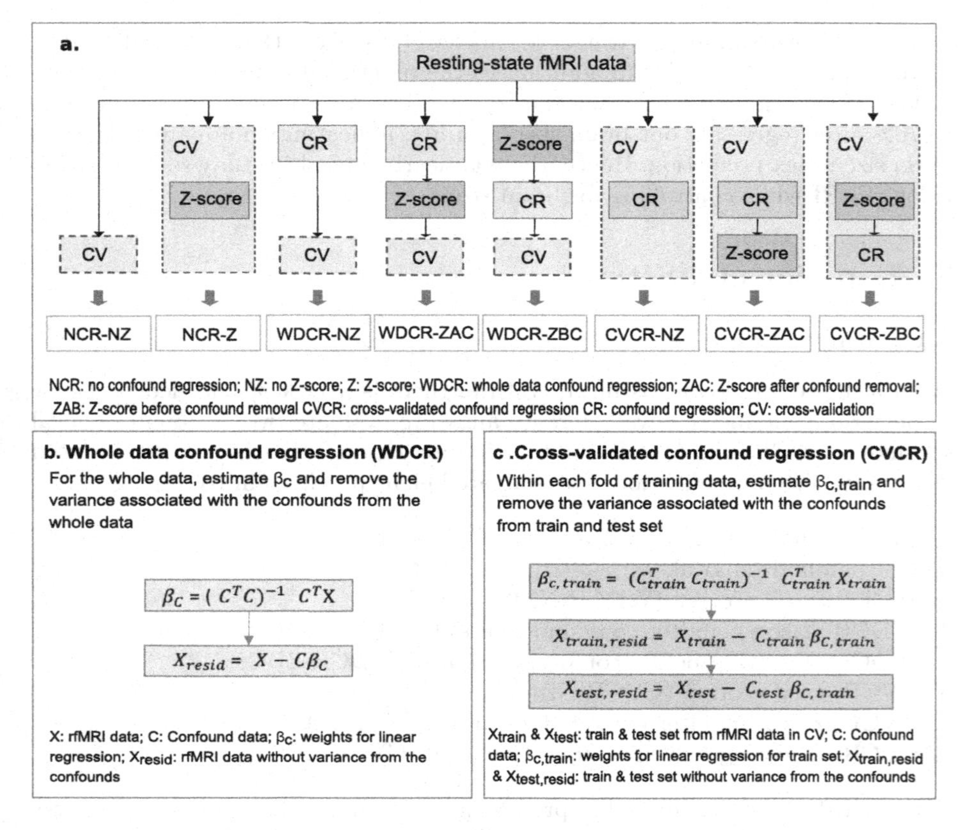

Fig. 2. a. The schematic diagram of various combinations of confound removal schemes and Z-score for confound removal evaluated in the study. b. Whole data confound regression (WDCR). c. Cross-validated confound regression (CVCR).

residual feature and the brain size was calculated to check for remaining univariate confounding effects. The adjusted r^2 of the multiple linear regression model predicting the brain size using residual features was used to check for remaining multivariate confounding effects.

In neuroimaging studies it is common that the data is acquired on different scanners [40] and there may exist demographic differences between samples. Such differences can lead to covariate shift [19] and by extension confound shift. An ideal model should generalize well despite such differences. To evaluate this (evaluation-2), we employed an additional sample (sample-3; N = 484) from a public dataset (eNKI, see Sect. 4.1) where demographics, scanner parameters and preprocessing are different than sample-1 and 2. We tested four ways to remove confounds from OOS data.

1. **Train-to-test**: The confound removal models from the train data were applied to the OOS data. This is the standard method.
2. **Test WDCR**: WDCR was performed on the OOS data.

3. **Test CVCR**: CVCR was performed on the OOS data, i.e. confound regression was performed within CV for OOS data and the residuals were retained.
4. **Train and test combined**: WDCR was performed on the combined train and OOS data. The data was then re-split into train and test.

Methods 2, 3 aimed to obtain confound-free OOS data, with the assumptions that confound-removed models can perform well on confound-removed OOS data as confounds are handled within a sample. Method 4 assumes that the confound removal linear models can capture variance from both train and OOS data. Note that 2, 3 and 4 can only be used with sufficiently large OOS data. WDCR models trained on sample-1 were used to predict the confound-removed OOS data. The sample-2 and sample-3 with similar and different properties to sample-1 respectively were the OOS datasets. Note that for method 1, 2 and 3 trained models (on sample-1) come from the above-mentioned pipelines used for evaluation-1.

3.2 Confound Regression

We tested two different versions of confound regression, WDCR and CVCR (Fig. 2b and c). In WDCR, using multiple linear regression we regressed out the confounds from each of the predictors from the entire dataset before the cross-validated procedure. Note that, this procedure uses information from the whole dataset leading to data-leakage. In CVCR, we regressed the confounds in a similar way to WDCR but the confound removal models were estimated on the training data and subsequently applied to both train and validation sets. In this way, there is no leakage from train to test.

3.3 Predictive Modelling

We used two prediction models, Ridge Regression and Partial Least Square regression. Ridge Regression (RR) uses a sum of the square penalty on the model parameters to reduces model complexity and prevent overfitting [15]. The balance between the fit and the penalty is defined using a hyper-parameter λ which we tuned in an inner CV loop. PLS Regression (PLS) performs dimensionality reduction and learning simultaneously, making it a popular choice when there are more features than observations, and/or when there is multicollinearity among the features. It has performed well in MRI-based estimations for cognitive, behavioural and demographic variables [8,45]. PLS searches for a set of latent vectors that performs a simultaneous decomposition of predictors and the target such that these components explain the maximum covariance between them [1]. These latent vectors are then used for prediction. The hyperparameter for the PLS is the number of latent variables which was tuned in an inner CV loop.

4 Data Samples and Features

4.1 Data Samples

This study included three samples. Sample 1 and 2 are two independent subsets of the data provided by the Human Connectome Project (HCP) S1200 release

[39]. Sample-1 contained 377 subjects (age range: 22–37, mean age: 28.6 years; 182 females), sample-2 comprised 54 subjects (age range: 22–36, mean age: 28.9 years; 28 females). As the HCP data contains siblings and twins, the samples were constructed such that there were no siblings within or across the two samples, to avoid biases due to any similarity in the FC of the siblings. Within each of the two samples, males and females were matched for age, and education. Resting-state blood oxygen level-dependent (BOLD) data comprised 1200 functional volumes per subject, acquired on a Siemens Skyra 3T scanner with the following parameters: TR = 720 ms, TE = 33.1 ms, flip angle = 52°, voxel size = $2 \times 2 \times 2$ mm^3, FoV = 208×180 mm^2, matrix = 104×90, slices = 72. Sample-3 was obtained from the Enhanced Nathan Kline Institute–Rockland Sample (eNKI-RS) [23] with 484 subjects (age range: 6–85, mean age: 41.9 years; 311 females). Images were acquired on a Siemens TimTrio 3T scanner using BOLD contrast with the following parameters: TR = 1400 ms, TE = 30 ms, flip angle = 65°, voxel size = $2 \times 2 \times 2$ mm^3, slices = 64. Subjects were asked to lie with eyes open, with "relaxed" fixation on a white cross (on a dark background), think of nothing in particular, and not to fall asleep. The CAT-12 toolbox (http://www.neuro.uni-jena.de/cat/) was used to calculate the brain size of each subject based on T1-weighted images. Note the stark differences between sample-1, 2 and sample-3 in terms of demographics as well as scanner parameters. This selection was made to elucidate the common scenario of data heterogeneity.

Two-sample t-test revealed significant sex differences in the brain size across all the samples ($p < 0.001$; Fig. 1b). This clearly demonstrates that brain size is a confound in sex prediction. There was no difference in age between sexes in sample-1 but significant differences was observed in sample-2 and 3 ($p < 0.001$). Age is not expected to be related to sex but was included as a control confound.

4.2 Pre-processing and Feature Extraction

After standard rfMRI pre-processing we extracted two types of features based on the voxel-wise time-series.

Preprocessing. The rfMRI data needs to be pre-processed so that the effects of motion in the scanner are removed as well as the brain of each subject is normalized to a standard brain template (e.g., MNI-152) so that they can be compared across subjects. For samples 1 and 2, the pre-processed, FIX-denoised and spatially normalized to the MNI-152 template data provided by the HCP S1200 release was used. There was no difference in the movement parameters (measured as mean framewise displacement) between males and females in both the samples. No further motion correction was performed. For sample-3, physical noise and effects of motion in the scanner were removed by using FIX (FMRIB's ICA-based Xnoiseifier, version 1.061 as implemented in FSL 5.0.9; [13,31]). Unique variance related to the identified artefactual independent components and 24 movement parameters [32] were then regressed from the data. The FIX-denoised data were further preprocessed using SPM8 (Wellcome Trust Centre for Neuroimaging, London) and in-house Matlab scripts for movement correction and spatial normalization to the MNI-152 template [3].

Regions of Interest (ROI). The Dosenbach atlas was used to extract 160 ROIs from the whole-brain data. These ROIs are spheres of 10 mm diameter, identified from a series of meta-analyses of task-related fMRI studies and broadly cover much of the cerebral cortex and cerebellum [11]. This atlas has been utilized in several brain network analyses including for sex prediction [5,45].

Feature Space 1: Regional Homogeneity (ReHo) measures the similarity of the time-series of a set of voxels and thus reflects the temporal synchrony of the regional BOLD signal [44]. ReHo for each subject and each of the 160 ROIs was calculated as the Kendall's coefficient of concordance between all the time-series of the voxels within a given ROI resulting in 160 features per subject.

Feature Space 2: Functional Connectivity (FC) is the correlation between the time-series of different brain regions [36]. For each subject, the time series of all the voxels within a ROI were averaged and FC was calculated as the Pearson's correlation coefficients between them for all pairs of ROI. These were then transformed using Fisher's Z-score. Each subject had a feature vector of length 12,720 after vectorization of the lower triangle of the 160 $\times$ 160 FC matrix.

5 Results

We compiled the results from two viewpoints. We first asked which of the pipelines incorporating confound removal provides more realistic generalization performance estimates. Then we assessed the efficacy of the confound removal schemes in a model deployment scenario with data heterogeneity.

5.1 Generalization Performance Estimates

CV is commonly used to estimate generalization performance. However, it is not without caveats [7]. Therefore, we compared CV performance of the pipelines with "true" OOS performance. In this case, the CV was performed on sample-1 and sample-2 was used as the OOS data. PLS generally performed better than RR, so in the following we focus on the PLS results.

As expected, the CV performance was highest without controlling for confounds (Table 1). AUC and F1-scores for sex prediction with ReHo were 0.838 and 0.754 and with FC were 0.874 and 0.787, respectively. Both the schemes WDCR and CVCR showed reduced performance in line with previous studies [25,35]. As brain size is highly correlated to sex, regressing it out from every feature can remove sex-specific information, resulting in a lower performance.

WDCR provided lower generalization estimates than CVCR, with the balanced accuracy dropping close to chance level with WDCR. One might expect higher generalization performance with WDCR as it causes data leakage from the train to the validation set violating the crucial assumption of independence in cross-validated analysis. However, in this case, it leads to worse performance. This might be because WDCR is performed on the whole dataset and hence is more aggressive in removing the confounding signal than CVCR leading to poorer performance. When the trained models were applied to OOS data, we

Table 1. Comparison of the pipelines using RR and PLS. Models were trained on sample-1 and out-of-sample/test performance was tested on sample-2.

CR	Z-score	Feat.	Ridge regression						Partial least squares					
			CV: Sample-1			Test: Sample-2			CV: Sample-1			Test: Sample-2		
			AUC	F1	Acc.	AUC	F1	Acc.	AUC	F1	Acc.	AUC	F1	Acc.
NCR	NZ	ReHo	0.750	0.667	0.662	0.751	0.690	0.688	0.776	0.714	0.712	0.808	0.759	0.760
		FC	0.857	0.763	0.757	0.823	0.728	0.725	0.874	0.787	0.785	0.835	0.762	0.761
NCR	Z	ReHo	0.829	0.749	0.746	0.832	0.759	0.758	0.838	0.754	0.751	0.860	0.778	0.776
		FC	0.860	0.772	0.768	0.841	0.765	0.762	0.860	0.781	0.779	0.813	0.765	0.762
WDCR	NZ	ReHo	0.477	0.490	0.490	0.511	0.500	0.500	0.476	0.494	0.494	0.685	0.647	0.647
		FC	0.466	0.488	0.496	0.607	0.500	0.500	0.417	0.454	0.455	0.685	0.661	0.654
	ZAC	ReHo	0.528	0.523	0.522	0.501	0.500	0.500	0.553	0.548	0.546	0.735	0.685	0.683
		FC	0.467	0.482	0.483	0.611	0.500	0.500	0.409	0.444	0.446	0.677	0.578	0.577
	ZBC	ReHo	0.528	0.528	0.526	0.501	0.500	0.500	0.553	0.546	0.545	0.735	0.685	0.683
		FC	0.456	0.476	0.478	0.611	0.500	0.500	0.407	0.444	0.445	0.677	0.578	0.577
CVCR	NZ	ReHo	0.552	0.522	0.519	0.511	0.500	0.500	0.569	0.553	0.553	0.685	0.647	0.647
		FC	0.516	0.500	0.500	0.607	0.500	0.500	0.595	0.576	0.575	0.685	0.661	0.654
	ZAC	ReHo	0.632	0.589	0.585	0.577	0.611	0.518	0.668	0.637	0.634	0.694	0.666	0.665
		FC	0.543	0.532	0.529	0.661	0.592	0.582	0.588	0.565	0.563	0.705	0.595	0.595
	ZBC	ReHo	0.634	0.591	0.587	0.577	0.611	0.518	0.669	0.635	0.633	0.703	0.666	0.665
		FC	0.547	0.532	0.529	0.662	0.592	0.582	0.586	0.564	0.563	0.705	0.595	0.595

found that OOS performance was higher than the CV estimates for most of the pipelines. This might happen if the OOS data is easier to classify. The OOS performance was closer to the generalization performance estimated with CVCR. This result suggests that CVCR is a better way to do confound removal in predictive analyses with neuroimaging data.

We then checked whether the confound removal was happening as expected. First, in a univariate way we correlated the residuals (new features) with the confounding variables. We found no significant correlation with both confound removal schemes indicating effective univariate removal of the confounding signal from the features. However, as multivariate effects might still be remaining, we used multiple linear regression to predict brain size from the residual features. With CVCR and WDCR, these models on the training sets revealed negative adjusted r^2. This indicates that there were no remaining linear multivariate confounding effects with both WDCR and CVCR. Thus the models trained with the residual features contained no information from the confounds.

These trends were similar for both ReHo and FC. Z-scoring improved the model performance with ReHo but not with FC. There was no effect of Z-scoring the features before (raw features) or after (residuals) confound removal.

5.2 Predictive Features

One of the main objectives of a decoding analysis is to identify predictive features (brain regions) explaining the variance in phenotype. As the confounding effect can impact predictive features selection, it is important to compare them with and without confound removal. The Z-scored feature weights (the absolute value) averaged across CV runs were used to select predictive features. We found that predictive features with and without confound removal were different (Fig. 3).

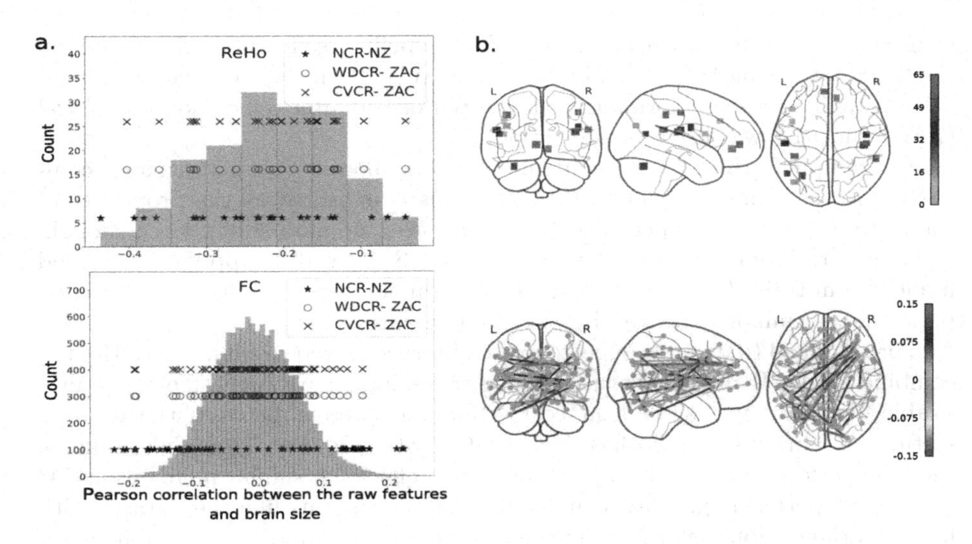

Fig. 3. a. Pearson correlation between the raw features and the brain size as histograms. The dots show the correlations of the selected features (jittered); 25 for ReHo (top) and 70 for FC (bottom) for NCR-NZ, WDCR-ZAC and CVCR-ZAC pipelines. b. Brain regions associated with the selected features; ReHo (top, relative weights), and FC (bottom), both with the CVCR-ZAC pipeline.

We compared 25 ReHo and 70 FC features with highest absolute weights from 3 pipelines, NCR-NZ, WDCR-ZAC and CVCR-ZAC (Fig. 3a). The features selected without confound removal had relatively higher positive or negative correlation with brain size. However, after confound removal (WDCR and CVCR), for FC the features with lower correlation were selected. This suggests that the features selected after confound removal represent the functional signal predictive of sex. We then identified features selected after confound removal (CVCR-ZAC) but not selected without confound removal (NCR-NZ) (Fig. 3b). With ReHo, selected regions were in dorsolateral prefrontal cortex, inferior parietal lobule, occipital, ventromedial prefrontal cortex, precentral gyrus, post insula, parietal, temporoparietal junction and inferior cerebellum, in line with a study identifying regions in the inferior parietal lobule and precentral gyrus [43]. In contrast, another study found sex differences in right hippocampus and amygdala [21]. We found important FC features widespread across the entire brain with strong interhemispheric connections. In contrast to the study by Zhang and colleagues [45] we did not find many intra-network FC in the DMN. Z-score feature normalization before or after confound removal did not affect selected features.

5.3 Out-of-Sample Performance

To study how a model deployment would work, especially in the presence of data heterogeneity common in neuroimaging studies, we tested four different ways to remove confounds from the OOS data including, applying confound models from train to OOS data using CVCR-ZAC pipeline, self-confound removal on the OOS data using WDCR and CVCR, and WDCR on the combined train and OOS data. The Z-score normalization was performed after the confound removal (ZAC) and PLS was used for prediction.

For sample-2, train-to-test confound removal showed best performance compared to other three methods (Table 2). This is expected as the properties of these two samples are expected to be similar (i.e., no data shift). Even though, residual correlations were observed in the OOS data after applying confound models from train data (Fig. 4a), the training models were confound-free so this performance cannot be driven by confounding effects.

For sample-3 (data shift expected), we observed mixed results. For ReHo, the combined WDCR model (learned on the train data) gave highest performance (Table 2b). However, significant correlation was present between the residual features and brain size in both train and OOS data (Fig. 4b). This might indicate that the performance is driven by confounding effects. A similar model using FC was lowest performing. With combined WDCR, it seems like the dataset with higher variance dominates leaving the other part correlated, indicating it might be suboptimal. Predictions on self-confound removed OOS data (sample-3) (Test WDCR and Test CVCR) were similar to when the confound models from sample-1 were applied (Table 2a). However, the OOS performance using ReHo dropped compared to CV while that of FC improved.

Table 2. Comparison of confound removal schemes on out-of-sample/test data. a. Confound models learned from the train data (sample-1) applied to test data (sample-2 and 3), WDCR and CVCR performed only on test data. b. WDCR on the combined train and test data.

a. Method	Feat.	CV: Sample-1			Test: Sample-2			Test: Sample-3		
		AUC	F1	Acc.	AUC	F1	Acc.	AUC	F1	Acc.
Train-to test:	ReHo	0.668	0.637	0.634	0.694	0.666	0.665	0.549	0.528	0.527
CVCR-ZAC	FC	0.588	0.565	0.563	0.705	0.595	0.595	0.637	0.628	0.619
Test WDCR:	ReHo	0.553	0.548	0.546	0.562	0.573	0.573	0.524	0.530	0.531
WDCR-ZAC	FC	0.409	0.444	0.446	0.632	0.576	0.576	0.635	0.592	0.595
Test CVCR:	ReHo	0.668	0.637	0.634	0.582	0.591	0.591	0.505	0.508	0.509
CVCR-ZAC	FC	0.588	0.565	0.563	0.603	0.578	0.577	0.634	0.597	0.601

b. Feat.	CV: Sample-1			Test: Sample-2			CV: Sample-1			Test: Sample-3		
	AUC	F1	Acc.	AUC	F1	Acc.	AUC	F1	Acc.	AUC	F1	Acc.
ReHo	0.533	0.538	0.538	0.580	0.558	0.560	0.870	0.788	0.786	0.614	0.577	0.502
FC	0.450	0.459	0.461	0.387	0.409	0.412	0.871	0.779	0.777	0.541	0.502	0.501

Taken together, we found that train-to-test application of confound removal models and self-confound removal to be better strategies but inconsistent across feature spaces. This raises questions regarding optimal confound removal strategies when data heterogeneity is present. Based on the results, we also speculate that covariate and confound shift is more pronounced in ReHo compared to FC.

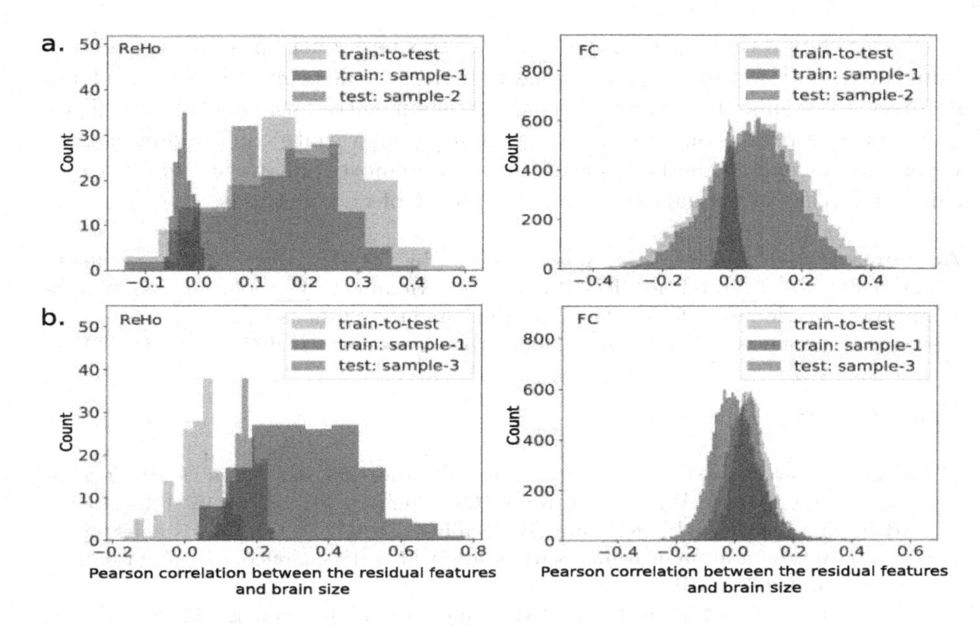

Fig. 4. Correlation between the residual features and brain size: for out-of-sample/test data when training confound removal models were applied (orange), and for train (purple) and test (green) data when combined train and test WDCR was performed. (Color figure online)

6 Conclusion

In this study, several confound removal pipelines were tested on the task of rfMRI data based sex classification. As expected, the two confound removal schemes (WDCR and CVCR) could effectively remove the signal corresponding to confounds leading to a substantial drop in prediction performance compared to without confound removal. Analyses on the residual features after WDCR and CVCR revealed that there were no remaining univariate and multivariate confounding effects. Thus, both these confound removed models should not have confound-related information encoded. We found CVCR to be a better method compared to WDCR as CVCR estimated generalization performance was closer to OOS performance. As WDCR leads to data leakage, one might expect it to be over-optimistic. However, our results point to the opposite. This is likely due to the aggressive confound removal. Our findings provide further corroboration to the idea of applying data analysis operations within the CV loop. In this work we focused on the sex prediction problem and whether our results apply to other problems remains to be seen.

The Z-score normalization of the features before or after confound removal did not affect model performance. We recommend to normalize after confound removal, as some learning algorithms might benefit from well-scaled features.

We also found that the OOS performance was best when the confound models from the train data were used, provided that the sample properties between train and test are similar but results were inconsistent with data shift. Although we used multiple regression to test for remaining multivariate confounding effects, we are not aware of a method that can directly remove multivariate effects. This calls for further investigations and development of new methods.

Acknowledgments. This study was supported by the Deutsche Forschungsgemeinschaft (DFG, PA 3634/1-1 and EI 816/21-1), the Helmholtz Portfolio Theme "Supercomputing and Modelling for the Human Brain" and the European Union [Horizon 2020 Research and Innovation Programme under Grant Agreement No. 945539 (HBP SGA3)].

References

1. Abdi, H.: Partial least squares regression and projection on latent structure regression (pls regression). Wiley Interdiscip. Rev. Comput. Stat. **2**(1), 97–106 (2010)
2. Adeli, E., Zhao, Q., Pfefferbaum, A., Sullivan, E.V., Fei-Fei, L., Niebles, J.C., et al.: Representation learning with statistical independence to mitigate bias. arXiv:1910.03676 (2019)
3. Ashburner, J., Friston, K.J.: Unified segmentation. Neuroimage **26**(3), 839–851 (2005)
4. Barnes, J., Ridgway, G.R., Bartlett, J., Henley, S.M., Lehmann, M., Hobbs, N., et al.: Head size, age and gender adjustment in mri studies: a necessary nuisance? Neuroimage **53**(4), 1244–1255 (2010)
5. Cao, M., Wang, J.H., Dai, Z.J., Cao, X.Y., Jiang, L.L., Fan, F.M., et al.: Topological organization of the human brain functional connectome across the lifespan. Developmental Cognitive Neurosci. **7**, 76–93 (2014)
6. Casanova, R., Whitlow, C., Wagner, B., Espeland, M., Maldjian, J.: Combining graph and machine learning methods to analyze differences in functional connectivity across sex. The Open Neuroimaging Journal **6**, 1 (2012)
7. Cawley, G.C., Talbot, N.L.: On over-fitting in model selection and subsequent selection bias in performance evaluation. J. Machine Learn. Res. **11**, 2079–2107 (2010)
8. Chen, C., Cao, X., Tian, L.: Partial least squares regression performs well in mri-based individualized estimations. Front. Neurosci. **13**, 1282 (2019)
9. Chen, J., Patil, K.R., Weis, S., Sim, K., Nickl-Jockschat, T., Zhou, J., et al.: Neurobiological divergence of the positive and negative schizophrenia subtypes identified on a new factor structure of psychopathology using non-negative factorization: An international machine learning study. Biol. Psychiatry **87**(3), 282–293 (2020)
10. Chyzhyk, D., Varoquaux, G., Thirion, B., Milham, M.: Controlling a confound in predictive models with a test set minimizing its effect. In: 2018 International Workshop on Pattern Recognition in Neuroimaging (PRNI), pp. 1–4. IEEE (2018)
11. Dosenbach, N.U., Nardos, B., Cohen, A.L., Fair, D.A., Power, J.D., Church, J.A., et al.: Prediction of individual brain maturity using FMRI. Science **329**(5997), 1358–1361 (2010)
12. Fox, M.D., Raichle, M.E.: Spontaneous fluctuations in brain activity observed with functional magnetic resonance imaging. Nat. Rev. Neurosci. **8**(9), 700–711 (2007)
13. Griffanti, L., Salimi-Khorshidi, G., Beckmann, C.F., Auerbach, E.J., Douaud, G., Sexton, C.E., et al.: Ica-based artefact removal and accelerated FMRI acquisition for improved resting state network imaging. Neuroimage **95**, 232–247 (2014)

14. Hahn, T., Nierenberg, A., Whitfield-Gabrieli, S.: Predictive analytics in mental health: applications, guidelines, challenges and perspectives. Molecular Psychiatry **22**(1), 37–43 (2017)
15. Hoerl, A.E., Kennard, R.W.: Ridge regression: biased estimation for nonorthogonal problems. Technometrics **42**(1), 80–86 (2000)
16. Kilbertus, N., Ball, P.J., Kusner, M.J., Weller, A., Silva, R.: The sensitivity of counterfactual fairness to unmeasured confounding. arXiv:1907.01040 (2019)
17. Kostro, D., Abdulkadir, A., Durr, A., Roos, R., Leavitt, B.R., Johnson, H., et al.: Correction of inter-scanner and within-subject variance in structural mri based automated diagnosing. NeuroImage **98**, 405–415 (2014)
18. Kusner, M.J., Loftus, J., Russell, C., Silva, R.: Counterfactual fairness. In: Advances in Neural Information Processing Systems, pp. 4066–4076 (2017)
19. Landeiro, V., Culotta, A.: Robust text classification under confounding shift. J. Artif. Intell. Res. **63**, 391–419 (2018)
20. Liem, F., Varoquaux, G., Kynast, J., Beyer, F., Masouleh, S.K., Huntenburg, J.M., et al.: Predicting brain-age from multimodal imaging data captures cognitive impairment. Neuroimage **148**, 179–188 (2017)
21. Lopez-Larson, M.P., Anderson, J.S., Ferguson, M.A., Yurgelun-Todd, D.: Local brain connectivity and associations with gender and age. Dev. Cogn. Neurosci. **1**(2), 187–197 (2011)
22. Miller, D.I., Halpern, D.F.: The new science of cognitive sex differences. Trends in Cognitive Sciences **18**(1), 37–45 (2014)
23. Nooner, K.B., Colcombe, S., Tobe, R., Mennes, M., Benedict, M., Moreno, A., et al.: The nki-rockland sample: a model for accelerating the pace of discovery science in psychiatry. Front. Neurosci. **6**, 152 (2012)
24. Pain, O., Dudbridge, F., Ronald, A.: Are your covariates under control? how normalization can re-introduce covariate effects. Euro. J. Hum. Genet. **26**(8), 1194–1201 (2018)
25. Pervaiz, U., Vidaurre, D., Woolrich, M.W., Smith, S.M.: Optimising network modelling methods for FMRI. NeuroImage **211**, 116604 (2020)
26. Picco, L., Subramaniam, M., Abdin, E., Vaingankar, J.A., Chong, S.A.: Gender differences in major depressive disorder: findings from the singapore mental health study. Singapore Med. J. **58**(11), 649 (2017)
27. Pourhoseingholi, M.A., Baghestani, A.R., Vahedi, M.: How to control confounding effects by statistical analysis. Gastroenterol Hepatol Bed Bench **5**(2), 79 (2012)
28. Rao, A., Monteiro, J.M., Mourao-Miranda, J., Initiative, A.D., et al.: Predictive modelling using neuroimaging data in the presence of confounds. NeuroImage **150**, 23–49 (2017)
29. Ritchie, S.J., Cox, S.R., Shen, X., Lombardo, M.V., Reus, L.M., Alloza, C., et al.: Sex differences in the adult human brain: evidence from 5216 uk biobank participants. Cerebral Cortex **28**(8), 2959–2975 (2018)
30. Ruigrok, A.N., et al.: A meta-analysis of sex differences in human brain structure. Neurosci. Biobehav. Rev. **39**, 34–50 (2014)
31. Salimi-Khorshidi, G., Douaud, G., Beckmann, C.F., Glasser, M.F., Griffanti, L., et al.: Automatic denoising of functional mri data: combining independent component analysis and hierarchical fusion of classifiers. Neuroimage **90**, 449–468 (2014)
32. Satterthwaite, T.D., Elliott, M.A., Gerraty, R.T., Ruparel, K., Loughead, J., Calkins, M.E., et al.: An improved framework for confound regression and filtering for control of motion artifact in the preprocessing of resting-state functional connectivity data. Neuroimage **64**, 240–256 (2013)

33. Seeman, M.V.: Psychopathology in women and men: focus on female hormones. Am. J. Psychiatry **154**(12), 1641–1647 (1997)
34. Smith, S.M., Beckmann, C.F., Andersson, J., Auerbach, E.J., Bijsterbosch, J., Douaud, G., et al.: Resting-state FMRI in the human connectome project. Neuroimage **80**, 144–168 (2013)
35. Snoek, L., Miletić, S., Scholte, H.S.: How to control for confounds in decoding analyses of neuroimaging data. NeuroImage **184**, 741–760 (2019)
36. Stephan, K., Friston, K., Squire, L.: Functional connectivity. Encyclopedia of Neuroscience, pp. 391–397 (2009)
37. Todd, M.T., Nystrom, L.E., Cohen, J.D.: Confounds in multivariate pattern analysis: theory and rule representation case study. Neuroimage **77**, 157–165 (2013)
38. Tripepi, G., Jager, K.J., Dekker, F.W., Zoccali, C.: Stratification for confounding-part 1: the mantel-haenszel formula. Nephron Clin. Pract. **116**(4), c317–c321 (2010)
39. Van Essen, D.C., Ugurbil, K., Auerbach, E., Barch, D., Behrens, T., Bucholz, R., Chang, A., et al.: The human connectome project: a data acquisition perspective. Neuroimage **62**(4), 2222–2231 (2012)
40. Wachinger, C., Becker, B.G., Rieckmann, A., Pölsterl, S.: Quantifying confounding bias in neuroimaging datasets with causal inference. In: Shen, D., Liu, T., Peters, T.M., Staib, L.H., Essert, C., Zhou, S., Yap, P.-T., Khan, A. (eds.) MICCAI 2019. LNCS, vol. 11767, pp. 484–492. Springer, Cham (2019). https://doi.org/10.1007/978-3-030-32251-9_53
41. Weis, S., Patil, K.R., Hoffstaedter, F., Nostro, A., Yeo, B.T., Eickhoff, S.B.: Sex classification by resting state brain connectivity. Cerebral Cortex **30**(2), 824–835 (2020)
42. Werling, D.M., Geschwind, D.H.: Sex differences in autism spectrum disorders. Current Opinion Neurol. **26**(2), 146 (2013)
43. Xu, C., Li, C., Wu, H., Wu, Y., Hu, S., Zhu, Y., et al.: Gender differences in cerebral regional homogeneity of adult healthy volunteers: a resting-state FMRI study. BioMed research international **2015** (2015)
44. Zang, Y., Jiang, T., Lu, Y., He, Y., Tian, L.: Regional homogeneity approach to FMRI data analysis. Neuroimage **22**(1), 394–400 (2004)
45. Zhang, C., Dougherty, C.C., Baum, S.A., White, T., Michael, A.M.: Functional connectivity predicts gender: evidence for gender differences in resting brain connectivity. Human Brain Mapp. **39**(4), 1765–1776 (2018)

Energy Consumption Forecasting Using a Stacked Nonparametric Bayesian Approach

Dilusha Weeraddana[1](✉), Nguyen Lu Dang Khoa[1], Lachlan O'Neil[2], Weihong Wang[1], and Chen Cai[1]

[1] Data61-The Commonwealth Scientific and Industrial Research Organisation (CSIRO), Eveleigh, Australia
dilusha.weeraddana@data61.csiro.au

[2] Energy-The Commonwealth Scientific and Industrial Research Organisation, Newcastle, Australia

Abstract. In this paper, the process of forecasting household energy consumption is studied within the framework of the nonparametric Gaussian Process (GP), using multiple short time series data. As we begin to use smart meter data to paint a clearer picture of residential electricity use, it becomes increasingly apparent that we must also construct a detailed picture and understanding of consumer's complex relationship with gas consumption. Both electricity and gas consumption patterns are highly dependent on various factors, and the intricate interplay of these factors is sophisticated. Moreover, since typical gas consumption data is low granularity with very few time points, naive application of conventional time-series forecasting techniques can lead to severe over-fitting. Given these considerations, we construct a stacked GP method where the predictive posteriors of each GP applied to each task are used in the prior and likelihood of the next level GP. We apply our model to a real-world dataset to forecast energy consumption in Australian households across several states. We compare intuitively appealing results against other commonly used machine learning techniques. Overall, the results indicate that the proposed stacked GP model outperforms other forecasting techniques that we tested, especially when we have a multiple short time-series instances.

Keywords: Nonparametric Bayesian · Gaussian process · Energy forecasting · Prediction interval · Sparse time series data

1 Introduction

1.1 Energy Forecasting Across Major Australian States

Accurately forecasting residential energy usage is critically important for informing network infrastructure spending, market decision making and policy development. With concerns not only over the total volume of consumption within

Y. Dong et al. (Eds.): ECML PKDD 2020, LNAI 12461, pp. 19–35, 2021.
https://doi.org/10.1007/978-3-030-67670-4_2

a time period, but also the maximum 'peak' simultaneous consumption, this becomes a complicated topic to explore, one which the Australian Energy Market Operator (AEMO) explores yearly in their Electricity Statement of Opportunities (ESOO) [7] and Gas Statement of Opportunities (GSOO) [6]. Forecasting energy consumption with high accuracy is a challenging task. The energy consumption time series is non-linear and non-stationary, with daily and weekly cycles. In addition, it is also noisy due to the large loads with unknown hours of operation, special events and holidays, extreme weather conditions and sudden weather changes.

Energy forecasting has been an active area of research. There are two main groups of approaches: statistical, such Autoregressive Integrated Moving Average (ARIMA), exponential smoothing [30] and Machine Learning, with Tree based models, Neural Networks and Bayesian methods being the most popular representative of this group [16,33].

In Australia, interval meters typically record electricity usage in half hour intervals. This provides a rich source of data to develop forecasting models. However, the majority of electricity meters in Australia are basic accumulation meters which simply accumulate the amount of energy consumed over a longer timeframe, typically on a quarterly basis [36,37]. When it comes to gas metering, almost all meters in Australia are basic meters. This low penetration rate of interval meters in the electricity and gas sectors hinders the industry's ability to forecast accurately.

Against this backdrop, this current study aims to develop a method which can be used to successfully predict residential energy consumption for individual households using readings from basic meters (and interval meters) and other available information such as weather or demographics.

Current research has calculated the average 10 year annual energy consumption growth for Australia to be 0.7% for the residential sector, up until the 2016-17 financial year (Table 2.3 in [19]). However, these estimates only consider aggregated load. There is still a poor understanding of changes in individual household energy consumption. This makes it difficult to build robust models which can capture unusual aggregate changes created by the chaotic nature of a system made up of millions of households. As such, the key objective of this work is to collaborate with leading Energy utilities and make use of machine learning techniques to identify future gas and electricity consumption for individual households. In this research, more than 2,500 households are studied across different states of Australia. Each of these households contain some form of electricity and/or gas metering data as well as linked demographics, building characteristics and appliance uptake information gathered through household surveying.

The long-term objective of this work is to better understand the complex nature of Australian consumer's behaviour and provide comprehensive information to underpin key decision making, helping to deliver an affordable and sustainable energy future for Australia.

1.2 Challenges and Related Work

A household's energy consumption depends on numerous factors such as the socio-demographic characteristics and behaviour of occupants, features of the dwelling, and climate/seasonal factors, among many others [2,19,20,22]. In addition, energy load series is highly non-linear and non-stationary, with several cycles that are nested into each other. All these factors combined together make the task of building accurate prediction models challenging. As such, the underlying function which governs the energy demand is fairly sophisticated, making it difficult to obtain a suitable model by looking only at the input data.

In practice, many time series scenarios are best considered as components of some vector-valued time series having not only serial dependence within each component series, but also interdependence between the other series of elements [15]. There are various statistical and machine learning techniques to handle time series predictions. Ensemble learners such as Random Forest and Gradient Boosting are known to be among the most competitive forms of solving time series predictive tasks [31]. The ARIMA (Autoregressive Integrated Moving Average) technique is another popular time series forecasting model which in the past, was mainly used for load forecasting. ARIMA models aim to describe the autocorrelation of the data as well as the error associated with the prediction of the previous time instance [17]. Neural networks have also received a lot of attention recently due to their ability to capture the non-linear relationship between the predictor variables and the target variable [1]. These methods focus on point forecasting (at time t the task is to predict the load value for time $t+d$, where d is the forecasting horizon). Predicting an interval of values with a certain probability instead of just predicting a single value gives more information about the variability of the target variable [34].

GPs are nonparametric Bayesian models which have been employed in a large number of fields for a diverse range of applications [4,32]. The work reported by Swastanto [40] highlights the inflexibility of using parametric models in time series forecasting, thus encouraging the use of nonparametric models. In the nonparametric model, training data is best fitted when constructing the mapping function, whilst maintaining some ability to generalise to unseen data. Nonparametric models use a flexible number of parameters, and the number of parameters are allowed to grow as it learns from more data [25]. Therefore, a nonparametric prediction method such as GP is more suitable [35] for applications such as energy demand forecasting. Such methods allow the data to speak for itself by requiring minimal assumptions on the data. Further to this, additive models are also popular nonparametric regression methods used for time series forecasting which combine adaptive properties of nonparametric methods and the usability of regression models [41].

Regression can face issues when there are too few training data points available to learn a good model. Trying to learn a flexible model with many parameters from a few data points may result in over-fitting where the model mistakes artefacts of the specific available samples as actual properties of the underlying distribution [12,13,42]. We encounter this issue when forecasting energy consumption across Australia, especially when only a few training data points are available for each household. The datasets provide sparse and often irregularly sampled time series instances. Though this is a challenge, there is potential to use the energy consumption data of similar households to help learn the consumption model of each individual household.

There has been a lot of work in recent years on time series prediction with GP [13,21,28]. Most of the frameworks developed on GPs have focused on using large datasets [11]. In the current study, however, we focus on the other frontier of GP: when the dataset consists of a considerably large number of short independent instances. The model that we propose takes inspiration from the construction of stacked Gaussian processes [10,29] and multi-tasking Gaussian Process learning [12], where they take the form of a hierarchical multi-layered Gaussian process, with a stacked kernel.

1.3 Contribution of Our Approach

We construct a simple yet effective Gaussian process framework in the domain of nonparametric supervised learning regression for time series forecasting. In the proposed stacked GP model, the predictive posteriors of each GP applied on each instance are embedded in the prior and likelihood of the next level stacked GP. Furthermore, kernel functions are designed to cater for the time-varying seasonal patterns, and shared across each instance. Therefore, with our framework, it is possible to extract the underlying structure from the data, even when using short time series data. Another strength of our model is its analytical simplicity, which provides a clear insight into how energy consumption relates to the time series, spatial disparity and other demographic data. Therefore, the model is capable of capturing the seasonal variation of energy consumption along with other related household data. Furthermore, the proposed model has a flexible modelling capability through kernel modification. Hence, it can model a time series with a broad range of complexity.

As our model is based on GP which is a special class of Bayesian probabilistic modelling, it provides a distribution instead of a point forecast. Therefore, we will have a point forecast which can be obtained by the mean of the distribution and its uncertainty (the variance of the distribution). As a result, for each point that is predicted with a GP, we are given the perceived uncertainty of that prediction.

2 Overview of Gaussian Process

GP sets an assumption that for every finite set of points x_i in X, the prior distribution of the vector $(f(x_1), ..., f(x_n))$ is multivariate Gaussian. Thus, the response function $f : X \rightarrow Y$ is 'a-priori' modelled by the sample path of a GP. Here, we assume some multivariate Gaussian distribution can be used to represent any observation in our dataset. This nonparametric model has a much less restrictive assumption over the function f, where we only assume that the function is sufficiently smooth. This smoothness of the function can be defined by tuning the hyper-parameters of a Gaussian process regression. Such a Gaussian distribution can be characterised by defining the mean vector and the covariance function.

We have n observations such that $\mathbf{y} = \{y_1, y_2, ..., y_n\}$ which correspond to set of inputs $\mathbf{X} = \{\mathbf{x_1}, \mathbf{x_2}, ..., \mathbf{x_n}\}$ where each $\mathbf{x_i}$ is a d-dimensional data point. Assuming that the noise or uncertainty when making an observation can be represented by a Gaussian noise model which has a zero mean and σ_n variance, any observation can be written as,

$$y_i = f(\mathbf{x_i}) + \mathcal{N}(0, \sigma_n^2). \tag{1}$$

Gaussian distributions are completely parameterized by their mean $m(\mathbf{x})$ and the covariance function $k(\mathbf{x_1}, \mathbf{x_2})$, defined as

$$m(\mathbf{x}) = \mathbb{E}[f(\mathbf{x})], \quad k(\mathbf{x_1}, \mathbf{x_2}) = cov\ (f(\mathbf{x_1}), f(\mathbf{x_2}))$$

Any collection of function values has a joint Gaussian distribution

$$[f(\mathbf{x_1}), f(\mathbf{x_1}) ...., f(\mathbf{x_n})]^\top \sim \mathcal{N}(\mu, K) \tag{2}$$

where the elements in the covariance matrix $\mathbf{K}(\mathbf{X}, \mathbf{X})$ are covariance functions between all training inputs, has entries $\mathbf{K}(\mathbf{X}, \mathbf{X})_{ij} = k(\mathbf{x}_i, \mathbf{x}_j)$, and the mean μ have entries $\mu_i = m(\mathbf{x}_i)$. The properties of the functions such as smoothness and periodicity are determined by the kernel function, which should be any symmetric and positive semi-definite function [43]. In contrast to regression methods where the goal is to find $f(\mathbf{x_i})$, in GP we try to predict y_i whose expectation value is equal to that of $f(\mathbf{x_i})$. Consider that the covariance function (k) is defined using some hyperparameters, θ. Training or learning a GP regression model means determining the optimum set of hyperparameters in the covariance functions.

Standard gradient-based techniques such as conjugate gradient or quasi-Newton methods can be used to determine the best parameters. The prediction for some test inputs $\mathbf{X}_*$ is given by mean $\mu_\mathbf{T}$ and the uncertainty of the prediction is captured by its variance $\mathbf{\Sigma_T}$ as shown in Eq. (3).

$$\begin{aligned} \mu_\mathbf{T} &= \mathbf{K}(\mathbf{X}_*, \mathbf{X})[\mathbf{K}(\mathbf{X}, \mathbf{X}) + \sigma_n^2\mathbf{I}]^{-1}\mathbf{y} \\ \mathbf{\Sigma_T} &= \mathbf{K}(\mathbf{X}_*, \mathbf{X}_*) - \mathbf{K}(\mathbf{X}_*, \mathbf{X})[\mathbf{K}(\mathbf{X}, \mathbf{X}) + \sigma_n^2\mathbf{I}]^{-1}\mathbf{K}(\mathbf{X}, \mathbf{X}_*) + \sigma_n^2\mathbf{I} \end{aligned} \tag{3}$$

3 Proposed Data Analytic Model for Energy Prediction

3.1 Feature Selection

Feature selection is the process of choosing a set of informative variables that are necessary and sufficient for an accurate prediction. Selection of a suitable set of features is one of the critical factors for successful prediction [24]. Factor analysis can measure the correlation between energy consumption based on the comprehensive data and a large range of factors (including environmental, demographic and other dwelling-specific factors) [26]. While a significant amount of literature exists [19,20] on the factors affecting household energy use, this step is critical to discern which of these causes would explain the most variance when forecasting. In our case, the features selected are the previous energy consumption (gas or electricity) of each dwelling, weather information such as air temperature (weather data was collected from Bureau of Meteorology [9]), HDD (heating degree day), and CDD (cooling degree day) [38]; and other demographic and dwelling information (income level, number of rooms etc.). By linking variables such as demographics and weather, we can aim to accurately forecast future energy use.

Data related to household energy consumption is held by numerous parties. It is also formatted to different standards and access is often restricted. In our previous research, we have conducted single and multi-factor analysis on various such data sets obtained from different energy bodies. However, in this paper we specifically focus on a dataset obtained from the Australian Energy Regulator's (AER's) 2017 Electricity Bill Benchmark's survey [8].

We have studied factor analysis extensively on the AER data and other datasets in terms of both gas and electricity consumption. A few examples for demographic and weather related factor analysis outcomes are illustrated in Fig. 1 (a)–(c). Figure 1 (a) clearly shows that there is a gradual increase in household gas consumption with respect to total household income per annum. Figure 1 (b) shows the association between household electricity consumption and distance from the sea for dwellings across suburbs in New South Wales (NSW) and Victoria (VIC). It appears that dwellings located further from the sea tend to have higher levels of electricity consumption. It can be observed that the gas consumption is higher during the winter period and gradually decreases with increases in the temperature. This effect is due predominantly due to gas space heating. However, the lower temperatures may also increase the amount of energy consumed by appliances such as gas hot water systems. The relationship between air temperature and gas consumption is illustrated in Fig. 1 (c) for a given dwelling. It can be observed that the gas consumption is higher during the winter season and gradually decreases with the increase of the temperature. This effect is due predominantly due to gas space heating. However, the lower temperatures also increase the amount of energy consumed by appliances such as gas hot water systems.

To this end, we have also identified some of the potential advantages of including these additional factors in our analysis to help explain variation in the energy consumption of different households across various Australian regions. We have the opportunity to explore similarities and differences across regions and measure the effect of various factors on household energy use. This will allow us to develop an improved framework for energy prediction.

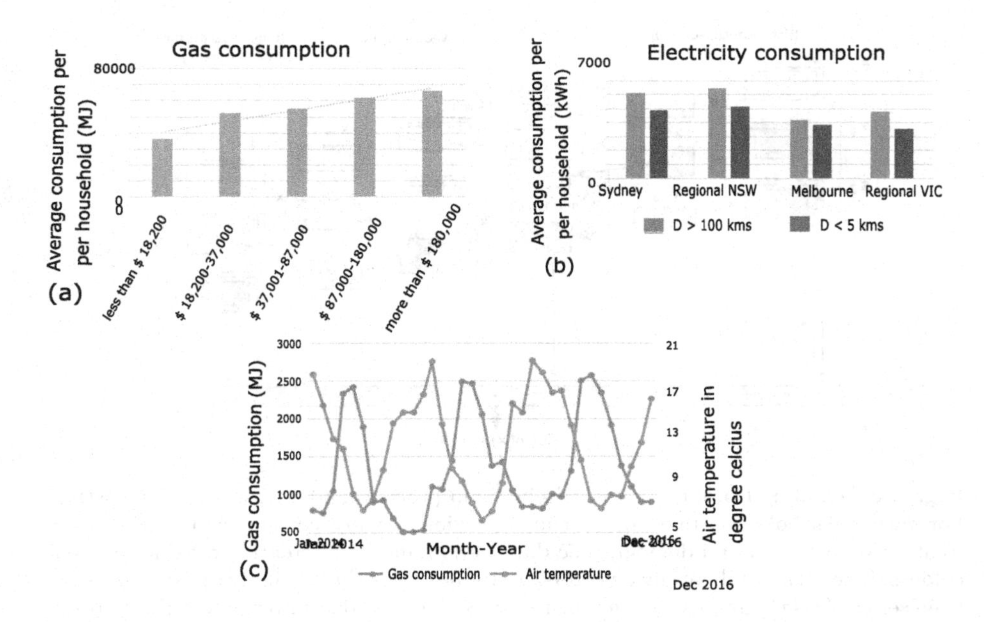

Fig. 1. (a) Average annual gas consumption for different levels of household income, (b) Average annual electricity consumption across regions in NSW and VIC based on distance to the sea, where D denotes distance from the dwelling to the sea in kiolmetres, (c) Relationship between air temperature and household gas consumption over time for a selected dwelling in Australia. In Australia, December-February is summer season and June-August is winter season.

3.2 Modelling for Energy Forecasting

We propose a stacked GP model which combines different models to produce a meta-model with a better predictive performance than the constituent parts. In the context of energy consumption forecasting, our meta-model fuses the seasonally varying pattern of the energy consumption of each household with the other information related to household consumption. Thus, our goal is to fuse data from multiple households with a GP to fully exploit the information contained in the covariates and model spatio-temporal correlations. Figure 2 illustrates the proposed stacked GP model for energy consumption forecasting, where the mean and variance vectors of the predictive posteriors of a set of time series GPs are embedded into the prior and likelihood of the ensemble GP as: $\mathrm{GP}(m(\mathbf{x}), k(\mathbf{x}, \mathbf{x}'))$.

We will begin by computing the posterior over all the random functions. As the likelihood and prior are Gaussian, the posterior over functions is also another Gaussian given by:

$$p(\mathbf{f_t}|\mathbf{T}, \mathbf{y_t}) \sim \mathcal{N}(\mathbf{f_t}|\bar{\mu}, \bar{\mathbf{\Sigma}}) \tag{4}$$

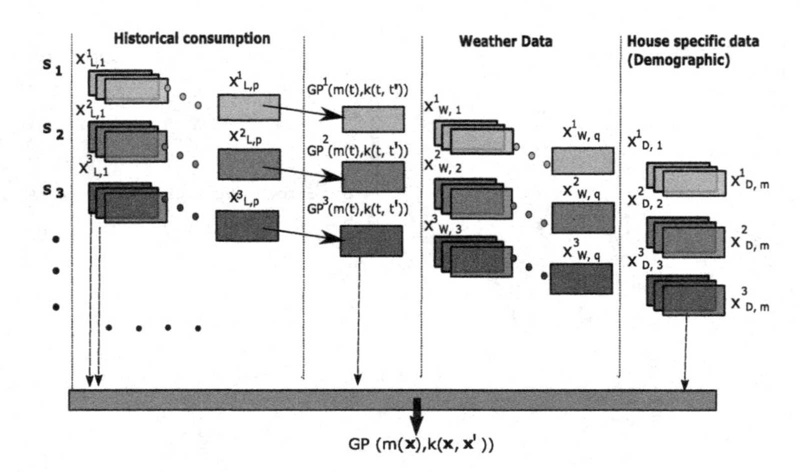

Fig. 2. Schematic of proposed stacked Gaussian process model for energy forecasting. For each household, we take into account previous energy consumption data ($x^i_{L,z}$), weather data ($x^i_{W,z}$) and demographic data ($x^i_{D,z}$) and is differentiated using different colours. The mean and variance vectors computed from the predictive posterior of the time series GP (x^i_t) applied to each instance, S_i is embedded into the feature vector of the proposed stacked GP model.

where

$$\begin{aligned} \bar{\mu} &= \mathbf{K}(\mathbf{T},\mathbf{T})[\mathbf{K}(\mathbf{T},\mathbf{T}) + \sigma_n^2\mathbf{I}]^{-1}\mathbf{y_t} \\ \bar{\mathbf{\Sigma}} &= \mathbf{K}(\mathbf{T},\mathbf{T}) - \mathbf{K}(\mathbf{T},\mathbf{T})[\mathbf{K}(\mathbf{T},\mathbf{T}) + \sigma_n^2\mathbf{I}]^{-1}\mathbf{K}(\mathbf{T},\mathbf{T}) + \sigma_n^2\mathbf{I} \end{aligned} \tag{5}$$

where $\mathbf{f_t}$ is a vector containing all the random functions evaluated at training data input vector, $\mathbf{f_t} = [\mathbf{f(t_1)}, \mathbf{f(t_2)}..., \mathbf{f(t_n)}]^\top$. $\mathbf{T}$ is the time vector and $\mathbf{y_t}$ denotes the training response variable (in our case energy consumption). To add some flexibility to capture time varying seasonality, a local kernel (squared-exponential kernel) is combined with a periodic kernel. This will allow us to model functions that are only locally periodic and change over time. $k(t,\bar{t})$ is given by

$$k(t,\bar{t}) = \sigma^2 \exp\Big(\frac{-2\sin^2(\pi|t-\bar{t}|/p)}{l^2}\Big) + \exp\Big(\frac{-(t-\bar{t})^2}{2l^2}\Big) \tag{6}$$

where σ, l, p denote hyper-parameters, t represents the time, and $*$ denotes the testing data. We can now predict a household's energy consumption at any given point in time by computing the predictive posterior:

$$p(\mathbf{y}_{*,\mathbf{t}}|\mathbf{T}_*, \mathbf{T}, \mathbf{y_t}) = \int p(\mathbf{y}_{*,\mathbf{t}}|\mathbf{T}_*, \mathbf{f_t}, \mathbf{T}) p(\mathbf{f_t}|\mathbf{T}, \mathbf{y_t}) \mathbf{df_t} \tag{7}$$

$$p(\mathbf{y}_{*,\mathbf{t}}|\mathbf{T}_*, \mathbf{T}, \mathbf{y_t}) \sim \mathcal{N}(\mathbf{y}_{*,\mathbf{t}}|\bar{\mu}_{*,\mathbf{t}}, \bar{\mathbf{\Sigma}}_{*,\mathbf{t}}) \tag{8}$$

This is applied for each task (denoted by i, where task is a sample time series drawn from one dwelling) and the mean and variance are calculated from the below formula:

$$\bar{\mu}^{i}_{*,t} = \mathbf{K}^{i}(\mathbf{T}_{*}, \mathbf{T})[\mathbf{K}^{i}(\mathbf{T}, \mathbf{T})]^{-1}\mathbf{y_t}^{i} \tag{9}$$

$$\bar{\Sigma}_{*,t} = \mathbf{K}(\mathbf{T}_{*}, \mathbf{T}_{*}) - \mathbf{K}(\mathbf{T}_{*}, \mathbf{T})[\mathbf{K}(\mathbf{T}, \mathbf{T}) + \sigma_n^2\mathbf{I}]^{-1}\mathbf{K}(\mathbf{T}, \mathbf{T}_{*}) + \sigma_n^2\mathbf{I} \tag{10}$$

The mean $\bar{\mu}^{i}_{*,t}$ and variance $\bar{\Sigma}^{i}_{*,t}$ obtained from the predictive posterior of each task are taken into the next stage,

$$(\bar{\mu}_{*,\mathbf{t}}, \bar{\boldsymbol{\Sigma}}_{*,\mathbf{t}}) = \{(\bar{\mu}^{1}_{*,t}, \bar{\Sigma}^{1}_{*,t})..,(\bar{\mu}^{i}_{*,t}, \bar{\Sigma}^{i}_{*,t})..,(\bar{\mu}^{n}_{*,t}, \bar{\Sigma}^{n}_{*,t})\} = \{(\bar{\mu}^{\mathbf{Tr}}_{*,\mathbf{t}}, \bar{\boldsymbol{\Sigma}}^{\mathbf{Tr}}_{*,\mathbf{t}}), (\bar{\mu}^{\mathbf{Te}}_{*,\mathbf{t}}, \bar{\boldsymbol{\Sigma}}^{\mathbf{Te}}_{*,\mathbf{t}})\}$$

$$\mathbf{X}^{\mathbf{Tr}}_{\mathbf{t}} = (\bar{\mu}^{\mathbf{Tr}}_{*,\mathbf{t}}, \bar{\boldsymbol{\Sigma}}^{\mathbf{Tr}}_{*,\mathbf{t}}) \text{ if MPE} < \tau, \quad \mathbf{X}^{\mathbf{Te}}_{\mathbf{t}} = (\bar{\mu}^{\mathbf{Te}}_{*,\mathbf{t}}, \bar{\boldsymbol{\Sigma}}^{\mathbf{Te}}_{*,\mathbf{t}}) \tag{11}$$

Tr and Te represent training and testing data. Only the tasks with MPE (Mean Percentage Error) $< \tau$ are considered for model training. τ can be learnt from the training data. $\mathbf{X}^{\mathbf{Tr}}_{\mathbf{t}}, \mathbf{X}^{\mathbf{Te}}_{\mathbf{t}}$ are then used in the prior and likelihood of the predictive posterior in the next stage.

At the next stage of our proposed stacked model, let's state that the function f_{st} is random and drawn from a multivariate Gaussian distribution, where i and j represent different tasks (or dwellings). As such:

$$\begin{bmatrix} f_{st}(\mathbf{x}_i) \\ f_{st}(\mathbf{x}_j) \\ \vdots \end{bmatrix} \sim \mathcal{N}\left(\begin{bmatrix} m_{st}(\mathbf{x}_i) \\ m_{st}(\mathbf{x}_j) \\ \vdots \end{bmatrix}, \begin{bmatrix} k_{st}(\mathbf{x}_i, \mathbf{x}_i) & k_{st}(\mathbf{x}_i, \mathbf{x}_j) & \cdots \\ k_{st}(\mathbf{x}_j, \mathbf{x}_i) & k_{st}(\mathbf{x}_j, \mathbf{x}_j) & \cdots \\ \vdots & \vdots & \ddots \end{bmatrix}\right) \tag{12}$$

where m and K are the mean and covariance function, respectively. Since the function f_{st} is random, the optimum f_{st} will be inferred via the Bayesian inference. Therefore, we can compute the predictive posterior over all the random functions considering all the tasks.

$$p(\mathbf{y}_{*,\mathbf{st}}|\mathbf{X}_{*}, \mathbf{X}, \mathbf{y}) = \int p(\mathbf{y}_{*,\mathbf{st}}|\mathbf{X}_{*}, \mathbf{f_{st}}, \mathbf{X})p(\mathbf{f_{st}}|\mathbf{X}, \mathbf{y})\mathbf{df_{st}} \tag{13}$$

here $\mathbf{y}_{*,\mathbf{st}}$ is the predictive posterior of the proposed stacked GP.
$\mathbf{X} = \{\mathbf{X}^{\mathbf{Tr}}_{\mathbf{L}}, \mathbf{X}^{\mathbf{Tr}}_{\mathbf{W}}, \mathbf{X}^{\mathbf{Tr}}_{\mathbf{D}}, \mathbf{X}^{\mathbf{Tr}}_{\mathbf{LT}}, \mathbf{X}^{\mathbf{Tr}}_{\mathbf{t}}\}$, representing training data for historical load ($\mathbf{X}^{\mathbf{Tr}}_{\mathbf{L}}$), weather ($\mathbf{X}^{\mathbf{Tr}}_{\mathbf{W}}$), demographic ($\mathbf{X}^{\mathbf{Tr}}_{\mathbf{D}}$), lag time ($\mathbf{X}^{\mathbf{Tr}}_{\mathbf{LT}}$) and mean predictive posterior of time series GP ($\mathbf{X}^{\mathbf{Tr}}_{\mathbf{t}}$) respectively. $\mathbf{X}_{*}$ denotes the testing data. The predictive distribution is again Gaussian, with a mean $\mu_{*,\mathbf{st}}$, and covariance $\boldsymbol{\Sigma}_{*,\mathbf{st}}$.

$$p(\mathbf{y}_{*,\mathbf{st}}|\mathbf{X}_{*}, \mathbf{X}, \mathbf{y}) \sim \mathcal{N}(\mathbf{y}_{*,\mathbf{st}}|\mu_{*,\mathbf{st}}, \boldsymbol{\Sigma}_{*,\mathbf{st}}) \tag{14}$$

where, $\mu_{*,\mathbf{st}}, \mathbf{\Sigma}_{*,\mathbf{st}}$ can be obtained from Eq.(3).

The following proposed kernel is used to model the inter-sample relationship.

$$k_{st}(i,j) = \prod_{z=1}^{z=p} k(x^i_{\mathrm{L},z}, x^j_{\mathrm{L},z}) \times \prod_{z=1}^{z=q} k(x^i_{\mathrm{W},z}, x^j_{\mathrm{W},z}) \\ \times \prod_{z=1}^{z=m} k(x^i_{\mathrm{D},z}, x^j_{\mathrm{D},z}) \times k(x^i_{\mathrm{LT}}, x^j_{\mathrm{LT}}) \times k(x^i_{\mathrm{t}}, x^j_{\mathrm{t}}) \tag{15}$$

Here, $k(x^i_{\mathrm{L},z}, x^j_{\mathrm{L},z}), k(x^i_{\mathrm{W},z}, x^j_{\mathrm{W},z}), k(x^i_{\mathrm{D},z}, x^j_{\mathrm{D},z}), k(x^i_{\mathrm{LT}}, x^j_{\mathrm{LT}})$, and $k(x^i_{\mathrm{t}}, x^j_{\mathrm{t}})$ denote covariates generated by historical consumption load, weather data, demographic data, time stamp of the lag time and the mean values obtained from the time series GP respectively. p is the number of months, q is the total number of weather data points (e.g.. $x^i_{\mathrm{L},z}$ denotes the energy consumption of household i in the zth month). These notations are illustrated in Fig. 2. Intuitively, if two tasks, $\mathbf{x}_i$ and $\mathbf{x}_j$ are similar, then $f_{st}(\mathbf{x}_i)$ and $f_{st}(\mathbf{x}_j)$ should also be similar, which explains why the function generated by a GP is smooth. Our assumption of this similarity or smoothness is encoded by the kernel function $k(\mathbf{x}_i, \mathbf{x}_j)$.

4 Experimental Setup

4.1 Data Description and Preparation

In our study, we have used a dataset obtained from the AER to demonstrate how our stacked GP model learns spatio-temporal patterns in the data. We investigate two cases: gas consumption and electricity consumption forecasting. Our primary focus is on gas use forecasting where we have short but multiple time series data (gas data is collected from basic meters, thus the dataset is sparse. To overcome this issue, we have drawn multiple short time series data for our experiments). Nevertheless, in order to further demonstrate the capability of our model, we have also applied the proposed stacked GP model on household electricity data.

1. **Case 1:** The AER gas consumption dataset consists of quarterly gas consumption data for approximately 2600 dwellings, with data spanning from year 2013–2016. Though this dataset consists of quarterly consumption readings, the duration of a quarter is not consistent across years nor within individual household. Thus, a measure of monthly gas usage was generated based on daily average usage. Thus, monthly gas usage was generated based on daily average usage. Table 1 shows the statistics for our experimental set-up, including the mean gas consumption and number of samples in the training and testing data split across major states (e.g.. NSW, VIC, QLD, ACT) and focusing only on the state of Victoria (VIC). Our reason for examining Victoria separately from other states (i.e. 'Overall' in Table 1) is due to AEMO's current interest in understanding VIC's energy consumption as a stepping stone to understanding Australia's consumption as a whole. This is in part

because smart meter data is in such high abundance in Victoria due to the state government's smart meter roll-out [39] and in part due to the high penetration of gas appliances in Victoria. We then split the overall population of the data into training and testing based on the year. The years 2013–2015 are used for training and the year 2016 is used for testing.

2. **Case 2:** AER smart meter electricity data consists of about 3100 households situated in VIC and NSW. The data spans from year 2015–2016. We generated monthly electricity consumption for each household based on half-hourly consumption data. Table 1 provides detailed statistics of the experimental set-up.

Evaluation Metrics. To measure predictive accuracy, we use two standard performance measures: Mean Absolute Error (MAE) and R-squared (R^2).

$$\mathrm{MAE} = \frac{1}{n}\sum_{i=1}^{n} |y_i - f_i| \tag{16}$$

$$\mathrm{R}^2 = 1 - \frac{\sum_{i=1}^{n}(y_i - f_i)^2}{\sum_{i=1}^{n}(y_i - \bar{y})^2} \tag{17}$$

where y_i and f_i are the actual and predicted gas or electricity consumption at time i, and n is the total number of predicted loads.

4.2 Baselines and Other Machine Learning Models Used for Comparison

We compare our proposed stacked GP method against the following commonly used machine learning techniques. We use the same feature set for all the methods.

1. Random Forest (RF): RF is a supervised learning algorithm which builds multiple decision trees to obtain more accurate and stable results [14].
2. Gradient Boosting (GB): GB uses a boosting technique which sequentially ensembles trees into a single strong learner [23].
3. Multivariate Long Short-Term Memory (LSTM): This is a recurrent neural network (RNN) architecture which consists of feedback connections and widely used in time series forecasting [27].
4. Auto-regressive (AR): AR models learn from a series of timed steps and takes measurements from previous actions [3].
5. Auto-regressive integrated moving average (ARIMA): ARIMA uses the regression error as a linear combination of error terms whose values occurred contemporaneously and at various times in the past [18].
6. Traditional time-series GP: Application of GP to a time series sequence, where predictor variable is a sequence of time instances and response variable is energy consumption (in our case). This method is also described from Eq. 4–7, in Sect. 3.2.

Table 1. Data statistics for gas and electricity consumption.

Training/Testing	Gas		Electricity	
	Region		Region	
	VIC	Overall	VIC	NSW
Training load mean	4100 MJ	3600 MJ	576 kWh	560 kWh
Testing load mean	4500 MJ	3700 MJ	409 kWh	527 kWh
Training samples	4586	7481	6031	3499
Testing samples	8882	15774	5650	2191

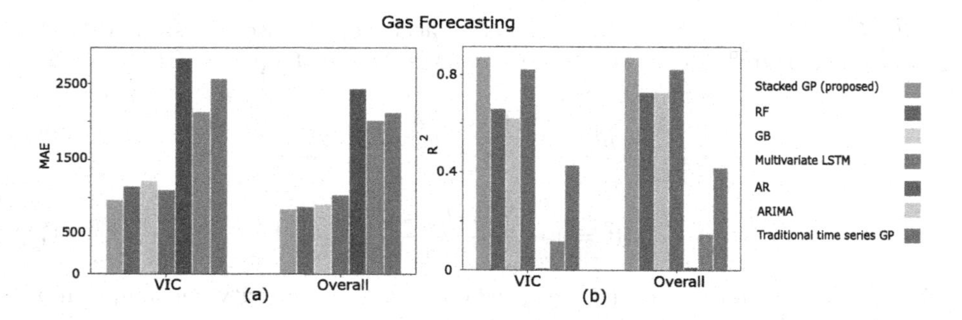

Fig. 3. Results: (a) MAE (Mean Absolute Percentage Error) and (b) R^2 values for gas consumption forecasting, first for the state of VIC and then overall across Australia. For both MAE and R^2 metrics, the proposed Stacked GP method produced the best results (i.e. smaller MAE and higher R^2) when assessing both Victoria only and Overall.

5 Results and Discussion

In energy forecasting, we are primarily interested in the predictive accuracy of the model rather than the parameter of the function. Therefore, instead of inferring the parameters to get the latent function of interests (as in parametric methods), here we infer the function f_{st} given in Eq. (13) directly, as if the function f_{st} is the 'parameters' of our model. We have inferred the function f_{st} from the data through the Bayesian inference to obtain a probability distribution.

We performed the time series GP prediction by applying a single GP on each data sample, as given in Eq. (4–9). In order to capture the periodic random functions that vary over time, we used an exponential kernel with a periodic kernel as given in Eq. (6). Training samples of MPE less than $\tau = 1$ are taken to the next stage and embedded in the prior and likelihood in the proposed stacked GP, as given in Eq. (13).

To evaluate the effectiveness of the proposed method, we compared its predictive performance with several machine learning based and naive techniques (Sect. 4.2). Grid search is used for hyper-parameter tuning for all the machine learning techniques. The lowest MAE is reported for the proposed stacked GP method (see Fig. 3 (a) and Fig. 4 (a)). In terms of the R^2 metric, the proposed

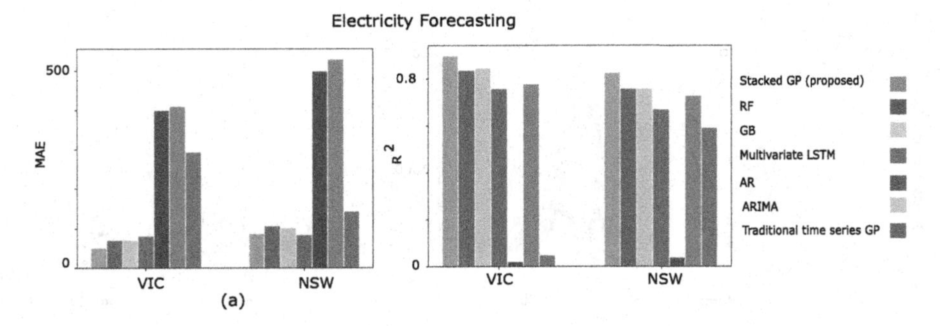

Fig. 4. Results: (a) MAE and (b) R^2 values for VIC and NSW electricity consumption forecasting. For both MAE and R^2 metrics, the proposed Stacked GP method produced the best results (i.e. smaller MAE and higher R^2) when assessing both VIC and NSW

method outperforms all the other techniques as shown in the Fig. 3 (a) and Fig. 4 (b).

Further to this, for illustrative purposes, we have depicted the average mean and variance of the proposed model output from all the dwellings across Australia (Fig. 5 (a)) and those located in VIC (Fig. 5 (b)), for gas consumption. These figures demonstrate the ability of the proposed model to capture the trend in household consumption patterns. The latest report from AEMO [5], states that the accuracy of AEMO's 2017–18 annual operational consumption forecast is approximately $R^2 = 0.80$. Though this R^2 value is lower than the one reported in this paper, AEMO's value provides forecasting accuracies for state-level energy consumption. This is in contrast to our research based on household-level energy forecasting. As such these may not be directly comparable.

It can also be observed that popular time series forecasting methods such as AR and ARIMA do not perform well in our case, as we have only limited amount of time series data for each task. The shortcomings of these two techniques are caused by their limited assumptions over the underlying function. The ARIMA model assumes that y_{t+1} is linear in the past data and past errors. These linear assumptions are inadequate to model a complex forecasting problem and thus a more powerful model is needed.

It is also important to note that the application of time series GP on individual households provides weak performance metrics due to the limited number of training data samples. This is illustrated in Fig. 5 (c) with the aid of a median performing dwelling, where the mean prediction and the variance of the prediction from the proposed stacked GP are compared against the output from the traditional time-series GP. The main motivation of this work is the handling of multiple short time series tasks which require several decisions in terms of how we describe the dynamics of the observed values of other related tasks. Settling on a single answer to these decisions or assumptions over the underlying functions may be risky in real world time series where one frequently observes changes in properties such as demographics or the weather. The proposed stacked GP

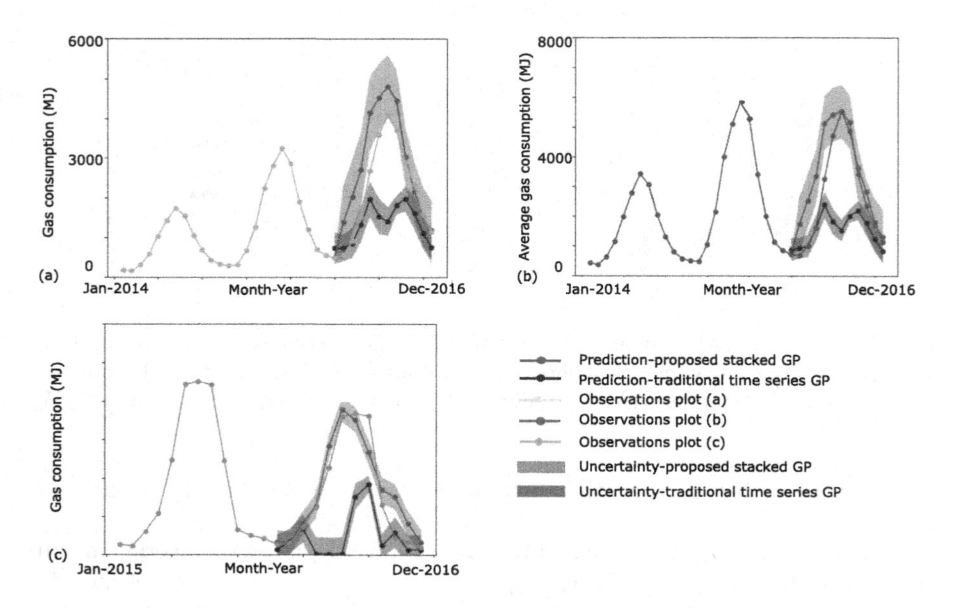

Fig. 5. Results comparison of the proposed method against the traditional time series GP. (a) Average overall gas consumption per dwelling in Australia, (b) Average overall gas consumption per dwelling in VIC, Australia. 2014–2015 data used for training and 2016 data used for model evaluation (in (a) and (b) monthly average of the mean and variance of proposed stacked GP output of dwellings across Australia and dwellings located only in VIC are considered), (c) Gas consumption model testing for a median performing dwelling.

method which uses mean and variance from time series GP and considers the inter-household relationship shows superior performance compared to the other techniques considered.

6 Conclusion

In this paper, we proposed a nonparametric Bayesian model which uses Gaussian Process (GP) with an ensemble learning approach to forecast energy consumption at a household level across major states in Australia. Our proposed stacked GP model inherits the advantages from both time series GP and multi-dimensional GP, without any prior knowledge about the underlying functional interplay among the latent temporal dynamics. This model was formulated based on the GP predictive distribution over the secondary tasks given the primary task. Here, the primary tasks are formulated as a set of time series functions and the secondary task models the inter-sample relationship using task related descriptors. Experiments on real-world datasets demonstrate that the proposed model is effective and robust to model spatio-temporal patterns in time series data and performs consistently better compared to other machine learning and time series forecasting techniques.

This is one of our ongoing predictive analytic projects in the energy domain, which is being carried-out in close collaboration with two leading government bodies in Australia. This highlights Australia's efforts in using machine learning to provide meaningful information that will unlock the mysteries of Australia's energy behaviour – and help to deliver an efficient energy future.

Acknowledgement. We sincerely thank AEMO for providing valuable feedback on this research and AER for providing access to the valuable dataset used herein. We would also like to thank Dr. Elisha Frederiks (Energy-CSIRO), Peter Goldthorpe (Energy-CSIRO) and Dr. Nariman Mahdavi (Energy-CSIRO) for their constructive and insightful feedback to the manuscript.

References

1. Abdel-Nasser, M., Mahmoud, K.: Accurate photovoltaic power forecasting models using deep LSTM-RNN. Neural Comput. Appl. **31**(7), 2727–2740 (2019)
2. Abrahamse, W., Steg, L.: How do socio-demographic and psychological factors relate to households' direct and indirect energy use and savings? J. Econ. Psychol. **30**(5), 711–720 (2009)
3. Akaike, H.: Fitting autoregressive models for prediction. Ann. Inst. Stat. Math. **21**(1), 243–247 (1969)
4. Ambrogioni, L., Maris, E.: Complex-valued gaussian process regression for time series analysis. Signal Process. **160**, 215–228 (2019)
5. Australian Energy Market Operator: Aemo forecasting accuracy report **2018**, December 2018 (2018). https://www.aemo.com.au/Electricity/National-Electricity-Market-NEM/Planning-and-forecasting/Forecasting-Accuracy-Reporting
6. Australian Energy Market Operator: 2019 Gas Statement of Opportunities, March 2019 (2019) https://www.aemo.com.au/Media-Centre/2019-Gas-Statement-of-Opportunities-released
7. Australian Energy Market Operator: 2019 NEM Electricity Statement of Opportunities, August 2019 (2019). https://www.aemo.com.au/Media-Centre/2019-NEM-Electricity-Statement-of-Opportunities
8. Australian Energy Regulator: Electricity and gas bill benchmarks for residential customers **2017** (2017)
9. Australian Government Bureau of Meteorology: Australian Government Bureau of Meteorology (October 2018) (2018) http://www.bom.gov.au/
10. Bhatt, S., Cameron, E., Flaxman, S.R., Weiss, D.J., Smith, D.L., Gething, P.W.: Improved prediction accuracy for disease risk mapping using gaussian process stacked generalization. J. Royal Soc. Interface **14**(134), 20170520 (2017)
11. Binois, M., Gramacy, R.B., Ludkovski, M.: Practical heteroscedastic gaussian process modeling for large simulation experiments. J. Comput. Graph. Stat. **27**(4), 808–821 (2018)
12. Bonilla, E.V., Chai, K.M., Williams, C.: Multi-task gaussian process prediction. In: Advances in Neural Information Processing Systems, pp. 153–160 (2008)
13. Brahim-Belhouari, S., Bermak, A.: Gaussian process for nonstationary time series prediction. Comput. Stat. Data Anal. **47**(4), 705–712 (2004)

14. Breiman, L.: Random forests. Machine Learn. **45**(1), 5–32 (2001)
15. Brockwell, P.J., Davis, R.A.: Introduction to Time Series and Forecasting. STS. Springer, Cham (2016). https://doi.org/10.1007/978-3-319-29854-2
16. Cheng, Y.Y., Chan, P.P., Qiu, Z.W.: Random forest based ensemble system for short term load forecasting. In: 2012 International Conference on Machine Learning and Cybernetics, vol. 1, pp. 52–56. IEEE (2012)
17. Contreras, J., Espinola, R., Nogales, F.J., Conejo, A.J.: Arima models to predict next-day electricity prices. IEEE Trans. Power Syst. **18**(3), 1014–1020 (2003)
18. Das, S.: Time Series Analysis. Princeton University Press, Princeton (1994)
19. Department of the Environment and Energy: Australian energy update 2018. Commonwealth of Australia **2018**, September 2018 (2018). https://www.energy.gov.au/publications/australian-energy-update-2018
20. Energy Rating: Report: Energy Use in the Australian Residential Sector 1986–2020, January 2008 (2018). http://www.energyrating.gov.au/document/report-energy-use-australian-residential-sector-1986-2020
21. Filip, S., Javeed, A., Trefethen, L.N.: Smooth random functions, random odes, and gaussian processes. SIAM Review **61**(1), 185–205 (2019)
22. Frederiks, E., Stenner, K., Hobman, E.: The socio-demographic and psychological predictors of residential energy consumption: a comprehensive review. Energies **8**(1), 573–609 (2015)
23. Friedman, J.H.: Stochastic gradient boosting. Comput. Stat. Data Anal. **38**(4), 367–378 (2002)
24. Guyon, I., Elisseeff, A.: An introduction to variable and feature selection. J. Machine Learn. Res. **3**, 1157–1182 (2003)
25. Hjort, N.L., Holmes, C., Müller, P., Walker, S.G.: Bayesian nonparametrics, vol. 28. Cambridge University Press (2010)
26. Jones, R.V., Fuertes, A., Lomas, K.J.: The socio-economic, dwelling and appliance related factors affecting electricity consumption in domestic buildings. Renew. Sustainable Energy Rev. **43**, 901–917 (2015)
27. Li, Y., Zhu, Z., Kong, D., Han, H., Zhao, Y.: EA-LSTM: Evolutionary attention-based LSTM for time series prediction. Knowledge-Based Systems (2019)
28. McDowell, I.C., Manandhar, D., Vockley, C.M., Schmid, A.K., Reddy, T.E., Engelhardt, B.E.: Clustering gene expression time series data using an infinite gaussian process mixture model. PLoS Comput. Biology **14**(1), e1005896 (2018)
29. Neumann, M., Kersting, K., Xu, Z., Schulz, D.: Stacked gaussian process learning. In: 2009 Ninth IEEE International Conference on Data Mining, pp. 387–396. IEEE (2009)
30. de Oliveira, E.M., Oliveira, F.L.C.: Forecasting mid-long term electric energy consumption through bagging arima and exponential smoothing methods. Energy **144**, 776–788 (2018)
31. Oliveira, M.R., Torgo, L.: Ensembles for time series forecasting (2014)
32. Pasolli, L., Melgani, F., Blanzieri, E.: Gaussian process regression for estimating chlorophyll concentration in subsurface waters from remote sensing data. IEEE Geosci. Remote Sensing Lett. **7**(3), 464–468 (2010)
33. Pole, A., West, M., Harrison, J.: Applied Bayesian Forecasting and Time Series Analysis. Chapman and Hall/CRC (2018)
34. Rana, M., Koprinska, I., Khosravi, A., Agelidis, V.G.: Prediction intervals for electricity load forecasting using neural networks. In: The 2013 International Joint Conference on Neural Networks (IJCNN), pp. 1–8. IEEE (2013)

35. Rasmussen, C.E.: Gaussian processes in machine learning. In: Bousquet, O., von Luxburg, U., Rätsch, G. (eds.) ML -2003. LNCS (LNAI), vol. 3176, pp. 63–71. Springer, Heidelberg (2004). https://doi.org/10.1007/978-3-540-28650-9_4
36. Renew Economy: Australia leads global energy disruption, but is it "smart" enough to stay in front? July 2019 (2019). https://www.smart-energy.com/industry-sectors/smart-meters/falling-behind-down-under/
37. Smart Energy International: Falling behind 'down under', June 2019 (2019). https://www.smart-energy.com/industry-sectors/smart-meters/falling-behind-down-under/
38. Spinoni, J., et al.: Changes of heating and cooling degree-days in europe from 1981 to 2100. Int. J. Climatol. **38**, e191–e208 (2018)
39. State Government of Victoria: Smart meters (2019). https://www.energy.vic.gov.au/electricity/smart-meters
40. Swastanto, B.A.: Gaussian process regression for long-term time series forecasting (2016)
41. Thouvenot, V., Pichavant, A., Goude, Y., Antoniadis, A., Poggi, J.M.: Electricity forecasting using multi-stage estimators of nonlinear additive models. IEEE Trans. Power Syst. **31**(5), 3665–3673 (2015)
42. Topa, H., Honkela, A.: Gaussian process modelling of multiple short time series. arXiv preprint arXiv:1210.2503 (2012)
43. Wilson, A., Adams, R.: Gaussian process kernels for pattern discovery and extrapolation. In: International Conference on Machine Learning, pp. 1067–1075 (2013)

Reconstructing the Past: Applying Deep Learning to Reconstruct Pottery from Thousands Shards

Keeyoung Kim[1,2](✉), JinSeok Hong[3], Sang-Hoon Rhee[4], and Simon S. Woo[5](✉)

[1] Stony Brook University/SUNY, Korea, Incheon, Republic of Korea
kykim@sunykorea.ac.kr
[2] Ingenio AI, Seoul, Republic of Korea
[3] Artificial Intelligence Research Institute, Seongnam, Republic of Korea
hjs@airi.kr
[4] Linkgenesis, Anyang, Republic of Korea
rhee@linkgenesis.co.kr
[5] Sungkyunkwan University, Suwon, Republic of Korea
swoo@g.skku.edu

Abstract. A great deal of time, patience, and effort are required to excavate pottery. For example, archaeologists dig hundreds to thousands of pottery shards from an excavation site. However, restoring pottery is a time-consuming and challenging process, requiring considerable amounts of expertise, experience, and time. Therefore, computer-assisted restoration methods are indispensable to assist the pottery restoration process. However, existing restoration approaches mostly resort to heuristic-based approaches, which are computationally expensive to match and align different shards together. It is often infeasible to handle and process a large number of shards to reconstruct pottery in 3D. In this paper, we propose a deep learning-based pottery restoration algorithm to classify a pottery shard to a specific pottery type and further predict the exact shard location in the pottery type. We use a novel 3D Convolutional Neural Networks and Skip-dense layers to achieve these objectives. Our model first processes a 3D point cloud data of each shard and predicts the shape of the pottery, which a shard possibly belongs to. We first apply Dynamic Graph CNN to effectively perform learning on 3D point clouds of shards and use Skip-dense layers for a classifier. In particular, we generate features from the 3D scanned point cloud of each shard using spatial transform and edge convolution, then classify shards into one of the pottery shape types using Skip-dense. We achieve 98.4% of classification accuracy over 5 different pottery types and 0.032 RMSE for shard location prediction.

Keywords: Pottery restoration · Point cloud · Skip-dense · Dynamic Graph CNN

Y. Dong et al. (Eds.): ECML PKDD 2020, LNAI 12461, pp. 36–51, 2021.
https://doi.org/10.1007/978-3-030-67670-4_3

1 Introduction

Pottery is one of the oldest human inventions dating back to 20,000 BC [19]. Therefore, pottery can reveal much about the culture and human life at the time when the pottery were made. Thus, archaeologists spend a significant amount of time and effort to excavate, restore, and research shards at an excavating site. Creating pottery often requires professional skills, and it can be improved through trial and error. Hence, many unsatisfactory pieces are often thrown away and discarded during the pottery making process. These discarded pottery are fragmented and further corroded, as time goes. Moreover, most of the pottery found in the field are rarely intact. Therefore, restoring a pottery from thousands of shards buried around the kiln site is a very challenging and time-consuming task, requiring significant expertise, even to archaeologists. While restoring a pottery from shards appears to be similar to solving 3D jigsaw puzzles, it is much more difficult than 3D puzzles for the following reasons: 1) conditions of the shards are not perfect and may have been worn out, 2) the total number of shards is unknown, and 3) there would be multiple pottery and tens of thousands shards need to be mapped to possibly different pottery. In addition, each shard must be cleaned first, before starting a restoration process. Therefore, it is much more challenging than solving a jigsaw puzzle due to the reasons mentioned above. Most of the current computer-based pottery restoration processes are often performed by a heuristic method [1], which directly collate or identify adjacent shards using thickness, curvature parameters, or fracture lines. During this process, many resources and significant manual human efforts are required.

To address these challenges, we propose a deep learning algorithm to assist in performing a pottery restoration process efficiently. We first use a 3D scanner to generate point cloud of sample shards, where the point cloud data is widely used to represent a 3D shape. Next, we use Dynamic Graph Convolutional Neural Network (DGCNN), which is one of the best deep learning models for point cloud data. DGCNN has been successfully applied to object classification, segmentation, scene segmentation, etc. After applying DGCNN, we use Skip-dense by Kim et al. [6] to classify where a pottery shard belongs to a specific pottery type as well as to predict the location of the shard in the pottery. Using our approach, we can predict the pottery type, and the exact location of the shards with high accuracy with 5 different test pottery type classes, trained with 125,000 shards. Our proposed deep learning-based method can significantly assist in restoring pottery with high accuracy, reducing manual human efforts and expertise. Our work clearly demonstrates that AI assisted technology can help better understand our cultural heritage from the past. Our main contributions are summarized as follows:

- We propose a novel pottery shard identification approach for pottery restoration process based on Skip-dense for 3D shard point cloud data for classification and prediction to restore the original pottery from thousands of shards.
- We compare our approach with four different deep learning baseline methods and show the effectiveness of our DGCNN approach.

2 Related Work

Existing Computer-Aided Restoration Algorithms. Currently, there are two main approaches to recover a pottery using a computer-based method as follows: (1) typological recognition and (2) vessel reconstruction methods [1]. Typological recognition method [1] often requires digitized shard footprint data and typological databases of potteries. Therefore, the main task is to search for pottery in the typological database and cross-examine those. However, this method is often difficult to classify the shard correctly, because shards can be so tiny or damaged that they are difficult to recognize. In addition, new classes of potteries may not be stored in a database. On the other hand, the Vessel reconstruction method [1] aims to assemble unknown pottery using digitized shards data, which can be similar to solving a 3D puzzle problem. In this process, several shard parameters are used such as thickness profile [16], the radius of curvature [1], handles, and holes, in addition to cross-sections of the shards and patterns for comparisons.

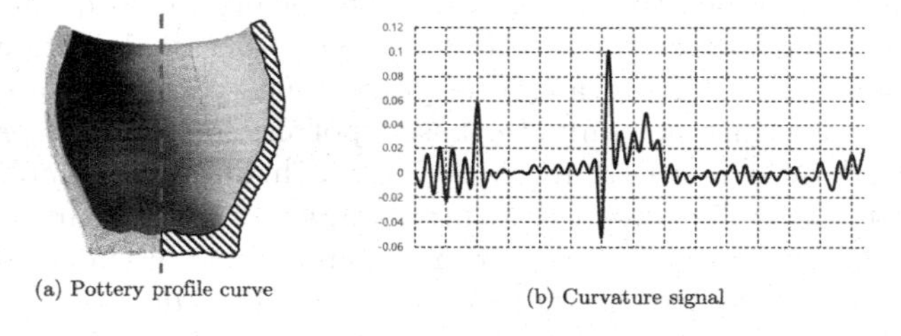

(a) Pottery profile curve (b) Curvature signal

Fig. 1. Profile curve and curvature signal: (a) Profile curve is the curve of the cross section of the pottery and (b) Curvature signal (signature) is the finite differences of profile curve

Main parameters used for the vessel reconstruction method are the profile curve and curvature signal, as shown in Fig. 1(a) and (b), respectively. The profile curve in Fig. 1(a) is a 2D planar curve generated from cutting a plane containing the axis of the symmetry of the pottery. For the pottery with axial symmetry, rotating the profile curve around the Y-axis can restore the shape of the circular 3D pottery because it is symmetric. Because this profile curve is a representative and unique feature of the pottery, it can be used as one of the criteria for the vessel reconstruction method. On the other hand, the curvature signal as shown in Fig. 1(b) is generated by extracting features from the profile curve using the finite difference method [7]. We can compare the profile curves between difference potteries or shards, using this curvature signal [1].

Both typological recognition and vessel reconstruction are popularly used to restore potteries. However, these methods are based on heuristics, requiring significant pre-processing to obtain the hand-crafted features using the unique property for potteries. To reduce the pre-processing effort, Makridis and Daras [8] applied machine learning in shard classification using bag-of-words features from color and local texture of shards. Comparing these features of potteries with the feature of each shard is time-consuming because these individual features have to be compared with many combinations of tens of thousand shards. To overcome these challenges with existing restoration methods, we propose a novel method to automatically identify and locate shards with a high accuracy using a deep learning algorithm to reduce time and efforts required to restore potteries.

3D Data Representation. Current computer graphics algorithms to represent 3D objects can be categorized into the following methods: (1) multi-view based method, (2) volumetric method, (3) point cloud or (4) triangle mesh based method, where we present these methods in Fig. 2.

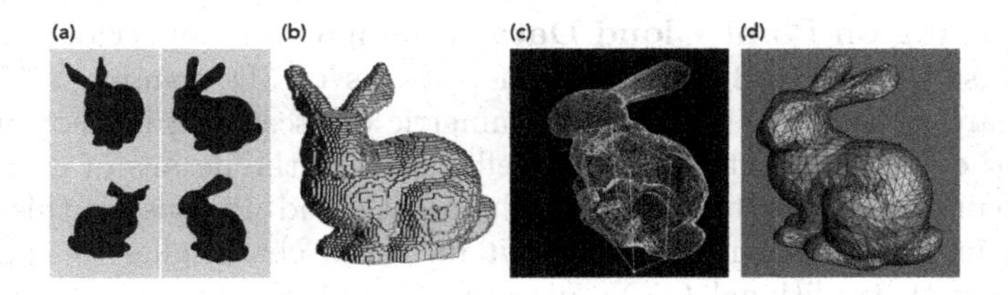

Fig. 2. 3D data representations [4]: (a) multi-view representation, (b) voxel grid, (c) point cloud, and (d) triangle mesh

Multi-view based method utilizes the multi-view representation as shown in Fig. 2(a), where a set of images are viewed at different viewpoints. Therefore, the multi-view based method is based on 2D CNNs, which handles multiple images [17]. The volumetric method converts an object into the voxel grid, as shown in Fig. 2(b), then it adapts 3D CNN [9,12,20]. The voxel is a good data representation to extend 2D CNN to 3D naturally. However, the voxel has a spare representation, which cannot capture the details of the model. Hence, the voxel is not the best 3D representation. As shown in Fig. 2(c), the point cloud method learns point cloud data from the objects which are represented as a set of coordinates in 3D. The Point cloud is widely used for 3D objects and scenes, and especially many 3D point cloud datasets nowadays can be easily obtained due to 3D scanners, depth cameras, and LIDAR. An example of obtaining 3D point cloud data is presented in Fig. 3. In addition, even triangle mesh data for 3D graphics, as shown in Fig. 2(d) can be easily converted into point cloud data by removing edges. However, point cloud data are not appropriate for CNNs, because the order of points in point cloud could not keep the locality because point cloud is a set of points without any order. Recently, several deep learning algorithms are proposed to address such problems in point cloud data as follows: PointNet, PointNet++, and DGCNN [11,13,18]. We further explain the details of these approaches in the next paragraph. In this work, we mainly chose point cloud method to model the potteries and shards in 3D using shards and pottery data from 3D scanners.

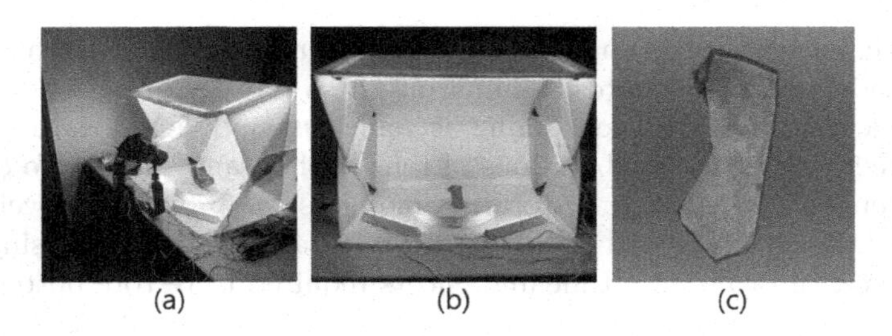

Fig. 3. Example of 3D scanning method for pottery shards: (a) photogrammetric scanning system in action, where it has a turntable, mini studio, and a camera, (b) frontal view of the mini-studio and the turntable, and (c) a scanned result of a shard

Deep Learning on Point Cloud Data. We can obtain point cloud data from 3D scanners, where Fig. 3 illustrates the process of 3D scanning. We scan a pottery shard sample using a photogrammetric 3D scanning method in a mini studio. Although the point cloud can well represent the details of a 3D object, there is no order between the points in the point cloud. Because of this unstructured and unordered nature of points, it has been challenging to apply point cloud data with traditional CNNs directly. *PointNet* [11] is the first work to successfully implement deep learning on a 3D point cloud data, where point cloud data is variant-to-permutation, and neural network operation cannot be directly applied. However, PointNet enables a point cloud permutation invariant by using a symmetric max pooling function. PointNet applies spatial transform to send the input point cloud to canonical space, and then apply convolution and max pooling, repeatably. From these, permutation invariant features can be successfully obtained. Subsequent papers such as *PointNet++* [13] and *Dynamic Graph Convolutional Neural Network (DGCNN)* [18] are introduced and provide a better performance by considering geometric features in the point cloud. In DGCNN, k-nearest neighbor (kNN) computation is added before the spatial transform and convolutions, using local geometric features. Gao and Geng [5] applied PointNet to classify terracotta warrior fragments according to each part of the body. In our work, we choose DGCNN for the world-first pottery shard identification deep learning algorithm. Even we improved its prediction accuracy combining DGCNN with Skip-dense.

3 Our Approach

3.1 Dataset Generation Method

In order to acquire training data, a large number of scanned shards and their relative position data are required. Since the point cloud data of such shards were not available, we first generated deep learning models to generate synthetic data using different methods as shown in Fig. 4.

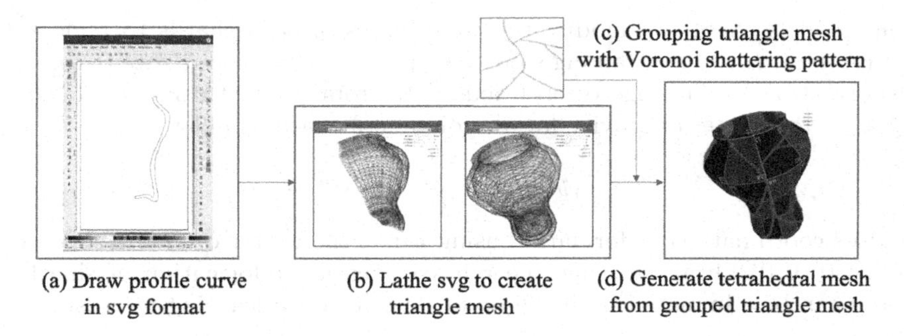

Fig. 4. Dataset generation pipeline: (a) Draw profile curve of a pottery in SVG vector graphics format, (b) Lathe the profile curve to get triangle mesh for the pottery, (c) Generate Voronoi shattering pattern texture randomly, map the texture on the triangle mesh to group triangle meshes, and (d) Using the triangle mesh and its groups to generate grouped tetrahedral to represent shards

Drawing Profile Curve. To create a pottery and its shards point cloud data, we first draw the profile curve of the target pottery prototype in vector graphics format as shown in Fig. 4(a). We used open source vector graphics tool, Inkscape [2] to generate a profile curve. We loaded a cross-section image of a pottery prototype into Inkscape and then drew the profile curve along the cross-section image.

Triangle Mesh Generation. We lathed the profile curves to create a 3D virtual pottery shape in the triangle mesh format, as shown in Fig. 4(b). We develop our own python code to lathe the profile curve, where we upload the code to a public repository (https://github.com/violeta7/lathe-and-shatter). Triangle meshes only cover inner surface and outer surface of the pottery, using tetrahedral meshes to fill inside of the model.

Grouping Triangle Mesh. Next, we generate a crack pattern texture using Voronoi shattering [15] as shown in Fig. 4(c). Although Voronoi shattering is simulated over the surface of a pottery, it generates quite realistic shattering patterns. This Voronoi shattering is popularly used in the game industry as well [3]. In this work, we map the shattering pattern texture on the triangle mesh and group all triangles by shards. In particular, we cluster each point of the triangle mesh if they are in the same shard on the texture.

Tetrahedral Mesh Generation. We use the open source tool TetGen [14], which is a mesh generator to partition a 3D geometry into tetrahedrons. Feeding triangle meshes and the group we create in the previous step, TetGen can generate tetrahedral meshes and its grouped tetrahedral meshes, as shown in Fig. 4(d). After this step, tetrahedral meshes fill inside of the model. In addition, because tetrahedral meshes are grouped, we can consider each group is a shard of the pottery. The (x, y, z) coordinates can be sampled from points in a tetrahedral group and they are modeled as a shard, which is the input data to our models. Next, we modify the coordinates of each shards to remove its location information for training step. The new (x', y', z') coordinate is obtained by subtracting the average of each axis data from (x, y, z) using Eq. 1, where $(x, \bar{y}, z)$ is the center of gravity for all points in a shard as follows:

$$(x - \bar{x}, y - \bar{y}, z - \bar{z}) \rightarrow (x', y', z') \quad (1)$$

After coordinate transformation using Eq. 1, the center of gravity of shards shift to $(0, 0, 0)$ always so that the relative location information of shards in pottery can be eliminated from the input data for deep learning models.

3.2 Our Proposed Model

We aim to classify shards into different pottery classes and predict a relative location of shards within the pottery using point cloud representation. In order to achieve these objectives, we use DGCNN to extract the shard features from the point cloud data. Further, we apply Skip-dense to classify pottery type classes and predict the shard location more precisely. Our complete single shard model is presented in Fig. 5, which is composed of (1) DGCNN part, which has Spatial transform and Edge convolution and (2) 10 Skip-dense layers and a Fully-connected (FC) layer.

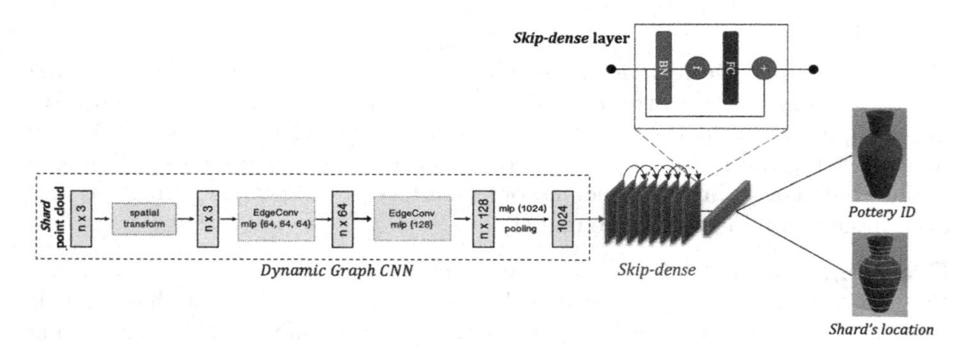

Fig. 5. Our proposed single shard model: Black box indicates feature dimension. Yellow dotted box is a Spatial Transform Block and Blue dotted boxes are Edge Convolution Blocks. Though our Single shard model is based on DGCNN, we apply Skip-dense layers to have more depth, where additional depth helps improve both pottery type classification and shard prediction performance. Skip-dense layer is a Fully-connected layer with skip-connection. Our single shard model stacks Skip-dense layer 10 times. (Color figure online)

Dynamic Graph Convolutional Neural Network. The Dynamic Graph Convolutional Neural Network (DGCNN) [18] has shown to achieve great object classification, part segmentation, and scene segmentation performance with point cloud. DGCNN is adapted to our proposed model to process a shard point cloud and outputs 1,024 size features, where DGCNN is illustrated in the first part of

Fig. 5. In particular, DGCNN consists of a spatial transform block and an edge convolution block, as shown in Fig. 6(a) and Fig. 6(b), respectively.

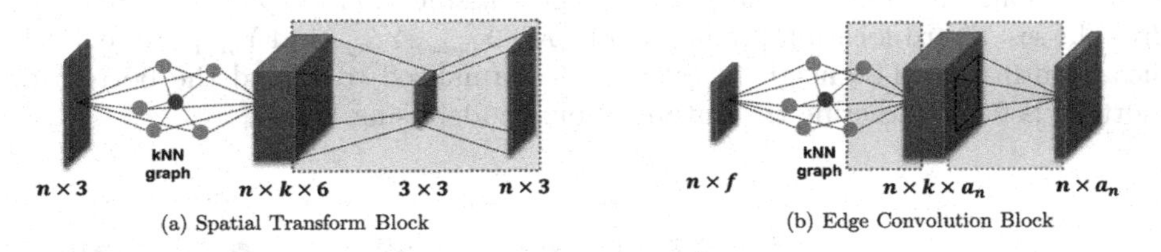

Fig. 6. (a) Spatial Transform Block: Blue boxes are feature maps and edges are layers. After computing kNN graph, there are two layers for matrix computation (edges in a red box) (b) Edge Convolution Block: After computing kNN graph, there are two layers: one for convolution (edges in a red box) and one for pooling (edges in blue box) (Color figure online)

The goal of a Spatial transform block is to apply the convolution filter by aligning point cloud data to canonical space as shown in Fig. 6(a) so that we can apply convolutions and pooling layers. The k-nearest neighbor (kNN) as shown in the left side of Fig. 6(a) determines the local features, and then the matrix computation is followed to estimate the geometric transformation matrix to align it with the canonical space.

The goal of Edge convolution block is to extract permutation invariant features, which is shown in Fig. 6(b). kNN as shown in the left graph in Fig. 6(b) first determines the local features again. Then, we apply a convolutional layer and max pooling layer. Max pooling layer works as a symmetric function, so that the pooling layer helps find permutation invariant features. Therefore, after two or three Edge convolution blocks, we can expect that the feature map from DGCNN can represent a shard component effectively.

Skip-Dense. We use the features of the shards from DGCNN as an input to the *Skip-dense* [6], which is stacked Fully-connected layers with skip-connections as shown in Fig. 5. We can also express Skip-dense using Eq. 2, where $\mathbf{I}_l$ is l-th layer's input of Skip-dense, $BN_{\gamma,\beta}$ is a batch normalization. γ and β are parameters for the batch normalization. Next, we have a ReLU activation function and a Fully-connected layer. $\mathbf{W}$ and $\mathbf{b}$ are parameters of the Fully-connected layer. Then, α is a coefficient to adjust the ratio for a skip connection:

$$\mathbf{I}_{l+1} = \mathbf{W} \cdot max(0, BN_{\gamma,\beta}(\mathbf{I}_l)) + \mathbf{b} + \alpha \times \mathbf{I}_l \tag{2}$$

We claim the Skip-dense can improve classification accuracy by enabling a deeper network optimization than a Fully-connected layer. In addition, a conventional Fully-connected layer is difficult to stack more than 2~3 layers. However, Skip-dense is capable of stacking up more layers. We compare the performance of Fully-connected layers and Skip-dense in the next section.

In our model, we stacks Skip-dense layers 10 times over DGCNN. Finally, we have a Fully-connected layer to generate outputs. Our model predicts both pottery type and normalized shard location (minimum, average (centroid), maximum) along the Y-axis. The pottery type classification output is softmax for five classes. Shard location prediction output, Y_{max}, Y_{avg}, and Y_{min} are real values from 0 to 1. The height of pottery is normalized to 1, and the bottom of pottery is 0. We describe the output of our model using Fig. 7.

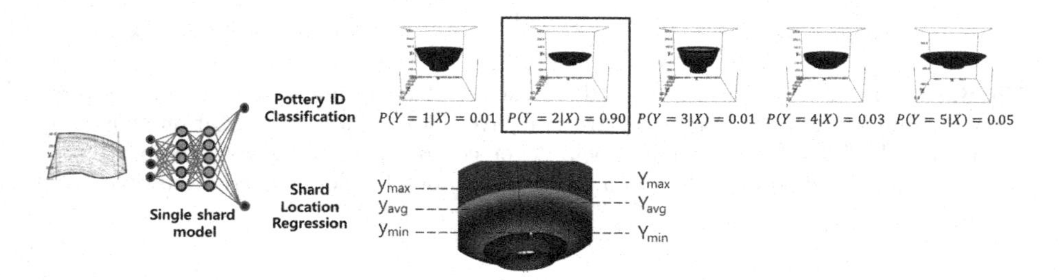

Fig. 7. Model outputs. Probabilities under the class types are softmax probability for each class. In this example, $P(Y = 2|X)$ (red boxed) is the most likely pottery type. Y_{max}, Y_{avg}, and Y_{min} (blue lines) are the correct answers and y_{max}, y_{avg}, and y_{min} (red-to-yellow gradation) are the predicted results. Y_{max}, Y_{avg}, and Y_{min} are normalized from 0 to 1, so that the rim of a pottery is always 1, and the bottom of a pottery is always 0. (Color figure online)

4 Experimental Setup

Dataset. In this work, we use five different pottery classes (types) for synthetic data generation, where these five pottery classes are obtained from Hanseo University Museum in South Korea. These potteries are the Korean Celadons which are made from 1150 to 1350 during the Goreyo dynasty. Our sample potteries include 2 Korea Celadon soup bowls, 1 rice bowl, and 2 dishes. Their shape and the shape of their shards are presented in Table 1.

We generate 2,000 different pottery samples for each pottery class, producing a total 10,000 samples of all five pottery classes. Typically, the number of shards for each pottery type ranges from 5 to 20. For each pottery class, we generate 25,000 shards for training, producing a total of 125,000 shards for five pottery classes. In addition, we randomly generate different Voronoi patterns to create a number of different shard patterns for each pottery class. The original point cloud data of the shards are sampled and saved as 2,048 (x, y, z) coordinates. And the specific pottery class (e.g., 1, 2, 3, 4, and 5) was assigned as classification label. The max, min, and average value in each coordinate of shards are provided as a location label. We train our model using 7,000 pottery shapes and evaluate with 3,000 synthetic shapes based on 5 different pottery classes. We use 87,500 shards for training and 37,500 shards for testing, respectively.

Table 1. Five different pottery classes (IDs), their restored shape, and generated shards

Class number (Pottery ID)	Restored shape	Shards
Class 1 (18-0702-02)	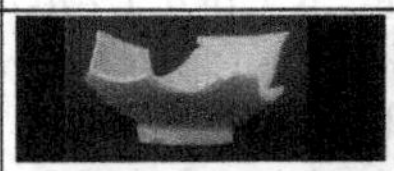	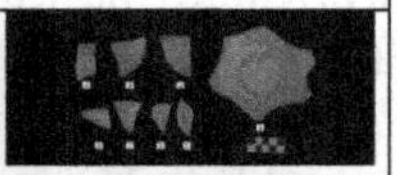
Class 2 (18-0702-03)	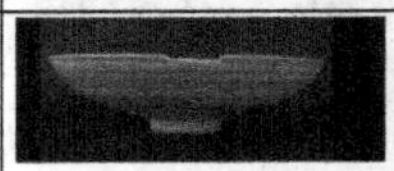	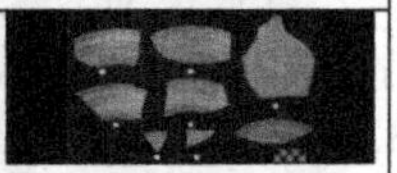
Class 3 (18-0702-07)	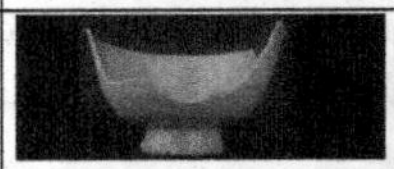	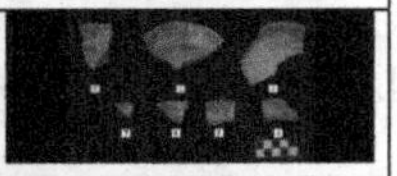
Class 4 (18-0702-08)	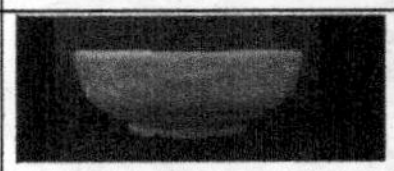	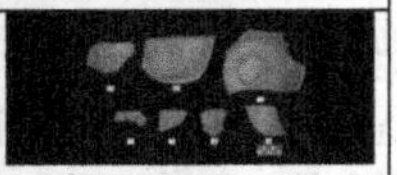
Class 5 (18-0702-09)	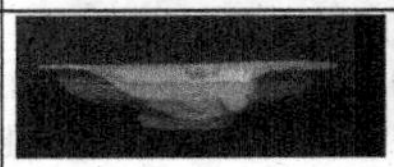	

Data Augmentation. To further increase the dataset from the original scanned shard data inputs and improve the overall performance, we perform the data augmentation by applying the rotation, jittering, scaling, and shifting of scanned shards. In our system, when we load point cloud data of a shard, data augmentation is automatically performed such as randomly rotating, scaling, shifting, etc.

Model Comparison. We compare our model with other different baseline models. The first model is DGCNN [18], which has Spatial transform block, two Edge convolution blocks, and 2 Fully-connected layers with 1024 nodes and 512 nodes. We also use two variations of DGCNN. DGCNN-1FC has one less Fully-connected layer than DGCNN. Therefore, it has one Spatial transform block, two Edge convolution blocks, and a Fully-connected layer. DGCNN-3FC has one more Fully-connected layer than DGCNN. Therefore, it has one Spatial transform block, two Edge convolution blocks, and three Fully-connected layers. Our proposed 'DGCNN-SD' has Skip-dense instead of Fully-connected layers of DGCNN so that it has one Spatial transform block, two Edge convolution blocks, and 10 Skip-dense layers.

Training Parameters, Loss Function, and Training Strategy. The scaling factor for skip-connection is 0.1, and dropout rate is 0.5 for our DGCNN-SD. We use 20 for batch size during training. Cross entropy is used as a loss function for pottery type classification. On the other hand, root mean square error (RMSE) is used as a loss function for pottery shard location prediction. In addition, L2-regularization loss term is used to prevent overfitting. We used the Adam optimizer for learning and suppressed the gradient explosion by using the gradient clipping [10]. We find out that pottery type classification was more challenging to train. For example, if we train the pottery type classification model from scratch, the pottery class classification result chooses one out of five classes many times so that its accuracy yields around 20%. This appears to be the fact that the model is trapped into a local optima easily and cannot recover from it.

In order to overcome this issue, we use *transfer learning* by utilizing pre-trained weights from the shard location prediction. We hypothesize that the features from shard location prediction could be helpful to solve the pottery type classification problem. Hence, to learn how to classify pottery types, we used trained features from the shard location prediction. Therefore, we build our loss function as Eq. 3, where $L_{Typ.}$ is the loss from pottery type classification, $L_{Loc.}$ is the loss from the shard location prediction, and $L_{Reg.}$ is the regularization loss from each classifier. $k_{Typ.}$, $k_{Loc.}$, and $k_{Reg.}$ are coefficients to balance their ratio.

$$L = k_{Typ.} \cdot L_{Typ.} + k_{Loc.} \cdot L_{Loc.} + k_{Reg.} \cdot L_{Reg.} \quad (3)$$

At first, we trained our model for 100 epochs with very small $k_{Typ.}$ such as 10^{-7} to focus on solving shard location prediction. After the well-trained shard location prediction model, we applied transfer learning by increasing $k_{Typ.}$ from one hundred times to one thousand times to tackle the pottery classification problem. Finally, when we reach $k_{Loc.}$ and $k_{Typ.}$ to 1:1 ratio, we achieve the model that can successfully classify both pottery type and shard location after the transfer learning for about 150 epochs in total.

5 Result

Prediction Loss and Classification Accuracy. We present the overall results of the shard location prediction error as well as the pottery ID classification accuracy in Table 2. We also show the confusion matrix and F1 score for pottery type classification in Table 3. As shown in Table 2, DGCNN-1FC, DGCNN, and our DGCNN-SD model achieve very good performance in the shard location prediction with RMSE around 0.03 to 0.04. This means that predicted location error is 3 to 4% on average. However, for DGCNN-3FC, three Fully-connected layers are difficult to optimize. Therefore, it has a large error rate, 0.095, compared to other models, as shown in Table 2.

In particular, for pottery type classification, we can observe that accuracy of the DGCNN-SD is the best among all the other DGCNN variant models with 98.4% as shown in Table 2. DGCNN is the second-best with 96.6%. DGCNN-1FC and DGCNN-3FC cannot achieve good performance in pottery type classification with 21.3% and 58.0%, respectively. For one Fully-connected layer model (DGCNN-1FC), its RMSE for shard location prediction was 0.04. However, it fails to predict pottery type prediction, because it predicts to one class for all shards as shown in blue in Table 3. We can observe that it predicts all test dataset into class 5. Therefore, its accuracy for pottery type classification is stuck around 20% as shown in Fig. 9 and its F1 score is 0.067.

For the DGCNN model, its RMSE for shard location prediction was 0.033, as shown in Table 2. DGCNN's performance is very close to the DGCNN-SD model, even achieving 96.6% after the pottery type classification phase as shown

Table 2. Results of shard location prediction and pottery type classification

		DGCNN-1FC (Baseline)	DGCNN-3FC	DGCNN	**(Ours)**
After shard location training phase	Shard location (RMSE)	0.040	0.095	0.033	**0.029**
After pottery ID training phase (final)	Shard location (RMSE)	0.204	0.200	0.127	**0.083**
	Pottery type (ACC)	21.3%	58.0%	96.6%	**98.4%**

in Table 2. Its F1 score is also very high, 0.966 as shown in Table 3. In three Fully-connected layer model (DGCNN-3FC), which has more layers than DGCNN, however, it is much more difficult to train due to a large number of parameters. Its shard location prediction performance is even worse than DGCNN-1FC, as shown in Table 2. This demonstrates that difficulty of adding more Fully-connected layers, where they have enormous parameters, and they are not easy to train because of the vanishing gradient issue. As shown in confusion matrix in Table 3, DGCNN-3FC's pottery type classification cannot differ from class 1, 2 and 3 (shown in red). On the other hand, our DGCNN-SD model with 10 Fully-connected layers, achieves the best RMSE prediction with 0.029 among all models, as shown in Table 2. Accuracy and F1 score of pottery class classification with our DGCNN-SD model are also the best among four models with 98.4% and 0.983, as shown in Table 2 and Table 3, respectively. In addition, even after the pottery type training phase, our DGCNN-SD model shows the best RMSE for the shard location prediction with 0.083, demonstrating that our DGCNN-SD model can be trained to predict both of the shard location and pottery type effectively.

Training Curves for Shard Location Prediction and Pottery Type Classification Phase. Figure 8 shows the training curve for the shard location prediction, where the X-axis is the epoch, and the Y-axis is the shard location prediction RMSE on the evaluation set. Red, purple, blue, and green lines represent DGCNN-SD, DGCNN, DGCNN-1FC, and DGCNN-3FC, respectively. We can observe that DGCNN-SD, DGCNN-1FC, and DGCNN models achieve great training performance, and DGCNN-SD is trained better than others as shown in the red curve in Fig. 8. Figure 9 presents the training curve for the pottery type classification phase, where the X-axis is the epoch, the main Y-axis is the pottery type classification accuracy, and the auxiliary Y-axis (on the right side) is the shard location prediction loss. Red, purple, blue, and green lines are pottery type classification accuracy for DGCNN-SD, DGCNN, DGCNN-1FC, and DGCNN-3FC, respectively. Cyan and orange lines are shard location prediction

Table 3. Confusion matrix and F1 score for each model

Confusion Matrix							F1 score
Predicted result for each model		Pottery type label					
		1	2	3	4	5	
DGCNN-1FC	1	0	0	0	0	0	0.067
	2	0	0	0	0	0	
	3	0	0	0	0	0	
	4	0	0	0	0	0	
	5	7293	7324	6972	7833	7438	
DGCNN-3FC	1	7189	7211	6894	1240	36	0.495
	2	0	0	0	0	0	
	3	0	0	0	0	0	
	4	104	113	78	6589	317	
	5	0	0	0	4	7085	
DGCNN	1	6983	57	286	23	4	0.966
	2	137	7166	234	64	9	
	3	73	31	6415	0	7	
	4	100	70	37	7746	105	
	5	0	0	0	0	7313	
DGCNN-SD	1	**7136**	**53**	**94**	**27**	**20**	**0.983**
	2	**54**	**7203**	**57**	**51**	**13**	
	3	**79**	**47**	**6800**	**13**	**3**	
	4	**18**	**15**	**21**	**7742**	**30**	
	5	**6**	**6**	**0**	**0**	**7372**	

loss for DGCNN-SD and DGCNN, respectively. Red vertical lines show that the time, when loss $k_{Typ.}$ of DGCNN-SD model and DGCNN model are increased.

Red dashed lines at 10 and 20 epochs are the instances when we increase $k_{Typ.}$ in DGCNN-SD model training. We can observe that increasing $k_{Typ.}$ provides the destructive influences to the model at 10 epoch and 20 epoch to our DGCNN-SD model. However, it recovers from it and finally becomes a much more stable model for pottery type classification. However, in the DGCNN model, we can observe that when we increase $k_{Typ.}$ at 97, 107, and 117 epochs (also denoted by red dashed vertical lines), classification accuracy is dramatically increasing. And the shard location prediction loss also increases at that time instances. Therefore, we can observe the trade-off in DGCNN between pottery type classification and shard location prediction. On the other hand, our DGCNN-SD model has a much weaker trade-off in prediction and classification. Therefore, our approach keeps better performance and balance in both pottery type classification and shard location prediction, even after the pottery type classification training phase. Consequently, the DGCNN-SD model achieves both better shard location

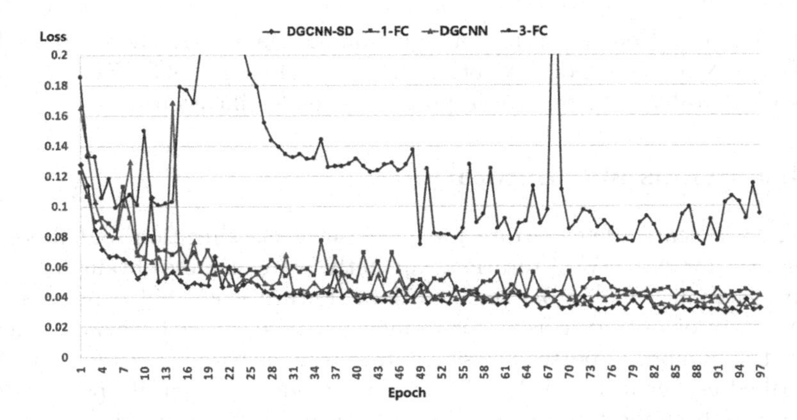

Fig. 8. Shard location training phase loss: All models except DGCNN-3FC trained well for shard location prediction. Among them, our DGCNN-SD is the best. (Color figure online)

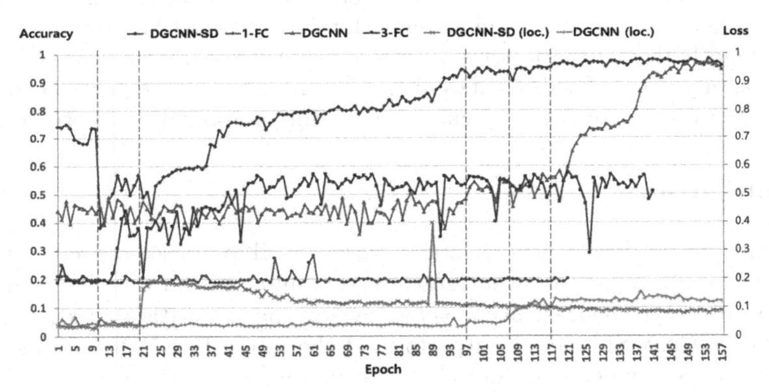

Fig. 9. Pottery type classification accuracy during training phase: Pottery type classification accuracy is plotted by main axis on the left side. Shard location prediction loss (loc.) is plotted by auxiliary axis on the right side. Although both DGCNN-SD and DGCNN achieve above 90% accuracy for pottery type classification, DGCNN-SD is better than DGCNN. Red dotted lines are the timing to increase $k_{Typ.}$. (Color figure online)

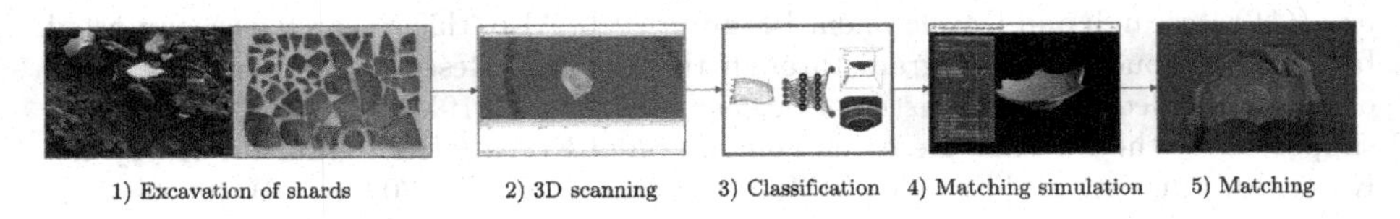

Fig. 10. Description of our method to be used for the real world pottery restoration environment

prediction loss and pottery type classification accuracy than those from all the other DGCNN baseline models. Both of DGCNN-1FC and DGCNN-3FC cannot achieve comparable performance in pottery type classification.

6 Discussions and Limitations

Although we clearly demonstrate the promise in classification and prediction performance for pottery restoration, currently we can only obtain five representative Korean Celadon 3D scanned data. Future work will explore a more diverse family of pottery classes for evaluation. Also, we believe we can easily increase the number of pottery classes and shards, using a 3D scanning method as described earlier in Fig. 3. In addition, our model can identify pottery types of a shard only among pre-defined classes. Future work is to explore a better joint classification and prediction model, which does not lock on specific pottery types from the beginning.

We are in the process of deploying our system on an excavation site with a large amount of shard scanning system along with the traditional existing vessel reconstruction method. Figure 10 shows the whole pottery restoration process from the excavation of pottery shards to matching them in reality. To achieve, we are planning to cover many more pottery classes, including non-axial symmetric ones.

7 Conclusion

We develop novel deep learning model based on Skip-dense to predict pottery type and shard location from the 3D scanned point cloud of a shard. We achieved 98.4% pottery type classification accuracy for shard location prediction. The proposed DGCNN-SD model outperformed the latest DGCNN baseline model in all aspects. We expect that our deep learning model can be used for restoring one of the valuable cultural heritages from centuries ago.

Acknowledgement. This research is supported by the Ministry of Science and ICT (MSIT), Korea, under the ICT Consilience Creative Program (IITP-2020-2011-1-00783) supervised by the Institute for Information and communication Technology Planning and evaluation (IITP), and the Ministry of Culture, Sports, and Tourism (MCST) and the Korea Creative Content Agency (KOCCA) in the Culture Technology (CT) Research and Development Program 2020. Also, this research was supported by Energy Cloud R&D Program through the National Research Foundation (NRF) of Korea Funded by the Ministry of Science, ICT (No. 2019M3F2A1072217) and was supported by the National Research Foundation of Korea (NRF) grant funded by the Korea government (MSIT) (No. 2017R1C1B5076474 and No. 2020R1C1C1006004).

References

1. Computational analysis of archaeological ceramic vessels and their fragments part 1–4. http://what-when-how.com/digital-imaging-for-cultural-heritage-preservation/

2. Inkscape. http://inkscape.org/
3. Unreal engine: Destructibles content examples. https://docs.unrealengine.com/en-US/Resources/ContentExamples/Destructables/index.html
4. Beyond the pixel plane: sensing and learning in 3D, August 2018. https://thegradient.pub/beyond-the-pixel-plane-sensing-and-learning-in-3d/
5. Gao, H., Geng, G.: Classification of 3D terracotta warrior fragments based on deep learning and template guidance. IEEE Access **8**, 4086–4098 (2020). https://doi.org/10.1109/ACCESS.2019.2962791
6. Kim, K., Seo, B., Rhee, S.H., Lee, S., Woo, S.S.: Deep learning for blast furnaces: skip-dense layers deep learning model to predict the remaining time to close tapholes for blast furnaces. In: Proceedings of the 28th ACM International Conference on Information and Knowledge Management, pp. 2733–2741, November 2019. https://doi.org/10.1145/3357384.3357803
7. Liszka, T., Orkisz, J.: The finite difference method at arbitrary irregular grids and its application in applied mechanics. Comput. Struct. **11**(1–2), 83–95 (1980)
8. Makridis, M., Daras, P.: Automatic classification of archaeological pottery sherds. J. Comput. Cult. Heritage **5**(4) (2013). https://doi.org/10.1145/2399180.2399183
9. Maturana, D., Scherer, S.: VoxNet: a 3d convolutional neural network for real-time object recognition. In: 2015 IEEE/RSJ International Conference on Intelligent Robots and Systems (IROS), pp. 922–928. IEEE (2015)
10. Pascanu, R., Mikolov, T., Bengio, Y.: On the difficulty of training recurrent neural networks. In: International Conference on Machine Learning, pp. 1310–1318 (2013)
11. Qi, C.R., Su, H., Mo, K., Guibas, L.J.: PointNet: deep learning on point sets for 3D classification and segmentation. In: Proceedings of the IEEE Conference on Computer Vision and Pattern Recognition, pp. 652–660 (2017)
12. Qi, C.R., Su, H., Nießner, M., Dai, A., Yan, M., Guibas, L.J.: Volumetric and multi-view CNNs for object classification on 3D data. In: Proceedings of the IEEE Conference on Computer Vision and Pattern Recognition, pp. 5648–5656 (2016)
13. Qi, C.R., Yi, L., Su, H., Guibas, L.J.: PointNet++: deep hierarchical feature learning on point sets in a metric space. In: Advances in Neural Information Processing Systems, pp. 5099–5108 (2017)
14. Si, H.: Tetgen, a delaunay-based quality tetrahedral mesh generator. ACM Trans. Math. Software **41**(2), 11:1–11:36 (2015). https://doi.org/10.1145/2629697
15. Smith, J., Witkin, A.P., Baraff, D.: Fast and controllable simulation of the shattering of brittle objects. Comput. Graphics Forum **20**(2), 81–90 (2001)
16. Stamatopoulos, M.I., Anagnostopoulos, C.N.: 3D digital reassembling of archaeological ceramic pottery fragments based on their thickness profile. arXiv preprint arXiv:1601.05824 (2016)
17. Su, H., Maji, S., Kalogerakis, E., Learned-Miller, E.: Multi-view convolutional neural networks for 3D shape recognition. In: Proceedings of the IEEE International Conference on Computer Vision. pp. 945–953 (2015)
18. Wang, Y., Sun, Y., Liu, Z., Sarma, S.E., Bronstein, M.M., Solomon, J.M.: Dynamic graph CNN for learning on point clouds. arXiv preprint arXiv:1801.07829 (2018)
19. Wu, X., et al.: Early pottery at 20,000 years ago in Xianrendong cave, China. Science (New York, N.Y.) **336**, 1696–700 (2012). https://doi.org/10.1126/science.1218643
20. Wu, Z., et al.: 3D shapenets: a deep representation for volumetric shapes. In: Proceedings of the IEEE Conference on Computer Vision and Pattern Recognition, pp. 1912–1920 (2015)

CrimeForecaster: Crime Prediction by Exploiting the Geographical Neighborhoods' Spatiotemporal Dependencies

Jiao Sun(✉), Mingxuan Yue, Zongyu Lin, Xiaochen Yang, Luciano Nocera, Gabriel Kahn, and Cyrus Shahabi

University of Southern California, Los Angeles, CA 90007, USA
{jiaosun,mingxuay,xiaochey,nocera,gabriel.kahn,shahabi}@usc.edu,
lin-zy17@mails.tsinghua.edu.cn

Abstract. Crime prediction in urban areas can improve the allocation of resources (e.g., police patrols) towards a safer society. Recently, researchers have been using deep learning frameworks for urban crime forecasting with better accuracies as compared to previous work. However, these studies typically partition a metropolitan area into synthetic regions, e.g., grids, which neglects the geographical semantics of a region, nor captures the spatial correlation across the regions, e.g., precincts, neighborhoods, blocks, and postal division. In this paper, we design and implement an end-to-end spatiotemporal deep learning framework, dubbed CrimeForecaster, which captures both the temporal recurrence and the spatial dependency simultaneously within and across regions. We model temporal dependencies by using the Gated Recurrent Network with Diffusion Convolution modules to capture the cross-region dependencies at the same time. Empirical experiments on two real-world datasets showcase the effectiveness of CrimeForecaster, where Crime-Forecaster outperforms the current state-of-the-art algorithm by up to 21%. We also collect and publish a ten-year crime dataset in Los Angeles for future use by the research community.

Keywords: Social good · Crime precition · Spatiotemporal learning

1 Introduction

Crime is a significant factor affecting the well-being of residents in urban communities. Understanding crime trends in metropolitan areas is important for both residents in general and for city officials to make decisions. For example, residents can use the crime information to choose where to live and how to protect their property. Law enforcement can use the crime information to allocate

Z. Lin—The work was conducted and completed while the author was an intern at USC.

Y. Dong et al. (Eds.): ECML PKDD 2020, LNAI 12461, pp. 52–67, 2021.
https://doi.org/10.1007/978-3-030-67670-4_4

resources to protect and serve residents, and journalists can write news stories that inform residents on crime in their neighborhoods.

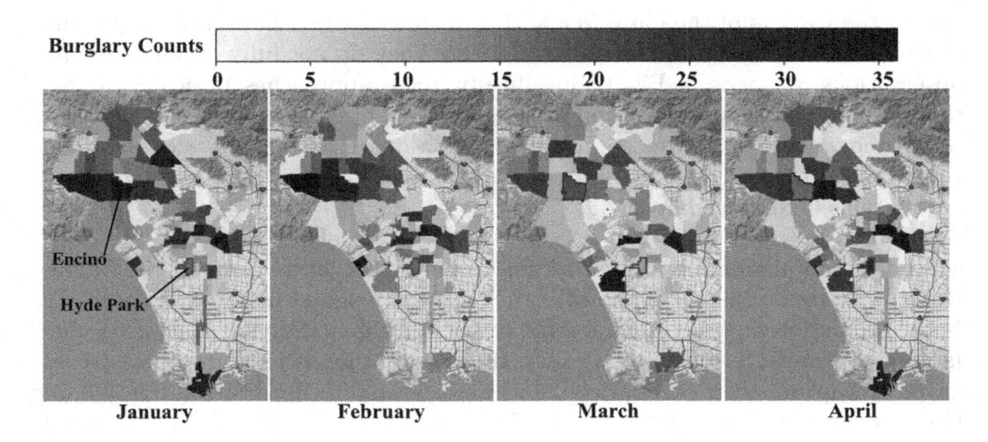

Fig. 1. Choropleth maps of burglaries for neighborhoods in Los Angeles from Jan. to Apr. 2018. We mark and highlight the Encino and Hyde Park neighborhoods.

Figure 1 shows a choropleth map of burglary counts in neighborhoods in Los Angeles over a four months period, to illustrate the spatial and temporal dynamics of crime data. The temporal dependencies are apparent in the figure as the number of burglaries in each neighborhood varies month to month. Monthly changes can be gradual (e.g., in Encino) or abrupt (e.g., in Hyde Park). In addition, we can find instances where month-to-month changes appear to be correlated for neighborhoods in the close proximity. This shows an implicit connection between spatial and temporal crime patterns. Clearly, these observations show the importance of using geographical information, i.e., neighborhoods, to understand the spatiotemporal dynamics of crime data. In fact, in the United States, neighborhood-level dynamics are very much dependent on socio-economic factors. One can therefore think of the geographical extent of a neighborhood as implicitly capturing the underlying socio-economic factors at play. From the standpoint of users who are interested in utilizing crime data, neighborhoods geographical extents are a natural representation to use, analyze and understand crime data at the neighborhood level.

Crime forecasting is especially important to crime data users, e.g., crime predictions can be used by law enforcement to manage resources. Researchers have extensively studied the predictive crime analytics. One approach to predict crime is to use the available *local information*, such as demographic data (e.g., population counts, education levels), data from the local municipality (e.g., property taxes, zoning information) and infrastructure (e.g., number of law enforcement facilities, number of police patrols). However, this data is updated rarely (e.g., the census data in the US is updated every ten years), and therefore is not suitable for making short to medium terms predictions. A separate body of work relies on crime data reported by law enforcement at the municipal level. For example, the current state-of-the work MiST [17] uses RNN to capture the temporal aspects and applies attention-based [26] methods on a regular grid to model the spatial aspects of the data. While this and other prior work incorporate the spatial and temporal properties of crime data they do not fully account for the observations we made in Fig. 1. Specifically, these approaches 1) deal with the temporal and spatial information separately, thereby neglecting the spatiotemporal dependencies; 2) use regular grid partitions that are not modeling the geographical extent of neighborhoods and therefore 3) do not model the crime correlations across neighborhoods.

To address these deficiencies, we propose CrimeForecaster, a new deep spatial-temporal learning framework for crime prediction. In essence, the framework incorporates the geographical extents of the neighborhoods to account for both the spatiotemporal dynamics and implicitly, the underlying socio-economic factors of crimes. Specifically, we map the neighborhoods as nodes in a spatial graph to capture their correlations, and we use the diffusion convolution [23] in Gated Recurrent Units (GRU) networks to deal with the spatial and temporal information simultaneously. The main contributions of our work are as follows.

- **Methodology.** We introduce a new end-to-end spatiotemporal learning framework dubbed CrimeForecaster that: 1) represents the geographical extents of neighborhoods and their correlations in a graph; 2) uses graph convolution to predict crimes. Compared to the current state-of-the-art crime prediction framework, CrimeForecaster achieves up to 21% improvement.
- **Evaluation.** We evaluate the performance of CrimeForecaster on two datasets: an existing dataset for the city of Chicago and our own Los Angeles dataset. We publish the CrimeForecaster's code and the dataset that we use in the experiments at Github[1].
- **Dataset.** We contribute a new crime dataset[2] for the entire region of Los Angeles County covering a ten-year period from 2010 to 2019.

The remainder of the paper is organized as follows: Sect. 2 introduces the dataset we collected and used for the experiments. We give the formal definition of our prediction task in Sect. 3. Section 4 introduces CrimeForecaster's components. We evaluate the performance of CrimeForecaster in Sect. 5 through two case studies. Finally, we review related studies in Sect. 6 and conclude in Sect. 7.

2 Datasets

In this section, we present our collected Los Angeles crime dataset and describe two datasets that we use for the experiment in Sect. 5.

Los Angeles County Crime Dataset. Collecting crime datasets at the city scale is challenging because municipalities publish their local data in different

[1] https://github.com/sunjiao123sun/CrimeForecaster.

[2] https://drive.google.com/open?id=1nbqr0gdp_bO2QaIRN6qPx-7Tvd5pKa0P.

formats. In collaboration with journalists from Crosstown [4], we collected crime data across 216 neighborhoods in Los Angeles County from January of 2010 to October of 2019, but the crime categories reported are not standardized. Thus, we consolidated crime categories into a set of 54 unique crime categories. We made the resulting Los Angeles County Crime dataset publicly available[3].

Table 1. Los Angeles City crime dataset statistics.

Category	Theft	Vehicle theft	Burglary	Fraud	Assault	Sexual offenses	Robbery	Vandalism
Counts	66,697	17,123	14,517	15,578	32,372	6,161	8,864	17,123

Table 2. Chicago crime dataset statistics.

Category	Theft	Criminal damage	Narcotics	Robbery	Assault	Deceptive practices	Burglary	Battery
Counts	56,695	28,589	21,607	9,632	16,692	14,085	13,103	48,824

Los Angeles City Crime Dataset. From the Los Angeles County dataset, we extracted a dataset that focuses on Los Angeles City for the latest complete year (i.e., 2018) and further aggregate crime into eight categories. For example, `theft, person` and `theft, coin machine` were two different categories, but we can aggregate them into a single `theft` category. Table 1 shows the statistics of the Los Angeles City dataset for all eight categories that we consider.

Chicago Crime Dataset. This dataset was published by *The City of Chicago* website [1]. It reflects reported incidents of crime (except for murders where data exists for each victim) that occurred in the City of Chicago in 2015. Table 2 shows the statistics for eight crime categories in this dataset.

Experiment Dataset. For the experiments in Sect. 5, we use the Chicago Crime Dataset (the same as the previous work [17] used in their experiments) and the Los Angeles Dataset. We further supplement two crime datasets with geographical data.

Geographical Data. We collect neighborhood boundaries and neighborhood connectivity information of the Los Angeles City from the Census Bureau [3] which comprises 133 neighborhoods. We also collect the geographical dataset for Chicago from the boundary widget of *the City of Chicago* [2].

3 Problem Definition

In this section, we formally define terminologies *neighorhoods*, *crime event* and our crime prediction task.

[3] https://drive.google.com/open?id=1nbqr0gdp_bO2QaIRN6qPx-7Tvd5pKa0P.

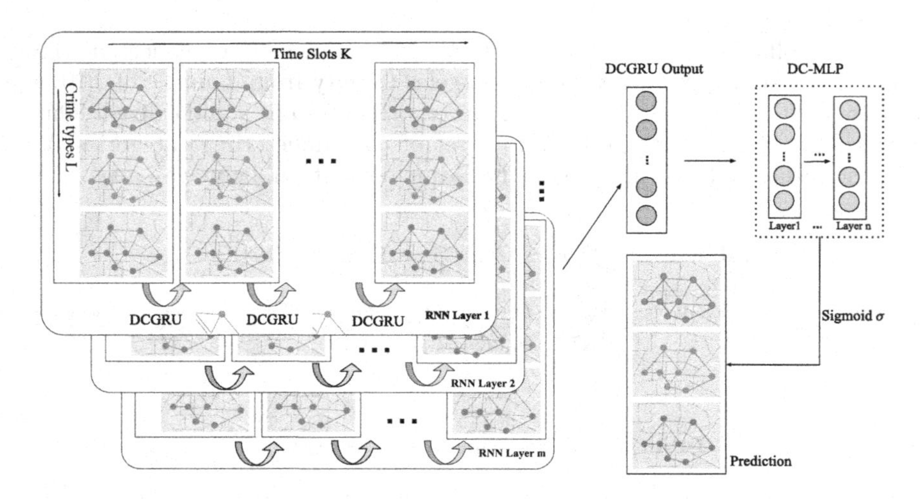

Fig. 2. CrimeForecaster Framework

Definition 1. *Neighborhoods.* As introduced in Sect. 2, we follow the definition of neighborhoods given by the government census data, then we can represent a city as I disjointed neighborhoods as $R = (r_1, r_2, \ldots, r_I)$.

Based on the definition and adjacency information of neighborhoods, we can further represent the city as a weighted undirect graph $\mathcal{G} = (\mathcal{V}, \mathcal{E}, W)$, where $\mathcal{V}$ is a set of nodes $|\mathcal{V}| = N$ and N is the number of neighborhoods in the city. $W \in \mathbb{R}^{N \times N}$ is a weighted adjacency matrix. For each element $W_{i,j} = exp(-\frac{\text{dist}(v_i, v_j)}{\theta^2})$. θ is the standard deviation of distances. $W_{i,j}$ represents the edge weight between neighborhood r_i and neighborhood r_j. $\text{dist}(v_i, v_j)$ is 0 if two neighborhoods are adjacent. If not, $\text{dist}(v_i, v_j)$ is the minimum distance between their boundaries.

Definition 2. *Crime Event.* Given a time window T, we split T as non-overlapping and consequent time slots $T = (t_1, t_2, \ldots, t_K)$, where K is the length of the time sequence. For a region r_i, we use $Y_i = (y_{i,1}^1, \ldots, y_{i,l}^k \ldots, y_{i,L}^K) \in \mathbb{R}^{L \times K}$ to denote all L types of crimes happened at the region during the past K slots' observations. We set each element $y_{i,l}^k$ to 1 if the crime type l happens at region r_i in the time slot t, and 0 if not.

CrimeForecaster Target. Given the historical records of crime events across the city R from time slot t_1 to t_K, we aim to learn a predictive framework which could infer the crime events of any category $l \in \{1, \ldots, L\}$ for region r_i at time t_{K+1}, based on all regions' previous time slots (i.e., temporal features) and their neighborhood relationships (i.e., spatial features). It means that given

$$\{Y_1, \ldots, Y_I\}, \text{where } Y_i = (\underbrace{y_{i,l}^1, \ldots, y_{i,L}^K}_{temporal}, \underbrace{\mathcal{G}}_{spatial}) \text{ and } i \in [1, \ldots, I],$$

and we want to infer the crime occurrence $y_{i,l}^{K+1}$ in the next time $K+1$.

4 Methodology

4.1 CrimeForecaster Framework Overview

Figure 2 shows the framework of CrimeForecaster. In this section, we will introduce how we tackle the crime prediction problem from the spatiotemporal perspective and introduce the details of each component in CrimeForecaster.

Spatial Perspective. Crime occurrences at one location are highly correlated with nearby locations [12,27], especially by the diffusion of neighborhoods [24]. To model and learn such diffusion process, we relate it to the walking transition in undirected graphs. Similar to the recent advances in Graph convolutional neural networks such as GCN [21], GraphSAGE [14] and DCRNN [22], we use graph convolutional operation to model the spatial dependency. For the traditional convolution neural network (CNN), it works for the grid-based data such as pixels in the image. The convolution operation scans across the image with a filter to extract the features. The Diffusion Convolution [22] extends this idea to the graph-structured data. Such convolutional operation is based on a diffusion matrix defined by the graph $\mathcal{G}$, and a learnable filter θ for different layers of neighbors. At each convolutional layer, each node collects messages from input signals of its neighbors according to this transition matrix weighted by parameter θ at different hops of neighbors. The diffusion process can be regarded as a finite Marcov process with trainable weights at each step. We define the maximum diffusion step as M. Formally, the diffusion convolution operation $\star_{\mathcal{G}}$ over all signals $X \in \mathbb{R}^{I \times P}$ of the graph $\mathcal{G}$ and a filter f_θ is

$$X_{:,p} \star_{\mathcal{G}} f_\theta = \sum_{m=0}^{M-1} (\theta_m (D^{-1}W)^m) X_{:,p} \quad \text{where} \quad p \in \{1, \ldots, P\} \tag{1}$$

D is the diagonal degree matrix of the neighborhood graph $\mathcal{G}$ we constructed, and $D^{-1}W$ represents one-step transition during the diffusion process, and the learnable transition matrix is $\sum_{m=0}^{M-1}(\theta_m (D^{-1}W)^m)$. Specifically, the i-th row of the weighted transition matrix represents the likelihood of diffusion from the neighborhood r_i within M-hop random walks. If X is the input, P equals to L, otherwise P is the number of hidden dimensions.

The graph diffusion convolution is a utilization of the inductive bias that the closer neighborhoods tend to have similar patterns and the crime diffusion behavior as we introduced in Sect. 1. Similar to classical convolutional operations, it allows sparse interactions and parameter sharing across the nodes in the graph, i.e., neighborhoods in the city. By using the diffusion convolution over the spatial graph of neighborhoods, we can capture the implicit spatial dependency of neighbors for different pattern transitions in the neural network, e.g., the gates in RNN modules that we will describe later.

Spatiotemporal Perspective. To better capture the temporal dynamics, we use RNN with DCGRU (Diffusion Convolutional Gated Recurrent Units) units [22] in CrimeForecaster, to encode the complex intra- and inter-region correlations across the previous time slots (from t_1 to t_K). DCGRU replaces the

Algorithm 1. Training procedure of CrimeForecaster

Require: the crime occurrence sequence $X \in \{0,1\}^{(N,K,I,L)}$, the ground-truth prediction $Y \in \{0,1\}^{(N,I,L)}$, the neighborhood graph $\mathcal{G}$, the max diffusion step M, number of RNN layers S

1: *Initialize all parameters*
2: **for** $x \in$ sample(X) **do**
3: // first RNN layer
4: $h^0(0) = h^0$ // initial state
5: **for** $t \leftarrow 1$ **to** K **do**
6: $h^t(0) = DCGRU(x^t, h^{t-1}(0), \mathcal{G}, N)$
7: **end for**
8: // stacked RNN layers
9: **for** $s \leftarrow 1$ **to** S **do**
10: $h^0(s) = h^0$ // initial state
11: **for** $t \leftarrow 1$ **to** K **do**
12: $h^t(s) = DCGRU(h^t(s-1), h^{t-1}(s), \mathcal{G}, N)$
13: **end for**
14: **end for**
15: $\hat{y} = \sigma(DC\text{-}MLP(h^K(S)))$
16: $\mathcal{L} =$ categorical_cross_entropy$(y, \hat{y})$ //as equation 4
17: update parameters over $\mathcal{L}$
18: **end for**

matrix multiplications in GRU with the diffusion convolution in Eq. 1. Compared to the traditional reccurent methods of using GRU or LSTM units, the diffusion convolutional layer in DCGRU can learn the representations from the whole graph, which enable the signal of a node to learn not only on its previous status but also its neighbors' previous status. Formally, the updating functions for hidden state h^t at t-th time step of input signal sequence in our RNN encoder with DCGRU units are as follows:

$$\begin{aligned} r^t &= \sigma(\sum_m^M \theta_r(m)(D^{-1}W)^m[x^t, h^{t-1}] + b_r) \\ u^t &= \sigma(\sum_m^M \theta_u(m)(D^{-1}W)^m[x^t, h^{t-1}] + b_u) \\ c^t &= tanh(\sum_m^M \theta_c(m)(D^{-1}W)^m[x^t, r^t \odot h^{t-1}] + b_c) \\ h^t &= u^t \odot h^t + (1-u^t) \odot c^t \end{aligned} \tag{2}$$

where x^t, h^t are the input and output of the DCGRU cell at time t, and r^t, u^t are the reset gate and update gate respectively. b_r, b_u and b_c are the bias terms. The gates also have separate filters θ_r, θ_u and θ_c enabling the resetting and updating to have different diffusion pattern from the neighbors. σ denotes the sigmoid function. $\odot$ denotes the Hadamard Product. As shown in Fig. 2, from the encoder-decoder perspective, CrimeForecaster's encoder consists of multiple stacked layers of RNNs, where each cell is a DCGRU. The encoder will capture both the spatial and temporal transitions across regions (neighborhoods) and time. The encoder outputs, i.e., the final hidden states of the graph, are delivered to the decoder for predicting the crime occurrences in the future.

Forecasting and Model Inference. After encoding the spatiotemporal history through the encoder, CrimeForecaster's decoder learns to predict the crime occurrences from the encoded hidden states. The decoder utilizes the diffusion convolutional operation on a Multilayer Perceptron (MLP) structure to map the hidden states to predicting outputs in a non-linear way. Formally we represent MLP with diffusion convolutional operations as follows:

$$\begin{aligned}\psi_1 &= \sigma(W_1 \star_{\mathcal{G}} \lambda_1 + b_1) \\ &\ldots \\ \psi_L &= \sigma(W_L \star_{\mathcal{G}} \lambda_L + b_L)\end{aligned} \tag{3}$$

In the final layer of the decoder, each output ψ_s corresponds to the crime occurrence probabilities of all neighborhoods for category l at the predicting time. Note that L is the number of hidden layers in MLP (in the last layer, it is the number of crime types). W is the weight matrix, and b is the bias term. σ is the sigmoid activation function as before. During the training process, we use binary cross entropy as the loss function:

$$\mathcal{L} = - \sum_{i \in \{1,\ldots,I\}, l \in \{1,\ldots,L\}} y_{i,l}^k \log \hat{y}_{i,l}^k + (1 - y_{i,l}^k) \log(1 - \hat{y}_{i,l}^k), \tag{4}$$

where $\hat{y}_i^{l,k}$ is the estimated probability of the l-th category crime happen at region r_i at the time k. In CrimeForecaster, we minimize the loss function by using the Adam optimizer [20] with learning rate decay strategy to learn the parameters of CrimeForecaster.

5 Experiment

Evaluation Metrics. We use Micro-F1 [13] and Macro-F1 [28] measures to evaluate the average prediction performance across all crime categories, similar to the previous work MiST [17]. We calculate them over all crime categories L:

$$\text{Macro-F1} = \frac{1}{L} \sum_{l=1}^{L} \frac{2 \cdot TP_l}{2 \cdot TP_l + FN_l + TP_l}$$

$$\text{Micro-F1} = \frac{2 \cdot \sum_{l=1}^{L} TP_l}{2 \cdot \sum_{l=1}^{L} TP_l + \sum_{l=1}^{J} FN_l + \sum_{l=1}^{L} FP_l}.$$

5.1 CrimeForecaster Experiment Data and Setup

We evaluate the performance of CrimeForecaster through two real-world datasets introduced in Sect. 2. In our experiments, we use the same "train-validation-test" setting as the previous work [17]. We chronologically split the dataset as 6.5 months for training, 0.5 month for the validation and 1 month for testing. We implement CrimeForecaster with Tensorflow [5] architecture. In our experiments, we set *learning rate* as 0.01, *the number of epochs* as 100 and the *batch size* as 64. For hyperparameters in CrimeForecaster, we use two stacked layers of RNNs. Within each RNN layer, we set 64 as the size of the hidden dimension. We use Adam optimizer [20] with learning rate annealing.

5.2 Performance Comparison

Table 3. Performance comparison of CrimeForecaster V.S. the state-of-the-art.

	Month	F1	SVR	LR	ARIMA	MLP	LSTM	GRU	MiST	MiST*	CF	
Chicago	Aug	Macro	0.342	0.427	0.392	0.434	0.436	0.439	0.567	0.565	**0.664**	17%
		Micro	0.582	0.608	0.487	0.512	0.499	0.511	0.570	0.596	**0.705**	24%
	Sep	Macro	0.329	0.424	0.398	0.437	0.410	0.444	0.568	0.567	**0.649**	14%
		Micro	0.606	0.609	0.484	0.506	0.490	0.519	0.575	0.573	**0.693**	21%
	Oct	Macro	0.335	0.416	0.388	0.435	0.445	0.435	0.568	0.578	**0.651**	15%
		Micro	0.640	0.600	0.476	0.521	0.499	0.504	0.573	0.580	**0.693**	21%
	Nov	Macro	0.330	0.421	0.367	0.445	0.428	0.444	0.577	0.570	**0.636**	10%
		Micro	0.594	0.603	0.452	0.492	0.497	0.504	0.581	0.571	**0.678**	17%
	Dec	Macro	0.333	0.422	0.370	0.450	0.435	0.436	0.554	0.565	**0.638**	15%
		Micro	0.626	0.600	0.453	0.517	0.498	0.499	0.565	0.575	**0.677**	20%
Los Angeles	Aug	Macro	0.212	0.341	0.304	0.385	0.385	0.376	0.520	0.504	**0.534**	3%
		Micro	0.457	0.493	0.371	0.467	0.454	0.429	0.551	0.539	**0.604**	10%
	Sep	Macro	0.201	0.347	0.296	0.390	0.382	0.385	0.513	0.498	**0.518**	1%
		Micro	0.475	0.494	0.367	0.459	0.445	0.444	0.551	0.539	**0.604**	10%
	Oct	Macro	0.199	0.344	0.301	0.385	0.381	0.372	0.512	0.499	**0.515**	1%
		Micro	0.530	0.486	0.372	0.450	0.504	0.426	0.547	0.537	**0.594**	9%
	Nov	Macro	0.205	0.341	0.309	0.379	0.378	0.380	0.510	0.500	**0.518**	2%
		Micro	0.490	0.493	0.380	0.425	0.460	0.445	0.545	0.539	**0.601**	10%
	Dec	Macro	0.208	0.342	0.301	0.380	0.380	0.370	0.505	0.493	**0.520**	3%
		Micro	0.529	0.488	0.373	0.435	0.448	0.428	0.541	0.533	**0.596**	10%

We compare CrimeForecaster with five kinds of baseline methods:

1) *the time-series forecasting method* (i.e., ARIMA). ARIMA [8] is a conventional time-series prediction model for predicting the future occurrences based on previous temporal records;

2) *conventional machine learning methods* (i.e., SVR, LR and Random Forest). Epsilon-Support Vector Regression (SVR) [6] is a supervised learning model for regression problems based on the RBF kernel function. We use the same input as CrimeForecaster except for the graph, i.e., use the historical crime occurrences as features and the predicting crimes as labels. Logistic Regression (LR) [16] is a statistical model which can forecast each region's crime occurrences. Please note that we apply Logistic Regression on the whole historical crime records, the same setting as CrimeForecaster, which is different from how researchers use it in the previous work such as MiST [17], where the authors only used the extracted temporal features (e.g., the day of a week or the month of a year) for the experiment. It means that the Logistic Regression in their setting cannot learn from the spatial perspective, which largely weakens the prediction ability of Logistic Regression model.

3) *neural network classifier* (Multi-layer Perceptron classifier, i.e., MLP). MLP [9] is a conventional deep neural network which learns the non-linearity from the historical distributions of crimes.

4) *recurrent neural network* (i.e., LSTM and GRU). Long Short-term Memory (LSTM) [11] and Gated Recurrent Unit (GRU) [7] are artificial recurrent network models with feedback connections. Researchers always use them to predict future event occurrences based on the historical time series data.

5) *spatiotemporal learning models for crime prediction and its variant* (i.e., MiST and MiST*). To the best of our knowledge, MiST [18] is the most state-of-the-art crime prediction model. It learns both the inter-region temporal and spatial correlations. Please note that we replace the synthetic partitions (i.e., grids) with the geographical neighborhoods, which got a significant performance improvement compared to the reported result in MiST when we use the same data setting of Chicago [17]. It also further proves the efficiency of using the census neighborhood partition over the grid partition. MiST* is a variant of MiST, where we replace the one-hot embedding vectors of locations with the ones learned from LINE [25], a classical network embedding method.

Comparison with Other Methods. Table 3 shows the crime prediction accuracy across categories in both Chicago and Los Angeles, where the last column shows the performance improvement of CrimeForecaster over MiST. We evaluate the performance of crime forecasting in terms of Macro-F1 and Micro-F1 for all methods. We summarized the following key observations:

CrimeForecaster performs significantly better than all other learning models. The most notable advancement occurs when predicting crimes in Chicago during September. CrimeForecaster reaches a 21% relative improvement over the state-of-art (i.e., MiST) in terms of Micro-F1. CrimeForecaster also persistently predicts more accurately than the best performed baseline (i.e., LR with the complete historical crime record). For example, when predicting crime in Chicago in September, CrimeForecaster outputs a Micro-F1 value 13.7% higher than LR. The improvement illuminates the benefits of using neighborhoods' information by doing spatial partitions based on geographical information compared to the manual partitions (e.g., grids).

Performance of CrimeForecaster over Different Months. We experiment with CrimeForecaster and all baseline models thoroughly across different training and test time periods across five months. We output the results of each month from August to December, and it is notable in Table 3 that CrimeForecaster consistently outputs the best result among all the methods over different months.

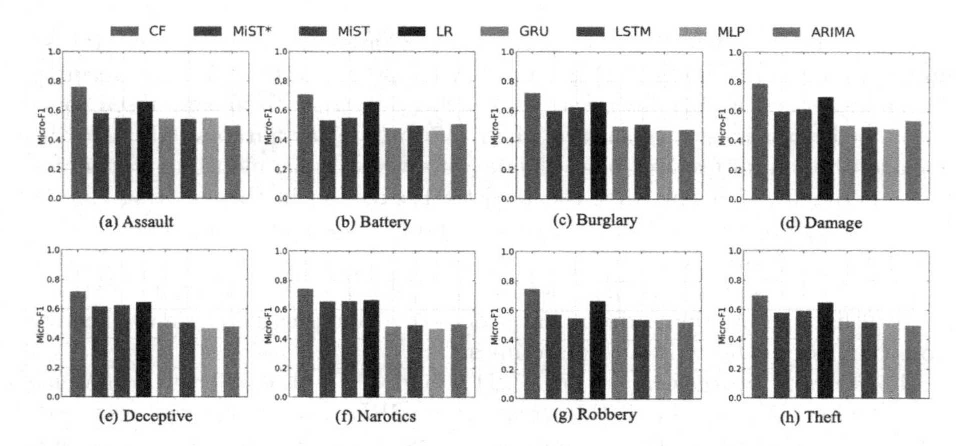

Fig. 3. Forecasting results for individual crime category in terms of Micro-F1 on the Chicago dataset.

Performance of CrimeForecaster Across Crime Categories. In addition to testing CrimeForecaster across different time periods, we also investigate CrimeForecaster's effectiveness over different crime categories. In Fig. 3 and Fig. 4, we show the performance comparisons for Los Angeles and Chicago's eight crime categories. We observed that CrimeForecaster achieves better performance across all crime categories. For example, CrimeForecaster improves the Micro-F1 score for forecasting assault over the best baseline (i.e., LR) by 19.43% on average and improve the forecasting for theft by 7.41%. When compared to MiST, CrimeForecaster improves forecasting assault for 38.4%, and theft by 15%. Additionally, we noticed that for certain crime categories in Los Angeles (i.e., burglary, fraud, vandalism, and vehicle theft), CrimeForecaster makes a very significant improvement over MiST (i.e., over 50%). For example, for forecasting burglary, CrimeForecaster outmatches MiST by 83.9% in Los Angeles, and for forecasting fraud, CrimeForecaster's accuracy exceeds MiST by 86.3%. In contrast, this kind of huge improvement does not happen to any crime categories in Chicago. We deduce this effect may be caused by the fact that the correlation between crimes can be recognized from a larger expansion in Los Angeles, while in Chicago, the correlated crimes are concentrated in a narrower range. Therefore, this result proves the superiority of keeping a neighborhood's original graph structure again.

5.3 Parameter Study

Temporal Sequence Length Influence. In order to know how CrimeForecaster performs with the change of the temporal sequence K, we vary K from 4 to 14. Figure 5a shows that CrimeForecaster performs better as the time sequence length K increases, and it tends to decrease when K reaches 10. Due to the space limit, here we only report our results on the Chicago crime data in August. Other

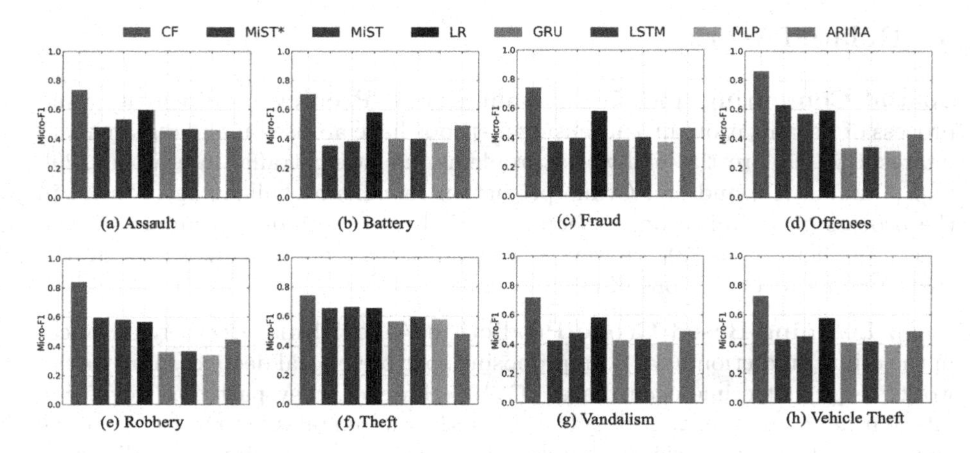

Fig. 4. Forecasting results for individual crime category in terms of Micro-F1 on the Los Angeles dataset.

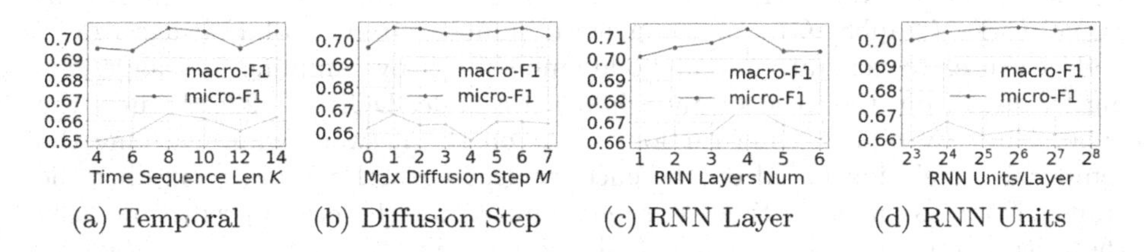

Fig. 5. The parameter study on Chicago august data.

months' data also show the similar trend. The experiment on Los Angeles dataset also shows the similar pattern.

CrimeForecaster Parameter Studies. CrimeForecaster handles the spatiotemporal information by using the graph structure in the diffusion process. *Max Diffusion Step M* determines which and how much the neighborhoods' information would influence. Figure 5b shows that both the Micro-F1 and Macro-F1 of CrimeForecaster gets a huge performance improvement with M increasing from zero to one, when it first perceives the signals from the neighborhoods. However, it starts to decrease when it reaches four. It is because although more neighborhoods provide more abundant information to use, further neighborhoods bring irrelevant or confusing information, which leads to the performance decrease. We also show the performance of CrimeForecaster under different settings of *the number of RNN layers* in Fig. 5c and *number of RNN units per layer* in Fig. 5d.

6 Related Work

Spatial Computing and Traditional Crime Prediction. With the rapid process of urbanization, urban sensing methods have aroused many researchers's interests to explore the real-world problems, including traffic forecasting [29], trip planning [15] and etc. Crime prediction remains a challenging problem in the process of urbanization. Researchers tried to use the demographical statistics to predict crimes [10]. However, the traditional method cannot capture the spatiotemporal correlations of crime occurrences.

Deep Learning-Based Crime Prediction Algorithms. Recently advances in the crime prediction have been proposing spatiotemporal deep learning frameworks to predict crime occurrences [17,19]. However, they partition the region into synthetic units (e.g., grids). Therefore, they were not able to incorporate the geographical extent of neighborhoods as well as the socio-econimical implications of neighborhoods.

Graph Neural Networks. There exists a great number of deep graph neural networks. GraphSAGE [14] is a general inductive framework that aggregates node features to generate decent node embeddings by sampling and aggregating features from the local neighborhood of a node. DCRNN [22] captures the dependency by bidirectional random walks on the network while learning the temporal correlation based on the encoder-decoder architecture to predict the traffic. However, none of them were customized for the crime prediction task; they either apply on the regression task [22] or do not take into account the spatiotemporal correlations [14,21]. Inspired by the advances in the graph neural networks, CrimeForecaster uses the graph structure across geographical regions to further improve the spatiotemporal learning on the crime prediction scenario.

7 Conclusion

In this work, we proposed CrimeForecaster, an end-to-end deep learning spatial-temporal framework for crime forecasting, which relies on the geographical information of the neighborhoods in a city to capture the complex spatiotemporal dependencies of crimes. CrimeForecaster represents neighborhoods in a graph and uses diffusion convolution over the graph to learn the spatiotemporal dependencies within and across neighborhoods. We evaluate CrimeForecaster on two real-world datasets: Chicago and our own Los Angeles dataset. We publish both the code and datasets. In both experiments, CrimeForecaster consistently outperforms state-of-the-art crime prediction models. CrimeForecaster outperforms the best state-of-the-art model by 21% for Micro-F1 in September in the Chicago dataset. In future work, we will explore other ways to construct the graph, e.g., by using the explicit demographic information of neighborhoods. We believe that CrimeForecaster will bring more attention to both the graph neural network application and the crime prediction community.

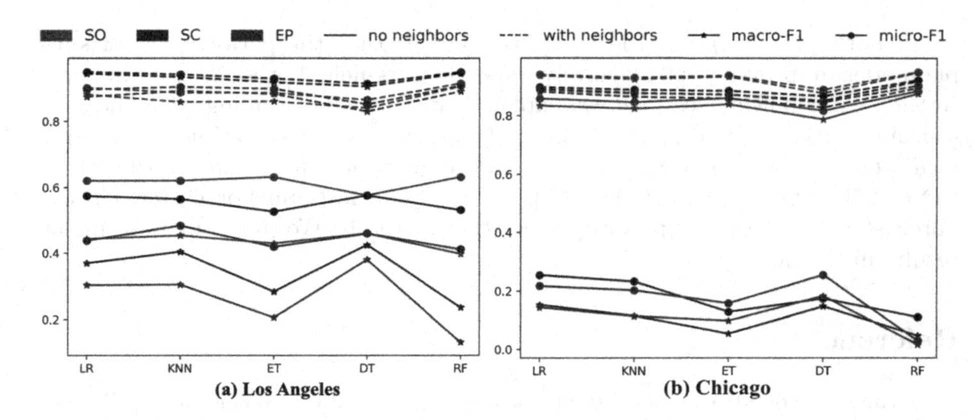

Fig. 6. Forecasting results for all crimes on three different neighborhoods in Los Angeles and Chicago. (Los Angeles: SO – Sherman Oaks; EP – Echo Park; SC – Studio City. Chicago: AG – Auburn Gresham; FP – Fuller Park; RD: RiverDale)

Acknowledgement. This work has been supported in part by Annenberg Leadership Initiative, the USC Integrated Media Systems Center, and unrestricted cash gifts from Microsoft and Google. The opinions, findings, and conclusions or recommendations expressed in this material are those of the authors and do not necessarily reflect the views of the sponsors.

Appendix A Impact of the Adjacent Neighborhoods' Information

We have observed that the spatial correlation would help with the crime prediction in Sect. 1. Here we want to know experimentally about if the neighborhood information helps to predict the crimes for a target individual neighborhood. Formally, for each neighborhood r_i, we would like to know if its neighborhoods' information helps to predict the crime event in the next time slot given the previous records. Therefore, we train the same classification models on two different feature settings: a) using the previous crime occurrences of r_i itself as features, and b) using the previous crime occurrences of both r_i itself and the aggregates of r_i's neighborhoods.

$$f\big(y_{i,l}^{K+1}|(y_{i,l}^{1},\ldots,y_{i,L}^{K})\big) \quad \text{V.S.} \quad f\big(y_{i,l}^{K+1}|\text{agg}(\{Y_{\mathcal{N}(1)},\ldots,Y_{\mathcal{N}(M)}\})\big)$$

where $Y_{\mathcal{N}(1)} = (y_{\mathcal{N}(1),l}^{1},\ldots,y_{\mathcal{N}(1),L}^{K})$, and L is the number of crime categories. Here f is a prediction function, which can be classification models, deep learning models and etc. $\mathcal{N}(m)$ denotes an adjacent neighbor of R_i, i.e., $W_{i,\mathcal{N}(m)} = 1$. In this experiment, we use classical machine learning classifiers: Logistic Regression (LR), K-Nearest Neighbor (KNN), Extra Tree (ET), Decision Tree (DT) and Random Forest (RF) classifiers.

In both Fig. 6 (*a*) and Fig. 6 (*b*), we can see that the performance of simple models improves a lot by simply appending its neighborhoods' information. In the next following sessions, we explored other ways of dealing with the spatial information to make the model applicable to predict the crime at multiple locations, such as proposing MiST* by adding network embedding learned from LINE [25] or using the diffusion graph convolution in CrimeForecaster. CrimeForecaster shows great superiority over other methods. We show our experiment results in the Sect. 5.

References

1. Crime data of 2015 in The City of Chicago. https://data.cityofchicago.org/Public-Safety/Crimes-2015/vwwp-7yr9
2. Boundary information of Chicago (2020). https://data.cityofchicago.org/widgets/bbvz-uum9
3. Census Bureau (2020). https://www.census.gov/
4. Crosstown Los Angeles (2020). https://xtown.la/
5. Abadi, M., et al.: Tensorflow: a system for large-scale machine learning. In: 12th {USENIX} Symposium on Operating Systems Design and Implementation ({OSDI} 16), pp. 265–283 (2016)
6. Chang, C.C., Lin, C.J.: LibSVM: a library for support vector machines. ACM Trans. Intell. Syst. Technol. (TIST) **2**(3), 1–27 (2011)
7. Chung, J., Gulcehre, C., Cho, K., Bengio, Y.: Empirical evaluation of gated recurrent neural networks on sequence modeling. arXiv preprint arXiv:1412.3555 (2014)
8. Contreras, J., Espinola, R., Nogales, F.J., Conejo, A.J.: Arima models to predict next-day electricity prices. IEEE Trans. Power Syst. **18**(3), 1014–1020 (2003)
9. Covington, P., Adams, J., Sargin, E.: Deep neural networks for YouTube recommendations. In: Proceedings of the 10th ACM Conference on Recommender Systems, pp. 191–198 (2016)
10. Featherstone, C.: Identifying vehicle descriptions in microblogging text with the aim of reducing or predicting crime. In: 2013 International Conference on Adaptive Science and Technology, pp. 1–8. IEEE (2013)
11. Gers, F.A., Schmidhuber, J., Cummins, F.: Learning to forget: Continual prediction with LSTM (1999)
12. Gorman, D.M., Speer, P.W., Gruenewald, P.J., Labouvie, E.W.: Spatial dynamics of alcohol availability, neighborhood structure and violent crime. J. Stud. Alcohol **62**(5), 628–636 (2001)
13. Grover, A., Leskovec, J.: node2vec: scalable feature learning for networks. In: Proceedings of the 22nd ACM SIGKDD International Conference on Knowledge Discovery and Data Mining, pp. 855–864 (2016)
14. Hamilton, W., Ying, Z., Leskovec, J.: Inductive representation learning on large graphs. In: Advances in Neural Information Processing Systems, pp. 1024–1034 (2017)
15. He, J., Qi, J., Ramamohanarao, K.: A joint context-aware embedding for trip recommendations. In: 2019 IEEE 35th International Conference on Data Engineering (ICDE), pp. 292–303. IEEE (2019)
16. Hosmer Jr., D.W., Lemeshow, S., Sturdivant, R.X.: Applied Logistic Regression, vol. 398. Wiley, Hoboken (2013)

17. Huang, C., Zhang, C., Zhao, J., Wu, X., Chawla, N.V., Yin, D.: Mist: a multiview and multimodal spatial-temporal learning framework for citywide abnormal event forecasting. In: WWW 2019 (2019)
18. Huang, C., Zhang, C., Zhao, J., Wu, X., Yin, D., Chawla, N.: Mist: a multiview and multimodal spatial-temporal learning framework for citywide abnormal event forecasting. In: The World Wide Web Conference, pp. 717–728 (2019)
19. Huang, C., Zhang, J., Zheng, Y., Chawla, N.V.: Deepcrime: attentive hierarchical recurrent networks for crime prediction. In: Proceedings of the 27th ACM International Conference on Information and Knowledge Management, pp. 1423–1432, CIKM 2018, Association for Computing Machinery, New York, NY, USA (2018). https://doi.org/10.1145/3269206.3271793
20. Kingma, D.P., Ba, J.: Adam: a method for stochastic optimization. arXiv preprint arXiv:1412.6980 (2014)
21. Kipf, T.N., Welling, M.: Semi-supervised classification with graph convolutional networks. arXiv preprint arXiv:1609.02907 (2016)
22. Li, Y., Yu, R., Shahabi, C., Liu, Y.: Diffusion convolutional recurrent neural network: data-driven traffic forecasting. arXiv preprint arXiv:1707.01926 (2017)
23. Li, Y., Yu, R., Shahabi, C., Liu, Y.: Diffusion convolutional recurrent neural network: Data-driven traffic forecasting. In: International Conference on Learning Representations (ICLR 2018) (2018)
24. Morenoff, J.D., Sampson, R.J.: Violent crime and the spatial dynamics of neighborhood transition: Chicago, 1970–1990. Social Forces **76**(1), 31–64 (1997)
25. Tang, J., Qu, M., Wang, M., Zhang, M., Yan, J., Mei, Q.: Line: large-scale information network embedding. In: Proceedings of the 24th International Conference on World Wide Web, pp. 1067–1077 (2015)
26. Vaswani, A., et al: Attention is all you need. In: Advances in Neural Information Processing Systems, pp. 5998–6008 (2017)
27. Wang, H., Kifer, D., Graif, C., Li, Z.: Crime rate inference with big data. In: Proceedings of the 22nd ACM SIGKDD International Conference on Knowledge Discovery and Data Mining, pp. 635–644 (2016)
28. Wang, J., Cong, G., Zhao, W.X., Li, X.: Mining user intents in Twitter: a semi-supervised approach to inferring intent categories for tweets. In: AAAI (2015)
29. Zhao, L., et al: T-GCN: a temporal graph convolutional network for traffic prediction. IEEE Trans. Intell. Transp. Syst. (2019)

PS3: Partition-Based Skew-Specialized Sampling for Batch Mode Active Learning in Imbalanced Text Data

Ricky Maulana Fajri(✉), Samaneh Khoshrou, Robert Peharz, and Mykola Pechenizkiy

Department of Mathematics and Computer Science, Eindhoven University of Technology, 5600 Eindhoven, MB, The Netherlands
{r.m.fajri,s.khoshrou,r.peharz,m.pechenizkiy}@tue.nl

Abstract. While social media has taken a fixed place in our daily life, its steadily growing prominence also exacerbates the problem of hostile contents and hate-speech. These destructive phenomena call for automatic hate-speech detection, which, however, is facing two major challenges, namely i) the dynamic nature of online content causing significant data-drift over time, and ii) a high class-skew, as hate-speech represents a relatively small fraction of the overall online content. The first challenge naturally calls for a batch mode active learning solution, which updates the detection system by querying human domain-experts to annotate meticulously selected batches of data instances. However, little prior work exists on batch mode active learning with high class-skew, and in particular for the problem of hate-speech detection. In this work, we propose a novel partition-based batch mode active learning framework to address this problem. Our framework falls into the so-called screening approach, which pre-selects a subset of most uncertain data items and then selects a representative set from this uncertainty space. To tackle the class-skew problem, we use a data-driven skew-specialized cluster representation, with a higher potential to "cherry pick" minority classes. In extensive experiments we demonstrate substantial improvements in terms of G-Means, and F1 measure, over several baseline approaches and multiple datasets, for highly imbalanced class ratios.

Keywords: Batch-mode active learning · Imbalance data · Hate-speech recognition

1 Introduction

Today, the digital revolution provides new platforms to communicate, share ideas and advertise products. It opens up an almost unlimited space for users to express themselves freely and (sometimes) anonymously. While freedom of speech is a fundamental and cherishable human right, spreading hate towards other members of the society is a violence of this liberty and cannot be tolerated. Due to

Y. Dong et al. (Eds.): ECML PKDD 2020, LNAI 12461, pp. 68–84, 2021.
https://doi.org/10.1007/978-3-030-67670-4_5

the massive amount of data, the use of AI-technologies to combat hate-speech seems to be a natural choice. However, despite the rapid advancement of this technology, AI is no silver bullet, and in many complex situations and ambiguous contents, human experts still need to take the lead. For example, hate-speech recognition on Facebook illustrates that a pure AI solution is clearly out of sight: Despite the long-term effort to develop a fully algorithmic-based hate-speech recognition system, the company has hired an "army" of over 10,000 human reviewers to address the problem.[1]

Given these circumstances, the question is: How can human and AI team up to solve such tasks? Current state-of-the-art calls for *active learning strategies*, which allow AI to interactively query human experts to label the most ambiguous instances. Active learning comes in two ways, namely sequential and batch-mode active learning. In sequential active learning, a single data item is selected at each iteration, which is annotated and subsequently used to update (re-train) the model. This strategy is rather costly, especially when combined with complex and expensive models, and puts quite some strain on human annotators. Therefore, in practical applications batch-mode active learning (BMAL) is typically preferred [17,20,24].

However, state-of-the-art BMAL is rather ill-suited for hate-speech recognition, as the problem is inherently class-skewed: It is somewhat ironic that aggressive users, whose targets are often minority groups, are a minority themselves among the pool of network users, leading to highly class-skewed data. While there exist several active learning approaches specifically designed to deal with imbalanced data [28,29], to the best of our knowledge, this paper is the first take to explore BMAL under heavy class-skew. Furthermore, much of the literature is mainly concerned with the accuracy of the prediction and not its deployment in real-world applications with respect to human limitations. In this work, we take the first steps towards designing an interactive framework to deal with the imbalance problem in the real-world, by taking the domain-expert characteristics into account. In fact, the budget of BMAL in our experiments is assessed from domain-experts acting in real working situations, and not determined by a hard-coded oracle.

Main Contributions: Motivated by the problem of hate-speech recognition, we propose *Partition-based Skew-Specialized Sampling* (PS3), a partition-based skew-specialized BMAL framework which is specifically designed to classify imbalanced text data. PS3 follows a *screening approach* [24] which, in each iteration, pre-selects a pool of uncertain samples of which a representative batch is then selected for annotation. In contrast to the *objective-driven* (or direct) approach, the screening approach generally scales much more gracefully to large data. Our selection process incorporates a preference for the minority class, leading to a more balanced query set. In particular, we use Gaussian mixture models to capture the diversity within the uncertain set, and select a representative number from each cluster to be annotated. Since the choice of representatives is

[1] http://theconversation.com/why-ai-cant-solve-everything-97022.

key in any partition-based active learning framework, we propose a novel data-driven partition representation, targeted at skewed data. We mainly focus on hate-speech detection as one of the most challenging problems in digital world, and conduct extensive experiments on multiple benchmark datasets. We design a set of experiments to formulate the real human expert handling ability to interact with the system, leading to an upper bound for the labeling budget.

The rest of the paper is organized as follows: in Sect. 2 we introduce the line of works related to this paper, Sect. 3 presents an overview of the learning framework, in Sect. 4 we discuss the experimental methodology, in Sect. 5 we experimentally investigate the different characteristics of our proposed approach compared with the baseline approaches, and finally, Sects. 6 concludes the paper's contributions and indicate the future directions.

2 Related Work

Active learning has been extensively studied in the literature, see e.g. [7,19, 22,28]. Most of the methods operate in a sequential manner, where "the most informative sample" is chosen for labeling and added to the training set, followed by re-training of the model. This procedure is iterated until most of the remaining unlabeled examples can be classified with "reasonably high confidence" [25,27]. However, re-training after each iteration is quite costly [13,15], which is the main rationale behind batch-mode active learning (BMAL), which selects a batch of informative instances rather than a single instance. Another incentive for the BMAL approach is to reduce the strain on the human annotator(s), as the delay between labeling single examples, caused by re-training the model, has an unnerving effect. In this study, we evaluated the annotation time for the task of hate-speech detection and calibrated the BMAL batch size accordingly. The central task for BMAL is to select, in each iteration, a batch of unlabeled samples, trading off between *uncertainty* and *representativeness* [24]. Uncertain samples are likely helpful to refine the decision boundary, while representativeness (according to some surrogate criterion) strives to capture the data structure of the unlabeled set. There are two main BMAL approaches aiming to achieve this trade-off: i) objective-driven methods, and ii) screening approaches. *Objective-driven methods* formulate this trade-off in a single objective, such that the most informative/representative batch of samples is found by solving an optimization problem [3,4,9,12,14,21,24]. These approaches are often theoretically well-grounded and have demonstrated good performance in practice. However, they typically do not scale well to big datasets [17].

On the other hand, *screening-based methods* first reduce the unlabeled space to a subset containing the most uncertain samples. Subsequently, they partition (cluster) this subset and select a representative sample from each cluster, in order to reduce the probability of picking correlated samples [6,20,27]. Some approaches also omit the screening stage and work with the entire unlabeled space [23]. Typically, the sample closest to the cluster center is selected as representative instance [6], but at times also the most uncertain [20] or the highest ranked sample [2]. In comparison to objective-driven BMAL approaches, screening/partition-based approaches scale gracefully to large datasets, since the main computational bottleneck is the clustering stage, for which many computationally attractive algorithms exist.

The underlying assumption is that homogeneous clusters in uncertainty space also reflect homogeneous clusters in data space, and that each cluster can be essentially shrunk into a single data point without scarifying performance. Our main insight in this paper is that, under strong class-skew, the *single* representative of each cluster might be biased towards the majority class. To this end, we devise a data-driven approach to select *several* representative samples per cluster, in particular "extreme" examples, in order to better capture the minority class. While classic sequential active learning has been explored for skewed data [16,28,29], BMAL approaches for class-skewed data are an unexplored area. Our screening-based framework is, to the best of our knowledge, the first attempt in this direction.

3 PS3: Partition-Based Skew-Specialized Sampling for Batch-Mode Active Learning

In this section, we present the Partition-based Skew-Specialized Sampling (PS3) for batch-mode active learning framework, which has been specifically designed to tackle the class-imbalance problem in real-world applications, where annotators work over limited hours daily. The framework is comprised of two main components: i) Batch-mode imbalance learning that mainly focuses on finding a compromise between exploration and exploitation to effectively cover the uncertain space subject to a predefined budget. ii) Human-in-the-Loop, that models the real-domain expert behavior in a working environment.

3.1 Batch-Mode Imbalance Learning

The main objective of PS3 is to develop a scalable batch-model framework for the class-imbalance problem. The success of batch mode active learning depends on selecting representative samples [24] as well as the batch size and total budget constraints [17]. The key question is how to find the most representative samples from both the minority and majority classes to cover the whole uncertain space given the limited budget. To achieve these goals, PS3 learning component consists of three folds: i) screening stage, ii) representative set, and iii) weighted model training.

First, we introduce the formal setting of BMAL. Let $X = \{x_1, x_2, ..., x_n\}$ denote a dataset of n instances, which consists of a *labeled* set L and an *unlabeled* set U, where $L \bigcup U = X$ and $L \bigcap U = \emptyset$. Every instance $x_i^L \in L$ is associated with a label y_i^L, which has been revealed by a domain expert d, whereas the labels associated with x_i^U are not known yet. PS3 interactively selects a batch B of samples, satisfying $B \subset U$ and $|B| = b$, where the *batch size* b is determined by *human labeling ability*. In this paper, we assume that all instances are of equal annotation cost. PS3 operates in T iterations, where in each iteration the learner selects b instances to be labeled by the domain expert, which are subsequently added to the labeled set L to update the classifier.

As mentioned in Sect. 2, most BMAL approaches select the batch B guided by a trade-off between *uncertainty* and *representativeness.* While *objective-driven* BMAL formulates this trade-off explicitly as an objective function to be optimized, *screening-based* BMAL operates in a two-stage manner, by first screening for a candidate subset of most uncertain samples, and subsequently selecting a representative set from the candidate set to be labeled. Our approach falls in the latter category, as screening-based BMAL methods generally scale better to big data.

Screening Stage. Let h be the classifier trained on the labeled set L from iteration $t-1$. One of the most prominent uncertainty criteria is the sample-wise *class-entropy*:

$$H(x) = -\sum_{y \in \mathcal{Y}} P(y|x,h) \log P(y|x,h), \tag{1}$$

where $\mathcal{Y}$ is the set of possible class-values and $P(y|x,h)$ is the probability of class y under model h, usually given by a soft-max output layer. The entropy is computed for each sample in the unlabeled set U and the samples with highest entropy are selected, forming a subset $US \subseteq U$. 10% of the most uncertain instances of U is sent for the partitioning stage, for example in Founta dataset US includes 500 instances out of 5000 instances of U, Infact we reduce the search space by a factor of 10. The screening stage is the "goal-directed" and "information-seeking" exploration stage, reducing the unlabeled space U to the *uncertainty space* US. Typically, the set US is too large to be annotated as a whole, especially when dealing with large-scale datasets. Therefore, it is further reduced by selecting representative examples from it.

Partitioning Stage. Almost all screening BMAL methods in the literature [6,20] apply clustering to partition the uncertainty space US into subgroups. The hope is that clustering captures the inherent structure of US, where clusters represent diverse modes of the data, while samples within each cluster belong to the same mode. Most paper apply k-means clustering, which strives for clusters modeled as isotropic balls. In this paper, we employ Gaussian mixture models trained with expectation-maximization, followed by a hard assignment of samples to their most likely cluster. Due to the large dimensionality of text data, we employ Gaussians with diagonal covariances. Thus, the resulting clusters correspond to axis-aligned ellipsoids, which allow to capture clusters whose features are expressed on different length-scales. In preliminary experiments we found that GMM-based clustering generally performed better than k-means clustering. After clustering, a representative set (one or multiple instances) is selected from each cluster.

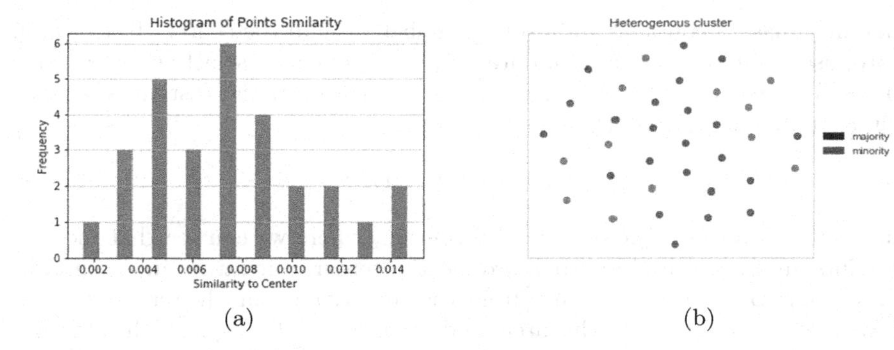

(a) (b)

Fig. 1. Analysis of clusters after partitioning-stage, taken from the Founta dataset (see Sect. 4). (a) Histogram of the distances to cluster center. (b) An example of a heterogeneous cluster. While a common assumption in BMAL is that clusters after the partitioning stage are homogeneous, a typical cluster contains samples from both the majority and minority classes. For visualization in 2 dimensions, we used t-SNE embedding [18].

Representative Set. After the partitioning stage, most state-of-the-art BMAL approaches select a *single* point (e.g. cluster centroid or most uncertain point [6, 20]) to represent the cluster. The basic assumption of these approaches is that clusters are roughly homogeneous with respect to the class value, and that this homogeneity is maintained in the uncertainty space. Under the homogeneity assumption, also class-skew would be roughly maintained and reducing each cluster to a single sample, using either of the selection mechanisms above, would not sacrifice any performance.

However, according to our own experiments and observations, this assumption is often violated, and clusters in the uncertainty space are often quite *heterogeneous* with respect to the class value. Figure 1(b) illustrates an examplar heterogeneous cluster of Founta dataset (for visualization purposes a 2-d projection is presented). One observes that the minority class rather spread out over cluster space. This effect is especially disadvantageous when dealing with high class-skew, as many clusters might be dominated by the majority class. In these cases, selecting cluster centroids as representatives will bias the selected samples towards the majority class. One might expect that selecting representatives according to uncertainty might be less prone to such bias, as the label uncertainty is likely uncorrelated with the class label, in which case the class ratio of the selected samples should be roughly equal to the overall class ratio. In this paper, however, we devise a selection mechanism which actively strives for samples from the minority class.

Tale of Tails Representation (T^2). As discussed above, selecting a single sample per cluster might bias the class proportion towards the majority class. Moreover, it is well known that the mode (center) of a Gaussian distribution is not a typical sample of the distribution, and that most of the Gaussian's probability mass is contained in a spherical shell around the center. To this end, we propose a simple yet effective data-driven approach to select representative samples from each cluster. We first take a histogram of the distances of each point x_i to its cluster center μ_k:

$$\forall x_i \in C_k, d_i = dist(x_i, \mu_k) \tag{2}$$

Figure 1 shows an example of such a histogram, where we can see that most of the points are not located in a ball around the center, but rather in an annulus with a fairly considerable distance from the center. For our cluster representation we select a sample from this area, and, in order to also capture the cluster's tails, we select two samples corresponding to the %5 and %95 quantiles, respectively. In order to select these three samples, we simply fit a Gaussian to the cluster distances (2), and select the three samples closest to the 5%, 50% and 95% quantiles of the fitted distribution. Our experiments demonstrate that our "Tale of Tails" (T^2) representation has a higher potential to pick samples of the minority class as representatives.

Weighted Model Training. Once a batch is obtained, a classifier (h) is trained or updated to reflect the latest knowledge. PS3 is a model agnostic framework, and any classifier can be applied. In this paper, we employ SVM as our base learner. In a highly skewed environment where the percentage of the minority class is extremely low (under 10%) [1], conventional active learning approaches tends to perform poorly [16], and the model's performance tends to fluctuate over the training iterations. One of the classic approaches to overcome the class imbalance is to increase the penalty for miss-classifying minority samples, using a weighing strategy. The weighing compensates for the class-skew and essentially moves the generated hyperplane towards the real classification boundary. Similar to [28], we employ weights given by

$$w_i = \begin{cases} \frac{|N^-|}{|N^-|+|N^+|}, & \text{if } y_i = 1 \text{ (hate-speech)}, \\ \frac{|N^+|}{|N^+|+|N^-|}, & \text{if } y_i = 0 \text{ (no hate-speech)}, \end{cases} \tag{3}$$

where N^- and N^+ are the number of non hate-speech samples and number of hate-speech samples, respectively. The weights are updated after each iteration, in order to reflect the characteristics of the latest labeled set.

3.2 Human in the Loop – Assessment of Labeling Effort for Hate-Speech Detection

Most studies on BMAL use an arbitrary batch size for their experiments. While BMAL should, in principle, work for any batch size, we aimed to mimic a realistic setting for hate-speech detection. To this end, we performed a small preliminary study to assess the labeling ability of domain experts. Our experiments were performed in our research lab in an academic setting. We asked 6 fluent English

Table 1. The characteristics of datasets used in our experiments

Name	#Instance	#Majority class	#Minority class	%Minority	Class
Davidson [5]	5233	3857	1376	26	Hate/No hate
Founta [8]	11863	11126	737	6	Hate/No hate
Gao and Huang [10]	1127	704	423	37	Hate/No hate
Golbeck [11]	20324	15042	5282	25	Hate/No hate
Waseem [26]	1811	1582	229	12	Hate/No hate
Amazon review[a]	20000	16000	4000	25	Positive/Negative
Amazon polarity[b]	35000	26250	8750	33	Positive/Negative
Fake news[c]	11887	10387	1500	14	Reliable/Un-reliable
Hotel review[d]	25000	20000	5000	25	Positive/Negative
Sentiment[e]	30000	25000	5000	20	Positive/Negative

[a] https://www.kaggle.com/shap/amazon-fine-food-review/.
[b] https://www.kaggle.com/bittlingmayer/amazonreviews.
[c] https://www.kaggle.com/c/fake-news/data.
[d] https://www.kaggle.com/jonathanoheix/sentiment-analysis-with-hotel-reviews/.
[e] https://www.kaggle.com/lakshi25npathi/imdb-dataset-of-5ok-movie-reviews/version/1.

speakers without English linguistics background to label our pilot set (dataset Founta, see below). The guidelines for the labeling procedure were provided to the annotators beforehand in order to ensure consistency with the ground-truth labels.[2] Every hour each domain expert (d) provided a vector $Y_d = [y_{d,1}, y_{d,2} \dots]$, where $\forall y_{d,i}\colon y_{d,i} \in Y_d, y_{d,i} = \{0, 1\}$ (0 and 1 refers to the majority and minority samples, respectively) for every instance $x_i \in Pilot$. The average length of Y_d over domain experts, denoted as B, defines the average number of queries a domain expert can handle hourly. We ran all the experiments for 8 h as it would represent a full working day. On average, the annotators labeled $2\frac{2}{3}$ sentences per minute. Therefore the BMAL budget was fixed to 180 instances per hour.

4 Experimental Methodology

4.1 Datasets

We conducted our experiments on 10 publicly available text datasets from various domains such as hate-speech recognition, sentiment analysis, and product review. We mainly focus on skewed text data as they represent a large group of real world challenges in data science. Table 1 illustrates the characteristics of datasets used in this paper.

Data Pre-processing and Parameter Tuning. We follow the preprocessing steps used in [5], which include lowercasing and stemming each of the

[2] Annotators had to label "hate-speech" ($y = 1$) vs. "no hate-speech" ($y = 0$). Hate-speech was defined as a statement expressing hate or extreme bias towards a particular group, in particular defined via religion, race, gender, or sexual orientation. Offensive or hateful expressions directed towards individuals, without reference to a group-defining property, did not count as hate-speech.

instances using the Natural Language Toolkit.[3] Furthermore, we removed hashtags, retweet, URLS and Mention from the original tweet data. For PS3, only hyper parameter k (i.e. the number of clusters) is computed by human handling ability (i.e. $B = 180$ samples per hour) as well as the number of representative points per cluster (3 points per cluster for T^2). Thus, k is set to $180/3 = 60$. For all BMAL methods, we employed support vector machines (SVMs) with RBF kernel as underlying model.

4.2 Baseline Methods

We compared PS3 with multiple screening BMAL approaches as well as one of the most recent objective-driven method, namely LBC, which outperforms multiple state-of-the-art objective-driven BMAL.

- **CBMAL (Certainty Based BMAL)** [20]. As one of the first screening-based methods, the points which the classifier is less certain about, are clustered together. Then, the most ambiguous point from each cluster is forwarded to be labeled by a hard-coded oracle.
- **MCLU (Most Informative Clustered Based Unit)** [6]. MCLU follows almost similar steps to cluster uncertain points, however instead of the most ambiguous samples, the most "informative" sample, usually cluster centroid, is chosen to represent the cluster.
- **Random strategy.** Following similar steps as above. The uncertain space is partitioned into a pre-set number of clusters, each of which are represented by a randomly selected instance.
- **LBC** [24]. As one of the most recent state-of-the-art *objective-driven* batch active learning, LBC uses a lower bounded certainty score of each unlabeled data. Then a large similarity matrix over all unlabeled space is formed and a random greedy algorithm is employed to find a candidate batch for labeling.
- **Batch-Rank** [2]. Batch-Rank, which lies at the intersection of screening and objective-driven methods, ranks instances according to the classifier's uncertainty and similarity to the pre-labeled instances. In particular, low confidence and high similarity correspond to high rank. At each iteration the batch of instances with highest rank instances is queried.

4.3 Evaluation Criteria

In conventional classification problem *accuracy* is a standard choice for performance evaluation. However, it fails to reflect the performance on the skewed datasets. In such scenarios G-Means and F1 measures are widely used in the literature. G-Means is defined as the geometric mean of sensitivity (true positive rate) and specificity (true negative rate), and F-1 as the harmonic mean of precision and recall:

$$G - Means = \sqrt{Sensitivity \times Specificity} \tag{4}$$

[3] nltk.org.

$$F1 = \frac{2 \times Precision \times Recall}{Precision + Recall} \tag{5}$$

Since in this paper we particularly focus on class-skewed problems, we refrain from reporting classification accuracies in this study.

5 Results and Discussion

We conducted two sets of experiments to address the following questions: How well does PS3 work compared to other state-of-the-art approaches? And, in particular, how well does PS3 work when facing highly skewed data? These questions are discussed in the following section.

5.1 Performance Evaluation

We randomly divided each dataset into 3 disjoint sets: labeled train (10%), unlabeled train (70%), and test (20%). To have a fair comparison in terms of the query number, k is set to 180 for all the baseline approaches. The average of 10 runs are reported. Figure 2 illustrates G-Means as a function of query number for PS3 compared with all baseline approaches. We observe that PS3 outperforms the other methods in most of the datasets, often by a considerable margin. Furthermore, looking closer at the instances chosen for labeling we find that PS3 is able to query more instances from the minority class, leading to a better classification accuracy, which is supported by our results in terms of F1 measure. Table 2 provides F1 values for PS3 as well as baseline approaches on different datasets.

To further assess the performance of PS3, we tested for significant differences in terms of G-Means score, using paired t-tests at significance level 0.05. The results (i.e. win/tie/loss of PS3 versus other methods) are presented in Table 4. PS3 achieves the highest number of wins over the baselines on all of the used datasets. Interestingly, PS3 is the "absolute winner" (32/0/0) for the most skewed dataset, *Founta.*

The second set of experiments was designed to investigate PS3's capabilities to treat imbalanced data. First, we processed each data set into new versions with different skewness level (i.e. containing different numbers of majority and minority samples), ranging from a balanced version to a highly skewed version, including only 1% of minority samples. Due to lack of sufficient data for a balanced version, we excluded Davidson, Gao & Huang, and Waseem, and conducted our experiments on the remaining 7 datasets. We performed the same data splitting strategy as in the previous experiment, leaving most of the samples in the unlabeled pool.

Figure 3 presents the performance of all the methods as a function of different imbalance ratio. The average G-Means performance over 10 runs demonstrates that almost all the methods are performing equally well in a balanced setting and the G-Means deviation is below 2%. In fact, for most of the datasets there is one baseline which marginally outperforms PS3 (see Fig. 4(a)). We also

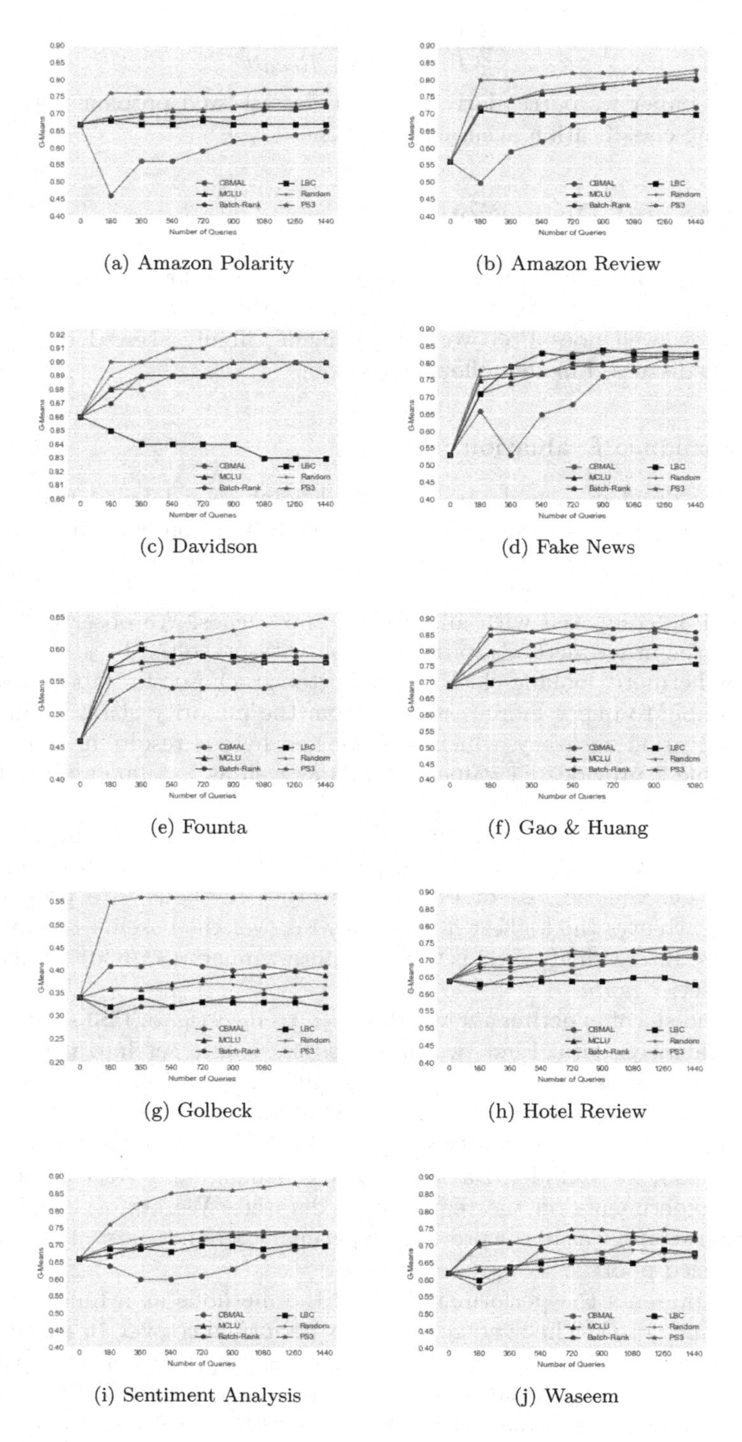

(a) Amazon Polarity (b) Amazon Review

(c) Davidson (d) Fake News

(e) Founta (f) Gao & Huang

(g) Golbeck (h) Hotel Review

(i) Sentiment Analysis (j) Waseem

Fig. 2. Comparison of the performance of PS3 with baseline methods on multiple datasets (G-Means against number of queries).

Table 2. F1 measure of various algorithms on multiple datasets

Dataset	Methods					
	CBMAL	MCLU	Batch rank	LBC	Random	PS3
Amazon polarity	0.43	0.63	0.61	0.59	0.60	**0.66**
Amazon review	0.55	0.65	0.64	0.64	0.59	**0.74**
Fake news	0.74	0.82	0.81	0.84	0.79	**0.86**
Davidson	0.85	0.87	0.83	0.65	0.88	**0.92**
Founta	0.60	0.59	0.55	0.63	0.60	**0.67**
Gao and Huang	0.84	0.83	**0.87**	0.75	0.76	0.85
Golbeck	0.43	0.42	0.34	0.36	0.38	**0.54**
Hotel review	0.68	**0.74**	0.66	0.58	0.63	0.71
Sentiment	0.64	0.69	0.66	0.65	0.69	**0.75**
Waseem	0.68	**0.74**	0.66	0.64	0.62	0.72

Table 3. Percentage of minority class selected by each method

Dataset	Ratio of minority class selected (%)					
	CBMAL	MCLU	Batch rank	LBC	Random	PS3
Amazon polarity	6%	17%	18%	18%	16%	**19%**
Amazon review	3%	8%	8%	8%	5%	**10%**
Davidson	14%	21%	19%	20%	17%	**27%**
Fake news	40%	26%	33%	17%	35%	**49%**
Founta	34%	32%	36%	33%	37%	**43%**
Gao and Huang	39%	34%	42%	38%	36%	**48%**
Golbeck	5%	11%	10%	11%	9%	**16%**
Hotel review	14%	**15%**	13%	14%	**15%**	**15%**
Sentiment	5%	14%	13%	12%	10%	**18%**
Waseem	50%	66%	62%	53%	65%	**77%**

observe that, *the more imbalanced the dataset is, the lower is G-Means*, which can be explained by the lack of training data for the minority class. However, PS3 shows a more stable performance under such circumstances. The improvement gets more tangible when facing more skewed versions of the datasets. In fact, PS3 is the absolute winner in the most imbalanced setting (see Fig. 4(b)), which is explained by higher potential of "Tale of the tails" representation to capture minorities. Table 3 compares the ratio (%) of minority instances (e.g. Hate samples) picked by PS3 against all other baseline approaches on different datasets. Attaining top rank among all other approaches on all the explored datasets confirms the exploration power of PS3 to "target" minorities.

Table 4. Win/tie/loss counts of PS3 versus other methods on different datasets

Dataset	PS3 versus					
	CBMAL	MCLU	Batch rank	LBC	Random	In all
Amazon polarity	8/0/0	8/0/0	8/0/0	8/0/0	8/0/0	40/0/0
Amazon review	8/0/0	8/0/0	8/0/0	8/0/0	8/0/0	40/0/0
Fake news	8/0/0	8/0/0	8/0/0	8/0/0	8/0/0	40/0/0
Davidson	8/0/0	8/0/0	6/2/0	8/0/0	7/1/0	37/3/0
Founta	7/1/0	8/0/0	8/0/0	8/0/0	8/0/0	39/1/0
Gao and Huang	5/0/0	5/0/0	2/1/2	5/0/0	5/0/0	22/1/2
Golbeck	8/0/0	8/0/0	8/0/0	8/0/0	8/0/0	40/0/0
Hotel review	8/0/0	3/3/2	8/0/0	8/0/0	8/0/0	35/3/2
Sentiment	8/0/0	8/0/0	8/0/0	8/0/0	8/0/0	40/0/0
Waseem	8/0/0	8/0/0	8/0/0	8/0/0	8/0/0	40/0/0
In All	76/1/0	72/3/2	72/3/2	77/0/0	76/1/0	**373/8/2**

Table 5. Computational time of each methods per one iteration in seconds

Dataset	Methods					
	CBMAL	MCLU	Batch rank	LBC	Random	PS3
Amazon polarity	25.60	24.23	26.62	656.23	24.20	**24.10**
Amazon review	15.53	16.23	17.23	489.02	**15.40**	**15.40**
Davidson	5.06	4.36	4.50	256.3	3.69	**3.4**
Fake news	7.89	7.40	7.12	458.40	7.12	**7.10**
Founta	7.01	7.58	7.20	456.78	7.18	**6.89**
Gao and Huang	**2.80**	2.88	2.81	150.23	2.89	2.90
Golbeck	8.89	9.59	8.56	487.20	9.03	**8.07**
Hotel review	21.20	22.34	21.43	468.89	**21.03**	21.10
Sentiment	26.56	27.88	26.50	486.88	**24.50**	26.68
Waseem	2.98	2.82	2.88	189.55	**2.744**	2.79

Reproducibility: All the code and data sets used in the experiments are available at "https://github.com/rmfajri/PS3".

5.2 Computational Time Analysis

The results above were evaluated in terms of performance over the number of queries, which is a natural approach to evaluate BMAL methods. Additionally, we measured actual runtimes of all methods, in order to demonstrate that PS3 also scales gracefully with limited resources. Table 5 compares PS3 and the baselines in terms of runtime (in seconds) required for one iteration on an Intel I7

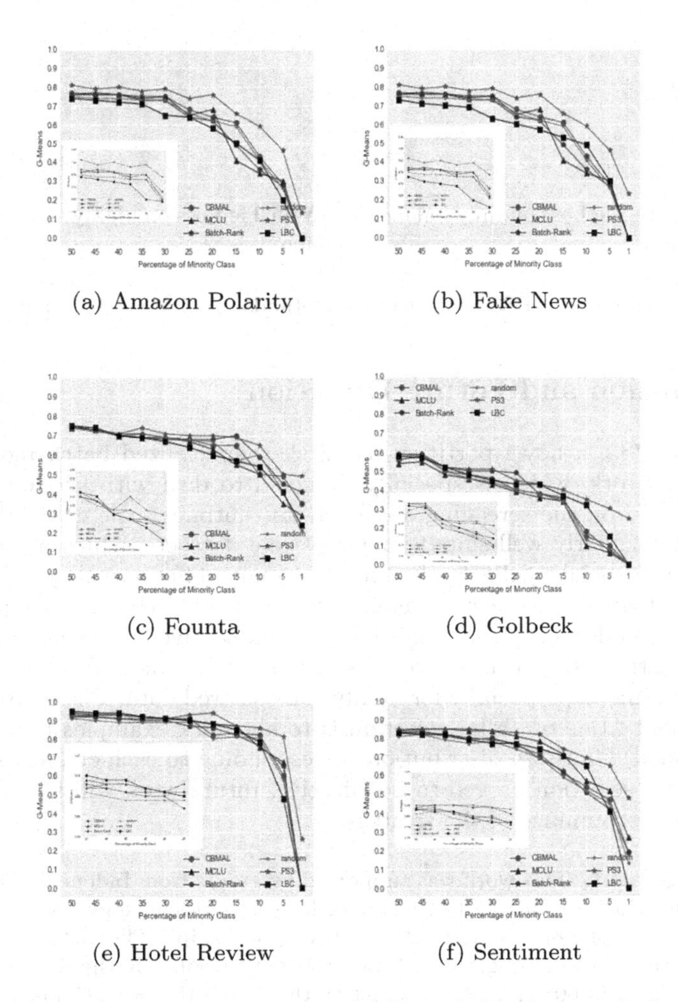

(a) Amazon Polarity (b) Fake News

(c) Founta (d) Golbeck

(e) Hotel Review (f) Sentiment

Fig. 3. Comparison of the performance of PS3 with baseline methods on different imbalance ratio.

CPU (4 cores) with 32 GB of Ram. From the table one can summarize that LBC, as a objective-driven approach, is a the most expensive, while PS3 constitutes top rank, attaining the shortest processing time for hate-speech and the second spot for other text datasets.

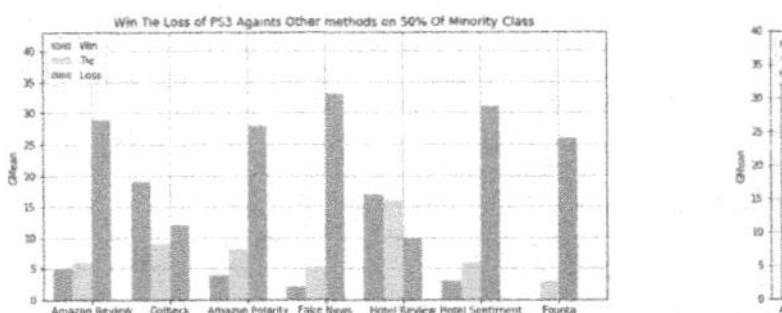

(a) Win Tie Loss on fully balanced version

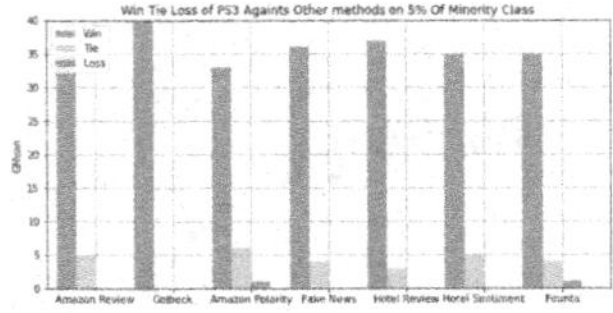

(b) Win Tie Loss on highly skewed version

Fig. 4. Win tie loss of PS3 against other methods on different imbalance ratio

6 Conclusion and Future Direction

We proposed PS3, a novel partition-based skew-specialized batch mode active learning framework, which is specially designed to deal with skewed datasets. Since it is based on the screening approach its runtimes scale gracefully, which make PS3 particularly well-suited for large-scale datasets, which are nowadays common in the social media landscape. The main difference to previous techniques is the strategy for selecting representatives from each cluster in the uncertainty space. While being based on a simple data-driven analysis, our experiments demonstrate that this strategy is very effective and led to large performance improvements, especially for highly class-skewed data. The main intuition of our so-called "Tale of Tails" selection is to also pick examples from the tails of the (assumed) Gaussian distribution corresponding to each cluster. While the "Tale of Tails" selection proved to be effective, future work will investigate the theoretical underpinnings of the strategy.

Acknowledgement. This work was supported by grants from Indonesia Endowment Fund for Education (LPDP) and Ministry of Research, Technology and Higher Education of the Republic of Indonesia (BUDI-LN Scholarship). The authors also would like to thank the research programme Commit2Data, specifically the RATE-Analytics project NWO628 003 001 (partly) financed by the Dutch Research Council.

References

1. Attenberg, J., Provost, F.J.: Why label when you can search?: alternatives to active learning for applying human resources to build classification models under extreme class imbalance. In: KDD (2010)
2. Cardoso, T.N.C., Silva, R.M., Canuto, S.D., Moro, M.M., Gonçalves, M.A.: Ranked batch-mode active learning. Inf. Sci. **379**, 313–337 (2017)
3. Chakraborty, S., Balasubramanian, V.N., Panchanathan, S.: Dynamic batch mode active learning. In: CVPR 2011, pp. 2649–2656 (2011)
4. Chakraborty, S., Balasubramanian, V.N., Panchanathan, S.: Adaptive batch mode active learning. IEEE TNNLS **26**, 1747–1760 (2015)
5. Davidson, T., Warmsley, D., Macy, M.W., Weber, I.: Automated hate speech detection and the problem of offensive language. In: ICWSM (2017)

6. Demir, B., Persello, C., Bruzzone, L.: Batch-mode active-learning methods for the interactive classification of remote sensing images. IEEE Trans. Geosci. Remote Sens. **49**, 1014–1031 (2011)
7. Elahi, M., Braunhofer, M., Ricci, F., Tkalcic, M.: Personality-based active learning for collaborative filtering recommender systems. In: Baldoni, M., Baroglio, C., Boella, G., Micalizio, R. (eds.) AI*IA 2013. LNCS (LNAI), vol. 8249, pp. 360–371. Springer, Cham (2013). https://doi.org/10.1007/978-3-319-03524-6_31
8. Founta, A.M., et al.: Large scale crowdsourcing and characterization of twitter abusive behavior. In: ICWSM (2018)
9. Gal, Y., Islam, R., Ghahramani, Z.: Deep Bayesian active learning with image data. In: ICML (2017)
10. Gao, L., Huang, R.: Detecting online hate speech using context aware models. In: RANLP (2017)
11. Golbeck, J., et al.: A large labeled corpus for online harassment research. In: WebSci (2017)
12. Guo, Y., Schuurmans, D.: Discriminative batch mode active learning. In: NIPS (2007)
13. Haußmann, M., Hamprecht, F.A., Kandemir, M.: Deep active learning with adaptive acquisition. In: IJCAI (2019)
14. Hoi, S.C.H., Jin, R., Lyu, M.R.: Batch mode active learning with applications to text categorization and image retrieval. IEEE TKDE **21**, 1233–1248 (2009)
15. Konyushkova, K., Sznitman, R., Fua, P.: Learning active learning from data. In: NIPS (2017)
16. Lin, C.H., Mausam, M., Weld, D.S.: Active learning with unbalanced classes and example-generation queries. In: HCOMP (2018)
17. Lourentzou, I., Gruhl, D., Welch, S.: Exploring the efficiency of batch active learning for human-in-the-loop relation extraction. In: WWW (2018)
18. Maaten, L.V.D., Hinton, G.: Visualizing data using t-SNE. J. Mach. Learn. Res. **9**, 2579–2605 (2008)
19. McCallum, A., Nigam, K.: Employing EM and pool-based active learning for text classification. In: ICML (1998)
20. Patra, S., Bruzzone, L.: A cluster-assumption based batch mode active learning technique. Pattern Recognit. Lett. **33**(9), 1042–1048 (2012)
21. Schohn, G., Cohn, D.: Less is more: active learning with support vector machines. In: ICML (2000)
22. Settles, B.: Active learning literature survey. Computer Sciences Technical report 1648, University of Wisconsin-Madison (2009)
23. Singla, A., Patra, S.: A fast partition-based batch-mode active learning technique using SVM classifier. Soft Comput. **22**(14), 4627–4637 (2018)
24. Wang, H., Zhou, R., Shen, Y.D.: Bounding uncertainty for active batch selection. In: AAAI (2019)
25. Wang, Z., Yan, S., Zhang, C.: Active learning with adaptive regularization. Pattern Recognit. **44**, 2375–2383 (2011)
26. Waseem, Z.: Are you a racist or am i seeing things? Annotator influence on hate speech detection on twitter. In: Proceedings of the First Workshop on NLP and Computational Social Science, pp. 138–142 (2016)
27. Xia, X., Protopapas, P., Doshi-Velez, F.: Cost-sensitive batch mode active learning: designing astronomical observation by optimizing telescope time and telescope choice. In: Proceedings of SIAM 2016, pp. 477–485 (2016)

28. Yu, H., Yang, X., Zheng, S., Sun, C.: Active learning from imbalanced data: a solution of online weighted extreme learning machine. IEEE Trans. Neural Netw. Learn. Syst. **30**, 1088–1103 (2019)
29. Zhang, X., Yang, T., Srinivasan, P.: Online asymmetric active learning with imbalanced data. In: KDD (2016)

An Uncertainty-Based Human-in-the-Loop System for Industrial Tool Wear Analysis

Alexander Treiss(✉), Jannis Walk(✉), and Niklas Kühl

Karlsruhe Institute of Technology, Kaiserstrasse 89, 76133 Karlsruhe, Germany
alexander.treiss@alumni.kit.edu, {walk,kuehl}@kit.edu

Abstract. Convolutional neural networks have shown to achieve superior performance on image segmentation tasks. However, convolutional neural networks, operating as black-box systems, generally do not provide a reliable measure about the confidence of their decisions. This leads to various problems in industrial settings, amongst others, inadequate levels of trust from users in the model's outputs as well as a non-compliance with current policy guidelines (e.g., EU AI Strategy). To address these issues, we use uncertainty measures based on Monte-Carlo dropout in the context of a human-in-the-loop system to increase the system's transparency and performance. In particular, we demonstrate the benefits described above on a real-world multi-class image segmentation task of wear analysis in the machining industry. Following previous work, we show that the quality of a prediction correlates with the model's uncertainty. Additionally, we demonstrate that a multiple linear regression using the model's uncertainties as independent variables significantly explains the quality of a prediction ($R^2 = 0.718$). Within the uncertainty-based human-in-the-loop system, the multiple regression aims at identifying failed predictions on an image-level. The system utilizes a human expert to label these failed predictions manually. A simulation study demonstrates that the uncertainty-based human-in-the-loop system increases performance for different levels of human involvement in comparison to a random-based human-in-the-loop system. To ensure generalizability, we show that the presented approach achieves similar results on the publicly available Cityscapes dataset.

Keywords: Human-in-the-loop · Image segmentation · Uncertainty · Deep learning

1 Introduction

Machining is an essential manufacturing process [1], which is applied in many industries, such as aerospace, automotive, and the energy and electronics industry. In general, machining describes the process of removing unwanted material from a workpiece [3]. Thereby, a cutting tool is moved in a relative motion to

Alexander Treiss and Jannis Walk contributed equally in a shared first authorship.

Y. Dong et al. (Eds.): ECML PKDD 2020, LNAI 12461, pp. 85–100, 2021.
https://doi.org/10.1007/978-3-030-67670-4_6

the workpiece to produce the desired shape [4]. Figure 1a displays an exemplary image of a machining process applying a *cutting tool insert*. Cutting tools are consumables because of the occurrence of wear on the tools, which ultimately results in unusable tools. In the following, we will briefly describe three common wear mechanisms, compare Figs. 1b–1d for exemplary images. *Flank wear* occurs due to friction between the tools flank surface and the workpiece [1]. It is unavoidable and thus the most commonly observed wear mechanism [29]. *Chipping* refers to a set of particles breaking off from the tool's cutting edge [1]. A *built-up edge* arises when workpiece material deposits on the cutting edge due to localized high temperatures and extreme pressures [3]. Chipping and built-up edge are less desirable than flank wear since they induce a more severe and sudden deformation of the tool's cutting edge, leading to a reduced surface quality on the workpiece. Ultimately, this can lead to an increase of scrap components. A visual inspection of cutting tools enables an analysis of different wear mechanisms and provides insights into the usage behavior of cutting tools. Tool manufacturers, as well as tool end-users, can later leverage these insights to optimize the utilization of tools and to identify promising directions for the development of the next tool generations. Furthermore, an automated visual inspection enables the application of tool condition monitoring within manufacturing processes [12]. These analytics-based services possess a high economic value. Research suggests that tool failures are responsible for 20% of production downtime in machining processes [21]. Furthermore, cutting tools and their replacement account for 3–12% of total production cost [6]. Due to the relevance of these analytics-based services, our industry partner Ceratizit Austria GmbH, a tool manufacturer, agreed to closely collaborate within the research and implementation of an automated visual inspection for tool wear analysis.

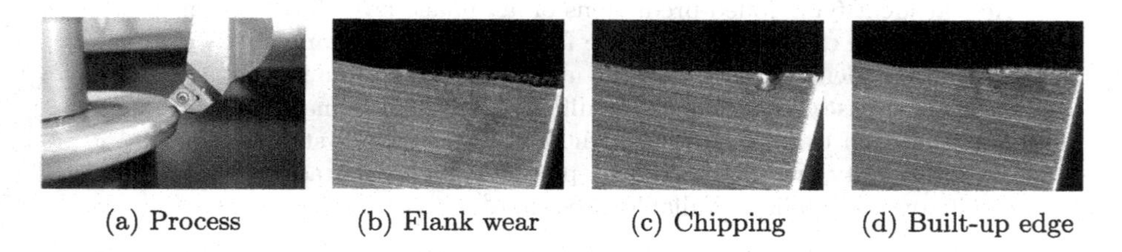

(a) Process (b) Flank wear (c) Chipping (d) Built-up edge

Fig. 1. Machining process and common wear mechanisms

Numerous studies have examined wear analysis in the machining industry to address the need for automated wear analysis. However, the majority of the research to date focuses on traditional computer vision techniques [30]. Since traditional computer vision approaches require the user to fine-tune a multitude of parameters [25], scalability often becomes an issue. Contrary to traditional computer vision approaches, deep learning-based approaches learn the required features themselves and can, therefore, be applied to different wear problems more efficiently. Another critical advantage of deep neural networks (DNNs) is their performance. In particular, the exploitation of convolutional neural networks (CNNs) has contributed to a performance increase in several computer vision tasks, e.g., classification and segmentation. CNNs are even able to surpass human-level performance in some of these settings [15], which demonstrates that a variety of visual tasks, previously performed by humans, can be automated using CNNs. Recently, Lutz et al. (2019) [23] published the first work utilizing a CNN for wear analysis on cutting tool inserts, reporting promising results. However, CNNs, functioning as black-box systems, generally do not provide a reliable measure about the confidence of their decisions. This shortcoming is critical because the trustworthiness of a model's output remains unclear for a human supervisor. Two scenarios [22] can unfold: First, the model's capabilities can be underestimated, resulting in a disuse of the system. Secondly, the human supervisor can overestimate the model's capabilities, leading to a misuse. The correct balance between trust and distrust constitutes one of the main barriers for a successful adoption of CNNs in many real-world use cases [9]. In particular, a measure of confidence is essential in safety-critical applications and in scenarios for which data is limited [20]. Limited amounts of data often occur in industrial problems, where resources and knowledge required to label and retrieve data are frequently restricted. In these settings, CNNs can occasionally produce sub-optimal results because they usually require a substantial amount of training data. Moreover, while performance can be high on average, within safety-critical applications, it is crucial to filter out any erroneous output. Lastly, due to the black-box property, CNNs are currently non-compliant with the Ethics Guidelines for Trustworthy Artificial Intelligence by the European Union [17]. In the future, these guidelines could translate into legislation that would limit the application of CNN-based systems in some industrial settings.

In this work, we address the need for CNN-based systems to output a confidence measure in an industrial environment. We consider the task of tool wear analysis using a unique, real-world data set from our industry partner Ceratizit. We employ an image segmentation algorithm based on the U-Net [26] for the pixel-wise classification of three different wear mechanisms on cutting tool inserts. To increase transparency and performance, we further enhance the tool wear analysis system with capabilities to function as an uncertainty-based human-in-the-loop system. The suggested system aims at classifying the quality of a prediction, enabling the incorporation of a human expert. In particular, we estimate the quality of a prediction using the model's uncertainty. As a foundation for uncertainty, we apply Monte-Carlo dropout (MC-dropout), which approximates a Bayesian Neural Network (BNN) [13]. The approximated BNN outputs a probability distribution over the outputs. Based on the probability distribution, we apply multiple measures that aim at capturing the uncertainty of an output. Subsequently, we show that for the use case of wear analysis, there exists a significant linear correlation between the uncertainty measures and the performance metric, the Dice coefficient. The linear relationship enables the utilization of the uncertainty measures as explanatory variables to predict the quality of a prediction in the absence of ground truth. We utilize these quality estimations in the following way: Predictions, which are estimated to be of high quality, are marked as successful by the system. These predictions are then passed on for automated analysis without any further human involvement. Otherwise, if an output is marked as failed, a human expert is requested to annotate the image manually. Hence, the system is introducing transparency by measuring the confidence of the predictions and is furthermore increasing performance with the selective use of a human expert.

Overall, we see the following contributions of our work. While research is carried out within the field of medical imaging, no previous study, to the best of our knowledge, has investigated the use of uncertainty estimates in order to predict segmentation quality in an industrial setting. We contribute by showing how an uncertainty-based assessment of segmentation quality can be utilized in an industrial task of tool wear analysis. While Lutz et al. (2019) [23] implement a CNN for tool wear analysis, we are additionally able to generate and leverage confidence estimates of the predictions. Besides the industrial relevance, our work also contributes on a more technical level. Most studies derive uncertainty estimates for binary classification problems, we are only aware of a study of Roy et al. (2018) [27], which focuses on the task of deriving uncertainty measures in a multi-class image segmentation problem. Therefore, we contribute by deriving uncertainty measures for two further multi-class image segmentation problems. Additionally, we demonstrate, that a multiple linear regression can be applied to estimate segmentation quality in these multi-class segmentation problems. Regarding the challenge of estimating segmentation quality using uncertainty measures, we are only aware of DeVries and Taylor (2018) [10], who use a DNN to predict segmentation quality. Within our use case, we rely on a multiple linear regression model, as it is interpretable, and also can be used in scarce data settings. Additionally, researchers press for more insights on human-in-the-loop systems [5], as successful designs are still scarce. Especially the allocation of (labeling) tasks between humans and machines is under-researched [9]. We contribute by implementing our human-in-the-loop system and evaluating it by a simulation study. To ensure generalizability, we assess the use of our approach on the publicly available Cityscapes [7] dataset for urban scene understanding.

2 Foundations and Related Work

First, we shortly introduce the motivation to use MC-dropout as an approach to estimate uncertainty. Subsequently, we present selected related studies, which focus on assessing the quality of a predicted segmentation by the use of uncertainty estimates.

There is a considerable body of literature growing around the theme of uncertainty estimation in DNNs. In classification tasks, a softmax output displays the probability of an output belonging to a particular class. Thus, softmax outputs are occasionally used to represent model uncertainty [16]. However, as illustrated by Gal (2015) [13], a model can be uncertain even with a high softmax output, indicating that softmax outputs do not represent model uncertainty accurately. Contrary to traditional machine learning approaches, a Bayesian perspective provides a more intuitive way of modeling uncertainty by generating a probability distribution over the outputs. However, inference in BNNs is challenging because the marginal probability can not be evaluated analytically, and, therefore, inference in BNNs is computationally intractable. Nevertheless, in a recent advance, Gal and Ghahramani (2015) [13] show that taking Monte Carlo samples from a DNN in which dropout is applied at inference time approximates a BNN. In a study by Kendall and Gal (2017) [20], the authors show that these approximated BNNs lead to an improvement in uncertainty calibration compared to non-Bayesian approaches. Since dropout exists already in many architectures for regularization purposes, MC-dropout presents a scalable and straightforward way of performing approximated Bayesian Inference using DNNs without the need to change the training paradigm.

For the human-in-the-loop system, we are particularly interested in the prediction of segmentation quality based on uncertainty estimates. To date, several studies on this particular topic have been conducted within the field of medical imaging. DeVries and Taylor (2018) [10] use MC-dropout as a source of uncertainty to predict segmentation quality within the task of skin lesions segmentation. A separate DNN is trained to predict segmentation quality based on the original input image, the prediction output, and the uncertainty estimate. While the subsequent DNN achieves promising results in predicting segmentation quality, the subsequent model lacks transparency and explainability itself. In particular, the subsequent model does not provide any information why a prediction failed. Furthermore, a DNN is only applicable if a considerable amount of data is available. Nair et al. (2020) [24] utilize MC-dropout to explore the use of different uncertainty measures for Multiple Sclerosis lesion segmentation in 3D MRI sequences. The authors show that for small lesion detection, performance increases by filtering out regions of high uncertainty. While the majority of studies focus on pixel-wise uncertainties, there is a need to aggregate uncertainty on whole segments of an image. These aggregated structure-wise uncertainty measures allow an uncertainty assessment on an image-level. The work of Roy et al. (2018) [27] introduces three structure-wise uncertainty measures, also based on MC-dropout, for brain segmentation. While the authors show that these uncertainty measures correlate with prediction accuracy, the work does not display how these uncertainty measures are applicable in a broader context, e.g., in a human-in-the-loop system.

3 Methodology

On the basis of the previously presented foundations, we now introduce our applied methodology. In Subsect. 3.1, we shortly introduce foundations regarding image segmentation. Subsequently, in Subsect. 3.2, we present the modified U-Net architecture, which we use to approximate a BNN. Subsect. 3.3 then describes the loss function and the evaluation metric, which we use to train and evaluate the modified U-Net. Lastly, Subsect. 3.4 depicts the computation of the uncertainty measures on which our human-in-the-loop system relies.

3.1 Image Segmentation

For our use case of wear analysis in the machining industry, information must be available on a pixel level to facilitate the assessment of location and size of wear for a given input image. An approach that provides this detailed information is called *image segmentation.* While image segmentation can be performed unsupervised, it is often exercised as a supervised learning problem [14]. In supervised learning problems, the task is generally to learn a function $f : X \rightarrow Y$ mapping some input values (X) to output values (Y). In the case of image segmentation, the concrete task is to approximate a function f, which takes an image x as input and produces a segmentation $\hat{y}$. The predicted segmentation assigns a category label $c \in C$ to each pixel $i \in N$ in the input image, where C denotes the possible classes and N the number of pixels in an input image. Therefore, the task of image segmentation is also referred to as pixel-wise classification. Consequently, the outputs must have the same height and width as the input image, the depth is defined by the number of possible classes.

3.2 Model: Dropout U-Net

We apply a modified U-Net architecture due to its ability to produce good results even with a small amount of labeled images [26]. To avoid overfitting and increase performance, we implement the following adaptions to the original U-Net architecture: We use an L2-regularization in every convolutional layer, and additionally, we reduce the number of feature maps in the model's architecture, starting with 32 feature maps instead of 64 feature maps in the first layer [2]. The number of feature maps in the remaining layers follows the suggested approach from the original U-Net architecture, which doubles the number of feature maps in the contracting path and then analogically halves the number of feature maps in the expansive path. To accelerate the learning process, we add batch normalization between each pair of convolutional layers [18]. We use a softmax activation function in the last layer of the model to obtain predictions in the range [0,1]. Therefore, the modified U-Net takes an input image x and produces softmax probabilities $p_{i,c} \in [0, 1]$ for each pixel $i \in N$ and class $c \in C$, compare Eq. (1).

$$p_{i,c} = f(x) \qquad \forall i \in N, \forall c \in C \tag{1}$$

As a source of uncertainty, we realize the human-in-the-loop system based on MC-dropout because of its implementation simplicity, while still being able to generate reasonable uncertainty estimates [13]. We employ dropout layers in the modified U-Net to enhance the model with the ability to approximate a BNN [13]. Units within the dropout layers have a probability of 0.5 to be multiplied with zero and therefore, to drop out. We follow the suggested approach by Kendall et al. (2015) [19] in the context of the Bayesian SegNet to use dropout layers at the five most inner decoder-encoder blocks. Hereinafter, we will refer to the modified U-Net, which applies dropout at inference time, as the *Dropout U-Net.* By the application of dropout at inference time, the Dropout U-Net constitutes a stochastic function f. For multiple stochastic forward passes T of an input image x, the Dropout U-Net generates a probability distribution for each $p_{i,c}$. To obtain a segmentation $\hat{y}$, we calculate the mean softmax probability first, as described in Eq. (2). Then, the segmentation $\hat{y}$ is derived by applying the argmax function over the possible classes of the softmax probabilities $p_{i,c}$, compare Eq. (3).

$$p_{i,c} = \frac{1}{T} \sum_{t=1}^{T} p_{i,c,t} \qquad \forall i \in N, \forall c \in C \tag{2}$$

$$\hat{y}_i = \underset{c \in C}{\operatorname{argmax}}(p_{i,c}) \qquad \forall i \in N \tag{3}$$

3.3 Loss Function and Performance Evaluation Metric

As a loss function, we use a *weighted category cross-entropy* loss. We weight each class with the inverse of its occurrence (pixels) in the training data due to class imbalance. This leads to an equal weighting between the classes in the loss function [8]. The weighted categorical cross-entropy loss is defined in Eq. (4); $g_{i,c}$ denotes the one-hot-encoded ground truth label, and w_c the computed weights for each class.

$$\mathcal{L}(p, g) = -\frac{1}{N} \sum_{i=1}^{N} \sum_{c=1}^{C} w_c \, g_{i,c} \, \log \, p_{i,c} \tag{4}$$

To reflect the quality of a predicted segmentation, we rely on the *Dice coefficient* [11] (DSC) as a performance evaluation metric. The Dice coefficient assesses the overlap, or intersection, between the model's outputs and the one-hot-encoded ground truth labels $g_{i,c}$. A full overlap between a prediction and a label is represented by a value of one. If there is no overlap, the Dice coefficient returns zero. Representing the outputs, one could use the softmax probabilities $p_{i,c}$ or the binarized one-hot encoded predictions $\hat{y}_{i,c}$. We use the binarized predictions $\hat{y}_{i,c}$ since these predictions represent the foundation for the build-on tool wear analysis. As suggested by Garcia et al. (2017) [14] for the related Jaccard-Coefficient, we compute the Dice coefficient for each class separately. To assess the segmentation quality of an input image, we compute the averaged Dice coefficient across all classes, leading to a *mean Dice coefficient*, defined in Eq. (5). To evaluate a model on the test set, we calculate the average across the mean Dice coefficients per image. Next to the Dice coefficient, we also compute the pixel accuracy as a performance measure. It defines the percentage of correctly classified pixels.

$$\textit{Mean Dice Coefficient} = \frac{2}{C} \sum_{c=1}^{C} \frac{\sum_{i=1}^{N} \hat{y}_{i,c} \, g_{i,c}}{\sum_{i=1}^{N} \hat{y}_{i,c} + \sum_{i=1}^{N} g_{i,c}} \tag{5}$$

3.4 Uncertainty Estimation

As a next step, we describe how uncertainty measures can be calculated using the probability distribution outputs of the Dropout U-Net. In general, the information-theory concept of entropy [28] displays the expected amount of information contained in the possible realizations of a probability distribution. Following previous work [20], we utilize the entropy as an uncertainty measure to reflect the uncertainty of each pixel in a predicted segmentation:

$$H(p_i) = -\sum_{c=1}^{C} p_{i,c} \log p_{i,c} \qquad \forall i \in N \tag{6}$$

The entropy displays its maximum if all classes have equal softmax probability and it reaches its minimum of zero if one class holds a probability of 1 while the other classes have a probability of 0. Therefore, the entropy reflects the uncertainty of a final output $\hat{y}_i$ by considering the model's outputs $p_{i,c}$ over all classes C. For the task of image segmentation, the entropy is available per pixel. However, for several applications, it is necessary to derive an uncertainty estimate on a higher aggregation level. For example, within the tool wear analysis, we want to decide on an image basis, whether a segmentation is successful or failed. One approach is to average the pixel-wise entropy values over an image to come up with an image-wise uncertainty estimate. Roy et al. (2018) [27] propose to calculate the average pixel uncertainty for each predicted class in a segmentation. We utilize this idea in the context of wear analysis and define the entropy per predicted class U_c in Eq. (7). Equation (8) defines the number of pixels for each predicted class. The entropy per predicted class provides information on an aggregated class-level and can be used to estimate uncertainty for each class in an input image. Since only the predictions are used, this uncertainty estimate is applicable in the absence of ground truth.

$$U_c = \frac{1}{S_c} \sum_{i=1}^{S_c} H(p_i) \qquad \forall c \in C \tag{7}$$

$$S_c = \{i \in N \mid \hat{y}_i = c\} \qquad \forall c \in C \tag{8}$$

4 Experiments

With the methodology at hand, the upcoming Subsect. 4.1 provides information about the two datasets, the preprocessing and training procedures. Subsequently, Subsect. 4.2 briefly describes the performance results in terms of the Dice coefficient. In Subsect. 4.3, we evaluate the uncertainty-based human-in-the-loop system in the following way: First, we illustrate the relation between uncertainty and segmentation quality using two exemplary predictions. Then, we quantitatively assess the relation between uncertainty and segmentation quality using the Bravais-Pearson correlation coefficient. Next, we fit a multiple linear regression on each test set, which uses the uncertainty measures as independent variables to explain segmentation quality. Lastly, we simulate the performance of the uncertainty-based human-in-the-loop system based on the multiple regression and compare it against a random-based human-in-the-loop system.

4.1 Datasets, Preprocessing and Training Procedure

The unique **Tool wear dataset** consists of 213 pixel-wise annotated images of cutting tool inserts, which were previously used by Ceratizit's customers in real manufacturing processes until their end-of-life. The labels are created as follows: The first 20 images are labeled jointly by two domain experts from Ceratizit. Afterward, labels are assigned individually, whereas at least two domain experts discuss unclear cases. The recording of the images is standardized to reduce the required amount of generalization of the learning algorithm. The images initially have a resolution of 1600 × 1200 pixels. As a preprocessing step, we cut the image to a shape of 1600 × 300, which lets us focus on the cutting edge where the wear occurs. For computational efficiency and as a requirement of the U-Net architecture, we resize the images to a shape of 1280 × 160 using a bilinear interpolation. We randomly split the dataset into 152 training, 10 validation and 51 test images. The model is trained for 200 epochs, using an Adam optimizer, a learning rate of 0.00001, an L2-regularization with an alpha of 0.01, and a batch size of 1. We choose the hyperparameters after running a brief hyperparameter search. The training is conducted on a Tesla V100-SXM2 GPU, with a training duration of approximately two hours. In the literature, the number of conducted forward passes ranges from 10 to 100 [10], we use 30 Monte Carlo forward passes to create the probability distribution over the outputs.

The **Cityscapes dataset** [7] is a large-scale dataset, which contains images of urban street scenes from 50 different cities. It can be used to assess the performance of vision-based approaches for urban scene understanding on a pixel-level. The Cityscapes dataset initially consists of 3475 pixel-wise annotated images for training and validation. Performance is usually specified through 1525 test set images for which ground truth labels are only available on the Cityscapes website [7]. Since we need ground truth labels to assess the uncertainty estimates, we use the 500 proposed validation images as the test set and randomly split the remaining 2975 images into 2725 training and 250 validation images. There are initially 30 different classes which belong to eight categories. We group classes belonging to the same category together, to create a similar problem setting between the Cityscapes dataset and the Tool wear dataset. Furthermore, we combine the categories 'void', 'object', 'human', and 'nature' to one class, which we consider in the following as the background class. The remaining categories are 'flat', 'construction', 'sky' and 'vehicle', ultimately resulting in five different classes. The original images have a resolution of 2048 × 1024 pixels. During training, an augmentation step flips the images horizontally with a probability of 0.5. Then, a subsequent computation randomly crops the images with a probability of 0.15 to an input size of 1024 × 512. Otherwise, the images are resized using a bilinear interpolation to the desired input shape of 1024 × 512. Following a brief hyperparameter search, we train the model for 15 epochs with a learning rate of 0.0001 using an Adam optimizer, a batch size of 10, and an L2 weight regularization with an alpha of 0.02. We train the model on a Tesla V100-SXM2 GPU for approximately 2.5 h. The Dropout U-Net uses five Monte Carlo forward passes, considering the higher computational complexity due to the more extensive test set and the large image size.

4.2 Performance Results

Next, we assess the quality of a predicted segmentation in terms of the Dice coefficient. Table 1 displays the performance results on each respective test set. The background class has the largest proportion of pixels (93.9%) in the tool wear dataset, followed by flank wear (4.7%), built-up edge (0.9%) and lastly, chipping (0.4%). We assume that the number of labeled pixels of a specific class is closely related to prediction performance. We find, that the model has particular difficulties segmenting chipping phenomena, which is expressed by a Dice coefficient of 0.244. We explain this lack of prediction performance by the punctual and minor occurrence of chipping phenomena within images, compare Fig. 2, and a characteristic of the Dice coefficient. In particular, the Dice coefficient per class drops to zero, if the model produces a false negative, meaning the model falsely predicts a wear mechanism, and if there is no wear for the corresponding class labeled in the image. This characteristic and the challenging task of classifying small chipping phenomena in an input image causes the Dice coefficient of the chipping class to drop to zero for several images. In contrast to chipping, the Dropout U-Net recognizes the background class well and is classifying flank wear and built-up edge considerably well. Compare the label (Fig. 2c) and the prediction (Fig. 2e) for an illustration. As illustrated on the left hand side of Fig. 2, the input image is predicted well as the intersection between label and prediction is high. The performance results of the Cityscapes dataset also indicate that the Dropout U-Net can generate a predicted segmentation for each class considerably well.

Table 1. Performance results

Tool wear dataset	
Class	Dice coefficient
Background	0.991
Flank wear	0.695
Chipping	0.244
Built-up edge	0.596
Mean DSC	0.631
Pixel accuracy	0.977

Cityscapes dataset	
Class	Dice coefficient
Background	0.929
Flat	0.693
Construction	0.830
Sky	0.769
Vehicle	0.773
Mean DSC	0.799
Pixel accuracy	0.875

4.3 Evaluation: Uncertainty-Based Human-in-the-Loop System

With the performance results at hand, we focus on assessing the uncertainty-based human-in-the-loop system. Figure 2 presents two preprocessed images, their corresponding human labels, their predictions, and the generated uncertainty maps for the Tool wear dataset. The uncertainty maps are generated using pixel-wise uncertainties based on the entropy, compare Eq. (6). Within the uncertainty maps, brighter pixels represent uncertain outputs, and darker pixels represent certain predictions. The uncertainty map (g) of the input image (1) displays uncertain outputs, indicated by brighter pixels, at the edge between classes. This behavior is often noticed within uncertainty observation in image segmentation tasks [19], as it reflects the ambiguity of defining precise class regions on a pixel-level. The most interesting aspect of Fig. 2 is the prediction (f) and the corresponding uncertainty map (h). The model falsely predicted several areas on the right-hand side as flank wear (red). However, the model also indicates high uncertainty for this particular area, indicated by brighter pixels in the corresponding uncertainty map (h).

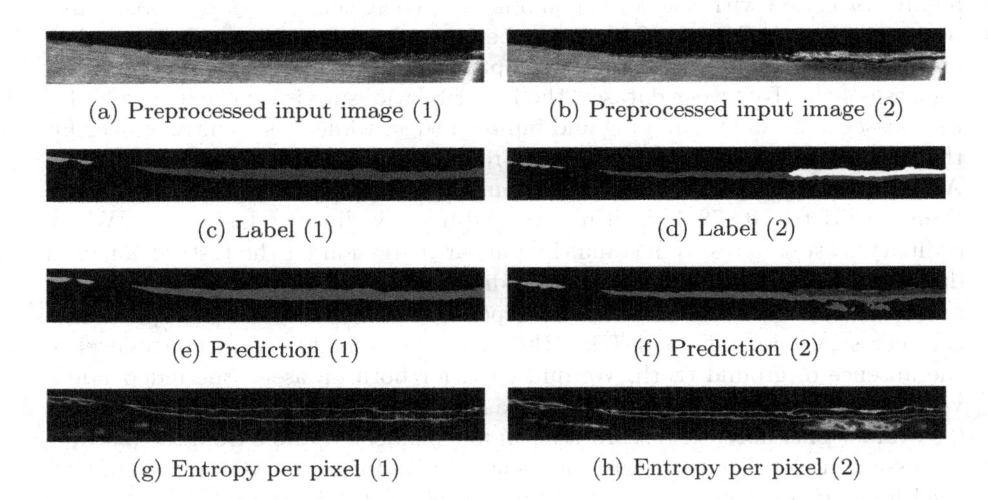

(a) Preprocessed input image (1) (b) Preprocessed input image (2)

(c) Label (1) (d) Label (2)

(e) Prediction (1) (f) Prediction (2)

(g) Entropy per pixel (1) (h) Entropy per pixel (2)

Fig. 2. Illustration of two images of the test set with their corresponding labels, predictions and uncertainty maps (best viewed in color). Color coding: Flank wear = dark grey/red, Chipping = light grey/green and Built-up edge = white/blue (Color figure online)

This relationship is essentially the foundation of the uncertainty-based human-in-the-loop system. The relationship between a pixel's uncertainty and the probability of being classified correctly enables the system to distinguish between images which the model segments successfully and images which the model segments poorly. As the goal is to distinguish segmentation quality on an image level, the average uncertainty of a predicted segmentation can be used as

Table 2. Correlation coefficients between the uncertainty per predicted class (U_c) and the Dice coefficients per class

Tool wear dataset	
Class	Correlation
Background	-0.656
Flank wear	-0.911
Chipping	-0.818
Built-up edge	-0.932

Cityscapes dataset	
Class	Correlation
Background	-0.208
Flat	-0.878
Construction	-0.754
Sky	-0.751
Vehicle	-0.858

an uncertainty measure for an image. However, we find that for the task of tool wear, there does not exist a significant correlation (-0.34) between the averaged entropy per prediction and the mean Dice coefficient on the 51 images of the test set. The same analysis yields a correlation of -0.57 for the Cityscapes dataset. However, we find that the mean entropy of a predicted class, U_c, is highly correlated with the corresponding Dice coefficient per class. Table 2 displays the Bravais-Pearson Correlation between the entropy of a predicted class and the corresponding Dice coefficient per class on the respective test sets. In the case of the Tool wear dataset, the linear relationship is especially strong for the classes flank wear, chipping and built-up edge, while it is slightly weaker for the background class. These results are reproducible on the Cityscapes dataset. As can be seen from Table 2, correlations, besides the background class, range from -0.751 to -0.878, indicating a strong negative linear relationship. We use ordinary least squares to fit a multiple linear regression on the test set for both datasets using the uncertainties per predicted class U_c as independent variables and the mean Dice coefficient as the dependent variable. Subsequently, the linear regression aims at quantifying the prediction quality on an image-level in the absence of ground truth. We find that, for both datasets, the independent variable 'uncertainty per background class' is statistically not significant at the 0.05 value. Therefore, we discard it as an independent variable from the multiple regression model. The remaining independent variables are significant at the 0.01 level for both datasets. The regression results yield a $R^2 = 0.718$ for the Tool wear dataset and a $R^2 = 0.655$ for the Cityscapes dataset. This indicates that the multiple linear regression can explain a substantial amount of variation of the mean Dice coefficient. The full regression results can be found in the online appendix (https://dsi.iism.kit.edu/downloads/ECML_HITL_Appendix.pdf).

In the following paragraph, we assess the use of the multiple linear regression in the context of a human-in-the-loop system by running a simulation. The multiple linear regression predicts the quality of a predicted segmentation in terms of the mean Dice coefficient, using the uncertainties per predicted class as independent variables. Then, in an iterative process, the input image, for which the prediction displays the lowest estimated mean Dice coefficient is forwarded for human annotation. Images, displaying a higher estimated Dice coefficient are retained by the system. Within the simulation, we assume a perfect human segmentation, and set the corresponding Dice coefficient of the forwarded image to one. The performance of the system is then calculated by combining the mean Dice coefficients of the retained images and the forwarded human annotated images. Figure 3 shows the simulation results for both datasets. The x-axis displays the number of images, which are forwarded to human annotation. The y-axis displays the performance of the system. We compare the performance of the human-in-the-loop system (blue line) against a random-based human-in-the-loop system (orange line). Contrary to the uncertainty-based system, the random-based system decides randomly, which images are forwarded to human annotation. To avoid overfitting, we use a split for each dataset as follows: For the Tool wear dataset, the multiple linear regression is fitted on 30 images, the remaining 21 predictions are then used for simulation. Within the Cityscapes dataset, we use 300 test set images to fit the regression and 200 images for simulation. For both datasets, the uncertainty-based human-in-the-loop system is able to achieve a better mean Dice coefficient using fewer human annotations than a random-based approach. This is due to the multiple regression, which identifies low-quality predictions and therefore enables the system to forward these predictions to human annotation first.

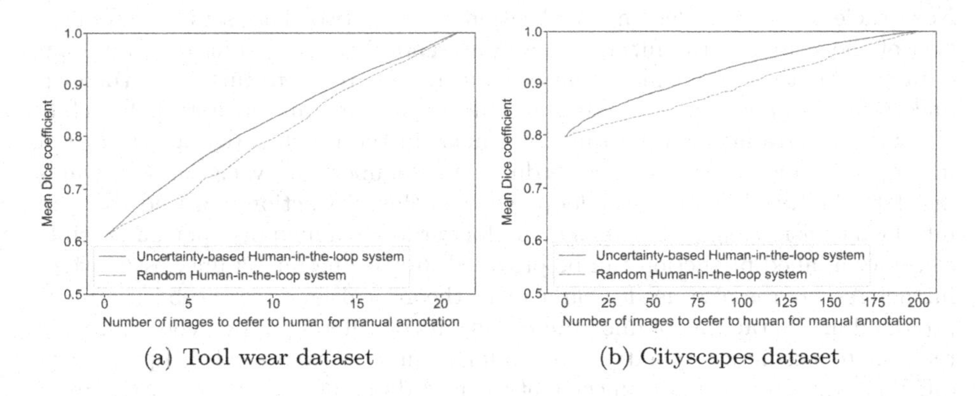

(a) Tool wear dataset

(b) Cityscapes dataset

Fig. 3. Simulation results (Color figure online)

5 Discussion and Outlook

In this work, we show the applicability and usefulness of an uncertainty-based human-in-the-loop system for the task of industrial tool wear analysis. The human-in-the-loop system addresses critical challenges regarding the adoption of CNNs in industry. In particular, it increases transparency by providing uncertainty measures which are correlated with segmentation performance. Additionally, it improves performance by incorporating a human expert for the annotation of images that are estimated to be of low quality.

Within the use case of tool wear analysis, we consider the task of segmenting three different wear mechanisms on cutting tool inserts. We apply and train a modified U-Net architecture on a real-world dataset of our industry partner Ceratizit, achieving good performance results. For the human-in-the-loop system, we enhance the existing tool wear analysis with the following capabilities: We implement MC-dropout and use the information-theory concept of entropy to compute pixel-wise uncertainties. Furthermore, we aggregate the pixel-wise uncertainties to compute class-wise uncertainty measures on an image-level. A multiple linear regression reveals that the class-wise uncertainties can be used as independent variables to explain a substantial amount of the mean Dice coefficient of an image. The multiple linear regression is then leveraged within the human-in-the-loop system to decide, whether a given segmentation should be forwarded to a human expert, or be retained in the system as a successful prediction. A simulation study demonstrates that the performance improves through the utilization of a human expert, which annotates estimated low-quality predictions. Furthermore, the system increases transparency by additionally issuing an estimate about the quality of a prediction.

We assess our system not only on our proprietary tool wear data set but also on the publicly available and substantially larger Cityscapes data set, confirming the generalizability of our approach to the task of urban scene understanding. Nevertheless, we consider the application to only two data sets as a limitation of the study. In the future, we aim at validating the system on additional datasets. Another promising avenue for future research is to further distinguish uncertainty into epistemic (model) and aleatoric (data) uncertainty [20]. While aleatoric uncertainty is due to inherent noise in the input data, e.g., a blurred image, epistemic uncertainty occurs due to model uncertainty, e.g., lack of training data. Within human-in-the-loop systems, this distinction can lead to more informed decision, e.g., when images are forwarded to a human expert, a possible cause for a failed prediction can be provided. Regarding uncertainty estimation, further research is also needed on a more theoretical level, to establish a more profound understanding of uncertainty outputs of different approaches and their relation to prediction quality. This could include a structured comparison of different ways to calculate uncertainty across different use cases. Lastly, from a human-centric machine learning standpoint, further research should assess, if the increased transparency of the human-in-the-loop system leads to a more calibrated level of trust from the user. While we found several indications in the literature, few studies have investigated this relationship in a systematic way.

We see a broad applicability of the uncertainty-based human-in-the-loop system in industrial applications. While we consider the task of image segmentation, the general observations should be relevant in a variety of supervised learning problems. A human-in-the-loop system can be beneficial for all types of automation tasks, in which human experts display superior performance than automated systems, but in which the automated system is more cost efficient. An example for such a system would be an industrial quality control system. Otherwise, we perceive limited potential for tasks, in which the performance of human experts is inferior compared to the performance of automated systems. This scenario would include many applications of time series forecasting. In these tasks, the estimated prediction quality could only be used to issue warnings whenever an output is likely to be faulty. Altogether, we believe that the uncertainty-based human-in-the-loop system represents an essential building block for the facilitation of a more widespread adoption of CNN-based systems in the industry.

Acknowledgments. We would like to thank Ceratizit Austria GmbH, in particular Adrian Weber for facilitating and supporting this research.

References

1. Altintas, Y.: Manufacturing Automation: Metal Cutting Mechanics, Machine Tool Vibrations, and CNC Design, p. 57. Cambridge University Press, Cambridge (2012)
2. Bishop, C.M.: Regularization and complexity control in feed-forward networks. Tech. rep. (1995)
3. Black, J.T.: Introduction to machining processes. In: ASM Handbooks: Volume 16: Machining. ASM International, 2nd edn. (1995)
4. Boothroyd, G.: Fundamentals of Metal Machining and Machine Tools, vol. 28. CRC Press, Boca Raton (1988)
5. Brynjolfsson, E., McAfee, A.: Race against the machine: how the digital revolution is accelerating innovation, driving productivity, and irreversibly transforming employment and the economy. Brynjolfsson and McAfee (2011)
6. Castejón, M., Alegre, E., Barreiro, J., Hernández, L.: On-line tool wear monitoring using geometric descriptors from digital images. Int. J. Mach. Tools Manuf. **47**(12–13), 1847–1853 (2007)
7. Cordts, M., et al.: The cityscapes dataset for semantic urban scene understanding. In: Proceedings of the IEEE Conference on Computer Vision and Pattern Recognition, pp. 3213–3223 (2016)
8. Crum, W.R., Camara, O., Hill, D.L.: Generalized overlap measures for evaluation and validation in medical image analysis. IEEE Trans. Med. Imaging **25**(11), 1451–1461 (2006)
9. Dellermann, D., Ebel, P., Söllner, M., Leimeister, J.M.: Hybrid intelligence. Bus. Inf. Syst. Eng. **61**(5), 637–643 (2019)
10. DeVries, T., Taylor, G.W.: Leveraging uncertainty estimates for predicting segmentation quality. arXiv preprint arXiv:1807.00502 (2018)
11. Dice, L.R.: Measures of the amount of ecologic association between species. Ecology **26**(3), 297–302 (1945)
12. Dutta, S., Pal, S., Mukhopadhyay, S., Sen, R.: Application of digital image processing in tool condition monitoring: a review. CIRP J. Manuf. Sci. Technol. **6**(3), 212–232 (2013)
13. Gal, Y., Ghahramani, Z.: Dropout as a Bayesian approximation: representing model uncertainty in deep learning. In: Proceedings of the 33rd International Conference on Machine Learning, vol. 48, pp. 1050–1059 (2016)
14. Garcia-Garcia, A., Orts-Escolano, S., Oprea, S., Villena-Martinez, V., Garcia-Rodriguez, J.: A review on deep learning techniques applied to semantic segmentation. arXiv preprint arXiv:1704.06857 (2017)

15. He, K., Zhang, X., Ren, S., Sun, J.: Delving deep into rectifiers: surpassing human-level performance on imagenet classification. In: Proceedings of the IEEE International Conference on Computer Vision, pp. 1026–1034 (2015)
16. Hendrycks, D., Gimpel, K.: A baseline for detecting misclassified and out-of-distribution examples in neural networks. arXiv preprint arXiv:1610.02136 (2016)
17. Hleg, AI.: Ethics guidelines for trustworthy AI. B-1049 Brussels (2019)
18. Ioffe, S., Szegedy, C.: Batch normalization: accelerating deep network training by reducing internal covariate shift. arXiv preprint arXiv:1502.03167 (2015)
19. Kendall, A., Badrinarayanan, V., Cipolla, R.: Bayesian SegNet: model uncertainty in deep convolutional encoder-decoder architectures for scene understanding. arXiv preprint arXiv:1511.02680 (2015)
20. Kendall, A., Gal, Y.: What uncertainties do we need in Bayesian deep learning for computer vision? In: Advances in Neural Information Processing Systems, pp. 5574–5584 (2017)
21. Kurada, S., Bradley, C.: A review of machine vision sensors for tool condition monitoring. Comput. Ind. **34**(1), 55–72 (1997)
22. Lee, J.D., See, K.A.: Trust in automation: designing for appropriate reliance. Hum. Factors **46**(1), 50–80 (2004)
23. Lutz, B., Kisskalt, D., Regulin, D., Reisch, R., Schiffler, A., Franke, J.: Evaluation of deep learning for semantic image segmentation in tool condition monitoring. In: Proceedings - 18th IEEE International Conference on Machine Learning and Applications, ICMLA 2019, pp. 2008–2013 (2019)
24. Nair, T., Precup, D., Arnold, D.L., Arbel, T.: Exploring uncertainty measures in deep networks for multiple sclerosis lesion detection and segmentation. Med. Image Anal. **59**, 101557 (2020)
25. O'Mahony, N., et al.: Deep learning vs. traditional computer vision. In: Arai, K., Kapoor, S. (eds.) CVC 2019. AISC, vol. 943, pp. 128–144. Springer, Cham (2020). https://doi.org/10.1007/978-3-030-17795-9_10
26. Ronneberger, O., Fischer, P., Brox, T.: U-Net: convolutional networks for biomedical image segmentation. In: Navab, N., Hornegger, J., Wells, W.M., Frangi, A.F. (eds.) MICCAI 2015. LNCS, vol. 9351, pp. 234–241. Springer, Cham (2015). https://doi.org/10.1007/978-3-319-24574-4_28
27. Roy, A.G., Conjeti, S., Navab, N., Wachinger, C.: Inherent brain segmentation quality control from fully ConvNet Monte Carlo sampling. In: Frangi, A.F., Schnabel, J.A., Davatzikos, C., Alberola-López, C., Fichtinger, G. (eds.) MICCAI 2018. LNCS, vol. 11070, pp. 664–672. Springer, Cham (2018). https://doi.org/10.1007/978-3-030-00928-1_75
28. Shannon, C.E.: A mathematical theory of communication. Bell Syst. Tech. J. **27**(3), 379–423 (1948)
29. Siddhpura, A., Paurobally, R.: A review of flank wear prediction methods for tool condition monitoring in a turning process. Int. J. Adv. Manuf. Technol. **65**(1–4), 371–393 (2013)
30. Walk, J., Kühl, N., Schäfer, J.: Towards leveraging end-of-life tools as an asset: value co-creation based on deep learning in the machining industry. In: Proceedings of the 53rd Hawaii International Conference on System Sciences, vol. 3, pp. 995–1004 (2020)

Filling Gaps in Micro-meteorological Data

Antoine Richard[1(✉)], Lior Fine[2,3], Offer Rozenstein[3,4], Josef Tanny[2,4], Matthieu Geist[5], and Cedric Pradalier[6]

[1] Georgia Institute of Technology, Atlanta, USA
antoine.richard@gatech.edu
[2] Institute of Soil, Water and Environmental Sciences, Agricultural Research Organization, Volcani Center, Rishon LeTsiyon, Israel
[3] The Robert H Smith Faculty of Agriculture, Food and Environment, The Hebrew University of Jerusalem, Jerusalem, Israel
[4] HIT – Holon Institute of Technology, Holon, Israel
[5] Google Research, Brain Team, Lorraine, France
[6] GeorgiaTech Lorraine – UMI2958 GT-CNRS, Metz, France

Abstract. Filling large data-gaps in Micro-Meteorological data has mostly been done using interpolation techniques based on a marginal distribution sampling. Those methods work well but need a large horizon of the previous events to achieve good results since they do not model the system but only rely on previously encountered iterations. In this paper, we propose to use multi-head deep attention networks to fill gaps in Micro-Meteorological Data. This methodology couples large-scale information extraction with modeling capabilities that cannot be achieved by interpolation-like techniques. Unlike Bidirectional RNNs, our architecture is not recurrent, it is simple to tune and our data efficiency is higher. We apply our architecture to real-life data and clearly show its applicability in agriculture, furthermore, we show that it could be used to solve related problems such as filling gaps in cyclic-multivariate-time-series.

Keywords: Evapo-transpiration · Gap-filling · Attention-models

1 Introduction

Gap filling is a crucial task in environmental science. In many fields such as precision agriculture or environmental monitoring, sophisticated sensor systems record data in remote areas over long periods of time. Often, those systems suffer from power breaks, sensor malfunction, or reduced data quality that requires binning some of the measurements. This implies that the data are frequently incomplete. As an example, approximately 30% of eddy-covariance flux measurements are missing or binned [6]. Setting up experiments to acquire environmental data is a time consuming and very tedious process, requiring highly

***.

Y. Dong et al. (Eds.): ECML PKDD 2020, LNAI 12461, pp. 101–117, 2021.
https://doi.org/10.1007/978-3-030-67670-4_7

skilled personnel, expensive equipment, and a long time to set up, process and analyze. Hence precisely filling gaps is of paramount importance to maximize the available data. Additionally, when data aggregation needs to be done, for example for calculating the seasonal water budget or the amount of daily evapotranspiration (ET) for irrigation purposes, it is necessary to complete the gaps in the time-series. In this paper, we aim to fill gaps in the ET measured by an Eddy-Covariance (EC) device on short-life-span crops (3 to 4 months): tomatoes, cotton, and wheat. The training data consists of a set of real meteorological variables: the net-radiation, the relative-humidity, the air-temperature, and the wind-speed, along with the measured ET. Our dataset (training and testing) is limited to those 3 crops with 2 independent recording seasons and sites for each crop, more details are provided in Sect. 4.1. Here, due to the warm climate resulting from the geographical location of the considered fields (middle-east), and the rapid development of the crops over the course of a short growing season, our system shows rapidly changing dynamics. Additionally, since we study seasonal crops, we cannot rely on site specific data recorded over multiple years as it is typically done when filling in EC measurements over forests.

Formally, the problem that we are studying consists of multiple variables that have dependencies with one or more periodic variables. Those periods are not necessarily known and may require a large time horizon to observe. We consider that the dependencies between the variables and the time cannot be modeled. To further focus our problem, we will only try to fill gaps in one variable, as shown in Fig. 1. This variable depends on the other remaining variables. Hence the other variables are always considered to be intact. Even-though this assumption can seem bold, neighboring weather stations can be used to fill missing meteorological variables if a sensor malfunction were to occur on the EC tower. To do so one may have to compensate for the bias in-between the EC measurements and the ones of the weather station.

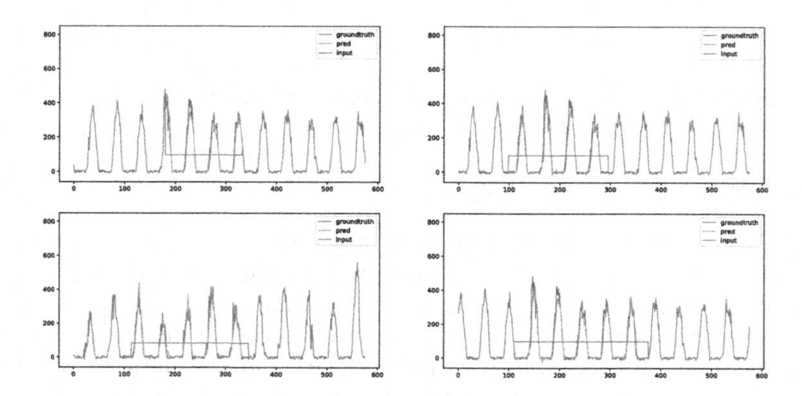

Fig. 1. Gap filling in a latent heat flux recording with gaps of 3 to 6 days (144 to 288 points). The results were obtained using our models for wheat in Saad, Israel. Blue: real sequence, orange: gap filled sequence using our method, green: sequence with gap. Other variables not shown. (Color figure online)

To tackle this problem, an intuitive approach would be to use Fourier's analysis. However, the problems to which we aim to apply our algorithms may consist of very long periods, at the scale of growing seasons (*i.e.* 2–6 months) or years. Furthermore, the measurements we are using often consist of only one growing season per location. Our measurements are performed in different locations and different times of the year. Since the beginning and the length of the growing season depend on the type of crop, the weather and other agronomic considerations, the measurements do not necessarily start at the same time, nor are they performed for the same duration. All those factors are the reason that techniques based on Fourier's analysis were not considered in this study.

The first consideration in the development of the algorithm to be described in Sect. 3 was the need for the irrigation research community that relies on evapotranspiration (loss of water from the soil by evaporation and plant transpiration) measurements to estimate the irrigation demand. Since the evapotranspiration depends on the air temperature, relative humidity, wind speed, and radiation, previous research in this field relied on the diurnal cycles of these variables to fill gaps in water vapor flux measurements from EC. To do so, they used Mean Diurnal Variation (MDV) [6] and Marginal Distribution Sampling (MDS) [15,18]. Those methods use neighboring full non-contaminated data to try and patch the missing data. Despite the very good performance of techniques such as Look-Up Tables (LUT) [13], they are often tailor-made solutions that only solve particular types of problems. Additionally, since they rely on neighboring data, they require a fairly large amount of non-corrupted neighboring data to work.

As far as we know, there is no general machine learning paper dealing with the gap filling problem: some of the work focused on making sense of the data despite the gaps [2], while the rest of the work was mostly using old architectures such as Multi-Layer Perceptron (MLP) [4]. However, the applicability of Recurrent Neural Networks (RNNs) to perform time-series-forecasting is not left to demonstrate, hence, we will present how those approaches could be used in the context of filling gaps. Yet, both MLPs and RNNs suffer from multiple drawbacks. The MLPs are not "context-aware"; thus, they cannot extract local trends, which leads to making them inapplicable in systems with rapidly changing dynamics like ours. As for the RNNs, they are context-aware, but they can suffer from vanishing gradient. There is also a limit to the reach and the amount of information that can be propagated. Additionally, they are hard to tune and expensive to train and infer, especially on long sequences. In this study, we will be considering sequences ranging from 200 to 600 elements.

To solve the aforementioned problems, we propose to rely on the most recent advances in Natural Language Processing (NLP) [5,17]. These models provide both the modeling capacities of the neural-networks but also the context understanding that offers the MDSs and RNNs without suffering from their respective limitations. Our model uses the multi-head attention layers defined in [17] with a modified positional encoding to account for the cyclicity of the data. We show that our approach outperforms current state of the art to fill gaps in evapo-transpiration data. And we also show that it can be applied to solve more general problems on a toy case.

In the end, this paper contributions can be summarized as follows:

- We derive a model for gap filling based on attention networks that account for the periodicity within the data.
- We apply our model to a real-world problem defining a new state-of-the-art performance despite a very limited training set.

2 Related Work

Reliably filling missing data-points in time-series is of the utmost importance in the field of environmental sciences. However, despite the importance of the task, this field still heavily relies on old methods.

Initially, the method used to replace missing data was LUT. It consists in using the previous occurrence that resembles most the point we are trying to recover. However, it has since been replaced by a smoother approach named MDS. In our problem, we consider that only the target data $Y \in \mathbb{R}$ is corrupted but that the remaining data $X \in \mathbb{R}^n$ is correct. We fit a probability distribution over $Y|X$. The missing values are recovered by sampling the marginal distribution over X and computing $P(Y|X)$. In practice, this is computed using K-Nearest-Neighbour (KNN). It amounts to performing a local weighted average over available values of $Y|X$. The weights are proportional to their distance from X, and only samples in the vicinity of the gap are used. The main drawback of those methods is that they solely rely on the neighboring points. Hence if the weighting window is too small, there may not be occurrences similar to the points to be filled. To avoid this problem, a large time horizon is required, as shown in [18]. However, when the horizon increases, the impact of non-observable variables (latent variables) may significantly deteriorate the filling quality. This is particularly true in systems with high paced dynamics, where the latent variables are responsible for amplitudes changes.

However, with the rise of deep neural-networks, those methods are being challenged by parametric approaches. Standard MLPs have been successfully used to fill gaps in slowly changing micro-meteorological time-series [4,14]. Those time-series have no seasonal trend or fixed seasonal trends and have large data-banks that range over multiple years. In those cases, MLPs are using the observation of the system X at a given point to predict the value of Y for that point. Hence it does not rely on neighboring values. Despite their positive results, the MLPs suffer from two significant drawbacks: they cannot natively embed temporal relations, and they cannot make long term relations. This means that if those networks were used for rapidly changing systems, they would not be able to extract local trends to re-scale their forecasts. Additionally, the use of MLP on those applications requires a large data-banks of previous years because those network leverages information like the day of the year. In our case this is not feasible as we do not have such information.

In the field of machine learning, some works have shown the applicability of RNNs to time-series forecasting. In general, Long Short-Term Memorys (LSTMs) [9] and Gated Recurrent Units (GRUs) [3] are common tools used in time-series forecasting. Unfortunately, to the best of our knowledge, in the case of gap filling, the machine learning literature has little to offer. This is most possibly due to the lack of interest of the community in this problem. Usually, when performing time-series forecasting, a sequence of chronologically ordered elements is embedded in a higher-dimensional space using a learned projection such as a linear transformation. It is then processed by a few stacked recurrent-layers, and, finally the results of the recurrent-layers are projected back into the desired shape. The recurrent nature of RNNs allows them to remember past information.

While these methods address the problems encountered with MLPs, they have limitations of their own. Indeed, LSTMs and GRUs suffer from limited memory capacity and range. Because they are recurrent architectures, the hidden-state of the network, also known as its memory, will go through n transformations, with n the number of elements in the sequence. This means that correlating elements at the beginning of the sequence with those at the end will be hard, especially since we consider processing sequences of 600 points. Additionally, properly initializing the RNNs' hidden state is tricky and leads to errors, especially in the case of a limited dataset. Finally, in the case of gap filling, it would make sense to use bidirectional RNNs [8]. Indeed standard RNN only leverages past information when bidirectional RNN can leverage past and future observations, making it a more appropriate choice. However, those architectures are not well suited for our task because the variable we are trying to fill gap in exhibits a cyclicity. We are confident that the RNN would be able to learn this cyclicity but other architectures can account for it natively if modified correctly. Directly accounting for the cyclicity should reduce the need for learning samples and make the overall learning process easier.

Recent advances in the field of NLP and Natural Language Generation (NLG) address those issues with the introduction of attention-based deep neural networks. Models such as Transformer [17], and now Bert [5], are the reference for many sequence-to-sequence (seq2seq) tasks in NLP. The mechanism at the root of these neural-networks is called self-attention. It offers the benefits of recurrent models without their downsides. Also, because the model is not iterative, it does not have to iterate through each of the sequence elements but can perform all the operation concurrently leveraging tensor-dot operations. This makes attention-based models much more efficient than their recurrent counterparts. Finally, these models can be modified to account for the cyclicity of the data. As they rely on a positional encoding to know where the different elements are located in the sequence, this encoding can modified to create a cyclicity. In the case of filling gaps in ET time-series we can use the time of the day. This results in all the values for a given time of the day to be located at the same position in the sequence from the perspective of the attention network. In comparison, a recurrent network iterates through the whole sequence and stores in its hidden state all that happened before to make its prediction. It is not aware that there might be a cyclic behavior unless it learns it. This consideration, associated with the one stated earlier on, explains why we did not consider RNNs for this study.

3 Filling Gaps

In order to leverage information from both past, present, and future, we chose to develop seq2seq attention-based models dedicated to filling gaps. Our architecture can be seen in Fig. 2.

From a formal point of view, our system can be formulated as in (1), where $x_i(t)$ is the set of observed variables such that $\forall i \in [1, n], \forall t \in [1, T]$, $x_i(t) \in \mathbb{R}$ and $z_i(t)$ is the set of non-observed variables such that $\forall i \in [1, m], \forall t \in [1, T]$, $z_i(t) \in \mathbb{R}$.

$$\begin{aligned} y(t) = f(x_1(t), .., x_n(t), z_1(t), .., z_m(t)) \in \mathbb{R} \\ \forall i \in [1, n], \exists \tau_i \in \mathbb{R} \, \text{s.t.} \, \mathrm{x_i(t)} = \mathrm{x_i(t} + \tau_\mathrm{i}) + \epsilon \end{aligned} \tag{1}$$

We will denote as x the "sequence of observations" of our system, and y the "sequence of targets" of our system. The goal of our model will be to recover the missing values from the target sequence using the sequence of observations and the sequence of targets with missing values. Our architecture leverages both past and future information using the attention mechanism.

3.1 Architecture

When building our model, we took inspiration from the canonical attention architecture: Transformer. An architecture like transformer does not iteratively process all the elements of a sequence but instead processes a whole sequence at once. Additionally, it features an encoder-decoder structure, where, the encoder is used to process the input sequence, and the decoder is used to generate the output sequence based on the processed input sequence. The encoder share a fairly similar structure where each element of the sequence is embedded using the same transformation and then a process called self attention is applied on the sequence. This process results in a sequence of similar length as the input sequence, and allows the network to make sequence wide correlations. Then a stack of shared feed-forward layer are applied on each element of the previously processed sequence. Complete details of the architecture is given in [17, sec 3.1].

We only relied on a single part of the transformer model: the encoder. This simplification of the encoder/decoder architecture effectively reduces the number of parameters within the model, reducing its complexity and making it easier to train. Similarly to transformer our architecture features multiple attention-heads which process embedded input. This embedding is of size d_{model}.

An attention-head (or Scaled Dot-Product Attention) computes the correlation for every combination of elements pairs in the sequence under the form of an attention matrix. This propagates information between two elements, even if they are far apart in the sequence. The exact implementation of this method is described in [17, sec 3.2]. Multi-head attention consists in using multiple attention-heads in parallel (with the same input) and then concatenate their output.

Initial results with a single head showed poor performances: the network was having a hard time separating the different variables. Indeed, in the case of the observation sequence, the model should look everywhere, but in the case of the target sequence, it should not pay attention to the gap. Hence, based on that observation, and in an effort to minimize the learning complexity, we chose to manually assign different variables to our network heads. We used three attention heads: one head processes the observation sequence, another head processes the target sequence, and finally, the last head processes a concatenation of observation and target sequences. Ideally, we would let the model learn how to separate the variables on its own, but, due to the limited training samples this was not feasible in the case of our dataset.

Furthermore, it is worth noting that those heads do not use a causal mask[1]. In attention heads, causal masks prevent the association of an element with other elements that happen later in the sequence. Removing it allows each attention head to look at the past, present, and future at the same time. This enables our model to extract local trends, and correlate information from elements far apart in the sequence in a way an RNN could not.

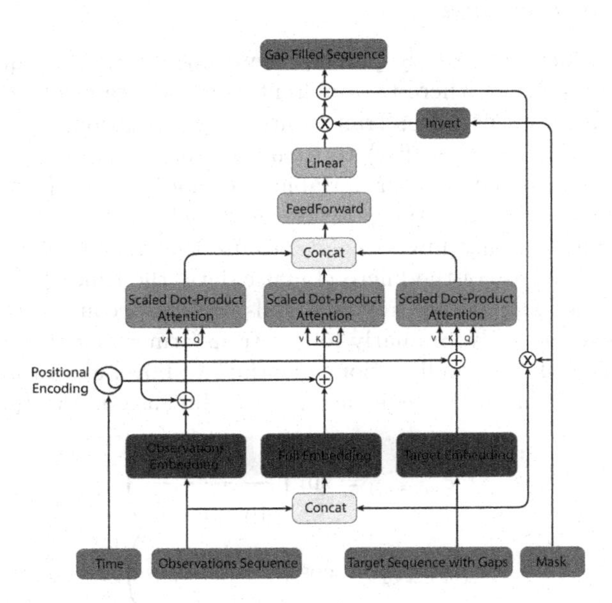

Fig. 2. Our architecture for gap filling

[1] Please note that this is only true in attention heads. We use a mask in the overall structure to copy the non-gap-points.

3.2 Feed Forward Layer and Copy Task

After applying the multi-head attention mechanism, the output heads are concatenated and passed to a feed-forward layer: two dense layers with Leaky-Rectified-Linear-Unit (Leaky-ReLU) [12] activation function, and a dropout layer. Similarly to [16], our models performed approximately 10% better when using Leaky-ReLUs over normal Rectified-Linear-Units (ReLUs). After the feed-forward layer, a dense layer is used to project back the output of our feed-forward layer to a one-dimensional sequence: the target sequence with its missing data filled.

Finally, we implemented a full skip-branch directly copying the original target value onto the output. This prevents the network from learning a complicated function, where part of the target sequence is copied, and the rest is changed to fill the gap. This is achieved by providing our network with a binary mask (a sequence of binary numbers). Where, True means that the data is to be copied, and False means the network should fill in the point. As of now, the detection of the gaps in the data is handled by the EddyPro software[2]. This software is one of the most prevalent in the field and features a wide panel of failure detection methods. Nevertheless, future work may focus on learning areas of the sequence that need correction, *i.e.* using a similar architecture to learn the sensor filtering.

3.3 Positional Encoding

Unlike RNNs, which recurrently process the elements of the sequence, attention models cannot know where the different elements are positioned inside the sequence. To alleviate this issue, Transformer uses a positional encoding, which gives a unique value (an identifier) to each element of the sequence. However, to account for the cyclicity of our variables we modified the positional encoding. When Transformer uses the position of the elements in the sequence to perform the positional encoding, we chose to use the value of the primary periodic variable: in the case of the micro-climatic data, the time of the day. Hence, the positional encoding becomes cyclic. This lets us account for the periodic time dependencies natively. Similarly to the Transformer model, we use sine and cosine functions with the modifications mentioned above, as can be seen in (2). $pos \in \mathbb{N}$ is the position in the sequence, i is the dimension, and $t(pos) \in \mathbb{N}$.

$$\begin{aligned} PE_{(pos,2i)} &= \sin\left(\frac{t(pos)}{10000^{\frac{2i}{d_{model}}}}\right) \\ PE_{(pos,2i+1)} &= \cos\left(\frac{t(pos)}{10000^{\frac{2i}{d_{model}}}}\right) \end{aligned} \tag{2}$$

Despite the appeal to learn the encoding [7], which should, in theory, allow the network to learn the frequencies that make most sense for the problem at hand,

[2] https://www.licor.com/env/products/eddy_covariance/software.html.

we chose not to. Firstly, based on the conclusion of [17], it seems that there are no benefits from using a learned positional encoding. Secondly, this lets us reduce the complexity of the learning process in regard to the limited amount of training examples at our disposal.

From a practical perspective, the positional encoding depends on the period from the different sequences of the batch, which have different time offsets. This means that it cannot be computed ahead of time. Thus to maximize performance, the positional encoding is computed directly within the network's graph.

3.4 Training

To train our models we used gap-free-data and generated artificial gaps inside them. Since gap-free-data almost do not exist in the real-world, we used linear interpolation to fill small gaps (1 to 2 points). In the case of larger gaps, we simply removed the days during which those occurred and did not took samples that overlapped with the gaps. The exact details of the dataset generation are given in Sect. 4.1. Please note we do not aim to fill those small gaps, this problem is considered trivial and we are only looking to fill continuous gaps larger than 24 points (or half a day) and up to 288 points.

We then had to put some values inside the gaps to tell our models where points are missing in the sequence. Unlike NLP, we cannot choose to set an unknown character to indicate the areas where we want to recover data as our variables are defined in $\mathbb{R}$. However, since the data is normalized to a zero-centered normal distribution with unit variance we can make the assumption that most of our values will lie within the $[-2, 2]$ interval. Hence, we could pick a value outside of this interval to fill our gaps. Yet, after some experimentation, we found that it was more reliable to fill the gaps with the mean of the non-corrupted points within that same sequence. Not only did it make the models predictions more reliable, but it also increased the convergence speed of our architectures.

Our networks were trained using the ADAM optimizer [10] (without using the advanced modification of transformer). As in most regression problems, we use an L_2-loss that we average over the whole sequence excluding the points outside of the gap.

4 Experiments

4.1 Datasets

We tested our model on two cases: a real use case from the field of agriculture and a toy case to demonstrate that our approach also scales to other related problems. For the application on real data, we aim to demonstrate that our methodology is capable of strong generalization by learning on a growing season in different crop fields and using this knowledge to fill gaps in situations that were not encountered before: a different crop at a different season at a different location.

Eddy Covariance (EC) Data. To evaluate our model on a real-world scenario, we chose to apply it to ET measurements. These measures of ET are acquired using an EC device. An EC tower measures latent heat flux (*i.e.* water vapor flux) from a crop. This can be used to infer the crop water consumption and hence its evapo-transpiration. Additionally, the tower also records the relative humidity in the air, along with the sun radiation, the wind speed and the air temperature. These 4 variables, which we will call meteorological data, are the only variables our neural network has access to, to reconstruct the missing point in the ET. Please note that sometimes the tower had sensors failures as well. In order to avoid having gaps in our input data, we used values from a nearby meteorological station to fill in the missing meteorological values. The files used to train and evaluate our model come from six different measurement campaigns[3] and feature two growing seasons in three different types of crops: processing tomatoes, cotton, and wheat. These recordings were acquired at different seasons: winter and summer, and in different regions in Israel: north and south. The direct consequence is that the amplitudes of the variables change significantly between the different recordings. Each recording has 5 variables, the latent heat flux (i.e. our target), the net radiation, the relative humidity, the air temperature, and the wind speed. Those files are recorded with a half-hourly rate over a period of 3 to 4 months. In total, this makes for about 3000 to 4000 continuous gap-free-points per recording.

To evaluate our networks on those crops, we could not train and test on each recording individually. This would result in too little data to train or evaluate our model properly. Also, it would mean that our model would be tuned for this specific recording, and would not be able to generalize to other crops, making our approach impractical. We verified this hypothesis using tomato crops and then chose to do 6 different train/test sets. To do so, we put all our recordings but one in the training and the remaining one in the test. We ran the 6 possible combinations and obtained 6 different datasets.

This dataset exhibits interesting behaviors when compared to similar problems, for instance in forestry. Here, the hot middle-eastern climate coupled to the spring season creates rapid changes in the plant canopy, increasing its leaf area index which in-turns increases the water it consumes. Ideally, we would include a vegetation variable, like the leaf-area index to our model, but this variable is very tedious to acquire, and in most cases is not measured. Using vegetation indices (e.g. NDVI - Normalized Difference vegetation index) derived from remote sensing data [11] could also be considered, but they often exhibits gaps and are not applicable to small fields. This is why in the end, we did not consider any vegetation variables in our experiments. On this dataset we only compare ourselves to REddyProc. As the MLP performed worsed than the REddyProc its results are not presented here.

Toy Problem. To test our architecture, we developed a small toy problem that features similar construction to our general problem. We create 3 variables

[3] Data available upon request.

$x_i(t) \in \mathbb{R}, t \in \mathbb{N}$ s.t. $x_i(t) = \alpha_i + \sin(t * \gamma + \tau_i) * \beta_i + \epsilon_i$ and γ is computed such that the periodicity of the variables is 48 points. Additionally, we create a latent variable (which will not be observed by our methods) $z_1(t) \in \mathbb{R}$ defined as $z_1 = \sin(t * \omega)$ where ω is set such that the periodicity of z_1 is 720 points. We chose 720 as this is larger than the maximum scope of our neural-networks. We then combine those variable to form (3).

$$y(t) = \left| x_1(t)^2 \times e^{x_2(t)} \times \log\left(x_3(t)\right) \right| \times (z_1(t) + 2) + \epsilon \tag{3}$$

This problem is interesting because the cyclicity of the positional encoding of our networks is set to 48 points and not 720 points. Hence, our model will have to adapt to the amplitude change created by z_1 and extract local trends to estimate the correct values. A total of 70,000 points were generated. On this problem, we compare ourselves to a KNN method with a maximum window of 300 points on each side of the gap. It is set to use up to 15 points as long as their L_∞-norm is below 5% error. If no points matching this condition are found, it then works as a LUT. We are also comparing ourselves to an MLP model that we acquired doing a grid-search for the optimal set of dense layers.

4.2 Evaluation

We chose to evaluate our approach on different sequences and gap sizes. To do so, we generated sequences with 3 different lengths: 192, 384, and 576 points, equivalent respectively to 4, 8, and 12 days on the real data. For each of these lengths, one model is learned. We will refer to sequences of 192 points as small (S), 384 points as medium (M), and 576 as large (L).

When considering our real data, the limited quantity of data-points was a problem. Even for small sequences, we only had 82 unique non-overlapping sequences. To increase that number, we generated sequences using a moving window with a stride of one. This allowed us to generate about 2500 sequences out of one recording. This high redundancy in our data explains why we aimed to minimize the network's parameters: to prevent overfitting. Additionally, we chose to generate gaps of random size, and at random positions when sampling batches during both training and testing. This further increases the quantity of available data and further mitigates the risk of overfitting. In the small sequences, the gaps ranged from 24 to 72 points; in the medium ones, the gaps ranged from 72 to 144 points, and in the large ones, the gaps ranged from 144 to 288. This is slightly higher than the usual 30% missing data in average in EC measurements.

To evaluate our approach on the real data, we compare ourselves to the most prevalent tool in the field: REddyProc [18]. This tool, developed by the Max-Planck Institute, is strictly dedicated to fill gaps within flux measurements by EC systems. Embedded as an R package, it relies on the MDS, and in some extreme cases on the MDV to recover the missing points. The main restriction of this tool comes from the techniques it uses. As it is based on MDS, a window of at least 7 days on each side of the gap is being used. Since the size of this window cannot be changed, the two approaches were compared using different sequence lengths to fill the gaps: Our network sees a much smaller horizon of points due to memory limitation of our GPUs. Also, using 14-day windows (7 on each side) and 6-day gaps would be equivalent to use sequences of 960 points, which we could not do with our data-recordings as it would result in too few sequences.

Finally, to evaluate the results of the different methods, we sample 100 gaps from the test set. On these gaps, we compute the Root Mean Squared Error (RMSE) and the Mean Bias Error (MBE) between the methods results and the ground-truth EC measurements. The MBE is the mean of all the errors on a given gap. For each gap the percentage of improvement is computed and the mean and standard deviation of the improvement is also reported. The metrics are computed per gap and then averaged. Additionally, we compute the standard deviation for each of the metrics. The objective of the MBE metric is to make sure that the model has a zero centered error. In irrigation, it is used to indicate the bias induced on the daily or seasonal sums of the water loss. Finally, to ensure the repeatability of our results, all our models are trained 3 times with a different optimizer seed, different test samples, and different training batch orders. The presented results are an average of the 3 runs.

4.3 Neural Networks and Training

When training on the small sequences, we used a batch size of 256, a learning rate of 0.0001, and dropout of 0.85. When training on medium sequences, the batch size was reduced to 128, and when training on large sequences, the batch size was set to 64. Going over those values resulted in tensors too large to be processed. For our model on the real data, the input embedding is of size 128, d_{model} is of size 512, the first dense layer after it has a size of 128, and the last dense layer as a size of 32.

We implemented our model in TensorFlow [1] 1.14 with a tensorboard front-end allowing us to visualize the evolution of the attention-heads over time along with the loss and accuracy. All experiments were carried out on an IBM Power Systems AC922 with 256 GB of RAM and 4 NVIDIA V100 16 GB GPU (using only a single GPU).

5 Results

5.1 Toy Case

On the toy problem, our approach outperforms both MLP and the KNN algorithm despite the larger view horizon of the KNN. As can be seen in Table 1, which summarizes the results for the different gap-size, our approach is consistently better than KNN and MLP across all metrics except for the MBE on large sequence sizes. This can be explain by the nature of the MBE metric: it is the mean of the error, thus one value can slightly change the overall result even more here since the error values are very small. Hence it is more important to focus

Table 1. RMSE and MBE of our model and KNN on the toy problem. Our problem easily bests the KNN approach.

Seq size	Methods	RMSE		MBE	
		Mean	Std	Mean	Std
S	Ours	**0.09**	**0.021**	**0.008**	**0.021**
	MLP	0.16	0.041	0.0016	0.044
	KNN	0.48	0.21	−0.016	0.21
M	Ours	**0.12**	**0.04**	**−0.004**	**0.026**
	MLP	0.25	0.08	−0.007	0.096
	KNN	0.54	0.21	0.011	0.23
L	Ours	**0.25**	**0.09**	−0.007	**0.033**
	MLP	0.41	0.09	−0.016	0.15
	KNN	0.69	0.19	**0.04**	0.34

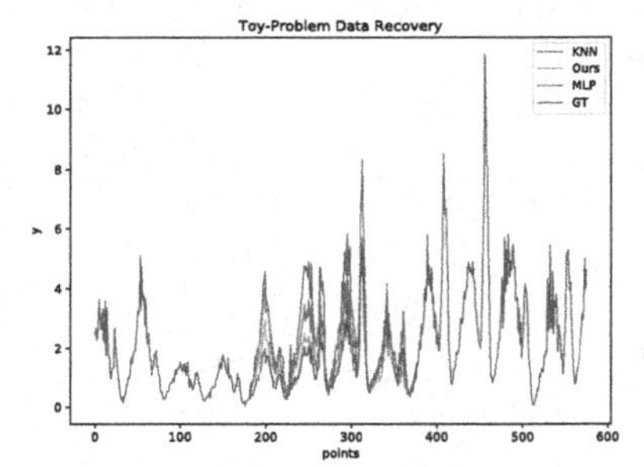

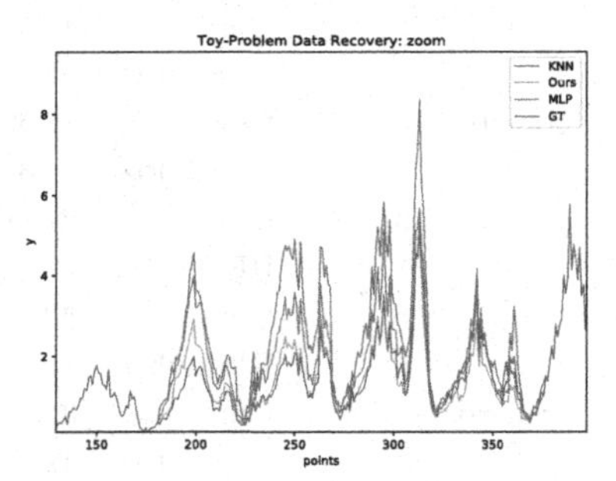

Fig. 3. Gap filling quality comparison of our model (NN), with KNN on the toy problem on large gaps. (Color figure online)

on the variance of the MBE as it depicts how is fluctuates over various gaps. Figure 3 compares the KNN method (in blue) to our architecture (in orange). We can see that our approach better fits the data. Our attention-based model is able to extract the local trend, whereas KNN is being tricked by the long term amplitude changes created by the latent variable z_1.

5.2 Evapotranspiration Data

Table 2 summarizes our models' performance on the EC datasets. Based on the RMSE results our model performs statistically better or as well as the reference method: REddyProc. Only on large gaps on Cotton2 does it performs slightly worse in average but the difference is not statistically significant. On an other hand, our model performs much better than the baseline on the Tomato crops. This is particularly visible on the Tomato 2 experiment, which yields an average improvement of 30% across all gaps. This is interesting as tomatoes are summer

Table 2. RMSE and MBE of our model and REddyProc applied on real EC data (lower RMSE and MBE values indicate better model performance). Values range from −50 to 800.

Crops	Seq size	Methods	RMSE		RMSE improvements	MBE	
			Mean	Std		Mean	Std
Cotton 1	S	Ours	**44.2**	**13.2**	+12% ± 17%	**−0.9**	**8.4**
		REddyProc	51.8	17.4		2.1	13.8
	M	Ours	**53.9**	**13.4**	+7% ± 14%	**−3.9**	13.9
		REddyProc	57.0	13.5		9.1	**12.2**
	L	Ours	**46.7**	**7.62**	+10% ± 10%	**−0.5**	7.4
		REddyProc	52.9	12.7		0.9	**5.7**
Cotton 2	S	Ours	**43.2**	**13.9**	+10% ± 15%	−0.7	**12.0**
		REddyProc	48.1	13.9		**0.1**	14.0
	M	Ours	**48.0**	**11.9**	+4% ± 22%	**0.1**	12.1
		REddyProc	51.0	12.0		-5.5	**9.7**
	L	Ours	49.5	14.6	−10% ± 49%	2.2	12.3
		REddyProc	**47**	**10.9**		**−1.1**	**8.6**
Tomato 1	S	Ours	**30.9**	**14.0**	+29% ± 19%	−10.5	**11.7**
		REddyProc	54.6	24.0		**−4.1**	19.3
	M	Ours	**31.0**	**4.2**	+35% ± 12%	−2.6	**4.8**
		REddyProc	49.9	14.9		**0.6**	10.7
	L	Ours	**44.6**	**9.1**	+35% ± 12%	−4.4	**5.4**
		REddyProc	71.9	23.8		**−3.3**	12.8
Tomato 2	S	Ours	**32.4**	**9.1**	+19% ± 25%	**−1.2**	**9.1**
		REddyProc	42.1	13.2		−3.4	16.6
	M	Ours	**36.4**	**6.88**	+14% ± 20%	**−1.6**	**8.5**
		REddyProc	44.2	12.34		−5.9	13.0
	L	Ours	**37.8**	**4.1**	+13% ± 15%	**1.0**	**5.9**
		REddyProc	44.4	9.2		−1.6	9.8
Wheat 1	S	Ours	**37.9**	**12.2**	+7% ± 30%	−8.9	**9.4**
		REddyProc	46.44	13.58		**−7.6**	22.5
	M	Ours	38.0	**8.1**	+0% ± 25%	**−4.3**	**4.8**
		REddyProc	**37.5**	11.4		6.5	5.8
	L	Ours	**37.9**	**4.6**	+0% ± 13%	−8.9	**4.2**
		REddyProc	38.7	6.6		**−1.7**	8.5
Wheat 2	S	Ours	**26.0**	**7.2**	+0% ± 30%	4.5	**6.3**
		REddyProc	28.42	13.6		**−1.8**	10.3
	M	Ours	**27.9**	**4.8**	+2% ± 21%	7.5	**7.4**
		REddyProc	29.9	9.1		**−0.8**	9.2
	L	Ours	**29.9**	**3.3**	+7% ± 17%	7.2	**4.9**
		REddyProc	31.9	7.0		**−2.8**	6.9

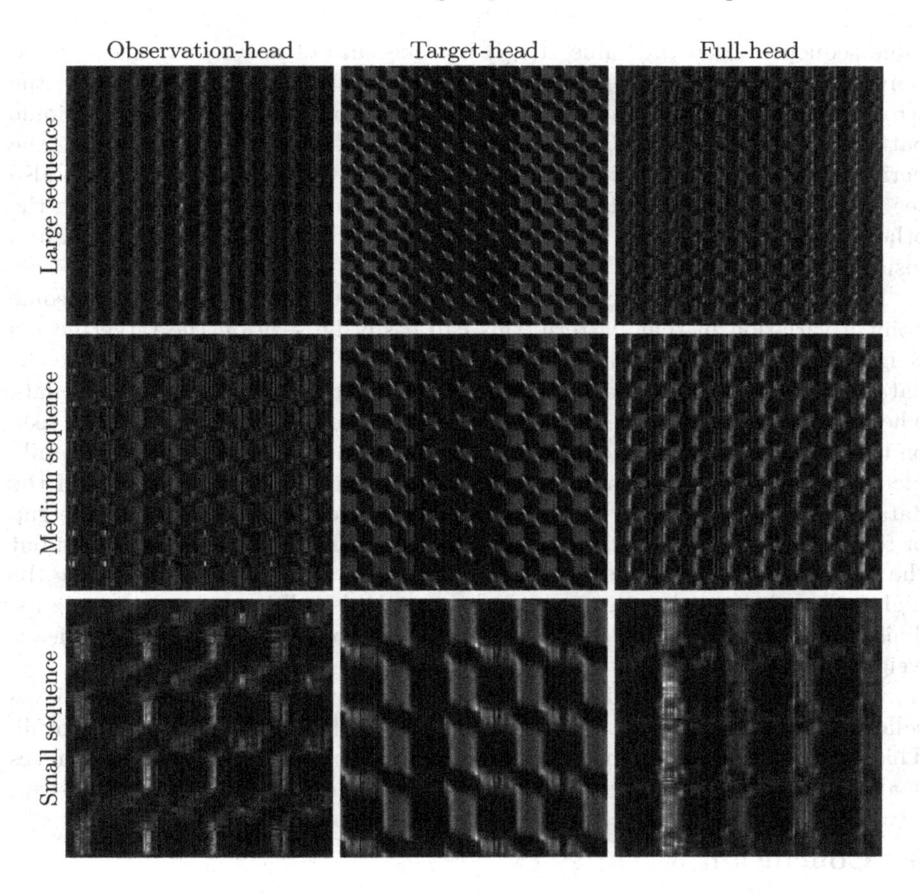

Fig. 4. Attention weights for the different sequence sizes on real data. The periodicity of the data was understood and leveraged by the attention mechanism.

crops with a quick canopy growth. This, particularities lead to faster ET dynamics than the one encountered in the other crops present in this dataset. The superior performance of our network on this crop shows that our approach performs well on system with high dynamics, which was our original goal. Additionally, the constant performance of our network demonstrates that, despite its smaller time horizon, our architecture is more reliable than REddyProc. Regarding the MBE our model performs as well as REddyProc. Overall, the MBE quantifies the irrigation bias but not the accuracy of the prediction, hence our prediction keeps similar performances.

Figure 4 shows examples of attention matrices for the different attention heads. Each of those matrices translates the cross-correlation between the elements of the same sequence. Let us define A, an attention matrix for a sequence of size k s.t $A \in R^{k \times k}$ and $i, j \in [1, k]$, the position of two elements inside that same sequence; then the value $A(i, j)$ is a measure of how strong the correlation between the elements i and j is. The brighter the pixels in the image, the stronger the correlation. The attention matrices of our networks present periodic patterns. The periodic patterns show that our architectures are leveraging the periodicity of the data to fill the missing values in our sequences. One can also see that on the target-heads the center of the matrix is less bright than on the other heads. This is due to the fact that the target heads avoid using the values inside the gaps.

If one looks at the last row of Fig. 4, one can see some patterns and at some point a variation in that pattern, this can easily be seen on the target-head's weights where a black band appears. This is where the gap is located inside the data. What this black band means, is that the network learns not to use data where there are gaps in order to fill the missing values. Similar things can be seen on the medium and large sequences (on the target head), but the gap is not fully black. Instead during the nights the network is still trying to make use of the data. This is probably due to the network having difficulties detecting the gap or to a lack of training data. On the other heads, similar pattern can be seen but the gap is not clearly visible. However, what is visible are the nights: during the nights the values are homogeneous and the variables all have the same weights. This is represented by this darker bands that can be seen in the different heads weights.

Finally, using this training-testing split, we show that our network achieves solid performances without even training on the dataset which we aim to fill. This demonstrates the strong generalization capacities of our method and makes it almost as convenient as the MDS since its application would be training-free.

6 Conclusion

A novel, data-driven, gap filling method that relies on multi-head attention is introduced in the context of evapotranspiration measurements of field crops. Our method performed better than the current state of the art when tested on both a toy problem and a real-world scenario using evapo-transpiration data. Furthermore, the data-efficiency of our method allows to achieve these results even with very small training datasets. Finally, the generality of this innovative approach is demonstrated in a real-world scenario where we learn a model on a set of crops and use the knowledge learned on those crops to successfully fill gaps on an other crop type. Future efforts will focus on creating a simple framework for the benefit of users outside the machine learning community to train and apply these models for any time-series.

Acknowledgements. We would like to thank Lionel Clavien from our partner Inno-Boost SA in Switzerland for providing us access to the server used for our experiments as well as some starting help on the platform. This work was supported by a grant from the Ministry of Science and Technology (MOST), Israel, under the France-Israel Maimonide Program, & Ministry of Europe and Foreign Affairs (MEAE), and the Ministry of Higher Education, Research and Innovation (MESRI) of France.

References

1. Abadi, M., Agarwal, A., Barham, P., et al.: TensorFlow: large-scale machine learning on heterogeneous systems (2015). http://tensorflow.org/. software available from tensorflow.org
2. Che, Z., Purushotham, S., Cho, K., Sontag, D., Liu, Y.: Recurrent neural networks for multivariate time series with missing values. Sci. Rep. **8**(1), 6085 (2018)
3. Cho, K., et al.: Learning phrase representations using RNN encoder-decoder for statistical machine translation. arXiv preprint arXiv:1406.1078 (2014)
4. Coutinho, E.R., Silva, R.M.d., Madeira, J.G.F., Coutinho, P.R.d.O.d., Boloy, R.A.M., Delgado, A.R.S., et al.: Application of artificial neural networks (ANNs) in the gap filling of meteorological time series. Revista Brasileira de Meteorologia **33**(2), 317–328 (2018)
5. Devlin, J., Chang, M.W., Lee, K., Toutanova, K.: Bert: pre-training of deep bidirectional transformers for language understanding. arXiv preprint arXiv:1810.04805 (2018)
6. Falge, E., et al.: Gap filling strategies for long term energy flux data sets. Agric. For. Meteorol. **107**(1), 71–77 (2001)
7. Gehring, J., Auli, M., Grangier, D., Yarats, D., Dauphin, Y.N.: Convolutional sequence to sequence learning. In: Proceedings of the 34th International Conference on Machine Learning, vol. 70, pp. 1243–1252. JMLR. org (2017)
8. Graves, A., Fernández, S., Schmidhuber, J.: Bidirectional LSTM networks for improved phoneme classification and recognition. In: Duch, W., Kacprzyk, J., Oja, E., Zadrożny, S. (eds.) ICANN 2005. LNCS, vol. 3697, pp. 799–804. Springer, Heidelberg (2005). https://doi.org/10.1007/11550907_126
9. Hochreiter, S., Schmidhuber, J.: Long short-term memory. Neural Comput. **9**(8), 1735–1780 (1997)
10. Kingma, D.P., Ba, J.: Adam: a method for stochastic optimization. arXiv preprint arXiv:1412.6980 (2014)
11. Lange, M., Dechant, B., Rebmann, C., Vohland, M., Cuntz, M., Doktor, D.: Validating MODIS and sentinel-2 NDVI products at a temperate deciduous forest site using two independent ground-based sensors. Sensors **17**(8), 1855 (2017)
12. Maas, A.L., Hannun, A.Y., Ng, A.Y.: Rectifier nonlinearities improve neural network acoustic models. In: Proceedings of ICML, vol. 30, p. 3 (2013)
13. Moffat, A.M., et al.: Comprehensive comparison of gap-filling techniques for eddy covariance net carbon fluxes. Agric. For. Meteorol. **147**(3–4), 209–232 (2007)
14. Papale, D., Valentini, R.: A new assessment of European forests carbon exchanges by eddy fluxes and artificial neural network spatialization. Glob. Change Biol. **9**(4), 525–535 (2003)
15. Reichstein, M., et al.: On the separation of net ecosystem exchange into assimilation and ecosystem respiration: review and improved algorithm. Glob. Change Biol. **11**(9), 1424–1439 (2005)
16. Richard, A., Mahé, A., Pradalier, C., Rozenstein, O., Geist, M.: A comprehensive benchmark of neural networks for system identification (2019)
17. Vaswani, A., et al.: Attention is all you need. In: Advances in Neural Information Processing Systems, pp. 5998–6008 (2017)
18. Wutzler, T., et al.: Basic and extensible post-processing of eddy covariance flux data with REddyProc. Biogeosciences **15**(16), 5015–5030 (2018)

Lagrangian Duality for Constrained Deep Learning

Ferdinando Fioretto[1(✉)], Pascal Van Hentenryck[2], Terrence W.K. Mak[2], Cuong Tran[1], Federico Baldo[3], and Michele Lombardi[3]

[1] Syracuse University, Syracuse, NY 13244, USA
{ffiorett,cutran}@syr.edu
[2] Georgia Institute of Technology, Atlanta, GA 30332, USA
{pvh,wmak}@isye.gatech.edu
[3] University of Bologna, 40126 Bologna, BO, Italy
{federico.baldo2,michele.lombardi2}@unibo.it

Abstract. This paper explores the potential of Lagrangian duality for learning applications that feature complex constraints. Such constraints arise in many science and engineering domains, where the task amounts to learning to predict solutions for constraint optimization problems which must be solved repeatedly and include hard physical and operational constraints. The paper also considers applications where the learning task must enforce constraints on the predictor itself, either because they are natural properties of the function to learn or because it is desirable from a societal standpoint to impose them.

This paper demonstrates experimentally that Lagrangian duality brings significant benefits for these applications. In energy domains, the combination of Lagrangian duality and deep learning can be used to obtain state of the art results to predict optimal power flows, in energy systems, and optimal compressor settings, in gas networks. In transprecision computing, Lagrangian duality can complement deep learning to impose monotonicity constraints on the predictor without sacrificing accuracy. Finally, Lagrangian duality can be used to enforce fairness constraints on a predictor and obtain state-of-the-art results when minimizing disparate treatments.

1 Introduction

Deep Neural Networks, in conjunction with progress in GPU technology and the availability of large data sets, have proven enormously successful at a wide array of tasks, including image classification [25], speech recognition [3], and natural language processing [10], to name but a few examples. More generally, deep learning has achieved significant success on a variety of regression and classification tasks. On the other hand, the application of deep learning to aid computationally challenging constrained optimization problems has been more sparse, but is receiving increasing attention, such as the efforts in jointly training prediction and optimization models [22,24,34] and incorporating optimization algorithms into differentiable systems [4,13,35].

Y. Dong et al. (Eds.): ECML PKDD 2020, LNAI 12461, pp. 118–135, 2021.
https://doi.org/10.1007/978-3-030-67670-4_8

This research originated in an attempt to apply deep learning to fundamentally different application areas: The learning of constrained optimization problems and, in particular, optimization problems with hard physical and engineering constraints. These constrained optimization problems arise in numerous contexts including in energy systems, mobility, resilience, and disaster management. Indeed, these applications must capture physical laws such as Ohm's law and Kirchhoff's law in electrical power systems, the Weymouth equation in gas networks, flow constraints in transportation models, and the Navier-Stoke's equations for shallow water in flood mitigation. Moreover, they often feature constraints that represent good engineering and operational practice to protect various devices. For instance, they may include thermal limits, voltage and pressure bounds, as well as generator and pump limitations, when the domain is that of energy systems. Direct applications of deep learning to these applications may result in predictions with severe constraint violations, as shown in Sect. 5.

There is thus a need to provide deep learning architectures with capabilities that would allow them to capture constraints directly. Such models can have a transformative impact in many engineering applications by providing high-quality solutions in real-time and be a cornerstone for large planning studies that run multi-year simulations. To this end, this paper proposes a *Lagrangian Dual Framework* (LDF) for Deep Learning that addresses the challenge of enforcing constraints during learning: Its key idea is to exploit Lagrangian duality, which is widely used to obtain tight bounds in optimization, during the training cycle of a deep learning model.

Interestingly, the proposed LDF can be applied to two distinct context: (1) constrained optimization problems, which are characterized by constraints modeling relations among features of each data sample, and (2) problems that require specific properties to hold on the predictor itself, called constrained predictor problems. For instance, energy optimization problems are example problems of the first class. These problems impose constraints that are specific to each data sample, such as, flow conservation constraints or thermal limits bounds. An example problem of the second class is *transprecision computing*, a technique that achieves energy savings by adjusting the precision of power-hungry algorithms. An important challenge in this area is to predict the error resulting from a loss in accuracy and the error should be monotonically decreasing with increases in accuracy. As a result, the learning task may impose constraints over different samples with their predictions used during training. Other applications in dataset-dependent constraint learning may impose fairness constraints on the predictor, e.g., a constraint ensuring equal opportunity [18] or no disparate impact [37] in a classifier that enforces a relation among multiple samples of the dataset.

This paper shows that the proposed LDF provides a versatile tool to address these constrained learning problems, it presents the theoretical foundations of the proposed framework, and demonstrates its practical potential on both constrained optimization and constrained predictors problems. The LDF is evaluated extensively on a variety of real benchmarks in power system optimization and gas compression optimization, that present hard engineering and operational constraints. Additionally, the proposed method is tested on several datasets that enforce non-discriminatory decisions and on a realistic transprecision computing application, that requires constraints to be enforced on the predictors themselves. The results present a dramatic improvement in the number of constraint violations reduction, and often result in substantial improvements in the prediction accuracy in energy optimization problems.

2 Preliminaries: Lagrangian Duality

Consider the optimization problem

$$\mathcal{O} = \operatorname*{argmin}_y f(y) \quad \text{subject to} \quad g_i(y) \leqslant 0 \;\; (\forall i \in [m]). \tag{1}$$

In *Lagrangian relaxation*, some or all the problem constraints are relaxed into the objective function using *Lagrangian multipliers* to capture the penalty induced by violating them. When all the constraints are relaxed, the *Lagrangian function* becomes

$$f_\lambda(y) = f(y) + \sum_{i=1}^{m} \lambda_i g_i(y) \tag{2}$$

where the terms $\lambda_i \geqslant 0$ describe the Lagrangian multipliers, and $\lambda = (\lambda_1, \ldots, \lambda_m)$ denotes the vector of all multipliers associated to the problem constraints. Note that, in this formulation, $g(y)$ can be positive or negative. An alternative formulation, used in augmented Lagrangian methods [20] and constraint programming [17], uses the following Lagrangian function

$$f_\lambda(y) = f(y) + \sum_{i=1}^{m} \lambda_i \max(0, g_i(y)) \tag{3}$$

where the expressions $\max(0, g_i(y))$ capture a quantification of the constraint violations. This paper abstracts the constraints formulations in (2) and (3) by using a function $\nu(\cdot)$ that returns either the constraint satisfiability or the violation degree of a constraint.

When using a Lagrangian function, the optimization problem becomes

$$LR_\lambda = \operatorname*{argmin}_y f_\lambda(y) \tag{4}$$

and it satisfies $f(LR_\lambda) \leqslant f(\mathcal{O})$. That is, the Lagrangian function is a lower bound for the original function. Finally, to obtain the strongest Lagrangian relaxation of $\mathcal{O}$, the *Lagrangian dual* can be used to find the best Lagrangian multipliers, i.e.,

$$LD = \operatorname*{argmax}_{\lambda \geqslant 0} f(LR_\lambda). \tag{5}$$

For various classes of problems, the Lagrangian dual is a strong approximation of $\mathcal{O}$. Moreover, its optimal solutions can often be translated into high-quality feasible solutions by a post-processing step, i.e., using a *proximal operator* that minimizes the changes to the Lagrangian dual solution while projecting it into the problem feasible region [32].

3 Learning Constrained Optimization Problems

This section describes how to use the Lagrangian dual framework for approximating constrained optimization problems in which constraints model relations among features of each data sample. Importantly, in the associated learning task, each data sample represents a different instantiation of a constrained optimization problem. The section first reviews two fundamental applications that serve as motivation.

3.1 Motivating Applications

Several energy systems require solving challenging (non-convex, non-linear) optimization problems in order to derive the best system operational controls to serve the energy demands of the customers. Power grid and gas pipeline systems are two examples of such applications. While these problems can be solved using effective optimization solvers, their resolution relies on *accurate* predictions of the energy demands. The increasing penetration of renewable energy sources, including those behind the meter (e.g., solar panels on roofs), has rendered accurate predictions more challenging. In turn, predictions need to be performed at minute time scales to ensure sufficient accuracy. Thus, finding optimal solutions for these underlying optimization problems in these reduced time scales becomes computationally challenging, opening opportunities for machine-learning approaches. The next paragraphs review two energy applications that motivate the proposed framework. An extended description of these models is provided in [15].

Optimal Power Flow. The *Optimal Power Flow* (OPF) problem determines the best generator dispatch ($y = S^g$) of minimal cost ($\mathcal{O} = \min_{S^g} \text{cost}(S^g)$) that meets the demands ($d = S^d$) while satisfying the physical and engineering constraints ($g(y)$) of the power system [8], where S^g and S^d denote the vectors (in the complex domain) of generator dispatches and power demands. Typical constraints include the non-linear non-convex AC power flow equations, Kirchhoff's current laws, voltage bounds, and thermal limits. The OPF problem is a fundamental building bock of many applications, including security-constrained OPFs [28]), optimal transmission switching [16], capacitor placement [6], and expansion planning [31] which are of fundamental importance for ensuring a reliable and efficient behavior of the energy system.

Optimal Gas Compressor Optimization. The Optimal Gas Compressor Optimization (OGC) problem aims at determining the best compression controls ($y = R$) with minimum compression costs ($\mathcal{O} = \min_R \text{cost}(R)$) to meet gas demands ($d = q^d$) while satisfying the physical and operational limits ($g(y)$) of the natural gas pipeline systems [19]. Therein, R and q^d are compressors control values and gas demands. Typical constraints include: the non-linear gas flow equations describing pressure losses, the flow balance equations, the non-linear non-convex compressor objective $\mathcal{O}$, and the pressure bounds. Similar to the

OPF problem, the OGC is a non-linear non-convex optimization problem with physical and engineering constraints and a fundamental building block for many gas systems.

The next section describes how to approximate OPFs and OGCs, by viewing them as parametric optimization problems, using the proposed Lagrangian dual framework.

3.2 The Learning Task

The learning task estimates a parametric version of problem (1), defined as

$$\mathcal{O}(d) = \operatorname*{argmin}_{y} f(y, d) \quad \text{subject to} \quad g_i(y, d) \leqslant 0 \ \ (\forall i \in [m]) \tag{6}$$

with a set of samples $D = \{(d_l, y_l = \mathcal{O}(d_l))\}_{l=1}^{n}$. More precisely, given a parametric model $\mathcal{M}[w]$ with weights w and a loss function $\mathcal{L}$, the learning task must solve the following optimization problem

$$w^* = \operatorname*{argmin}_{w} \sum_{l=1}^{n} \mathcal{L}(\mathcal{M}[w](d_l), y_l) \tag{7a}$$

$$\text{subject to} \ \ g_i(\mathcal{M}[w](d_l), d_l) \leqslant 0 \ \ (\forall i \in [m], l \in [n]) \tag{7b}$$

to obtain the approximation $\widehat{\mathcal{O}} = \mathcal{M}[w^*]$ of $\mathcal{O}$.

The main difficulty lies in the constraints $g_i(y, d) \leqslant 0$, which can represent physical and operational limits, as mentioned in the motivating applications. Observe that the model weights must be chosen so that the constraints are satisfied for all samples, which makes the learning particularly challenging. A naive approach to the learning task is thus likely to result in predictors that significantly violate these constraints, as demonstrated in Sect. 5, producing a model that would not be useful in practice.

3.3 Lagrangian Dual Framework for Constrained Optimization Problems

To learn constrained optimization problems, the paper proposes a *Lagrangian dual framework* (LDF) to the learning task. The framework relies on the notion of *Augmented Lagrangian* [20] used for solving constrained optimization problems [17].

In more details, LDF exploits a Lagrangian dual approach in the learning task to approximate the minimizer $\mathcal{O}$. Given multipliers $\lambda = (\lambda_1, \ldots, \lambda_m)$, consider the Lagrangian loss function

$$\mathcal{L}_{\lambda}(\hat{y}_l, y_l, d_l) = \mathcal{L}(\hat{y}_l, y_l) + \sum_{i=1}^{m} \lambda_i \, \nu \left(g_i(\hat{y}_l, d_l)\right),$$

Algorithm 1: LDF for Constrained Optimization Problems

input: $D = (d_l, y_l)_{l=1}^n$: Training data;
$\alpha, s = (s_0, s_1, \ldots)$: Optimizer and Lagrangian step sizes.
1 $\lambda_i^0 \leftarrow 0 \;\; \forall i \in [m]$
2 **for** *epoch* $k = 0, 1, \ldots$ **do**
3 **foreach** $(y_l, d_l) \in D$ **do**
4 $\hat{y}_l \leftarrow \mathcal{M}[w(\lambda^k)](d_l)$
5 $w(\lambda^{k+1}) \leftarrow w(\lambda^k) - \alpha \nabla_w \mathcal{L}_{\lambda^k}(\hat{y}_l, y_l, d_l)$
6 $\lambda_i^{k+1} \leftarrow \lambda_i^k + s_k \sum_{l=1}^n \nu_i \left(g_i(\hat{y}_l, d_l)\right) \;\; \forall i \in [m]$

where $\hat{y}_l = \mathcal{M}[w](d_l)$ represents the model prediction. For multipliers λ, solving the optimization problem

$$w^*(\lambda) = \operatorname*{argmin}_w \sum_{l=1}^n \mathcal{L}_\lambda(\mathcal{M}[w](d_l), y_l, d_l) \tag{8}$$

produces an approximation $\widehat{\mathcal{O}}_\lambda = \mathcal{M}[w^*(\lambda)]$ of $\mathcal{O}$. The Lagrangian dual computes the optimal multipliers, i.e.,

$$\lambda^* = \operatorname*{argmax}_\lambda \min_w \sum_{l=1}^n \mathcal{L}_\lambda(\mathcal{M}[w](d_l), y_l, d_l) \tag{9}$$

to obtain $\widehat{\mathcal{O}}^* = \mathcal{M}[w^*(\lambda^*)]$, i.e., the strongest Lagrangian relaxation of $\mathcal{O}$.

Learning $\widehat{\mathcal{O}}^*$ relies on an iterative scheme that interleaves the learning of a number of Lagrangian relaxations (for various multipliers) with a subgradient method to learn the best multipliers. The LDF, described in Eqs. (8) and (9), is summarized in Algorithm 1. Given the input dataset D, the optimizer step size $\alpha > 0$, and a Lagrangian step size s_k, the Lagrangian multipliers are initialized in line 1. The training is performed for a fixed number of epochs, and each epoch k optimizes the model weights w of the optimizer $\mathcal{M}[w(\lambda^k)]$ using the Lagrangian multipliers λ^k associated with current epoch (lines 3–5). Finally, after each epoch, the Lagrangian multipliers are updated according to a *dual ascent* rule [7] (line 6).

4 Learning Constrained Predictors

This section describes how to use the Lagrangian dual framework for problems in which constraints are not sample-independent, but enforcing global properties between different samples in the dataset and the predictor outputs. It starts with two motivating applications.

4.1 Motivating Applications

Several applications require to enforce constraints on the learning process itself to attain desirable properties of the predictor. These constraints impose conditions on subsets of the samples that must be satisfied. For instance, assume that there is a partial order $\preceq$ on the optimization inputs and the following property holds:

$$d_1 \preceq d_2 \Rightarrow f(\mathcal{O}(d_1), d_1) \leqslant f(\mathcal{O}(d_2), d_2).$$

The predictor should ideally satisfy these constraints as well:

$$d_1 \preceq d_2 \Rightarrow f(\widehat{\mathcal{O}}(d_1), d_1) \leqslant f(\widehat{\mathcal{O}}(d_2), d_2).$$

Transprecision Computing. Transprecision computing is the idea of reducing energy consumption by reducing the precision (a.k.a. number of bits) of the variables involved in a computation [27]. It is especially important in low-power embedded platforms, which arise in many contexts such as smart wearable and autonomous vehicles. Increasing precision typically reduces the error of the target algorithm. However, it also increases the energy consumption, which is a function of the maximal number of used bits. The objective is to design a *configuration* d_l, i.e., a mapping from input computation to the precision for the variables involved in the computation. The sought configuration should balance *precision* and *energy consumption*, given a bound to the error produced by the loss in precision when the highest precision configuration is adopted.

However, given a configuration, computing the corresponding error can be very time-consuming and the task considered in this paper seeks to learn a mapping between configurations and error. This learning task is non-trivial, since the solution space precision-error is non-smooth and non-linear [27]. The samples (d_l, y_l) in the dataset represent, respectively, a configuration d_l and its associated error y_l obtained by running the configuration d_l for a given computation. The problem $\mathcal{O}(d_l)$ specifies the error obtained when using configuration d_l.

Importantly, transcomputing expects a *monotonic* behavior: Higher precision configurations should generate more accurate results (i.e., a smaller error). Therefore, the structure of the problem imposes the learning task to require a dominance relation $\preceq$ between instances of the dataset. More precisely, $d_2 \preceq d_2$ holds if

$$\forall i \in [N] : \; x_{1_i} \leqslant x_{2_i}$$

where N is the number of variables involved in the computation and x_{1_i}, x_{2_i} are the precision values for the variables in d_1 and d_2 respectively.

Fair Classifier. The second motivating application considers the task of building a classifier that satisfies *disparate impact* [37] with respect to a protected attribute d^s and outcome y. A binary classifier does not suffer from disparate impact if

$$\Pr(\hat{y} = 1 \mid d^s = 0) = \Pr(\hat{y} = 1 \mid d^s = 1). \qquad (10)$$

For outcome $y = 1$, the constraint above requires the predictor $\hat{y}$ to have equal *predicted positive rates* across the different sensitive classes: $d^s = 0$ and $d^s = 1$, in the binary task example above. For $y = 0$, the constraint enforces equal *predicted negative rates.* Disparate impact constraints the predicted positive (or negative) rates to be similar across all sensitive attributes. To construct an estimator that minimizes the disparate impact, the paper considers $|\mathcal{D}_s| = 2$ estimators $\mathcal{M}_0$ and $\mathcal{M}_1$, each associated with a dataset partition $D_{|s_i} = \{(d_l, y_l) | d_l^s = s_i\}$ that marginalizes for a particular (combination of) value(s) of the protected feature(s), in addition to the classical estimator $\mathcal{M}$ that is trained over the entire dataset D. Thus, the learning process is defined by the following objective:

$$\min_{w, w_0, w_1} \mathcal{L}\left(\mathcal{M}[w](D)\right) + \sum_{i=0}^{1} \mathcal{L}\left(\mathcal{M}_i[w_i](D_{|s_i})\right) \tag{11a}$$

$$\text{such that} \left| \frac{\sum_{x_i \in D_{s_0}} I(\hat{y}_i = 1)}{|D_{s_0}|} = \frac{\sum_{x_i \in D_{s_1}} I(\hat{y}_i = 1)}{|D_{s_1}|} \right|, \tag{11b}$$

where I is the indicator function. It enforces a constraint on the output of the classifiers $\mathcal{M}_0$, trained on data D_{s_0} to be equivalent to that of the output of the classifier $\mathcal{M}_1$, trained on the dataset D_{s_1}, when their predicted outcome is positive.

The next section will specify how to encode such type of constraints as well as how to express and enforce dominance relations in the proposed constrained learning framework.

4.2 The Learning Task

Consider a set $\mathcal{S} = \{S_1, \ldots, S_q\}$ where S_i is a subset of the inputs that must satisfy the associated constraint

$$h_i(\{\mathcal{O}(d_l)\}_{l \in S_i}, \boldsymbol{d}_{S_i}),$$

where $\boldsymbol{d}_{S_i} = \{d_l\}_{l \in S_i}$, and denote $\boldsymbol{y}_{S_i} = \{y_l\}_{l \in S_i}$.

In this context, the learning task is defined by the following optimization problem

$$\underset{w}{\text{argmin}} \sum_{l=1}^{n} \mathcal{L}(\mathcal{M}[w](d_l), y_l) \tag{12a}$$

$$\text{subject to } g_i\left(\mathcal{M}[w](d_l), d_l\right) \leqslant 0 \quad (\forall i \in [m], l \in [n]) \tag{12b}$$

$$h_i\left(\{\mathcal{M}[w](d_l)\}_{l \in S_i}, \boldsymbol{d}_{S_i}\right) \quad (\forall i \in [q]). \tag{12c}$$

4.3 Lagrangian Dual Framework for Constrained Predictors

To approximate Problem (12), the learning task considers Lagrangian loss functions, for subset of the inputs $S_i \in \mathcal{S}$, of the form

$$\mathcal{L}_{\mu,\lambda}(\tilde{\boldsymbol{y}}_{S_i}, \boldsymbol{y}_{S_i}, \boldsymbol{d}_{S_i}) = \sum_{l \in S_i} \mathcal{L}_{\lambda}(\tilde{y}_l, y_l, d_l) + \sum_{i=1}^{q} \mu_i \, \nu\left(h_i(\tilde{\boldsymbol{y}}_{S_i}, \boldsymbol{d}_i)\right), \tag{13}$$

where $\tilde{y}_l = \mathcal{M}[w](d_l)$ and $\tilde{\boldsymbol{y}}_{S_i} = \{\mathcal{M}[w](d_l)_{l \in S_i}\}$. It learns approximations of the Lagrangian relaxations $\widehat{\mathcal{O}}_{\lambda,\mu}$ of the form

$$w^*(\mu, \lambda) = \underset{w}{\operatorname{argmin}} \sum_{i=1}^{q} \mathcal{L}_{\mu,\lambda}(\{\mathcal{M}[w](d_l)\}_{l \in S_i}, \boldsymbol{y}_{S_i}, \boldsymbol{d}_{S_i}), \tag{14}$$

as well as the Lagrangian duals of Eq. (14) of the form

$$\lambda^*(\mu) = \underset{\lambda}{\operatorname{argmax}} \min_{w} \sum_{i=1}^{q} \mathcal{L}_{\mu,\lambda}(\{\mathcal{M}[w](d_l)\}_{l \in S_i}, \boldsymbol{y}_{S_i}, \boldsymbol{d}_{S_i}), \tag{15}$$

and, finally, the Lagrangian dual of the Lagrangian duals (Eq. (15)) as

$$\mu^* = \underset{\mu}{\operatorname{argmax}} \max_{\lambda} \min_{w} \sum_{i=1}^{q} \mathcal{L}_{\mu,\lambda}(\{\mathcal{M}[w](d_l)\}_{l \in S_i}, \boldsymbol{y}_{S_i}, \boldsymbol{d}_{S_i}) \tag{16}$$

to obtain the best estimator $\widehat{\mathcal{O}}^* = \mathcal{M}[w^*]$, where

$$w^* = \underset{w}{\operatorname{argmin}} \sum_{i=1}^{q} \mathcal{L}_{\mu^*,\lambda^*(\mu^*)}(\{\mathcal{M}[w](d_l)\}_{l \in S_i}, \boldsymbol{y}_{S_i}, \boldsymbol{d}_{S_i}).$$

The Lagrangian dual framework for constrained predictors, described in Eqs. (14)–(16), is summarized in Algorithm 2. The learning algorithm interleaves the learning of the Lagrangian duals with the subgradient optimization of the multipliers μ. Given the input dataset D, a set $\mathcal{S}$ of subsets of inputs, the optimizer step size $\alpha > 0$, and Lagrangian step sizes s_k, and t_k, the Lagrangian multipliers are initialized in lines 1 and 2. The training is performed for a fixed number of epochs, and each epoch k optimizes the model weights w of the optimizer $\mathcal{M}$ using the Lagrangian multipliers λ^k and μ^k associated with current epoch k, denoted $\mathcal{M}[w(\lambda^k, \mu^k)]$ in the algorithm (lines 4–6). Similarly to Algorithm 1, the Lagrangian multipliers λ_i for the dual variables are updated after each epoch, on line 7). Finally, the algorithm updates the multipliers μ_i associated to the Lagrangian duals of the Lagrangian duals (line 8).

5 Experiments

This section evaluates the proposed LDF on constrained optimization problems for energy and gas networks and on constrained learning problems–that enforce constraints on the predictors–for applications in transprecision computing and fairness.

5.1 Constrained Optimization Problems

Data Set Generation. The experiments examine the proposed models on a variety of power networks from the NESTA library [9] and natural gas benchmarks from [26] and GasLib [33]. The ground truth data are constructed as

Algorithm 2: LDF for Constrained Predictor Problems

input: $D = (d_l, y_l)_{l=1}^n, \mathcal{S} = \{S_1, \ldots, S_n\}$ Training data and data partitions;
$\alpha, s = (s_0, s_1, \ldots), t = (t_0, t_1, \ldots)$: Optimizer and Lagrangian step sizes.

1 $\lambda_i^0 \leftarrow 0 \;\; \forall i \in [m]$
2 $\mu_i^0 \leftarrow 0 \;\; \forall i \in [q]$
3 **for** *epoch* $k = 0, 1, \ldots$ **do**
4   **foreach** $S_i \in \mathcal{S}$ **do**
5     $\hat{\boldsymbol{y}}_{S_i} \leftarrow \{\mathcal{M}[w(\lambda^k, \mu^k)](d_l)\}_{l \in S_i}$
6     $w(\lambda^{k+1}, \mu^{k+1}) \leftarrow w(\lambda^k, \mu^k) - \alpha \nabla_w \mathcal{L}_{\lambda^k, \mu^k}(\hat{\boldsymbol{y}}_{S_i}, \boldsymbol{y}_{S_i}, \boldsymbol{d}_{S_i})$
7   $\lambda_i^{k+1} \leftarrow \lambda_i^k + s_k \sum_{l=1}^n \nu_i\left(g_i(\hat{y}_l, d_l)\right) \;\; \forall i \in [m]$
8   $\mu_i^{k+1} \leftarrow \mu_i^k + t_k \nu_i\left(h(\hat{\boldsymbol{y}}_{S_i}, \boldsymbol{d}_{S_i})\right) \;\; \forall i \in [q]$

follows: For each power and gas network, different benchmarks are generated by altering the amount of nominal demands $d = S^d$ (for power networks) and $d = q^d$ (for gas networks) within a ±20% range. The resulting 4000 demand vectors are used to generate solutions to the OPF and OGC problems. Increasing loads causes heavily congestions to the system, rendering the computation of optimal solutions challenging. A network value, that constitutes a dataset entry $(d_l, y_l = \mathcal{O}(d))$, is a feasible solution obtained by solving the AC-OPF problem [8], for electricity networks, or the OGC problem, for gas networks [19]. The experiments use a 80/20 train-test split and results are reported on the test set.

Learning Models. The experiments use a baseline ReLU network $\mathcal{M}$, with 5 layers which minimizes the Mean Squared Error (MSE) loss $\mathcal{L}$ to predict to active power $\hat{p}$, voltage magnitude $\hat{v}$, and voltage angle $\hat{\theta}$, for energy networks, and compression ratios $\hat{R}$, pressure $\hat{p}$, and gas flows $\hat{q}$, for gas networks.

This baseline model is compared with a model $\mathcal{M}_{\text{C}}$ that exploits the problem constraints and minimizes the loss: $\mathcal{L} + \lambda\nu(\cdot)$, with multiplier values λ fixed to **1**. Finally, $\mathcal{M}_{\text{C}}^{\text{D}}$ extends model $\mathcal{M}_{\text{C}}$ by learning the Lagrangian multipliers using the LDF introduced in Sect. 3.3. The constrained learning model for power systems also exploits the hot-start techniques used in [14], with states differing by at most 1%. Experiments using larger percentages (up to 3%) showed similar trends. The training uses the Adam optimizer with learning rate ($\alpha = 10^{-3}$) and was performed for 80 epochs using batch sizes $b = 64$. Finally, the Lagrangian step size ρ is set to 10^{-4}. Extensive additional information about the network structure, the optimization model (OPF and OGC), the learning loss functions and constraints, as well as additional experimental analysis is provided in [15].

Prediction Errors. Table 1 and 2 report the average L_1-distance and the *prediction errors* between a subset of predicted variables y (marked with $\hat{y}$) on both the power and gas benchmarks and their original ground-truth quantities. The error for y is reported in percentage as $100\frac{\|\hat{y}-y\|_1}{\|y\|_1}$ and the gain (in parenthesis) reports the ratio between the error obtained by the baseline model accuracy and the constrained models.

Table 1. Mean Prediction Errors (%) and accuracy gain (%) on OPF Benchmarks.

Test case	Type	$\mathcal{M}$	$\mathcal{M}_{\rm C}$		$\mathcal{M}_{\rm C}^{\rm D}$	
		err (%)	err (%)	gain	err (%)	gain
30_ieee	$\hat{p}$	3.3465	0.3052	(10.96)	**0.0055**	(608.4)
	$\hat{v}$	14.699	0.3130	(46.96)	**0.0070**	(2099)
	$\hat{\theta}$	4.3130	0.0580	(74.36)	**0.0041**	(1052)
	$\tilde{p}^f$	27.213	0.2030	(134.1)	**0.0620**	(438.9)
118_ieee	$\hat{p}$	0.2150	0.0380	(5.658)	**0.0340**	(6.323)
	$\hat{v}$	7.1520	0.1170	(61.12)	**0.0290**	(246.6)
	$\hat{\theta}$	4.2600	1.2750	(3.341)	**0.2070**	(20.58)
	$\tilde{p}^f$	38.863	0.6640	(58.53)	**0.4550**	(85.41)
300_ieee	$\hat{p}$	0.0838	0.0174	(4.816)	**0.0126**	(6.651)
	$\hat{v}$	28.025	3.1130	(9.002)	**0.0610**	(459.4)
	$\hat{\theta}$	12.137	7.2330	(1.678)	**2.5670**	(4.728)
	$\tilde{p}^f$	125.47	26.905	(4.663)	**1.1360**	(110.4)

For the power networks, the models focus on predicting the active generation dispatches $\hat{p}^g = Re(S^g)$, voltage magnitudes $\hat{v}$, voltage angles $\hat{\theta}$, and the active transmission line (including transformers) flows $\tilde{p}^f$. Power flows $\tilde{p}^f$ are not directly predicted but computed from the predicted quantities through Ohm's laws[1]. For the gas networks, the models focus on predicting compression ratios $\hat{R}$, pressure values $\hat{p}$, and gas flows $\hat{q}$. The best results are highlighted in bold.

A clear trend appears: The prediction errors decrease with the increase in model complexity. In particular, model $\mathcal{M}_{\rm C}$, which exploits the problem constraints, predicts voltage quantities and power flows that are up to two order of magnitude more precise than those predicted by $\mathcal{M}$, for OPF problems. The prediction errors on OGC benchmarks, instead remain of the same order of magnitude as those obtained by the baseline model $\mathcal{M}$, albeit the accuracy of the prediction increases consistently when adopting the constrained models. This can be explained by the fact that the gas networks behave largely monotonically in compressor costs for varying loads. *Finally, the LDF that finds the best multipliers ($\mathcal{M}_{\rm C}^{\rm D}$) consistently improves the baseline model on OGC benchmarks, and further improves $\mathcal{M}_{\rm C}$ predictions by an additional order of magnitude, for OPF problems.*[2]

Measuring the Constraint Violations. This section simulates the prediction results in an operational environment, by measuring the minimum required adjustments in order to satisfy the operational limits and the physical constraints in the energy domains studied. Given the predictions $\hat{y}$ returned by a model, the

[1] See more details in [15].

[2] The accuracy gains appear more pronounced on OPF problems since the baseline model $\mathcal{M}$ produces already extremely accurate results for OGC benchmarks.

Table 2. Mean Prediction Errors (%) and accuracy gain (%) on OGC Benchmarks.

Test case	Type	$\mathcal{M}$	$\mathcal{M}_C$		$\mathcal{M}_C^D$	
		err (%)	err (%)	gain	err (%)	gain
24-pipe	$\hat{R}$	0.0052	0.0079	(0.658)	**0.0025**	(2.080)
	$\hat{p}$	**0.0057**	0.0068	(0.838)	**0.0057**	(1.000)
	$\hat{q}$	0.0029	0.0592	(0.049)	**0.0007**	(4.142)
40-pipe	$\hat{R}$	0.0009	0.0103	(0.087)	**0.0006**	(1.833)
	$\hat{p}$	0.0011	0.0025	(0.240)	**0.0006**	(1.500)
	$\hat{q}$	0.0006	0.0329	(0.033)	**0.0004**	(1.500)
135-pipe	$\hat{R}$	0.0206	0.0317	(0.650)	**0.0199**	(1.307)
	$\hat{p}$	0.0260	**0.0209**	(1.067)	0.0225	(0.916)
	$\hat{q}$	0.0223	0.0572	(0.455)	**0.0222**	(1.005)

experiments compute a projection $\bar{y}$ of $\hat{y}$ into the feasible region and reports the minimal distance $\|\bar{y} - \hat{y}\|_2$ of the predictions from the satisfiable solution. This step is executed on all the predicted control variables: generator dispatch and voltage set points, for power systems, and compression ratios, for gas systems. Table 3 reports the minimum distance (normalized in percentage) required to satisfy the operational limits and physical constraints, and the best results are highlighted in bold. These results provide a proxy to evaluate the degree of constraint violations of a model. Notice that the adjustment required decrease with the increase in model complexity. *The results show that the LDF can drastically reduce the effort required by a post-processing step to satisfy the problem constraints.*

Table 3. Average distances (in percentage) for the active power p^g, voltage magnitude v, and compressor ratios R of the simulated solutions w.r.t. the corresponding predictions.

Test case	Type	$\mathcal{M}$	$\mathcal{M}_C$		$\mathcal{M}_C^D$	
		violation (%)	violation (%)	gain	violation (%)	gain
30_ieee	p^g	2.0793	0.1815	(11.45)	**0.0007**	(2970.0)
	v	83.138	0.0944	(880.7)	**0.0037**	(22469)
118_ieee	p^g	0.1071	0.0043	(24.91)	**0.0038**	(28.184)
	v	3.4391	0.0956	(35.97)	**0.0866**	(39.712)
300_ieee	p^g	0.0447	0.0091	(4.912)	**0.0084**	(5.3214)
	v	31.698	0.2383	(133.0)	**0.1994**	(158.97)
24-pipe	R	0.1012	0.1033	(0.978)	**0.0897**	(1.1282)
40-pipe	R	0.0303	0.0277	(1.094)	**0.0207**	(1.4638)
135-pipe	R	0.0322	0.0264	(1.219)	**0.0005**	(64.4)

5.2 Constrained Predictor Problems

This section examines the LDF for constrained predictor problems discussed in Sect. 4 on transprecision computing and fairness application domains.

Transprecision Computing. The benchmark considers training a neural network to predict the error of transprecision configurations. The monotonicity property is expressed as a constraint exploiting the relation of dominance among configurations of the training set, i.e. $\nu_i = \max(0, \mathcal{M}(x_1) - \mathcal{M}(x_2))$ if $x_1 \preceq x_2$ for every pair (x_1, x_2) in the dataset. This approach is particularly suited for instances of training with scarce data points with a high rate of violated constraints, since it guides the learning process towards a more general approximation of the target function. In order to explore different scenarios the experiments use 5 different training sets of increasing size, i.e. 200, 400, 600, 800, and 1000. The test set size is fixed to 1000 samples. The data sets are constructed by generating random configuration (d_i) and computing errors (y) by measuring the performance loss obtained when running the configuration d_i on the target algorithm. Ten disjoint training sets are constructed so that the violation constraint ratio is 0.5, while the test set was fixed.

Table 4 illustrates the average results comparing a model $(\mathcal{M})$ that minimizes the Mean Absolute Error (MAE) prediction error, one $(\mathcal{M}_C)$ that include the Lagrangian loss functions $\mathcal{L}_\lambda$ associated to each constraint and where all weights λ are fixed to value 1.0, and the proposed model $(\mathcal{M}_C^D)$ that uses the LDF to find the optimal Lagrangian weights. All prediction models are implemented as classical feed-forward neural network with 3 hidden layers and 10 units and minimize the MAE as loss function. The training uses 150 epochs, Lagrangian step sizes $t_k = 10^{-3}$ and learning rate 10^{-3}. The table also shows the average number of constraint violations (VC) and the sum of the magnitudes of the violated constraints (SMVC), i.e., $\sum_{x_i,x_j \in \mathcal{D}; x_i \preceq x_j \wedge \mathcal{M}(x_i) > \mathcal{M}(x_j)} |\mathcal{M}(x_i) - \mathcal{M}(x_j)|$.

The table clearly illustrates the positive effect of adding the constraints within the LDF on reducing the number of constraint violations. Notice that model $\mathcal{M}_C$, that weights all the constraints violations equally, produces a degradation of both the MAE score and the number and magnitude of the constraint violations, when compared to the baseline model $(\mathcal{M})$. The benefit of using the LDF is substantial in both reducing the number of constraint violations and in retaining a high model precision (i.e., a low MAE score). The most significant contribution was obtained on training sets with fewer data points, confirming that *exploiting the Lagrangian Duals of the Constraint Violations can be an important tool for constrained learning.*

Fairness Constraints. The benchmark considers building a classifier that minimizes *disparate treatment* [37]. The paper considers the disparate DT index, introduced by Aghaei et al. [2], to quantify the disparate impact in a dataset. Given a dataset of samples $D = (x_i, y_i)_{i \in [n]}$, this index is defined as:

$$\mathrm{DT}(D) = \left| \sum_{x_i \in D_{s_0}} I(\hat{y}_i = 1)/|D_{s_0}| - \sum_{x_i \in D_{s_1}} I(\hat{y}_i = 1)/|D_{s_1}| \right|,$$

Table 4. Mean Absolute Error (MAE), number of constraints violations (VC), and sum of absolute magnitude of violated constraints (SMVC). Best results are highlighted in bold.

n_{tr}	$\mathcal{M}$			$\mathcal{M}_C$			$\mathcal{M}_D^s$		
	MAE	VC	SMVC	MAE	VC	SMVC	MAE	VC	SMVC
200	0.1902	9.6	0.2229	0.1919	35.8	0.4748	**0.1883**	**7.4**	**0.1872**
400	0.1765	4.5	0.0804	0.1999	19.4	0.2149	**0.1763**	**2.6**	**0.0369**
600	**0.1687**	2.5	0.0397	0.2022	9.1	0.0683	0.1723	**1.7**	**0.0224**
800	**0.1672**	3.0	0.0600	0.2007	8.5	0.0746	0.1704	**0.6**	**0.0131**
1000	**0.1640**	0.4	0.0048	0.2012	5.7	0.0511	0.1642	**0.5**	**0.0043**

where I is the characteristic function and $\hat{y}_i$ is the predicted outcome for sample x_i. The idea is to use a locally weighted average to estimate the conditional expectation. The higher is the DT score for a dataset, the more it suffers from disparate treatment, with DT $= 0$ meaning that the dataset does not suffer from disparate treatment.

Since the *DT* constraint introduced in Eq. (11b) is not differentiable with the respect to the model parameters, the paper uses an expectation matching constraints between the predictors for the protected classes, defined as:

$$\left| E_{x \sim D_{s_0}}[\mathcal{M}_0(x) \mid z(x) = 0] - E_{x \sim D_{s_1}}[\mathcal{M}_1(x) \mid z(x) = 1] \right| = 0. \quad (17)$$

The effect of the Lagrangian Dual framework on reducing disparate treatment was evaluated on three datasets: The *Adult* dataset [23], containing 30,000 samples and 23 features, in which the prediction task is that of assessing whether an individual earns more than $50K$ per year and the protected attribute is *race*. The *Default* of Taiwanese credit card users [36], containing 45,000 samples and 13 features, in which the task is to predict whether an individual will default and the protected attribute is *gender*. Finally, the *Bank* dataset [37], containing 41,188 samples, each with 20 attributes, where the task is to predict whether an individual has subscribed or not and the protected attribute is *age*. The experiments use a 80/20 train/test split and executes a 5-fold cross-validation to evaluate the accuracy and the fairness score (DT) of the predictors.

Table 5 illustrates the results comparing model $\mathcal{M}$ that minimizes the Binary Cross Entropy (BCE) loss, model $\mathcal{M}_C$ that includes the Lagrangian loss functions $\mathcal{L}_\lambda$ associated with each constraint and where all λ are fixed to value 1.0, and the proposed model $\mathcal{M}_C^D$ that uses the LDF to find the optimal Lagrangian weights. All prediction models use a classical feed-forward neural network with 3 layers and 10 hidden units. The training uses 100 epochs, Lagrangian step size $s_k = 10^{-4}$ and learning rate 10^{-3}. The models are also compared against a state-of-the-art fair classifier which enforces fairness by limiting the covariance between the loss function and the sensitive variable [37]. While [37] focuses on logistic regression, the model is implemented as a neural network with the same hyper parameters of model $\mathcal{M}$. The table clearly shows the effect of the Lagrangian constraints on reducing the DT score. Not only such reduction attains state-of-the-art results on the DT score, but it also comes at a much more contained cost of accuracy degradation.

Table 5. Classification accuracy (Acc.) and fairness score (DT)

Dataset	$\mathcal{M}$		$\mathcal{M}_C$		$\mathcal{M}_C^D$		Zafar'19	
	Acc.	DT	Acc.	DT	Acc.	DT	Acc.	DT
Adult	**0.8423**	0.1853	0.8333	0.0627	0.8335	**0.0545**	0.7328	0.1037
Default	0.8160	0.0162	**0.8182**	0.0216	0.8166	**0.0101**	0.6304	0.0109
Bank	**0.8257**	0.4465	0.7744	0.4515	0.8135	0.1216	0.7860	**0.0363**

6 Related Work

The application of Deep Learning to constrained optimization problems is receiving increasing attention. Approaches which embed optimization components in neural networks include [22,24,34]. These approaches typically rely on problems exhibiting properties like convexity or submodularity. Another line of work leverages explicit optimization algorithms as a differentiable layer into neural networks [4,13,35]. A further collection of works interpret constrained optimization as a two-player game, in which one player optimizes the objective function and a second player attempt at satisfying the problem constraints [1,21,30]. For instance Agarwal et al. [1], proposes a best-response algorithm applied to fair classification for a class of linear fairness constraints. To study generalization performance of training algorithms that learn to satisfy the problem constraints, Cotter et al. [11] propose a two-players game approach in which one player optimizes the model parameters on the training data and the other player optimizes the constraints on a validation set. Arora et al. [5] proposed the use of a multiplicative rule to iteratively changing the weights of different distributions to maintaining some properties and discuss the applicability of the approach to a constraint satisfaction domain. The closest related work is that of Nandwani et al. [29]. It discusses a primal-dual formulation for deep learning models with hard logical rules. It however, does not concern with the task of predicting solutions for constraint optimization problems and focuses on natural language processing problems. A different strategy for minimizing empirical risk subject to a set of constraints is that of using projected stochastic gradient descent (PSGD). Cotter et al. [12] proposed an extension of PSGD that stay close to the feasible region while applying constraint probabilistically at each iteration of the learning cycle.

Different from these proposals, this paper proposes a framework that exploits key ideas in Lagrangian duality to encourage the satisfaction of generic constraints within a neural network learning cycle and apply to both sample dependent constraints (as in the case of energy problems) and dataset dependent constraints (as in the case of transprecision computing and fairness problems). This paper builds on the recent results that were dedicated to learning and optimization in power systems [14].

7 Conclusions

This paper proposed a Lagrangian dual framework to encourage the satisfaction of constraints in deep learning. It was motivated by a desire to learn parametric constrained optimization problems that feature complex physical and engineering constraints. The paper showed how to exploit Lagrangian duality for deep learning to obtain predictors that minimize constraint violations. The proposed framework can be applied to constrained optimization problems, in which the constraints model relations among features of each data sample, and to constrain predictors in which the constraints enforce global properties over multiple dataset samples and the predictor outputs.

The Lagrangian dual framework for deep learning was evaluated on a collection of realistic energy networks, by enforcing non-discriminatory decisions on a variety of datasets, and on a transprecision computing application. The results demonstrated the effectiveness of the proposed method that dramatically decreases constraint violations committed by the predictors and, in some applications, as in those in energy optimization, increases the prediction accuracy by up to two orders of magnitude.

References

1. Agarwal, A., Beygelzimer, A., Dudík, M., Langford, J., Wallach, H.: A reductions approach to fair classification. arXiv preprint arXiv:1803.02453 (2018)
2. Aghaei, S., Azizi, M.J., Vayanos, P.: Learning optimal and fair decision trees for non-discriminative decision-making. In: AAAI, pp. 1418–1426 (2019)
3. Amodei, D., et al.: Deep speech 2: end-to-end speech recognition in English and mandarin. In: ICML, pp. 173–182 (2016)
4. Amos, B., Kolter, J.Z.: Optnet: differentiable optimization as a layer in neural networks. In: ICML, pp. 136–145. JMLR. org (2017)
5. Arora, S., Hazan, E., Kale, S.: The multiplicative weights update method: a meta-algorithm and applications. Theory Comput. **8**(1), 121–164 (2012)
6. Baran, M.E., Wu, F.F.: Optimal capacitor placement on radial distribution systems. IEEE TPD **4**(1), 725–734 (1989)
7. Boyd, S., Parikh, N., Chu, E., Peleato, B., Eckstein, J., et al.: Distributed optimization and statistical learning via the alternating direction method of multipliers. Foundations Trends® Mach. Learn. **3**(1), 1–122 (2011)
8. Chowdhury, B.H., Rahman, S.: A review of recent advances in economic dispatch. IEEE Trans. Power Syst. **5**(4), 1248–1259 (1990)
9. Coffrin, C., Gordon, D., Scott, P.: NESTA, the NICTA energy system test case archive. CoRR abs/1411.0359 (2014). http://arxiv.org/abs/1411.0359
10. Collobert, R., Weston, J.: A unified architecture for natural language processing: Deep neural networks with multitask learning. In: ICML, pp. 160–167 (2008)

11. Cotter, A., et al.: Training well-generalizing classifiers for fairness metrics and other data-dependent constraints. arXiv preprint arXiv:1807.00028 (2018)
12. Cotter, A., Gupta, M., Pfeifer, J.: A light touch for heavily constrained SGD. In: Conference on Learning Theory, pp. 729–771 (2016)
13. Donti, P., Amos, B., Kolter, J.Z.: Task-based end-to-end model learning in stochastic optimization. In: NIPS, pp. 5484–5494 (2017)
14. Fioretto, F., Mak, T., Van Hentenryck, P.: Predicting AC optimal power flows: combining deep learning and Lagrangian dual methods. In: AAAI, p. 630 (2020)
15. Fioretto, F., Mak, T.W.K., Baldo, F., Lombardi, M., Van Hentenryck, P.: A lagrangian dual framework for deep neural networks with constraints. CoRR, arXiv:2001.09394 (2020)
16. Fisher, E.B., O'Neill, R.P., Ferris, M.C.: Optimal transmission switching. IEEE Trans. Power Syst. **23**(3), 1346–1355 (2008)
17. Fontaine, D., Laurent, M., Van Hentenryck, P.: Constraint-based Lagrangian relaxation. In: O'Sullivan, B. (ed.) Principles and Practice of Constraint Programming. CP 2014. LNCS, vol. 8656, pp. 324–339. Springer, Cham (2014). https://doi.org/10.1007/978-3-319-10428-7_25
18. Hardt, M., Price, E., Price, E., Srebro, N.: Equality of opportunity in supervised learning. In: Lee, D.D., Sugiyama, M., Luxburg, U.V., Guyon, I., Garnett, R. (eds.) Advances in Neural Information Processing Systems, vol. 29, pp. 3315–3323 (2016)
19. Herty, M., Mohring, J., Sachers, V.: A new model for gas flow in pipe networks. Math. Methods Appl. Sci. **33**(7), 845–855 (2010)
20. Hestenes, M.R.: Multiplier and gradient methods. J. Optim. Theory Appl. **4**(5), 303–320 (1969). https://doi.org/10.1007/BF0092767
21. Kearns, M., Neel, S., Roth, A., Wu, Z.S.: Preventing fairness gerrymandering: auditing and learning for subgroup fairness. arXiv preprint arXiv:1711.05144 (2017)
22. Khalil, E., Dai, H., Zhang, Y., Dilkina, B., Song, L.: Learning combinatorial optimization algorithms over graphs. In: NIPS, pp. 6348–6358 (2017)
23. Kohavi, R.: Scaling up the accuracy of Naive-Bayes classifiers: a decision-tree hybrid. KDD **96**, 202–207 (1996)
24. Kool, W., Van Hoof, H., Welling, M.: Attention, learn to solve routing problems! arXiv preprint arXiv:1803.08475 (2018)
25. Krizhevsky, A., Sutskever, I., Hinton, G.E.: Imagenet classification with deep convolutional neural networks. In: NIPS, pp. 1097–1105 (2012)
26. Mak, T.W.K., Hentenryck, P.V., Zlotnik, A., Bent, R.: Dynamic compressor optimization in natural gas pipeline systems. INFORMS J. Comput. **31**(1), 40–65 (2019)
27. Malossi, A.C.I., Schaffner, M., et al.: The transprecision computing paradigm: concept, design, and applications. In: Design, Automation & Test in Europe Conference & Exhibition (DATE), 2018, pp. 1105–1110. IEEE (2018)
28. Monticelli, A., Pereira, M., Granville, S.: Security-constrained optimal power flow with post-contingency corrective rescheduling. IEEE TPS **2**(1), 175–180 (1987)
29. Nandwani, Y., Pathak, A., Singla, P.M.: A primal dual formulation for deep learning with constraints. In: NIPS (2019)
30. Narasimhan, H.: Learning with complex loss functions and constraints. In: International Conference on Artificial Intelligence and Statistics, pp. 1646–1654 (2018)
31. Verma, S.N., Mukherjee, V.: Transmission expansion planning: a review. In: International Conference on Energy Efficient Technologies for Sustainability, pp. 350–355 (2016)

32. Parikh, N., Boyd, S., et al.: Proximal algorithms. Foundations Trends® Optim. **1**(3), 127–239 (2014)
33. Pfetsch, M., et al.: Validation of nominations in gas network optimization: models, methods, and solutions. Optim. Methods Softw. **30**(1), 15–53 (2015)
34. Vinyals, O., Fortunato, M., Jaitly, N.: Pointer networks. In: NIPS, pp. 2692–2700 (2015)
35. Wilder, B., Dilkina, B., Tambe, M.: Melding the data-decisions pipeline: decision-focused learning for combinatorial optimization. In: AAAI, vol. 33, pp. 658–1665 (2019)
36. Yeh, I.C., Lien, C.H.: The comparisons of data mining techniques for the predictive accuracy of probability of default of credit card clients. Expert Syst. Appl. **36**(2), 2473–2480 (2009)
37. Zafar, M.B., Valera, I., Gomez-Rodriguez, M., Gummadi, K.P.: Fairness constraints: a flexible approach for fair classification. JMLR **20**(75), 1–42 (2019)

Applied Data Science: Healthcare

Few-Shot Microscopy Image Cell Segmentation

Youssef Dawoud[1(✉)], Julia Hornauer[1], Gustavo Carneiro[2], and Vasileios Belagiannis[1]

[1] Universität Ulm, Ulm, Germany
{youssef.dawoud,julia.hornauer,vasileios.belagiannis}@uni-ulm.de
[2] The University of Adelaide, Adelaide, Australia
gustavo.carneiro@adelaide.edu.au

Abstract. Automatic cell segmentation in microscopy images works well with the support of deep neural networks trained with full supervision. Collecting and annotating images, though, is not a sustainable solution for every new microscopy database and cell type. Instead, we assume that we can access a plethora of annotated image data sets from different domains (sources) and a limited number of annotated image data sets from the domain of interest (target), where each domain denotes not only different image appearance but also a different type of cell segmentation problem. We pose this problem as meta-learning where the goal is to learn a generic and adaptable few-shot learning model from the available source domain data sets and cell segmentation tasks. The model can be afterwards fine-tuned on the few annotated images of the target domain that contains different image appearance and different cell type. In our meta-learning training, we propose the combination of three objective functions to segment the cells, move the segmentation results away from the classification boundary using cross-domain tasks, and learn an invariant representation between tasks of the source domains. Our experiments on five public databases show promising results from 1- to 10-shot meta-learning using standard segmentation neural network architectures.

Keywords: Cell segmentation · Microscopy image · Few-shot learning

1 Introduction

Microscopy image analysis involves many procedures including cell counting, detection and segmentation [37]. Cell segmentation is particularly important for studying the cell morphology to identify the shape, structure, and cell size. Manually segmenting cells from microscopy images is a time-consuming and costly process. For that reason, many methods have been developed to automate the process of cell segmentation, as well as counting and detection.

Electronic supplementary material The online version of this chapter (https://doi.org/10.1007/978-3-030-67670-4_9) contains supplementary material, which is available to authorized users.

Y. Dong et al. (Eds.): ECML PKDD 2020, LNAI 12461, pp. 139–154, 2021.
https://doi.org/10.1007/978-3-030-67670-4_9

Although reliable cell segmentation algorithms exist for more than a decade [34,37], only recently they have shown good generalization with the support of deep neural networks [6,36]. Current approaches in cell segmentation deliver promising results based on encoder - decoder network architectures trained with supervised learning [36]. However, collecting and annotating large amounts of images is practically inviable for every new microscopy problem. Furthermore, new problems may contain different types of cells, where a segmentation model, pre-trained on different data sets, may not deliver good performance.

To address this limitation, methods based on domain generalization [12,19], domain adaptation [33] and few-shot learning [28] have been developed. In these approaches, it is generally assumed that we have access to a collection of annotated images from different domains (source domains), no access to the target domain (domain generalisation), access to a large data set of un-annotated images (domain adaptation), or access to a limited number (less than 10) of annotated images from the target domain (few-shot learning). Domain generalisation and adaptation generally involve the same tasks in the source and target domains, such as the same cell type segmentation problem with images coming from different sites and having different appearances. However, our setup is different because we consider that the source and target domains consist of different types of cell segmentation problems. Each domain consists of different cell types such as mitochondria and nuclei. In this way, we form a typical real-life scenario where a variety of microscopy images, containing different cell structures, are leveraged from various resources to learn a cell segmentation model. Therefore, the challenge in our setup is to cope with different image and cell appearances, as well as different types of cell for each domain, as illustrated in Fig. 1. In such setup, we argue that few-shot learning is more appropriate, where we aim to learn a generic and adaptable few-shot learning model from the available source domain data sets. This model can be afterwards fine-tuned on the few annotated images of the new target domain. Such problem can be formulated as an optimization-based meta-learning approach [11,14,27].

In this paper, we present a new few-shot meta-learning cell segmentation model. For meta-training the model, we propose the combination of three loss functions to 1. impose pixel-level segmentation supervision, 2. move the segmentation predictions away from the classification boundary using cross-domain tasks and 3. learn an invariant representation between tasks. In our evaluations on five microscopy data sets, we demonstrate promising results compared to the related work on settings from 1- to 10-shot learning by employing standard network architectures [30,36]. To the best of our knowledge, this is the first work to introduce few-shot task generalisation using meta-learning and to apply few-shot learning to microscopy image cell segmentation.

2 Related Work

Cell segmentation is a well-established problem [37] that is often combined with cell counting [1,10] and classification [38]. We discuss the prior work that is relevant to our approach, as well as the related work on few-shot learning.

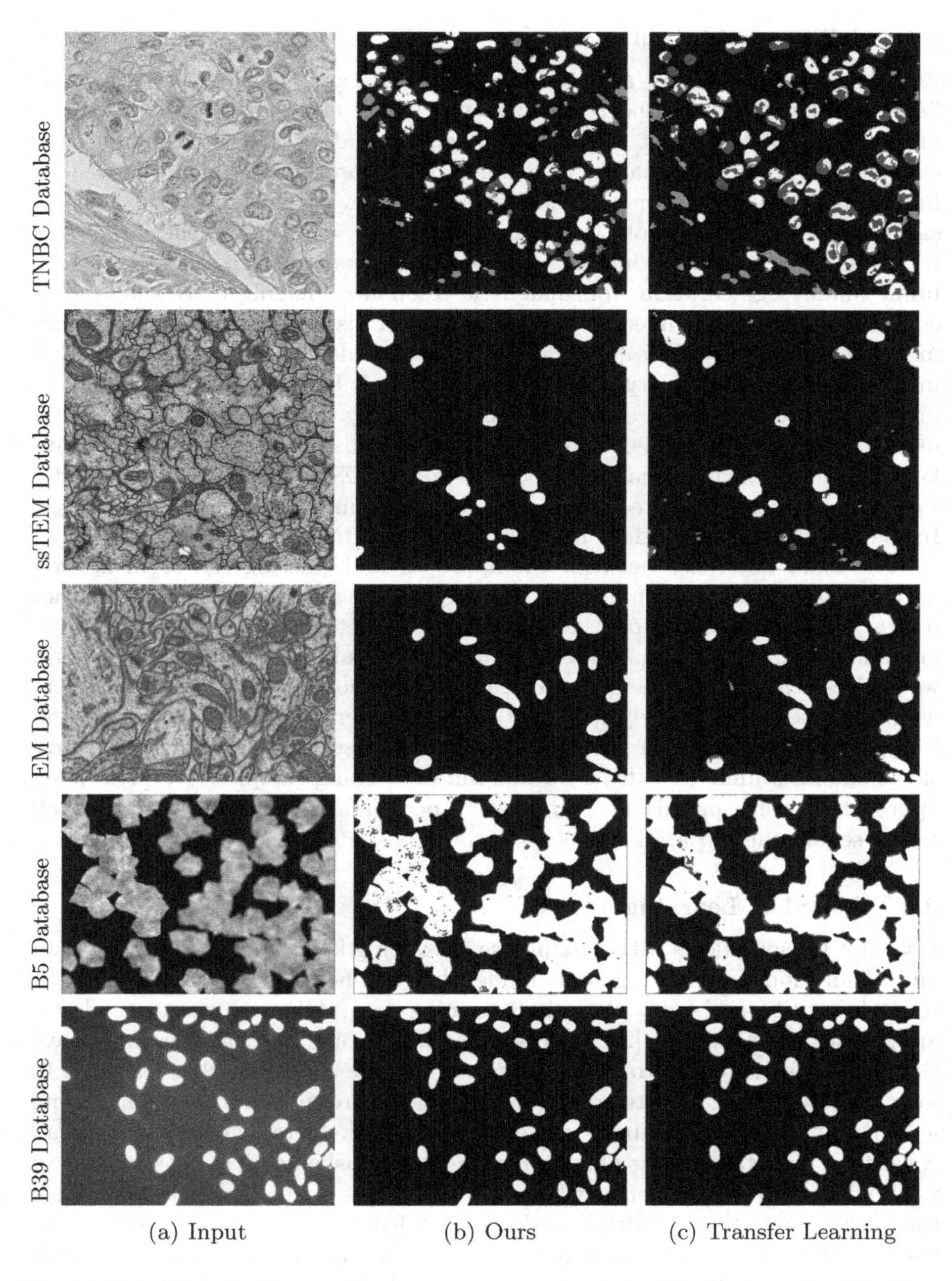

Fig. 1. Visual Result. We visually compare our approach to transfer learning using the U-Net architecture. Ours refers to meta-training with all objectives, namely $\mathcal{L}_{BCE} + \mathcal{L}_{ER} + \mathcal{L}_{D}$. The red color corresponds to false positive, the green color to false negative, the black color to true negative, and the white color to true positive. Best viewed in color. (Color figure online)

2.1 Cell Segmentation

Automatic cell detection and counting in microscopy images has been earlier studied with the support of image processing and computer vision techniques [13,21,34]. In past few years, advances in deep neural networks have changed the type of approaches developed for microscopy images. In particular, fully convolutional neural networks (FCNs) [20] allow to make predictions on the same or similar spatial resolution of the input image. FCN approaches have been widely adapted in medical imaging with applications to nuclei segmentation [25], brain tumor segmentation from magnetic resonance imaging (MRI) [3] and of course segmentation in microscopy images [6]. For instance, histology image segmentation relies on FCNs [7] to perform mitosis detection. Among the FCN models, U-Net [30] is a very popular architecture. It has been developed for segmenting neuronal structures in electron microscopy images, but it is presently used for any kind of medical imaging that demands spatial predictions. Similarly, the fully convolutional regression network (FCRN) [36] is another encoder - decoder architecture for cell segmentation and counting in microscopy images. In the evaluation, we consider both U-Net and FCRN architectures.

The main difference between our approach and existing cell segmentation approaches lies in the training algorithm. While the aforementioned approaches deliver promising results on the examined data sets, they all require a large amount of annotated data and a fully supervised training process. In this work, we address the problem as a more realistic few-shot learning problem, where we have access to relatively large data sets of different domain and of several types of cell segmentation problems, but the data set for the target segmentation domain is limited with just a handful of training samples. We present an approach to reach promising segmentation performance regardless of the small target segmentation data set.

2.2 Few-Shot Learning

Few-shot learning deals with making use of existing knowledge, in terms of data and annotation, to build a generic model that can be adapted to different (but related) target problems with limited training data. Although there are classic approaches from the past [23,29], deep neural networks and meta-learning have significantly contributed to improve the state of the art of few-shot learning. In meta-learning, the goal is to train a model to learn from a number of tasks using a limited number of training samples per task [31]. Meta-learning, in general, consists of a meta-training phase where multiple tasks adapt a base learner to work with different problems (where each task uses a small training set), then the classifiers of those multiple tasks are pooled together to update the base learner. After the meta-training process converges, the model of the base learner is fine-tuned to a limited number of annotated data sampled from the target problem. Meta-learning approaches can be categorized in metric-learning [32,35], model- [8,24] and optimization-based learning [14,27,28]. The optimization meta-learning approaches function with gradient-based learning, which is less complex to implement and computationally efficient. For that reason, we rely on Reptile [27], an optimization-based approach, to develop our algorithm.

3 Few-Shot Cell Segmentation

In this section, we define the few-shot segmentation problem. Afterwards, we pose cell segmentation from microscopy images as few-shot meta-learning approach.

3.1 Problem Definition

Let $\mathcal{S} = \{\mathcal{S}_1, \mathcal{S}_2, \ldots, \mathcal{S}_{|\mathcal{S}|}\}$ be a collection of $|\mathcal{S}|$ microscopy cell data sets. Each data set $\mathcal{S}_m$ is defined as $\mathcal{S}_m = \{(\mathbf{x}, \mathbf{y})_k\}_{k=1}^{|\mathcal{S}_m|}$, where $(\mathbf{x}, \mathbf{y})_k$ is a pair of the microscopy cell image $\mathbf{x} : \Omega \rightarrow \mathbb{R}$ (Ω denotes the image lattice with dimensions $H \times W$) and the respective ground-truth segmentation mask $\mathbf{y} : \Omega \rightarrow \{0, 1\}$; and $|\mathcal{S}_m|$ is the number of samples in $\mathcal{S}_m$. Note that each data set corresponds to a different task and domain, each representing a new type of image and segmentation task. All data sets in $\mathcal{S}$ compose the source data sets. We further assume another data set, which we refer to as target, defined by $\mathcal{T} = \{(\mathbf{x}, \mathbf{y})_k\}_{k=1}^{|\mathcal{T}|}$, where the number of training images with ground-truth segmentation masks is limited, e.g. between 1 and 10 training images and segmentation mask pairs. Also, we assume that the target data set comes from a different task and domain.

Our goal is to perform cell segmentation in the images belonging to the target data set $\mathcal{T}$ through a segmentation model $\mathbf{y} = f(\mathbf{x}; \theta)$, where θ denotes the model parameter (i.e. the weights of the deep neural network). However, the limited number of annotated training images prohibits us from learning a typical fully-supervised data-driven model. This is a common problem in real-life when working with cell segmentation problems, where annotating new data sets, i.e. the target data sets, does not represent a sustainable solution. To address this limitation, we propose to learn a generic and adaptable model from the source data sets in $\mathcal{S}$, which is then fine-tuned on the limited annotated images of the target data set $\mathcal{T}$.

3.2 Few-Shot Meta-learning Approach

We propose to learn a generic and adaptable few-shot learning model with gradient-based meta-learning [14,27]. The learning process is characterized by a sequence of episodes, where an episode consists of the meta-training and meta-update steps. In meta-training, the model parameter θ (this is known as the meta-parameter) initialises the segmentation model, with $\theta_m = \theta$ in $\mathbf{y} = f(\mathbf{x}; \theta_m)$ (defined in Sect. 3.1) for each task $m \in \{1, ..., |\mathcal{S}|\}$, where each task is modeled with a training set $\tilde{\mathcal{S}}_m \subset \mathcal{S}_m$. Next, the model meta-parameter θ is meta-updated from the learned task parameters $\{\theta_m\}_{m=1}^{|\mathcal{S}|}$.

The segmentation model that uses the meta parameter θ, defined as $\mathbf{y} = f(\mathbf{x}; \theta)$, is denoted by the base learner. To train this base learner, we propose three objective functions that account for segmentation supervision, and two regularisation terms. In our implementation, we rely on the Reptile algorithm [27] to develop our approach because of its simplicity and efficiency. Our approach is described in Algorithm 1. Next, we present each part of the approach.

Algorithm 1: Gradient-Based Meta-Learning for Cell Segmentation

1: Input: source domain data sets $\mathcal{S} = \{\mathcal{S}_1, \mathcal{S}_2, \ldots, \mathcal{S}_{|\mathcal{S}|}\}$, with $\mathcal{S}_m = \{(\mathbf{x}, \mathbf{y})_k\}_{k=1}^{|\mathcal{S}_m|}$
2: Input: target domain data set $\mathcal{T}$, with $\mathcal{T} = \{(\mathbf{x}, \mathbf{y})_k\}_{k=1}^{|\mathcal{T}|}$
3: Initialize θ, α, β, where $0 < \alpha, \beta < 1$
4: **for** iteration $= 1, 2, \ldots$ **do**
5: **for** $m = 1, 2, \ldots, |\mathcal{S}|$ **do**
6: Sample K image and segmentation samples to form $\tilde{\mathcal{S}_m} = \{(\mathbf{x}, \mathbf{y})_k\}_{k=1}^{K} \subset \mathcal{S}_m$ from current task $\mathcal{S}_m$
7: Randomly choose a different task $n \in \{1, ..., |\mathcal{S}|\}$ such that $n \neq m$
8: Randomly choose a different task $p \in \{1, ..., |\mathcal{S}|\}$ such that $p \neq m$
9: Sample K image samples from task n to form $\tilde{\mathcal{S}_n} = \{(\mathbf{x})_k\}_{k=1}^{K} \subset \mathcal{S}_n$
10: Sample K image samples from task p to form $\tilde{\mathcal{S}_p} = \{(\mathbf{x})_k\}_{k=1}^{K} \subset \mathcal{S}_p$
11: Compute gradient $\theta'_m = g(\mathcal{L}_{BCE}(\theta, \tilde{\mathcal{S}_m}) + \alpha\mathcal{L}_{ER}(\theta, \tilde{\mathcal{S}_n}) + \beta\mathcal{L}_D(\theta, \tilde{\mathcal{S}_m}, \tilde{\mathcal{S}_p}))$
12: **end for**
13: Meta-update $\theta \leftarrow \theta + \epsilon \frac{1}{|\mathcal{S}|} \sum_{m=1}^{|\mathcal{S}|} (\theta'_m - \theta)$
14: **end for**
15: Produce few-shot training set from target by sampling K samples to form $\tilde{\mathcal{T}} = \{(\mathbf{x}, \mathbf{y})_k\}_{k=1}^{K} \subset \mathcal{T}$
16: Fine-tune $f(\mathbf{x}; \theta)$ with $\tilde{\mathcal{T}}$ using $\mathcal{L}_{BCE}(\theta, \tilde{\mathcal{T}})$
17: Test $f(\mathbf{x}; \theta)$ with testing set $\hat{\mathcal{T}} = \mathcal{T} \setminus \tilde{\mathcal{T}}$.

3.3 Meta-learning Algorithm

During meta-training, i.e. lines 5 to 12 in Algorithm 1, $|\mathcal{S}|$ tasks are generated by sampling K images from each source domain in $\mathcal{S}$. Hence, a task is represented by a subset of K images and segmentation maps from $\mathcal{S}_m$. In our experiments, we work from $K \in \{1, ..., 10\}$-shot learning problems. After sampling a task, the baser learner is trained with the three objective functions. The first objective function consists of the standard binary cross entropy loss $\mathcal{L}_{BCE}(.)$ that uses the images and segmentation masks of the sampled task. The second loss function is based on the entropy regularization $\mathcal{L}_{ER}(.)$ [16] that moves the segmentation results away from the classification boundary using a task $n \neq m$ without the segmentation maps from task n – such regularisation makes the segmentation results for task n more confident, i.e. more binary like. The third objective loss function consists of extracting an invariant representation between tasks [12] by enforcing the learning of a common feature representation across different tasks with the knowledge distillation loss $\mathcal{L}_D(.)$ [17]. The use of the entropy regularisation and knowledge distillation losses in few-shot meta-learning represent the main technical contribution of our approach. The learned task parameters θ'_m are optimized using the three objective functions above during meta-training, as follows:

$$\theta'_m = \arg\min_{\theta} \mathbb{E}_{\tilde{\mathcal{S}}_m, \tilde{\mathcal{S}}_n, \tilde{\mathcal{S}}_p}[\mathcal{L}(\theta, \tilde{\mathcal{S}}_m, \tilde{\mathcal{S}}_n, \tilde{\mathcal{S}}_p)], \quad (1)$$

where the loss $\mathcal{L}(\theta, \tilde{\mathcal{S}}_m, \tilde{\mathcal{S}}_n, \tilde{\mathcal{S}}_p) = \mathcal{L}_{BCE}(\theta, \tilde{\mathcal{S}}_m) + \alpha\mathcal{L}_{ER}(\theta, \tilde{\mathcal{S}}_n) + \beta\mathcal{L}_D(\theta, \tilde{\mathcal{S}}_m, \tilde{\mathcal{S}}_p)$ is a combination of the three objectives that depend on K-shot source domain training sets $\tilde{\mathcal{S}}_{\{m,n,p\}} \subset \mathcal{S}_{\{m,n,p\}}$ (with $m \neq n$, $m \neq p$ and $\{m, n, p\} \subset \{1, ..., |\mathcal{S}|\}$)

and $|\tilde{\mathcal{S}}_{\{m,n,p\}}| = K$. In Sect. 3.4, we present in detail the objective functions. Finally, the parameters of the base learner are learned with stochastic gradient descent and back-propagation.

At last, the meta-update step takes place after iterating over $|\mathcal{S}|$ tasks during meta-training. The meta-update, i.e. line 11 in Algorithm 1, updates the model parameter θ using the following rule:

$$\theta \leftarrow \theta + \epsilon \frac{1}{|\mathcal{S}|} \sum_{m=1}^{|\mathcal{S}|} (\theta'_m - \theta), \tag{2}$$

where ϵ is the step-size for the parameter update. The episodic training takes place based on the meta-training and meta-update steps until convergence.

3.4 Task Objective Functions

We design three objective functions for the meta-training stage to update the parameters of the base learner.

Binary Cross Entropy Loss. Every sampled task includes pairs of input image and segmentation mask. We rely on the pixel-wise binary cross entropy (BCE) as the main objective to learn predicting the ground-truth mask. In addition, we weight the pixel contribution since we often observe unbalanced pixel ratio between the foreground and background pixels. Given the foreground probability prediction $\mathbf{y}' = f(\mathbf{x}; \theta)$ of the input image $\mathbf{x}$ and the segmentation mask $\mathbf{y}$ that belong to the K-shot training set $\tilde{\mathcal{S}}_m$ for task m, the pixel-wise BCE loss is given by:

$$\begin{aligned} \mathcal{L}_{BCE}(\theta, \tilde{\mathcal{S}}_m) = -\frac{1}{|\tilde{\mathcal{S}}_m|} \sum_{(\mathbf{x},\mathbf{y}) \in \tilde{\mathcal{S}}_m} \sum_{\omega \in \Omega} [\mathbf{w}\mathbf{y}(\omega) \log(\mathbf{y}'(\omega)) \\ + (1 - \mathbf{y}(\omega)) \log(1 - \mathbf{y}'(\omega))] , \end{aligned} \tag{3}$$

where $\omega \in \Omega$ denotes to the spatial pixel position in the image lattice Ω and $\mathbf{w}$ is the weighting factor of the foreground class probability $\mathbf{y}(\omega)$ which equals to the ratio of background to foreground classes in $\mathbf{y}$. This is the standard loss function for segmentation-related problems, which we also employ for our binary problem [5].

Entropy Regularization. The BCE loss can easily result in over-fitting the base-learner to the task m. We propose to use Shannon's entropy regularization on a different task loss to prevent this behavior. More specifically, while minimizing the BCE loss on a sampled task of one source domain, e.g. task m, we sample a second task $n \neq m$ from a different domain, e.g. task n; and seek to minimize Shannon's entropy loss for task n without using that segmentation masks of that task. As a result, while minimizing the BCE loss for task m, we are also aiming to make confident predictions for task n by minimizing Shannon's entropy. The regularization is defined as:

$$\mathcal{L}_{ER}(\theta, \tilde{\mathcal{S}}_n) = \frac{1}{|\tilde{\mathcal{S}}_n|} \sum_{(\mathbf{x}) \in \tilde{\mathcal{S}}_n} \sum_{\omega \in \Omega} \left[\mathcal{H}(\mathbf{y}'(\omega)) \right], \tag{4}$$

where Shannon's entropy is $\mathcal{H}(\mathbf{y}'(\omega)) = -\mathbf{y}'(\omega) \log(\mathbf{y}'(\omega))$ for the foreground pixel probability $\mathbf{y}'(\omega)$, and $\mathbf{y}' = f(\mathbf{x}; \theta)$. Our motivation for the entropy regularizer originates from the field of semi-supervised learning [16]. The same loss has been recently applied to few-shot classification [9].

Distillation Loss. Although we deal with different source domains and cell segmentation problems, we can argue that microscopy images of cells must have common features in terms of cell morphology (shape and texture) regardless of the cell type. By learning from as many source data sets as possible, we aim to have a common representation that addresses the target data set to some extent. We explore this idea during meta-training by constraining the base learner to learn a common representation between different source data sets. To that end, two tasks from two source data sets m, p ($m \neq p$, $m, p \in \{1, ..., |\mathcal{S}|\}$) are sampled in every meta-training iteration. Then, the Euclidean distance between the representations of two images (one from each task) from the lth layer of the segmentation model is minimised. This idea is common in knowledge distillation [17] and network compression [2] where the student network learns to predict the same representation as the teacher network. Recently, the idea of distillation between tasks has been also used for few-shot learning [12]. We sample an image $\mathbf{x}^{(m)}$ from data set m and $\mathbf{x}^{(p)}$ from the data set p. Then, we define the distillation loss as:

$$\mathcal{L}_D(\theta, \tilde{\mathcal{S}}_m, \tilde{\mathcal{S}}_p) = \frac{1}{|\tilde{\mathcal{S}}_m||\tilde{\mathcal{S}}_p|} \sum_{(\mathbf{x}^{(m)}) \in \tilde{\mathcal{S}}_m} \sum_{(\mathbf{x}^{(p)}) \in \tilde{\mathcal{S}}_p} \left[f^l(\mathbf{x}^{(m)}; \theta) - f^l(\mathbf{x}^{(p)}; \theta) \right]^2, \tag{5}$$

where $f^l(\mathbf{x}^{(m)}; \theta)$ and $f^l(\mathbf{x}^{(p)}; \theta)$ correspond to the l-th layer activation maps of the base learner for images $\mathbf{x}^{(m)}$ and $\mathbf{x}^{(p)}$, respectively. Furthermore, the lth layer feature of the base learner representation is the latent code of an encoder - decoder architecture [30,36].

3.5 Fine-Tuning

After the meta-learning process is finished, the segmentation model is fine-tuned using a K-shot subset of the target data set, denoted by $\tilde{\mathcal{T}} \subset \mathcal{T}$, with $|\tilde{\mathcal{T}}| = K$ (see line 16 in Algorithm 1). This fine-tuning process is achieved with the following optimisation:

$$\theta^* = \arg\min_{\theta} [\mathcal{L}_{BCE}(\theta, \tilde{\mathcal{T}})]. \tag{6}$$

Here, we only need the binary cross entropy loss $\mathcal{L}_{BCE}$ for fine-tuning. We also rely on Adam optimization and back-propagation for fine-tuning, though we use different hyper-parameters as we later report in the implementation details (Sect. 4.1). At the end, our model with the updated parameters θ^* is evaluated on the target test set $\hat{\mathcal{T}} = \mathcal{T} \setminus \tilde{\mathcal{T}}$ (see line 17 in Algorithm 1).

4 Experiments

We evaluate our approach on five microscopy image data sets using two standard encoder - decoder network architectures for image segmentation [30,36]. All evaluations take place for 1-, 3-, 5-, 7- and 10-shot learning, where we also compare with transfer-learning. In transfer learning the model is trained on all available source data sets and then fine-tuned on the few shots available from the target data set. This work is the first to explore few-shot microscopy image cell segmentation, so we propose the assessment protocol and data sets.

4.1 Implementation Details

We implement two encoder - decoder network architectures. The first architecture is the fully convolutional regression network (FCRN) from [36]. We moderately modify the decoder by using transposed convolutions instead of bi-linear up-sampling as we observed better performance. We also rely on sigmoid activation functions for making binary predictions, instead of the heat-map predictions because of the better performance too. The second architecture is the well-established U-Net [30]. We empirically adapted the original architecture to a lightweight variant, where the number of layers are reduced from 23 to 12. We trained them in meta-training with Adam optimizer with learning rate 0.001 and weight decay of 0.0005. In meta-learning, we set ϵ from Eq. (2) to 1.0. Both networks contain batch-normalization. However, batch-normalization is known to face challenges in gradient-based meta-learning due to the task-based learning [4]. The problem is on learning global scale and bias parameters of batch-normalization based on the tasks. For that reason, we only make use of the mean and variance during meta-training. On fine-tuning, we observed that the scale and bias parameters can be easily learned for the FCRN architecture. For U-Net, we do not rely on the scale and bias parameters for meta-learning. To allow a fair comparison with transfer-learning, we follow the same learning protocol of batch normalization parameters. During fine-tuning, we rely on the Adam optimizer with 20 epochs, learning rate 0.0001, and weight decay 0.0005. The same parameters are used for transfer learning. Our implementation and evaluation protocol is publicly available[1].

4.2 Microscopy Image Databases

We selected five data sets of different cell domains, i.e. $|\mathcal{S}| = 5$. First, we rely on the Broad Bioimage Benchmark Collection (BBBC), which is a collection of various microscopy cell image data sets [18]. We use BBBC005 and BBBC039 from BBBC, which we refer to as B5 and B39. B5 contains 1200 fluorescent synthetic stain cells, while B39 has 200 fluorescent nuclei cells. Second, the Serial Section Transmission Electron Microscopy (ssTEM) [15] database has 165 images of mitochondria cells in neural tissues. Next, the Electron Microscopy (EM) data

[1] https://github.com/Yussef93/FewShotCellSegmentation.

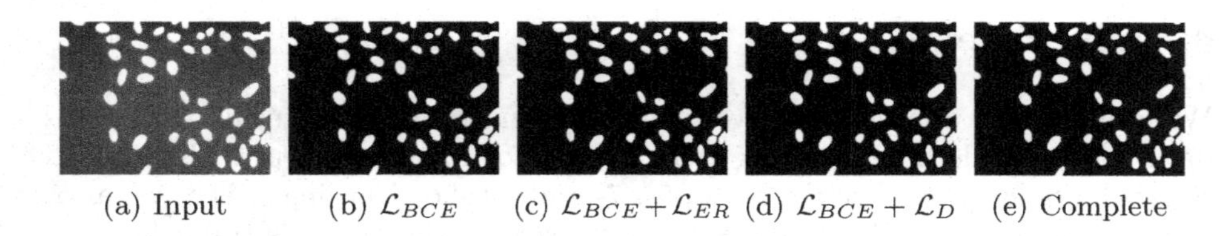

(a) Input (b) $\mathcal{L}_{BCE}$ (c) $\mathcal{L}_{BCE}+\mathcal{L}_{ER}$ (d) $\mathcal{L}_{BCE}+\mathcal{L}_{D}$ (e) Complete

Fig. 2. Results of Our Objectives: We visually compare the effect of our objective functions in meta-training. The complete figure (e) refers to meta-training with all objectives, namely $\mathcal{L}_{BCE}+\mathcal{L}_{ER}+\mathcal{L}_{D}$. The green color corresponds to false negative, the white color to true positive and black to the true negative. Best viewed in color. (Color figure online)

set [22] contains 165 electron microscopy images of mitochondria and synapses cells. Finally, the Triple Negative Breast Cancer (TNBC) database [26] has 50 histology images of breast biopsy. We summarize the cell type, image resolution and number of images for data sets in Table 1. Moreover, we crop the training images to 256×256 pixels, while during testing, the images are used in full resolution.

Table 1. Microscopy Image Databases. We present the details of the five data sets upon which we conduct our experiments.

Data set	B5 [18]	B39 [18]	ssTEM [15]	EM [22]	TNBC [26]
Cell Type	Synthetic stain	Nuclei	Mitochondria	Mitochondria	Nuclei
Resolution	696×520	696×520	1024×1024	768×1024	512×512
# of Samples	1200	200	20	165	50

4.3 Assessment Protocol

We rely on the mean intersection over union (IoU) as the evaluation metric for all experiments. This is a standard performance measure in image segmentation [11, 20]. We conduct a leave-one-dataset-out cross-validation [12], which is a common domain generalization experimental setup. This means that we use the four microscopy image data sets for meta-training and the remaining unseen data set for fine-tuning. From the remaining data set, we randomly select K samples for fine-tuning. Since the selection of the K-shot images can vary significantly the final result, we repeat the random selection of the K-shot samples ten times and report the mean and standard deviation over these ten experiments. The same evaluation is performed for 1-,3-,5-,7- and 10-shot learning. Similarly, we evaluate the transfer learning approach as well.

Objectives Analysis: We examine the impact of each objective, as presented in Sect. 3.4, to the final segmentation result. In particular, we first meta-train our model only with the binary cross entropy $\mathcal{L}_{BCE}$ from Eq. 3. Second, we meta-train the binary cross entropy $\mathcal{L}_{BCE}$ jointly with entropy regularization $\mathcal{L}_{ER}$ from Eq. 4. Next, we meta-train the binary cross entropy and distillation loss $\mathcal{L}_D$ from Eq. 5. At last, we meta-train with all loss functions together, where we weight $\mathcal{L}_{ER}$ with $\alpha = 0.01$ and $\mathcal{L}_D$ with $\beta = 0.01$. We also noticed that the values of these two hyper-parameters depend on the complete loss function. As a result, we empirically set $\alpha = 0.1$ when the complete loss is composed of $\mathcal{L}_{ER}$ and $\mathcal{L}_{BCE}$; and keep $\beta = 0.01$ when the complete loss is $\mathcal{L}_D$ and $\mathcal{L}_{BCE}$. We later analyze the sensitivity of the hyper-parameter in Sect. 4.4. Overall, the values have been found with grid search. A visual comparison of our objective is shown in Fig. 2.

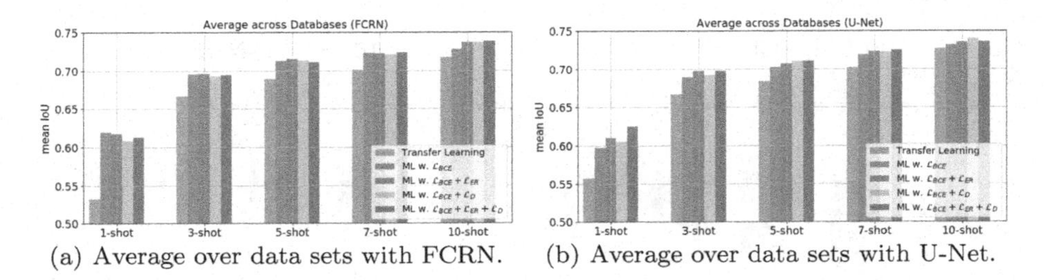

(a) Average over data sets with FCRN. (b) Average over data sets with U-Net.

Fig. 3. Mean intersection over union (IoU) comparison using all datasets. We compare all loss function combinations and transfer learning for all data sets using the FCRN architecture in (a) and U-Net architecture in (b).

4.4 Results Discussion

We present the results of our approach with different combinations of the objective function as described in Sect. 4.3, as well as the complete model as presented in Algorithm 1. The quantitative evaluation is summarized in Fig. 3 and Fig. 4. In Fig. 3 we present the mean intersection over union (IoU) over all data sets while in Fig. 4 we present the mean IoU and the standard deviation over our ten random selections of the K-shot samples for each data set individually. We also provide a visual comparison with transfer learning in Fig. 1.

At first, relying only on the binary cross entropy loss $\mathcal{L}_{BCE}$ for meta-training represents the baseline result for our approach. Adding the entropy regularization $\mathcal{L}_{ER}$ has a positive impact on some K-shot learning experiments. This can be seen in Fig. 4, which depicts meta-training with $\mathcal{L}_{BCE}$ and meta-training with $\mathcal{L}_{BCE} + \mathcal{L}_{ER}$. Similarly, the use of the distillation loss $\mathcal{L}_D$ together with the binary cross entropy, i.e. $\mathcal{L}_{BCE} + \mathcal{L}_D$ in Fig. 4, generally improves the mean IoU. The complete loss function $\mathcal{L}_{BCE} + \mathcal{L}_{ER} + \mathcal{L}_D$ gives the best results on average. This is easier to be noticed in Fig. 3(a) and Fig. 3(b) where we compare the different objective combinations; and the results show that the complete

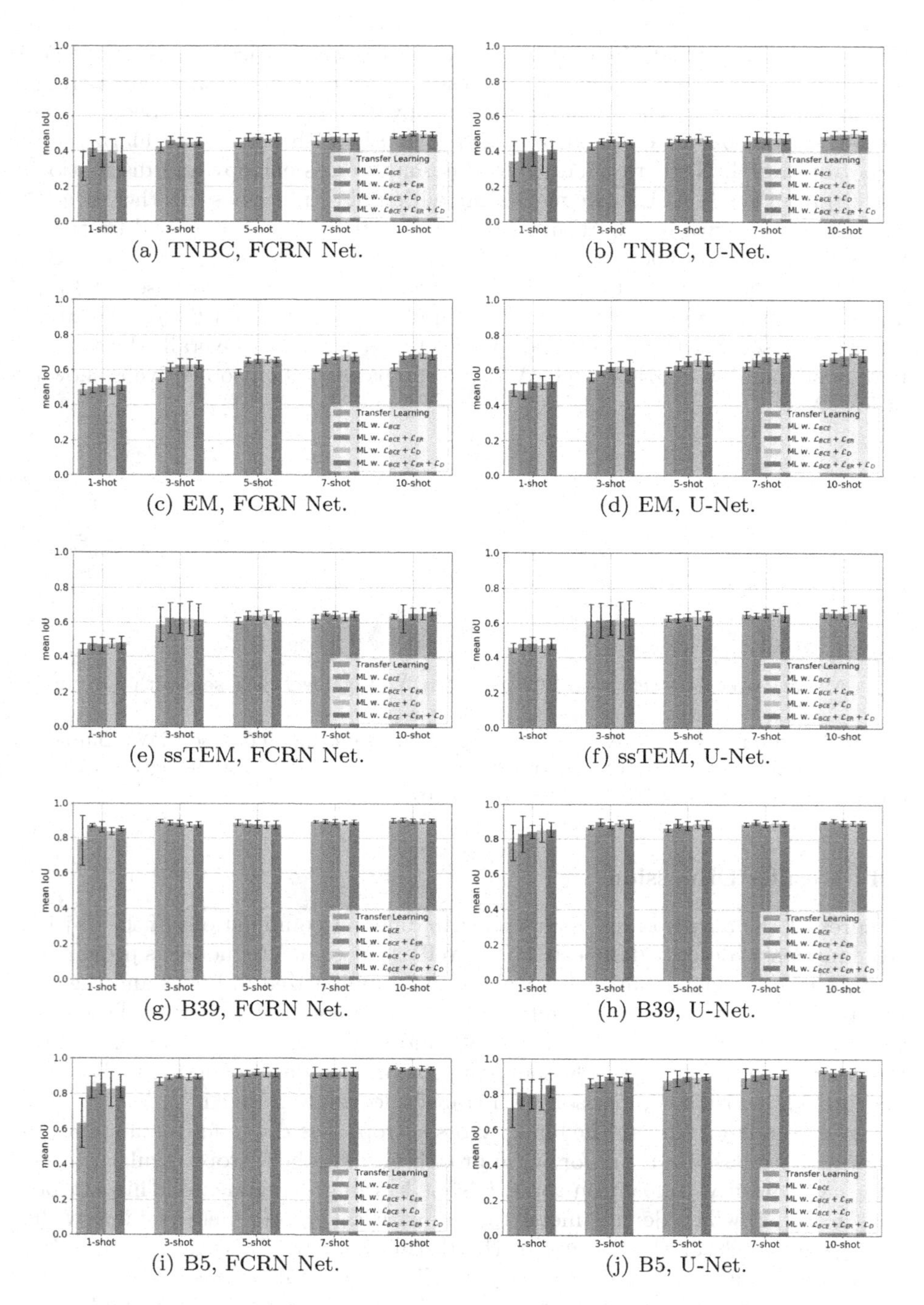

(a) TNBC, FCRN Net. (b) TNBC, U-Net.

(c) EM, FCRN Net. (d) EM, U-Net.

(e) ssTEM, FCRN Net. (f) ssTEM, U-Net.

(g) B39, FCRN Net. (h) B39, U-Net.

(i) B5, FCRN Net. (j) B5, U-Net.

Fig. 4. Mean intersection over union (IoU) and standard deviation result comparison. We include of all loss functions and transfer learning the FCRN architecture (*left* column) and U-Net architecture (*right* column). ML stands for meta-learning.

loss produces the best results for most K shot problems. In addition, it is clear (Fig. 3) that our contributions, the $\mathcal{L}_{ER}$ and $\mathcal{L}_D$ objectives, have a positive impact on the outcome when more K-shot samples are added.

Comparing the FCRN with U-Net architectures in Fig. 3(a) and Fig. 3(b), we observe that U-Net delivers slightly better results for the complete loss combination. Nevertheless, the overall behavior is similar for both architectures. The standard binary cross entropy without our regularization and distillation objectives is already more accurate than transfer learning. Figure 4 also shows that the transfer learning performance can be comparable to ours (complete model) for 10-shot learning in some cases, e.g. Fig. 4(g) and Fig. 4(h) for FCRN and U-Net. On 10-shot learning, transfer-learning benefits from the larger number of training data, resulting in performance closer to meta-learning. Other than the better performance, we also make two consistent observations when comparing to transfer learning. First, we argue that the main reason for the better performance is the parameter update rule. Transfer-learning averages gradients over shuffled samples, while meta-learning relies on task-based learning and then on averaging over the tasks. Second, meta-learning shows faster convergence. In particular, meta-learning converges 40x faster for FCRN and around 60x faster for U-Net. This is an important advantage when working with multiple training sets. The disadvantage of meta-learning is the selection of the hyper-parameters. Our experience in the selection of the optimizer, including its hyper-parameters, as well as the selection of α and β demonstrates that we need to be careful in this process. However, the optimizer and hyper-parameters are consistent for all tasks of cell segmentation. On the other hand, transfer learning involves only the selection of the optimizer.

We can conclude the following points from the experiments. Meta-learning yields an overall stronger performance than transfer-learning as we average across our data sets, in addition, the new loss functions boosts the meta-learning performance in many cases, as shown in Fig. 3(a) and 3(b). Both approaches, have low standard deviation when we add more shots in the fine-tuning step, as shown in Fig. 4. The same overall behavior is observed regardless the network architecture. Moreover, both architectures exhibit similar mean IoU. Besides, we notice that our regularization loss functions are not necessary for results around 90% or more. In this case, only the additional shots make an impact. Finally, we notice the difference in performance across the data sets. For example, the mean IoU in TNBC target domain at 10-shot learning is within 50% range while in EM we are able to reach around 70% range. The reason could be that the source domains are more correlated with the EM target domain in terms of features than TNBC. Overall, our approach produces promising results when having at least 5-shots, i.e. annotated images, from the target domain for fine-tuning. This is a reasonably low amount of required to annotation to reach good generalization.

5 Conclusion

We have presented a few-shot meta-learning approach for microscopy image cell segmentation. The experiments show that our approach enables the learning of a generic and adaptable few-shot learning model from the available annotated images and cell segmentation problems of the source data sets. We fine-tune this model on the target data set using a limited annotated images. In the context of meta-training, we proposed the combination of three objectives to segment and regularize the predictions based on cross-domain tasks. Our experiments on five microscopy data sets show promising results from 1- to 10-shot meta-learning. As future work, we will include more data sets in our study to explore their correlation and impact on the target data set. Also, we plan to pose our problem in the context of domain generalization where the target data sets lacks of annotation.

Acknowledgments. This work was partially funded by Deutsche Forschungsgemeinschaft (DFG), Research Training Group GRK 2203: Micro- and nano-scale sensor technologies for the lung (PULMOSENS), and the Australian Research Council through grant FT190100525. G.C. acknowledges the support by the Alexander von Humboldt-Stiftung for the renewed research stay sponsorship.

References

1. Arteta, C., Lempitsky, V., Zisserman, A.: Counting in the Wild. In: Leibe, B., Matas, J., Sebe, N., Welling, M. (eds.) ECCV 2016. LNCS, vol. 9911, pp. 483–498. Springer, Cham (2016). https://doi.org/10.1007/978-3-319-46478-7_30
2. Belagiannis, V., Farshad, A., Galasso, F.: Adversarial network compression. In: Leal-Taixé, L., Roth, S. (eds.) ECCV 2018. LNCS, vol. 11132, pp. 431–449. Springer, Cham (2019). https://doi.org/10.1007/978-3-030-11018-5_37
3. de Brebisson, A., Montana, G.: Deep neural networks for anatomical brain segmentation. In: Proceedings of the IEEE Conference on Computer Vision and Pattern Recognition Workshops, pp. 20–28 (2015)
4. Bronskill, J., Gordon, J., Requeima, J., Nowozin, S., Turner, R.E.: Tasknorm: ethinking batch normalization for meta-learning. arXiv preprint arXiv:2003.03284 (2020)
5. Chen, L.C., Papandreou, G., Kokkinos, I., Murphy, K., Yuille, A.L.: Deeplab: semantic image segmentation with deep convolutional nets, atrous convolution, and fully connected crfs. IEEE Trans. Pattern Anal. Mach. Intell. **40**(4), 834–848 (2017)
6. Ciresan, D., Giusti, A., Gambardella, L.M., Schmidhuber, J.: Deep neural networks segment neuronal membranes in electron microscopy images. In: Advances in Neural Information Processing Systems, pp. 2843–2851 (2012)
7. Cireşan, D.C., Giusti, A., Gambardella, L.M., Schmidhuber, J.: Mitosis detection in breast cancer histology images with deep neural networks. In: Mori, K., Sakuma, I., Sato, Y., Barillot, C., Navab, N. (eds.) MICCAI 2013. LNCS, vol. 8150, pp. 411–418. Springer, Heidelberg (2013). https://doi.org/10.1007/978-3-642-40763-5_51
8. Clavera, I., Rothfuss, J., Schulman, J., Fujita, Y., Asfour, T., Abbeel, P.: Model-based reinforcement learning via meta-policy optimization. arXiv preprint arXiv:1809.05214 (2018)

9. Dhillon, G.S., Chaudhari, P., Ravichandran, A., Soatto, S.: A baseline for few-shot image classification. arXiv preprint arXiv:1909.02729 (2019)
10. Dijkstra, K., van de Loosdrecht, J., Schomaker, L.R.B., Wiering, M.A.: Centroid-Net: a deep neural network for joint object localization and counting. In: Brefeld, U., Curry, E., Daly, E., MacNamee, B., Marascu, A., Pinelli, F., Berlingerio, M., Hurley, N. (eds.) ECML PKDD 2018. LNCS (LNAI), vol. 11053, pp. 585–601. Springer, Cham (2019). https://doi.org/10.1007/978-3-030-10997-4_36
11. Dong, N., Xing, E.: Few-shot semantic segmentation with prototype learning. In: BMVC, vol. 3 (2018)
12. Dou, Q., de Castro, D.C., Kamnitsas, K., Glocker, B.: Domain generalization via model-agnostic learning of semantic features. In: Advances in Neural Information Processing Systems, pp. 6447–6458 (2019)
13. Faustino, G.M., Gattass, M., Rehen, S., de Lucena, C.J.: Automatic embryonic stem cells detection and counting method in fluorescence microscopy images. In: 2009 IEEE International Symposium on Biomedical Imaging: From Nano to Macro, pp. 799–802. IEEE (2009)
14. Finn, C., Abbeel, P., Levine, S.: Model-agnostic meta-learning for fast adaptation of deep networks. In: Proceedings of the 34th International Conference on Machine Learning-Volume 70, pp. 1126–1135. JMLR. org (2017)
15. Gerhard, S., Funke, J., Martel, J., Cardona, A., Fetter, R.: Segmented anisotropic ss TEM dataset of neural tissue. figshare (2013)
16. Grandvalet, Y., Bengio, Y.: Semi-supervised learning by entropy minimization. In: Advances in Neural Information Processing Systems, pp. 529–536 (2005)
17. Hinton, G., Vinyals, O., Dean, J.: Distilling the knowledge in a neural network. arXiv preprint arXiv:1503.02531 (2015)
18. Lehmussola, A., Ruusuvuori, P., Selinummi, J., Huttunen, H., Yli-Harja, O.: Computational framework for simulating fluorescence microscope images with cell populations. IEEE Trans. Med. Imaging **26**(7), 1010–1016 (2007)
19. Li, D., Yang, Y., Song, Y.Z., Hospedales, T.M.: Learning to generalize: meta-learning for domain generalization. In: 32nd AAAI Conference on Artificial Intelligence (2018)
20. Long, J., Shelhamer, E., Darrell, T.: Fully convolutional networks for semantic segmentation. In: Proceedings of the IEEE Conference on Computer Vision and Pattern Recognition, pp. 3431–3440 (2015)
21. Lu, Z., Carneiro, G., Bradley, A.P.: An improved joint optimization of multiple level set functions for the segmentation of overlapping cervical cells. IEEE Trans. Image Process. **24**(4), 1261–1272 (2015)
22. Lucchi, A., Li, Y., Fua, P.: Learning for structured prediction using approximate subgradient descent with working sets. In: Proceedings of the IEEE Conference on Computer Vision and Pattern Recognition, pp. 1987–1994 (2013)
23. Mensink, T., Verbeek, J., Perronnin, F., Csurka, G.: Metric learning for large scale image classification: generalizing to new classes at near-zero cost. In: Fitzgibbon, A., Lazebnik, S., Perona, P., Sato, Y., Schmid, C. (eds.) ECCV 2012. LNCS, vol. 7573, pp. 488–501. Springer, Heidelberg (2012). https://doi.org/10.1007/978-3-642-33709-3_35
24. Munkhdalai, T., Yu, H.: Meta networks. In: Proceedings of the 34th International Conference on Machine Learning-Volume 70, pp. 2554–2563 (2017). JMLR. org
25. Naylor, P., Laé, M., Reyal, F., Walter, T.: Nuclei segmentation in histopathology images using deep neural networks. In: 2017 IEEE 14th International Symposium on Biomedical Imaging (ISBI 2017), pp. 933–936. IEEE (2017)

26. Naylor, P., Laé, M., Reyal, F., Walter, T.: Segmentation of nuclei in histopathology images by deep regression of the distance map. IEEE Trans. Med. Imaging **38**(2), 448–459 (2018)
27. Nichol, A., Achiam, J., Schulman, J.: On first-order meta-learning algorithms. arXiv preprint arXiv:1803.02999 (2018)
28. Ravi, S., Larochelle, H.: Optimization as a model for few-shot learning (2016)
29. Rohrbach, M., Ebert, S., Schiele, B.: Transfer learning in a transductive setting. In: Advances in Neural Information Processing Systems, pp. 46–54 (2013)
30. Ronneberger, O., Fischer, P., Brox, T.: U-Net: convolutional networks for biomedical image segmentation. In: Navab, N., Hornegger, J., Wells, W.M., Frangi, A.F. (eds.) MICCAI 2015. LNCS, vol. 9351, pp. 234–241. Springer, Cham (2015). https://doi.org/10.1007/978-3-319-24574-4_28
31. Schmidhuber, J.: Learning to control fast-weight memories: an alternative to dynamic recurrent networks. Neural Comput. **4**(1), 131–139 (1992)
32. Snell, J., Swersky, K., Zemel, R.: Prototypical networks for few-shot learning. In: Advances in Neural Information Processing Systems, pp. 4077–4087 (2017)
33. Tzeng, E., Hoffman, J., Saenko, K., Darrell, T.: Adversarial discriminative domain adaptation. In: Proceedings of the IEEE Conference on Computer Vision and Pattern Recognition, pp. 7167–7176 (2017)
34. Wählby, C., Sintorn, I.M., Erlandsson, F., Borgefors, G., Bengtsson, E.: Combining intensity, edge and shape information for 2D and 3D segmentation of cell nuclei in tissue sections. J. Microsc. **215**(1), 67–76 (2004)
35. Vinyals, O., Blundell, C., Lillicrap, T., Wierstra, D., et al.: Matching networks for one shot learning. In: Advances in Neural Information Processing Systems, pp. 3630–3638 (2016)
36. Xie, W., Noble, J.A., Zisserman, A.: Microscopy cell counting and detection with fully convolutional regression networks. Comput. Methods Biomech. Biomed. Eng. Imaging Vis. **6**(3), 283–292 (2018)
37. Xing, F., Yang, L.: Robust nucleus/cell detection and segmentation in digital pathology and microscopy images: a comprehensive review. IEEE Rev. Biomed. Eng. **9**, 234–263 (2016)
38. Zhang, X., Wang, H., Collins, T.J., Luo, Z., Li, M.: Classifying stem cell differentiation images by information distance. In: Flach, P.A., De Bie, T., Cristianini, N. (eds.) ECML PKDD 2012. LNCS (LNAI), vol. 7523, pp. 269–282. Springer, Heidelberg (2012). https://doi.org/10.1007/978-3-642-33460-3_23

Deep Reinforcement Learning for Large-Scale Epidemic Control

Pieter J.K. Libin[1,2,3](✉), Arno Moonens[1], Timothy Verstraeten[1], Fabian Perez-Sanjines[1], Niel Hens[3], Philippe Lemey[2], and Ann Nowé[1]

[1] Vrije Universiteit Brussel, Brussels, Belgium
{pieter.libin,arno.moonens,timothy.verstraeten,fperezsa,ann.nowe}@vub.be
[2] KU Leuven, Leuven, Belgium
philippe.lemey@kuleuven.be
[3] Hasselt University, Hasselt, Belgium
niel.hens@uhasselt.be

Abstract. Epidemics of infectious diseases are an important threat to public health and global economies. Yet, the development of prevention strategies remains a challenging process, as epidemics are non-linear and complex processes. For this reason, we investigate a deep reinforcement learning approach to automatically learn prevention strategies in the context of pandemic influenza. Firstly, we construct a new epidemiological meta-population model, with 379 patches (one for each administrative district in Great Britain), that adequately captures the infection process of pandemic influenza. Our model balances complexity and computational efficiency such that the use of reinforcement learning techniques becomes attainable. Secondly, we set up a ground truth such that we can evaluate the performance of the "Proximal Policy Optimization" algorithm to learn in a single district of this epidemiological model. Finally, we consider a large-scale problem, by conducting an experiment where we aim to learn a joint policy to control the districts in a community of 11 tightly coupled districts, for which no ground truth can be established. This experiment shows that deep reinforcement learning can be used to learn mitigation policies in complex epidemiological models with a large state and action space.

Keywords: Multi-agent systems · Epidemic control · Pandemic influenza · Deep reinforcement learning

1 Introduction

Epidemics of infectious diseases are an important threat to public health and global economies. The most efficient way to combat epidemics is through prevention. To develop prevention strategies and to implement them as efficiently as possible, a good understanding of the complex dynamics that underlie these epidemics is essential. To properly understand these dynamics, and to study emergency scenarios, epidemiological models are necessary. Such models enable

Y. Dong et al. (Eds.): ECML PKDD 2020, LNAI 12461, pp. 155–170, 2021.
https://doi.org/10.1007/978-3-030-67670-4_10

us to make predictions and to study the effect of prevention strategies in simulation [18]. The development of prevention strategies, which need to fulfil distinct criteria (i.a., prevalence, mortality, morbidity, cost), remains a challenging process. For this reason, we investigate a deep reinforcement learning (RL) approach to automatically learn prevention strategies in an epidemiological model. The use of model-free deep reinforcement learning is particularly interesting, as it allows us to set up a learning environment in a complex epidemiological setting (i.e., large state space and non-linear dependencies) while imposing few assumptions on the policies to be learned [22]. In this work, we conduct our experiments in the context of pandemic influenza, where we aim to learn optimal school closure policies to mitigate the epidemic.

Pandemic preparedness is important, as influenza pandemics have made many victims in the (recent) past [23] and the ongoing COVID-19 epidemic is yet another reminder of this fact [36]. Contrary to seasonal influenza epidemics, an influenza pandemic is caused by a newly emerging virus strain that can become pandemic by spreading rapidly among naive human hosts (i.e., human hosts with no prior immunity) worldwide [23]. This means that at the start of the pandemic no vaccine will be available and it will take several months before vaccine production can commence [30]. For this reason, learning optimal strategies of non-therapeutic intervention measures, such as school closure policies, is of great importance to mitigate pandemics [20]. To meet this objective, we consider a reinforcement learning approach. However, as the state-of-the-art of reinforcement learning techniques require many interactions with the environment in order to converge, our first contribution entails a realistic epidemiological model that still has a favourable computational performance.

Specifically, we construct a meta-population model that consists of a set of 379 interconnected patches, where each patch corresponds to an administrative region in Great Britain and is internally represented by an age-structured stochastic compartmental model. To conduct our experiments, we establish a Markov decision process with a state space that directly corresponds to our epidemiological model, an action space that allows us to open and close schools on a weekly basis, a transition function that follows the epidemiological model's dynamics, and a reward function that is targeted to the objective of reducing the attack rate (i.e., the proportion of the population that was infected). In this work, we will use "Proximal Policy Optimization" (PPO) [26] to learn the school closure policies.

First, we set up an experiment in an epidemiological model that covers a single administrative district. This setting enables us to specify a ground truth that allows us to empirically assess the performance of the policies learned by PPO. In this analysis, we consider different values for the basic reproductive number R_0[1] and the population composition (i.e., proportion of adults, children, elderly, adolescents) of the district. Both parameters induce a significant change of the epidemic model's dynamics. Through these experiments, we demonstrate

[1] The number of infections that is, on average, generated by one single infected individual that is placed in an otherwise fully susceptible population.

the potential of deep reinforcement learning algorithms to learn policies in the context of complex epidemiological models, opening the prospect to learn in even more complex stochastic models with large action spaces. In this regard, we consider a large scale setting where we examine whether there is an advantage to consider the collaboration between districts when designing school closure policies.

2 Related Work

The closing of schools is an effective way to limit the spread of an influenza pandemic [20]. For this reason, the use of school closures as a mitigation strategy has been explored in a variety of modelling studies (exhaustive list in Supplementary Information, SI), of which the work presented in [8] is the most recent and comprehensive study.

The concept to learn dynamic policies by formulating the decision problem as a Markov decision process (MDP) was first introduced in [32]. The proposed technique was used to investigate dynamic tuberculosis case-finding policies in HIV/tuberculosis co-epidemics [33]. Later, the technique was extended towards a method to include cost-effectiveness in the analysis [34], and applied to investigate mitigation policies (i.e., school closures and vaccines) in the context of pandemic influenza in a simplified epidemiological model. The work presented in [32,34] uses a policy iteration algorithm to solve the MDP. To scale this approach to larger problem settings, we explore the use of on-line reinforcement learning techniques (e.g., TD-learning, policy gradient). Note that the "Deep Q-networks" algorithm was recently used to investigate culling and vaccination in farms in a simple individual-based model to delay the spread of viruses in a cattle population [24]. Furthermore, recently, reinforcement learning, with a neural network as function approximator, was used to address the graph protection problem, to aim to restrain the epidemic propagation in a network [31]. However, to our best knowledge, the work presented in this manuscript is the first attempt to use deep reinforcement learning algorithms directly on a complex meta-population model. Furthermore, we experimentally validate the performance of these algorithms using a ground truth, in a variety of model settings (i.e., different census compositions and different R_0's). This is the first validation of this kind and it demonstrates the potential of on-line deep reinforcement learning techniques in the context of epidemic decision making. Finally, we present a novel approach to investigate how intervention policies can be improved by enabling collaboration between different geographic districts, by formulating the setting as a multi-agent problem, and by solving it using deep reinforcement learning algorithms. Next to stateful reinforcement learning, the use of multi-armed bandits has recently been explored to assist decision makers to efficiently select the optimal prevention strategy [17,18].

3 Epidemiological Model

We construct a meta-population model that consists of 379 patches, where each patch represents one administrative region in Great Britain. Great Britain consists of three countries with the following administrative regions: 325 districts in England, 22 unitary authorities in Wales and 32 council areas in Scotland. Each patch consists of a stochastic age-structured compartmental model, which we describe in Subsect. 3.1, and the different patches are connected via a mobility model, as detailed in Subsect. 3.2. In Subsect. 3.3 we discuss how we validate and calibrate the model. We analyse the model's computational complexity and discuss the model's performance in SI.

3.1 Intra-patch Model

We consider a stochastic compartmental SEIR model from which we sample trajectories. We first describe the model in terms of ordinary differential equations (i.e., a deterministic representation) that we then transform to stochastic differential equations [1] to make a stochastic evaluation possible (background on compartmental models in SI). An SEIR model divides the population in a Susceptible, Exposed, Infected and Recovered compartment, and is commonly used to model influenza epidemics [4]. More specifically, we consider an age-structured SEIR model (see Fig. 1 for a visualization) with a set of n disjoint age groups [4,7]. This model is formally described by this system of ordinary differential equations (ODEs), defined for each age group i:

$$\begin{aligned}\frac{\mathrm{d}S_i}{\mathrm{d}t} &= -\phi_i(t)S_i(t)\\ \frac{\mathrm{d}E_i}{\mathrm{d}t} &= \phi_i(t)S_i(t) - \zeta E_i(t)\\ \frac{\mathrm{d}I_i}{\mathrm{d}t} &= \zeta E_i(t) - \gamma I_i(t)\\ \frac{\mathrm{d}R_i}{\mathrm{d}t} &= \gamma I_i(t).\end{aligned} \tag{1}$$

Every susceptible individual in age group i is subject to an age-specific and time-dependent force of infection:

$$\phi_i(t) = \sum_{j=1}^{n} \beta M_{ij}(t) \frac{I_j(t)}{N_j(t)}, \tag{2}$$

which depends on:

- The probability of transmission β when a contact occurs.
- The time-dependent contact matrix M, where $M_{ij}(t)$ is the average frequency of contacts that an individual in age group i has with an individual in age group j.

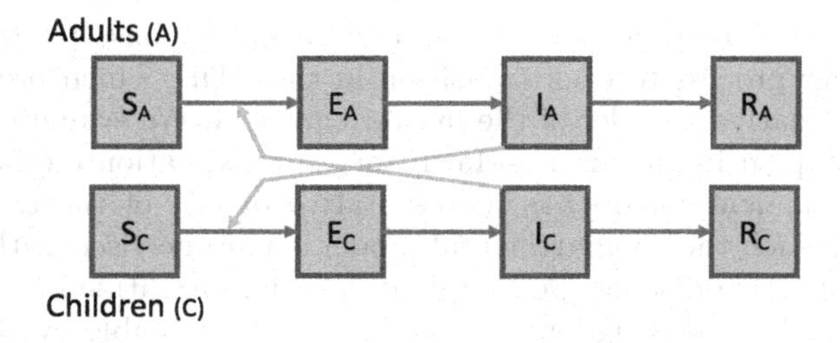

Fig. 1. We depict an age-structured SEIR model that considers two age groups (i.e., adults and children). This model consists of two SEIR models, one for each age group, that are connected to represent mixing between the age groups (yellow arrows). Note that it is also possible to mix within the age groups. Note that we use two age groups in this figure to provide a clear visualization of the model. In our actual model, we consider four different age groups. (Color figure online)

- The frequency at which contacts with infected individuals (in age group j) occur: $I_j(t)/N_j(t)$.

Once exposed, individuals move to the infected state according to the latency rate ζ. Individuals recover from infection (i.e., get better or die) at a recovery rate γ.

We omit vital dynamics (i.e., births and deaths that are not caused by the epidemic) in this SEIR model, as the epidemic's time scale is short and we therefore assume that births and deaths will have a limited influence on the epidemic process [28]. Therefore, at any time:

$$N_i(t) = S_i(t) + E_i(t) + I_i(t) + R_i(t), \tag{3}$$

where the total population size N_i corresponds to age-specific census data. Our model considers 4 age groups: children (0–4 years), adolescents (5–18 years), adults (19–64 years) and elderly (65 years and older).

Note that the contact frequency $M_{ij}(t)$ is time-dependent, in order to model school closures, i.e., a different contact matrix is used for school term and school holiday. Following [4], we consider *conversational contacts*, i.e., contacts for which physical touch is not required. As we aim to model the effectiveness of school closure interventions, we use the United Kingdom contact matrices presented in [4], which encodes a contact matrix for both school term and school holiday. These contact matrices are the result of an internet-based social contact survey completed by a cohort of participants [4]. The contact matrices encode for the same age groups as mentioned before: children, adolescents, adults and elderly.

We defined the SEIR model in terms of a system of ODEs which implies a deterministic evaluation of the system. However, for predictions, stochastic models are preferred, as they to account for stochastic variation and allow us to quantify uncertainty [13]. In order to sample trajectories from this set of differential equations, we transform the system of ODEs to a system of stochastic differential equations (SDEs), using the transformation procedure presented in

[1] (details in SI). This procedure introduces stochastic noise to the system by adding a Wiener process to each transition in the ODE, which provides a reasonable approximative sampler of the master equation. We evaluate the SDE at discrete time steps using the Euler-Maruyama approximation method [1].

Each compartmental model is representative of one of the administrative districts and as such the compartmental model is parametrised with the census data of the respective district, i.e., population counts stratified by age groups. We use the 2011 United Kingdom census data made available by NOMIS[2]. We present more details on the census data in SI.

3.2 Inter-patch Model

Our model, that is comprised of a set of connected SEIR patches, is inspired by the recent BBC pandemic model [15]. The BBC pandemic model was in its turn motivated by the model presented in [9].

At each time step, our model checks whether a patch p becomes infected. This is modulated by the patch's force of infection, which combines the potential of the infected patches in the system, weighted by a mobility model, that represents the commuting of adults between the different patches:

$$\mathring{\phi}_p(t) = \sum_{p' \in \mathcal{P}} \mathcal{M}_{p'p} \cdot \beta \cdot \left(S_p^{\mathrm{A}}(t)\right)^{\mu} \cdot \mathcal{I}_{p'}(t), \tag{4}$$

where $\mathcal{P}$ is the set of patches in the model, $\mathcal{M}_{p'p}$ is the mobility flux between patch p' and p, β is the probability of transmission on a contact, $S_p^{\mathrm{A}}(t)$ is the susceptible population of adults in patch p at time t and its contribution is modulated by parameter μ (range in $[0, 1]$), and $\mathcal{I}_{p'}(t)$ is the infectious potential of patch p' at time t. We define this infectious potential as,

$$\mathcal{I}_{p'}(t) = I_{p'}^{\mathrm{A}}(t) \cdot M_{\mathrm{AA}}, \tag{5}$$

where $I_{p'}^{\mathrm{A}}(t)$ is the size at time t of the infectious adult population in patch p' and M_{AA} is the average number of contacts between adults, as specified in the contact matrix (see Subsect. 3.1).

$\mathcal{M}$ is a matrix based on the mobility dataset provided by NOMIS[3]. This dataset describes the amount of commuting between the districts in Great Britain.

In general, this inter-patch model is constructed from first principles i.e., census data, a mobility model, the number of infected individuals and the transmission potential of the virus. However, for the parameter μ that modulates the contribution of the susceptibles in the receptive patch (while it is commonly used in literature [5,9,14]) no such intuition is readily available. Therefore, this parameter is typically fitted to match the properties of the epidemic that is

[2] https://www.nomisweb.co.uk.

[3] We use the NOMIS WU03UK dataset that was released in 2011.

under investigation [5,9,14]. We will calibrate this parameter such that it can be used for a range of R_0 values, as detailed in the next sub-section.

Given this time-dependent force of infection, we model the event that a patch becomes infected with a non-homogeneous Poisson process [29]. As the process' intensity depends on how the model (i.e., the set of all patches) evolves, we cannot sample the time at which a patch becomes infected a priori. Therefore, we determine this time of infection using the time scale transformation algorithm [3]. Details about this procedure can be found in SI. Following [15], we assume that a patch will become infected only once.

By using this time scale transformation algorithm and evaluating the stochastic differential equation at discrete time steps, we produced a model with favourable performance, i.e., in our experiments we can run about 2 simulation runs per second on a MacBook Pro (CPU: 2,3 GHz Intel Core i5). We analyse the model's computational complexity in SI.

3.3 Calibration and Validation

Our objective is to construct a model that is representative for contemporary Great Britain with respect to population census and mobility trends. This model will be used to study school closure intervention strategies for future influenza pandemics. While in many studies [5,9,14] a model is created specifically to fit one epidemic case, we aim for a model that is robust with respect to different epidemic parameters, most importantly R_0, the basic reproduction number.

To validate our model according to these goals, we conduct two experiments. In the first experiment, we compare our patch model to an SEIR compartmental model that uses the same contact matrix and age structure, but with homogeneous spatial mixing (i.e., no spatial structure). While we do not expect our model to behave exactly like the compartmental model, as the patches and the mobility network that connects them induces a different dynamic, we do observe similar trends with respect to the epidemic curve and peak day. This experiment also enables us to calibrate the μ parameter. We present a detailed description of this analysis and report the results in SI. In the second experiment we show that our model is able to reproduce the trends that were observed during the 2009 influenza pandemic, commonly known as the swine-origin influenza pandemic, i.e., A(H1N1)v2009, that originated in Mexico. The 2009 influenza pandemic in Great Britain is an interesting case to validate our model for two reasons. Firstly, the pandemic occurred quite recently and thus our model's census and mobility scheme are a good fit, as both the datasets on which we base our census and mobility model were released in 2011. Secondly, due to the time when the virus entered Great Britain, the summer holiday started 11 weeks after the emergence of the epidemic. The timing of the holidays had a severe impact on the progress of the epidemic and resulted in a epidemic curve with two peaks. This characteristic epidemic curve enables us to demonstrate the predictive power of our age-structured contact model with support for school closures. In Fig. 2, we show a set of model realisations in conjunction with the symptomatic case data, which shows that we were able to closely match the epidemic trends observed during the British pandemic in 2009 (qualitative model fit, details on this case study in SI). Note that our model reports the number of infections while the British Health Protection Agency only recorded symptomatic cases. Therefore we scale the epidemic curve with a factor of $\frac{1}{4}$. This large number of asymptomatic cases produced by our model is in line with earlier serological surveys [21] and with previous modelling studies [16].

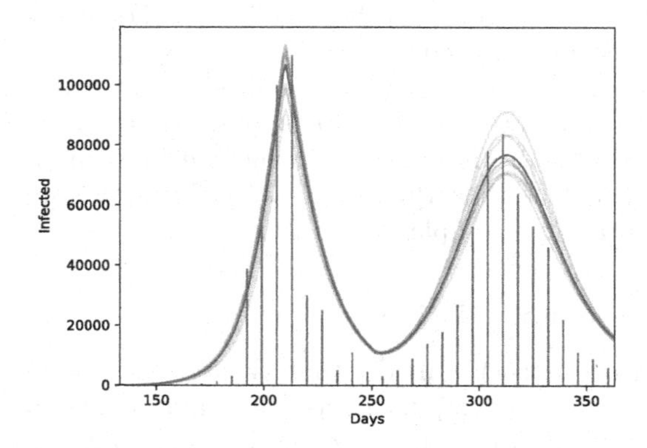

Fig. 2. show that our model (blue epidemic curves) is able to match the trends observed in the British pandemic of 2009 (the vertical bars represent the number of infected individuals that was recorded during the epidemic). We show 10 stochastic trajectories. (Color figure online)

4 Learning Environment

In order to apply reinforcement learning, we construct a Markov Decision process (background in SI) based on the epidemiological model that we introduced in the previous section. This epidemiological model consists of patches that correspond to administrative regions.

We have an agent for each patch that we attempt to control, and for each agent we have an action space $\mathcal{A} = \{\text{open}, \text{close}\}$ that allows us to open and close schools for one week. Each agent has a predefined budget $\mathscr{b}$ of school closure actions it can execute. Once this budget is depleted, executing a close action will default to executing an open action. We refer to the remaining budget at time t as $\mathscr{b}^{(t)}$. In the epidemiological model, when schools are closed we use a contact matrix that is representative for school holidays and when schools are open we use a contact matrix that is representative for school term (details in Sect. 3.1).

For each patch, we consider a state space that combines the state of the SEIR model and the remaining budget of school closures $\mathscr{b}_p^{(t)}$. For the SEIR model, we have 16 state variables (i.e., $\mathbb{R}^{16}$), as we have an SEIR model (4 state variables) for each of the four age groups. The remaining school closure budget is encoded as an integer, resulting in a combined state space of 17 variables. We refer to the state space of one patch p, that thus combines the epidemiological states and the budget, as $\mathcal{S}_p$. The state space $\mathcal{S}$ of the MDP corresponds to the aggregation of the state space of each patch that we attempt to control:

$$\bigtimes_{p \in \mathcal{P}^c} \mathcal{S}_p, \tag{6}$$

where $\mathcal{P}^c \subseteq \mathcal{P}$ is the set of patches that we control.

The transition probability function $T(\mathbf{s}' \mid \mathbf{s}, \mathbf{a})$ stochastically determines the state of the epidemic in the next week, taking into account the school closure actions that were chosen, using the epidemiological dynamics as defined in the previous section.

To reduce the attack rate, we consider an immediate reward function that quantifies the negative loss in susceptibles over one simulated week:

$$R_{\mathrm{AR}}(\mathbf{s}, \mathbf{a}, \mathbf{s}') = -(S(\mathbf{s}) - S(\mathbf{s}')), \tag{7}$$

where $S(.)$ is the function that determines the total number of susceptible individuals given the state of the epidemiological model.

In this work, we use the Proximal Policy Optimization algorithm (PPO), as policy gradient algorithms tend to be more suitable to deal with large action spaces [26], such as the action space of the multi-district setting that we will investigate in Sect. 6 (background on policy gradient and PPO in SI). PPO requires both a policy and value network. The policy network accepts the state of the epidemiological model as input (details in Sect. 4) and the output of the network contains 1 unit, which is passed through a sigmoid activation function. This output thus represents the probability of keeping the schools open in the district. Every hidden layer in the PPO network uses the hyperbolic tangent activation function. The value network has the same architecture as the policy network, with the exception that the output is not passed through an activation function. We will refer to this setting throughout this work as the single-district PPO agent.

PPO's hyper-parameters are tuned (hyper-parameter values in SI) on a single-district (i.e., the Greenwich district) learning environment with $R_0 = 1.8$. To this end, we performed a hyper-parameter sweep using Latin hypercube sampling ($n = 1000$) [27].

We conduct two kinds of experiments: in the context of a single district and in the context of the Great Britain model that combines all 379 districts. We consider two values for the reproductive number, i.e., $R_0 = \{1.8, 2.4\}$, to investigate the effect of distinct reproductive numbers. $R_0 = 1.8$ represents an epidemic with moderate transmission potential [6] and $R_0 = 2.4$ represents an epidemic with high transmission potential [19]. We investigate the effect of different school closure budgets, i.e., $\mathscr{b} = \{2, 4, 6\}$ weeks. The epidemic is simulated for a fixed number of weeks, chosen beforehand, to ensure that there is enough time for the epidemic to fade out after its peak. Following [2], we use a latent period of one day ($\zeta = \frac{1}{1}$) and an infectious period of 1.8 days ($\gamma = \frac{1}{1.8}$). Given the contact matrix $M_{ij}(t)$, the latency rate ζ, the recovery rate γ, we can compute β for an R_0, as specified in SI.

The source code of the model and experiments is available on GitHub[4].

5 PPO Versus Ground Truth

We now establish a ground truth for different population compositions, i.e., the proportion of the different age groups in a population. We will use this ground truth to empirically validate that PPO converges to the appropriate policy.

To establish this ground truth[5], first consider that when we deal with a single district, we can approach the 'average' behaviour of the model by removing the stochastic terms from the differential equations. Hence, for a particular parameter configuration (i.e., district, R_0, γ, ζ), the model will always produce the same epidemic curve. This means that the state space of this deterministic epidemic model directly corresponds to the time of the epidemic. For an epidemic that spans w weeks, we can formulate a school closure policy as a binary number with w digits, where the digit at position i signifies whether schools should be open (1) or closed (0) during the i-th week. For short-lived epidemics, such as influenza epidemics, we can enumerate these policies and evaluate them in our model (i.e., using exhaustive policy search). Note that, in the epidemiological models that we consider, the epidemic spans no more than 25 weeks, and thus exhaustive search is possible.

In this analysis, we consider different values for the basic reproductive number R_0 and the population composition of the district, both parameters that induce a significant change of the epidemic model's dynamics. To this end, we select 10 districts that are representative of the population heterogeneity in Great Britain: one district that is representative for the average of this census distribution and a set of nine districts that is representative for the diversity in this census distribution. Details on this selection procedure can be found in SI.

To evaluate PPO with respect to the ground truth, we repeat the experiment for which we established a ground truth (i.e., $R_0 \in \{1.8, 2.4\}$, 10 districts and $\mathscr{b} \in \{2, 4, 6\}$) and learn a policy using PPO, in the stochastic epidemic model. For each experimental setting (i.e., the combination of a district, an R_0 value, and a school closure budget $\mathscr{b}$), we run PPO 5 times (5 trials), to asses the variance of the learning performance. Each PPO trial is run for 10^4 episodes of 43 weeks. We show the learning curves, i.e., total reward at the end of the episode, for the district that is representative for the average of the census distribution (i.e., the Barnsley district in England), with $R_0 = 2.4$ in Fig. 3, for the other settings we report similar learning curves in SI.

To compare each of the learned policies to its ground truth (one for each district), we take the learned policy and apply it 1000 times in the stochastic

[4] https://github.com/plibin-vub/epi-rl.

[5] Note that this is a proxy to the ground truth, as we use a deterministic version of the model.

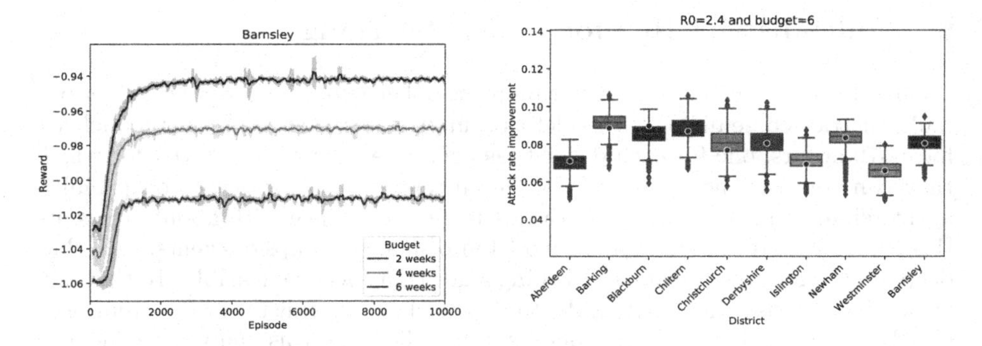

Fig. 3. [Left panel] PPO learning curves for the Barnsley district with $R_0 = 2.4$ for the three school closure budgets $\mathscr{b} = \{2, 4, 6\}$. [Right panel] We compare the PPO results to the ground truth for $R_0 = 2.4$ and $\mathscr{b} = 6$. Per district, we show a box plot that denotes the outcome distribution that was obtained by simulating the policy learned by PPO 1000 times. On top of this box plot, we show the ground truth, as a blue dot. (Color figure online)

model, which results in a distribution over model outcomes (i.e., attack rate improvement: the difference between the attack rate produced by the model and the baseline when no schools are closed). We then compare this distribution to the attack rate improvement that was recorded for the ground truth. We show these results, for the setting with a school closure budget of 6 weeks and $R_0 = 2.4$, in Fig. 3, and for the other settings in SI. These results show that PPO learns a policy that matches the ground truth for all districts and combinations of R_0 and $\mathscr{b}$.

Note that for these experiments, we use the same hyper-parameters for PPO that were introduced in Sect. 4. This demonstrates that, for different values of R_0 and for different census compositions (which induce a significant change in dynamics in the epidemic model) these hyper-parameters work well. This indicates that these hyper-parameters are adequate to be used for different variations of the model.

In this section, we compare to the ground truth (that has been found through an exhaustive policy search) to a policy learned by PPO, a deep reinforcement learning algorithm. This allows us to empirically validate that PPO converges to the optimal policy. This experimental validation is important, as it demonstrates the potential of deep reinforcement learning algorithms to learn policies in the context of complex epidemiological models. This indicates that it is possible to learn in even more complex stochastic models with large action spaces, for which it is impossible to compute the ground truth. In Sect. 6, we investigate such a setting, where we aim to learn a joint policy for a set of districts.

6 Multi-district Reinforcement Learning

To investigate the collaborative nature of school closure policies, we consider our multi-district epidemiological model as a multi-agent system. In our model, we have 379 agents, one for each district, as agents represent the district for which they can control school closure. As the current state-of-the-art of deep multi-agent reinforcement learning algorithms is limited to deal with about 10 agents [11], we thus need to partition our model into smaller groups of agents, such that deep multi-agent reinforcement learning algorithms become feasible. To this end, we analyse the mobility matrix $\mathcal{M}$ to detect clusters of districts that represent closely connected communities (details in SI). Through this analysis we identify a community with 11 districts, to which we will refer as the Cornwall-Devon community, as it is comprised of the Cornwall and Devon regions.

We now examine whether there is an advantage to consider the collaboration between districts when designing school closure policies. We conduct an experiment in our epidemiological model with 379 districts, and attempt to learn a joint policy to control the districts in the Cornwall-Devon community. To this end, we assign an agent to each of the 11 districts of the Cornwall-Devon community, and use a reinforcement learning approach to learn a joint policy. We compare this joint policy to a non-collaborative policy (i.e., aggregated independent learners).

We refer to the state space of one patch p as $\mathcal{S}_p$, as detailed in Sect. 4. The state space $\mathcal{S}$ of the MDP corresponds to the aggregation of the state space of the set of patches $\mathcal{P}^c$ that we attempt to control. In this experiment, $\mathcal{P}^c$ corresponds to the 11 districts in the Cornwall-Devon community.

In order to learn a joint policy, we need to consider an action space that combines the actions for each district $p \in \mathcal{P}^c$ that we attempt to control. This results in a joint action space with a size that is exponential with respect to the number of agents. To approach this problem, we use a PPO super-agent that controls multiple districts simultaneously, to learn a joint policy. To this end, we use a custom policy network that gets as input the combined model state of each district $p \in \mathcal{P}^c$ (Eq. 6), and as a result, the input layer has $17 \cdot |\mathcal{P}^c|$ input units. In contrast to the single-district PPO, that was introduced in Sect. 4, the output layer of the policy network of this agent has a unit for each district that we attempt to control. Again, each output unit is passed through a sigmoid activation function, and hence corresponds to the probability of closing the schools in the associated district. Similar to the single-district PPO, each hidden layer uses the hyperbolic tangent activation function. The value network has the same architecture for the input layers and hidden layers, but only has a single output unit that represents the value for the given state. We will refer to this agent as *multi-district PPO*.

We conduct experiments for $R_0 = 1.8$ (i.e., moderate transmission potential) and $R_0 = 2.4$ (i.e., high transmission potential), and we consider a school closure budget of 6 weeks, i.e., $\mathscr{b} = 6$. We run multi-district PPO 5 times, to assess the variance of the learning signal, for 10^5 episodes of 43 weeks, and we show the learning curves in Fig. 4. These learning curves demonstrate a stable and steady

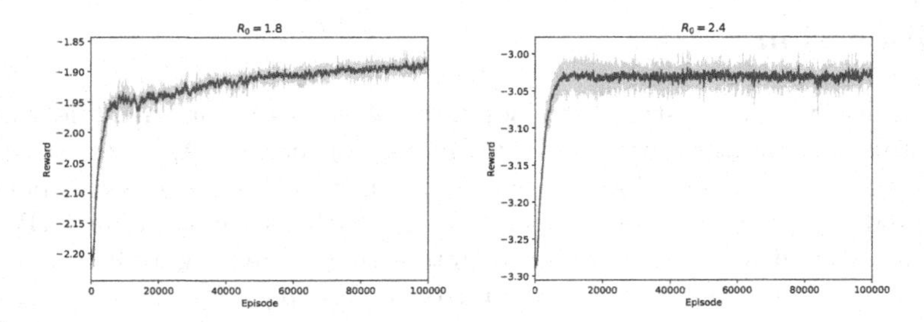

Fig. 4. We show the reward curves for multi-district PPO for $R_0 = 1.8$ (left panel) and $R_0 = 2.4$ (right panel). The reward curves are visualized using a rolling window of 100 steps. The shaded area shows the standard deviation of the reward signal, over 5 multi-district PPO runs.

learning process. For $R_0 = 1.8$ the reward curve is still increasing, while for $R_0 = 2.4$ the reward curve indicates that the learning process has converged.

To investigate whether these *joint policies* provide a collaborative advantage, we compare it to the aggregation of single district policies, to which we will refer as the *aggregated policy*. To construct this aggregated policy, we learn a distinct school closure policy for each of the 11 districts in the Cornwall-Devon community, using PPO, following the same procedure as in Sect. 5. To evaluate this aggregated policy, we execute the distinct policies simultaneously. For the districts that are not controlled (both for the joint and aggregated policy) we keep the schools open for all time steps. For both $R_0 = 1.8$ and $R_0 = 2.4$, respectively, we simulate the joint and the aggregated policy 1000 times, and we show the attack rate improvement distribution in Fig. 5. These results corroborate that there is a collaborative advantage when devising school closures policies, for both $R_0 = 1.8$ and $R_0 = 2.4$. However, the improvement is most significant for $R_0 = 1.8$. We conjecture that this difference is due to the fact that there is less flexibility when the transmission potential of the epidemic is higher, since there is less time to act. Although, we observe an improvement when a joint policy is learned, it remains challenging to interpret deep multi-agent policies, and we discuss in Sect. 7 possible directions for future work with respect to multi-agent reinforcement learning.

In this analysis, where we have a limited number of actions per agent, the use of multi-district PPO proved to be successful. However, the use of more advanced multi-agent reinforcement learning methods is warranted when a more complex action space is considered or a larger number of districts needs to be controlled. For this reason, we also investigated the recently introduced QMIX [25] algorithm, but the resulting learning curve proved to be quite unstable (shown in SI).

We conducted our experiments in the setting of school closures, and our findings are of direct relevance with respect to the mitigation of pandemic influenza.

7 Discussion

In this work, we demonstrate the potential of deep reinforcement learning in the context of complex stochastic epidemiological models. As few assumptions are made on the epidemiological model, our new technique has the potential to be used for other epidemiological settings, such as the ongoing COVID-19 pandemic. For future work, it would be interesting to investigate how well these algorithms scale to even larger state and/or action spaces.

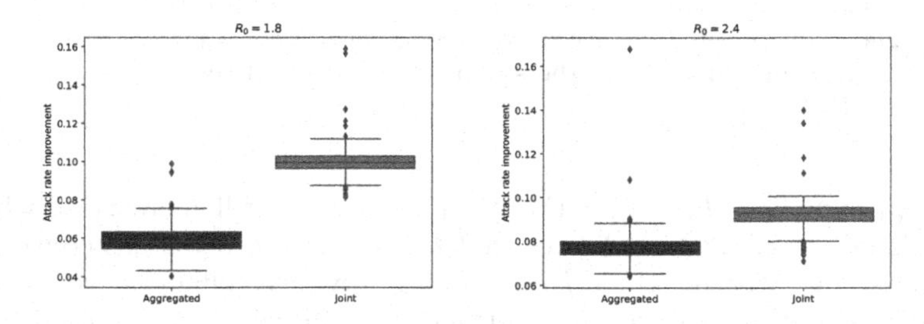

Fig. 5. We compare the simulation results of the aggregated policy (blue) and the joint policy (orange) for $R_0 = 1.8$ (left panel) and $R_0 = 2.4$ (right panel). For both distributions (i.e., aggregated versus joint), we show a box plot that denotes the outcome distribution that was obtained by simulating the respective policy 1000 times. (Color figure online)

Another important concern is to scale these reinforcement learning methods to individual-based epidemiological models, as such models can be easily configured to approach a variety of research scenarios, i.a., vaccine allocation, telecommuting, antiviral drug allocation. However, the computational burden that is associated with individual-based models complicates the use of reinforcement learning methods [35]. To this end, it would be interesting to devise methods to automatically learn a surrogate model from the individual-based model, such that the reinforcement learning agent can learn in this computationally leaner surrogate model.

While we show that deep reinforcement learning algorithms can be used to learn optimal mitigation strategies, the interpretation of such policies remains challenging [10]. This is especially the case for the multi-district setting we considered, where state and time do not match, and the infection onset of the patches is highly stochastic. To this end, further research into explainable reinforcement learning, both in a single-agent and multi-agent setting, is warranted. In this regard, we believe it to be interesting future work to compare the learnt policies to baselines policies, such as reactive school closure policies [12].

Acknowledgements. Pieter Libin and Timothy Verstraeten were supported by a PhD grant of the FWO (Fonds Wetenschappelijk Onderzoek - Vlaanderen). This research acknowledges funding from the Flemish Government (AI Research Program) and from the EpiPose project (H2020/101003688). We thank the anonymous reviewers for their insightful comments.

References

1. Allen, E.J., Allen, L.J., Arciniega, A., Greenwood, P.E.: Construction of equivalent stochastic differential equation models. Stochas. Anal. Appl. **26**(2), 274–297 (2008)
2. Baguelin, M., Van Hoek, A.J., Jit, M., Flasche, S., White, P.J., Edmunds, W.J.: Vaccination against pandemic influenza a/h1n1v in England: a real-time economic evaluation. Vaccine **28**(12), 2370–2384 (2010)
3. Cinlar, E.: Introduction to Stochastic Processes. Courier Corporation, North Chelmsford (2013)
4. Eames, K.T., Tilston, N.L., Brooks-Pollock, E., Edmunds, W.J.: Measured dynamic social contact patterns explain the spread of h1n1v influenza. PLoS Comput. Biol. **8**(3), e1002425 (2012)
5. Eggo, R.M., Cauchemez, S., Ferguson, N.M.: Spatial dynamics of the 1918 influenza pandemic in England, wales and the united states. J. R. Soc. Interface **8**(55), 233–243 (2010)
6. Ferguson, N.M., Cummings, D.A., Fraser, C., Cajka, J.C., Cooley, P.C., Burke, D.S.: Strategies for mitigating an influenza pandemic. Nature **442**(7101), 448 (2006)
7. Fumanelli, L., Ajelli, M., Manfredi, P., Vespignani, A., Merler, S.: Inferring the structure of social contacts from demographic data in the analysis of infectious diseases spread. PLoS Comput. Biol. **8**(9), e1002673 (2012)
8. Germann, T.C., et al.: School dismissal as a pandemic influenza response: when, where and for how long? Epidemics **28**, 100348 (2019)
9. Gog, J.R., et al.: Spatial transmission of 2009 pandemic influenza in the US. PLoS Comput. Biol. **10**(6), e1003635 (2014)
10. Gunning, D., Aha, D.W.: Darpa's explainable artificial intelligence program. AI Mag. **40**(2), 44–58 (2019)
11. Hernandez-Leal, P., Kartal, B., Taylor, M.E.: A survey and critique of multiagent deep reinforcement learning. Auton. Agent. Multi Agent Syst. **33**(6), 750–797 (2019). https://doi.org/10.1007/s10458-019-09421-1
12. House, T., et al.: Modelling the impact of local reactive school closures on critical care provision during an influenza pandemic. Proc. Roy. Soc. B **278**(1719), 2753–2760 (2011)
13. King, A.A., Domenech de Cellès, M., Magpantay, F.M., Rohani, P.: Avoidable errors in the modelling of outbreaks of emerging pathogens, with special reference to Ebola. Proc. Roy. Soc. B **282**(1806), 20150347 (2015)
14. Kissler, S.M., et al.: Geographic transmission hubs of the 2009 influenza pandemic in the United States. Epidemics **26**, 86–94 (2019)
15. Klepac, P., Kissler, S., Gog, J.: Contagion! the BBC four pandemic-the model behind the documentary. Epidemics **24**, 49–59 (2018)
16. Kubiak, R.J., McLean, A.R.: Why was the 2009 influenza pandemic in England so small? PLoS ONE **7**(2), e30223 (2012)
17. Libin, P., Verstraeten, T., Roijers, D.M., Wang, W., Theys, K., Nowe, A.: Bayesian anytime m-top exploration. In: 2019 IEEE 31st ICTAI, pp. 1422–1428. IEEE (2019)

18. Libin, P.J.K., et al.: Bayesian best-arm identification for selecting influenza mitigation strategies. In: Brefeld, U., et al. (eds.) ECML PKDD 2018. LNCS (LNAI), vol. 11053, pp. 456–471. Springer, Cham (2019). https://doi.org/10.1007/978-3-030-10997-4_28
19. Longini, I.M., et al.: Containing pandemic influenza at the source. Science **309**(5737), 1083–1087 (2005)
20. Markel, H., et al.: Nonpharmaceutical interventions implemented by US cities during the 1918–1919 influenza pandemic. JAMA **298**(6), 644–654 (2007)
21. Miller, E., Hoschler, K., Hardelid, P., Stanford, E., Andrews, N., Zambon, M.: Incidence of 2009 pandemic influenza a h1n1 infection in England: a cross-sectional serological study. Lancet **375**(9720), 1100–1108 (2010)
22. Mnih, V., et al.: Human-level control through deep reinforcement learning. Nature **518**(7540), 529 (2015)
23. Paules, C., Subbarao, K.: Influenza. The Lancet **390**, 697–708 (2017)
24. Probert, W.J., et al.: Context matters: using reinforcement learning to develop human-readable, state-dependent outbreak response policies. Philos. Trans. Roy. Soc. B **374**(1776), 20180277 (2019)
25. Rashid, T., Samvelyan, M., Schroeder, C., Farquhar, G., Foerster, J., Whiteson, S.: QMIX: monotonic value function factorisation for deep multi-agent reinforcement learning. In: ICML, vol. 80, pp. 4295–4304, 10–15 July 2018
26. Schulman, J., Wolski, F., Dhariwal, P., Radford, A., Klimov, O.: Proximal policy optimization algorithms. arXiv preprint arXiv:1707.06347 (2017)
27. Stein, M.: Large sample properties of simulations using Latin hypercube sampling. Technometrics **29**(2), 143–151 (1987)
28. Towers, S., Feng, Z.: Social contact patterns and control strategies for influenza in the elderly. Math. Biosci. **240**(2), 241–249 (2012)
29. Wang, L., Wu, J.T.: Characterizing the dynamics underlying global spread of epidemics. Nat. Commun. **9**(1), 218 (2018)
30. Webby, R.J., Webster, R.G.: Are we ready for pandemic influenza? Science **302**(5650), 1519–1522 (2003)
31. Wijayanto, A.W., Murata, T.: Effective and scalable methods for graph protection strategies against epidemics on dynamic networks. Appl. Netw. Sci. **4**(1), 1–31 (2019). https://doi.org/10.1007/s41109-019-0122-7
32. Yaesoubi, R., Cohen, T.: Dynamic health policies for controlling the spread of emerging infections: influenza as an example. PLoS ONE **6**(9), e24043 (2011)
33. Yaesoubi, R., Cohen, T.: Identifying dynamic tuberculosis case-finding policies for HIV/TB coepidemics. Proc. Natl. Acad. Sci. **110**(23), 9457–9462 (2013)
34. Yaesoubi, R., Cohen, T.: Identifying cost-effective dynamic policies to control epidemics. Stat. Med. **35**(28), 5189–5209 (2016)
35. Yu, Y.: Towards sample efficient reinforcement learning. In: IJCAI, pp. 5739–5743 (2018)
36. Zhu, N., et al.: A novel coronavirus from patients with pneumonia in China, 2019. New England Journal of Medicine (2020)

GLUECK: Growth Pattern Learning for Unsupervised Extraction of Cancer Kinetics

Cristian Axenie[1,2](✉) and Daria Kurz[3]

[1] Audi Konfuzius-Institut Ingolstadt Lab, Esplanade 10, 85049 Ingolstadt, Germany
cristian.axenie@audi-konfuzius-institut-ingolstadt.de
[2] Technische Hochschule Ingolstadt, Esplanade 10, 85049 Ingolstadt, Germany
[3] Interdisziplinäres Brustzentrum, Helios Klinikum München West, Steinerweg 5, 81241 Munich, Germany
daria.kurz@helios-gesundheit.de

Abstract. Neoplastic processes are described by complex and heterogeneous dynamics. The interaction of neoplastic cells with their environment describes tumor growth and is critical for the initiation of cancer invasion. Despite the large spectrum of tumor growth models, there is no clear guidance on how to choose the most appropriate model for a particular cancer and how this will impact its subsequent use in therapy planning. Such models need parametrization that is dependent on tumor biology and hardly generalize to other tumor types and their variability. Moreover, the datasets are small in size due to the limited or expensive measurement methods. Alleviating the limitations that incomplete biological descriptions, the diversity of tumor types, and the small size of the data bring to mechanistic models, we introduce Growth pattern Learning for Unsupervised Extraction of Cancer Kinetics (GLUECK) a novel, data-driven model based on a neural network capable of unsupervised learning of cancer growth curves. Employing mechanisms of competition, cooperation, and correlation in neural networks, GLUECK learns the temporal evolution of the input data along with the underlying distribution of the input space. We demonstrate the superior accuracy of GLUECK, against four typically used tumor growth models, in extracting growth curves from a four clinical tumor datasets. Our experiments show that, without any modification, GLUECK can learn the underlying growth curves being versatile between and within tumor types.

Keywords: Unsupervised learning · Neural networks · Cancer kinetics · Tumor growth

1 Background

Ideally, if we would have understood all tumor biology from neoplastic cells to metastatic cancer, we could build a model that would anticipate the development curve of a tumor into what's to come given its present state. Of course,

Y. Dong et al. (Eds.): ECML PKDD 2020, LNAI 12461, pp. 171–186, 2021.
https://doi.org/10.1007/978-3-030-67670-4_11

we cannot do that because our knowledge is sadly incomplete. Contributing to this is the fact that repeated measures of growing tumors - required for any detailed study of cancer kinetics - are very difficult to obtain. Tumor growth can typically be described in three frameworks: 1) in vitro [14]; 2) exploratory in vivo frameworks (animal models with implanted tumors, [6,26]); or 3) clinical tumor growth using non intrusive imaging strategies, [9]. But, even in the best scenarios, obtaining accurate size measures of the irregular, three-dimensional masses without influencing the physiology of either tumor or host remains problematic. Nevertheless, such an assessment is crucial as it can subsequently guide cancer therapy. Mostly due to such limitations, theorists sought "growth laws" that seek to represent general tumor kinetics without reference to any particular tumor type.

1.1 Tumor Growth and Its Implications

Extensive biological studies have been devoted to tumor growth kinetics, addressing both mass and volume evolution. Such studies postulated that, principles of tumor growth might result from general growth laws, typically amenable to differential equations [12]. These models have a twofold objective: 1) testing growth hypotheses or theories by assessing their descriptive power against experimental data and 2) estimating the prior or future course of tumor progression. Such an assessment can be used as a personalized clinical prediction tool [3], or in order to determine the efficacy of a therapy [7]. With GLUECK we try to address the two objectives by exploiting the benefits of learning and generalization in neural networks.

1.2 Mechanistic Models of Tumor Growth

Different tumor growth patterns have been identified experimentally and clinically and modelled over the years. A large number of ordinary differential equations (ODE) tumor growth models [12] have been proposed and used to make predictions about the efficacy of cancer treatments. The first theory that formulated a consistent growth law implied that tumors grow exponentially, a concept that obviously stems from the "uncontrolled proliferation" view of cancer. However, this hypothesis is wrong. Assuming that a tumor increases exponentially, then a one-dimensional estimate of its size (e.g. diameter) will also increase. In other words, both volume and "diameter" log plots over time are linear, if and only if the tumor is exponentially increasing. But, detailed studies of tumor growth in animal models [23] reveal that the tumor "diameter" appears to be linear, not its logarithm, suggesting a decreasing volumetric growth rate over time, Fig. 1. There probably is no single answer to the question, which is the best model to fit cancer data [17]. In our analysis we chose four of the most representative and typically used scalar growth models, namely Logistic, von Bertalanffy, Gompertz and Holling, described in Table 1 and depicted in Fig. 2. Despite their ubiquitous use, in either original form or embedded in more complex models [18], the aforementioned scalar tumor growth models are confined due to: a)

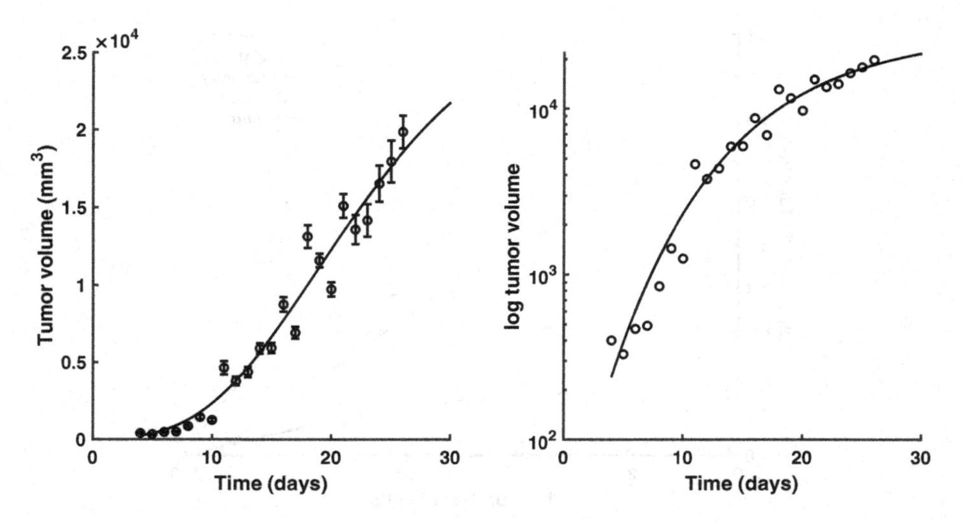

Fig. 1. Growth kinetics of Fortner Plasmacytoma 1 tumors. Points represent mean volume of subcutaneous tumor implants in mice, error bars represent ±1 standard error of the mean at each point. Data from [23]

Table 1. Overview of tumor growth models in our study. Parameters: N - cell population size (or volume/mass thorough conversion [15]), α - growth rate, β - cell death rate, λ - nutrient limited proliferation rate, k - carrying capacity of cells.

Model	Equation
Logistic [25]	$\frac{dN}{dt} = \alpha N - \beta N^2$
Bertalanffy [27]	$\frac{dN}{dt} = \alpha N^{\lambda} - \beta N$
Gompertz [13]	$\frac{dN}{dt} = N(\beta - \alpha \ln N)$
Holling [2]	$\frac{dN}{dt} = \frac{\alpha N}{k+N} - \beta N$

the requirement of a precise biological description (i.e. values for α, β, λ and k correspond to biophysical processes); b) incapacity to describe the diversity of tumor types (i.e. each is priming on a type of tumor), and c) the small amount and irregular sampling of the data (e.g. 15 measurements with at days 6, 10, 13, 17, 19, 21, 24, 28, 31, 34, 38, 41, 45, 48 for breast cancer growth in [26]).

Although such competing canonical tumor growth models are commonly used, how to decide which of the models to use for which tumor types is still an open question. In an attempt to answer this questions, the work in [22] built a broad catalogue of growth laws and recommendations on the best fit and parameter ranges. Now, assuming that a set of models was selected for a treatment, the next question is how to quantify their differences in predictions, for instance, in the presence and absence of chemotherapy? Addressing this final question, the study in [19] showed that tumor growth model selection can lead to as much

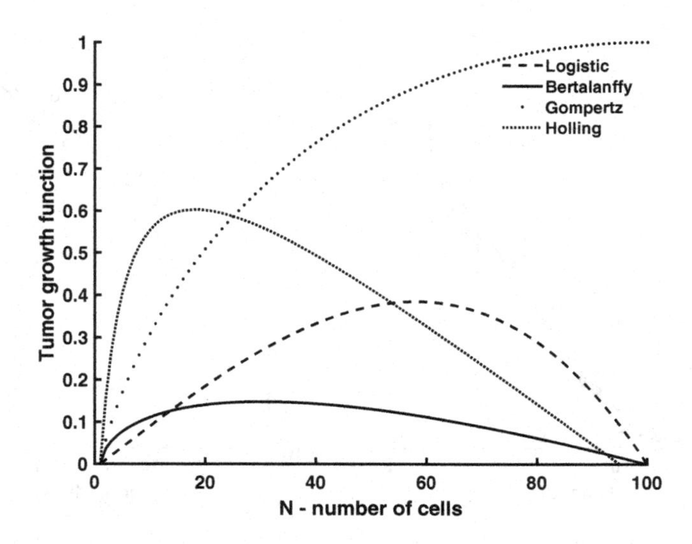

Fig. 2. Comparison of four tumor growth rate functions.

as twelve-fold change in estimated outcomes, and that the model that best fits experimental data might not be the best model for future growth. To cope with these inherent limitations, predictive models of tumor growth were developed.

1.3 Predictive Models of Tumor Growth

Several researchers have tried to find the "right" ODE growth law by fitting different models onto a limited number of experimental tumor growth data sets [22]. All in all, the findings are somewhat inconclusive, with findings indicating that the choice of growth model is at least partially dependent on the type of tumor [5]. Addressing this challenge multiple studies "augmented" traditional growth models with predictive capabilities wrapping their parameter estimation into probabilistic or machine learning frameworks. We comparatively review three most relevant approaches.

Building upon the basic growth models (i.e. adapted Gompertz model on tumor phases) and vascular endothelial growth factor (VEGF), the work in [11] trained and validated a model using published in vivo measurements of xenograft tumor volume. The model employed Nonlinear Least Squares (NLS) optimization together with a global sensitivity analysis to determine which of the four tumor growth kinetic parameters impacts the predicted tumor volume most significantly. The major disadvantage of the method is that NLS requires iterative optimization to compute the parameter estimates. Initial conditions must be reasonably close to the unknown parameter estimates or the optimization procedure may not converge. Moreover, NLS has strong sensitivity to outliers. GLUECK alleviates the need of expensive optimization through a learning mechanism on multiple timescales, that is guaranteed to converge [4]. Instead of selecting best model parameters that describe a certain dataset, GLUECK uses a data-driven learning and adaptation to the data distribution (see Methods) implicitly achieving this.

Using growth simulations on functional Magnetic Resonance Imaging (fMRI) and Partial Differential Equations (PDE)-constrained optimization, the work in [1] used discrete adjoint and tangent linear models to identify the set of parameters that minimize an objective functional of tumor cell density and tissue displacement at a specific observation time point. Conversely, GLUECK does not depend on expensive imaging data and a precise PDE modelling of the problem, rather it exploits the data to extract the underlying growth pattern through simple and efficient operations in neural networks (see Methods).

Finally, the work in [24] performed a quantitative analysis of tumor growth kinetics using nonlinear mixed-effects modeling on traditional mechanistic models. Using Bayesian inference, they inferred tumor volume from few (caliper and fluorescence) measurements. The authors proved the superior predictive power of the Gompertz model when combined with Bayesian estimation. Such work motivated the need for GLUECK. This is mainly because GLUECK: 1) departs completely from the limitations of a mechanistic model of tumor growth (e.g. Gompertz) and 2) it alleviates the difficulty of choosing the priors, the computational cost at scale, and the incorporation of the posterior in the actual prediction - that typically affect Bayesian methods.

1.4 Peculiarities of Tumor Growth Data

Before we introduce GLUECK we strengthen our motivation for this work by iterating through the peculiarities of tumor growth data.
Tumor growth data:

- is **small**, only a few data points with, typically, days level granularity, [21].
- is **unevenly sampled**, with irregular spacing among measurements [26].
- data has **high variability** between and within tumor types (e.g. breast versus lung cancer, [5] and each type of treatment (i.e. administered drug) modifies the growth curve [11].
- is **heterogeneous** and sometimes **expensive to obtain** (e.g. volume assessed from bio-markers, fMRI [1], fluorescence imaging [20], flow cytometry, or calipers [6]).
- poses **challenges in selecting the best model** [1,11,19].
- **determines cancer treatment planning** [22].

2 Materials and Methods

In the current section we describe the underlying functionality of GLUECK along with the datasets used in our experiments.

2.1 Introducing GLUECK

GLUECK is an unsupervised learning system based on Self-Organizing Maps (SOM) [16] and Hebbian Learning (HL) [8] as main ingredients for extracting underlying relations among correlated timeseries. In order to introduce GLUECK, we provide a simple example in Fig. 3. Here, we consider data from a cubic growth law (3rd powerlaw) describing the impact of sequential dose density over a 150 weeks horizon in adjuvant chemotherapy of breast cancer [10]. In this simple example, the two input timeseries (i.e. the cancer cell number and the irregular measurement index over the weeks) follow a cubic dependency (cmp. Fig. 3a). When presented the data, GLUECK has no prior information about timeseries' distributions and their generating processes, but learns the underlying (i.e. hidden) relation directly from the input data in an unsupervised manner.

Core Model. The input SOMs (i.e. 1D lattice networks with N neurons) produce a discretized representation of the input space. They are responsible to extract the distribution of the incoming timeseries data (i.e. 1D tumor size/volume), depicted in Fig. 3a, and encode timeseries samples in a distributed activity pattern, as shown in Fig. 3b. This activity pattern is generated such that the closest preferred value of a neuron to the input sample will be strongly activated and will decay, proportional with distance, for neighbouring units. This accounts for the basic quantization capability of SOM and augments it with a dimension corresponding to latent representation resource allocation (i.e. number of neurons allocated to represent the input space). The SOM specialises to represent a certain (preferred) value in the input space and learns its sensitivity, by updating its tuning curves shape. Given an input sample $s^p(k)$ from one timeseries at time step k, the network follows the processing stages depicted in Fig. 4. For each i-th neuron in the p-th input SOM, with preferred value $w^p_{in,i}$ and tuning curve size $\xi^p_i(k)$, the elicited neural activation is given by

$$a^p_i(k) = \frac{1}{\sqrt{2\pi}\xi^p_i(k)} e^{\frac{-(s^p(k) - w^p_{in,i}(k))^2}{2\xi^p_i(k)^2}}. \tag{1}$$

The winning neuron of the p-th population, $b^p(k)$, is the one which elicits the highest activation given the timeseries sample at time k

$$b^p(k) = \underset{i}{\text{argmax}}\ a^p_i(k). \tag{2}$$

The competition for highest activation in the SOM is followed by cooperation in representing the input space. Hence, given the winning neuron, $b^p(k)$, the cooperation kernel,

$$h^p_{b,i}(k) = e^{\frac{-||r_i - r_b||^2}{2\sigma(k)^2}}. \tag{3}$$

allows neighbouring neurons (i.e. found at position r_i in the network) to precisely represent the input sample given their location in the neighbourhood $\sigma(k)$ of the

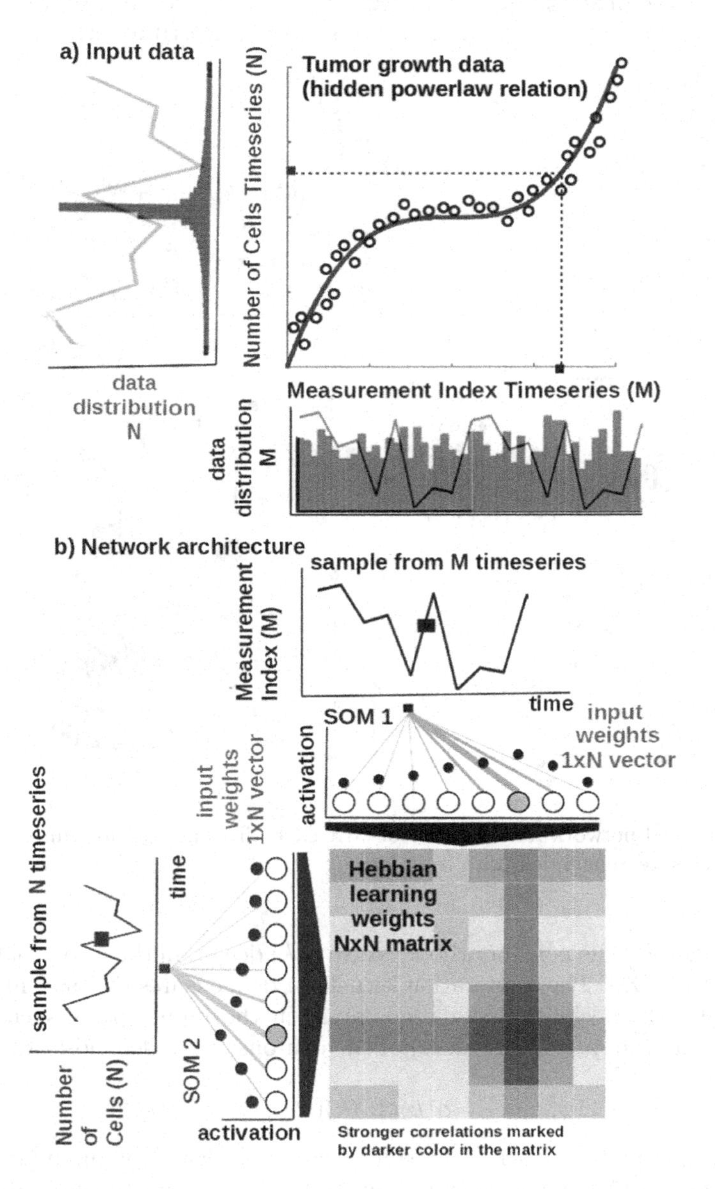

Fig. 3. Basic functionality of GLUECK. a) Tumor growth data resembling a non-linear relation and its distribution - relation is hidden in the timeseries (i.e. number of cells vs. measurement index). Data from [10]. b) Basic architecture of GLUECK: 1D SOM networks with N neurons encoding the timeseries (i.e. number of cells vs. measurement index), and a NxN Hebbian connection matrix coupling the two 1D SOMs that will eventually encode the relation.

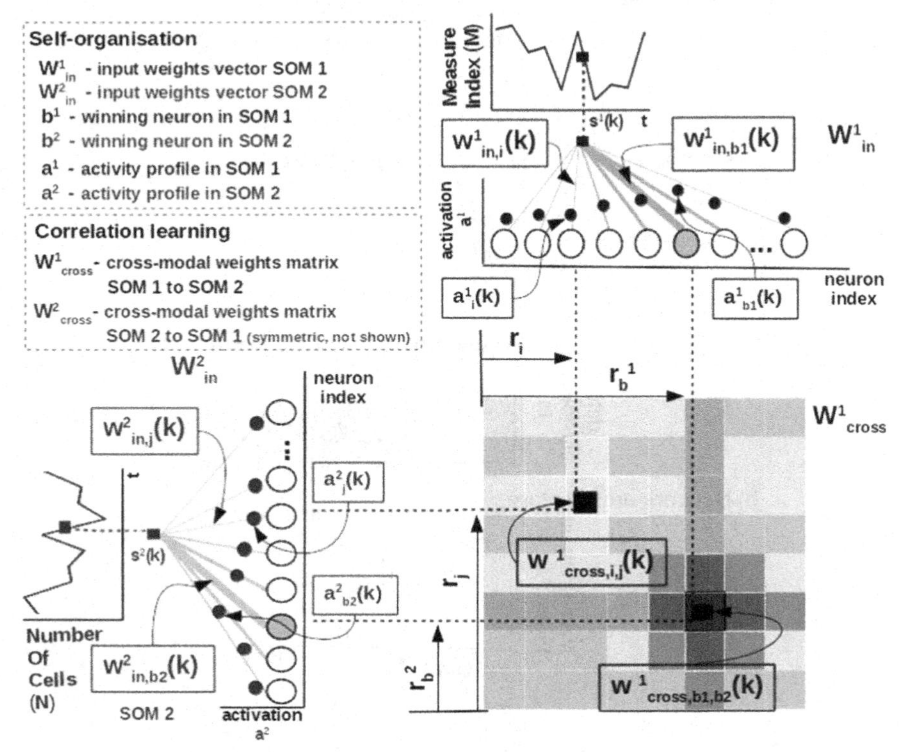

Fig. 4. Detailed network functionality of GLUECK, instantiated for tumor number of cells vs. measurement index data from [10].

winning neuron. The neighbourhood width $\sigma(k)$ decays in time, to avoid twisting effects in the SOM. The cooperation kernel in Eq. 3, ensures that specific neurons in the network specialise on different areas in the input space, such that the input weights (i.e. preferred values) of the neurons are pulled closer to the input sample,

$$\Delta w_{in,i}^{p}(k) = \alpha(k) h_{b,i}^{p}(k)(s^{p}(k) - w_{in,i}^{p}(k)). \tag{4}$$

This corresponds to updating the tuning curves width ξ_i^p as modulated by the spatial location of the neuron in the network, the distance to the input sample, the cooperation kernel size, and a decaying learning rate $\alpha(k)$,

$$\Delta \xi_i^p(k) = \alpha(k) h_{b,i}^{p}(k)((s^{p}(k) - w_{in,i}^{p}(k))^2 - \xi_i^p(k)^2). \tag{5}$$

As an illustration of the process, let's consider learned tuning curves shapes for 5 neurons in the input SOMs (i.e. neurons 1, 6, 13, 40, 45) encoding the breast cancer cubic tumor growth law, depicted in Fig. 5. We observe that higher input

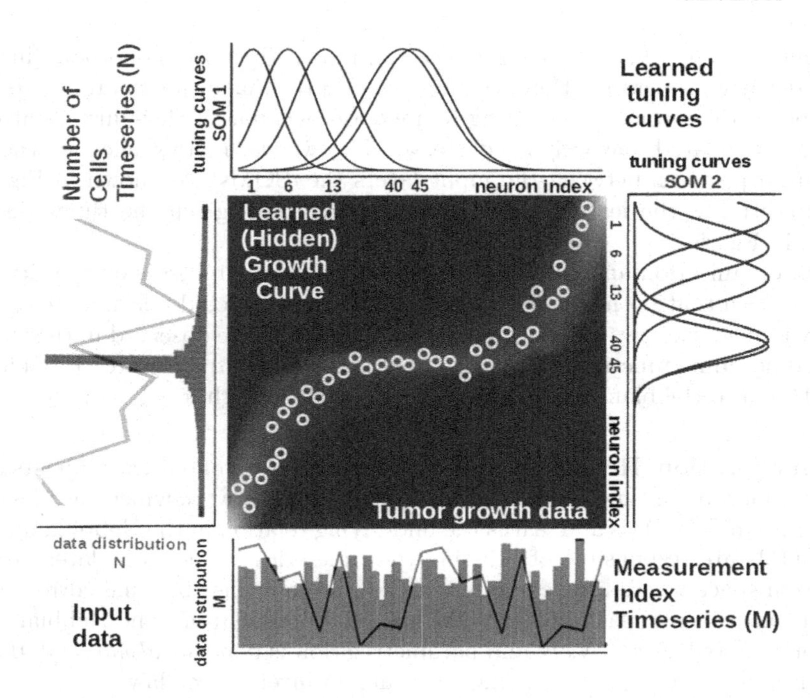

Fig. 5. Extracted timeseries relation describing the growth law and data statistics for the data in Fig. 3a depicting a cubic breast cancer tumor growth law among number of cells and irregular measurement over 150 weeks from [10]. Timeseries overlay on the data distribution and corresponding model encoding tuning curves shapes.

probability distributions are represented by dense and sharp tuning curves (e.g. neuron 1, 6, 13 in SOM1), whereas lower or uniform probability distributions are represented by more sparse and wide tuning curves (e.g. neuron 40, 45 in SOM1). Using this mechanism, the network optimally allocates resources (i.e. neurons). A higher amount of neurons to areas in the input space, which need a finer resolution; and a lower amount for more coarsely represented areas. Neurons in the two SOMs are then linked by a fully (all-to-all) connected matrix of synaptic connections, where the weights in the matrix are computed using Hebbian learning. The connections between uncorrelated (or weakly correlated) neurons in each population (i.e. w_{cross}) are suppressed (i.e. darker color) while correlated neurons connections are enhanced (i.e. brighter color), as depicted in Fig. 4. Formally, the connection weight $w^{p}_{cross,i,j}$ between neurons i, j in the different input SOMs are updated with a Hebbian learning rule as follows:

$$\Delta w^{p}_{cross,i,j}(k) = \eta(k)(a^{p}_{i}(k) - \overline{a}^{p}_{i}(k))(a^{q}_{j}(k) - \overline{a}^{q}_{j}(k)), \tag{6}$$

where

$$\overline{a}^{p}_{i}(k) = (1 - \beta(k))\overline{a}^{p}_{i}(k-1) + \beta(k)a^{p}_{i}(k), \tag{7}$$

is a "momentum" like exponential decay and $\eta(k)$, $\beta(k)$ are monotonic (inverse time) decaying functions. Hebbian learning ensures that when neurons fire synchronously their connection strengths increase, whereas if their firing patterns are anti-correlated the weights decrease. The weight matrix encodes the co-activation patterns between the input layers (i.e. SOMs), as shown in Fig. 3b, and, eventually, the learned growth law (i.e. relation) given the timeseries, as shown in Fig. 5.

Self-organisation and Hebbian correlation learning processes evolve simultaneously, such that both the representation and the extracted relation are continuously refined, as new samples are presented. This can be observed in the encoding and decoding functions where the input activations are projected though w_{in} (Eq. 1) to the Hebbian matrix and then decoded through w_{cross} (Eq. 8).

Parametrization, Read-Out and Validation. In all of our experiments data from tumor growth timeseries is fed to the GLUECK network which encodes each timeseries in the SOMs and learns the underlying relation in the Hebbian matrix. The SOMs are responsible of bringing the timeseries in the same latent representation space where they can interact (i.e. through their internal correlation). In our experiments, each of the SOM has $N = 100$ neurons, the Hebbian connection matrix has size NxN and parametrization is done as: $alpha = [0.01, 0.1]$ decaying, $\eta = 0.9$, $\sigma = \frac{N}{2}$ decaying following an inverse time law.

We use as decoding mechanism an optimisation method that recovers the real-world value given the self-calculated bounds of the input timeseries. The bounds are obtained as minimum and maximum of a cost function of the distance between the current preferred value of the winning neuron (i.e. the value in the input which is closest to the weight vector of the neuron in Euclidian distance) and the input sample at the SOM level. Depending on the position of the winning neuron in the SOM, the decoded/recovered value $y(t)$ from the SOM neurons weights is computed as:

$$y(t) = \begin{cases} w_{in,i}^p + d_i^p & \text{if } i \geq \frac{N}{2} \\ w_{in,i}^p - d_i^p & \text{if } i < \frac{N}{2} \end{cases}$$

where, $d_i^p = \sqrt{2\xi_i^p(k)^2 log(\sqrt{2\pi} a_i^p(k) \xi_i^p(k)^2)}$ for the most active neuron with index i in the SOM, a preferred value $w_{in,i}^p$ and $\xi_i^k(k)$ tuning curve size and $\Delta a_i^p(k) = w_{cross,i,j}^p(k) a_j^q(k)$. The activation $a_i^p(k)$ is computed by projecting one input through SOM q and subsequently through the Hebbian matrix to describe the paired activity (i.e. at the other SOM p, Eq. 8).

$$\Delta a_i^p(k) = w_{cross,i,j}^p(k) a_j^q(k) \tag{8}$$

where $w_{cross,i,j}^p(k) = rot90(w_{cross,i,j}^q(k))$ and $rot90$ is a clockwise rotation. The validation of the encoded timeseries in the SOMs was performed by recovering the probability density from the number and width of the tuning curves, whereas the validation of the learnt growth curve, stored in the Hebbian matrix, was validated by evaluating typical timeseries metrics, respectively.

2.2 Datasets

In our experiments we used publicly available tumor growth datasets (see Table 2), with real clinical tumor volume measurements, for breast cancer (datasets 1 and 2) and other cancers (e.g. lung, leukemia - datasets 3 and 4, respectively). This choice is to probe and demonstrate transfer capabilities of the models to tumor growth patterns induced by other cancer types.

Table 2. Description of the datasets used in the experiments.

Experimental dataset setup					
Dataset	Cancer type	Data type	Data points	Data freq.	Source
1	Breast (MDA-MB-231)	Fluorescence imaging	7	2x/week	[20]
2	Breast (MDA-MB-435)	Digital caliper	14	2x/week	[26]
3	Lung	Caliper	10	7x/week	[6]
4	Leukemia	Microscopy	23	7x/week	[23]

2.3 Procedures

In order to reproduce the experiments, the MATLAB ® code and copies of all the datasets are available on GitLab[1] Each of the four mechanistic tumor growth models (i.e. Logistic, Bertalanffy, Gompertz, Holling) and GLUECK were presented the tumor growth data in each of the four datasets. When a dataset contained multiple trials, a random one was chosen. Note that the validation of the model was performed by feeding the trained model with data from other trials/patients.

Mechanistic Models Setup. Each of the four tumor growth models was implemented as ordinary differential equation (ODE) and integrated over the dataset length. We used a solver based on a modified Rosenbrock formula of order 2 that evaluates the Jacobian during each step of the integration. To provide initial values and the best parameters (i.e. $\alpha, \beta, \lambda, k$) for each of the four models the Nelder-Mead simplex direct search (i.e. derivative-free minimum of unconstrained multi-variable functions) was used, with a termination tolerance of $10e^{-6}$ and upper bounded to 500 iterations. Finally, fitting was performed by minimizing the sum of squared residuals (SSR).

GLUECK Setup. For GLUECK the data was normalized before training and de-normalized for the evaluation. The system was comprised of two input SOMs, each with $N = 50$ neurons, encoding the volume data and the irregular sampling time sequence, respectively. Both input density learning and cross-modal learning cycles were bound to 100 epochs. The full parametrization details of GLUECK are given in Parametrization, read-out and validation subsection.

[1] https://gitlab.com/akii-microlab/ecml-2020-glueck-codebase.

3 Results

In the current section we describe the experimental results, discuss the findings and also evaluate an instantiation of GLUECK learning capabilities to predict surgical tumor size. All of the five models were evaluated through multiple metrics on each of the four datasets. Note that we consider the clinical time-frame for the evaluation (as specified in the dataset of origin) and compare on the prediction capabilities. This choice of different tumor types (i.e. breast, lung, leukemia) is to probe and demonstrate between and within tumor type prediction versatility. Assessing the distribution of the measurement error as a function of the caliper-measured volumes of subcutaneous tumors [5] suggested the following model for the standard deviation of the error σ_i at each time point i,

$$\sigma_i = \begin{cases} \sigma(y_m^i)^\alpha, & \text{if } y_m^i \geq y^i \\ \sigma(y^i)^\alpha, & \text{if } y_m^i < y^i \end{cases}$$

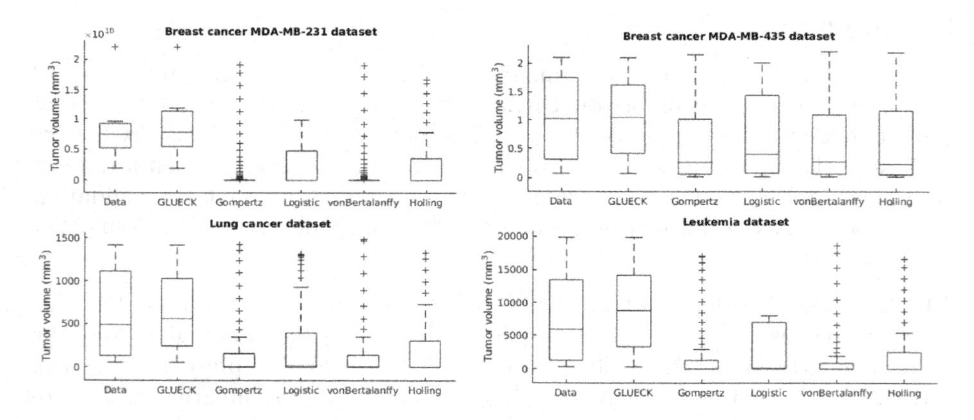

Fig. 6. Evaluation of the tumor growth models: summary statistics.

Table 3. Evaluation metrics for tumor growth models. We consider: N - number of measurements, σ - standard deviation of data, p - number of parameters of the model.

Metric	Equation
SSE	$\Sigma_{i=1}^{N}(\frac{y^i - y_m^i}{\frac{\sigma_i}{\sigma}})$
RMSE	$\sqrt{\frac{SSE}{N-p}}$
sMAPE	$\frac{1}{N}\Sigma_{i=1}^{N}(2\frac{\lvert y^i - y_m^i\rvert}{(\lvert y^i\rvert + \lvert y_m^i\rvert)})$
AIC	$Nln(\frac{SSE}{N}) + 2p$
BIC	$Nln(\frac{SSE}{N}) + ln(N)p$

Table 4. Evaluation of the tumor growth models.

Evaluation metrics (smaller value is better)						
Dataset/Model	SSE	RMSE	sMAPE	AIC	BIC	Rank[a]
Breast[b] *cancer* [20]						
Logistic	7009.6	37.4423	1.7088	52.3639	52.2557	**2**
Bertalanffy	8004.9	44.7350	1.7088	55.2933	55.1310	**5**
Gompertz	7971.8	39.9294	1.7088	53.2643	53.1561	**4**
Holling	6639.1	40.7403	1.4855	53.9837	53.8215	**3**
GLUECK	119.3	4.1285	0.0768	19.8508	19.8508	**1**
Breast[c] *cancer* [26]						
Logistic	0.2936	0.1713	0.1437	−40.5269	−39.5571	**4**
Bertalanffy	0.2315	0.1604	0.1437	−41.3780	−39.9233	**2**
Gompertz	0.3175	0.1782	0.1437	−39.5853	−38.6155	**5**
Holling	0.2699	0.1732	0.1512	−39.5351	−38.0804	**3**
GLUECK	0.0977	0.0902	0.0763	−57.7261	−57.7261	**1**
Lung cancer [6]						
Logistic	44.5261	2.2243	1.5684	19.3800	20.1758	**2**
Bertalanffy	54.1147	2.6008	1.5684	23.5253	24.7190	**5**
Gompertz	53.2475	2.4324	1.5684	21.3476	22.1434	**4**
Holling	50.6671	2.5166	1.5361	22.8012	23.9949	**3**
GLUECK	3.6903	0.5792	0.2121	−12.0140	−12.0140	**1**
Leukemia [23]						
Logistic	223.7271	3.2640	1.6368	56.3235	58.5944	**2**
Bertalanffy	273.6770	3.6992	1.6368	62.9585	66.3649	**5**
Gompertz	259.9277	3.5182	1.6368	59.7729	62.0439	**4**
Holling	248.5784	3.5255	1.6001	60.7461	64.1526	**3**
GLUECK	35.2541	1.2381	0.3232	9.8230	9.8230	**1**

[a] Calculated as best in 3/5 metrics.
[b] MDA-MB-231 cell line.
[c] MDA-MB-435 cell line.

This formulation shows that, when overestimating ($y_m \geq y$), the measurement error is α sub-proportional and, when underestimating ($y_m < y$), the error made is the same as the measured data points. In the following experiments we consider $\alpha = 0.84$ and $\sigma = 0.21$, as a good trade-off of error penalty and enhancement. This interpretation of the notion of measurement error is used in the metrics employed for prediction performance (i.e. Sum of Squared Errors (SSE), Root Mean Squared Error (RMSE), Symmetric Mean Absolute Percentage Error (sMAPE)) and goodness-of-fit and parsimony (i.e. Akaike Information Criterion (AIC) and Bayesian Information Criterion (BIC)), as shown in Table 3. Important to note that the evaluation metrics also take into account the number of parameters p each tumor growth model has, as in Table 1: Logistic $p = 2$, Bertalannfy $p = 3$, Gompertz $p = 2$, Holling $p = 3$. GLUECK is a data-driven approach that does not use any biological assumptions or dynamical description of the process, rather it infers the underlying statistics of the data and uses it in predicting it without supervision. Note that the GLUECK hyper-parameters were unchanged during the experiments (i.e. the same network size, learning epochs etc.) and described in the Materials and methods. Due to its inherent learning capabilities GLUECK provides overall better accuracy between and within tumor type growth curve prediction. Assessing both summary statistics in Fig. 6 and the broad evaluation in Table 4 we can conclude that a data-driven approach for tumor growth extraction can overcome the limitations that incomplete biological descriptions, the diversity of tumor types, and the small size of the data bring to mechanistic models.

4 Conclusion

Notwithstanding, tumor growth models have had a profound influence on modern chemotherapy, especially dose management, multi-drug strategies and adjuvant chemotherapy. Yet, the selection of the best model and its parametrization is not always straightforward. To tackle this challenge GLUECK comes as a support tool that could assist oncologists in extracting tumor growth curves from the data. Using unsupervised learning, GLUECK overcomes the limitations that incomplete biological descriptions, the diversity of tumor types, and the small size of the data bring on the basic, scalar growth models (e.g. Logistic, Bertalanffy, Gompertz, Holling). GLUECK exploits the temporal evolution of timeseries of growth data along with its distribution and obtains superior accuracy over basic models in extracting growth curves from different clinical tumor datasets. Without changes either to structure or parameters GLUECK's versatility has been demonstrated in between cancer types (i.e. breast vs. lung vs. leukemia) and within cancer type (i.e. breast MDA-MB-231 and MDA-MB-435 cell lines) tumor growth curve extraction.

References

1. Abler, D., Büchler, P., Rockne, R.C.: Towards model-based characterization of biomechanical tumor growth phenotypes. In: Bebis, G., Benos, T., Chen, K., Jahn, K., Lima, E. (eds.) ISMCO 2019. LNCS, vol. 11826, pp. 75–86. Springer, Cham (2019). https://doi.org/10.1007/978-3-030-35210-3_6
2. Agrawal, T., Saleem, M., Sahu, S.: Optimal control of the dynamics of a tumor growth model with hollings' type-ii functional response. Comput. Appl. Math. **33**(3), 591–606 (2014)
3. Baldock, A., et al.: From patient-specific mathematical neuro-oncology to precision medicine. Front. Oncol. **3**, 62 (2013)
4. Benaïm, M., Fort, J.C., Pagès, G.: Convergence of the one-dimensional kohonen algorithm. Adv. Appl. Probab. **30**(3), 850–869 (1998)
5. Benzekry, S., et al.: Classical mathematical models for description and prediction of experimental tumor growth. PLoS Comput. Biol. **10**(8), e1003800 (2014)

6. Benzekry, S., Lamont, C., Weremowicz, J., Beheshti, A., Hlatky, L., Hahnfeldt, P.: Tumor growth kinetics of subcutaneously implanted Lewis Lung carcinomacells, December 2019. https://doi.org/10.5281/zenodo.3572401
7. Bernard, A., Kimko, H., Mital, D., Poggesi, I.: Mathematical modeling of tumor growth and tumor growth inhibition in oncology drug development. Expert Opin. Drug Metab. Toxicol. **8**(9), 1057–1069 (2012)
8. Chen, Z., Haykin, S., Eggermont, J.J., Becker, S.: Correlative learning: abasis for brain and adaptive systems, vol. 49. Wiley (2008)
9. Christensen, J., Vonwil, D., Shastri, V.P.: Non-invasive in vivo imaging and quantification of tumor growth and metastasis in rats using cells expressing far-red fluorescence protein. PloS one **10**(7), e0132725 (2015)
10. Comen, E., Gilewski, T.A., Norton, L.: Tumor growth kinetics. Holland-Frei Cancer Medicine, pp. 1–11 (2016)
11. Gaddy, T.D., Wu, Q., Arnheim, A.D., Finley, S.D.: Mechanistic modeling quantifies the influence of tumor growth kinetics on the response to anti-angiogenic treatment. PLoS Comput. Biol. **13**(12), e1005874 (2017)
12. Gerlee, P.: The model muddle: in search of tumor growth laws. Cancer Res. **73**(8), 2407–2411 (2013)
13. Gompertz, B.: On the nature of the function expressive of the law of human mortality, and on a new mode of determining the value of life contingencies. in a letter to francis baily, ESQ. FRS & C. Philos. Trans. Royal Soc. London (115), 513–583 (1825)
14. Katt, M.E., Placone, A.L., Wong, A.D., Xu, Z.S., Searson, P.C.: In vitro tumor models: advantages, disadvantages, variables, and selecting the right platform. Front. Bioeng. Biotechnol. **4**, 12 (2016)
15. Kisfalvi, K., Eibl, G., Sinnett-Smith, J., Rozengurt, E.: Metformin disrupts crosstalk between g protein–coupled receptor and insulin receptor signaling systems and inhibits pancreatic cancer growth. Cancer Res. **69**(16), 6539–6545 (2009)
16. Kohonen, T.: Self-organized formation of topologically correct feature maps. Biol. Cybern. **43**(1), 59–69 (1982)
17. Kuang, Y., Nagy, J.D., Eikenberry, S.E.: Introduction to Mathematical Oncology, vol. 59. CRC Press, Boca Raton (2016)
18. Kühleitner, M., Brunner, N., Nowak, W.G., Renner-Martin, K., Scheicher, K.: Best fitting tumor growth models of the von bertalanffy-püttertype. BMC Cancer **19**(1), 683 (2019)
19. Murphy, H., Jaafari, H., Dobrovolny, H.M.: Differences in predictions of ode models of tumor growth: a cautionary example. BMC Cancer **16**(1), 163 (2016)
20. Rodallec, A., Giacometti, S., Ciccolini, J., Fanciullino, R.: Tumor growth kinetics of human MDA-MB-231 cells transfected with dTomato lentivirus, December 2019. https://doi.org/10.5281/zenodo.3593919
21. Roland, C.L., et al.: Inhibition of vascular endothelial growth factor reduces angiogenesis and modulates immune cell infiltration of orthotopic breast cancer xenografts. Mol. Cancer Ther. **8**(7), 1761–1771 (2009).https://doi.org/10.1158/1535-7163.MCT-09-0280, https://mct.aacrjournals.org/content/8/7/1761
22. Sarapata, E.A., de Pillis, L.: A comparison and catalog of intrinsic tumor growth models. Bull. Math. Biol. **76**(8), 2010–2024 (2014)
23. Simpson-Herren, L., Lloyd, H.H.: Kinetic parameters and growth curves for experimental tumor systems. Cancer Chemother Rep. **54**(3), 143–174 (1970)

24. Vaghi, C., et al.: Population modeling of tumor growth curves, the reduced gompertz model and prediction of the age of a tumor. In: Bebis, G., Benos, T., Chen, K., Jahn, K., Lima, E. (eds.) ISMCO 2019. LNCS, vol. 11826, pp. 87–97. Springer, Cham (2019). https://doi.org/10.1007/978-3-030-35210-3_7
25. Verhulst, P.F.: Notice sur la loi que la population suit dans son accroissement. Corresp. Math. Phys. **10**, 113–126 (1838)
26. Volk, L.D., Flister, M.J., Chihade, D., Desai, N., Trieu, V., Ran, S.: Synergy of nab-paclitaxel and bevacizumab in eradicating large orthotopic breast tumors and preexisting metastases. Neoplasia **13**(4), 327-IN14 (2011)
27. Von Bertalanffy, L.: Quantitative laws in metabolism and growth. Q. Rev. Biol. **32**(3), 217–231 (1957)

Automated Integration of Genomic Metadata with Sequence-to-Sequence Models

Giuseppe Cannizzaro, Michele Leone, Anna Bernasconi, Arif Canakoglu, and Mark J. Carman(✉)

Dipartimento di Elettronica, Informazione e Bioingegneria, Politecnico di Milano, Via Ponzio 34/5, 20133 Milan, Italy
giuseppe.cannizzaro@mail.polimi.it, {michele.leone, anna.bernasconi,arif.canakoglu,mark.carman}@polimi.it

Abstract. While exponential growth in public genomic data can afford great insights into biological processes underlying diseases, a lack of structured metadata often impedes its timely discovery for analysis. In the Gene Expression Omnibus, for example, descriptions of genomic samples lack structure, with different terminology (such as "breast cancer", "breast tumor", and "malignant neoplasm of breast") used to express the same concept. To remedy this, we learn models to extract salient information from this textual metadata. Rather than treating the problem as classification or named entity recognition, we model it as machine translation, leveraging state-of-the-art sequence-to-sequence (seq2seq) models to directly map unstructured input into a structured text format. The application of such models greatly simplifies training and allows for imputation of output fields that are implied but never explicitly mentioned in the input text.

We experiment with two types of seq2seq models: an LSTM with attention and a transformer (in particular GPT-2), noting that the latter outperforms both the former and also a multi-label classification approach based on a similar transformer architecture (RoBERTa). The GPT-2 model showed a surprising ability to predict attributes with a large set of possible values, often inferring the correct value for unmentioned attributes. The models were evaluated in both homogeneous and heterogenous training/testing environments, indicating the efficacy of the transformer-based seq2seq approach for real data integration applications.

Keywords: Genomics · High-throughput sequencing · Metadata integration · Deep Learning · Translation models · Natural language processing

1 Introduction

Technologies for DNA sequencing have made incredible steps in the last decade, producing rapidly expanding quantities of various types of genomic data with

Y. Dong et al. (Eds.): ECML PKDD 2020, LNAI 12461, pp. 187–203, 2021.
https://doi.org/10.1007/978-3-030-67670-4_12

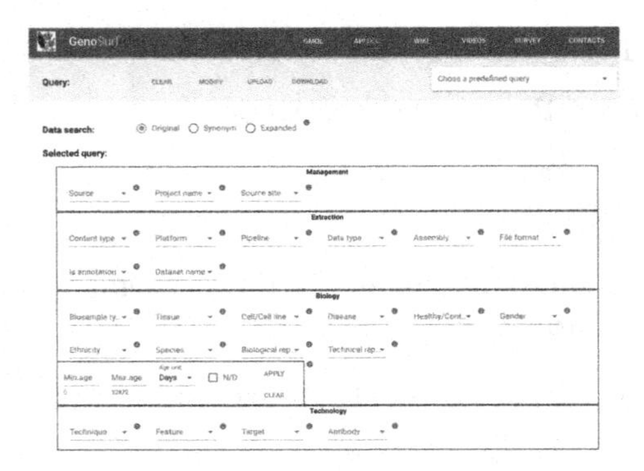

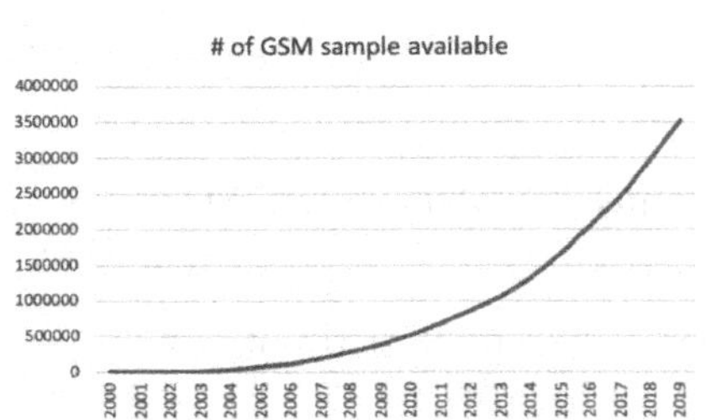

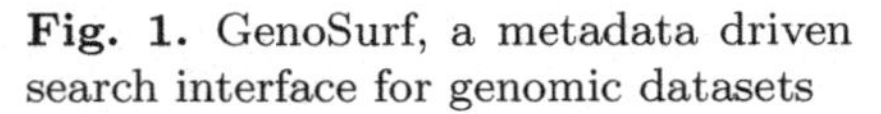

Fig. 1. GenoSurf, a metadata driven search interface for genomic datasets

Fig. 2. Growth over time of # samples available in the GEO database

ever lower costs[1] and faster production times. Biologists and bioinformaticians need access to such datasets for their everyday work, and open data is available through various platforms. Unfortunately, each platform enforces its own data model and formats, and this heterogeneity can hinder data analysis. There is need to integrate resources [4] to prevent scientists from missing relevant data or wasting time on data preparation. Metadata-driven search engines such as GenoSurf [7] attempt to do this by allowing users to search for genomic samples with given characteristics using a structured interface (see Fig. 1). GenoSurf integrates metadata schemas from important genomic sources (ENCODE [8], TCGA [31], 1000 Genomes [27], and Roadmap Epigenomics [18]), but misses samples from the largest public genomic repository, the Gene Expression Omnibus (GEO) [2].

The data in GEO is of fundamental importance to the scientific community for understanding various biological processes, including species divergence, protein evolution and complex disease. The number of samples in the database is growing exponentially (see Fig. 2), and while tools for retrieving information from GEO datasets exist[2], large-scale analysis is complicated due to heterogeneity in the data processing across studies and most importantly in the metadata describing each experiment. When submitting data to the GEO repository, scientists enter experiment descriptions in a spreadsheet (see Fig. 3) where they can provide unstructured information and create arbitrary fields that need not adhere to any predefined dictionary. The validity of the metadata is not checked at any point during the upload process[3], thus the metadata associated with gene expression data, usually does not match with standard class/relation identifiers from

[1] Companies currently offer complete genome sequencing for under 600USD (e.g. https://www.veritasgenetics.com/myGenome) with costs expected to fall.

[2] NCBI E-utilities [17] provide a federated search engine supporting information on experimental protocols, but lack functionality regarding characteristics of the sample, such as species of origin, age, gender, tissue, mutations, disease state, etc.

[3] Information regarding the submission of high-throughput sequences is provided at https://www.ncbi.nlm.nih.gov/geo/info/seq.html.

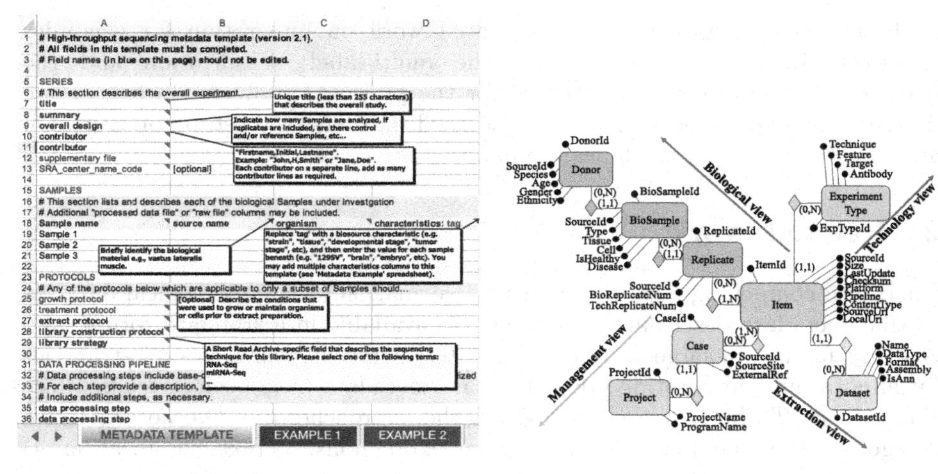

Fig. 3. Spreadsheet for describing data when submitting dataset to GEO

Fig. 4. Genomic conceptual model

specialized biomedical ontologies. The resulting free-text experiment descriptions suffer from redundancy, inconsistency, and incompleteness [32].

In this paper, we develop automated machine learning methods for extracting structured information from the heterogeneous GEO metadata. Our aim is to populate a structured database with attributes according to the Genomic Conceptual Model (GCM) [5], which recognizes a limited set of concepts[4] (shown in Fig. 4), that are supported by most genomic data sources. GCM provides a common language for genomic dataset integration pipelines (see META-BASE framework [3]), fuelling user search-interfaces such as GenoSurf.

The main contributions of this paper are the following:

1) We provide a novel formulation of the metadata integration problem as a machine translation (MT) problem, which has a number of benefits over both a named-entity recognition (NER) based approach (since there is no requirement for annotating input training sequences), and over a multi-label classification based approach (since the same model architecture can be used regardless of the target attributes to be extracted).
2) We provide experimental evidence demonstrating the effectiveness of the transformer-based translation models over simpler attention based seq2seq models and over the classification based approach using a similar transformer architecture. Experiments are performed in both homogeneous and heterogenous training/testing environments, indicating the ability of the seq2seq model to impute values often unobserved in the input, and the efficacy of the approach for real data integration applications.

[4] The data model centers on the *item* of experimental data, with views describing *biological* elements, *technology* used, *management* aspects, and *extraction* parameters.

In the next section we discuss related work on integration for genomics resources. In Sect. 3, we overview the multi-label classification and the translation-based approaches studied for the proposed problem. In Sect. 4, we describe the experiments, including the used datasets, the setup configuration and results. Section 5 concludes the paper.

2 Related Work

There is a compelling need to structure information in large biological datasets so that metadata describing experiments is available in a standard format and is ready for use in large-scale analysis [16]. In recent years, several strategies for annotating and curating GEO database metadata have been developed (see Wang et al. [29] for a survey). We group the approaches into five non-exclusive categories: 1) manual curation, 2) regular expressions, 3) text classification, 4) named-entity recognition, and 5) imputation from gene expression.

Manual Curation: Structured methods for authoring and curating metadata have been promoted by numerous authors [14,20,24]. Moreover, a number of biological metadata repositories (e.g. RNASeqMetaDB [13], SFMetaDB [19] and CREEDS [30]) manually annotate their datasets, guaranteeing high accuracy. This option is however, highly time-consuming and hardly practicable as the volume and diversity of biological data grows.

Regular Expressions: The use of regular expressions for extracting structured metadata fields from unstructured text is common [12]. This simple technique is limited, however, to matching patterns that are: *expressible*, yet identifiers for biological entities often do not follow any particular pattern (e.g., IMR90, HeLa-S3, GM19130); *explicit*, i.e. matching the cell line K562 cannot produce the implied sex, age, or disease information[5]; and *unique*, i.e. it cannot discern between multiple string matches in the same document.

Text Classification: Machine learning techniques can be used to predict the value of metadata fields based on unstructured input text. Posch et al. [25] proposed a framework for predicting structured metadata from unstructured text using tf-idf and topics modeling based features. The limitations of the classification approach include that a separate model needs to be trained for each attribute to extract and that values of the attribute need to be known in advance.

Named-Entity Recognition: NER models are often used to extract knowledge from free text data including medical literature. They work by identifying spans within an input text that mention named entities, and then assigning these spans into predefined categories (such as *Cell Line*, *Tissue*, *Sex*, etc.). By learning their parameters and making use of the entire input sentence as context, these systems overcome the limitations of simple regular expression based approaches. In particular, certain works [11,16] have employed NER to map free textual description

[5] K562 is a widely known cell line originally extracted from tissue belonging to a 53 year-old woman affected by myeloid leukemia.

into existing concepts from curated specialized ontologies that are well-accepted by the biomedical community [6] to improve the integrated use of heterogeneous datasets. In practice, training NER models can be difficult, since the training sequences must be labelled on an individual word level. This is especially time consuming in the genomics domain, where biomedical fields require specific and technical labels. Moreover, there is no way to make use of publicly available curated datasets that overlap with GEO to synthesize training data, since the information they contain applies to the entire GEO sample as a whole. A further drawback of NER methods, is that they can produce false positives with high frequency, due to misleading information in samples' descriptions (e.g., presence of pathologies in the family of an healthy patient).

Imputation from Gene Expression: The automated label extraction (ALE) platform [12], trains ML models based on high-quality annotated gene expression profiles (leveraging on text-extraction approaches based on regular expression and string matching). However, the information is limited to a small set of patient's characteristics (i.e., gender, age, tissues). Authors in [10] also predict sample labels using gene expression data; a model is built and evaluated for both biological phenotypes (sex, tissue, sample source) and experimental conditions (sequencing strategy). The approach is applied on repositories alternative to GEO (i.e., training from TCGA samples, testing on GTEx [22] and SRA).

Each of the aforementioned approaches to genomic metadata extraction have their limitations. As we will discuss in the next section, many (or all) of these limitations can be overcome by making use of a translation (a.k.a. sequence-to-sequence) modeling approach. To the best of our knowledge, no previous work has applied this approach to the problem of automating the integration of experiment metadata before.

3 Approaches

We now discuss two different approaches that we applied to the metadata extraction problem. Both leverage recent advances in Deep Learning for text analysis. The first approach builds a multi-label classifier to predict metadata attribute values using a deep embedding, and will serve as our baseline for later experiments. The second makes use of a novel translation-based approach where powerful sequence-to-sequence models are leveraged to solve the metadata extraction problem in a more elegant and extensible fashion.

3.1 Multi-label Classification Approach

To model the metadata extraction problem using a classification approach, we can simply turn the attribute-value prediction problem into a *multi-label classification problem*, by treating each possible value for each attribute as a separate class to be predicted. An alternative would be to model the task as a *multi-task multi-class classification problem*, where each attribute is associated with its own softmax function (thereby constraining that each attribute must appear in the

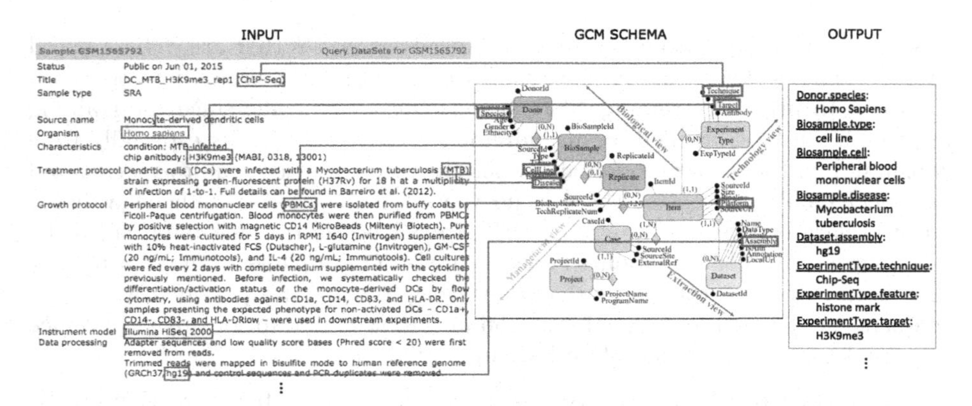

Fig. 5. Example mapping task: from GEO sample GSM1565792 input text, into GCM attributes, to finally produce output key-value pairs

output and must take on a single – possibly *unknown* – value). For simplicity and extensibility purposes we choose instead to model the task as a single multi-label classification problem where each attribute-value is associated with its own sigmoid function. We then use a post-processing to select the most likely attribute value for each attribute. We note that the classification approach requires that each attribute have a finite set of values and each value must be known at training time. It is most suitable for extracting attributes with a relatively small number of possible values, and does not accommodate the situation where an attribute needs to take on multiple values at once (e.g. because a single GEO sample contains data for multiple cell lines).

RoBERTa: To build an embedding from the input text that can provide the feature space for the classifier, we make use of RoBERTa [21], a variation on the BERT [9] language model. These self-attention based transformer models [28] have recently shown state-of-the-art performance for all kinds of text classification tasks owing to the pre-training of the language model in an unsupervised fashion on large text corpora. To build the multi-label classifier, a dense feed-forward layer is place on top of the transformer stack. The last layer presents a number of neurons equal to the total number of attribute-values, (the target attributes are one-hot-encoded for the multi-label model).

3.2 Translation-Based Approach

We treat the problem of extracting metadata from unstructured text as a *translation task*, where instead of translating input text into another language, we translate it into a well structured list of extracted attributes. By approaching the problem in this fashion, many strengths of translation models can be exploited: i) translation models do not expect translated text to follow string patterns; ii) translation models do not use a lookup approach (i.e., they can disambiguate correctly input words whenever the text contains multiple possible choices for

Table 1. Size of Encoder and Decoder networks

Network	Layer	Size
Encoder	Embedding	256
	LSTM	512
Decoder	Embedding	256
	LSTM	512
	Tan_h	512
	Dense	Vocab_size

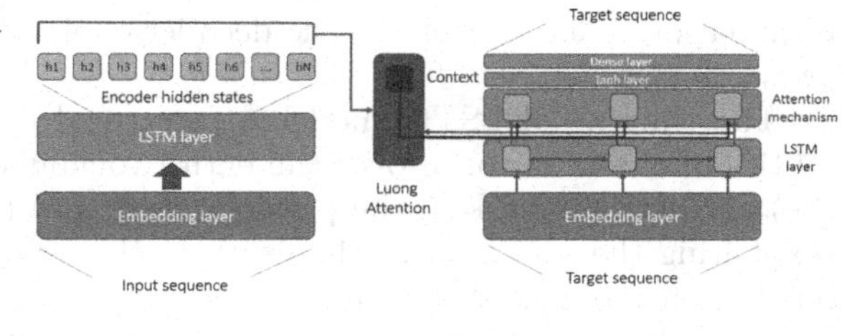

Fig. 6. Encoder-Decoder structure with attention mechanism

a certain concept); iii) translation models do not expect a fixed number of output values; iv) translation models can extract hidden information from the text context.

Input and Output Formats: Each training sample is composed of *input-output* pairs, where *input* corresponds to the textual description of a biological sample and *output* is a list of attribute-value pairs. Figure 5 shows an example translation task: on the left, a metadata record from GEO repository describing a human biological sample, in the middle the target schema, and on the right the resulting output pairs. The text output produced by the translation model should be human and machine readable, so we used a dash-separated list of "key: value" pairs, `Cell Line: HeLa-s3 - Cell Type: Epithelium - Tissue: Cervix`.

We now discuss the Encoder-Decoder LSTM and OpenAI GPT-2 architectures employed as translation models in our experiments.

Encoder-Decoder LSTM: This model is composed of two LSTM networks, an encoder and a decoder, and exploits a Luong attention [23] mechanism. The *encoder* is composed of an embedding layer plus an LSTM layer, which provides hidden states to feed the attention mechanism for the decoding phase. The *decoder* is composed of an embedding layer, an LSTM layer and 2 dense layers as shown in Fig. 6. We report the number of neurons of each layer of the encoder and decoder network in Table 1. The size of the dense layer depends on the vocabulary size (and is thus determined by the tokenizer). The two dense layers are needed for the attention mechanism: The output of the LSTM layer is concatenated with the context vector, thus doubling the size of the vectors coming from the LSTM layer. The first dense layer re-shapes the LSTM output to the same size as the LSTM, while the second maps the output of the first dense layer to the size of the vocabulary. The vocabulary token with the highest probability is then predicted for each time step.

The embedding layer of the Encoder is fed with a tokenized version of the input text and is executed once for each sample (batch of items). The decoding phase takes place iteratively, thus the output is generated token-by-token. At each i time step (corresponds to a single token), the embedding layer of the decoder is fed with a tokenized version of the output text starting from the *start* token (<*start*>) reaching the i-th token. The decoder is trained to generate the i+*1*-th token until the entire sequence has been generated, producing a termination token (<*end*>). The decoder exploits the attention mechanism, as shown in Fig. 6.

The training of LSTM model is performed by learning conditioned probabilities of the next token over the entire vocabulary, given the (embedding for the) current input token, the previous state and the sequence up to that point (exploiting the attention mechanism). Each token is a tensor that represents a one-hot-encoding over the entire vocabulary.

To evaluate LSTM performance, we generated the output sequences for the input strings in the test set as follows: the model encodes each input sequence, the decoder generates the predicted probabilities over the entire vocabulary given the input and the <start> token. The most probable output token is then selected and concatenated to the input sequence after the <start> token. After that point, the decoder generates a prediction given the input and the generated sequence; the procedure goes on iteratively until the termination token (<end>) is generated. In the unlikely case of a generation process that does not end (because the termination character is never generated), production is stopped when the output sequence reaches the maximum output length in the training set.

OpenAI GPT-2 is a more powerful sequence-to-sequence pre-trained language model [26], whose structure is based on Transformer Decoders [28]. Text generation is done in a similar fashion as encoder-decoder, i.e., a generation token-by-token. Differently from LSTM models, the text generation phase is not preceded by an encoding phase. This means that the model is not trained on input-output pairs; instead, it is trained on single sequences. Thus, we prepared sentences composed of both input and output, separated with the "=" character and terminated by '$' (e.g., `[Input sentence] = Cell line: HeLa-S3 - Cell Type: Epithelium - Tissue Type: Cervix - Factor: DNase $`).

GPT-2 training is performed by learning conditioned probabilities of the next token over the entire vocabulary, given only the sequence of previous tokens. As with the LSTM, each token is a tensor that represents a one-hot-encoding over the entire vocabulary. To evaluate GPT-2 performance, we generated the output sequences for the input strings in the test set employing a similar approach as with the Encoder-Decoder LSTM model. GPT-2 outputs the probabilities over the entire vocabulary for a given input sequence which terminates with "=". The output token with the highest probability is then concatenated to the input sequence. The model then outputs the probabilities given the new input sequence (which is composed of the input sequence used at previous step concatenated to the generated token); the process goes on until the termination token ($) is generated.

In both described translation models, after the entire sequence is generated, the tokenized output sequence is decoded back to text.

4 Experiments

Our experiments aim to evaluate and compare results of two seq2seq models: a simple Encoder-Decoder model using a Long Short-Term Memory (LSTM) layer

Table 2. Cistrome attributes: percentage of 'None' and count of distinct values

Attributes	% 'None'	#distinct values
Cell line	52	519
Cell type	19	152
Tissue type	29	82
Factor	0	1252

Table 3. ENCODE attributes: percentage of 'None' and count of distinct values

Attributes	% 'None'	#distinct values
Age	1	169
Age units	32	6
Assay name	0	26
Assay type	0	9
Biosample term name	0	9
Classification	1	6
Ethnicity	74	15
Genome assembly	16	11
Health status	53	65
Investigated as	48	22
Life stage	1	17
Organism	1	5
Project	0	3
Sex	1	10
Target of assay	48	344

with Luong attention [23], and the OpenAI Generative Pretrained Transformer 2 (GPT-2) Language Model [26], which makes use of transformer decoder cells and has been proved to perform very well in NLP tasks, in particular in those regarding text generation. As our baseline, we used the RoBERTa multi-label classification model [21]. In the following, we first describe the data that we use in the experiments, we then detail the experiment setup and finally report on the obtained results. The code used in the experiments is publicly available[6].

4.1 Datasets: GEO, Cistrome and ENCODE

We make use of data from GEO, Cistrome and ENCODE for our experiments.

GEO: Input text descriptions are taken from the GEOmetadb database [34]. We extracted the *Title*, *Characteristics_ch1*, and *Description* fields, which include information about the biological sample from the `gsm` table. We format the input by alternating a field name with its content and separating each pair with the dash "–" character, e.g., `Title: [...] - Characteristics: [...] - Description: [...]`. In this way, we allow the model to learn possible information patterns, for example, information regarding "Cell Line" is often included in the "Title" section. We pre-processed the input text by replacing special characters (i.e., !@#$&̂*[]? \—'~_+") with spaces and by removing "\n" and "\t".

[6] https://github.com/DEIB-GECO/GEO-metadata-translator.

Cistrome: The Cistrome Data Browser [33] provides a collection of publicly available data derived from the GEO Database. More specifically, it contains ChIP-seq and chromatin accessibility experiments, two techniques used to analyze protein interactions with DNA and physically accessible DNA areas, respectively. Importantly, the samples in Cistrome have been manually curated and annotated with the *cell line*, *cell type*, *tissue type*, and *factor name*. We downloaded in total 44,843 metadata entries from Cistrome Data Browser[7] with the four mentioned attributes. As indicated in Table 2, three of the fields contain many "None" values, but these should not be interpreted as missing, since they actually indicate that the specific sample does not carry that kind of information.

ENCODE: The Encyclopedia of DNA Elements [8] is a public genomic repository of datasets related to functional DNA sequences and to the regulatory elements that control gene expression. The ENCODE Consortium exploited manual curation to collect and organize metadata for the DNA sequences [15], making the repository one of the most complete and accurate genomic archives from the point of view of data description. We downloaded 16,732 metadata entries from ENCODE web portal[8], by requesting the fields listed in Table 3 for each experiment sample. The free text input related to each sample, was retrieved by either: (i) exploiting a reference to the GEO GSM (only available for 6,233 entries) or (ii) by concatenating the additional ENCODE fields *Summary*, *Description* and *Biosample Description*.

4.2 Experimental Setup

We designed three experiments to validate our proposal. Experiment 1 and 2 allow to compare performances of the three analyzed models on two different datasets: Cistrome (with input from GEO) and ENCODE (with input both from GEO and ENCODE itself). Experiment 3, instead, tested the performance of the best proposed model on randomly chosen instances from GEO.

The Transformer library from HuggingFace[9] was used for the GPT-2 model, while the SimpleTransformers library[10] was used for the RoBERTa model. The LSTM encoder-decoder was built with Tensorflow [1] version 2.1 using the Keras API. For the LSTM model, we performed the tokenization process using the default Keras tokenizer, setting the API parameters to convert all characters into lower case, using empty space as a word separator, and disabling character-level tokenization. We added a space before and after the following characters: opening/closing parenthesis, dashes, and underscores[11]. We also removed equal signs. For the LSTM models, the resulting vocabulary had a size of 36,107 for Experiment 1 and 17,880 for Experiment 2.

[7] http://cistrome.org/db/#/bdown.
[8] https://www.encodeproject.org/.
[9] https://github.com/huggingface/transformers.
[10] https://github.com/ThilinaRajapakse/simpletransformers.
[11] Pre-processing was motivated by the fact that important character ngrams often appear in sequences separated by special characters, e.g., "RH_RRE2_14028".

Table 4. Setup of the three different models for each experiment (BPE = Byte Pair Encoding; LR = learning rate)

Model	Batch size	Loss function	Tokenizer	Optimizer	LR	beta_1	beta_2	Epsilon
RoBERTa	10	Cross entropy	BPE	Adam	2e−4	0.9	0.999	1e−6
LSTM	64	Sparse cross entropy	Keras	Adam	1e−3	0.9	0.999	1e−7
GPT-2	5	Cross entropy	BPE	Adam	1e−3	0.9	0.999	1e−6

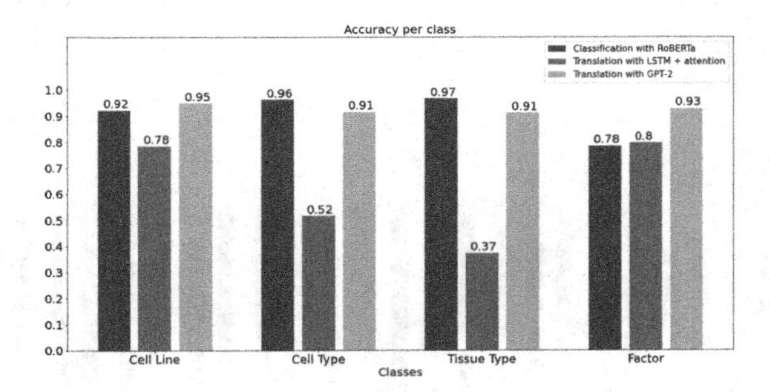

Fig. 7. Experiment 1: per-class accuracy for the three models on Cistrome data.

RoBERTa and GPT-2 were trained using a Tesla P100-PCIE-16GB GPU, while the LSTM model was trained on Google Colaboratory[12] with GPU accelerator. Table 4 lists the configurations for the systems. All models were subject to early stopping method to avoid over-fitting.

For Experiment 1 and 2, data was split into training set (80%), validation set (10%) and test set (10%). Some text cleaning and padding processes were adopted: Encoder-Decoder requires input-output pairs that are padded to the maximum length of concatenation of input and output; GPT-2 requires single sentences that are padded to a maximum length of 500 characters. We excluded sentences exceeding the maximum length.

4.3 Experiments 1 and 2

We evaluated the performances of LSTM (with attention mechanism) and GPT-2 seq2seq models against RoBERTa, using samples from Cistrome (Experiment 1) and samples from ENCODE (Experiment 2).

In both experiments, overall GPT-2 outperforms both Encoder-Decoder LSTM and RoBERTa, as it can be observed in Tables 5 and 6. Results divided by class are shown in Fig. 7 for Experiment 1 and in Fig. 8 for Experiment 2.

Experiment 1 Considerations. From Fig. 7, RoBERTa seems to perform better for classes that contain a low number of distinct values, i.e. *cell type* and *tissue type* (which contain 380 and 249 possible values). Instead, for *cell line* and *factor* (both with more than a thousand possible values) GPT-2 outperforms

[12] https://colab.research.google.com/.

Table 5. Experiment 1: overall accuracy, precision, and recall. Precision and recall are weighted by the number of occurrences of each attribute value.

Model	# Epochs	Accuracy	Precision	Recall
RoBERTa	69	0.90	0.89	0.91
LSTM + Attention	15	0.62	0.65	0.62
GPT-2	47	0.93	0.93	0.93

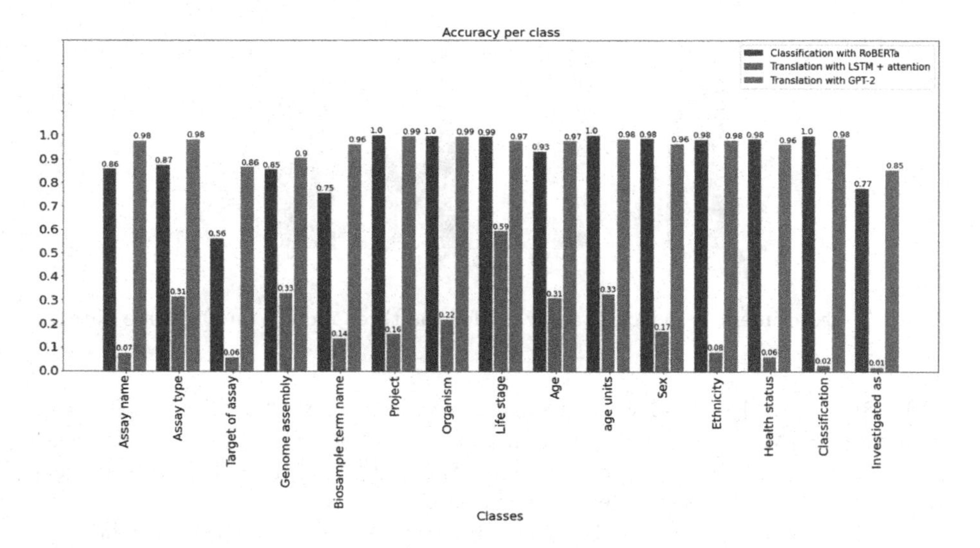

Fig. 8. Experiment 2: per-class accuracy of the three models on ENCODE data.

RoBERTa. The number of "None" values is taken into consideration (Table 2), the classes *cell line*, *cell type* and *tissue type* present a relevant percentage of "None", the weighted precision and recall analysis, however, shows high scores, despite the unbalance of values count; this implies that the models were able to correctly classify samples which lack of labels for certain classes.

Experiment 2 Considerations. From Fig. 8, we appreciate a similar behaviour as in Experiment 1, i.e., translation models perform better for attributes with larger amount of distinct values. The attributes *target of assay* and *biosample term name* present the highest number of distinct values and GPT-2 far exceeded RoBERTa in terms of accuracy. Instead, this experiment highlights how the LSTM model with attention does not perform well for a larger amount of target attributes, at least with the tested model size. The labels *health status* and *ethnicity* presented several "None" values (74% and 53%), but both RoBERTa and GPT-2 were able to predict correctly almost the totality of samples, producing results with high weighted precision and weighted recall.

Table 6. Experiment 2: overall accuracy, precision, and recall. Precision and recall are weighted by the number of occurrences of each attribute value.

Model	# Epochs	Accuracy	Precision	Recall
RoBERTa	71	0.90	0.89	0.90
LSTM + Attention	22	0.19	0.19	0.19
GPT-2	48	0.96	0.96	0.96

Table 7. Experiment 3: Results of prediction of 200 manually labelled samples for ENCODE class *biosample term name.*

Condition	Accuracy	Precision	Recall
Label **present in** the input	0.83	0.70	0.68
Label **absent from** the input	0.062	0.038	0.038

Previous works aimed to extract a restricted set of labels (such as *age* and *sex*) with unsatisfactory results; they often limited the target *age unit* to "years" or "months" and the target *sex* to only "Male" and "Female". A lot of different scenarios for the input text made it impossible – for previous work – to extract correctly the target attributes (for example cases for which the information needs to be inferred, or when the experiment presents multiple cells, consequently multiple ages and multiple sex). This experiment shows that a our proposed translation approach can outperform state-of-the-art approaches, additionally handling a different number of non-standard cases.

4.4 Experiment 3: Randomly Chosen GEO Instances

In this experiment we study the behaviour of GPT-2 on a realistic scenario involving randomly chosen samples from GEO. These samples were not selected based on presence in a database (Cistrome or ENCODE) and thus provide a realistic test scenario for the proposed use-case of the system. No reference labels are available for the randomly selected set of 200 input descriptions, so each instance was manually checked to provide ground-truth labels. The system was trained using both the Cistrome and ENCODE datasets (10 epochs of training on the former followed by 17 epochs of training on the latter). We note the heterogeneity between the training and test examples for this experiment.

Table 7 reports performance for one exemplar class, the *biosample term name*, which defines the tissue or cell line analyzed in the experimental sample. In order to understand the ability of the system to impute values even when the desired output label is not explicitly present in the input, the evaluation metrics (accuracy, precision, and recall) are computed under two different conditions: the true label is *present in* and *absent from* the input. *Biosample term name* class contains a large number of heterogeneous values: some are only represented by acronyms, e.g., "HeLaS3" that is a cell line; others are more verbose, e.g.,

Table 8. Examples of GPT-2 translations of randomly selected GEO descriptions

Input sentence	Output sentence
DNase-seq on 91 day old male fetal human kidney tissue left kidney female embryo (59 d) and male embryo (91 day)	Assay name: DNase-seq - Assay type: DNA accessibility - Target of assay: None - Genome assembly: GRCh38 - Biosample term name: left kidney - Project: ENCODE - Organism: Homo sapiens - Life stage: embryonic - Age: 91, 59 - Age units: day - Sex: male, female - Ethnicity: None - Health status: None - Classification: tissue - Investigated as: None
RNA Bind-n-Seq (RBNS) pulldown experiment against PUM1	Assay name: RNA Bind-n-Seq - Assay type: RNA binding - Target of assay: PUM1 - Genome assembly: hg19 - Biosample term name: cell-free sample - Project: ENCODE - Organism: None - Life stage: None - Age: None - Age units: None - Sex: None - Ethnicity: None - Health status: None - Classification: None - Investigated as: RNA binding protein
ChIP-seq against eGFP-efl-1 in L1 larva worms. strain YL418 (unc-119(ed3)III; vrIs65 [pGES-1::EFL-1::GFP FLAG:EFL-1 3'UTR, unc-119 (+)]) whole organism L1 larva genetically modified using bombardment	Assay name: ChIP-seq - Assay type: DNA binding - Target of assay: efl-1 - Genome assembly: ce11, ce10 - Biosample term name: whole organism - Project: modENCODE - Organism: Caenorhabditis elegans - Life stage: L1 larva - Age: unknown - Age units: None - Sex: mixed - Ethnicity: None - Health status: None - Classification: whole organisms - Investigated as: cofactor, transcription factor
datatype: DnaseSeq; datatype description: DNaseI HS Sequencing; labversion: baseAlignCounts.pl v 1, fseq v 1.84, iff_generic_male, p-value cutoff: 0.05; replicate: 1,2; cell: Colo829; cell organism: human; cell description: malignant melanoma; cell karyotype: cancer; cell sex: M; labversion: baseAlignCounts.pl v 1 Duke human Colo829 DNase-seq COLO829 malignant melanoma	Assay name: DNase-seq - Assay type: DNA accessibility - Target of assay: None - Genome assembly: GRCh38, hg19 - Biosample term name: COLO829 - Project: ENCODE - Organism: Homo sapiens - Life stage: adult - Age: 45 - Age units: year - Sex: male - Ethnicity: Caucasian - Health status: metastatic melanoma - Classification: cell line - Investigated as: None

"Peripheral blood mononuclear cells"; others indicate tissues of provenance, e.g., "synovial membrane". Accuracy, precision and recall are very promising when the model is able to exploit the information contained in the input text (i.e., condition "label **present in** the input"), while they are low in the opposite case; the model still correctly predicts information that is not contained in the input, in some few cases.

In addition to performance on the *biosample term name* attribute, we noted interesting findings regarding the other classes (examples of input and output shown in Table 8). In particular we found cases in which the output contains a label that is:

1. *unseen* in training data, e.g., no sample contained *target of assay*: *MYC-1*.
2. *absent* from input description, e.g., for the input *"HNRNPK ChIP-seq in K562 K562 HNRNPK ChIP-seq in K562"* the output correctly contained: *Organism: Homo sapiens - Age: 53 - Age units: year - Sex: female - Health status: chronic myelogenous leukemia (CML)*, etc.
3. *multi-valued*: e.g., a particular GEO record contained samples from both male and female donors[13], and the output correctly noted both genders: *"Sex: male, female ..."*.
4. *reordered* with respect to the input, e.g., an input containing *"Tfh2_3 cell type: Tfh2 CD4+ T cell; ..."* correctly produced the output "*Biosample term name: CD4-positive Tfh2*".

5 Conclusions and Future Work

In this paper we targeted the problem of extracting useful metadata from free-text descriptions of genomic data samples. Rather than treating the problem as classification or named entity recognition, we model it as machine translation, leveraging state-of-the-art sequence-to-sequence (seq2seq) models to directly map unstructured input into a structured text format. The application of such models greatly simplifies training and allows for imputation of output fields that are implied but never explicitly mentioned in the input text.

We experimented with two types of seq2seq models: an LSTM with attention and GPT-2 (a transformer based language model). We compared the seq2seq models with a multi-label classification based approach using the RoBERTa transformer-based embedding. The GPT-2 model outperforms both the LSTM and the classifier. It demonstrated the ability to predict high-arity attributes and to infer the correct value even for attributes that were not explicitly mentioned in (but were implied by0 the input text. The models were evaluated in both homogeneous and heterogenous training/testing environments, indicating the efficacy of the transformer-based seq2seq approach for real data integration applications.

[13] The input in this case was: *microRNA profile of case NPC362656 survival status (1-death,0-survival): 0; gender (1-male,2-female): 1; age (years): 56; ...*

A goal for future work is to apply the technique to other genomic and biomedical databases, and to develop a crowdsourcing-based online training framework that can allow us to scale up performance for a production system.

Acknowledgments. This research is funded by the ERC Advanced Grant 693174 GeCo.

References

1. Abadi, M., Agarwal, A., Barham, P., et al.: Tensorflow: Large-scale machine learning on heterogeneous distributed systems. arXiv preprint arXiv:1603.04467 (2016)
2. Barrett, T., Wilhite, S.E., Ledoux, P., et al.: NCBI GEO: archive for functional genomics data sets-update. Nucleic Acids Res. **41**(D1), D991–D995 (2012)
3. Bernasconi, A., Canakoglu, A., Masseroli, M., et al.: META-BASE: a novel architecture for large-scale genomic metadata integration. IEEE/ACM Trans. Comput. Biol. Bioinform. https://doi.org/10.1109/TCBB.2020.2998954
4. Bernasconi, A., Canakoglu, A., Masseroli, M., et al.: The road towards data integration in human genomics: players, steps and interactions. Briefings in Bioinform. **22**(1), 30–44 (2021). https://doi.org/10.1093/bib/bbaa080
5. Bernasconi, A., Ceri, S., Campi, A., Masseroli, M.: Conceptual modeling for genomics: building an integrated repository of open data. In: Mayr, H.C., Guizzardi, G., Ma, H., Pastor, O. (eds.) ER 2017. LNCS, vol. 10650, pp. 325–339. Springer, Cham (2017). https://doi.org/10.1007/978-3-319-69904-2_26
6. Bodenreider, O.: Biomedical ontologies in action: role in knowledge management, data integration and decision support. Yearbook of Medical Informatics, p. 67 (2008)
7. Canakoglu, A., Bernasconi, A., Colombo, A., et al.: GenoSurf: metadata drivensemantic search system for integrated genomic datasets. Database **2019** (2019)
8. Davis, C.A., Hitz, B.C., Sloan, C.A., et al.: The encyclopedia of DNA elements (ENCODE): data portal update. Nucleic Acids Res. **46**(D1), D794–D801 (2017)
9. Devlin, J., Chang, M.W., Lee, K., et al.: BERT: Pre-training of deep bidirectional transformers for language understanding. In: Proceedings of the 2019 Conference of the North American Chapter of the Association for Computational Linguistics: Human Language Technologies, vol. 1 (Long and Short Papers), pp. 4171–4186 (2019)
10. Ellis, S.E., Collado-Torres, L., Jaffe, A., et al.: Improving the value of public RNA-seq expression data by phenotype prediction. Nucleic Acids Res. **46**(9), e54–e54 (2018)
11. Galeota, E., Kishore, K., Pelizzola, M.: Ontology-driven integrative analysis of omics data through onassis. Sci. Rep. **10**(1), 1–9 (2020)
12. Giles, C.B., Brown, C.A., Ripperger, M., et al.: ALE: automated label extraction from GEO metadata. BMC Bioinform. **18**(14), 509 (2017)
13. Guo, Z., Tzvetkova, B., Bassik, J.M., et al.: RNASeqMetaDB: a database and web server for navigating metadata of publicly available mouse RNA-Seq datasets. Bioinformatics **31**(24), 4038–4040 (2015)
14. Hadley, D., Pan, J., El-Sayed, O., et al.: Precision annotation of digital samples in NCBI's Gene Expression Omnibus. Sci. Data **4**, 170125 (2017)
15. Hong, E.L., Sloan, C.A., Chan, E.T., et al.: Principles of metadata organization at the ENCODE data coordination center. Database **2016** (2016)

16. Huang, C.C., Lu, Z.: Community challenges in biomedical text mining over 10 years: success, failure and the future. Briefings Bioinform. **17**(1), 132–144 (2016)
17. Kans, J.: Entrez direct: E-utilities on the unix command line. In: Entrez Programming Utilities Help [Internet]. National Center for Biotechnology Information (US) (2020)
18. Kundaje, A., Meuleman, W., Ernst, J., et al.: Integrative analysis of 111 reference human epigenomes. Nature **518**(7539), 317 (2015)
19. Li, J., Tseng, C.S., Federico, A., et al.: SFMetaDB: a comprehensive annotation of mouse RNA splicing factor RNA-Seq datasets. Database **2017** (2017)
20. Li, Z., Li, J., Yu, P.: GEOMetaCuration: a web-based application for accurate manual curation of Gene Expression Omnibus metadata. Database J. Biol. Databases Curation 2018 (2018)
21. Liu, Y., Ott, M., Goyal, N., et al.: RoBERTa: a robustly optimized bert pretraining approach. arXiv preprint arXiv:1907.11692 (2019)
22. Lonsdale, J., Thomas, J., Salvatore, M., et al.: The genotype-tissue expression (GTEx) project. Nat. Genet. **45**(6), 580 (2013)
23. Luong, T., Pham, H., Manning, C.D.: Effective approaches to attention-based neural machine translation. In: Proceedings of the 2015 Conference on Empirical Methods in Natural Language Processing, pp. 1412–1421 (2015)
24. Musen, M.A., Sansone, S.A., Cheung, K.H., et al.: CEDAR: semantic web technology to support open science. In: Companion Proceedings of the The Web Conference 2018, pp. 427–428. International World Wide Web Conferences Steering Committee (2018)
25. Posch, L., Panahiazar, M., Dumontier, M., et al.: Predicting structured metadata from unstructured metadata. Database **2016** (2016)
26. Radford, A., Wu, J., Child, R., et al.: Language models are unsupervised multitask learners. OpenAI Blog. **1**(8), 9 (2019)
27. Genomes Project Consortium: A global reference for human genetic variation. Nature **526**(7571), 68 (2015)
28. Vaswani, A., Shazeer, N., Parmar, N., et al.: Attention is all you need. In: Advances in Neural Information Processing Systems, pp. 5998–6008 (2017)
29. Wang, Z., Lachmann, A., Ma'ayan, A.: Mining data and metadata from the Gene Expression Omnibus. Biophys. Rev. **11**(1), 103–110 (2019)
30. Wang, Z., Monteiro, C.D., Jagodnik, K.M., et al.: Extraction and analysis of signatures from the Gene Expression Omnibus by the crowd. Nature Commun. **7**(1), 1–11 (2016)
31. Weinstein, J.N., Collisson, E.A., Mills, G.B., et al.: The cancer genome atlas pan-cancer analysis project. Nat. Genet. **45**(10), 1113 (2013)
32. Zaveri, A., Hu, W., Dumontier, M.: MetaCrowd: crowdsourcing biomedical metadata quality assessment. Hum. Comput. **6**(1), 98–112 (2019)
33. Zheng, R., Wan, C., Mei, S., et al.: Cistrome Data Browser: expanded datasets and new tools for gene regulatory analysis. Nucleic Acids Res. **47**(D1), D729–D735 (2018)
34. Zhu, Y., Davis, S., Stephens, R., et al.: GEOmetadb: powerful alternative search engine for the Gene Expression Omnibus. Bioinformatics **24**(23), 2798–2800 (2008)

Explaining End-to-End ECG Automated Diagnosis Using Contextual Features

Derick M. Oliveira[1(✉)], Antônio H. Ribeiro[1], João A.O. Pedrosa[1], Gabriela M.M. Paixão[1,2], Antonio Luiz P. Ribeiro[1,2], and Wagner Meira Jr.[1]

[1] Universidade Federal de Minas Gerais, Belo Horizonte, Minas Gerais, Brazil
{derickmath,antoniohorta,joao.pedrosa,meira}@dcc.ufmg.br, gabrielamiana@ufmg.br, tom@hc.ufmg.br

[2] Telehealth Center from Hospital das Clínicas da UFMG, Belo Horizonte, Minas Gerais, Brazil

Abstract. We propose a new method to generate explanations for end-to-end classification models. The explanations consist of meaningful features to the user, namely contextual features. We instantiate our approach in the scenario of automated electrocardiogram (ECG) diagnosis and analyze the explanations generated in terms of interpretability and robustness. The proposed method uses a noise-insertion strategy to quantify the impact of intervals and segments of the ECG signals on the automated classification outcome. These intervals and segments and their impact on the diagnosis are common place to cardiologists, and their usage in explanations enables a better understanding of the outcomes and also the identification of sources of mistakes. The proposed method is particularly effective and useful for modern deep learning models that take raw data as input. We demonstrate our method by explaining diagnoses generated by a deep convolutional neural network.

Keywords: Explainability · Machine learning · Cardiology

1 Introduction

The evolution of artificial intelligence and related technologies have the potential to increase the clinical importance of automated diagnosis tools. The deployment of theses tools, however, is challenging, since its outcome may be used to make important clinical decisions. Models should not only provide accurate predictions, but also evidence that supports the predictions, so that they can be audited and double-checked by an expert. Several methods have been proposed to generate explanations to complex machine learning models [15,28,29], the achieved solutions, however, are usually not tailored for the needs of the physicians and do not take any medical background into consideration. Our claim in this work is that explanations must be based on features that are meaningful to the final users, i.e., the physicians. We call those *contextual* features (Fig. 1).

Y. Dong et al. (Eds.): ECML PKDD 2020, LNAI 12461, pp. 204–219, 2021.
https://doi.org/10.1007/978-3-030-67670-4_13

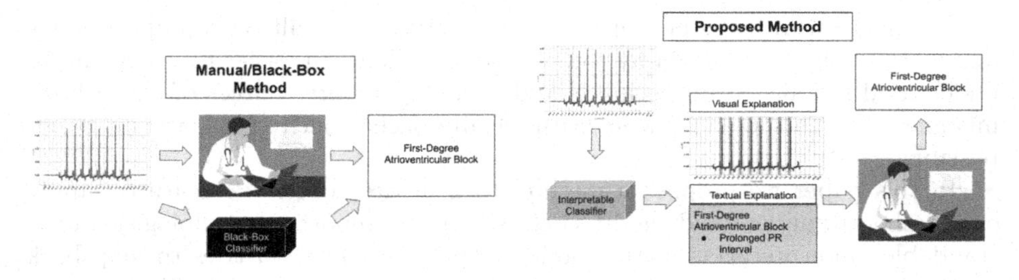

Fig. 1. This illustration shows the difference between the traditional methodology used in ECG classification and our proposal. In the leftmost figure we show the traditional methodology where the input sample, an ECG, is given to a physician or a black-box classification method and then a diagnosis is generated. In contrast, the rightmost figure illustrates the use of an explanation of the classifier outcome, that is also provided to a physician, who uses this information to more accurate and consistent diagnoses.

In the last decades, the usage of deep learning application demonstrated the ability to provide extremely accurate diagnosis [4,5,20]. However, these *black-box* models apply several layers of transformation to the input data and do not support an easy interpretation of the outcome. There are examples of how small perturbations in black-box models may affect the results [9], and, even when they potentially transform the clinical practice [12,21,33,36], there are challenges to be addressed. An interesting example is the neural network model approved for medical use in the European market for melanoma recognition, which suffers significantly when fake surgical markers are given as input, suggesting the model is using these unwanted features for diagnostics [38].

In this work we generate explanations based on *contextual features* for electrocardiogram diagnosis. Cardiovascular diseases are the leading cause of death in the world [27] and the electrocardiogram (ECG) is a major exam for screening cardiovascular diseases. ECG is a tool capable of identifying abnormalities in the heart beat from electrical frequencies that are detected by sensors scattered throughout the patient's body [27].

Classical methods for automated ECG analysis, such as the University of Glasgow ECG analysis program [16], employ a two-step approach: they first extract the main features of the ECG signal using traditional signal processing techniques and then use these features as inputs to a classifier. The features employed by these traditional methods are based on physician background. Deep learning presents an alternative to this approach, since the raw signal itself is used as an input to the classifier, which learns from examples to extract the features. This paradigm is called "end-to-end" learning. Some recent works on the use of these techniques in ECG analysis show the immense potential of this technology, outperforming not only classical methods [32], but also actual physicians in detecting rhythms and abnormalities in the ECG signal both in a single lead [11] and in a 12-lead setup [25,26].

In the classical two-step approach, the models are built on top of measures and features that are known by the physicians, making it easier to verify and to understand the algorithm decisions and, also, to identify sources of algorithmic mistakes. In "end-to-end" deep learning approaches such transparency is not possible.

In this paper we propose a method to generate contextual features, apply our method to understand the ECG classifiers and report a "cardiologist understandable" interpretation of the model output. The method tries to map back the classification decision into features that doctors understand. We explore a noise-insertion method and study its application to the deep neural network classifier proposed in [26]. To the best of our knowledge, this is the first work that generates explanations tailored to ECG classification algorithms and the information that physicians need.

2 Related Works

Several papers on interpretability have been published in the recent years [2,3,8, 10,13,15,24,28–30,37,39]. In this section, we focus on *post-hoc* methods, which try to interpret predictive models after training.

We start by discussing some works which propose visualizations that highlight the most relevant regions for the classifier decision [6,30,34]. These techniques are more common for image classification, where the user is interested on the image regions that are the most relevant to the classifier, and are typically proposed in the context of deep learning model architectures [30]. The relevance of this class of methods to the interpretation of the automated ECG classification is limited, since the visual explanation provided by the method usually does not allow easy interpretations and make no attempt to match the physician background.

Another class of *post-hoc* interpretation methods provide noise-based explanations to classification outcome. These explanation methods insert noise on the features used by the classifier and compute the importance of each feature to the model outcomes [28]. These models may provide local [28,29,31] or global explanations [15] based on features.

Some methods are based on data transformations that enable tabular-like explanations, such as those proposed by [28], where *super-pixels* act like tabular data, which provide an aggregated view of the image that work as model features. Other works target specifically time-series data [7,22] using symbolic representation of time series to achieve an interpretable model. Notice that it is not clear whether these transformations guarantee meaningful explanations for end users, that is, the nature of the explanations is not enough for ensuring their efficacy.

One limitation of all methods presented in this section is that they do not enforce that the features used in the explanations are meaningful to the end user. In this work we propose a method that generates explanations based on contextual features, which are meaningful to experts.

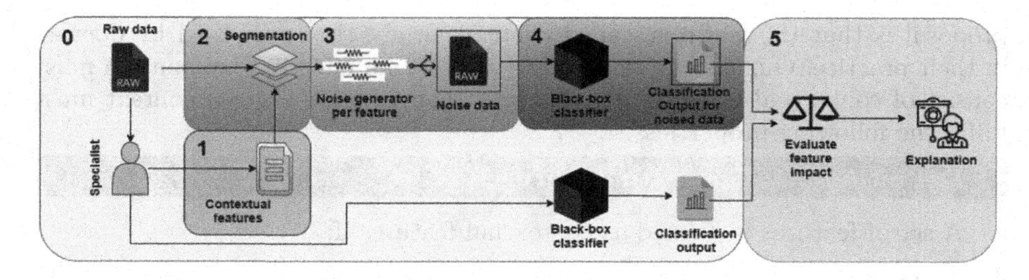

Fig. 2. General procedure of the proposed explanation method. Each step is identified by a different color and a different number in the sequence of steps.

3 Methodology

In this work, we represent a black-box model by a function $f : I \rightarrow K$, which maps the input space I to the classes of interest K, and we denote $f(x) = \hat{y}$. This map f can be obtained by (approximately) minimizing the discrepancy between predicted values $\hat{y}$ and observed values y, i.e., the loss $l(y, \hat{y})$, along with all training examples. We focus on end-to-end models, for which the input x is raw data.

Let F be a feature space, which contains features that are meaningful to the end-user. We state that the features in F are key to the classification. Our application has focused on problems where the classes we are interested in are defined by the features from F.

The methodology of our method is illustrated in Fig. 2, where each block represents one step to generate an explanation and defines the contextual features of the problem. The first step is to determine the *contextual* features used in the explanation. We enforce that, for this step, an expert is necessary: the feature space must be capable to explain the output classes, at least considering the current knowledge about the classification problem. Notice that we may still discover new features that compose the explanation outcomes as well, as illustrated in Sect. 3.1.

After defining the contextual features, we can identify their occurrence in the raw data (step number two), namely *segmentation step*, which consists of extracting the contextual features from the raw data. In automated ECG analysis, this step is equivalent to segmenting the ECG signal, finding peaks and wave locations. The noise insertion step is a Monte Carlo process that generates N perturbed versions of the input x. The noise is concentrated on the signal-region related to one contextual feature (cf. Sect. 3.1) and the noise parameters are set with the help of an expert, so that the final perturbed feature is still realistic.

The classification outcomes of noise-inserted inputs are compared with the original ones to estimate the impact of each contextual feature on the model performance. We use the mean difference of the values as the comparison metric.

For building an interpretable model, we first must determine an explanation, that is, a set of instantiated features. As mentioned, a key difference in our proposal is that the features, called contextual, are the same used by doctors in their practice. Our premise in this work is that an effective explanation must consist of contextual features. The features that describe a given context must fulfill the following properties:

Definition 3.1: Contextual features

A set of features is defined as contextual features if:

- They must be well known to the context specialist;
- The set of features must cover all possible classes;
- The set of features may be updated as we determine new relevant features.

3.1 Case Study Method

Our case study is the ECG classification where x consists of the points of the ECG 12-lead exam, that is, $x \in I \subset \mathbb{R}^{NxM}$ where N is the number of samples per lead (signal channels) and $M = 12$ is the number of leads. The feature space comprises the segments depicted in Fig. 3. Information about these segments is meaningful for the cardiologists and can be used as diagnosis criteria. For example, let us consider the *Left Bundle Branch Block* (LBBB), a well-known diagnosis based on an ECG, which is characterized by the following criteria [18]:

1. QRS duration greater than 120 ms;
2. Absence of Q wave in leads DI, V5 and V6;
3. Monomorphic R wave in DI, V5 and V6;
4. ST and T wave displacement opposite to the major deflection of the QRS complex.

Notice that these criteria are examples of features that a physician would expect in an explanation.

Based on Definition 3, we need to define the contextual features for the ECG application, based on background employed by doctors when analyzing an ECG. Usually, the ECG reports 10-second time series generated by 12 leads (*DI, DII, DIII, AVR, AVF, AVL, V1, V2, V3, V4, V5, V6*), where each lead refers to the heart readings from a different location [23]. The leads support the diagnosis of a large spectrum of diseases and are analyzed in groups or individually, all of them contributing somehow to the diagnosis of some disease.

We state that the elements in 3.1 are interpretable features since they are used by physicians to analyze an ECG report, as described in [18].

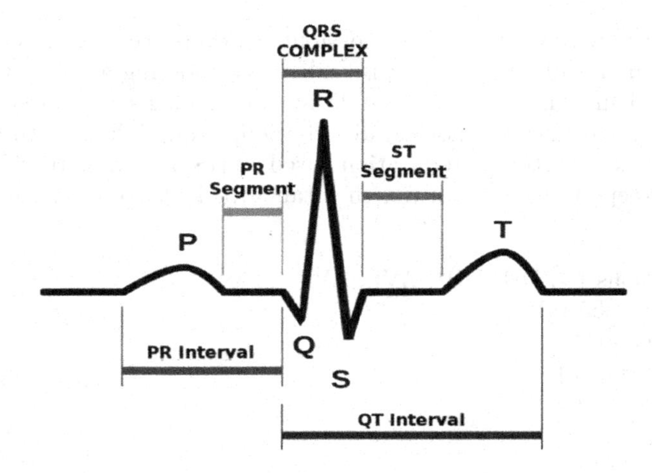

Fig. 3. Illustration of how the ECG is divided into intervals and segments.

Claim 3.1: Explanation for ECG

An explanation for electrocardiogram classifiers must be presented in terms of elements used by cardiologists, namely: (1) Interval and segments and axis of the ECG; (2) Atrial and ventricular rate; (3) ECG rhythm; and (4) ECG morphology.

The intervals and segments of an electrocardiogram are divided into 5 main waves, namely *P, Q, R, S and T*, and, along with these 5 waves, there are 4 segments and intervals, namely *TP, ST, QT and PR*, and the *QRS complex*, which is the union of the waves *Q*, *R*, and *S*, as the name indicates. The scope of all intervals and segments is depicted in Fig. 3, except for the *TP Interval*, for better visualization. The rate of ECG may be measured based on the segmentation of the electrocardiogram. For example, the *ventricular rate* is the frequency of the *R waves* in the ECG.

This set of features fulfill the requirements of Definition 3: (1) all features are defined by experts; (2) all possible diagnosis may be explained, at least partially, using this set; and, (3) new features may arise by combining these data. The discovery of new features is possible once we did not constrain the possible waves or segments and the morphology of the ECG. New intervals may be included in this definition, such as the *U* wave.

3.2 Segmentation-Based Noise Insertion

The noise insertion procedures were designed and implemented with the help of a cardiologist in order to respect the morphology of the waves, segments, intervals, *QRS* complex and the atrial/ventricular rate, and, as a result, we minimize the possibility to output a disturbed signal that is not realistic. Since we cannot insert an arbitrary random noise in the Monte Carlo process, we insert a specific perturbation in x to build a $\tilde{x}$ that is realistic considering an actual ECG, i.e., the noise level in $\tilde{x}$ must be acceptable by a cardiologist as a real ECG. We derive our randomness by a normal distribution, using different parameters for each segment axis, rhythm and duration based on [18,35]. All perturbations have zero mean except the Axis feature, the standard deviations used in this work are:

- for derivations DI, DII, DIII, AVL, AVF:
  - QRS = 1.55 mV
  - T = 0.95 mV
- for derivations V1 – V6:
  - QRS = 1.70 mV
  - T = 1.10 mV
- for all derivations:
  - P = 0.30 mV
  - Duration = 150(bpm)
  - Axis mean = 90° and std = 30°

Our methodology is currently constrained by the segmentation accuracy, making it hard to include the morphology feature in the explanations. This feature may be implemented through the contrast of each element to its normal morphology, replicate the pattern of interest and assess its impact. However, we need a more accurate segmentation, especially w.r.t. *P Wave*.

In order to assess the proper usage of the interpretable features, we also introduce a *random* feature, which inserts noise in random intervals of the ECG with variable levels of impact. We chose the normal distribution to generate the random feature.

In summary, the key idea of our model is to determine the interpretable features, to insert noise in those features, and to compute the impact of each one on the classifier result by a Monte Carlo approximation, while assuring that the method did not introduce an outlier $\tilde{x}$ that may impact the outcome significantly. The proposed methodology is summarized next:

1. Identification of the electrocardiogram waves. This step is known as ECG segmentation, and several previous works addressed this issue. Our empirical analysis identified the algorithm [19] as the best segment identifier so far.
2. Determination of the actual outcome of the sample.
3. For each segment of the ECG, we insert noise by changing its shape. The noise criteria were defined together with cardiologists.
4. We repeat this process N times to evaluate whether the affected feature changes the outcome and such change is not caused by an outlier $\tilde{x}$.

Steps **1** and **2** pre-process the input data, preparing the function f to be analyzed. In these steps, we identify our interpretable features using, for instance, a segmentation method such as [19], and compute the outcome value $\hat{y}$ generated by the classifier, respectively. Step **3** inserts noise on the detected features, and calculates the impact of the feature through a loss function $loss(\hat{y}, \tilde{y})$. In this work, we use the L_0 as the loss function. The final impact is the average of all simulated impacts.

The segmentation accuracy is a critical issue for our proposal, since it is well known that it is a challenging and open problem, and it is beyond the scope of this work to address it. In order to maximize the information provided by the explanations, we use the aforementioned segmentation algorithm, since it is used in many scenarios where the aim is to generate interpretable classifiers [17]. On the other hand, the lack of segmentation accuracy may compromise the explanations we generate.

In summary, our work targets this trade-off between realistic and accurate explanations, as we discuss in Sect. 4. Our implementation can be found at our GitHub repository[1].

4 Contextual Features for a Convolutional Network

In order to illustrate the methods proposed in this work, we apply them to the automated ECG classification system described in [26] (Table 1).

4.1 Model Description

Table 1. Classifier comparison to the average performance of: i) 4th year cardiology resident (*cardio.*); ii) 3rd year emergency resident (*emerg.*); and, iii) 5th year medical students (*stud.*). (PPV = positive predictive value). Reprinted from [26].

	Precision (PPV)				Recall (Sensitivity)				Specificity				F1 Score			
	DNN	Cardio.	Emerg.	Stud.	DNN	Cardio.	Emerg.	Stud.	DNN	Cardio.	Emerg.	Stud.	DNN	Cardio.	Emerg.	Stud.
1dAVb	0.867	0.905	0.639	0.605	0.929	0.679	0.821	0.929	0.995	0.997	0.984	0.979	**0.897**	0.776	0.719	0.732
RBBB	0.895	0.868	0.963	0.914	1.000	0.971	0.765	0.941	0.995	0.994	0.999	0.996	**0.944**	0.917	0.852	0.928
LBBB	1.000	1.000	0.963	0.931	1.000	0.900	0.867	0.900	1.000	1.000	0.999	0.997	**1.000**	0.947	0.912	0.915
SB	0.833	0.833	0.824	0.750	0.938	0.938	0.875	0.750	0.996	0.996	0.996	0.995	**0.882**	**0.882**	0.848	0.750
AF	1.000	0.769	0.800	0.571	0.769	0.769	0.615	0.923	1.000	0.996	0.998	0.989	**0.870**	0.769	0.696	0.706
ST	0.947	0.968	0.946	0.912	0.973	0.811	0.946	0.838	0.997	0.999	0.997	0.996	**0.960**	0.882	0.946	0.873

This system is based on the deep convolutional neural network depicted in Fig. 4, which contains 9 convolutional layers and more than 6 million trainable parameters. This neural network was trained using a dataset that consists of 2,322,513 ECG records from 1,676,384 different patients from 811 counties in the state of Minas Gerais/Brazil, acquired through the Telehealth Network of Minas Gerais (TNMG) [1]. 98% of the dataset was used for training and 2% for hyperparameter tuning. The resulting model is capable of classifying the 6 abnormalities in Fig. 5: (1) 1st degree AtrioVentricular block (1dAVb); (2) Right Bundle Branch Block (RBBB); (3) Left Bundle Branch Block (LBBB); (4) Sinus Bradycardia

[1] https://github.com/DerickMatheus/ECG-interpretation.

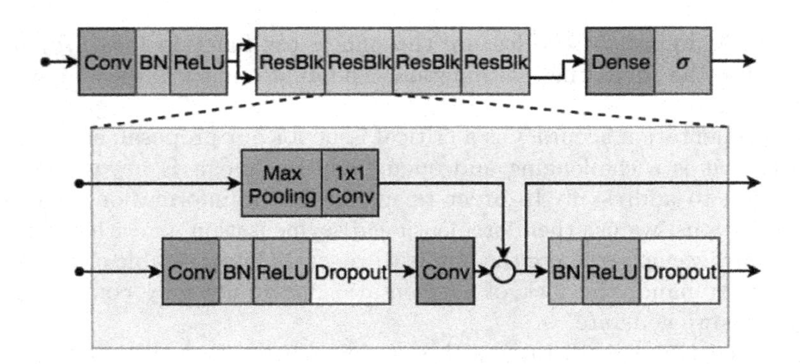

Fig. 4. The uni-dimensional residual neural network architecture used for ECG classification. Reprinted from [26].

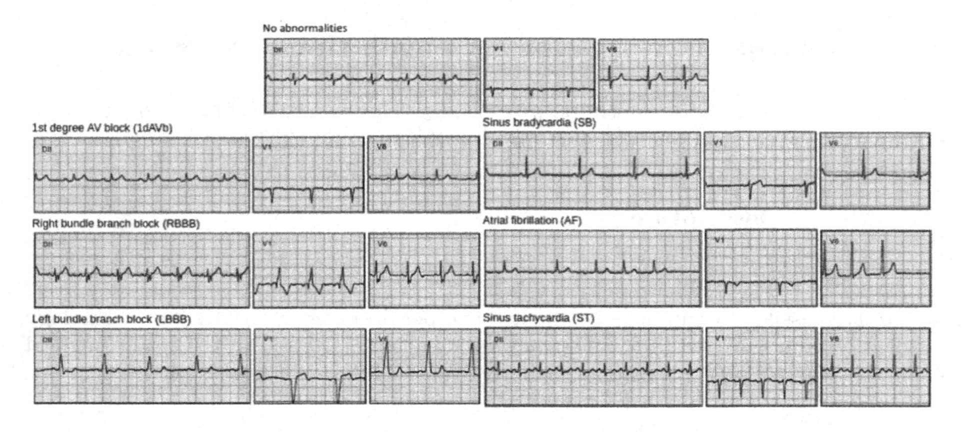

Fig. 5. A list of all the abnormalities the model classifies. We show only 3 representative leads (DII, V1 and V6). Reprinted from [26].

(SB); (5) Atrial Fibrillation (AF); and (6) Sinus Tachycardia (ST). The model presented in this work has better accuracy (on a separate test set) than cardiology residents with more than 9 years of medical training and is openly available[2] on Zenodo platform. The test dataset used in [25,26] is also openly available[3] and consists of 827 exams with diagnosis agreed by two cardiologists. We used this test set and model to validate our method.

While highly accurate, this end-to-end model returns a diagnosis that is hard for the cardiologist to interpret and our method aims to provide a way to extract interpretation from its classification, making it more useful in a practical scenario. The only variable for our explanation method is the number of

[2] https://doi.org/10.5281/zenodo.3625017.
[3] https://doi.org/10.5281/zenodo.3625006.

simulations. In order to minimize the likelihood of random results, we performed a large number of simulations, 499.

4.2 Model Evaluation

Table 2. Relevance of each interpretable feature and random noise used by each class in the [26] classifier, A/V Rate is the variation of atria or ventricular rate. We displayed only the random feature but test with 5 types of random noise, but none of them show any impact. Thus we display only one column with this result. The column #Exams presents the number of exams used in the experiment for each diagnosis.

	#Exams	Measurements					A/V rate	Rhythm	Random
		P wave	T wave	PR interval	QT interval	QRS complex			
1dAVb	30	0.00	0.00	0.30	0.26	0.78	0.00	0.26	0.00
RBBB	38	0.00	0.21	0.00	0.00	0.52	0.48	0.64	0.00
LBBB	30	0.00	0.07	0.00	0.00	0.86	0.14	0.25	0.00
SB	18	0.00	0.00	0.00	0.00	0.00	0.86	1.00	0.00
AF	10	0.00	0.00	0.30	0.00	0.20	0.70	1.00	0.00
ST	38	0.00	0.00	0.00	0.00	0.03	0.69	1.00	0.00

In order to analyze our method efficacy, we also assess the explanation robustness. We define that a robust explanation must report similar feature impacts for all samples from the same class, that is, we did not expect to find significantly different explanations for different occurrences of the same disease. To assess robustness, we compute the number of times each feature appears as an explanation on the test set, and we show the compiled result in Table 2. We observe that a common set of features explains each class, confirming the robustness of our explanations. We may also observe that there is a small variation on the results, e.g., the *QRS* complex explains the *1dAVb* in some cases. These errors are associated with segmentation inaccuracies and shared importance among features, e.g., A/V rate affects the duration of the *QRS* complex. It is important to highlight that the feature *A/V rate* is related to all others, and the variation on the ECG frequency modifies the duration of all segments. Thus, *A/V rate* can be used as duration criterion for other features.

In Table 2, we show how frequently random features impact significantly each class. We tested five levels of noise in random intervals, the level of noise is generated from a normal distribution and variable variance, in this case 15%, 30%, 50%, 100% and 200% of the signal variance. As we may expect, the random features did not impact significantly the classifier performance, regardless the impact level employed. As a consequence, we summarize all results associated with random features in a single column in Table 2.

Note that a feature does impact explanations if it affects a large number of tests, then being an explanation for the classifier as reported by our method. The result presented by our model to the end user is the impact that each interpretable feature has on the classifier, as discussed in [15]. An explanation consists of both a visual and a textual explanation. Each visual explanation is a horizontal bar graph where each bar is associated with a feature, its length represents the impact, and the error bar at the right end of the colored bar represents the impact standard deviation considering samples from a given disease. Contextual features that do not impact any of the samples are omitted. The red dotted line is the usage threshold of the feature, as proposed by [25,26]. The textual explanation is an automatically generated text that *reads* the visual explanation for the end user. Figure 6 shows some examples of visual and textual explanations for the various diagnosis.

Next, we illustrate our methodology on actual cases, and organize our discussion according to the effectiveness of our methodology in generating correct, incorrect, and dubious explanations.

Correct Explanations. We were able to explain correctly three classes of diseases: *SB*, *ST*, and *1dAVb*. Analyzing Table 2, the explanation given by our method for *SB* and *ST* is the feature A/V rate. Another correct explanation is for *1dAVb*, where the most significant feature is the duration of the *PR* interval. All cases match real diagnoses [18].

Several samples of *LBBB* and *RBBB* show a correct explanation, as we can see on Fig. 6, where the *QRS* complex is an actual explanation. Figure 6 shows the A/V rate for all classes since we did not disregard changes in the duration of A/V rate. Such procedure would be necessary because, when we disturb the A/V rate, we affect the other features w.r.t. their duration. We decided to keep the A/V rate in the explanation to enforce the importance of feature duration.

Incorrect Explanations. By analyzing the explanations for *RBBB*, we realize that our explanation is incorrect. Considering the real criteria for *RBBB* and our explanation features, our explanation should indicate the *QRS* duration as explanation feature, but, in Table 2, almost 50% of the exams did not present the *QRS* complex as part of an explanation. In Fig. 6, we present an instance of a correct explanation for *RBBB*, where we can see the A/V rate and the *QRS complex*, which indicates that the mean duration of the complex is the diagnosis explanation.

Notice that this incorrect explanation does not necessarily represent a classifier error, since our method did not use all of our possible explanation features (as defined in Claim 3.1), and thus the classifier may use a different feature. On the other hand, this explanation reveals that, for *RBBB*, the classifier did not use information about the *QRS* complex, which is not commonplace among cardiologists, leading us to consider it an incorrect explanation.

Dubious Explanations. The dubious explanation was for the *AF* diagnosis, as discussed next. In medical practice, the main criteria for characterizing *AF* are: (1) fast atrial rate and (2) absence of *P* wave. One issue here is that segmenting the P wave is challenging. Analyzing Table 2, we can observe that the *A/V rate*

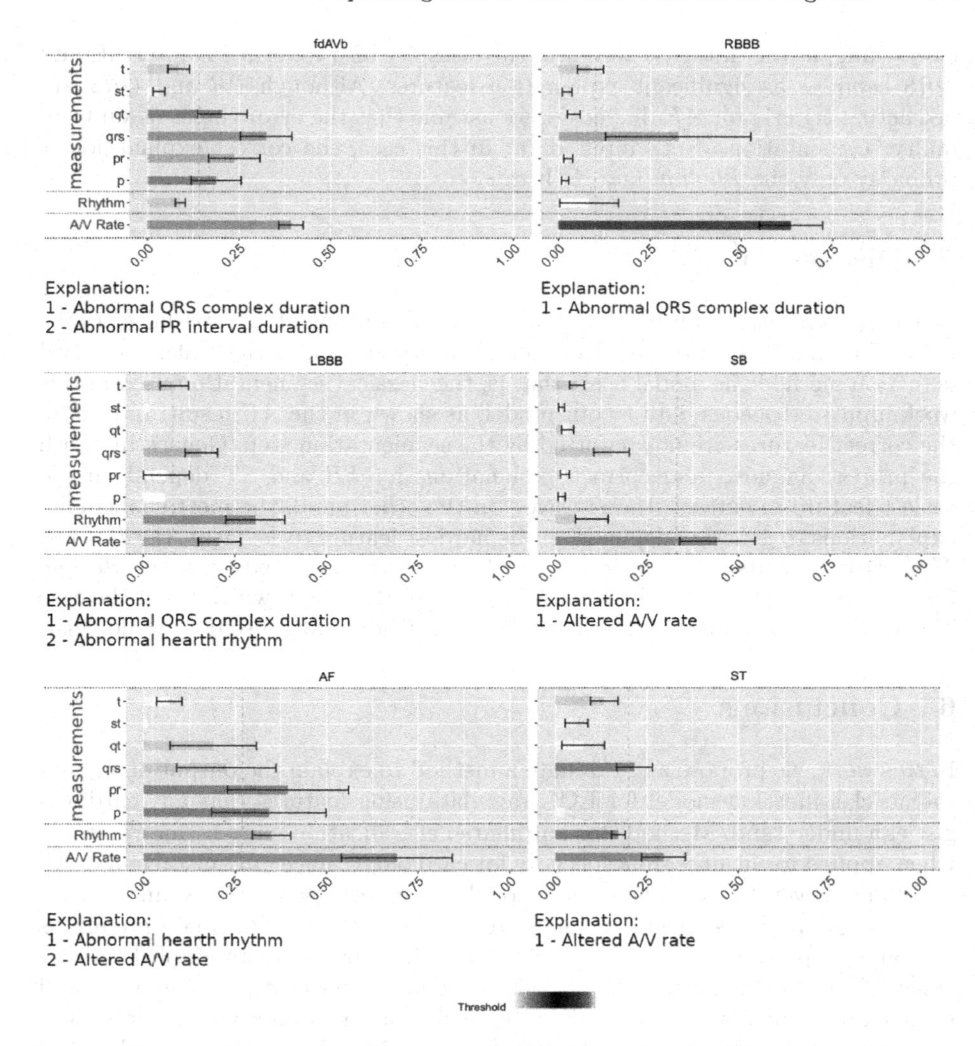

Fig. 6. Explanations examples for each class of ECG disease. Each explanation has a visual and a textual component. The visual component is a horizontal bar graph where each bar represents a feature. The colored bar is the mean value of the impact of the associated feature on the classifier and the error bar at the right end is the standard deviation. We consider an explanation as significant when the mean and the standard deviation are above the threshold vertical dotted line. The textual explanation is presented below each graph.

and *Rhythm* do impact the classification in almost all cases, which are correct explanations. On the other hand, as in the *LBBB* case, other explanation features are not significant. We compare explanations for the true and false positive outcomes to better understand the observed phenomenon. In the false positive cases, the classifier emphasized more the A/V rate than in the true positive case, while, in the first case scenario, the classifier also reports *PR* intervals and *QRS complex* as significant explanation features. Although *PR* and *QRS* are not actual criteria for *AF* diagnoses, we assume that the error comes again from faulty segmentation of the input data. In this case, the correct explanation is probably only the *Rhythm* and *A/V rate.*

5 Discussion

Our main limitation is related to how we locate and insert contextual feature's noise. The main issue of our noise-based interpretation is the feature interval detection method the model used, that is, the classical segmentation used in this work inputs erroneous data to our model, as shown in the *AF* result. In Sect. 4, the correct features are not reported by the segmentation step. Generating both the proper diagnosis and correct explanations for *SB* and *ST* depend on the easiest feature to extract, the *R* wave. On the other hand, the most challenging explanations to generate depends on the hardest feature to segment, the *P* wave. This error may also be seen in classics ECG classifiers based on *segmentation*, but, in these applications, the error is inherent to the result, which is not the case in end-to-end methods. All of our results were validated by a senior cardiologist.

6 Conclusions

In this work, we propose and evaluate a method to explain the output of an end-to-end classification model for ECG raw data using features that any cardiologist can understand. Methods for automated classification must be interpretable when applied for healthcare, especially for cardiology, since any mistake may kill a patient. Several works show how models may be biased [14,28] and consistently make mistakes, even with high precision in test sets. Our main premise is that an interpretation must consist of features that are understandable by a specialist. As far as the authors know, this is the first work that proposes a method to generate explanations for an ECG classifier, using contextual features, that is, features that are understandable by any cardiologist. Our proposed model is based on contextual features that contribute to a better explanation of the results of a black-box classifier to a physician.

In order to improve our method, we need to improve the segmentation method. In particular, we expect that better segmentation will support more precise explanations and eliminate cases such as the *AF* class example, where the segmentation found a *P* wave even if one criteria for AF is its absence.

We also want to perform a larger-scale experiment with cardiologists, by providing the correct ECG along with the classification explanation and measure the aggregated value of our method. Such experiment should consider all features defined in Claim 3.1.

Finally, we expect to implement our method in other scenarios and contexts, validating our method robustness and generality.

Acknowledgement. The authors would like to thank FAPEMIG, CNPq and CAPES for their financial support. This work was also partially funded by projects MASWeb, EUBra-BIGSEA, INCT-Cyber, ATMOSPHERE and by the Google Research Awards for Latin America program.

References

1. Alkmim, M.B., et al.: Improving patient access to specialized health care: the telehealth network of Minas Gerais, Brazil. Bull. World Health Organ. **90**(5), 373–378 (2012). https://doi.org/10.2471/BLT.11.099408
2. Alvarez-Melis, D., Jaakkola, T.S.: Towards robust interpretability with self-explaining neural networks. arXiv preprint arXiv:1806.07538 (2018)
3. Bai, T., Zhang, S., Egleston, B.L., Vucetic, S.: Interpretable representation learning for healthcare via capturing disease progression through time. In: Proceedings of the 24th ACM SIGKDD International Conference on Knowledge Discovery & Data Mining, pp. 43–51. ACM (2018)
4. Bejnordi, B.E., et al.: Diagnostic assessment of deep learning algorithms for detection of lymph node metastases in women with breast cancer. JAMA **318**(22), 2199 (2017). https://doi.org/10.1001/jama.2017.14585
5. De Fauw, J., et al.: Clinically applicable deep learning for diagnosis and referral in retinal disease. Nat. Med. **24**(9), 1342–1350 (2018). https://doi.org/10.1038/s41591-018-0107-6
6. Erhan, D., Bengio, Y., Courville, A., Vincent, P.: Visualizing higher-layer features of a deep network. Univ. Montreal **1341**(3), 1 (2009)
7. Fawaz, H.I., Forestier, G., Weber, J., Idoumghar, L., Muller, P.A.: Deep learning for time series classification: a review. Data Min. Knowl. Disc. **33**(4), 917–963 (2019)
8. Fong, R.C., Vedaldi, A.: Interpretable explanations of black boxes by meaningful perturbation. In: Proceedings of the IEEE International Conference on Computer Vision, pp. 3429–3437 (2017)
9. Goodfellow, I.J., Shlens, J., Szegedy, C.: Explaining and Harnessing Adversarial Examples. arXiv:1412.6572 (2014)
10. Guidotti, R., Monreale, A., Ruggieri, S., Turini, F., Giannotti, F., Pedreschi, D.: A survey of methods for explaining black box models. ACM Comput. Surv. (CSUR) **51**(5), 1–42 (2018)
11. Hannun, A.Y., et al.: Cardiologist-level arrhythmia detection and classification in ambulatory electrocardiograms using a deep neural network. Nat. Med. **25**(1), 65–69 (2019). https://doi.org/10.1038/s41591-018-0268-3
12. Hinton, G.: Deep learning—a technology with the potential to transform health care. JAMA **320**(11), 1101–1102 (2018). https://doi.org/10.1001/jama.2018.11100
13. Ignatiev, A., Narodytska, N., Marques-Silva, J.: On relating explanations and adversarial examples. In: Advances in Neural Information Processing Systems, pp. 15857–15867 (2019)
14. Lipton, Z.C.: The doctor just won't accept that! arXiv preprint arXiv:1711.08037 (2017)
15. Lundberg, S.M., Lee, S.I.: A unified approach to interpreting model predictions. In: Advances in Neural Information Processing Systems, pp. 4765–4774 (2017)
16. Macfarlane, P.W., Devine, B., Clark, E.: The university of glasgow (Uni-G) ECG analysis program. In: Computers in Cardiology, pp. 451–454 (2005). https://doi.org/10.1109/CIC.2005.1588134

17. Macfarlane, P., Devine, B., Clark, E.: The university of glasgow (uni-g) ECG analysis program. In: Computers in Cardiology, 2005, pp. 451–454. IEEE (2005)
18. Macfarlane, P., van Oosterom, A., Pahlm, O., Kligfield, P., Janse, M., Camm, J.: Comprehensive electrocardiology. Springer, Heidleberg (2010). https://doi.org/10.1007/978-1-84882-046-3, 978-1-84882-046-3
19. Makowski, D.: Neurokit: A python toolbox for statistics and neurophysiological signal processing (eeg, eda, ecg, emg...). Memory and Cognition Lab' Day, 01 November, Paris, France (2016)
20. McKinney, S.M., et al.: International evaluation of an AI system for breast cancer screening. Nature **577**(7788), 89–94 (2020). https://doi.org/10.1038/s41586-019-1799-6
21. Naylor, C.: On the prospects for a (deep) learning health care system. JAMA **320**(11), 1099–1100 (2018). https://doi.org/10.1001/jama.2018.11103
22. Nguyen, T.L., Gsponer, S., Ilie, I., Ifrim, G.: Interpretable time series classification using all-subsequence learning and symbolic representations in time and frequency domains. arXiv preprint arXiv:1808.04022 (2018)
23. Pepine, C.J.: Complete cardiology in a heartbeat. https://www.healio.com/cardiology/learn-the-heart
24. Petsiuk, V., Das, A., Saenko, K.: Rise: randomized input sampling for explanation of black-box models. arXiv preprint arXiv:1806.07421 (2018)
25. Ribeiro, A.H., et al.: Automatic diagnosis of the 12-lead ECG using a deep neural network. Nat. Commun. **11**(1), 1–9 (2020)
26. Ribeiro, A.H., et al.: Automatic diagnosis of the short-duration 12-Lead ECG using a deep neural network: the CODE study. arXiv (2019)
27. Ribeiro, A.L.P., Duncan, B.B., Brant, L.C., Lotufo, P.A., Mill, J.G., Barreto, S.M.: Cardiovascular health in Brazil: trends and perspectives. Circulation **133**(4), 422–433 (2016)
28. Ribeiro, M.T., Singh, S., Guestrin, C.: Why should i trust you?: Explaining the predictions of any classifier. In: Proceedings of the 22nd ACM SIGKDD international conference on knowledge discovery and data mining. pp. 1135–1144. ACM (2016)
29. Ribeiro, M.T., Singh, S., Guestrin, C.: Anchors: Hhigh-precision model-agnostic explanations. In: Thirty-Second AAAI Conference on Artificial Intelligence (2018)
30. Selvaraju, R.R., Cogswell, M., Das, A., Vedantam, R., Parikh, D., Batra, D.: Grad-cam: visual explanations from deep networks via gradient-based localization. In: Proceedings of the IEEE International Conference on Computer Vision, pp. 618–626 (2017)
31. Shrikumar, A., Greenside, P., Kundaje, A.: Learning important features through propagating activation differences. In: Proceedings of the 34th International Conference on Machine Learning, vol. 70, pp. 3145–3153. JMLR. org (2017)
32. Smith, S.W., et al.: A deep neural network learning algorithm outperforms a conventional algorithm for emergency department electrocardiogram interpretation.J. Electrocardiol. **52**, 88–95 (2019)
33. Stead, W.W.: Clinical implications and challenges of artificial intelligence and deep learning. JAMA **320**(11), 1107–1108 (2018). https://doi.org/10.1001/jama.2018.11029
34. Sundararajan, M., Taly, A., Yan, Q.: Axiomatic attribution for deep networks. In: Proceedings of the 34th International Conference on Machine Learning, vol. 70, pp. 3319–3328. JMLR. org (2017)

35. Surawicz, B., Childers, R., Deal, B.J., Gettes, L.S.: Aha/accf/hrs recommendations for the standardization and interpretation of the electrocardiogram: part iii: intraventricular conduction disturbances a scientific statement from the american heart association electrocardiography and arrhythmias committee, council on clinical cardiology; the american college of cardiology foundation; and the heart rhythm society endorsed by the international society for computerized electrocardiology. J. Am. Coll. Cardiol. **53**(11), 976–981 (2009)
36. Topol, E.: Deep Medicine: How Artificial Intelligence Can Make Healthcare Human Again. Hachette UK, London (2019)
37. Ventura, F., Cerquitelli, T., Giacalone, F.: Black-box model explained through an assessment of its interpretable features. In: Benczúr, A., et al. (eds.) ADBIS 2018. CCIS, vol. 909, pp. 138–149. Springer, Cham (2018). https://doi.org/10.1007/978-3-030-00063-9_15
38. Winkler, J.K., et al.: Association between surgical skin markings in dermoscopic images and diagnostic performance of a deep learning convolutional neural network for melanoma recognition. JAMA Dermatology (2019). https://doi.org/10.1001/jamadermatol.2019.1735
39. Zhang, X., Solar-Lezama, A., Singh, R.: Interpreting neural network judgments via minimal, stable, and symbolic corrections. In: Advances in Neural Information Processing Systems, pp. 4874–4885 (2018)

Applied Data Science: E-Commerce and Finance

A Deep Reinforcement Learning Framework for Optimal Trade Execution

Siyu Lin(✉) and Peter A. Beling

Department of Engineering Systems and Environment, University of Virginia, Charlottesville, VA, USA
{sl5tb,pb3a}@virginia.edu

Abstract. In this article, we propose a deep reinforcement learning based framework to learn to minimize trade execution costs by splitting a sell order into child orders and execute them sequentially over a fixed period. The framework is based on a variant of the Deep Q-Network (DQN) algorithm that integrates the Double DQN, Dueling Network, and Noisy Nets. In contrast to previous research work, which uses implementation shortfall as the immediate rewards, we use a shaped reward structure, and we also incorporate the zero-ending inventory constraint into the DQN algorithm by slightly modifying the Q-function updates relative to standard Q-learning at the final step.

We demonstrate that the DQN based optimal trade execution framework (1) converges fast during the training phase, (2) outperforms TWAP, VWAP, AC and 2 DQN algorithms during the backtesting on 14 US equities, and also (3) improves the stability by incorporating the zero ending inventory constraint.

Keywords: Deep reinforcement learning · Optimal trade execution · Shaped reward structure · Zero ending inventory constraint · US equities

1 Introduction

Algorithmic trading involves the use of computer programs, algorithms, and advanced mathematical models to make trading decisions and transactions in the financial market. It has become prevalent in major financial markets since the late 1990s and has dominated the modern electronic trading markets. Optimal trading execution is one of the most crucial problems in the realm of algorithmic trading, as the profitability of many trading strategies depends on the effectiveness of the trade execution. Optimal trade execution problem concerns about how to best trade a set of stock shares at a minimal cost. Mostly, the trading agent has to trade off between two scenarios: 1) trade quickly with the risk of encountering a large amount of loss due to limited liquidity with certainty, and 2) trade slowly with the risk of adverse market movement.

Y. Dong et al. (Eds.): ECML PKDD 2020, LNAI 12461, pp. 223–240, 2021.
https://doi.org/10.1007/978-3-030-67670-4_14

Bertsimas and Lo are the pioneers in the realm of optimal trade execution. They use a dynamic programming approach to find an explicit closed-form solution by minimizing trade execution costs of large transactions over a fixed trading period [3]. Huberman, Stanzl [10] and Almgren, Chriss [11] extend their work by introducing transaction costs, more complex price impact functions, risk aversion parameters. The closed-form analytical solutions, however, have strong assumptions on the underlying price movement or distributions. In addition to the closed-form solutions, the time-weighted average price (TWAP) strategy and volume-weighted average price (VWAP) strategy are prevalent among practitioners in financial markets [12]. The TWAP and VWAP strategies have very few assumptions; however, both strategies are not able to learn from historical data. In the article, we choose TWAP, VWAP, and AC model as the baseline models as they are optimal under certain assumptions and are widely used in the financial industry.

Reinforcement learning (RL) seems to be a natural choice for the optimal trade execution problem, as it enables the trading agent to interact with the market and to learn from its experiences. The Q-learning, a reinforcement learning technique, does not require a model of the market and hence has fewer assumptions on the price dynamics than the closed-form solutions. Nevmyvaka, Feng, and Kearns have published the first large-scale empirical application of reinforcement learning to optimal trade execution problems [13]. Hendricks and Wilcox propose to combine the Almgren and Chriss model (AC) and RL algorithm and to create a hybrid framework mapping the states to the proportion of the AC-suggested trading volumes [16]. Despite the advantages mentioned above, traditional RL techniques are limited by the curse of dimensionality. This limitation has hindered its application in financial trading due to the high dimensions and the complexity of the underlying dynamics of the financial market.

The Deep Q-Network (DQN) [14], which combines the deep neural network and the Q-learning, can address the curse of dimensionality challenge faced by Q-learning. Deep neural network (DNN) is known as the universal function approximator and is capable of extracting intricate nonlinear patterns from raw dataset [15]. The Deep Reinforcement Learning (DRL) algorithm has achieved many successes, including Alpha Go [18] and autonomous driving. However, only a few research articles have been published to explore its capability in financial trading problems until recently. Bacoyannis et al. from J.P. Morgan have reviewed and discussed the application of deep reinforcement learning in electronic trading and some challenging problems that they have encountered from the perspective of financial practioners [28]. Ning et al. have claimed their research work to be the first to adapt and modify the Deep Q-Learning to the optimal trade execution [23].

An important constraint in the execution problem is that all the shares must be executed by the end of the trading period. In the real world business, the brokers receive contracts or directives from their clients to execute a certain amount of shares within a certain time period. For the brokers, the zero-ending inventory constraint is mandatory. However, previous research [13,16,23] are all based on a generic Q-learning algorithm which fails to address the zero-ending inventory constraint. We will discuss the flaws when applying the generic Q-learning algorithm and propose a modified algorithm for the zero-ending inventory constraint.

In this article, we perform an extensive empirical application of Deep Q-Network to the optimal trade execution problem leveraging NYSE market microstructure data. Our main contributions are: 1) We propose a modified DQN algorithm to address the zero ending inventory constraint. 2) We design a shaped reward structure to standardize the rewards among different stocks and under various market scenarios, which facilitates the convergence of the DQN algorithm and is robust to noise in market data. 3) We perform an extensive experiment to demonstrate the advantages of DQN method over TWAP, VWAP, and AC model as well as Ning et al.'s algorithm and its variant. 4) We carefully design the framework by integrating OpenAI Gym's RL framework [19] and Ray's distributed RLlib [30].

The paper is organized as follows. Section 2 reviews the concepts of limit order book and market microstructure. Section 3 revisits the basics of DQN algorithm, presents our DQN based optimal trade execution framework and describes the experimental methodology and settings. Section 4 describes the data sources, discusses and reports empirical results. Section 5 concludes the paper and proposes future work.

2 Limit Order Book and Market Microstructure

In this section, we provide a brief introduction to the limit order book (LOB) and market microstructure to help understand the problem formation, evaluation results, and analysis in later sections. The LOB has dominated the modern financial market and major stock exchanges such as NASDAQ and NYSE. A limit order book records the quantities and the prices that a buyer or seller is willing to buy or sell for the particular security. When there is a cross between the bid and ask price, i.e., bid price higher than or equal to ask price, the match engine executes the trade and remove the crossed bid/ask orders from the order book.

In Fig. 1, we demonstrate an example of the GOOG order book. The bid and ask orders are listed on the left and right-hand sides and are sorted descending (bid) and ascending (ask) by price, respectively. The top bid and ask orders are known as the best bid and ask. The difference between the best bid and ask prices is the bid-ask spread and is closely related to market liquidity. Suppose a sell order of 300 comes at the price of $576.74,

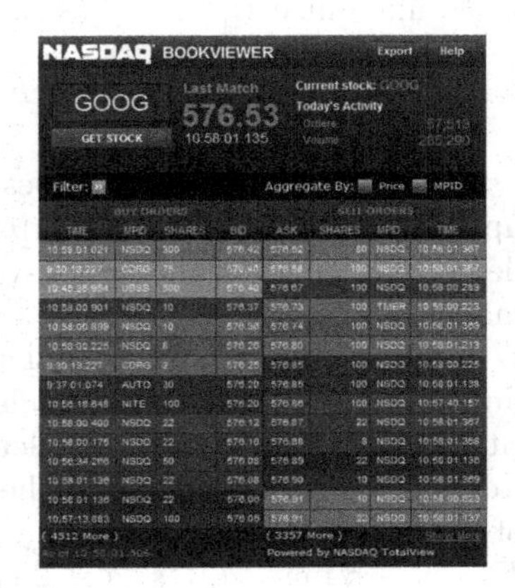

Fig. 1. A snapshot of GOOG order book

the order book is updated, and the share at \$576.74 becomes $100+300=400$. However, when a bid comes at \$576.74, the existing 100 shares are executed first due to First In First Out (FIFO) policies.

In this article, we use the NYSE daily millisecond Trades & Quotes (TAQ) data to reconstruct the limit order book. After order book reconstruction, we develop a market environment/simulator by leveraging OpenAI Gym's framework [19] to read in the constructed historical order book information sequentially and respond to the actions that RL agent takes based on the rules described above.

3 A DQN Formulation to Optimal Trade Execution

The nonlinear function approximation is known to cause instability or even divergence in RL [20]. The DQN algorithm addresses the instability issue by experience replay. The experience replay breaks down the correlations among samples and improves the stability of the DQN algorithm [17]. DeepMind has introduced the Target Q, which freezes the Q functions for a certain amount of steps instead of updating it in each iteration, to further improve the stability [14].

3.1 Preliminaries

Similar to Q-learning, the goal of the DQN agent is to maximize cumulative discounted rewards by making sequential decisions while interacting with the environment. The main difference is that DQN is using a deep neural network to approximate the Q function. At each time step, we would like to obtain the optimal Q function, which obeys the *Bellman equation*. The rationale behind *Bellman equation* is straightforward: if the optimal action-value function $Q^*(s', a')$ was completely known at next step, the optimal strategy at current step would be to maximize $E[r + \gamma Q^*(s', a')]$, where γ is the discounted factor [17]

$$Q^*(s, a) = E_{s' \sim \varepsilon}[r + \gamma \max_{a'} Q^*(s', a')|s, a] \tag{1}$$

The Q-learning algorithm estimates the action-value function by iteratively updating $Q_{i+1}(s, a) = E_{s' \sim \varepsilon}[r + \gamma \max_{a'} Q_i(s', a')|s, a]$. It has already been demonstrated that the action-value Q_i would eventually converge to the optimal action-value Q^* as $i \to \infty$ [21]. The Q-learning iteratively updates the Q table to obtain an optimal Q table. However, it suffers from the curse of dimensionality and is not scalable to large scale problems. The DQN algorithm has addressed this challenge faced by Q-learning. It trains a neural network model to approximate the optimal Q function by minimizing a sequence of loss function $L_i(\theta_i) = E_{s,a \sim \rho(\cdot)}[(y_i - Q(s, a; \theta_i))^2]$ iteratively, where $y_i = E_{s' \sim \varepsilon}[r + \gamma \max_{a'} Q^*(s', a'; \theta_{i-1})|s, a]$ is the target function and $\rho(s, a)$ refers to the probability distribution of states s and actions a. In the DQN algorithm, the target function is usually freezed for a while when optimizing the loss function to prevent the instability caused by the frequently shift in target function.

The gradient could be obtained by differentiating the loss function $\nabla_{\theta_i} L_i(\theta_i) = E_{s,a\sim\rho(\cdot);s'\sim\varepsilon}[((r + \gamma \max_{a'} Q^*(s', a'; \theta_{i-1}))_i - Q(s, a; \theta_i))\nabla_{\theta_i} Q(s, a; \theta_i)]$.

The model weights could be estimated by optimizing the loss function through stochastic gradient descent algorithms. In addition to the capability of handling high dimensional problems, the DQN algorithm is also a *model-free* RL algorithm and has no assumption about the dynamics of the environment. It learns about the optimal policy by exploring the state-action space. The learned policy maps the state space to action space, and is an agent's strategy.

3.2 Problem Formulation

The Q-learning algorithm is a *model-free* technique that has no assumptions on the environment. However, the curse of dimensionality has limited its application in high dimensional problems. Deep Q-Network seems to be a natural choice because it is capable of handling the high dimensional problems while inheriting Q-learning's ability to learn from its experiences. Given the rich amount of information available in the limit order book, we believe that the Deep Q-Network could be more suitable to utilize those high dimensional market microstructure information. In this section, we provide the DQN formulation for the optimal trade execution problem and describe the state, action, reward, and the algorithm used in the experiment.

States. The state is a vector to describe the current status of the environment. In the settings of optimal trade execution, the states consist of 1) Public state: market microstructure variables including top 5 bid/ask prices and associated quantities, bid/ask spread, 2) Private state: remaining inventory and elapsed time, 3) Derived state: we also derive features based on historical limit order book (LOB) states to account for the temporal component in the environment [1]. The derived features could be grouped into three categories: 1) Volatility in VWAP price; 2) Percentage of positive change in VWAP price; 3) Trends in VWAP price. More specifically, we derive the features based on the past 6, 12, and 24 steps (each step is a 5-second interval) respectively. Additionally, we use several trading volumes: 0.5*TWAP, 1*TWAP, and 1.5*TWAP[2] to compute the VWAP price. It is straightforward to derive features in 1). For 2), we record the steps that the current VWAP price increases compared with the previous step and compute the percentage of the positive changes. For 3), we calculate the difference between the current VWAP prices and the average VWAP prices in the past 6, 12, and 24 steps.

Actions. In this article, we choose different numbers of shares to trade based on the liquidity of the stock market. As the purpose of the research is to evaluate

[1] For example, the differences of immediate rewards between current time period and arrival time at various volume levels, manually crafted indicators to flag specific market scenarios (i.e., regime shift, a significant trend in price changes, and so on).

[2] TWAP represents the trading volume of the TWAP strategy in one step.

the capability of the proposed framework to balance the liquidity risk and timing risk, we choose the total number of shares to ensure that the TWAP orders[3] are able to consume at least the 2nd best bid price on average. The total shares to trade for each stock are illustrated in Table 1. In the optimal trade execution framework, we set the range of actions from 0 to 2TWAP and discretize the action space into 20 equally distributed grids. Hence, we have 21 available actions including: 0, 0.1TWAP, 0.2TWAP, ..., 1.9TWAP, 2TWAP. The learned policy maps the state to the 21 actions.

Table 1. The number of shares to trade for each stock in the article

Ticker	Trading shares	Ticker	Trading shares
FB	6000	GS	300
GOOG	300	CRM	1200
NVDA	600	BA	300
MSCI	300	MCD	600
TSLA	300	PEP	1800
PYPL	2400	TWLO	600
QCOM	7200	WMT	3600

Rewards. The reward structure is the key to the DQN algorithm and should be carefully designed to reflect the goal of the DQN agent. Otherwise, the DQN agent cannot learn the optimal policy as expected. In contrast to previous work [13,16,23], we use a shaped reward structure rather than the implementation shortfall[4], a commonly used metric to measure execution gain/loss, as the immediate reward received after execution at each non-terminal step.

There is a common misconception that the implementation shortfall (IS) is a direct optimization goal. In fact, the IS compares the model performance to an idealized policy that assumes infinite liquidity at the arrival price. In the real world, brokers often use TWAP and VWAP as benchmarks. IS is usually used as the optimization goal because TWAP and VWAP prices could only be computed until the end of the trading horizon and cannot be used for real-time optimization. Essentially, IS is just a surrogate reward function, but not a direct optimization goal. Hence, a reward structure is good as long as it can improve model performance.

Additionally, there are some shortcomings of IS reward: 1) it's noisy making learning hard; 2) it's not stationary across scenarios: in the scenario that price is going up, all actions might receive positive rewards while some actions might be bad decisions; in the scenario that price is going down, all actions

[3] TWAP order $= \frac{\text{Total \# of shares to trade}}{\text{Total \# of periods}}$.

[4] Implementation Shortfall=arrival price×traded volume - executed price×traded volume.

might receive negative rewards while we may still want the agent to trade constantly or even more aggressively to prevent an even larger loss at the end. The shaped reward structure attempts to remove the impact of trends to standardize the reward signals and is demonstrated as follows. It is worth mentioning that the shaped reward structure below is only roughly tuned on Facebook's training data. Although it might not be optimal for all the stocks, it generalizes reasonably well on the rest equities in the article as demonstrated in Sect. 4.

1. To penalize divergence from TWAP: $\mathrm{Reward} = -\frac{\mathrm{abs(Action-TWAP)}}{\mathrm{0.1TWAP}}$
2. To encourage the DQN agent to trade more if Implementation Shortfall (IS) of Action is higher than that of TWAP: Reward = 1 if $\mathrm{IS_{Action}} > \mathrm{IS_{TWAP}}$ and Action > TWAP
3. To reward or penalize by comparing the average IS per share at Action with Action − 0.2TWAP and Action + 0.2TWAP: $\mathrm{Reward} = \mathrm{I}\frac{\max(0,\mathrm{Action-TWAP})}{\mathrm{0.2TWAP}}$

$$\mathrm{I} = \begin{cases} 1 & \text{if } \mathrm{AvgIS_+ > AvgIS_- + abs(AvgIS_-)} \text{ or } \mathrm{AvgIS_+ > AvgIS_- + 1} \\ -1 & \text{if } \mathrm{AvgIS_+ < AvgIS_- - abs(AvgIS_-)} \text{ or } \mathrm{AvgIS_+ < AvgIS_- - 1} \\ 0.5 & \text{if } \mathrm{AvgIS_+ > AvgIS_- + 0.1abs(AvgIS_-)} \text{ or } \mathrm{AvgIS_+ > AvgIS_- + 0.1} \\ -0.5 & \text{if } \mathrm{AvgIS_+ < AvgIS_- - 0.1abs(AvgIS_-)} \text{ or } \mathrm{AvgIS_+ < AvgIS_- - 0.1} \\ 0 & \text{otherwise} \end{cases}$$

$\mathrm{AvgIS_+ = AvgIS(Action) - AvgIS(Action + 0.2TWAP)}$
$\mathrm{AvgIS_- = AvgIS(Action - 0.2TWAP) - AvgIS(Action))}$[5]

4. To encourage DQN agent to trade more when IS is positive and IS(Action + 0.1 TWAP) <= 0: $\mathrm{Reward} = 1 + \frac{\max(0,\mathrm{Action-TWAP})}{\mathrm{0.2TWAP}}$
5. To encourage DQN agent to trade as much as possible as long as IS is zero, the Action >= 0.8TWAP and IS(Action + 0.1TWAP) < 0: $\mathrm{Reward} = 0.5 + \frac{0.5\max(0,\mathrm{Action-0.8TWAP})}{\mathrm{0.2TWAP}}$
6. To reward or penalize by comparing the $\mathrm{IS_t}$[6] with $\mathrm{IS_{t-1}}$ and $\mathrm{IS_1}$:

$$\mathrm{Reward} = \begin{cases} \mathrm{I}(1 + \frac{\mathrm{Action-TWAP}}{\mathrm{0.2TWAP}}) \\ \mathrm{I}(1 - \frac{\mathrm{Action-TWAP}}{\mathrm{0.2TWAP}}) \end{cases}$$

$$\mathrm{I} = \begin{cases} 1 & \text{if } \mathrm{IS_t > IS_{t-1} + abs(IS_{t-1})} \text{ or } \mathrm{IS_t > IS_{t-1} + 1} \text{ or } \mathrm{IS_t > IS_1 + abs(IS_1)} \\ & \text{or } \mathrm{IS_t > IS_1 + 1} \text{ or } \mathrm{IS_t < IS_{t-1} - abs(IS_{t-1})} \text{ or } \mathrm{IS_t < IS_{t-1} - 1} \\ & \text{or } \mathrm{IS_t < IS_1 - abs(IS_1)} \text{ or } \mathrm{IS_t < IS_1 - 1} \\ 0.5 & \text{if } \mathrm{IS_t > IS_{t-1} + 0.1abs(IS_{t-1})} \text{ or } \mathrm{IS_t > IS_{t-1} + 0.1} \\ & \text{or } \mathrm{IS_t > IS_1 + 0.1abs(IS_1)} \text{ or } \mathrm{IS_t > IS_1 + 0.1} \\ & \text{or } \mathrm{IS_t < IS_{t-1} - 0.1abs(IS_{t-1})} \text{ or } \mathrm{IS_t < IS_{t-1} - 0.1} \\ & \text{or } \mathrm{IS_t < IS_1 - 0.1abs(IS_1)} \text{ or } \mathrm{IS_t < IS_1 - 0.1} \\ 0 & \text{otherwise} \end{cases}$$

[5] AvgIS is the average IS.
[6] $\mathrm{IS_t}$ = IS at time t.

7. The reward at the final step is different. Generally, we encourage the DQN agent to trade as soon as possible while not going too deep into the market depth. Hence, we penalize the DQN agent if remaining inventory at final step is larger than TWAP order: Reward $= -\frac{40*\text{abs(Action}-\text{TWAP)}}{\text{TWAP}}$. Additionally, we have a minor modification on the final state to incorporate the zero ending inventory constraint into the DQN algorithm. Please refer to Sect. 3.2.4 for details.

As the scale of implementation shortfall varies significantly at different time steps and for various stocks, we use shaped rewards as described above, similar to clipped rewards [14], to standardize the rewards. The shaped rewards are robust to noise in the implementation shortfall and have facilitated the learning process.

The rationale for the shaped rewards is to ensure that the DQN outperforms the TWAP baseline. Rules 1 and 7 above are straightforward. Rule 2 is to incent the DQN agent to trade more than TWAP when trading more can produce more rewards than TWAP. Rule 3 is to incent the DQN agent to minimize the execution cost per share for the selected action as well, as minimizing the total execution costs is equivalent to minimizing the average execution costs per share. Rule 3 could prevent the DQN agent from being myopic and obtaining more rewards at the current step by sacrificing rewards in the future. Rules 4 and 5 encourage the DQN agent to trade more whenever IS >= 0, since the non-negative IS is usually favorite to us and we would like to trade as much as possible in such scenarios. Rule 6 accounts for the trends by comparing rewards obtained with those at step 1 or step t-1. For example, in a downward trend, we would like the agent to trade more at earlier steps to prevent an even larger loss at the end. Moreover, 0.1TWAP in the above rules represents the minimal trading unit. The other coefficients in the rules are chosen either to reflect the significance of certain actions (0.5 in Rule 5 simply represents our preference for a positive reward/IS) or to restrict the action comparison to a certain range. The coefficients above are only roughly tuned on FB training data. In general, the rules above are to incent the DQN agent to stay with the TWAP strategy unless there is a good reason to diverge from it (for example, we would like the agent to trade more when the stock market becomes much more favorable to us).

Zero Ending Inventory Constraint. The major difference in our approach compared with previous research (Nevmyvaka et al. 2006, Hendricks et al. 2014 and Ning et al. 2018) is that our approach combines the last two steps (time $T-1$ and T) and estimates the Q-function for them together. We estimate the Q-function for only $T-1$ steps, while the authors in the aforementioned three articles estimate the Q-function for all T steps. The intuition of our approach is that at time $T-1$, we will know exactly what a_T is after we determine a_{T-1}. Hence, we make only $T-1$ decisions even though there are T steps, and we cannot estimate Q-function for time T alone. To rigorously demonstrate the intuition, let's rewrite the total reward below. γ is the discount factor, r_t , v_t , s_t , and a_t are the immediate reward, number of remaining shares, LOB states and number of shares to trade at time t respectively. At time T, we are forced to liquidate all remaining shares, and we have $a_T = v_T = v_{T-1} - a_{T-1}$. By Markov property, we have $s_T = f(v_{T-1}, s_{T-1}, a_{T-1})$.

$$\text{Total reward} = \sum_{t=1}^{T} \gamma^{t-1} r_t(v_t, s_t, a_t)$$
$$= r_1(v_1, s_1, a_1) + \ldots + \gamma^{T-2} r_{T-1}(v_{T-1}, s_{T-1}, a_{T-1}) + \gamma^{T-1} r_T(v_T, s_T, a_T)$$
$$= r_1(v_1, s_1, a_1) + \ldots + \gamma^{T-2} r_{T-1}(v_{T-1}, s_{T-1}, a_{T-1}) + \gamma^{T-1} r'_T(v_{T-1}, s_{T-1}, a_{T-1})$$

Since r'_T is also based on v_{T-1}, s_{T-1}, and a_{T-1}, we define the Q-function below given an optimal policy $\pi \in \Pi$.

$$Q_t(v_t, s_t, a_t) = \begin{cases} \mathrm{E}[r_t(v_t, s_t, a_t) + \gamma \max_{a_{t+1}} Q_{t+1}(v^{\pi}_{t+1}, s^{\pi}_{t+1}, a^{\pi}_{t+1}))], & t = 1, \ldots, T-2 \\ \mathrm{E}[r_{T-1}(v_{T-1}, s_{T-1}, a_{T-1}) + \gamma r'_T(v_{T-1}, s_{T-1}, a_{T-1})], & t = T-1 \end{cases}$$

However, if we estimate Q-function for time $T-1$ and T separately, we have the Q-function below.

$$Q_t(v_t, s_t, a_t) = \begin{cases} \mathrm{E}[r_t(v_t, s_t, a_t) + \gamma \max_{a_{t+1}} Q_{t+1}(v^{\pi}_{t+1}, s^{\pi}_{t+1}, a^{\pi}_{t+1}))], & t = 1, \ldots, T-1 \\ \mathrm{E}[r'_T(v_{T-1}, s_{T-1}, a_{T-1})], & t = T \end{cases}$$

After maximizing Q_T, we obtain the optimal action a^*_{T-1} and the optimal action-value Q^*_T. When we try to maximize Q_{T-1}: $Q^*_{T-1} = \max_{a_{T-1}} E[r_{T-1}(v_{T-1}, s_{T-1},\ a_{T-1}) + \gamma Q^*_T]$, we will optimize over the same action a_{T-1} again. If the optimal actions a^*_{T-1} estimated by the two functions are not the same, it will cause instability and even divergence as demonstrated in our article. This applies to Nevmyvaka et al. 2006's method as well. Even if their algorithm works backward from the final period, it still optimizes over the same action a_{T-1} twice. Ning et al. 2018's method of adding a quadratic penalty to all immediate rewards does not address the zero-ending issue fundamentally, as we still have to optimize Q-function over the same action a_{T-1} twice.

$$Q_t(v_t, s_t, a_t) = \begin{cases} \mathrm{E}[(r_t(v_t, s_t, a_t) - f_t(a_t)) + \gamma \max_{a_{t+1}} Q_{t+1}(v^{\pi}_{t+1}, s^{\pi}_{t+1}, a^{\pi}_{t+1}))], & t = 1, \ldots, T-1 \\ \mathrm{E}[r'_T(v_{T-1}, s_{T-1}, a_{T-1}) - f_T(a_{T-1})], & t = T \end{cases}$$

Assumptions. The most important assumption in our experiment is that the actions that DQN agent take have only a temporary market impact, and the market is resilient and will bounce back to the equilibrium level at the next time step. This assumption also suggests that the DQN agent's actions do not affect the behaviors of other market participants. The market resilience assumption is the core assumption of this article and also all previous research applying RL for optimal trade execution problems [13,16,23]. The reason is that we are training and testing on the historical data and cannot account for the permanent market impact. However, the equities we choose in the article are liquid, and the actions are relatively small compared with the market volumes. Therefore, the assumption should be reasonable. Secondly, we ignore the commissions and exchange fees as our research is primarily aimed at institutional investors, and those fees are relatively small fractions and are negligible. This is also the practice of previous research [13,16,23]. Thirdly, we apply a quadratic penalty if the trading volume exceeds the available volumes. Fourthly, the remaining unexecuted volumes will be liquidated all at once at the last time step to ensure the execution of all the shares. Fifthly, we also assume direct and fast access to exchanges with no order arrival delays. Finally, if multiple actions result in the same reward, we choose the maximum action (trading volumes). The rationale is that we would like to trade as quickly as possible while not encountering too much loss. Most of the assumptions are also the core assumptions in previous research [13,16,23] because we need a high-fidelity market simulation environments or data collected by implementing the DQN algorithm in the real market rather than historical data to account for these factors such as order delays, permanent market impact, and agent interactions, etc.

3.3 DQN Architecture and Extensions

The vanilla DQN algorithm has addressed the instability issues of using the nonlinear function approximation by techniques such as experience replay and the target network. In this article, we also incorporate several other techniques to improve its stability and performance further. The architectures are chosen in Ray's Tune platform by comparing model performances of all combinations of these extensions on Facebook data only, which is equivalent to ablation studies. A brief description of the DQN architecture and the techniques we use are below.

Network Architecture. In the proposed DQN algorithm, we use a fully connected feedforward neural network with two hidden layers, 128 hidden nodes in each hidden layer, and ReLU activation function in each hidden node. The input layer has 51 nodes including private attributes such as remaining inventory and time elapsed as well as derived attributes based on public market information. The output has 21 nodes with a linear function as the activation function. The Adam optimizer is chosen for weights optimization.

Double DQN. It is well known that the maximization step tends to overestimate the Q function. van Hasselt [24] addresses this issue by decoupling the action selection from its evaluation. In 2016, van Hasselt et al. [25] had successfully integrated this technique with DQN and had demonstrated that the double DQN could reduce the harmful overestimation and improve the performance of DQN. The only change is in the loss function below.

$$\nabla_{\theta_i} L_i(\theta_i) = E_{s,a\sim\rho(\cdot);s'\sim\varepsilon}[((r + \gamma Q(s', \max_{a'} Q^*(s', a'; \theta_{i-1})))_i - Q(s, a; \theta_i))\nabla_{\theta_i} Q(s, a; \theta_i)] \tag{2}$$

Dueling Networks. The dueling network architecture consists of two streams of computations with one stream representing state values and the other one representing action advantages. The two streams are combined by an aggregator to output an estimate of the Q function. Wang et al. have demonstrated the dueling architecture's capability to learn the state-value function more efficiently [26].

Noisy Nets. Noisy nets are proposed to address the limitations of the ϵ-greedy exploration policy in vanilla DQN. A linear layer with a deterministic and noisy stream, as demonstrated in Eq. 3, replaces the standard linear $y = b + Wx$. The noisy net enables the network to learn to ignore the noisy stream and enhance the effectiveness of exploration [27].

$$y = (b + Wx) + (b_{noisy} \odot \epsilon^b + (W_{noisy} \odot \epsilon^w)x) \tag{3}$$

3.4 Experimental Methodology and Settings

In our experiments, we apply the DQN algorithm on several stocks including Facebook (FB), Google (GOOG), Nvidia (NVDA), Msci (MSCI), Tesla (TSLA), PayPal (PYPL), Qualcomm (QCOM), Goldman Sachs (GS), Salesforce.com (CRM), Boeing (BA), Mcdonald's Corp (MCD), PepsiCo (PEP), Twilio (TWLO), and Walmart (WMT) which cover both liquid and illiquid stocks and also high tech, financial and retail industries. We tune the hyperparameters on FB only and apply the same neural network architecture to the remaining stocks due to limited computing resources. The experiment follows the steps below.

1. We obtain one-year millisecond Trade and Quote (TAQ) data of 14 stocks above from WRDS and reconstruct it into the limit order book. Then, we split the data into training (January-September) and test sets (October-December). We set the execution horizon to be 1 min, and the minimum trading interval to be 5 s.
2. The hyperparameters[7] are tuned on FB only due to the limited computing resources. After fine-tuning, we apply the DQN architecture and hyperparameters to the rest stocks.
3. Upon the completion of training, we check the average episode shaped rewards' progression and compare it with the actual rewards (IS). Then, we apply the learned policies to the testing data and compare the average episode rewards and the distribution of rewards against the baseline models.

[7] Please refer to the appendix for the chosen hyperparameters.

4 Experimental Results

In this section, we present the DQN's performance and compare it with the baseline models. The DQN converges very fast and has significantly outperformed the baseline models during the backtesting. In our experiment, we adopt DeepMind's framework to assess the stability in the training phase and the performance evaluation in the backtesting [17].

4.1 Data Sources

We use the NYSE daily millisecond TAQ data from January 1st, 2018 to December 31st, 2018, downloaded from WRDS. The TAQ data is used to reconstruct the limit order book. Only the top 5 price levels from both seller and buyer sides are kept and aggregated at 5 s, which is the minimum trading interval.

4.2 Training and Stability

Assessing the stability and the model performance in the training phase is straightforward in supervised learning by evaluating the training and testing samples. However, it is challenging to evaluate and track the RL agent's progress during training, since we usually use the average episode rewards gained by the agent over multiple episodes as the evaluation metric to track the agent's learning progress. The average episode reward is usually very noisy since the updates on the weights of the Q function can seriously change the distribution of states that the DQN agent visits.

In Fig. 2, we observe that our proposed DQN agent (orange) converges very fast in less than 200,000 steps. Additionally, its learning curves are much more stable than Ning et al.'s algorithm (blue) and its variant (red). Not only does Ning et al.'s algorithm demonstrate great oscillations on FB, PYPL, QCOM, NVDA, WMT, and CRM, but also it diverges on MSCI and PEP. Its variant has incorporated the zero-ending constraint, and is more stable. However, it still shows great oscillations on FB, PYPL, PEP, QCOM, WMT, and CRM, and diverges on BA. In contrast, our proposed DQN agent is much more stable, which is a desired feature in financial trading.

4.3 Main Evaluation and Backtesting

To evaluate the performance of the trained DQN algorithm, we apply it to the test samples from October 2018 to December 2018 for all the stocks and compare its implementation shortfalls with three commonly used baseline models in the industry: TWAP, VWAP, and AC model as well as the DQN algorithm proposed by Ning et al. [23] and an enhanced version of Ning et al.'s algorithm by $\Delta IS = IS_{DQN} - IS_{TWAP}$, $\Delta IS = IS_{DQN} - IS_{VWAP}$ and $IS_{DQN} - IS_{AC}$ (in US dollars) for each execution horizon. For a fair comparison with the AC model,

Table 2. Model Performances Comparison. Mean is based on ΔIS (in US dollars) and STD is based on IS (in US dollars)

Model name	Mean													
	BA	CRM	FB	GOOG	GS	MCD	MSCI	NVDA	PEP	PYPL	QCOM	TSLA	TWLO	WMT
TWAP	0.00	0.00	0.00	0.00	0.00	0.00	0.00	0.00	0.00	0.00	0.00	0.00	0.00	0.00
VWAP	49.66	−141.42	−4,040.78	−65.41	−12.58	−7.89	34.94	−31.46	−77.54	−459.28	−6,737.10	−65.04	−41.27	−1,376.78
AC	−0.78	−3.34	−40.26	−0.92	−0.16	−0.13	−0.34	−1.32	−0.34	−8.88	−127.87	−1.21	−2.96	−2.24
DQN (Ning2018)	91.35	16.65	−545.49	**183.63**	**42.41**	**104.18**	−770.09	70.33	−15,074.82	−30.81	−1,056.94	46.03	**49.77**	140.32
DQN (Ning2018 + ZeroConstraint)	−1,547.46	**56.48**	−2,086.12	160.86	39.76	88.78	**75.59**	**103.38**	86.90	**138.13**	**255.77**	**48.62**	45.38	−24.62
DQN + Shaped Reward + ZeroConstraint	**170.23**	**55.05**	**332.06**	117.65	16.20	34.47	24.23	33.62	**214.06**	63.46	98.87	8.44	31.19	**159.38**
Model name	Standard deviation													
	BA	CRM	FB	GOOG	GS	MCD	MSCI	NVDA	PEP	PYPL	QCOM	TSLA	TWLO	WMT
TWAP	335.05	885.62	**4,154.47**	707.03	92.74	221.72	159.14	478.18	1,058.16	681.91	**2,913.72**	**203.00**	196.88	1,442.92
VWAP	503.64	1,366.18	14,936.29	1,394.18	190.89	420.70	220.25	726.23	2,389.84	2,587.93	27,131.03	462.24	341.32	6,699.03
AC	335.80	894.99	4,517.32	707.40	93.19	222.10	159.59	479.65	1,058.48	776.66	4,724.56	210.49	215.42	1,456.07
DQN (Ning2018)	506.22	1,086.15	5,298.78	962.25	**64.98**	**183.17**	641.79	629.95	5,082.65	1,242.17	6,890.66	239.12	204.75	1,758.85
DQN (Ning2018 + ZeroConstraint)	2,042.13	890.35	6,811.13	953.61	72.21	224.04	158.66	577.01	1,397.35	844.30	3,426.35	220.14	208.52	2,292.27
DQN + Shaped Reward + ZeroConstraint	**156.77**	**744.70**	**4,168.56**	**647.71**	76.11	**183.40**	**147.54**	**434.66**	**948.46**	**581.74**	**3,032.75**	213.79	**178.03**	**1,368.99**
Model name	GLR													
	BA	CRM	FB	GOOG	GS	MCD	MSCI	NVDA	PEP	PYPL	QCOM	TSLA	TWLO	WMT
TWAP	–	–	–	–	–	–	–	–	–	–	–	–	–	–
VWAP	0.59	0.46	0.16	0.46	0.46	0.49	0.72	0.51	0.40	0.22	0.08	0.40	0.41	0.17
AC	0.04	0.02	0.00	0.10	0.16	0.11	0.15	0.06	0.18	0.00	0.01	0.10	0.05	0.02
DQN (Ning2018)	0.69	0.66	0.57	0.78	**2.96**	**2.27**	0.08	0.73	0.01	0.44	0.33	1.40	**1.16**	**1.14**
DQN (Ning2018 + ZeroConstraint)	0.18	**1.35**	0.21	0.71	2.37	1.50	**1.63**	0.89	0.79	**1.20**	**1.89**	**1.86**	**1.16**	0.63
DQN + Shaped Reward + ZeroConstraint	**2.07**	1.20	**1.00**	**1.14**	1.25	1.25	1.13	**1.10**	**1.18**	**1.16**	1.01	1.08	**1.14**	**1.13**

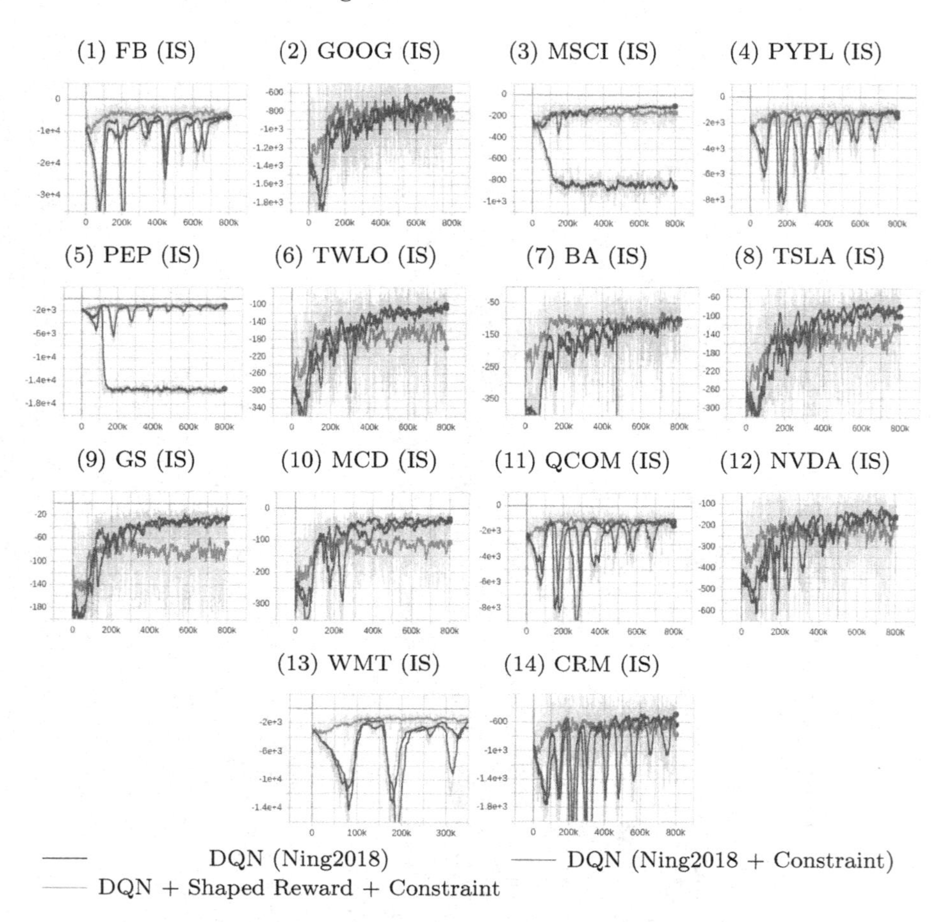

Fig. 2. Training curves tracking the DQN agent's average implementation shortfalls in US dollars (y-axis) against steps (x-axis): Each point is the average IS per episode for 800k steps on the 14 stocks (Color figure online)

we set the permanent price impact parameter of the AC model to 0. We report the mean, standard deviation, and gain-loss ratio (GLR)

$$\mathrm{GLR} = \frac{\mathrm{E}[\Delta\mathrm{IS}|\Delta\mathrm{IS} > 0]}{\mathrm{E}[-\Delta\mathrm{IS}|\Delta\mathrm{IS} < 0]} \tag{4}$$

The statistical results for all the stocks are summarized above in Table 2. We observe that our proposed DQN algorithm outperforms the TWAP and the AC model on all the 14 stocks and outperforms VWAP on all stocks except MSCI. By comparing the three DQN algorithms, we observe that the our proposed DQN algorithm have better overall performances than the other two in terms of Mean, Standard Deviation, and GLR. Although Ning et al.'s DQN algorithm has performed well on several stocks, it diverges on 3 stocks and has very large standard divergences on 5 stocks which are highlighted in red. In the real-world trading, such divergence is very dangerous and should be avoided. Its enhanced version is more stable, but still diverges on 2 stocks. In contrast, our proposed DQN algorithm is much more stable with no divergence and smaller standard deviations on most stocks suggesting that it is able to handle outliers and volatility better than the other two DQN algorithms.

Table 3. Hyperparameters used in the article

Hyperparameter	Value	Description
Minibatch size	16	Size of a batched sampled from replay buffer for training
Replay memory size	10000	Size of the replay buffer
Target network update frequency	2000	Update the target network every 'target network update freq' steps
Sample batch size	4	Update the replay buffer with this many samples at once
Discount factor	0.99	Discount factor gamma used in the Q-learning update
Timesteps per iteration	100	Number of env steps to optimize for before returning
Learning rate	0.0001	The learning rate used by Adam optimizer
Initial exploration	0.15	Fraction of entire training period over which the exploration rate is annealed
Final exploration	0.05	Final value of random action probability
Final exploration frame	125000	Final step of exploration
Replay start size	5000	The step that learning starts and a uniform random policy is run before this step
Number of hidden layers	2 for both main and noisy networks	All of main, dueling and noisy networks have 2 hidden layers
Number of hidden nodes	[128,128] for main network and [32,32] for noisy network	The main and dueling networks have 128 hidden nodes for each hidden layer; while noisy network has 32 hidden nodes for each hidden layer
Number of input/output nodes	Input: 51; output: 21	The input layer has 51 nodes while the output layer has 21 nodes
Activation functions	Relu for hidden layers and linear for output layer	For the hidden layers, we use Relu activation functions and a linear function for the output layer

We exclude the methods proposed by Nevmyvaka et al. [13] and Hendricks and Wilcox [16] from the performance comparison for several reasons: 1) Traditional Reinforcement Learning is not scalable to high dimensional problems. 2) Both articles are using market data quite a while ago. Nevmyvaka et al.'s method was tested on three NASDAQ stocks, AMZN, NVDA, and QCOM, before 2006. Hendricks and Wilcox's method was tested on three South Africa's stocks, SBK,

AGL, and SAB, in 2012. It's difficult to evaluate the effectiveness of their methods, since the dynamics of market microstructure have already changed dramatically in the past few years.

5 Conclusion and Future Work

In this article, we propose an adaptive trade execution framework based on a modified DQN algorithm which incorporates the zero-ending inventory constraint. The experiment results suggest that the DQN algorithm is a promising technique for the optimal trade execution and has demonstrated significant advantages over TWAP, VWAP, AC model as well as Ning et al.'s algorithm and its variant. We use shaped rewards as the reward signals to encourage the DQN agent to conduct correct behaviors and has demonstrated its advantage over the commonly used implementation shortfall.

The current experiment relies on the assumption that the DQN agent's actions are independent of other market participants' actions. It will be exciting if we could model the interactions of multiple DQN agents and their collective decisions in the market.

A Hyperparameters

We fine-tuned the hyperparameters on FB only, and we did not perform an exhaustive grid search on the hyperparameter space, but rather to draw random samples from the hyperparameter space due to limited computing resources. Finally, we choose the hyparameters listed in Table 3.

References

1. Abelson, H., Sussman, G.-J., Sussman, J.: Structure and Interpretation of Computer Programs. MIT Press, Cambridge (1985)
2. Baumgartner, R., Gottlob, G., Flesca, S.: Visual information extraction with Lixto. In: Proceedings of the 27th International Conference on Very Large Databases, pp. 119–128. Morgan Kaufmann, Rome (2001)
3. Bertsimas, D., Lo, A.-W.: Optimal control of execution costs. J. Finan. Mark. **1**(1), 1–50 (1998)
4. Brachman, R.-J., Schmolze, J.-G.: An overview of the KL-ONE knowledge representation system. Cogn. Sci. **9**(2), 171–216 (1985)
5. Gottlob, G.: Complexity results for nonmonotonic logics. J. Logic Comput. **2**(3), 397–425 (1992)
6. Gottlob, G., Leone, N., Scarcello, F.: Hypertree decompositions and tractable queries. J. Comput. Syst. Sci. **64**(3), 579–627 (2002)
7. Levesque, H.-J.: Foundations of a functional approach to knowledge representation. Artif. Intell **23**(2), 155–212 (1984)
8. Levesque, H.-J.: A logic of implicit and explicit belief. In: Proceedings of the Fourth National Conference on Artificial Intelligence, pp. 198–202. American Association for Artificial Intelligence, Austin (1984)

9. Nebel, B.: On the compilability and expressive power of propositional planning formalisms. J. Artif. Intell. Res. **12**, 271–315 (2000)
10. Huberman, G., Stanzl, W.: Optimal liquidity trading. Rev. Finan. **9**(2), 165–200 (2005)
11. Almgren, R., Chriss, N.: Optimal execution of portfolio transactions. J. Risk **3**, 5–40 (2000)
12. Berkowitz, S.-A., Logue, D.-E., Noser Jr., E.-A.: The total cost of transactions on the NYSE. J. Finan. **43**(1), 97–112 (1988)
13. Nevmyvaka, Y., Feng, Y., Kearns, M.: Reinforcement learning for optimal trade execution. In: Proceedings of the 23rd International Conference on Machine Learning, pp. 673–680. Association for Computing Machinery, Pittsburgh (2006)
14. Mnih, V., et al.: Human-level control through deep reinforcement learning. Nature **518**, 529–533 (2015)
15. Hornik, K., Stinchcombe, M., White, H.: Multilayer feedforward networks are universal approximators. Neural Netw. **2**(5), 359–366 (1989)
16. Hendricks, D., Wilcox, D.: A reinforcement learning extension to the Almgren-Chriss framework for optimal trade execution. In: Proceedings from IEEE Conference on Computational Intelligence for Financial Economics and Engineering, pp. 457–464. IEEE, London (2014)
17. Mnih, V., et al.: Playing atari with deep reinforcement learning. arXiv preprint arXiv:1312.5602 (2013)
18. Silver, D., et al.: Mastering the game of Go without human knowledge. Nature **550**, 354–359 (2017)
19. Brockman, G., et al.: OpenAI Gym. arXiv preprint arXiv:1606.01540 (2016)
20. Tsitsiklis, J.-N., Van Roy, B.: An analysis of temporal-difference learning with function approximation. IEEE Trans. Autom. Control **42**(5), 674–690 (1997)
21. Sutton, R., Barto, A.: Reinforcement Learning: An Introduction. MIT Press, Cambridge (1998)
22. Lundberg, S.-M., Lee, S.-I.: A unified approach to interpreting model predictions. In: Advances in Neural Information Processing Systems, Long Beach, CA, pp. 4768–4777 (2017)
23. Ning, B., Ling, F.-H.-T., Jaimungal, S.: Double Deep Q-Learning for Optimal Execution. arXiv:1812.06600 (2018)
24. van Hasselt, H.: Double Q-learning. In: Advances in Neural Information Processing Systems, Vancouver, British Columbia, Canada, pp. 2613–2621 (2010)
25. van Hasselt, H., Guez, A., Silver, D.: deep reinforcement learning with double Q-learning. In: Proceedings of the 30th AAAI Conference on Artificial Intelligence, Phoenix, Arizona, pp. 2094–2100 (2016)
26. Wang, Z., Schaul, T., Hessel, M., van Hasselt, H., Lanctot, M., de Freitas, N.: Dueling network architectures for deep reinforcement learning. In: Proceedings of the 33rd International Conference on Machine Learning, pp. 1995–2003. JMLR, New York (2016)
27. Fortunato, M., et al.: Noisy networks for exploration. In: International Conference on Learning Representations, Vancouver, British Columbia, Canada (2018)
28. Bacoyannis, V., Glukhov, V., Jin, T., Kochems, J., Song, D.-R.: Idiosyncrasies and challenges of data driven learning in electronic trading. In: NIPS 2018 Workshop on Challenges and Opportunities for AI in Financial Services: The Impact of Fairness, Montréal, Canada (2018)

29. Liaw, R., Liang, E., Nishihara, R., Moritz, P., Gonzalez, J.-E., Stoica, I.: Tune: a research platform for distributed model selection and training. arXiv preprint arXiv:1807.05118 (2018)
30. Liang, E., et al.: RLlib: abstractions for distributed reinforcement learning. In: Proceedings of the 35th International Conference on Machine Learning, Stockholm, Sweden (2018)

Detecting and Predicting Evidences of Insider Trading in the Brazilian Market

Filipe Lauar(✉) and Cristiano Arbex Valle

Department of Computer Science, UFMG - Universidade Federal de Minas Gerais, Av. Antônio Carlos, Belo Horizonte 6627, Brazil
filipe0lauar@gmail.com, arbex@dcc.ufmg.br

Abstract. Insider trading is known to negatively impact market risk and is considered a crime in many countries. The rate of enforcement however varies greatly. In Brazil especially very few legal cases have been pursued and a dataset of previous cases is, to the best of our knowledge, nonexistent. In this work, we consider the Brazilian market and deal with two problems. Firstly we propose a methodology for creating a dataset of evidences of insider trading. This requires both identifying impactful news events and suspicious negotiations that preceded these events. Secondly, we use our dataset in an attempt to recognise suspicious negotiations before relevant events are disclosed. We believe this work can potentially help funds in reducing risk exposure (suspicious trades may indicate undisclosed impactful news events) and enforcement agencies in focusing limited investigation resources. We employed a Machine Learning approach based on features from both spot and options markets. In our computational experiments we show that our approach consistently outperforms random predictors, which were developed due to lack of other related works in literature.

Keywords: Insider trading · Fraud detection · Data science

1 Introduction

Insider trading refers to negotiations of securities of a public company by individuals (insiders) with access to related nonpublic information. Traders with access to solely public information are always at a disadvantage when negotiating with insiders, and thus insider trading is generally seen as a source of increased overall risk and reduced participation in financial markets [2].

Non surprisingly, in most countries this practice is considered a crime. The crime occurs when insiders use such information to trade securities for the benefit of themselves or others, and it can be punished with fines and/or imprisonment. Bhattacharya and Daouk [4] revealed existing insider trading laws in 87 out of 103 countries, but enforcement had taken place in only 38 of those. The authors reported evidence that the equity cost in a country decreases significantly after

F. Lauar and C. Arbex Valle—Funded by CNPQ grant 420729/2018-6.

Y. Dong et al. (Eds.): ECML PKDD 2020, LNAI 12461, pp. 241–256, 2021.
https://doi.org/10.1007/978-3-030-67670-4_15

the first prosecution. Dent [7] argued about possible disastrous effects if insider trading were to be legalised.

The Securities and Exchange Commission (SEC), the American regulatory agency for financial markets, requires every insider to report their trades within 10 days. According to the Insider Trading Sanctions Act of 1984, the SEC may bring legal action against any person that is believed to have bought or sold securities while in possession of nonpublic information. In 2017 and 2018, nearly 100 cases of insider trading were identified by the SEC [21].

In Brazil, up to 2019 very few cases had been identified and prosecuted by the Brazilian Regulatory Agency (*Comissão de Valores Mobiliários*, CVM). Despite its legal repercussions, actual punishment is rare due to difficulty in detection and due to being hard to establish the causal link between the damage and the fault of the agent [20].

In this work, we deal with two problems regarding the Brazilian market. The first consists in creating a nontrivial dataset of possible evidences of insider trading. Due to lack of available data and successful prosecutions in Brazil, we had to prepare this dataset from the ground up.

For insider trading to be characterised, non-public impactful information must be disclosed, generally through the publication of news articles. Therefore the first step consisted in identifying a set of relevant news events which are believed to have caused abrupt stock price movement. The next step consisted in classifying as possible evidence "abnormal" or "suspicious" negotiations (or set of negotiations) that happened prior to the publication of these events. We developed a strict (albeit arbitrary) methodology for both classifying an event as relevant and trades as "abnormal" or "suspicious", that is, outside the normal trading patterns of the stock (in this paper we use trades and negotiations interchangeably). The methodologies are inspired by rules followed by other scientific studies and documents from regulatory agencies. We analysed data from both the spot and options markets.

We considered a universe of 54 stocks traded in the Brazilian stock market during 2017 and 2018. The datasets for each year were created separately. For 2017, much of the work had to be done manually. We then used the 2017 dataset as basis for a machine learning (ML) approach intended to help reducing the workload required to create the 2018 dataset. Our method identified 395 possible evidences of insider trading in 2017 and 2018, while only one related prosecution was carried forward in Brazil during those same years.

The second problem consists in attempting to recognise suspicious negotiations before relevant events are disclosed. We employed a ML approach based on features from both spot and options markets. Given market derived features and abnormal market movements, is it possible to predict whether impactful news events likely to impact stock prices are about to be disclosed so that these trades configure possible insider trading?

Successfully predicting impactful events may bring considerable financial benefit to investment funds - for example, by reducing risk exposure before such events increase market volatility. At the same time, recognising possible insider trading before the information is made public could help regulatory agencies in anticipating crimes and facilitating law enforcement.

Our proposed methodology is certainly subjective and biased to some extent. We cannot verify the dataset since that would require the outcome of successful and unsuccessful prosecutions as well as discarded investigations by the CVM, which we do not possess. We emphasise that the evidences in our dataset did not necessarily constitute actual crimes. We believe however that our methodology could help enforcement agencies focus limited investigation resources in fewer suspicious cases. Moreover, we are making our dataset and code available so that other researchers can help improve them. We invite the reader to verify the supplementary material accompanying this paper[1].

The rules derived to classify trades as abnormal are especially arbitrary and may not be the most appropriate. We however show that our computational results consistently outperform random predictors. We believe our approach may be a good starting point for further improvement.

The remainder of this paper is organised as follows. In Sect. 2, we give a brief overview of previous works found in literature. In Sect. 3, we discuss, in detail, the methodology and dataset proposed. In Sect. 4 we discuss the problem of identifying abnormal trades before the related events are disclosed, and we present our concluding remarks in Sect. 5.

2 Literature Review

Easley *et al.* (2002) [9] investigated the link between private information and asset returns and how it affects the dynamics of markets. They derived an explicit measure of the probability of private information-based trading for an individual stock and estimated this probability to NYSE stocks. The authors concluded that their measure was the predominant factor explaining returns. Martins *et al.* (2013) [15] used the same model for the Brazilian stock market and found similar results.

Mitchell and Netter (1993) [18] presented five detailed cases of securities fraud identified by the SEC. These cases showed how the SEC used empirical evidence derived from event studies based on the Efficient Market Hypothesis [10] to establish materiality and calculate damages in securities fraud cases. Minenna (2000, 2003) [16,17] argued that the SEC econometric approach could not be applied to all insider trading schemes due to the required assumption of statistically significant results. The author adapted the procedure for the Italian market with a probabilistic approach. At the time, the procedure was implemented by the Italian Securities and Exchange Comission.

Ahern (2018) [1] studied whether standard measures of illiquidity are able to capture the occurrence of insider trading. The author used data from SEC and the American Department of Justice for 410 insider trading cases between 1996 and 2013. Most cases occurred up to 20 days before the disclosure of the event.

[1] The dataset can be found at https://github.com/filipelauar/Insider-trading/, with detailed explanations given in the supplementary material.

Kulkarni *et al.* (2017) [14] used data from the SEC in building relationship graphs to capture complex patterns. The authors believe that a generative approach may be useful in learning the underlying distributions of insider events.

Goldberg *et al.* (2003) [12] developed the SONAR (Securities Observation, News Analysis & Regulation), an artificial intelligence based system whose aim is to monitor trades at the NASDAQ over-the-counter and futures market and identify possible instances of insider trading. The automated alert system improved the precision of guesses from 0.58% to 3.73%. The SONAR system first looks for news events and then looks backwards to find trading anomalies.

Donoho (2004) [8] analysed options market data and both structured and unstructured news events hoping to find traces of insider trading. The author emphasises the difficulty of the problem, and report three algorithms that were able to slightly outperform random classifiers.

Islam *et al.* (2018) [13] attempted to predict insider trading activities before the disclosure of news events. The authors used a LSTM recurrent neural network to predict stock volume based on the stock time series and used that information to detect anomalous patterns. The authors reported positive results when comparing their approach to real litigation cases from the SEC.

3 Dataset Preparation

In this section we detail the process developed to create a dataset of possible evidences of insider trading. Two key pieces of information are necessary: identifying impactful news events and "abnormal" negotiations prior to these events.

Our dataset spans 2017 and 2018. We however prepared data for each year separately. Due to lack of prior information, creating the 2017 dataset required heavy manual labour. For the 2018 dataset, on the other hand, we could make use of 2017 data to generate an initial version through a ML approach. We used 2017 data as training and validation sets in an attempt to identify both days with relevant events and possible evidences of insider trading. From this initial version we performed manual checks to discard data we believed was classified incorrectly.

Due to the laborious manual work involved, it is highly likely that some relevant news events may have been omitted and that the impact of some events may have been overestimated. It is also highly unlikely that we identified all possible evidences of insider trading and that all evidences we identified did indeed constitute crimes. We however emphasise that the goal is not to generate flawless data, but to help enforcement agencies focus on fewer promising cases. Moreover, we are making our dataset publicly available for future improvements.

3.1 Impactful Events in 2017

The first step consists in compiling a list of news events categorised as "impactful". There are several data providers which use natural language processing and ML to analyse sentiment in news events and their relevance to companies.

This results in preprocessed news data which can potentially be used as input in prediction models of financial data. However, to the best of our knowledge news events classified as impactful and of high relevance by such tools do not necessarily translate into large/extreme returns once such news are published. For more information we refer the reader to [19].

We therefore adopted an inverted logic: we started by looking for abnormal daily returns. For each stock in our asset universe we analysed their daily return distributions and compiled a list of extreme returns. We filter all days whose return is within the 5% most extreme in each tail of the distribution. We then manually looked for news events which may have been the trigger for such returns. If any is found, we classify it as a day with an impactful event.

Increased trading volume and large intraday movements (even if the final daily return is not extreme) could also be used for classifying preceding news events as impactful. However, we opted not to use them since a hypothetical insider negotiation prior to a day with these features may not necessarily result in financial benefit for the perpetrator.

The first 20 trading days in 2017 were ignored since this is the time frame prior to the publication of an event in which we look for evidences of insider trading (as will be described later). This leaves 222 remaining trading days in 2017. Considering our previously mentioned universe of 54 stocks, there are a total of 11988 trading days. Out of these, we classified 1119 days as days with an extreme return. This was further reduced by taking into account the known effect of volatility clustering [6] to discard abnormal returns following other abnormal returns, except when distinct relevant news events are found in consecutive days. Clearly some subjective judgment is necessary in defining the return threshold, and we hope to reduce any personal bias by making our dataset publicly available.

Ahern [1] classified the events for 410 insider trading cases in the United States, including 312 different companies. These events were distributed in the following categories: merges and acquisitions (52%), earnings announcements (28%), drug regulations (9%) and various other announcements about operations, securities issuance and financial distress (11%). Similarly we considered the following subjects: mergers and acquisitions, quarterly results, announcements and any kind of breaking news that affects the company. We considered news events published in traditional Brazilian and foreign media outlets.

Trading days in 2017 were divided in days with "Good events" (84 days), "Bad events" (87 days) and "No events" (11817 days, including days with no extreme returns). Non surprisingly the dataset is highly unbalanced, with only 1.43% days classified with events, corresponding to approximately 3 days per stock per year.

3.2 Classifying Possible Evidences of Insider Trading

Classifying negotiations as suspicious is nontrivial and also prone to subjectiveness and bias. Nevertheless we propose an analytical set of rules, some defined arbitrarily, partly based on analysis of previous works. We studied regulations

[11] and litigation releases [22] in both SEC and CVM documents [20]. We reviewed two relevant works in literature: two cases studied in [8] in the options market and five other cases detailed in [18] in both spot and options markets. We also reviewed the set of rules used in [13] as features in their LTSM approach. The rules shown below were either taken from these cases or based on our own observation of the data.

We begin by calculating some features from tick market data (obtained from B3, the Brazilian stock market [3]). Let N denote the set of stocks in our asset universe and T denote the set of trading days in 2017. Let S^{it} be the set of $|S^{it}|$ individual negotiations for stock $i \in N$ in day $t \in T$. For $s \in S^{it}$, we define x_s as the number of shares traded in negotiation s. Let σ_i, $i \in N$, be the standard deviation of x_s $\forall s \in S^{it}, t \in T$. Consider also M^t as the set of all single milliseconds during open market hours in day t. Let $\beta_{sm} = 1$ if negotiation s occurred in millisecond $m \in M^t$, zero otherwise.

For the spot market, we calculated the following features:

$$\forall t \in T, i \in N \qquad \max_{s \in S^{it}} x_s \tag{1}$$

$$\forall t \in T, i \in N \qquad \max_{m \in M^{it}} \sum_{s \in S^{it}} \beta_{sm} x_s \tag{2}$$

$$\forall t \in T, i \in N \qquad \sum_{s \in S^{it}} \{x_s : x_s > 10\sigma_i\} \tag{3}$$

For a given day and asset, Feature (1) represents the largest negotiation and Feature (2) is the largest volume traded in any single millisecond. Feature (3) is the sum of shares traded in all negotiations whose size is greater than $10\sigma_i$.

Let S_C^{it} and S_P^{it} be respectively the sets of $|S_C^{it}|$ call and $|S_P^{it}|$ put options negotiations for stock i in day t, with x_s for both $s \in S_C^{it}$ and $s \in S_P^{it}$ redefined accordingly. For the options market, we calculated:

$$\forall t \in T, i \in N \qquad \sum_{s \in S_C^{it}} x_s \tag{4}$$

$$\forall t \in T, i \in N \qquad \sum_{s \in S_P^{it}} x_s \tag{5}$$

$$\forall t \in T, i \in N \qquad \max_{s \in S_C^{it}} x_s \tag{6}$$

$$\forall t \in T, i \in N \qquad \max_{s \in S_P^{it}} x_s \tag{7}$$

$$\forall t \in T, i \in N \qquad \sum_{s \in S_C^{it}} x_s - \sum_{s \in S_P^{it}} x_s \tag{8}$$

$$\forall t \in T, i \in N \qquad \max_{s \in S_C^{it}} x_s - \max_{s \in S_P^{it}} x_s \tag{9}$$

For a day and asset, Features (4) and (6) are the sum of call and put options negotiated, while Features (6) and (7) give the largest respective negotiations.

The difference between the sum of calls and sum of puts is given in (8) and the difference between the maximum call and put negotiations is given in (9).

Different assets may have very distinct trading patterns, especially regarding liquidity. We observed that the range of values for the features vary greatly, which would prevent common rules applicable to all stocks. In order to have a single set of rules we applied a normalisation technique.

Possible evidences of insider trading are characterised by outliers in the values of features, so we initially adopted two traditional normalisation techniques that are known to preserve them, z-score and min-max. However, we observed that trading patterns for different stocks vary greatly. Some have such large outliers that the absolute vast majority of normalised trades would be within one standard deviation of the mean, while for others with smaller outliers trades would be more evenly distributed. This hinders the ability to find common sets of rules applicable for different stocks, and thus we opted to normalise the data using cumulative distribution functions (CDF) - which loses outlier information but keeps all information in the same scale.

For each pair (asset, feature) we sort its feature values and calculate their (discrete) CDF. We exclude days where the corresponding feature value is zero, instead we set their CDF values to 0.5. This was necessary due to liquidity of some assets: a few have no options negotiated in most trading days. In these cases if we had assigned them a zero in the normalised feature several regular negotiations would be wrongly classified as outliers.

Having all features normalised, we classified trades as abnormal according to the following rules:

- Features may only be considered as abnormal if they happen up to 20 days before the publication of a previously identified news event. We chose this value as Ahern (2018) [1] showed that most cases of insider trading identified by the SEC occurred within this interval.
- Put options anomalies before good events and call options anomalies before bad events are discarded.
- If the current daily return lies within the 5% lowest or 5% highest, we consider it to be a "volatile" day (although we do not attempt to measure volatility directly) and no feature value is classified as abnormal. This is irrespective of there having been a relevant news event prior to this day. If the current daily return lies between the 5% and 15% lowest or between the 85% and 95% highest, we do not consider any feature as abnormal if the succeeding relevant news event was published 10+ days afterwards.
- Features equal or above the 95th percentile are considered abnormal if there is a relevant event within the next 3 succeeding days. This threshold is raised to 96% (6 days), 97% (10 days) and 98% (20 days).
- If, for day t and asset i, $|S^{it}|$ (total number of trades) and $\sum_{s \in S^{it}} x_s$ (total volume of trades) are both within their respective 99th percentiles, then Features (1) and (3) are discarded if their values are below the 97th percentile.

Clearly these rules are arbitrary and not necessarily the most appropriate. We derived them both based on previous works and by empirically observing data patterns, and there is no way to confirm whether features classified as abnormal actually imply insider trades unless all cases are either successfully prosecuted by regulatory agencies or discarded/acquitted after formal investigation. We emphasise however that our intention here is to simply choose a subset of promising trades such that regulatory agencies may focus limited investigation resources. Work in this dataset is ongoing and there are a few exceptions to these rules which were subjectively thought to constitute evidence of insider trading.

After applying the methodology above we found possible evidence of insider trading in 89 out of the 171 events (≈52%). Since we considered 20 days prior to the news event being published, for some events we found abnormal trades more than once. Among 3098 days prior to a relevant event, we found possible evidence of insider negotiations in 146 of them (94 in the spot market, 45 in the options market and 7 in both).

3.3 Expanding the Dataset for 2018

In order to expand the dataset to include data from 2018 we applied a ML approach hoping to reduce the workload. We separated the 2017 data in a training set (up to 10/07) and a validation set (spanning 11/07 to 28/12). We employed XGBoost [5] as the prediction algorithm for all experiments reported in this paper. The corresponding R code is made available as supplementary material.

The first step is again identifying relevant events. The output is a binary value indicating whether there was a relevant news event or not in a given day. We employed as input all features given in the previous section (considering the same day as the new event) as well as the number of negotiations, total volume of shares traded, asset return for days t to $t-5$ and the relative spread between the highest and lowest prices ((high − low)/high). As we move forward in predicting days in 2018 we adopted a rolling window approach for updating σ_i and the normalisation of features, excluding the oldest day and adding a new one. We tuned the hyper-parameters to optimize the Area Under the ROC Curve (AUC) and to obtain 90% recall in the validation set.

With this approach, our algorithm predicted 674 days with possible relevant events in 2018. We manually searched these days for events and obtained the confusion matrix in Table 1.

Table 1. Confusion matrix for the relevant events in 2018

Predicted/Actual	Negative	Positive
Negative	1005	33
Positive	458	216

We found relevant news events in 216 out of the 674 days originally predicted to contain news. By performing a broader search we found 33 days with relevant events that were not identified by our algorithm. Clearly a manual verification of the data is naturally flawed and work on the dataset is ongoing. Nevertheless this approach reduced the manual analysis from an original 1712 days with a precision of ≈32% and a recall of ≈87%. The total distribution for 2017 and 2018 is shown in Table 2.

Table 2. Relevant events distribution for 2017 and 2018

Class	2017	2018	Total
Good events	84	142	226
Bad events	87	107	194
No events	11817	12981	24798
Total	11988	13230	25218

In order to find possible evidences of insider trading in 2018, we employed a similar ML approach. The major difference is that we use the results obtained earlier (relevant events in 2018) to create three additional features: for each trading day, we calculate the number of days until the next relevant event as a categorical feature (up to 20, otherwise "no event"), the stock return in the day of the next event (if the event exists, otherwise 0.5) and the number of days after the last relevant event (again, up to 20 or "no event"). The latter is intended to capture volatility clusters.

Our approach detected 616 days with possible evidences of insider trading in 2018. We verified all these days with the same methodology presented in Sect. 3.2. We also checked other days for any information that might have been overlooked. The results are shown in Table 3.

Table 3. Confusion matrix for evidence of insider trading in 2018

Predicted/actual	Negative	Positive
Negative	3411	20
Positive	387	229

Our approach obtained a precision of ≈37% and a recall of ≈92%. It is difficult to judge the quality of these results as we found no comparable work in scientific literature and our set of rules is unavoidably arbitrary. SONAR [12] is the closest approach we found, but it attempts to classify negotiations instead of days. Moreover SONAR is based on a different dataset to which we currently do not have access. It remains as future work to test our approach in other datasets.

We found possible evidence of insider trading in 144 out of 249 events (≈58%). Among 4047 days prior to a relevant event, we found possible evidence of insider negotiations in 249 of them (180 in the spot market, 54 in the options market and 15 in both). The total distribution for 2017 and 2018 is shown in Table 4. Figure 1 shows the full distribution of evidences according to the number of days before the relevant event is disclosed. In the figure we can see a decaying trend with evidences near the event being more common. This is similar to conclusions obtained in [1].

Table 4. Insider trading evidences distribution for 2017 and 2018

Class	2017	2018	Total
Days with evidences	146	249	395
Days without evidences	12922	12981	25903
Total	13068	13230	26298
Events with insider trading evidences	89	144	233
Events without insider trading evidences	82	105	181
Total	171	249	420

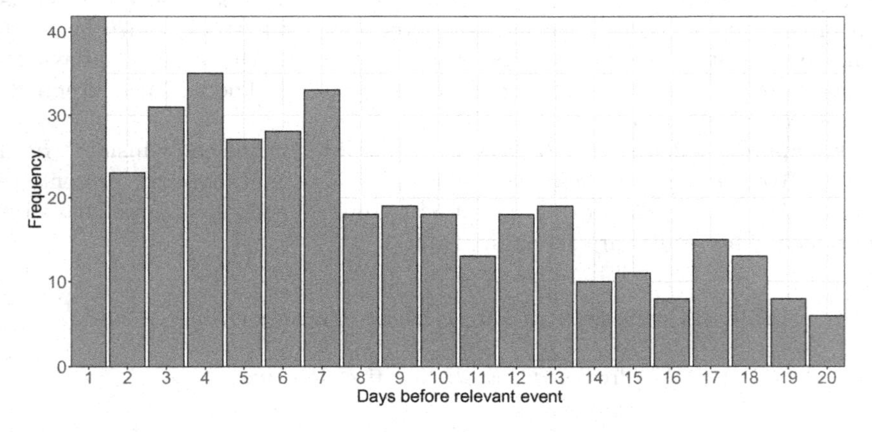

Fig. 1. Distribution of possible evidences of insider trading according to the number of days before the relevant event

4 Recognising Suspicious Trades Before Events Unfold

The main motivation behind the dataset is to answer the following question. Given market derived features and abnormal market movements, is it possible to predict whether impactful news events are about to be disclosed and so these trades configure possible insider trading? We envision two important applications for this problem. The first regards regulatory agencies being able to track traders before the information is made public. The second is helping investment funds manage risk exposure by anticipating market volatility.

For such we employ a ML approach where the output is a binary variable indicating the existence, for a given day and stock, of evidences of insider trading. A data point is classified as positive if it contains previously recognised abnormal trades according to our set of rules **and** if there is a related impactful event up to 20 days thereafter.

We apply here a similar approach as described in Sect. 3.3. However there are important differences between the problems. In that section we filtered out days in which there were no impactful news event in the upcoming 20 days (here we used **all** days). We also used as additional features the set of succeeding impactful events (the approach in Sect. 3.3 was aware that impactful news events would be disclosed). Here we attempt to identify suspicious trades without the knowledge that impactful events will eventually be disclosed. Finally, we make no use of our set of rules to validate whether a day should or not be classified as having suspicious negotiations. We simply employ the features described as input to our prediction algorithm.

Another important remark is that the dataset is highly imbalanced. As shown in Table 4, only 395 out of 26298 days had trades that were classified as possible insider trading evidence (less than 2%). We define as evaluation criteria metrics that are generally considered appropriate for imbalanced datasets. More specifically, we consider variations of the F1-score, which balances precision and recall, depending on the application mentioned above.

4.1 Monitoring Negotiations

In this application we consider that the more days with abnormal trades preceding relevant events we identify, the more potential enforcement agencies have to enforce existing laws. Here we assume that regulatory agencies have limited, but enough manpower or intelligent systems to monitor several trades classified as suspicious. We are interested in getting as much "strong" evidence (true positives) as possible (a high recall). At the same time, we would like to ideally save resources by preventing too many false positives (type I errors, low precision). Nevertheless we assume type I errors are less of a concern than type II errors (false negatives) from an enforcement agency point of view.

We therefore propose as evaluation metric the following modified F1-score:

$$\mathrm{F1}^R = \frac{2 \times \mathrm{Precision} \times \mathrm{Recall}^2}{\mathrm{Precision} + \mathrm{Recall}} \tag{10}$$

A similar approach in literature is the SONAR system [12]. SONAR alerts regulators whenever a suspicious trade has been detected. The main difference is that we attempt to generate alerts when trades happen, while SONAR generates alerts after relevant news events are disclosed (a somewhat similar problem to the one discussed in Sect. 3.3). Notice also that in this problem we are not interested in **when** the related news event will be disclosed, only **if** there will be an event eventually disclosed.

We divided our dataset in three subsets respecting temporality. The training and validation sets correspond to, respectively, 52% and 48% of 2017 data. The test set corresponds to the entire 2018 data. We optimised the hyper-parameters using only the training and validation samples. The R code for this model is also available as supplementary material.

Since we are unfamiliar with other similar methods in literature, we compare our algorithm to two dummy predictors: stratified and uniform. The stratified predictor respects the validation set class distribution, while the uniform predictor gives the same weight to both classes. Another common dummy classifier is to always predict the most frequent class. However, in our case that would mean both precision and recall would have values of zero.

For each data point, the output of XGBoost is a value between 0 and 1, which is rounded to one of the classes according to a classification threshold. The threshold is not set *a priori*, rather it is decided by the algorithm itself. Table 5 presents evaluation metrics considering the threshold that optimised the $F1^R$-score in the validation set. We include four metrics for the training, validation and test sets. The last two columns present the same (average) metrics for 100 executions of both the stratified and uniform predictors.

Table 5. Selected evaluation metrics for predicting insider trading evidences

Metric	Training set	Validation set	Test set	Stratified	Uniform
Precision	0.217	0.122	0.184	0.018	0.019
Recall	0.717	0.750	0.687	0.010	0.503
F1-score	0.334	0.211	0.291	0.013	0.037
$F1^R$-score	0.240	0.158	0.200	<0.001	0.018

The results shown in Table 5 substantially improve those from the dummy classifiers. The best value for the $F1^R$-score in the validation set was 0.158, obtained with a classification threshold of ≈0.074. The results displayed in the table for the test set are based on the same classification threshold.

Figure 2 shows how precision and recall change as we vary the threshold. The right vertical line corresponds to the best validation set threshold of 0.074. For reference, the best threshold for the test set (in regards to optimising Eq. (10)) would have been relatively similar, at 0.061 (the left vertical line).

4.2 Predicting Relevant Events

In this application we are interested in predicting whether impactful news events will be disclosed within the next days. It is essentially the same problem as in Sect. 4.1, but with one important difference: the output in 4.1 is positive if there is an impactful news event within the next 20 days **and** there exists abnormal negotiations classified as evidence of insider trading today. Here the output is

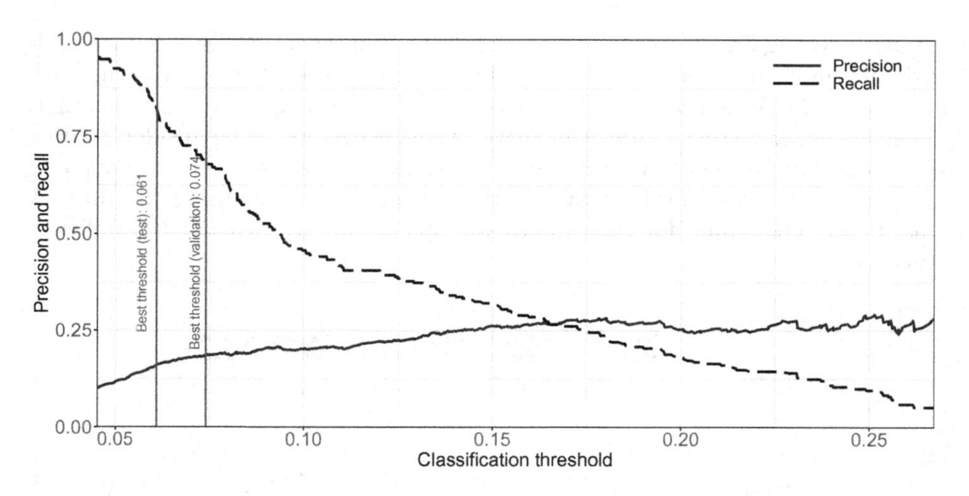

Fig. 2. Precision and Recall as the classification threshold varies

positive if there is a relevant news event within the next days, irrespective of there being evidence of insider trading today.

From an investment fund point of view, the more relevant news it is able to predict before they happen, the more volatility risk it can avoid. We assume the fund would like to be aware of as many impactful news events as possible. False positives mean that the fund may unnecessarily take conservative positions, perhaps missing promising return opportunities, while false negatives mean that the fund will not be hedged when the impactful news event happens. Once again, we aim for a balance between precision and recall, but this time we assume that type II errors (false negatives) are less of a concern than type I errors (false positives) - as most other funds would not be able to predict the event and would be similarly exposed to the incurred volatility risk.

Given these assumptions, we propose the following modified F1-score:

$$F1^P = \frac{2 \times \text{Precision}^2 \times \text{Recall}}{\text{Precision} + \text{Recall}} \tag{11}$$

In our dataset we classify days as having relevant events irrespective if there were evidences of insider trading before - as the reader may recall we used returns to decide whether a news event was relevant. For this problem however we only consider news events for which insider trading evidences were also identified. This is based on an assumption that news events which have no prior insider trading are unpredictable, and thus would bring noise to the dataset.

Our model hopes to answer the following question: "will there be an impactful news event within the next d days?". This is naturally a hard and ambitious problem. To simplify it a bit further we do not try to predict whether news events are "good" or "bad", but simply whether they will happen or not. As implied by the problem statement, we also do not try to predict the exact day of the event, rather we try to predict it within a time window.

We once again compare our approach to dummy classifiers, both stratified and uniform. Here we run 20 different instances of the problem, with $d = 1, \ldots, 20$. When $d = 15$, for instance, we are trying to predict whether, given a day and stock, impactful news events with preceding insider trading evidences will happen during the next 15 days. Figure 3 shows a summary of precision and recall obtained by the three approaches (including the dummy classifiers). The values for the dummy approaches are averages of 100 different runs each.

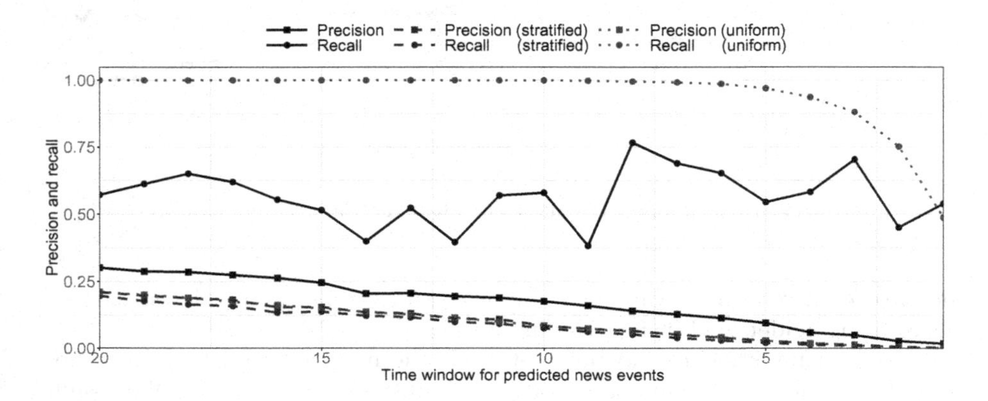

Fig. 3. Precision and recall for experiments with $d = 20, \ldots, 1$

From $d = 20$ to $d = 8$, the uniform dummy predictor achieved a recall of 1 - that is, all impactful news events were "predicted". Suppose today is day $t = 0$ and a news event takes place on $t = 5$. Then that news event is predicted if the classifier assigns a positive prediction for any one of days $t = 0, \ldots, 4$ - that is, a news event can be predicted any day prior to being disclosed. Since the uniform predictor classifies on average half the data points as 1 it is very likely that for large enough d it will not miss any news event. Its precision, however, is consistently lower than that of our approach. If we predict on $t = 0$ that there will be a news event and there is at least one event during days $t = 1, \ldots, d$, we count it as a correct guess. The stratified classifier has lower values of recall and precision in all cases.

Table 6. Selected evaluation metrics for predicting news events with $d = 15$

Metric	Training set	Validation set	Test set	Stratified	Uniform
Precision	0.666	0.167	0.245	0.155	0.149
Recall	0.325	0.628	0.514	0.137	1.000
F1-score	0.437	0.264	0.332	0.146	0.260
$F1^P$-score	0.291	0.044	0.081	0.022	0.039

Table 6 shows results for $d = 15$. The $F1^P$-score is higher in our test set then both dummy classifiers (0.081 as opposed to 0.022 and 0.039), even though the recall from the uniform predictor was at its maximum. Still, we believe the recall obtained in our test set was substantial, at 0.514, when considering the higher precision obtained (0.245 as opposed to 0.155 and 0.149).

Clearly this is overall a very hard problem. However we believe that our approach is promising as it consistently obtained better performance than dummy classifiers. We were careful to only use available market data as input features, but we are aware that the output is our own classification of relevant news events. We hope in the future to test our approach with other known insider trading datasets as well as improved versions of our own dataset.

5 Conclusion and Future Work

In this work, we studied insider trading in the Brazilian market. We proposed a methodology to create a dataset of possible insider trading evidences. We identified relevant news events which are believed to have impacted stock returns, and then we derived a set of rules to classify abnormal negotiations (or sets of negotiations) preceding those news events as possible evidence of insider trading. Our dataset spans 2017 and 2018. We created the 2017 dataset from the ground up, partly manually. We then used a ML approach to extend this dataset to 2018 in the hope of reducing the associated workload.

Having the dataset, we used it to tackle two problems: predicting whether a day had evidences of insider trading and predicting impactful news events before they are disclosed. Both are essentially the same problem but with subtle differences. We compared our approach to dummy classifiers and in both cases we obtained superior results. The problems are non-surprisingly hard but we believe our approach is promising. We are making our code and dataset publicly available in the hope that they are both improved by other researchers.

As future work, we intend to perform statistical tests to verify whether our results are significantly better than random predictors. We also plan to both refine the dataset and expand it to contain more recent data. We will try to improve our results by deriving new features and studying other ML algorithms that may be more suitable for these problems. Finally, since our dataset is highly unbalanced with few positive classifications, we intend to employ generative adversarial networks for both expanding it and (hopefully) improving our results.

Acknowledgments. We would like to thank Dr. Humberto Brandão for his essential contributions.

References

1. Ahern, K.R.: Do proxies for informed trading measure informed trading? Evidence from illegal insider trades. Technical report, University of Southern California - Marshall School of Business; National Bureau of Economic Research (NBER) (2018)

2. Astorino, E.S.: Insider trading networks in Brazil. Ph.D. thesis, Universidade de São Paulo (2017)
3. B3: Brasil, Bolsa, Balcão (2019). https://www.b3.com.br. Accessed 28 Mar 2020
4. Bhattacharya, U., Daouk, H.: The world price of insider trading. J. Financ. **57**(1), 75–108 (2002)
5. Chen, T., Guestrin, C.: XGBoost: a scalable tree boosting system. In: Proceedings of the 22nd ACM SIGKDD International Conference on Knowledge Discovery and Data Mining, pp. 785–794. ACM (2016). https://doi.org/10.1145/2939672.2939785
6. Conti, R.: Empirical properties of asset returns: stylized facts and statistical issues. Quant. Financ. **1**(2), 223–236 (2001)
7. Dent, G.W.: Why legalized insider trading would be a disaster. Del. J. Corp. Law **38**, 247–274 (2013)
8. Donoho, S.: Early detection of insider trading in option markets. In: Proceedings of the Tenth ACM SIGKDD International Conference on Knowledge Discovery and Data Mining, pp. 420–429 (2004)
9. Easley, D., Hvidkjaer, S., O'hara, M.: Is information risk a determinant of asset returns? J. Financ. **57**(5), 2185–2221 (2002)
10. Fama, E.F.: Efficient capital markets: a review of theory and empirical work. J. Financ. **25**(2), 383–417 (1970)
11. Foster, L.H.: SEC - Insider trading investigations (2018). https://www.sec.gov/about/offices/oia/oia_enforce/foster.pdf. Accessed 28 Mar 2020
12. Goldberg, H.G., Kirkland, J.D., Lee, D., Shyr, P., Thakker, D.: The NASD securities observation, new analysis and regulation system (SONAR). In: Proceedings of the 15th Conference on Innovative Applications of Artificial Intelligence, pp. 11–18 (2003)
13. Islam, S.R., Ghafoor, S.K., Eberle, W.: Mining illegal insider trading of stocks: a proactive approach. In: 2018 IEEE International Conference on Big Data (Big Data), pp. 1397–1406. IEEE (2018)
14. Kulkarni, A., Mani, P., Domeniconi, C.: Network-based anomaly detection for insider trading. arXiv preprint arXiv:1702.05809 (2017)
15. Martins, O.S., Paulo, E., Albuquerque, P.: Negociação com informação privilegiada na BMF&Bovespa e seu reflexo no retorno das ações (Informed trading and stock returns in the BMF&BOVESPA). Revista de Administração de Empresas FGV-EAESP **53**(4), 350–362 (2013)
16. Minenna, M.: A supervisory perspective on insider trading: estimating the value of the information. CONSOB, Quaderni di Finanza **45**(2000), 1–32 (2000)
17. Minenna, M.: The detection of market abuse on financial markets: a quantitative approach. CONSOB, Quaderni di finanza **1**(54), 1–53 (2003)
18. Mitchell, M.L., Netter, J.M.: The role of financial economics in securities fraud cases: applications at the securities and exchange commission. Bus. Lawyer **49**, 545–590 (1993)
19. Mitra, G., Yu, X.: The Handbook of Sentiment Analysis in Finance. Albury Books, New York (2016)
20. Parente, N.S.: Aspectos jurídicos do insider trading (1978). http://www.cvm.gov.br/export/sites/cvm/menu/acesso_informacao/serieshistoricas/estudos/anexos/Aspectos-Juridicos-do-insider-trading-NJP.pdf. Accessed 28 Mar 2020
21. SEC: Annual Report (2018). https://www.sec.gov/files/enforcement-annual-report-2018.pdf. Accessed 28 March 2020
22. U.S. Securities and Exchange Commission - SEC: Litigation releases. https://www.sec.gov/litigation/litreleases.shtml (May 2019)

Mend the Learning Approach, Not the Data: Insights for Ranking E-Commerce Products

Muhammad Umer Anwaar[1,3(✉)], Dmytro Rybalko[2], and Martin Kleinsteuber[1,3]

[1] Technische Universität München, München, Germany
umer.anwaar@tum.de, martin.kleinsteuber@mercateo.com
[2] IBM, Moscow, Russia
dmitriy.rybalko@ibm.com
[3] Mercateo, München, Germany

Abstract. Improved search quality enhances users' satisfaction, which directly impacts sales growth of an E-Commerce (E-Com) platform. Traditional Learning to Rank (LTR) algorithms require relevance judgments on products. In E-Com, getting such judgments poses an immense challenge. In the literature, it is proposed to employ user feedback (such as clicks, add-to-basket (AtB) clicks and orders) to generate relevance judgments. It is done in two steps: first, query-product pair data are aggregated from the logs and then order rate etc. are calculated for each pair in the logs. In this paper, we advocate counterfactual risk minimization (CRM) approach which circumvents the need of relevance judgements, data aggregation and is better suited for learning from logged data, i.e. contextual bandit feedback. Due to unavailability of public E-Com LTR dataset, we provide *Mercateo dataset* from our platform. It contains more than 10 million AtB click logs and 1 million order logs from a catalogue of about 3.5 million products associated with 3060 queries. To the best of our knowledge, this is the first work which examines effectiveness of CRM approach in learning ranking model from real-world logged data. Our empirical evaluation shows that our CRM approach learns effectively from logged data and beats a strong baseline ranker (λ-MART) by a huge margin. Our method outperforms full-information loss (e.g. cross-entropy) on various deep neural network models. These findings demonstrate that by adopting CRM approach, E-Com platforms can get better product search quality compared to full-information approach.

Keywords: Information retrieval · Ranking and preference learning · Learning to Rank · E-commerce search · Implicit feedback · Counterfactual risk minimization · Dataset · Mining data logs

Y. Dong et al. (Eds.): ECML PKDD 2020, LNAI 12461, pp. 257–272, 2021.
https://doi.org/10.1007/978-3-030-67670-4_16

1 Introduction

The E-Com industry is growing fast, with a projected global sales of 4.8 trillion USD in 2021[1]. Virtually every E-Com platform leverages machine learning (ML) techniques to increase their users' satisfaction and optimize business value for the company. Optimal ranking of search results plays a vital role in achieving these goals. The successful application of traditional ML algorithms, such as λ-MART [6] and AdaRank [31], requires hand-crafted features. This greatly reduces their applicability in commercial settings due to large amount of diverse products. Deep learning (DL) is a well established framework for automatically learning relevant features from raw data and has also inspired research in LTR [19,20]. But DL needs large-scale data for training a model. Fortunately, such data is readily available in almost every E-Com platform in the form of search, clicks, add-to-basket (AtB) clicks and orders logs. We will refer to this data as *log data* in the remainder of the paper.

As log data is abundantly and cheaply available, it is promising to devise learning algorithms which can learn effective ranking models from it. In contrast to online or active LTR methods [12,24], learning from log data avoids intrusive interactions in the live E-Com platform. This is highly desirable in practice because it avoids badly affecting users' experience. However, how to get relevance judgments (RJs) for training (offline) supervised LTR algorithms is a significant challenge in learning from log data. RJs for benchmark LTR datasets are performed either by experts or crowd sourcing [22]. Several studies [10,23] have shown that crowd-sourcing is not a reliable technique for getting RJs on products of E-Com platform. Getting RJs for millions of products from domain experts is prohibitively expensive. This has created a gap in the application of DL research for improving E-Com search quality.

In this paper, we aim to bridge this gap and improve product search with the practical constraints of a commercial setting. Santu et al. [23] have made an attempt to overcome this issue. They aggregate query-product pairs from the logs and calculate user feedback rate (e.g. order rate) for these pairs. RJs are done based on this rate (for details, see Sect. 3.3).

Such method of getting RJs ignores the fact that log data is in the form of so-called *contextual-bandit* feedback. This means that we have access to *only* those feedback signals which were generated in response to a limited set of actions taken by the ranking system (logging policy). For instance, we do not know how the user would have responded to the search results if another set of products was shown. That is why traditional supervised learning approach, where information about all possible actions is assumed, is not well-suited for learning from log data. We refer to the traditional approach as full-information (Full-Info) approach and its loss (such as cross-entropy and hinge) as Full-Info loss. This partial information (contextual-bandit feedback) challenge requires us to devise more efficient ways of utilizing the information contained in the log data. To address this challenge effectively, the learning problem should be reformulated.

---

[1] https://www.statista.com/statistics/379046/.

Due to these reasons, we advocate employing a counterfactual risk minimization (CRM) approach [27] and adapting the LTR algorithm to learn directly from the log data. CRM loss requires the knowledge of the logging policy, the actions taken by the logging policy and users' feedback on these actions (e.g. AtB clicks, orders etc.). All this information is contained in the log data. In simple terms, logging policy means the ranker which was used by our E-Com platform and the log data is *generated* by users' interactions with its actions. We propose that CRM approach is better suited for learning from such logged contextual-bandit feedback, as it does not require full-information about all actions and their rewards. Moreover, it also circumvents the need to aggregate the log data and generate RJs.

The contributions of this work are summarized as follows:

- We construct and publish a novel LTR dataset from a real-world E-Com platform.
- We adapt the CRM learning approach for E-Com LTR which enables learning directly from the log data.
- We conduct extensive experiments on the *Mercateo dataset* demonstrating the effectiveness of the CRM approach in comparison to the Full-Info approach.

2 Related Work

Over the past two decades, researchers have done significant amount of work on improving ranking for web search. They also investigated the problem of getting RJs for URLs via click logs. [14,15] used eye-tracking studies to devise a set of preference rules for interpreting the click logs. For instance, rule "Click > Skip Next" means that if user clicks on URL_A at position i and skips URL_B at position i + 1 then URL_A is preferable to URL_B. Similarly, rule "Click > Click Above" means that if both are clicked then URL_B is preferred over URL_A. Such rules generate pairwise preference judgments between URLs. [1] proposed a probabilistic interpretation of the click logs based on "Click > Skip Next" and "Click > Skip Above" rules. The problem with such judgments is that they tend to learn to reverse the list of results. A click at the lowest position is preferred over all other clicks, while a click at the top position is preferred to nothing else. Recently, researchers [4,13] have tried to overcome these issues but their focus is on improving web search with clicks and they do not take into account the unique challenges and feedback signals of E-Com search.

Another active research direction is to devise user models for estimating how users examine the list of results [2,17,21] and utilize these models to improve the automatic generation of RJs. In web search, such models are theoretically motivated and useful, however they are formulated in such a way that E-Com search can not benefit directly from them. Despite being quite noisy, clicks are extremely important user feedback for web search. But in E-Com search, we have far less noisier and more informative signals like orders, AtB clicks, revenue etc. Moreover, these approaches have to consider user intent modeling (exploratory, informative etc.) and position bias. In order to utilize click logs, models of users' click behavior or some preference rules are necessary. Due to these issues, we exclude click signals in this work. In the presence of more valuable signals this trade-off is beneficial.

In contrast to web search, there has been little work on learning effective ranking models for E-Com search. Santu et al. [23] have done a systematic study of applying traditional LTR algorithms on E-Com search. They studied the query-attribute sparsity problem, the effects of popularity based features and the reliability of relevance judgments via crowd-sourcing. They compared traditional ranking algorithms and reported that λ-MART showed best performance. They did not include DNN models in their work.

In this paper, we compare two approaches to overcome the hurdle of getting RJs for E-Com. The traditional approach of supervised learning (Full-Info) advocated by [23] and our CRM based approach of treating the log data as contextual-bandit feedback. Recently, [16] have shown that CRM loss is able to achieve predictive accuracy comparable to cross-entropy loss on object recognition tasks in computer vision. They replace cross-entropy loss with a self-normalized inverse propensity (SNIPS) estimator [28] that enables learning from contextual-bandit feedback. It is worth noting that [16] conducted all their experiments on simulated contextual-bandit feedback on the CIFAR10 dataset. According to authors' best knowledge, this is the first work which applies SNIPS estimator on actual logged bandit feedback and verify its effectiveness in solving a real-world problem i.e. E-Com LTR.

Another major issue which has inhibited research in LTR for E-Com is the unavailability of public E-Com LTR dataset. Some researchers [3,5,23] have conducted experiments on E-Com datasets, but they did not publish the dataset due to data confidentiality reasons. We think that such a dataset is critical in advancing the research of designing effective and robust LTR algorithms for E-Com search. We discuss this in detail in Sect. 3.1.

3 E-Com Dataset for LTR

In this section, we present the dataset constructed from our Business-to-Business (B2B) E-Com platform. This dataset contains information of queries from actual users, actions taken by the logging policy, the probability of these actions and feedback of users in response to these actions. The *Mercateo dataset* is publicly available to facilitate research on ranking products of E-Com platform. The dataset can be accessed at: https://github.com/ecom-research/CRM-LTR.

3.1 Need of a New Dataset

Unique Feedback Signals. E-Com platforms tackle diverse needs and preferences of their users while maximizing their profit. In practice, this translates to certain unique problems and opportunities which are irrelevant in comparable tasks like ranking documents or URLs. For instance, when users interact with the E-Com platform, they generate multiple implicit feedback signals e.g. clicks, AtB clicks, orders, reviews etc. These signals can serve as a proxy for users' satisfaction with the search results and business value for the E-Com platform.

Unavailablity of Public E-Com LTR Datasets. To the best of our knowledge, there is no publicly available E-Com LTR dataset. The principal reason is that such datasets may contain confidential or proprietary information, and E-Com platforms are unwilling to take this risk. For instance, [3,5,23] worked on LTR for E-Com search with *Walmart and Amazon* datasets but they did not publish it. There are some public E-Com datasets for problems like clustering or recommendation systems [9,26], but they have very few features (e.g. 8 features in [7]) and few products to be considered viable for the LTR task.

Logged Dataset. Although CRM loss does not need any RJs for training, it still requires logged contextual bandit feedback. Thus, we also publish the real-world logged dataset, which will prove instrumental in advancing the research on learning directly from logs.

Table 1. Statistics for the *Mercateo dataset*

Total # of queries	3060
# of queries in [Train/Dev/Test] set	[1836/612/612]
# of products in dataset	3507965
Avg. # of products per query [Train/Dev/Test]	[1106/1218/1192]

3.2 Scope of the Dataset

For a given query, E-Com search can be divided into two broad steps. The first step retrieves potentially relevant products from the product catalogs and the second step ranks these retrieved products in such a fashion which optimizes users' satisfaction and business value for the E-Com platform [18]. Our proprietary algorithm performs the first part effectively. Thus, the scope of this dataset is limited to overcoming the challenges faced in optimizing the second part of the E-Com search. Another important thing to consider is that most of our users (buyers) are small and medium-sized businesses (SMBs). On our platform, we employ user-specific information to personalise the final ranking based on the contracts of buyers with suppliers. Such user information is proprietary and it is against our business interests to publish it even after anonymising. Therefore, we exclude user information in this dataset.

User Feedback Signals. Although there are plenty of feedback signals logged by E-Com platforms, we considered only two common feedback signals in our dataset. One is AtB clicks on the results shown in response to a query and the second is the order signal on the same list of results. AtB clicks serve as implicit feedback on search quality, i.e. the user found the product interesting enough to first click it on the search result page and then click again to put it in the basket (cart). Orders, though more sparse than AtB clicks, are less noisy and correlate strongly with users' satisfaction with E-Com search results. They also serve as proxy to business value for the E-Com platform. The *Mercateo dataset* contains more than 10 million and 1 million AtB click and order log entries respectively. Table 1 shows some statistics about the dataset.

3.3 Dataset Construction

Data Sources. We collected data from sources commonly available in E-Com platforms. Namely: (i) title of the products, (ii) AtB click data from logs, (iii) search logs (containing information about the products displayed on the search result page in response to the query) and (iv) order logs.

Selecting the Queries. Most queries included in this dataset are those which were challenging for our current ranking algorithm. Queries were subsampled from the search logs, keeping in mind the following considerations: (i) ensure statistical significance of the learning outcomes (especially for dev and test sets, i.e. must have been searched at least 1000 times) and (ii) include queries with variable *specificity*. We say a query has low *specificity*, if the query entered by the user has a broad range of products associated with it. This can be ensured simply by looking at the number of *potentially relevant products* returned by our proprietary algorithm. For instance, for a very specific query like *pink painting brush* (20) products are returned, while for broad and common queries like *iphone* (354) or *beamer* (1,282) products are returned.

How to Get Relevance Judgments (RJs) on the Products for Supervised Dataset? Our CRM based approach does not require RJs for learning but RJs are required for Full-Info loss. These RJs are also needed for a fair comparison of CRM approach with Full-Info approach. But how do we get them for millions of products? RJs for benchmark datasets [6,22] are performed by human judges; who can be domain experts or crowd-source workers. Unfortunately it is too expensive to ask experts to judge products in an E-Com setting.

The problems associated with getting RJs via crowd-sourcing have been analyzed extensively by [10] using Amazon Mechanical Turk. They found that users of E-Com platforms have a very complex utility function and their criteria of relevance may depend on product's value for money, brand, warranty etc. Crowd-source workers fail to capture all these aspects of relevance. Hence, crowd-sourcing is an unreliable method for getting RJs. This was also confirmed by another study [23].

An alternative approach is to generate RJs from multiple feedback signals present in user interaction logs and historical sales data of products. Major benefits of this approach are: (i) such RJs are closer to the notion of relevance in E-Com search and quantify relevance as a proxy of user satisfaction and business value, (ii) it costs less time and money, as compared to other two approaches, and (iii) large-scale E-Com datasets can be constructed, as it only requires data generated by users' interaction with the E-Com website. Such data is abundantly available in almost every E-Com platform. One drawback of this approach is that the quality of these RJs may not be as good as human expert judgments. Based on the analysis of available choices for the supervised setting, generating RJs from logs in such a manner is justified. [23] also advocated this approach.

Calculating Relevance Judgments (Labels). For a given query q, we now show the steps taken to calculate relevance judgments on the set of products, $\mathcal{P}_q$, associated with it:

1. We calculate the visibility of a product, i.e. the number of times a product p was shown to the users. We remove the products from our dataset with visibility less than 50.
2. We then convert AtB click (order) signal to *relevance rate* (RR) by dividing the AtB clicks (orders) for product p, with its visibility.
3. Next, we normalize RR with the maximum value of RR for that query q to get normalized relevance rate (NRR).

We publish NRR for all query-product pairs. This allows researchers the flexibility in computing different types of relevance judgments (e.g. binary, graded, continuous etc.). We follow [23] and compute the 5-point graded relevance judgment, $l(p, q)$, by the following formula:

$$l(p, q) = ceil[4 \cdot NRR(p, q)] \quad (1)$$

Dataset Format. Although there are plenty of feedback signals logged by E-Com platforms, we considered only two common feedback signals in our dataset. One is AtB clicks on the results shown in response to a query and the second is the order signals on the same list of results. For each query, we retain all products that were AtB clicked (ordered) and negatively sample remaining products from the search logs, i.e. products which were shown to the user but not AtB clicked (ordered). A supervised train set is also published along with AtB click and order labels. The queries and the products are the same for both the supervised train and the logged training sets. Test and development (dev) sets are published *only* with supervisory labels. The split of queries among train, dev and test sets is done randomly with 60%, 20% and 20% of total number of queries. There is no overlap of queries in training, dev and test sets. Raw text can not be published because our sellers have the proprietary rights on the product title and description. We publish 100-dimensional GloVe word embeddings trained on the corpus comprising of queries and product titles. For each query-product pair, we also provide some proprietary features which can be employed as *dense features* in DNNs. These features contain information about price, delivery time, profit margin etc. of the product. Specific details can not be shared due to confidentiality reasons. Further details can be found on the github repository.

4 Problem Formulation

In Full-Info setting, e.g. pointwise LTR, a query-product pair and its RJ defines the training instance. The task of the ranker is to predict RJ for a given query-product pair. In our CRM setting, we say that a query-product pair defines the context. Additional information such as users' model for personalized ranking, product description etc., can also be incorporated into the context. But for reasons discussed in Sect. 3, we limit the context to search query, product title and dense features. The ranking system (logging policy) can only take binary actions given the context. That is, whether the product is to be displayed among the top-k positions or not. These actions incur reward (loss) based on feedback signals from the user. It is to be noted that the top-k positions are extremely important for real-world LTR systems. In E-Com platforms, k is usually not the same for all queries. It depends on the query specificity determined by our proprietary algorithm. For instance, too broad a query like *copy paper* signals informational user intent. Thus, k has a big value for such queries in comparison to specific queries like *pink painting brush.*

Formally, let context $c \in \mathcal{C}$ be word embeddings of the search query, word embeddings of the product title and dense features. Let $a \in \mathcal{A}$ denote the action taken by the logging policy. The probability of a given the context c is determined by the logging policy denoted as $\pi_0(a|c)$ running on the E-Com platform. For a sample i, this probability is also known as propensity p_i, i.e., $p_i = \pi_0(a_i|c_i)$. Based on the user feedback, a binary loss denoted as δ_i is incurred. If $a = 1$ is selected by π_0 for a context (query-product pair) and the user AtB clicked (ordered), it implies $\delta = 0$. This is *positive feedback*, i.e., decision of π_0 is correct and product is relevant to the user. On the other hand, if $a = 0$ and the user went beyond top-k results and AtB clicked (ordered) that product, we set $\delta = 1$. This is *negative feedback*, i.e., the product was relevant and π_0 made a wrong decision to not show the product in top-k results. The logged data is a collection of n tuples: $D = [(c_1, a_1, p_1, \delta_1), ..., (c_n, a_n, p_n, \delta_n)]$.

The goal of counterfactual risk minimization is to learn an unbiased stochastic policy π_w from logged data, which can be interpreted as a conditional distribution $\pi_w(A|c)$ over actions $a \in \mathcal{A}$. This conditional distribution can be modeled by a DNN, $f_w(\cdot)$, with a softmax output layer:

$$\pi_w(a|c) = \frac{exp(f_w(c, a))}{\sum_{a' \in \mathcal{A}} exp(f_w(c, a'))}. \tag{2}$$

Swaminathan et al. [28] proposed SNIPS as an efficient estimator to counterfactual risk. For details, we refer to [16]. We use the SNIPS estimator in all our experiments:

$$\hat{R}_{SNIPS}(\pi_w) = \frac{\sum_{i=1}^{n} \delta_i \frac{\pi_w(a_i|c_i)}{\pi_0(a_i|c_i)}}{\sum_{i=1}^{n} \frac{\pi_w(a_i|c_i)}{\pi_0(a_i|c_i)}}. \tag{3}$$

5 Experiments

We evaluate our proposed method on the real-world *Mercateo dataset.* We aim to answer the following research questions:

- **RQ1:** How does our CRM-based method perform compared to the Full-Info approach and λ-MART?
- **RQ2:** How does the performance of our method progress with more bandit feedback?
- **RQ3:** How does the DNN architecture affect the performance of CRM and Full-Info Loss?

5.1 Experimental Setup

In our experiments, we use mini-batch Adam and select the model which yields best performance on the dev set. Further implementation details can be found in the code[2]. For evaluation, we use the standard *trec_eval*[3] tool and several popular evaluation metrics.

5.2 Comparison of CRM and Full-Info Approaches (RQ1)

In this experiment, we selected a simple yet powerful CNN model [25], which we refer to as S-CNN. It utilizes convolutional filters for ranking pairs of short texts. We compare the performance of CRM loss with Full-Info (cross-entropy) loss on S-CNN model. All network parameters are kept the same for fair comparison.

We also report the performance of two strong baseline rankers, namely logging policy and λ-MART [30]. It has been recently shown that λ-MART outperforms traditional LTR algorithms on an E-Com dataset [23]. For λ-MART, we use the open source *RankLib* toolkit[4] with default parameters. λ-MART takes dense features engineered by domain experts as input. The logging policy also employs the same dense features. Predictions of this policy are used by our CRM method for actions a_i and $\pi_0(a_i|c_i)$.

Tables 2 and 3 summarize the results of the performance comparison on the test set with graded AtB click label and order label respectively. First, we note that S-CNN performs significantly better than the logging policy. In terms of MAP, the performance improvement against logging policy is 7.8% with Full-Info (cross-entropy) loss and 27.4% with CRM loss for target label AtB clicks. For target label orders, it is 32.6% with Full-Info loss and 54.6% with CRM loss.

Second we observe that the λ-MART model performs worse on all metrics. It performs significantly worse than even logging policy. Particularly, performance degradation is huge on NDCG@10 in comparison to MAP. This huge difference in metrics suggest that λ-MART has failed to learn graded relevance. Since MAP

[2] Available at: https://github.com/ecom-research/CRM-LTR.
[3] Available at: https://github.com/usnistgov/trec_eval.
[4] Available at: https://sourceforge.net/p/lemur/wiki/RankLib/.

treats all relevance other than zero as one, whereas NDCG metric is sensitive to graded relevance. These results also suggest that a deep learning approach can significantly outperform λ-MART, given enough training data. This is consistent with findings of [20]. One can argue that λ-MART has access to only few hand-crafted features, so its performance can be improved by adding more features. But feature engineering requires domain expertise and is a time-consuming process. Moreover, our logging policy employs exactly the same features and is performing quite better than λ-MART.

Table 2. Performance comparison for target label: AtB clicks. Significant degradation with respect to our implementation (p-value ≤ 0.05)

Ranker (loss)	MAP	MRR	P@5	P@10	NDCG@5	NDCG@10
Logging policy	0.4704	0.6123	0.4686	0.4613	0.2052	0.2537
λ-MART	0.3825	0.4472	0.2972	0.3119	0.0917	0.1164
S-CNN (full-info)	0.5074	0.8036	0.6552	0.6261	0.3362	0.3835
S-CNN (CRM) - ours	**0.5993**	**0.8391**	**0.7346**	**0.7093**	**0.4332**	**0.4964**

Third we note that our CRM based approach outperforms Full-Info loss on all metrics by a huge margin. Concretely, on NDCG@10, the performance gain of our CRM method over Full-Info loss is 29.4% for AtB click labels and 14.9% for order labels. These results show that our CRM approach not only learns effectively from logged data but also significantly outperforms the models trained in supervised fashion on aggregated data, i.e. Full-Info approach.

Table 3. Performance comparison for target label: orders. Significant degradation with respect to our implementation (p-value ≤ 0.05)

Ranker (loss)	MAP	MRR	P@5	P@10	NDCG@5	NDCG@10
Logging policy	0.2057	0.3225	0.1693	0.1717	0.0945	0.1250
λ-MART	0.1519	0.1772	0.0747	0.0884	0.0322	0.0479
S-CNN (full-info)	0.2728	0.4601	**0.2869**	0.2562	0.1720	0.1973
S-CNN (CRM) - ours	**0.3181**	**0.4609**	0.2841	**0.2791**	**0.1854**	**0.2266**

5.3 Learning Progress with Increasing Number of Bandit Feedback (RQ2)

It is quite insightful to measure how the performance of our CRM model changes with more bandit feedback (e.g. AtB clicks). To visualize this, we pause the training of S-CNN after a certain number of bandit feedback samples, evaluate the model on the test set and then resume training.

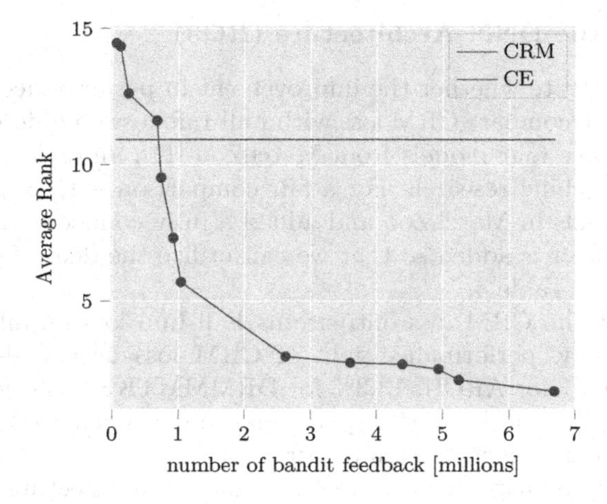

Fig. 1. Average rank of relevant product in AtB click test set

In Fig. 1 and Fig. 2, each green dot corresponds to these intermediate model evaluations. We plot average rank and average Discounted Cumulative Gain (DCG) scores of relevant products for all queries in AtB clicks test set. These curves show that there is a constant improvement in average rank of relevant (AtB clicked) product as more bandit feedback is processed. Similarly, average DCG values rise monotonically with increasing number of AtB click feedback. Due to space constraints, we omit the plot for orders but they follow a similar pattern. The red line in these figures correspond to Full-Info S-CNN model trained on complete training set. This experiment demonstrates the ability of the CRM approach to learn efficiently with increasing bandit feedback.

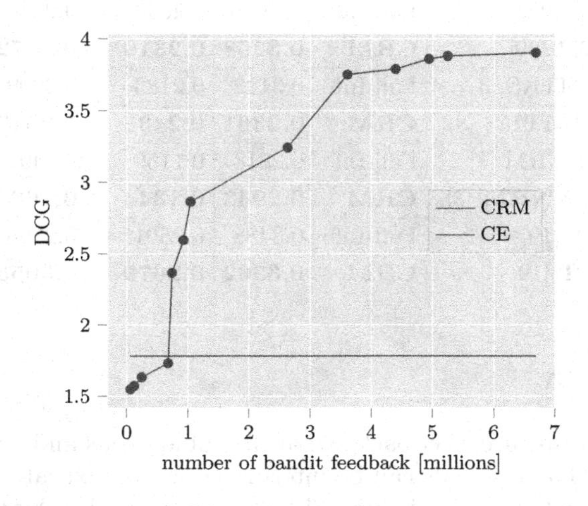

Fig. 2. Average DCG of relevant product in AtB click test set

5.4 Effect of the DNN Architecture (RQ3)

In order to investigate whether the improvement in performance is architecture agnostic or not, we compare CRM loss with Full-Info loss on different DNN architectures. We choose four models from MatchZoo [11], an open-source codebase for deep text matching research. For a fair comparison with S-CNN model, we modified the models in MatchZoo and added a fully connected layer before the last layer. This layer is added so that we can utilize the dense features. Results are summarized in Table 4.

We note that the CRM loss outperfoms Full-Info loss on all of these DNN models. Specifically, performance gain of CRM loss over Full-Info loss w.r.t NDCG@10 is 10.5% for ARCII, 13.3% for DRMMMTKS, 7.4% for ConvKNRM and 6.6% for DUET. It shows that performance gains achieved by our method are not limited to any specific architecture.

We also observe that the deep learning model architecture has significant impact on the performance. For instance, the best performing model with Full-Info loss, DUET has 27.8% improvement over S-CNN w.r.t MAP. For CRM loss, the best performing model DUET has 21% improvement over the worst performing model ConvKNRM w.r.t MAP.

The results of this experiment support our claim that adapting the learning approach to learn directly from logged data is beneficial as compared to modifying the data to fit the supervised (Full-Info) learning approach.

Table 4. Comparison of deep learning models for target label: orders.

Ranker	Loss	MAP	NDCG@5	NDCG@10
S-CNN [25]	Full-Info	0.2728	0.1720	0.1973
S-CNN [25]	**CRM**	**0.3181**	**0.1854**	**0.2266**
ARCII [29]	Full-Info	0.2891	0.2044	0.2238
ARCII [29]	**CRM**	**0.3208**	**0.2319**	**0.2472**
DRMMTKS [32]	Full-Info	0.3112	0.2183	0.2309
DRMMTKS [32]	**CRM**	**0.3361**	**0.2436**	**0.2617**
ConvKNRM [8]	Full-Info	0.2818	0.1159	0.1494
ConvKNRM [8]	**CRM**	**0.2942**	**0.1344**	**0.1604**
DUET [19]	Full-Info	0.3488	0.2501	0.2866
DUET [19]	**CRM**	**0.3562**	**0.2679**	**0.3055**

6 Conclusion

In E-Com platforms, user feedback signals are ubiquitous and are usually available in log files. These signals can be interpreted as contextual-bandit feedback, i.e. partial information which is limited to the actions taken by the logging policy and users' response to the actions. In order to learn effective ranking of the products from such logged data, we propose to employ counterfactual risk minimization approach. Our experiments on *Mercateo dataset* have shown that CRM approach outperforms traditional supervised (full-information) approach on several DNN models. On *Mercateo dataset*, it shows empirically that reformulating the LTR problem to utilize the information contained in log files is a better approach than artificial adapting of data for the supervised learning algorithm.

Acknowledgments. We would like to thank Alan Schelten, Till Brychcy and Rudolf Sailer for insightful discussions which helped in improving the quality of this work. This work has been supported by the Bavarian Ministry of Economic Affairs, Regional Development and Energy through the *WoWNet* project IUK-1902-003// IUK625/002.

A Comparison of Counterfactual Risk Estimators

We compare the performance of SNIPS estimator with two baseline estimators for counterfactual risk. We conduct the experiments on AtB click training data of *Mercateo dataset*. The inverse porpensity scoring (IPS) estimator is calculated by:

$$\hat{R}_{IPS}(\pi_w) = \frac{1}{n}\sum_{i=1}^{n} \delta_i \frac{\pi_w(a_i|c_i)}{\pi_0(a_i|c_i)}. \quad (4)$$

Second estimator is an empirical average (EA) estimator defined as follows:

$$\hat{R}_{EA}(\pi_w) = \sum_{(c,a)\in(\mathcal{C},\mathcal{A})} \overline{\delta}(c,a)\pi_w(a|c), \quad (5)$$

where $\overline{\delta}(c,a)$ is the empirical average of the losses for a given context and action pair. The results for these estimators are provided in Table 5. Compared to SNIPS both IPS and EA perform significantly worse on all evaluated metrics. The results confirm the importance of equivariance of the counterfactual estimator and show the advantages of SNIPS estimator.

Table 5. Results on *Mercateo dataset* with AtB click relevance for IPS and empirical average estimators

Estimator	MAP	MRR	NDCG@5	NDCG@10
CNN (CRM) - SNIPS	**0.5993**	**0.8391**	**0.4332**	**0.4964**
CNN (CRM) - IPS	0.4229	0.7139	0.3426	0.3703
CNN (CRM) - EA	0.2320	0.3512	0.2083	0.2253

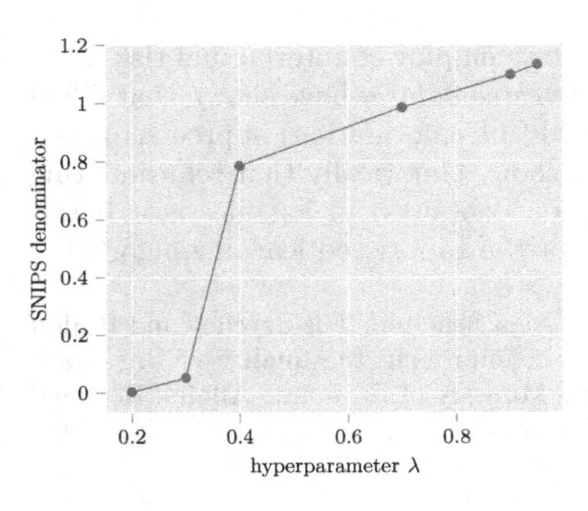

Fig. 3. SNIPS denominator vs λ on order logs (training set)

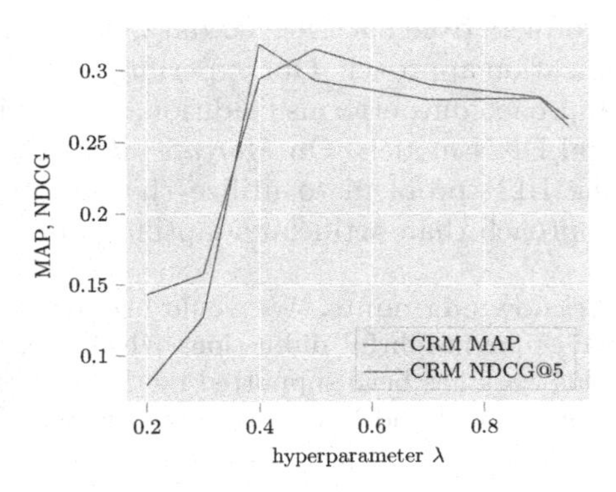

Fig. 4. Performance on orders test set of rankers trained with different λ

B Choosing Hyperparameter λ

One major drawback of SNIPS estimator is that, being a ratio estimator, it is not possible to perform its direct stochastic optimization [16]. In particular, given the success of stochastic gradient descent (SGD) training of deep neural networks in related applications, this is quite disadvantageous as one can not employ SGD for training.

To overcome this limitation, Joachims et al. [16] fixed the value of denominator in Eq. 3. They denote the denominator by S and solve multiple constrained optimization problems for different values of S. Each of these problems can be reformulated using lagrangian of the constrained optimization problem as:

$$\hat{w}_j = \underset{w}{\operatorname{argmin}} \frac{1}{n} \sum_{i=1}^{n} (\delta_i - \lambda_j) \frac{\pi_w(a_i|c_i)}{\pi_0(a_i|c_i)} \tag{6}$$

where λ_j corresponds to a fixed denominator S_j.

The main difficulty in applying the CRM method to learn from logged data is the need to choose hyperparameter λ. We discuss below our heuristics of selecting it. We also evaluate the dependence of λ on SNIPS denominator S, which can be used to guide the search for λ. To achieve good performance with CRM loss, one has to tune hyperparameter $\lambda \in [0, 1]$. Instead of doing a grid search, we follow a smarter way to find a suitable λ. Building on the observations proposed in [16], we can guide the search of λ based on value of SNIPS denominator S. It was shown in [16] that the value of S increases monotonically, if λ is increased. Secondly, it is straightforward to note that expectation of S is 1. This implies that, with increasing number of bandit feedback, the optimal value for λ should be selected such that its corresponding S value concentrates around 1. In our experiments, we first select some random $\lambda \in [0, 1]$ and train the model for two epochs with this λ. We then calculate S for the trained model; if S is greater than 1, we decrease λ by 10%, otherwise we increase it by 10%. The final value of λ is decided based on best performance on validation set.

In Fig. 3, we plot the values of denominator S on order logs (training set) of *Mercateo dataset* for different values of hyperparameter λ. On the figure below, Fig. 4, we also plot performance on orders test set, in terms of MAP and NDCG@5 scores, of different rankers for these values of hyperparameter λ. It is to be noted that the values of SNIPS denominator S monotonically increase with increasing λ. The MAP and NDCG@5 reach its highest value for $\lambda = 0.4$, but decrease only slightly with increasing values of λ. Furthermore, it can also be seen from these two figures that the λ values with good performance on test set have corresponding SNIPS denominator values close to 1.

References

1. Agrawal, R., Halverson, A., Kenthapadi, K., Mishra, N., Tsaparas, P.: Generating labels from clicks. In: WSDM 2009, pp. 172–181. ACM (2009). https://doi.org/10.1145/1498759.1498824
2. Bendersky, M., Wang, X., Najork, M., Metzler, D.: Learning with sparse and biased feedback for personal search. In: JCAI 2018, pp. 5219–5223. AAAI Press (2018)
3. Bi, K., Teo, C.H., Dattatreya, Y., Mohan, V., Croft, W.B.: Leverage implicit feedback for context-aware product search. In: eCOM@SIGIR (2019)
4. Borisov, A., Kiseleva, J., Markov, I., de Rijke, M.: Calibration: a simple way to improve click models. In: CIKM 2018 (2018)
5. Brenner, E.P., Zhao, J., Kutiyanawala, A., Yan, Z.: End-to-end neural ranking for ecommerce product search. In: SIGIR eCom, vol. 18 (2018)
6. Chapelle, O., Chang, Y.: Yahoo! Learning to rank challenge overview. In: Proceedings of the Learning to Rank Challenge, pp. 1–24 (2011)
7. Chen, D.: Data mining for the online retail industry: a case study of RFM model-based customer segmentation using data mining. J. Database Market. Customer Strategy Manag. **19**(3), 197–208 (2012). https://doi.org/10.1057/dbm.2012.17
8. Dai, Z., Xiong, C., Callan, J., Liu, Z.: Convolutional neural networks for soft-matching N-grams in ad-hoc search. In: WSDM 2018, pp. 126–134. ACM, New York (2018). https://doi.org/10.1145/3159652.3159659, http://doi.acm.org/10.1145/3159652.3159659
9. Dheeru, D., Taniskidou, E.: UCI machine learning repository (2017)
10. Alonso, O., et al.: Relevance criteria for e-commerce: a crowdsourcing-based experimental analysis. In: SIGIR 2009, pp. 760–761. ACM (2009)
11. Guo, J., Fan, Y., Ji, X., Cheng, X.: MatchZoo: a learning, practicing, and developing system for neural text matching. In: SIGIR 2019 (2019). https://doi.org/10.1145/3331184.3331403, http://doi.acm.org/10.1145/3331184.3331403
12. Hu, Y., Da, Q., Zeng, A., Yu, Y., Xu, Y.: Reinforcement learning to rank in e-commerce search engine: formalization, analysis, and application. In: KDD 2018, NY, USA (2018). https://doi.org/10.1145/3219819.3219846, http://doi.acm.org/10.1145/3219819.3219846
13. Jiang, S., et al.: Learning query and document relevance from a web-scale click graph. In: SIGIR 2016 (2016)
14. Joachims, T.: Optimizing search engines using clickthrough data. In: KDD 2002. ACM (2002). https://doi.org/10.1145/775047.775067

15. Joachims, T., Granka, L., Pan, B., Hembrooke, H., Radlinski, F., Gay, G.: Evaluating the accuracy of implicit feedback from clicks and query reformulations in web search. ACM Trans. Inf. Syst. **25**(2), 7-es (2007). https://doi.org/10.1145/1229179.1229181
16. Joachims, T., Swaminathan, A., Rijke, M.d.: Deep learning with logged Bandit feedback. In: ICLR 2018, May 2018
17. Joachims, T., Swaminathan, A., Schnabel, T.: Unbiased learning-to-rank with biased feedback. In: WSDM 2017. ACM (2017). https://doi.org/10.1145/3018661.3018699
18. Lucchese, C., Nardini, F.M., Orlando, S., Perego, R., Tonellotto, N.: Speeding up document ranking with rank-based features. In: SIGIR 2015, NY, USA (2015). https://doi.org/10.1145/2766462.2767776
19. Mitra, B., Diaz, F., Craswell, N.: Learning to match using local and distributed representations of text for web search. CoRR (2016)
20. Pang, L., Lan, Y., Guo, J., Xu, J., Xu, J., Cheng, X.: DeepRank: a new deep architecture for relevance ranking in information retrieval. CoRR abs/1710.05649 (2017)
21. Qi, Y., Wu, Q., Wang, H., Tang, J., Sun, M.: Bandit learning with implicit feedback. In: NIPS 2018, pp. 7287–7297. Curran Associates Inc., Red Hook (2018)
22. Qin, T., Liu, T.Y., Xu, J., Li, H.: LETOR: a benchmark collection for research on learning to rank for information retrieval. Inf. Retrieval **13**(4), 346–374 (2010). https://doi.org/10.1007/s10791-009-9123-y
23. Santu, S.K.K., Sondhi, P., Zhai, C.: On application of learning to rank for e-commerce search. In: SIGIR 2017 (2017)
24. Schuth, A., Hofmann, K., Whiteson, S., de Rijke, M.: Lerot: an online learning to rank framework. In: Proceedings of the 2013 Workshop on Living Labs for Information Retrieval Evaluation, pp. 23–26. ACM (2013)
25. Severyn, A., Moschitti, A.: Learning to rank short text pairs with convolutional deep neural networks. In: SIGIR 2015, pp. 373–382. ACM, New York (2015)
26. Sidana, S., Laclau, C., Amini, M.R., Vandelle, G., Bois-Crettez, A.: KASANDR: a large-scale dataset with implicit feedback for recommendation. In: SIGIR 2017, pp. 1245–1248 (2017)
27. Swaminathan, A., Joachims, T.: Batch learning from logged bandit feedback through counterfactual risk minimization. JMLR **16**, 1731–1755 (2015)
28. Swaminathan, A., Joachims, T.: The self-normalized estimator for counterfactual learning. In: NIPS 2015, pp. 3231–3239. MIT Press, Cambridge (2015)
29. Wan, S., Lan, Y., Guo, J., Xu, J., Pang, L., Cheng, X.: A deep architecture for semantic matching with multiple positional sentence representations. CoRR abs/1511.08277 (2015). http://arxiv.org/abs/1511.08277
30. Wu, Q., Burges, C.J., Svore, K.M., Gao, J.: Adapting boosting for information retrieval measures. Inf. Retrieval **13**(3), 254–270 (2010)
31. Xu, J., Li, H.: AdaRank: a boosting algorithm for information retrieval. In: SIGIR 2007, pp. 391–398. ACM, New York (2007)
32. Yang, Z., et al.: A deep top-K relevance matching model for ad-hoc retrieval. In: Zhang, S., Liu, T.-Y., Li, X., Guo, J., Li, C. (eds.) CCIR 2018. LNCS, vol. 11168, pp. 16–27. Springer, Cham (2018). https://doi.org/10.1007/978-3-030-01012-6_2

Multi-future Merchant Transaction Prediction

Chin-Chia Michael Yeh(✉), Zhongfang Zhuang(✉), Wei Zhang, and Liang Wang

Visa Research, Palo Alto, CA 94306, USA
{miyeh,zzhuang,wzhan,liawang}@visa.com

Abstract. The multivariate time series generated from merchant transaction history can provide critical insights for payment processing companies. The capability of predicting merchants' future is crucial for fraud detection and recommendation systems. Conventionally, this problem is formulated to predict one multivariate time series under the *multi-horizon* setting. However, real-world applications often require more than one future trend prediction considering the uncertainties, where more than one multivariate time series needs to be predicted. This problem is called *multi-future prediction*. In this work, we combine the two research directions and propose to study this new problem: *multi-future*, *multi-horizon* and *multivariate* time series prediction. This problem is crucial as it has broad use cases in the financial industry to reduce the risk while improving user experience by providing alternative futures. This problem is also challenging as now we not only need to capture the patterns and insights from the past but also train a model that has a strong inference capability to project multiple possible outcomes. To solve this problem, we propose a new model using convolutional neural networks and a simple yet effective encoder-decoder structure to learn the time series pattern from multiple perspectives. We use experiments on real-world merchant transaction data to demonstrate the effectiveness of our proposed model. We also provide extensive discussions on different model design choices in our experimental section.

Keywords: Multivariate time series · Multi-future · Multi-horizon

1 Introduction

The advances in digital payment systems in recent years have enabled billions of payment transactions to be processed every second. Merchant transaction history is a prevalent data type of payment processing systems. From airlines to book stores, the aggregation process for several critical features (e.g., the total amount of money spent in a book store between 5–6 pm) happens on an hourly-base, and the patterns of the transactions are monitored. Monitoring such patterns is essential to applications such as fraud detection (i.e., by observing

Y. Dong et al. (Eds.): ECML PKDD 2020, LNAI 12461, pp. 273–289, 2021.
https://doi.org/10.1007/978-3-030-67670-4_17

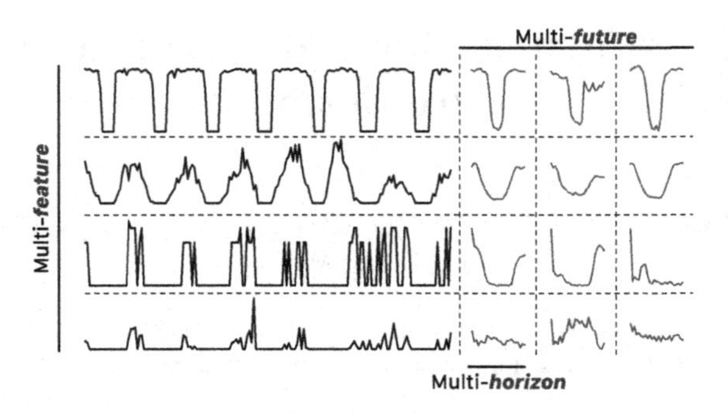

Fig. 1. Multi-*future* Multi-*horizon* Multi*variate* time series prediction.

the deviation from the *regular* trends) and shopping recommendation (i.e., by observing similar transaction histories).

One crucial step to build these applications is *estimating* every merchant's *future*, where each feature is predicted hourly in a "rolling" fashion. However, the constant high computation cost makes such an approach unrealistic in real-world scenarios. Instead, multi-horizon prediction, where the goal is to predict multiple time steps *at one time* instead of only one time step in the future, is a preferred approach in this scenario.

Moreover, predicting only one *future* in the financial industry may not be ideal as multiple factors can have a significant impact on the volume of merchant transactions. For example, severe weather may or may not have a significant impact on restaurants' business. If a model only predicts one trend signaling lower transaction volume, and there are still a large number of orders, it may cause a higher false-positive rate of credit card declines. Therefore, predicting multiple possible features is a more realistic approach. This problem is called *multi-future time series prediction*.

We depict the overall scope of this problem in Fig. 1. For instance, the first feature in Fig. 1 is the per hour transaction volume. The first and third possible future shows the business remains a similar trend as the previous history, while the second possible future shows jitter patterns. By knowing these two alternative patterns for the future transaction volume, downstream systems could prepare alternative plans for each of the possible futures.

Time series prediction is a well-studied problem [1]. Recent works on time series prediction [16,18] take a new look at the multi-horizon time series prediction problem from the neural network perspective. Specifically, Taieb and Atiya [16] proposes to train a neural network with a target consisting of multi-steps into the future; Wen *et al.* [18] approaches this problem from sequence-to-sequence perspective. However, none of those above work deals with the multi-future prediction problem. The closest problem formulation solved by prior work is described in [17], where multiple possible future trajectories of vehicles are predicted, but the system proposed is designed specifically for modeling driving behaviors, which is fundamentally different from the multivariate time series in the financial industry. Predicting multiple possible futures on multivariate time series in a multi-horizon setting remains challenging in the real-world applications as the real-world data may exhibit frequent patterns that discourage the prediction of various futures (Table 1).

In this work, we study the problem of *multi-future, multi-horizon multivariate time series prediction*. To tackle this challenge, we propose a novel model with two sub-networks: a *shape* sub-network responsible for learning the time series shape patterns, and a *scale* sub-network responsible for learning the magnitude and offset of the time series. Each sub-network is built using a simple yet effective encoder-decoder stack with layers (e.g., Linear Layers, Convolutional Neural Networks, Max/Average Pooling Layers) that have lower computation overhead considering the real-world application. We summarize our contributions as follows:

- We analyze the problem of multi-future prediction for multi-horizon multivariate time series data.
- We propose and design a novel architecture for learning both the shape and scale of the time series data.
- We conduct our experiments on real-world merchant transactions and demonstrate the effectiveness of our proposed model.

2 Notations and Problem Definition

In this section, we present several important definitions in this work. The most fundamental definitions are *time series* and *multivariate time series*.

Definition 1 (Time Series). A *time series* is an ordered set of real-valued numbers. For a time series $\boldsymbol{\tau}$ of length n, $\boldsymbol{\tau}$ is defined as $[t_1, \cdots, t_n] \in \mathbb{R}^n$.

Definition 2 (Multivariate Time Series). A *multivariate time series* is a set of co-evolving time series, denoted as $\mathbf{T}$. For a multivariate time series $\mathbf{T}$ of length n with d features, $\mathbf{T}$ is defined as $[\boldsymbol{\tau}_1, \cdots, \boldsymbol{\tau}_d] \in \mathbb{R}^{n \times d}$, where each $\boldsymbol{\tau}_i \in \mathbf{T}$ denotes a co-evolving time series in the set.

Moreover, given a multivariate time series $\mathbf{T}$, we use $\mathbf{T}[i]$ to denote the values at time i, and $\mathbf{T}[i : j]$ to denote the values of the multivariate time series from time i to time j.

With the basic notation defined, we are ready to define the problem we are solving in this work: *multi-future, multi-horizon, multivariate* time series prediction.

Definition 3 (Multi-future Time Series Set). A multi-future time series set $\widehat{\mathbb{T}}$ is a set of f *predicted* multivariate time series:

$$\widehat{\mathbb{T}} = \left[\widehat{\mathbf{T}}_1, \cdots, \widehat{\mathbf{T}}_f\right] \tag{1}$$

where each multivariate time series $\widehat{\mathbf{T}}_i \in \widehat{\mathbb{T}}$ is the prediction of a possible future.

Table 1. Important notations.

Notation	Meaning
$\boldsymbol{\tau} = [t_1, \cdots, t_n]$	Time series of length n
$\mathbf{T} = [\boldsymbol{\tau}_1, \cdots, \boldsymbol{\tau}_d]$	Multivariate time series
$\widehat{\mathbb{T}} = \left[\widehat{\mathbf{T}}_1, \cdots, \widehat{\mathbf{T}}_f\right]$	Set of f *predicted* time series
$\mathbf{T}[i]$	The i-th time step of the multivariate time series $\mathbf{T}$
$\mathbf{T}[i:j]$	The multivariate time series between time i and j of the multivariate time series $\mathbf{T}$
$\bar{\mathbf{T}}$	Multivariate time series ground truth
M	Multi-future time series prediction model
$\boldsymbol{h}$	Feature vector output by a encoder
$\mathbf{S}$	Shape bank (a matrix)
$\boldsymbol{r}$	Activation vector
$\widehat{\boldsymbol{\alpha}}_{i,j}^{(\mathrm{sp})}$	The shape prediction for i-th future for the j-th feature
$\widehat{\boldsymbol{A}}_{i}^{(\mathrm{sp})}$	The multivariate shape prediction for the i-th future
$\mu_{i,j}$	The offset of the i-th future, j-th feature
$\sigma_{i,j}$	The magnitude of the i-th future, j-th feature
i_{oc}	Oracle future index

Definition 4 (Multi-future Time Series Prediction Model). Given the current time index as i, a multi-future time series prediction model M predicts f possible futures for the next n_h time horizon using the multivariate time series $\mathbf{T}$ from the past n_p time points:

$$\widehat{\mathbb{T}} \leftarrow \mathsf{M}\left(\mathbf{T}\left[i - n_p + 1 : i\right]\right) \tag{2}$$

Definition 5. The goal of *multi-future multi-horizon multivariate time series prediction problem* is to train a multi-future time series prediction model M that minimizes the following performance measurement:

$$\sum_{i} \min_{1 \le j \le f} \mathtt{error}(\widehat{\mathbf{T}}_j, \mathbf{T}[i+1 : i+n_h]) \tag{3}$$

where $\mathtt{error}(\cdot)$ is a function computing error-based performance measurement. One possible choice for $\mathtt{error}(\cdot)$ is Root Mean Squared Error (RMSE).

3 Model Architecture

We use Fig. 2 to illustrate the overall architecture of the proposed model. The proposed model has two sub-networks: a **shape** sub-network and a **scale** sub-network. Both sub-networks use the input multivariate time series $\mathbf{T}$ in *parallel*; then, the output of each sub-network is combined to form the prediction.

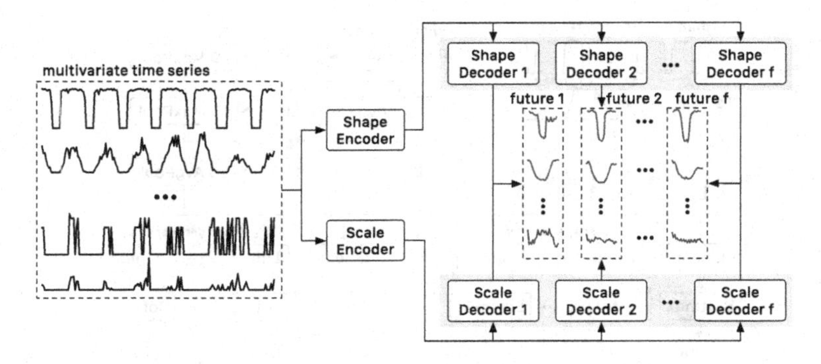

Fig. 2. The proposed model.

Each sub-network has an encoder and a set of decoders (i.e., an ensemble of decoders). We refer to the encoder and decoders of the shape sub-network as **shape encoder** and **shape decoder**. Similarly, we refer to the encoder and decoders of the scale sub-network as **scale encoder** and **scale decoder**.

While there is only one encoder, the number of decoders in each ensemble corresponds to the number of possible futures to predict. Specifically, each decoder in each ensemble is associated with a possible future. For example, shape decoder i and scale decoder i are both associated with the i-th future $\widehat{\mathbf{T}}_i$.

3.1 Shape Sub-network

We illustrate our proposed **shape sub-network** in Fig. 3. The shape sub-network consists of a single shape encoder and an ensemble of shape decoders. The purpose of the shape encoder is to capture relevant information for synthesizing the shape of the predicted time series.

Encoder. The encoder is built by stacking 1D convolution layer (denoted as `Conv`), rectified linear unit activation function (denoted as `ReLU`), 1D max pooling layers (denoted as `MaxPool`), and a 1D average pooling layer (denoted as `AvgPool`). Each stack is composed of a `Conv`, a `ReLU` and a `MaxPool` or `AvgPool`. That is, we use `AvgPool` instead of `MaxPool` in the last block to summarize the input time series along the temporal direction. For an encoder with l layers, the encoder can be expressed as:

$$\boldsymbol{h}_1 = \texttt{MaxPool}\left(\texttt{ReLU}\left(\texttt{Conv}\left(\mathbf{T}\right)\right)\right)$$
$$\dots$$
$$\boldsymbol{h} = \texttt{AvgPool}\left(\texttt{ReLU}\left(\texttt{Conv}\left(\boldsymbol{h}_{l-1}\right)\right)\right)$$

The encoder processes the input multivariate time series $\mathbf{T} \in \mathbb{R}^{n_p \times d}$ into a fixed-size vector representation $\boldsymbol{h} \in \mathbb{R}^{d_h}$, where d_h is the dimension of the representation. In our particular implementation, for all `Conv` layers, the receptive

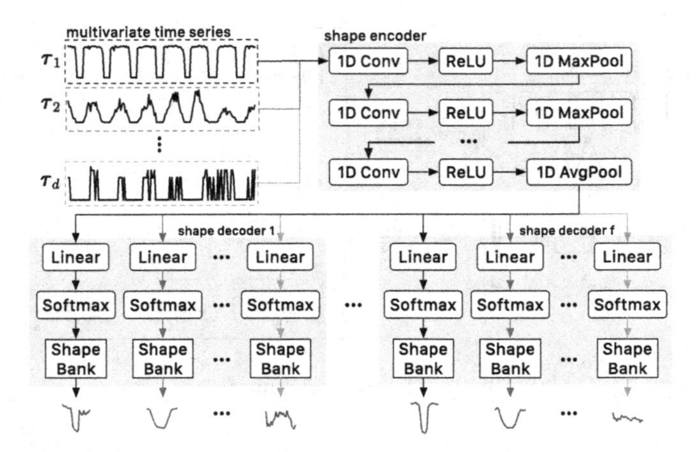

Fig. 3. The shape sub-network contains a single encoder and a decoder ensemble.

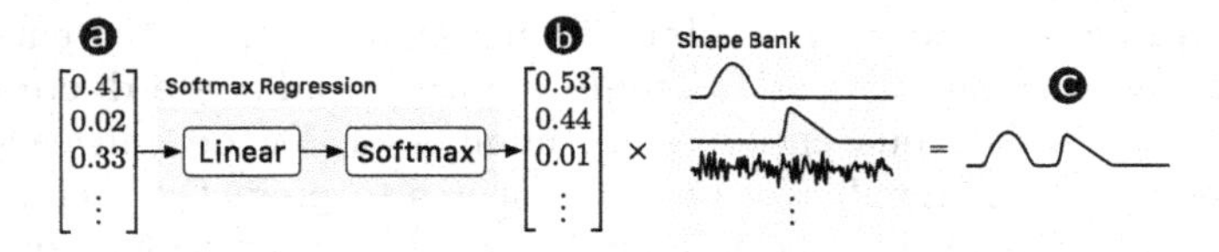

Fig. 4. Synthesizing process for the shape decoder for a dimension.

field size is set to 3, and the number of channels is set to 64. For the all `MaxPool`, the window size and stride are both set to 2.

Since our particular parameter settings for the layers would reduce the length of the time series by half each time the input passing trough a block, there are $\lfloor \log_2 n_p \rfloor$ blocks being used to process the input time series. Because $\log_2 n_p$ is not guaranteed to be an integer, we use an adaptive pooling layer (or global pooling layer) for the last pooling layer (i.e., `AvgPool`) to make sure there is only one pooling window and the pooling window covers the whole intermediate output of the previous layer.

Given a multivariate time series $\mathbf{T}$, our shape encoder would output the corresponding hidden representation vector $\boldsymbol{h}$ of size 64.

Decoder. Each decoder uses a shallow design with two major components: a *shape bank* and a *softmax regression.* In our model, each decoder is responsible for predicting one possible *futures.* For example, there will be three decoders if three different futures are to be predicted.

The shape bank stores a set of shape *templates* for synthesizing the prediction's shape for a specific feature dimension. Thus, in every shape decoder, the number of shape banks is the same as the number of feature dimensions. Moreover, each shape template has the same length as the output time series. Given there are a total of d feature dimensions in the multivariate time series, n_h as

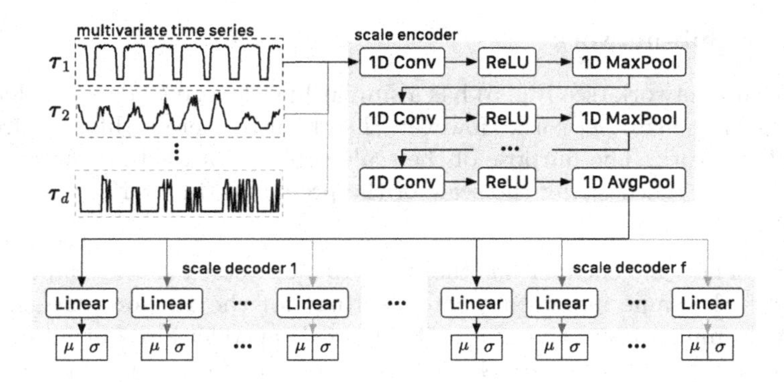

Fig. 5. The scale sub-network contains a single encoder and a decoder ensemble.

the time horizon (i.e., length of the shape template) and a total of n_s shape templates to be learned, each of the d shape bank is implemented as a matrix of $\mathbb{R}^{n_s \times n_h}$. Note, the shape banks can be either supplied by the user and/or refine/learned during the training process.

There are also d `softmax` regression models, where each of them corresponds to one of the d shape banks. Each of the `softmax` regression models dictates how the templates stored in the corresponding shape bank are combined to form the shape prediction.

Given the hidden representation $\boldsymbol{h}$ produced by the shape encoder and respective shape bank $\mathbf{S} \in \mathbb{R}^{n_s \times n_h}$ with n_s templates for the j-th feature dimension, the respective `softmax` regression model for the j-th feature dimension outputs the activation vector $\boldsymbol{r} \in \mathbb{R}^{n_s}$ where each value $r_k \in \boldsymbol{r}$ is the weighting for each shape template $\mathbf{s}_k \in \mathbf{S}$. Thereafter, the shape prediction $\widehat{\boldsymbol{\alpha}}_j^{(\mathrm{sp})}$ for the j-th feature dimension is computed as follows:

$$\widehat{\boldsymbol{\alpha}}_j^{(\mathrm{sp})} = \boldsymbol{r}\mathbf{S} \tag{4}$$

Since $\widehat{\boldsymbol{\alpha}}_j^{(\mathrm{sp})}$ represents only one future for the j-th feature dimension, we further denote $\widehat{\boldsymbol{\alpha}}_{i,j}^{(\mathrm{sp})}$ as the prediction of the i-th future for the j-th feature dimension. To simplify the notation in later section, we use $\widehat{\boldsymbol{A}}_i^{(\mathrm{sp})} \in \mathbb{R}^{d \times n_h}$ to denote the multivariate shape prediction for the i-th future. Figure 4 shows an example of a forward pass for synthesizing the shape prediction for one of the dimensions where ⓐ is the hidden representation output by the shape encoder, ⓑ is the activation vector computed by the `softmax` regression model, and ⓒ is the shape prediction. The example presented in Fig. 4 shows another benefit of our shape decoder design: the shape decoder is an interpretable model. By showing the activation vector and the associated shape template to the user, the user can understand both the synthesizing process and the relationship between the input time series and the relevant shape template in the shape bank.

3.2 Scale Sub-network

The scale sub-network (see Fig. 5) has a similar high-level encoder-decoder structure as the shape sub-network. However, the objective here is different from the shape sub-network. The purpose of the scale sub-network is to capture relevant information for synthesizing the scale of the predicted time series.

Encoder. The scale encoder has the same `1D Conv` stack as the shape encoder, and we use the same hyper-parameter settings for the `1D Conv` layers. Therefore, given a multivariate time series $\mathbf{T}$, the scale encoder also outputs a hidden representation vector $\boldsymbol{h}$ of size 64. Depending on the data, it is possible to have a simplified model with fewer trainable parameters for the scale encoder as the scale information could be relatively simpler comparing to the shape information. However, using the same architecture for the shape encoder and the scale encoder has proven the most effective on our merchant transaction data. Searching for a better encoder architecture for both the scale and the shape encoders is an interesting future direction.

Decoder. For the scale decoder, we also use a shallow model for the prediction process, which is similar to the shape decoder. However, instead of the `softmax` regression-based model, we use a linear model to predict the scale (i.e., offset μ and magnitude σ) of the output time series for each dimension. Similar to the notation we used for the shape prediction, we use $\mu_{i,j}$ and $\sigma_{i,j}$ to denote the μ and σ for i-th future and j-th feature dimension.

To form the i-th future prediction $\widehat{\mathbf{T}}_i$, we combine scale sub-network's output (i.e., $\mu_{i,j}$ and $\sigma_{i,j}$) with shape sub-network's output (i.e., $\widehat{\boldsymbol{\alpha}}_{i,j}^{(\mathrm{sp})}$) using the following equation:

$$\widehat{\mathbf{T}}_i = \begin{bmatrix} \mu_{i,1}\widehat{\boldsymbol{\alpha}}_{i,1}^{(\mathrm{sp})} + \sigma_{i,1} \\ \vdots \\ \mu_{i,d}\widehat{\boldsymbol{\alpha}}_{i,d}^{(\mathrm{sp})} + \sigma_{i,d} \end{bmatrix} \tag{5}$$

Can We Estimate the Probability of Each Possible Future? Such a problem can be solved with the concept of expert classifier [13]. An expert classifier is a model that predicts the most likely outcome out of all of the possible futures returned by the main model given an input. We can use an architecture similar to the shape encoder for expert classifiers by adding an additional linear and softmax layer to predict the most likely outcome. The expert classifier can be optimized by 1) feeding an input through the main network and stored the best-fitted outcome id as the ground truth label and 2) use the cross-entropy loss with the ground truth label and the input to optimize the expert classifier.

4 Training Algorithm

To ensure that each sub-network captures the corresponding information under multi-future prediction setting, we use the loss function shown in Eq. 6:

$$\mathcal{L}(\mathbf{T}, \bar{\mathbf{T}}) = \texttt{RMSE}(\bar{\mathbf{T}}, \widehat{\mathbf{T}}_{i_{\text{oc}}}) + \gamma\texttt{NRMSE}(\bar{\mathbf{T}}, \widehat{\boldsymbol{A}}_{i_{\text{oc}}}^{(\text{sp})}) \tag{6}$$

where $\mathbf{T} \in \mathbb{R}^{n_p \times d}$ is the input multivariate time series, $\bar{\mathbf{T}} \in \mathbb{R}^{n_h \times d}$ is the ground truth time series (i.e., the multivariate time series for the next n_h time steps), $\texttt{RMSE}(\cdot)$ is a function computes Root Mean Squared Error, $\texttt{NRMSE}(\cdot)$ is a function computes the Normalized Root Mean Square Error, $\widehat{\mathbf{T}}_{i_{\text{oc}}}$ is the prediction for i_{oc}-th future, $\widehat{\boldsymbol{A}}_{i_{\text{oc}}}^{(\text{sp})}$ is the shape prediction for i_{oc}-th future, i_{oc} is the oracle future index and $\gamma \in \mathbb{R}$ is a hyper-parameter balancing the $\texttt{RMSE}(\cdot)$ and $\texttt{NRMSE}(\cdot)$ terms. We set γ to 1 in all our experiments. The oracle future index i_{oc} is computed using Eq. 7, and it is only determined based on the shape prediction.

$$i_{\text{oc}} = \underset{1 \le j \le f}{\arg\min}\, \texttt{NRMSE}(\bar{\mathbf{T}}, \widehat{\boldsymbol{A}}_{j}^{(\text{sp})}) \tag{7}$$

The loss is computed by aggregating the RMSE and the NRMSE between the best-predicted future and the ground truth. Only the shape prediction determines the best-predicted future because shape prediction is the harder problem comparing to the scale prediction problem. Specifically, the $\texttt{NRMSE}(\bar{\mathbf{T}}, \widehat{\boldsymbol{A}}_{i_{\text{oc}}}^{(\text{sp})})$ term is computed by first z-normalized [12] each dimension of the ground truth $\bar{\mathbf{T}}$, then compute the RMSE between the normalized ground truth and the shape prediction $\widehat{\boldsymbol{A}}_{i_{\text{oc}}}^{(\text{sp})}$. Because both $\texttt{RMSE}(\bar{\mathbf{T}}, \widehat{\mathbf{T}}_{i_{o}c})$ and $\texttt{NRMSE}(\bar{\mathbf{T}}, \widehat{\boldsymbol{A}}_{i_{\text{oc}}}^{(\text{sp})})$ are computed using the best possible future in the set of multiple predicted future, we also refer to them as oracle RMSE and oracle NRMSE; both are special cases of the oracle loss function [7,10].

We use Algorithm 1 to training the model. The main input to the algorithm is the training data $\mathbf{T} \in \mathbb{R}^{n \times d}$, and the outputs are the trained model M. First, M is initialized in `Line 2`. The main training loop starts at `Line 3`. At the beginning of each iteration, a mini-batch consist of the input $\mathbf{T}_{\text{batch}}$ and the ground truth $\bar{\mathbf{T}}_{\text{batch}}$ is sampled from $\mathbf{T}$ as shown in `Line 4`. The input $\mathbf{T}_{\text{batch}}$ is a tensor of size (n_b, n_p, d), and the corresponding ground truth $\bar{\mathbf{T}}_{\text{batch}}$ is a tensor of size (n_b, n_h, d) where n_b is the batch size. For example, if $\mathbf{T}_{\text{batch}}[i, :, :]$ is sampled from $\mathbf{T}[j+1 : j+n_p, :]$, $\bar{\mathbf{T}}_{\text{batch}}[i, :, :]$ will contain $\mathbf{T}[j+n_p+1 : j+n_p+n_h, :]$. Next, from `Line 5` to `Line 8`, the loss for each instance in the mini-batch is computed using Eq. 6 and Eq. 7. The oracle future index is determined using Eq. 7 in `Line 7` and the loss is computed using Eq. 6 in `Line 8`. The loss for each instance within the mini-batch is aggregated together. Last, in the iteration, the model M is updated using the gradient computed using the loss at `Line 8`. The trained model M is returned at `Line 9` after the model is converged.

Algorithm 1. The Training Algorithm

```
   Input: multivariate time series T
   Output: model M
 1 function TRAIN(T)
 2     InitializeModel (M)
 3     for i ← 0 to n_iter do
 4         T_batch, T̄_batch ← GetMiniBatch(T, n_p, n_h)
 5         loss ← 0
 6         for each T_i, T̄_i ∈ T_batch, T̄_batch do
 7             i_oc ← GetOracleFutureIndex(T_i, T̄_i, M)        ▷ Use Eq. 7.
 8             loss ← ComputeLoss (T_i, T̄_i, M, i_oc)          ▷ Use Eq. 6.
 9         M ← UpdateModel(M, loss)
10     return M
```

5 Experiments

In this section, we aim to demonstrate the effectiveness of the proposed method by comparing it to the alternatives under both single-future prediction setup and multi-future prediction setup. For single-future experiments, we use `RMSE` and `NRMSE` as the performance measurement. For multi-future experiments, we use oracle `RMSE` and oracle `NRMSE` as the performance measurement. While `RMSE` gives us the measure of the deviation of prediction from the ground truth in raw value, it does not measure the deviation in terms of shape. To give us a complete picture of the proposed method's ability in prediction, we also choose to include `NRMSE` to differentiate different method's ability to predict the correct shape. In the financial industry, both the raw values and the shape of the future trend are essential for decision making. All the deep learning-based methods are implemented in PyTorch, and we use Adam optimizer [9] with the default parameters setting for optimization.

5.1 Description of the Datasets

We have organized four different datasets where each dataset consists of merchants from one of the following categories: department store, restaurants, sports facility, and medical services, denoted as Cat.1 $\sim$ Cat.4, respectively.

For each category, we randomly select 2,000 merchants located within California, United States. The time series datasets consist of four features, and each is produced by computing the hourly aggregation of the following statistics: number of approved transactions, number of unique cards, a sum of the transaction amount, and rate of the approved transaction. The training data consists of time series data from November 1, 2018, to November 23, 2018; the test data consists of a time series from November 24, 2018, to November 30, 2018.

As mentioned in Sect. 1, the goal of the system is to predict the next 24 h given the last 168 h (i.e., seven days). We predict every 24 h in the test data by supplying the latest 168 h to the system. For example, the transaction data of 168 h in `Week-10` is used to predict the values of 24 h on the Monday of `Week-11`.

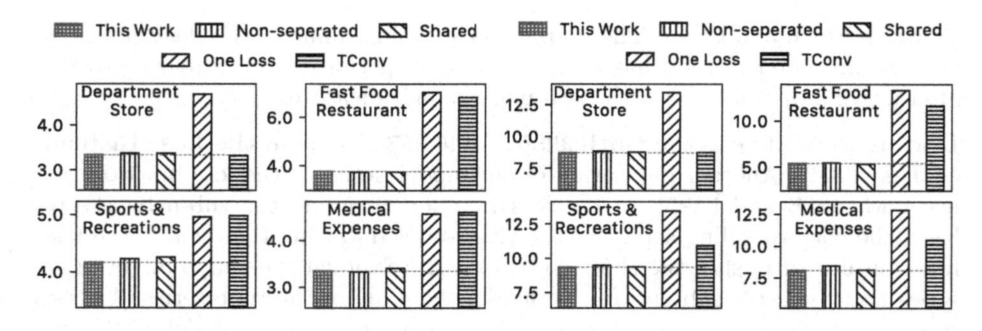

Fig. 6. Performance results in RMSE.

Fig. 7. Performance results in NRMSE.

5.2 Evaluation of Architecture Design Choice

In this section, we focus on evaluate the architecture design of our model. As design choice explored in the above questions is agnostic to the number of future to predict, we evaluate both the designs implemented in the proposed model architecture and the alternatives under a single-future prediction setting. Specifically, our experiments focus on answering the following questions:

Does It Benefit from Having Dedicated Shape Sub-network and Scale Sub-network? Instead of using dedicated encoder-decoder sub-network to model different aspects of the time series data (i.e., shape and scale), one could simply just using one encoder-decoder to model time series data. To evaluate the effectiveness of the dual sub-network design, we compare the proposed model to an alternative model where it just consists of an encoder with the shape-encoder architecture and a decoder with the shape-decoder architecture. As the alternative does not use separate sub-networks, we refer to this alternative as *non-separated model.* In Fig. 6 and Fig. 7, the performance of the proposed method (i.e.,) and the alternative (i.e.,) are presented. The proposed method has a higher or equal performance comparing to a non-separated model in all datasets with both RMSE and NRMSE, especially when the performance is measured with NRMSE.

Can We Share the Encoder in Shape/Scale Sub-network? As demonstrated by the experiments answering the last question, it is beneficial to have dedicated sub-network for modeling different aspects of time series data. However, does it require a dedicated encoder for the proposed model to function effectively, or is it possible to share the encoder for both shape and scale sub-network? To answer this question, we have implemented an alternative model structure where there is only one encoder, and the output of the encoder h_l is feed to a dedicated shape decoder and scale decoder. Because the alternative model shared the encoder, we call this alternative model *shared model.* The performance for the proposed model (i.e.,) and the shared model (i.e.,) is shown in Fig. 6 and Fig. 7. The overall conclusion is similar to the last question, and the proposed method has a higher or equal performance comparing to the alternative in all datasets with both performance measures.

Does it Benefit to Have Both `RMSE` and `NRMSE` Terms in the Loss Instead of Just `RMSE`? Another innovation of our model design is the loss function: we use both `RMSE` and `NRMSE` in the loss function to guide the sub-networks to learn the corresponding aspect of the time series data. In other words, the loss function design pushes the shape sub-network to model the shape of the time series and the scale sub-network to model the scale of the time series. To test this hypothesis, we use an alternative loss to train the same model where the alternative loss only consists of the `RMSE` term. Since this alternative only uses one `RMSE` in the loss, we call this alternative method *one loss*. In Fig. 6 and Fig. 7, we use ▦ to denote the purposed model train using the loss function with both `RMSE` and `NRMSE`; we use ▨ to denote the method trained only with `RMSE` loss. The design of the loss has a noticeable impact on the model's performance, the amount of improvement obtained by using the proposed loss function ranging from 15% to 59% compared to the alternative. It is crucial to use the proposed loss function when training model with the proposed architecture.

How Does the Shallow Shape Bank Design Comparing to the Commonly Seen Deep Transposed Convolutional Network? One alternative design for the decoder is the transposed convolutional network [11], which is relatively deep comparing to the shape bank design adopted in the proposed model. To compare our shallow design with the transposed convolutional network design, we implement a transposed convolutional (`TConv`) shape decoder as follows. Given an activation vector $\boldsymbol{r}$, the shape prediction $\widehat{\boldsymbol{\alpha}}_j^{(\mathrm{sp})}$ for the j-th feature dimension is computed by:

$$
\begin{aligned}
\boldsymbol{h}_0 &= \mathtt{ReLU}\left(\mathtt{Linear}\left(\boldsymbol{r}\right)\right)\\
\boldsymbol{h}_1 &= \mathtt{Upsample}\left(\mathtt{ReLU}\left(\mathtt{TConv}\left(\boldsymbol{h}_0\right)\right)\right)\\
&\ldots\\
\boldsymbol{h}_l &= \mathtt{Upsample}\left(\mathtt{ReLU}\left(\mathtt{TConv}\left(\boldsymbol{h}_{l-1}\right)\right)\right)\\
\widehat{\boldsymbol{\alpha}}_j^{(\mathrm{sp})} &= \mathtt{Conv}\left(\boldsymbol{h}_l\right)
\end{aligned}
$$

Particularly, the `Linear` layer has 64 channels. For both `TConv` and `Conv` layer, the receptive field size is set to 3, and the number of channels is set to 64. Aside from the last `Upsample` layer, we use the upsampling factor of 2. For the last `Upsample` layer, we set the output size to 24. We set l to 5 for generating a 24 sized time series. The performance of the shallow shape bank design (i.e., ▦) and the deep `TConv` design (i.e., ▤) is shown in Fig. 6 and Fig. 7. Aside from the department store dataset, the shape bank design achieves an improvement over the `TConv` design ranging from 14% to 54% for different datasets and performance measurements. As different datasets consist of different patterns, the `TConv` decoder design only capable of synthesizing the patterns may

Table 2. Comparison between the proposed method and the alternatives.

	RMSE					NRMSE				
	Cat.1	Cat.2	Cat.3	Cat.4	Avg.	Cat.1	Cat.2	Cat.3	Cat.4	Avg.
Nearest neighbor	5.9784	5.7083	8.6402	5.4327	6.4399	11.0063	6.8305	11.2740	9.5840	9.6737
Linear regression	3.6999	3.7094	4.3324	3.7532	3.8737	10.5669	5.5393	10.2107	9.9300	9.0617
Random forest	3.6918	4.2933	4.8370	3.9903	4.2031	8.9944	6.0096	10.5885	9.2405	8.7082
RNN (LSTM)	3.8138	3.7908	4.3121	3.3720	3.8222	9.9693	5.7273	10.2029	8.9879	8.7219
RNN (GRU)	3.8311	3.7381	4.3193	3.4631	3.8379	10.5512	5.6134	10.2579	9.1511	8.8934
This work	3.3390	3.7720	4.1736	3.3530	3.6594	8.7137	5.3977	9.3612	8.0980	7.8927

appear in the department store dataset while struggle on synthesizing the patterns appears in other datasets. Nevertheless, the proposed shape bank decoder has superb performance across the board comparing to the `TConv` decoder.

How Does the Proposed Architecture Comparing to Other Methods? Besides of evaluating each design choice we mode for the proposed model, it is also important to compare the proposed model to both the baseline model and the state of the art model [18] for multi-horizon time series prediction. We compared the proposed model with the following alternative methods:

- Nearest Neighbor. This method predicts the future by searching the nearest neighbor of the current time series from history. Once located the nearest neighbor of the training data, the following 24 h of this nearest neighbor is used as the prediction [12].
- Linear Regression. To apply this method to our problem, we first flatten the input multivariate time series into a feature vector, then we train 96 liner models using the flatten vector where each model is responsible for predicting one feature at one time step. The time series has four features, and we are predicting the next 24 time steps. We use $L2$ regularization when training the linear models.
- Random Forest. An ensemble method that trains a set of decision trees via bootstrap aggregating and random subspace method [2,8]. Similar to the Linear Regression model, we flatten the input multivariate time series into a feature vector, and we formulate the multi-horizon prediction problem as a multi-output regression problem before applying off-the-shelf random forest implementation.
- Recurrent Neural Network. We use two different types of RNNs: Long Short-Term Memory (LSTM) and Gated Recurrent Unit (GRU). For each type of RNN, we use two layers to encode the input time series and a Multi-layer Perceptron (MLP) to predict the time series for the next 24 h. This is the best performing architecture for multi-horizon time series prediction, according to [18].

The experiment result is summarized in Table 2. Out of the three non-deep learning-based methods, linear regression has the best performance based on

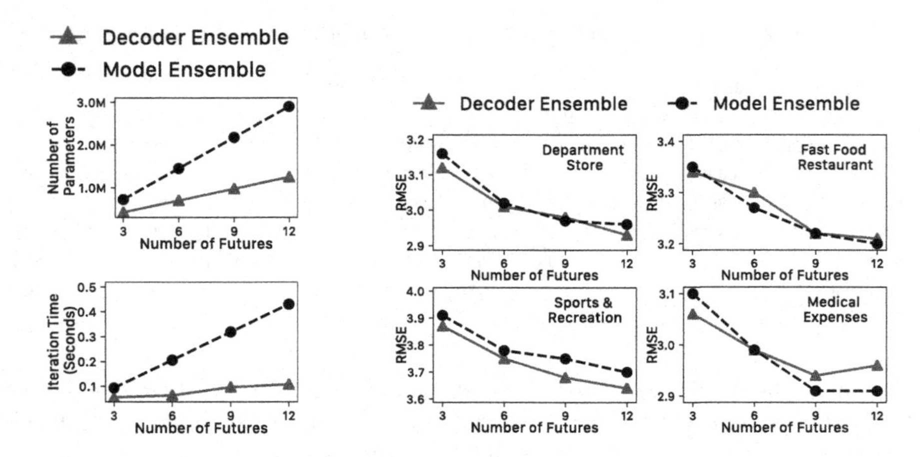

Fig. 8. Scalability comparisons.

Fig. 9. Accuracy comparisons.

averaged RMSE. When considering the averaged NRMSE, random forest outperforms the other two methods. Although the random forest has worse RMSE comparing to linear regression, it is more capable of modeling the shape of time series. When considering the alternative deep learning-based approaches, the RNNs achieve a superb RMSE comparing to linear regression and a similar NRMSE comparing to the random forest. Such observation confirms that the model presented in [18] is also the best off-the-shelf method for modeling time series from transaction data. Lastly, by comparing the proposed method to the alternatives, the proposed method demonstrates its superior performance in predicting both the raw values and the shape of the future time series.

5.3 Evaluation of the Multi-future Learning Scheme

Traditionally, multi-future learning is achieved by training a deep learning ensemble where each model within the ensemble is responsible for making a possible prediction [10,13]. In this section, we showcase the benefit of our multi-future learning scheme (i.e., decoder ensemble) comparing to the existing scheme (i.e., model ensemble) [10,13]. We apply both multi-future learning schemes to the proposed model using the same training algorithm presented in Sect. 4. We only consider the proposed model because the proposed model has superb performance comparing to the alternatives, as demonstrated in Sect. 5.2. To apply the model ensemble scheme to our model, we use multiple encoders, each corresponding to a decoder, to form the ensemble. To showcase the difference in terms of the scalability and accuracy for each method, we present the comparison for the number of parameters, average run-time per iteration, and oracle RMSE for each of the dataset under a different number of futures settings in Fig. 8 and Fig. 9. Regardless of the multi-future learning scheme, the number of parameters and training time grown linearly with the number of future (see Fig. 8); the

RMSEs of different datasets improves as the number of future (see Fig. 9). When comparing the different multi-future learning schemes, the proposed model can achieve comparable RMSEs with 168% and 443% improvement in training time for 3 and 12 futures, respectively comparing to the existing scheme. The number of parameters is only just 59%, and 42% relative to the model learned using an existing scheme when the number of futures is 3 and 12, respectively. The NRMSE figures are omitted due to space limitation, but the conclusion remains the same. In conclusion, the proposed multi-future learning scheme has better scalability comparing to the existing scheme with respect to the number of futures without sacrificing accuracy.

6 Related Work

Time Series Prediction is a well-studied problem dated back to [1]. Recent work utilizes the capability of neural networks to tackle this known problem: Taieb and Atiya [16] train a neural network with multiple steps in the future as the target; Wen *et al.* [18] approach from the sequence-to-sequence learning perspective. Another recent work [14] jointly trains a global matrix factorization model and a local temporal convolution network to model both global and local property of the time series. Although the motivation for their model design is different from ours, their final design shares some similarities with our shape subnetwork design. The major difference between our model and the one proposed by Sen *et al.* [14] is that they use a shared set of basic time series (i.e., shape bank) across all dimensions as they assume that all the dimensions have very similar behavior. However, as shown in Fig. 1, such an assumption does not hold for the time series generated from transaction records. Our method also focuses on the separation between shape and scale information, which is not considered in the model presented in [14]. Additional neural network-based methods are presented in [3,4,6,15] and [5] presents a comprehensive tutorial. Nevertheless, none of these works attempt to produce a multi-future prediction for time series, and they only focus on predicting one determined future given a known set of time series. Such fact limits the applicability of the methods mentioned above in real-world merchant transaction time series data.

Multi-future Learning or multiple-choice learning has attracted more attention recently. Notable work includes: [7], where multiple hypotheses are generated for prediction tasks that incorporate user interactions or successive components; [10] extended the learning algorithm presented in [7] for deep learning models; [13] uses a hierarchical incrementally growing CNN to count the number of people in a picture, the CNN grows following a binary tree structure throughout the training process, and an additional classifier is trained to route each test image through the network. All the works mentioned above focus on computer vision applications; therefore, their methods cannot be directly applied to our problem. In terms of problem formulation, the closest work to us is [17]; they are also solving time series prediction problems. However, since their model is designed specifically for modeling vehicular behaviors and only focuses on predicting the very next time step, we could not directly apply their method to our problem. In conclusion, these works are not tailored to tackle the challenges in time series prediction nor in multi-horizon multivariate time series prediction.

7 Conclusion

In this work, we identified the problem of multi-future multi-horizon prediction on multivariate time series data for merchant transactions. We design the model with the consideration of learning not only the numerical values but also the shape, the magnitude, and the offsets of the time series data. Our proposed model is flexible as it now predicts multiple possible futures of the merchant transactions. We conduct experimental evaluations on real-world merchant transaction data to demonstrate its effectiveness.

References

1. Box, G., et al.: Time series analysis forecasting and control holden-day: San francisco. BoxTime Series Analysis: Forecasting and Control Holden Day 1970 (1970)
2. Breiman, L.: Random forests. Mach. Learn. **45**(1), 5–32 (2001)
3. Cerqueira, V., Torgo, L., Pinto, F., Soares, C.: Arbitrated ensemble for time series forecasting. In: Ceci, M., Hollmén, J., Todorovski, L., Vens, C., Džeroski, S. (eds.) ECML PKDD 2017. LNCS (LNAI), vol. 10535, pp. 478–494. Springer, Cham (2017). https://doi.org/10.1007/978-3-319-71246-8_29
4. De Stefani, J., Caelen, O., Hattab, D., Le Borgne, Y.-A., Bontempi, G.: A Multivariate and Multi-step Ahead Machine Learning Approach to Traditional and Cryptocurrencies Volatility Forecasting. In: Alzate, C., et al. (eds.) MIDAS/PAP -2018. LNCS (LNAI), vol. 11054, pp. 7–22. Springer, Cham (2019). https://doi.org/10.1007/978-3-030-13463-1_1
5. Faloutsos, C., et al.: Forecasting big time series: theory and practice. In: ACM SIGKDD (2019)
6. Fan, C., et al.: Multi-horizon time series forecasting with temporal attention learning. In: ACM SIGKDD (2019)
7. Guzman-Rivera, A., et al.: Multiple choice learning: learning to produce multiple structured outputs. In: NeurIPS (2012)
8. Ho, T.K.: Random decision forests. In: Proceedings of 3rd International Conference on Document Analysis and Recognition, vol. 1, pp. 278–282. IEEE (1995)
9. Kingma, D.P., Ba, J.: Adam: a method for stochastic optimization. arXiv preprint arXiv:1412.6980 (2014)
10. Lee, S., et al.: Stochastic multiple choice learning for training diverse deep ensembles. In: NeurIPS (2016)
11. Radford, A., et al.: Unsupervised representation learning with deep convolutional generative adversarial networks. arXiv preprint arXiv:1511.06434 (2015)
12. Rakthanmanon, T., et al.: Searching and mining trillions of time series subsequences under dynamic time warping. In: ACM SIGKDD (2012)
13. Sam, B., et al.: Divide and grow: capturing huge diversity in crowd images with incrementally growing CNN. In: IEEE CVPR (2018)

14. Sen, R., et al.: Think globally, act locally: a deep neural network approach to high-dimensional time series forecasting. In: NeurIPS (2019)
15. Shih, S.-Y., Sun, F.-K., Lee, H.: Temporal pattern attention for multivariate time series forecasting. Mach. Learn. **108**(8), 1421–1441 (2019). https://doi.org/10.1007/s10994-019-05815-0
16. Taieb, S.B., et al.: A bias and variance analysis for multistep-ahead time series forecasting. IEEE Trans. Neural Netw. Learn. Syst. **27**(1), 62–76 (2015)
17. Tang, C., et al.: Multiple futures prediction. In: NeurIPS (2019)
18. Wen, R., et al.: A multi-horizon quantile recurrent forecaster. arXiv preprint arXiv:1711.11053 (2017)

Think Out of the Package: Recommending Package Types for E-Commerce Shipments

Karthik S. Gurumoorthy(✉), Subhajit Sanyal, and Vineet Chaoji

India Machine Learning, Amazon, Bangalore, India
{gurumoor,subhajs,vchaoji}@amazon.com

Abstract. Manifold of product attributes such as dimensions, weight, fragility and liquid content determine the package type used by e-commerce companies to ship products. Sub-optimal package types lead to damaged shipments, incurring huge damage related costs and adversely impacting the company's reputation for safe delivery. Items can be shipped in more protective packages to reduce damage costs, however this increases the shipment costs due to expensive packaging and higher transportation costs. In this work, we propose a multi-stage approach that trades-off between shipment and damage costs for each product, and accurately assigns the optimal package type using a scalable, computationally efficient linear time algorithm. A simple binary search algorithm is presented to determine the hyper-parameter that balances between the shipment and damage costs. Our approach when applied to choosing package type for Amazon shipments, leads to significant cost savings of tens of millions of dollars in emerging marketplaces, by decreasing both the overall shipment cost and the number of in-transit damages. Our algorithm is live and deployed in the production system, where package types for more than 130,000 products have been modified based on the model's recommendation, realizing a reduction in damage rate of 24%. Overall, the proposed approach has also helped reduce the carbon footprint.

Keywords: Package type selection · Ordinal enforcement · Discrete optimization · Constrained-unconstrained formulation equivalence · Trade off parameter selection

1 Introduction

E-commerce companies like Amazon uses several different package types to ship products from warehouses to the customer's doorstep. These package types vary in the extent of protection offered to the product during transit. Generally, robust package types that provide more protection to the product, resulting in reduced number of package related damages, cost more at the time of shipping due to high material and transportation costs, and vice versa. For instance as shown in Fig. 1,

Y. Dong et al. (Eds.): ECML PKDD 2020, LNAI 12461, pp. 290–305, 2021.
https://doi.org/10.1007/978-3-030-67670-4_18

Amazon has the following different package type options listed in increasing order of protection afforded to the product: (i) No Additional Packaging (NAP), (ii) Polybags: polythene bags small (PS) and special (PL), (iii) Jiffy mailer (JM), (iv) Custom pack (CP), (v) Corrugated T-folder box (T), (vi) Corrugated box with variable height (V), (vii) Corrugated carton box (C). Each package type comes in multiple sizes like small, medium, large and extra-large. The combination of packaging type and size is assigned a barcode, e.g. PS6 to PS9 for small Augmenting the training data polybags. When an item is ready to be packed, the chosen packaging material and size are used to ship the product.

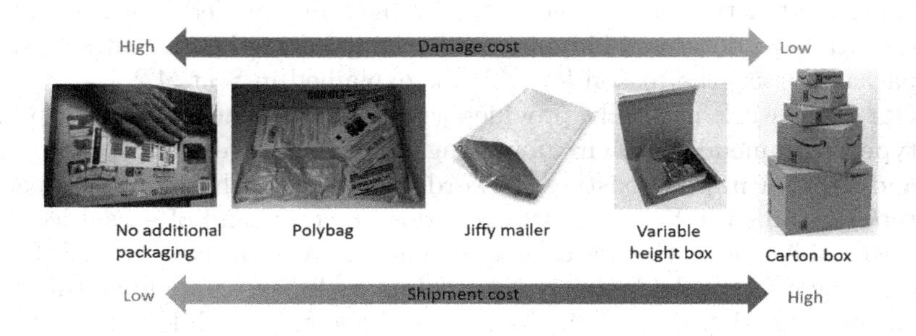

Fig. 1. Different package types

Packaging related damages can happen during transit or handling by an associate during shipment from the warehouse to the customer. As damages result in degraded customer experience, an extra compensatory amount is often paid to the customer over and above the product's price. As these damaged products need to be sent back to the warehouse, there is an additional return shipment cost. Such damages adversely affect the customer relationship since the company's reputation for reliable delivery is impacted. For instance, customers who are dissatisfied with time-critical purchases (e.g., during festivities), may hesitate to buy products in the future. Even the company's relationship with the sellers get impacted, particularly if new products suffer repeated damages over multiple shipments, since the first few customer experiences are critical to the seller's long term success on the platform. The sum of all these costs associated with damages including the product price, reshipment cost, customer compensation cost etc. will henceforth be referred to as damage cost C_{damage}.

To reduce the cost of damages, items can be packed in more protective packaging. However, more protective packaging (e.g. corrugated box (C)) costs more in terms of packaging materials and transportation costs, which could increase the shipping cost that customers have to pay, or costs that the company bears in case of free shipping. It also generates packaging waste at the customer's end which needs to be disposed off additionally. Hence the problem that needs to be addressed is: *"What is the right package type say, between polybag, jiffy mailer or different corrugated boxes that should be used for shipping a product with the best trade off between shipping cost and damage cost ?"* Once the package type is chosen, the smallest container (size) of that package type that could fit the product snugly would be used for actual shipment. This reduces the shipment volume and hence the shipping cost C_{ship}.

1.1 Contributions

Below, we list our main contributions in this work:

(i) We propose a two-stage approach to recommend the correct package type for products resulting in significant savings, primarily from decreased packaging related transit damages. Interestingly, our model's recommendation also leads to decreased shipping cost compared to the current selection of package types, the reason for which is explained in Sect. 4.2.
(ii) Our framework naturally provides a scalable mechanism for the package type recommendation, circumventing manual intervention at every stage and deprecating the existing keyword based approach of mapping package type explained in Sect. 2.1, which is slow, reactive and often subjective.
(iii) We establish novel theoretical connections between the constrained (Ivanov) and unconstrained (Tikhonov) formulations for our unique setting where the optimization variable is discrete, and show that while the constrained formulation is $NP-$complete, the unconstrained formulation enjoys a linear time solution. Though such connections based on the Lagrange dual formulation are known when the optimization variable is continuous [13], the proof methodology employed in our work to derive similar equivalences when the optimization variable is discrete requires fundamentally new insights into the solution space. To the best of our knowledge, this connection is unknown and not exploited before.
(iv) Our understanding of the solution space further enables us to consistently choose the hyper-parameter for the unconstrained formulation using a simple binary search type algorithm, which optimally provides the trade-off between the different cost parameters.

To summarize, we provide a scalable approach for choosing the best package type for products and also present an efficient algorithm to select the hyper-parameter involved in the optimization.

2 Related Work

2.1 Existing Packaging Selection Process

The decision to choose the package type for a product is currently based on a Keyword Based Approach (KBA), where the products are mapped to package types based on whether their title contains a predefined set of positive and negative keywords. Positive keywords work as enablers to ship products in inferior packaging types like polybags or NAP. Examples for positive keywords are helmet, diapers, mosquito net, bag pack, laptop sleeve, bedsheet, cushion, etc.

with the assumption being that such products will have low in-transit damages due to inferior packaging. Negative keywords on the other hand prohibit opting for polybag. Examples for negative keywords are bone-china, detergent, harpic, protein supplements, etc. After manually analyzing the product titles, a suitable package type is identified. Another approach that is closely followed is the selection of package types using the historical data on damages. Here, the damage rates of products are collected on a monthly basis and based on set guardrails, the packaging rules are modified for products with high damages. In addition to being a slow, manual process, this is a reactive approach which does not work for many new products or products whose attributes have been modified recently.

2.2 Why Not Ordinal Regression?

As the different packaging options can be graded in terms of their robustness, the package type forms an ordinal variable with implicit relative ordering between them. This observation naturally surfaces the following question: *"Is predicting the optimal package type for a product just an instance of ordinal regression?"*. Though the answer appears to be a *yes*, the problem lies in the lack of training data. The current assignment of package type to a product is known to be sub-optimal for most of the products w.r.t. trade off between shipping and damage costs. The ideal setting would demand that we have enough samples for every <product, package type> pair, so that one could assign true package type as the target label and perform ordinal regression on product features. This model could then be leveraged to predict package type for new products. Such an exercise would incur significant cost especially at the scale at which e-commerce company like Amazon operates and hence is practically infeasible. The lack of such ground truth data precludes us from performing ordinal regression analysis. We allude to this fact in Sect. 4.2.

2.3 Comparison with Standard Machine Learning Approaches for Package Planning

The work in [8] shows the adoption of machine learning (ML) in the manufacturing industry for automated package planning. Given a training data with well defined labels of package type to be used for product parts, the goal in these applications is to train a supervised ML model based on product characteristics, which are later used to predict package type for unseen products. Our current work differs from these approaches on the following factors:
(i) As described above, we do not have any ground truth data to learn a supervised ML model that directly predicts the package type given the product features. We have training label only at the shipment level that informs whether a product shipped in particular package type was damaged (1) or not (0).
(ii) There is natural ordering between the package types that should be enforced in any learning algorithm.

Given the above two constraints, we are not aware of any learning based framework that automatically chooses the best package type in linear time and is scalable to millions of products.

A majority of the work in logistics is around space optimization, which is broadly related to bin *packing* algorithms [11,12], not to be confused with the *packaging* type selection problem. The aim of the former is to identify those set of products, each of a specific volume, that should be loaded together in a container, in a specific orientation, so that the number of container used is minimum. The bin packing problem has no notion of choosing the best package type for each product. Our work also has very little connection with the box size optimization problem [18], where the goal is to determine the best box sizes that should be used to ship the products, so that the total shipment volume across all product is minimum. We do not optimize for the different sizes of the packages in the current work, but rather determine which package type is best suited for a product. Likewise, we do not forecast packaging demand like the methods developed in [1].

3 Two-Stage Approach for Optimal Package Selection

3.1 Stage 1: Estimating the Transit Damage Probability of a Product Given a Package Type

In this stage, we build a model to solve the following problem, *"Given a product and a package type, what is the probability that a shipment of the product with that package type is likely incur costs due to damages?"*. These damage probabilities are computed for every <product, package type> pair as a product may never have been shipped using a particular (say hitherto unknown optimal) package type to directly retrieve it from the shipment data. In short, our model predicts $p(d|i, j)$ where i refers the product, j refers to the package type and d is a variable indicating damage in transit, with $d = 0$ denoting no damage and $d = 1$ specifying a damage in transit.

For modeling, we considered historical shipment data where for every shipment we have a binary flag a.k.a. the target label indicating whether the shipment resulted in package related damage. We built this model using various metadata associated with the product as predictor variables. The following enumerate a sample set of attributes: product title, category, subcategory, product dimensions, weight, hazardous flag (indicating if product pertains to hazardous materials), fragile flag (denoting whether the product is fragile), liquid flag (representing if the product contains liquids), % air in shipment computed as the difference between the package volume and the product volume, etc. Based on the above set of features we trained a model to predict the probability that the shipment using the particular package type will incur a damage.

Maintaining Ordinal Relationship Between Different Package Types. The notion of graded robustness between package types correlates with the cost

of the packaging material where the cost of packaging goes up if we opt for a more robust package type and vice versa. As there exists an ordinal relationship among the various package types, i.e. they can be ordered in terms of their associated robustness, we need to impart this notion to our model while estimating the damage probabilities. Let m and n be the number of products and package type respectively and without loss of generality let the package type j_k be inferior to j_{k+1} represented by the ordinal relationship $j_1 \leq j_2 \leq \cdots \leq j_n$. During modeling we need to ensure that $p(d|i, j_{k+1}) \leq p(d|i, j_k)$ for all products i. In other words, the prediction function needs to be *rank monotonic* [9] where the rank denotes the robustness of the package type. Note that we require the predictions to satisfy the ranking relationship only between the different package types associated for a given product and not across two different products. We achieve rank monotonicity by two means: (a) Augmenting the training data, and (b) Proper representation of the package type feature and imposing lower bound constraints on the corresponding model coefficients.
Firstly, we append the modeling data as follows:

(i) For every damaged shipment, we create additional shipments with the same product and other inferior (less robust) package types and consider them to be damaged as well. This is to incorporate the notion that if a shipment of a product gets damaged with a particular package type, it is likely to get damaged in package types which are inferior in terms of robustness.
(ii) Likewise, for every shipment without any packaging related damages, we artificially introduce more shipments with the same product and other superior (more robust) package types and consider them to be not damaged as well. This is to incorporate the notion that if a shipment of a product does not get damaged with a particular package type, it is unlikely to get damaged in superior package types.

Appending the data set has an added advantage of creating many more samples for the positive damaged class (label = 1), as typically very few shipments, less than 0.6%, incur packaging related damages. This in turn reduces the model variance as even the positive class is well represented. Secondly, expressing the damage probability values in terms of the sigmoid function, namely $p(d|i, j) = \frac{1}{1+\exp(-f(\mathbf{z}_i, j))}$ where $\mathbf{z}_i$ denote the rest of input features barring the package type, we represent $f(.)$ as $f(\mathbf{z}_i, j) = g(\mathbf{z}_i) + \beta_j$. Here $\{\beta_j\}_{j=1}^n$ are the n model coefficients corresponding to each package type. Ensuring rank monotonicity is equivalent to constraining $\beta_k \geq \beta_{k+1}$. Expressing $\beta_k = \beta_{k+1} + \epsilon_k$, we enforce that $\epsilon_k \geq 0, \forall k \in \{1, 2, \ldots, n-1\}$. In the event that $g(.)$ is linear, i.e., $g(\mathbf{z}_i) = \mathbf{w}^T \mathbf{z}_i$ as the case with Logistic Regression classifier, then for each package type j_k, we append the feature vector $\mathbf{z}_i$ to create $\tilde{\mathbf{z}}_{ik} = [\mathbf{z}_i, \mathbf{p}_k]$ where $\mathbf{p}_k = [\underbrace{0, 0, \ldots, 0}_{k-1}, \underbrace{1, \ldots, 1}_{n-k+1}]$, augment the model coefficient vector $\mathbf{w}$ to $\tilde{\mathbf{w}} = [\mathbf{w}, \epsilon_1, \ldots, \epsilon_{n-1}, \beta_n]$, and express $f(\mathbf{z}_i, j_k) = \tilde{\mathbf{w}}^T \tilde{\mathbf{z}}_{ik}$. The vector $\tilde{\mathbf{w}}$ is determined as part of the model training process under the constraint that $\epsilon_k \geq 0, \forall k$.

3.2 Stage 2: Identifying the Optimal Package Type for Each Product

Optimally assigning the packaging type for each product involves finding the right balance between adopting a robust packaging and incurring more material and transport costs, and settling for an inferior option with a higher probability of in-transit damages leading to increased damage costs. We formulate this trade-off as an optimization problem. Given a packaging type assignment j for a product i, the packing material cost $m(i, j)$ and the transportation cost $s(i, j)$ are known and readily available. The quantity $s(i, j)$ is known as the *bill weight* and is proportional to the package volume. The net shipping cost, $C_{ship}(i, j) = m(i, j) + s(i, j)$. The total shipment cost, T_{ship}, computed over all the products equals: $T_{ship} = \sum_{i} C_{ship}(i, j) * s_{vel}(i)$, where $s_{vel}(i)$ is the sales velocity —number of units sold in a specified period— of the product i. Further, if a product i associated with the package type j is damaged in transit, we incur a net damage cost $C_{damage}(i)$. This damaged cost depends only on the product and independent of the package type used in the shipment. Using the in-transit damage probability $p(d|i, j)$ determined in stage 1 (Sect. 3.1), we estimate the damage cost as: $T_{damage} = \sum_i p(d|i, j) * s_{vel}(i) * C_{damage}(i)$.

Let us denote the current package type assignment of product i by j^{cur}. According to the current package type assignment, the total cost due to in-transit damages is: $T^{cur}_{damage} = \sum_{i} p(d|i, j^{cur}) * s_{vel}(i) * C_{damage}(i)$. The objective of the optimization is to determine the package types such that T_{ship} is minimized and at the same time T_{damage} is not largely different from T^{cur}_{damage} i.e., $T_{damage} \leq \gamma * T^{cur}_{damage}$, where $\gamma \geq 0$ sets the allowable tolerance w.r.t. T^{cur}_{damage} and is determined by business requirements.

Mathematical Formulation. Let the variable x_{ij} indicating whether a product i is to be shipped using the package type j, be the $<i, j>$ entry of the binary matrix X. These variables have to satisfy the following constraints, namely: $x_{ij} \in \{0, 1\}, \forall i, j$ and $\sum_j x_{ij} = 1, \forall i$. The first constraint states that a product is either shipped in a particular type ($x_{ij} = 1$) or not ($x_{ij} = 0$). The second constraint specify that a product should be shipped using one and only one package type. In additional to the aforesaid binary constraints, we also need to specify infeasible conditions that preclude certain products to be shipped via certain modes of packaging. For instance, liquid, fragile and hazardous products can neither be recommended polybags nor be shipped without any packaging if they are not currently shipped in these package options. We enforce these infeasibility constraints by creating a mask matrix M where we set $M_{ij} = 1$ if product i cannot be shipped in package type j and $M_{ij} = 0$ otherwise. By imposing the constraint $\sum_{i,j} M_{ij} * x_{ij} = 0$, the optimization algorithm will be coerced to set $x_{ij} = 0$ whenever $M_{ij} = 1$, thereby meeting our infeasibility requirements. Letting $S_{ij} = C_{ship}(i, j) * s_{vel}(i)$ to be the net shipment cost when product i is sent in package j, $D_{ij} = p(d|i, j) * s_{vel}(i) * C_{damage}(i, j)$ as the net damage cost when the shipment experiences an in-transit damage due to the packaging, $T = \gamma * T^{cur}_{damage}$, our objective can be mathematically expressed as:

$$\min_X \sum_{i,j} S_{ij} * x_{ij} \quad s.t. \quad \sum_{i,j} D_{ij} * x_{ij} \leq T \tag{3.1}$$
$$\text{where, } x_{ij} \in \{0,1\}, \sum_j x_{ij} = 1, \forall i \text{ and } \sum_{i,j} M_{ij} * x_{ij} = 0.$$

However, computing the optimal solution for X based on the Integer Programming (IP) formulation in Eq. (3.1) is computationally expensive as it is a known NP-complete problem [14]. The IP formulation is definitely not scalable and is of very limited use for our setting. Hence, we *do not* compute the solution for X by solving Eq. (3.1). We present the IP objective with the only intent of mathematically formulating and motivating our optimization problem. The direct minimization of the shipping cost, while enforcing that overall damage cost does not exceed the constant T, makes the setting easier to understand. We abstain from solving for X based on this IP objective.

A closer look into the constraints on X reveals that, the constraints are only intra-product, i.e., across different packaging options for a given product and there are no inter-product constraints. This insight enables us to derive an equivalent formulation for Eq. (3.1) whose solution, as we demonstrate, can be obtained via a simple linear time algorithm in $O(mn)$. To this end, let $S(X) = \sum_{i,j} S_{ij} * x_{ij}$, $D(X) = \sum_{i,j} D_{ij} * x_{ij}$, and consider the formulation:

$$\min_X E(X) = S(X) + \lambda D(X) \text{ s.t.}, \tag{3.2}$$
$$x_{ij} \in \{0,1\}, \sum_j x_{ij} = 1, \forall i \text{ and } \sum_{i,j} M_{ij} * x_{ij} = 0,$$

where the hyper-parameter λ is a single globally specified constant *independent* of the products and the package types. The constrained formulation in Eq. (3.1) is known as the Ivanov formulation [6] and the objective in Eq. (3.2) is referred to as the Tikhonov formulation [17]. The equivalences between the two are specifically known for Support Vector Machines [3,13] where the optimization variable, the weight vector $\mathbf{w}$, is continuous and is based on the Lagrange dual formulation. This approach does not work in our discrete setting where X is binary valued. We need to establish this equivalence without invoking the Lagrange formulation and hence our proof methodology is substantially different.

Let X_λ and X_T be the optimal solutions for the hyper-parameters λ and T in Tikhonov and Ivanov formulations respectively. Under mild conditions on the shipment and damage cost values, we prove that these two formulations are equivalent in the sense that for every T in Ivanov, $\exists$ a value of λ in Tikhonov such that both the formulations have the exact same optimal solution in X. To this end, we have the following lemmas:

Lemma 1. *The value of the objective function $E(X_\lambda)$ at the optimal solution X_λ strictly increases with λ.*

Lemma 2. *The overall damage cost $D(X_\lambda)$* [shipment cost $S(X_\lambda)$] *at the optimal solution X_λ is a non-increasing* [non-decreasing] *function of λ, i.e., if $\lambda_1 \leq \lambda_2$ then $D(X_{\lambda_1}) \geq D(X_{\lambda_2})$ $[S(X_{\lambda_1}) \leq S(X_{\lambda_2})]$. Further, if $X_{\lambda_1} \neq X_{\lambda_2}$ then we get the strict inequality, namely $D(X_{\lambda_1}) > D(X_{\lambda_2})$ $[S(X_{\lambda_1}) < S(X_{\lambda_2})]$.*

Lemma 2 states that $D(X_\lambda)$ is a piece-wise constant function of λ whose value decreases when the optimal solution changes. The length of the constant portion equals the range of λ having the same optimal solution. Further, $D(X_\lambda)$ is discontinuous and points of discontinuity occurs at those values of λ for which there are two different optimal solutions in X_λ. Lack of space precludes us from giving the details of the proof. We establish the equivalence through the following theorems.

Theorem 3. *For every λ in E (3.2), $\exists$ $T(\gamma)$ in Eq. (3.1) such that $X_\lambda = X_T$.*

We define a quantity Δ to equal the largest change between the two values of $D(X_\lambda)$ at the points of discontinuity. For our specific D matrix, $\Delta \leq \max_i \left[\max_{j, M_{ij}=0} D_{ij} - \min_{j, M_{ij}=0} D_{ij} \right]$. Armed with this definition, we now prove a mildly weaker equivalence in the opposite direction.

Theorem 4. *For every $T = \gamma * T^{cur}_{damage}$ in Eq. (3.1) for which the optimal solution X_T exists, one can find a $T^* \in [T, T + \Delta)$ such that for this value of T^*, $\exists \lambda$ in Eq. (3.2) satisfying $X_\lambda = X_{T^*}$.*

Linear Time Algorithm. The primary advantage of this equivalence is that the Tikhonov formulation in Eq. (3.2) enjoys a linear time algorithm compared to the Ivanov problem in Eq. (3.1) which is NP-complete. To see this, note that the constraints in the variables x_{ij} are only across the different package types j given a product i and there are no interaction terms between any two different products. Hence the optimization problem can be decoupled between the products and reduced to finding the optimal solution *independently* for each product agnostic to others. For each product i, define the vector $\mathbf{x}_i = [x_{i1}, x_{i2}, \ldots, x_{in}]$ and consider the optimization problem:

$$\min_{\mathbf{x}_i} \sum_j [S_{ij} + \lambda D_{ij}] x_{ij} \text{ s.t., } x_{ij} \in \{0,1\}, \sum_j x_{ij} = 1 \text{ and } \sum_j M_{ij} * x_{ij} = 0. \tag{3.3}$$

Among all the package types where $M_{ij} = 0$, the minimum occurs at that value of $j = j_i^*$ where the quantity $S_{ij_i^*} + \lambda D_{ij_i^*}$ takes the least value. In other words, define $j_i^* = \underset{j, M_{ij}=0}{\operatorname{argmin}} [S_{ij} + \lambda D_{ij}]$. Then $x_{ij_i^*} = 1$ and $x_{ik} = 0, \forall k \neq j_i^*$ is the optimal solution. As it only involves a search over the n values, its time complexity is $O(n)$ for each product. Hence the optimal solution X_λ across all the m products can be determined in $O(mn)$.

Selection of the Hyper-parameter λ. It is often easier to specify a bound on the overall damage cost $D(X_\lambda)$ via the tolerance constraint $T = \gamma * T^{cur}_{damage}$ in

Algorithm 1. Algorithm to determine λ given T and stopping criteria ρ

function DETERMINELAMBDA(T, ρ)
  Set: $\lambda_{min} = 0$, $\lambda_{max} =$ chosen high value, $\lambda_{mid} = \frac{\lambda_{min}+\lambda_{max}}{2}$, stoppingCriteria = False
  do
    Set: $\lambda = \lambda_{mid}$
    Determine: Optimal solution X_λ using the linear time algorithm.
    if $D(X_\lambda) < T$ **then**
      $\lambda_{max} = \lambda_{mid}$
    else
      $\lambda_{min} = \lambda_{mid}$
    end if
    Recompute: $\lambda_{mid} = \frac{\lambda_{min}+\lambda_{max}}{2}$
    if ($|\lambda_{mid} - \lambda| \leq \rho$) or ($D(X_\lambda) == T$) **then**
      Set: stoppingCriteria = True
    end if
  while (stoppingCriteria==False)
  return λ
end function

the Ivanov formulation in Eq. (3.1), as it is driven by business requirements such as customer satisfaction, impact of damages on downstream purchase behavior, etc. However, knowledge of γ alone is of little value as the Ivanov formulation being NP-complete, is computationally expensive to solve and we rightly refrain from doing so. Instead, we determine the corresponding λ through an efficient algorithm and then solve the Tikhonov formulation in Eq. (3.2) in linear time as explained in Sect. 3.2. Although no closed form expression exists relating the two, the non-increasing characteristic of $D(X_\lambda)$ in Lemma 2 can be leveraged to design a binary search algorithm for λ, as described in Algorithm 1. The crux of our method is to repeatedly bisect the interval for the search space of λ and then choose the subinterval containing the λ. The technique is very similar to the bisection method used to find the roots of continuous functions [4]. The user input ρ in Algorithm 1 is the stopping criteria on the minimum required change in λ values between successive iterations for the while loop to be executed. The number of iterations is inversely proportional to the magnitude of ρ.

Package Prediction for New Products. The definition of the net shipping and damage cost matrices includes the sales velocity term $s_{vel}(i)$, as the total shipment and damage costs across all products explicitly depend on the individual quantities of products sold. Hence the optimization problem in Eq. (3.1) deliberately makes use of the sales velocity term folded into the S_{ij} and D_{ij} matrix entries. However, for new products, the sales velocity is unknown and needs to be forecasted; which is generally very difficult and at most times noisy [10]. The lack of this term seems to preclude the new products from being part of the optimization in Eq. (3.1). However, the equivalent Tikhonov formulation in Eq. (3.2) comes to our rescue. Closely looking into the product-wise optimization problem in Eq. (3.3), note that $s_{vel}(i)$ appears in the same form (linearly) in both the S_{ij} and D_{ij} quantities and also *does not* depend on the package type j. Hence it can be factored out and dropped from the optimization altogether. The equivalent formulation has revealed a key insight that once λ is appropriately chosen, the optimal solution is *independent* of the sales velocity. Hence for all new products l, we only need to compute quantities $\{S_{lj}, D_{lj}\}_{j=1}^{n}$ without factoring in $s_{vel}(i)$ and choose that package type j_l^* with the least value of $S_{lj_l^*} + \lambda D_{lj_l^*}$ among the package types where $M_{lj} = 0$.

4 Experimental Results

Our training data for stage 1, where we predict the damage probabilities, consists of shipments during a 3 month period in 2019. We augmented the data with artificially induced inferior and superior packaging types and their corresponding 1 and 0 target values. We opted for the Logistic Regression classifier to learn and predict the damage probabilities $p(d|i,j)$, as it enables us to interpret and explain the predictions. Importantly, its linearity (post the link function) endows the model with the notion of ordinal relationship between packages by appending the sample features $\mathbf{z}_i$ with the package related features $\mathbf{p}_k$ as elaborated in Sect. 3.1. Though our classifier is linear in the feature space, we introduced non-linearity through various polynomial transformations of the input features and having interactions between the product and the package features to create new (non-linear) features. In more than 100 *million* augmented training shipments, only 0.7% shipments belonging to class 1 incurred damages due to packaging related issues. We counter this huge class imbalance by specifying class specific weight values of $1-\tau$ and τ to classes 1 and 0 respectively to the cross-entropy loss function where we set $\tau = 0.007$. We assessed the performance of our model on a test data consisting of 8 *million* shipments for about 600,000 products, out of which only 0.6% shipments incurred packaging related damages. The shipments in the test data occurred in a different time period w.r.t. training data. After augmenting the test data with artificially induced inferior and superior packaging types, the models performance on the area under the curve (AUC) metric was **0.902**.

4.1 Calibration

Since we are interested in estimating the actual probability of damage rather than binary classification, the estimated raw damage probabilities $p(d|i,j)$ need to be calibrated to reflect the true damage probabilities in the shipment data. This is more so, as we introduced class specific weights during training. We used Isotonic Regression [2] to learn the calibration function. It yielded the smallest average *log−loss* (log−loss = 0.0347) compared to the implicit calibration via the closed form expression derived in [7] (Eq. (28)) for binary Logistic Regression models (log−loss = 0.0379), and Platt Scaling [15] (log-loss = 0.0349). The log-loss for each shipment equals:$-y\log(p_{cal}) - (1-y)\log(1-p_{cal})$, where y is the actual label and p_{cal} is the calibrated damage probability. All these calibration methods significantly reduces the uncalibrated average log loss of 0.4763. To assess the correctness of post-calibrated values, for each package type we bucketed its shipments into 20 quantiles based on their calibrated values. For each quantile, we computed the absolute difference between the actual damage rate and the average of the calibrated values, weighted these absolute differences proportional to the number of shipments in each quantile, and then summed them. Figure 2 shows the summed, weighted absolute differences for each package type, for different calibration methods. Observe that Isotonic Regression has the lowest values across multiple package types. *Such low difference values highlight the estimation accuracy of our post-calibrated damage probability values.*

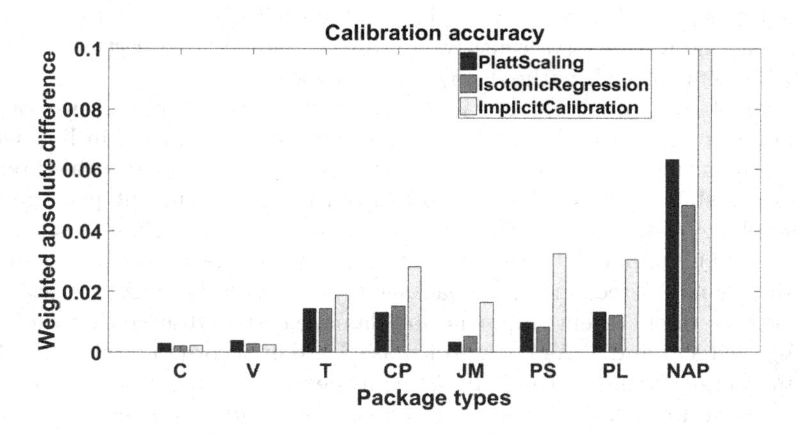

Fig. 2. Weighted absolute difference between estimated and true damage rates

4.2 Package Type Recommendation

For a dataset of about 250, 000 products in more than 10 categories with active purchase history in Amazon, we determined their raw damage probability for all possible package type options and then calibrated them using Isotonic Regression. Table 1 shows the relative average damage probabilities computed across the products for each package type. The damage probability for shipment without packaging (NAP) is set to 1 and the values for other package types are scaled relatively. The business sensitive nature of these damage probabilities precludes us from disclosing their actual estimated values. Observe that our model has indeed learned the implicit ordering between the package types, where the superior package types like C and V have the lowest values and inferior package types like PS and PL have the highest. The predicted damage probabilities are then fed into our optimization algorithm that proposes optimal packaging type for all the products.

For each <product, package type> tuple, we identified the smallest size of that package type that could fit the product snugly. This reduces the shipment volume and also the shipment cost. Recall that by setting entries $M_{ij} = 1$ in the mask matrix M, we can prevent the optimization from choosing the package type j for product i. We set $M_{ij} = 1$ for the following cases based on business rules: (i) products which due to its large size and volume cannot be shipped even in the largest container of certain package types, equivalent to setting the corresponding $S_{ij} = \infty$, (ii) liquid products from being shipped in JM, PS, PL or NAP; restricting fragile products from being sent in T, CP, JM, PS, PL or NAP; disallowing hazardous products to be shipped in PS, PL or NAP if these products are not currently shipped in these package types (the latter condition is required as these flags can sometimes be erroneously set), (iii) inferior package types compared to the current selection i.e., $j < j^{cur}$ for products (with active purchase history) having high damages in the current package type, (iv) superior package types $j > j^{cur}$ if $S_{ij} > S_{ij^{cur}}$ for products with very low damages in their current packaging type, (v) sensitive products belonging to certain categories from being sent in NAP without any packaging etc.

To corroborate our theoretical results in Lemmas 1 and 2, we ran the Tikhonov formulation in Eq. (3.2) for different values of λ, each in linear-time, and plot the results in Fig. 3. The values of $S(X_\lambda)$ and $D(X_\lambda)$ are scaled relative to the total shipment and damage costs from using the current package type, respectively. A value greater (lesser) than 1 indicates that these costs will be higher (lower) compared to the current levels when the products are shipped based on the model recommended package types. Similarly, $E(X_\lambda)$ is scaled relative to the sum of current shipment and damage costs. Observe that the trends of $E(X_\lambda)$, $D(X_\lambda)$ and $S(X_\lambda)$ as we increase λ are in accordance with the claims made in Lemmas 1 and 2. To verify the equivalence relations between the Ivanov and Tikhovov formulations stated in Theorems 3 and 4, we implemented the Integer Programming for Ivanov by setting $\gamma = 1.0$ using the CVXPY package [5]. We then determined the value of corresponding λ by executing our binary search method (Algorithm 1) for $\rho = 0.001$ and $\lambda_{max} = 1000$. The algorithm met the stopping criteria in 19 iterations and returned with $\lambda = 0.13387$. The identical results for $(\lambda = 0.13387, \gamma = 1.0)$ in columns VII and VIII of Table 1 is a testimony to this equivalence relationship. We validated this equivalence for other values of γ using our binary search algorithm and obtained similar results. In Fig. 3, note that though the net damage cost $D(X_\lambda)$ for $\lambda = 0.13387$ (and for $\gamma = 1.0$) exactly matches the cost value computed from using the current package types (ratio = 1 marked in horizontal red dotted line), the shipping cost $S(X_\lambda)$ is smaller than the current shipment cost (ratio = 0.843 marked in horizontal green dotted line). In other words, we are able to reduce the shipping cost from the current value without further increasing the damage cost. This again points to the fact that the existing product to package type mappings are sub-optimal, preventing us from pursuing the path of ordinal regression as explained in Sect. 2.2.

For each package type in Table 1, we show the ratio of number of products mapped to that package type by our algorithm and the number of products currently assigned to the package type, for different λ values. For instance if 100 products are currently shipped in package type C and our model recommends using C for 120 products, the ratio will equal 1.2. A number greater (lesser) than

1 denotes higher (lesser) recommendation of that package type compared to the current usage. Note that as we increase λ giving more importance to damage cost, the ratio for superior package types such as C and V steadily increases, and this trend is reversed for inferior packaging options such as PL and NAP. This shift is as expected since the damage rate and the damage cost decrease at higher λ values. In Table 2 we show the ratio between the number of products recommended to be sent in a particular package type computed at $\lambda = 1.5$ and the number of products currently shipped in these package types for different product categories. The value 0/0 means no product of that category is currently shipped in the specific package type and our model does not recommend it either. The true counts are confidential and cannot be disclosed. Observe that for liquid, fragile and hazardous products, the ratio is less than 1 for inferior package types such as JM, PS, PL and NAP, indicating that our method recommends lesser usage of these options for these kinds of products. Notably, many electronics products in column VI with high damage probability are moved to the most superior C package type, further contributing to the decreased damage rate of 24% as observed in Sect. 4.3, and cost savings of tens of millions of dollars even in emerging marketplaces.

Table 1. Relative avg. calibrated damage probabilities and change in product mappings for package types

I	II	III	IV	V	VI	VII	VIII
	Package type	*Relative avg. damage probability*	*Ratio for* $\lambda = 0.5$	*Ratio for* $\lambda = 1$	*Ratio for* $\lambda = 1.5$	*Ratio for* $\lambda = 0.13387$	*Ratio for* $\gamma = 1$
Superior	Carton box (C)	0.022	0.915	1.176	1.314	0.499	0.499
Package type	Variable height (V)	0.027	0.566	0.676	0.740	0.392	0.392
↓	T-folder (T)	0.043	1.227	1.282	1.298	0.898	0.898
	Custom pack (CP)	0.112	2.420	2.438	2.418	2.174	2.174
	Jiffy mailer (JM)	0.174	0.408	0.503	0.539	0.236	0.236
	Small polybag (PS)	0.447	1.586	1.360	1.238	1.955	1.955
Inferior	Special polybag (PL)	0.448	1.043	0.906	0.798	0.973	0.973
Package type	No packaging (NAP)	1.0	1.748	1.144	0.940	3.635	3.635

4.3 Impact Analysis from Actual Shipment Data

The numbers quoted below are excerpts from the actual shipment data, where for 130,000 products contributing to 21% shipments, their current package type was changed to the model's recommendation. We used the proposed package type obtained for $\lambda = 1.5$ ($\gamma = 0.69$), thus giving higher weight to reducing damage costs. The rationale being that receiving damage products negatively affects the customer trust in e-commerce companies and could affect their downstream purchase behavior. When these shipments were compared against those where the original package type was used, we observed the following significant positive

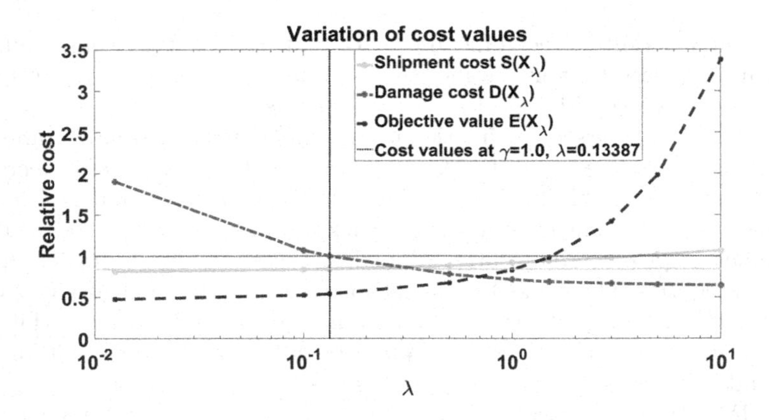

Fig. 3. Variation of relative cost values with λ

impacts: (i) Decrease in damage rate by **24%**, (ii) Decrease in transportation cost per shipment by 5%, (iii) Salability of products undelivered to customer because of transit damages improved by 3.5%. The only negative impact was that the material cost of the shipping supplies increased by 2%, as many products were moved to superior package types to reduce damages.

Table 2. Relative change in product mappings across different categories

I	II	III	IV	V	VI	VII	VIII
	Package type	*Liquid products*	*Fragile products*	*Hazardous products*	*Electronics category*	*Kitchen category*	*Beauty category*
Superior	Carton box(C)	0.929	1.794	1.975	9.525	1.214	0.927
Package type	Variable height(V)	0.754	1.047	0.915	1.017	0.859	0.478
↓	T-folder(T)	1.496	1.286	1.281	2.302	1.427	1.221
	Custom pack(CP)	0/0	1.383	1.081	1.873	2.855	0.818
	Jiffy mailer(JM)	0.519	0.486	0.488	0.392	0.695	0.928
	Small polybag(PS)	0.000	0.821	0.843	1.203	1.263	1.720
Inferior	Special polybag(PL)	0/0	0.541	0.500	0.863	1.124	4.000
Package type	No packaging(NAP)	0/0	0.598	0.806	0.737	0.701	2.000

5 Conclusion and Future Work

We presented a two-stage approach to recommend optimal packaging type for products, where we first estimated the calibrated damage probabilities for every <product, package type> tuple and then fed them into our linear-time optimization algorithm to select the best type. The binary search algorithm efficiently computes the trade-off parameter λ given the value γ in the Ivanov formulation.

In many scenarios, the extent of damages depend on the distance shipped, the air/ground mode of transportation used, the quality of roads along the route, the handling by the courier partners, the location of the warehouses or even the time of year as during monsoon seasons, more protection against water or moisture may be needed for some products. Going forward, we would like to lay emphasis on predicting the optimal packaging type based not only on the product, but using several aforementioned additional factors relating to a specific shipment of an item to a customer. Additionally, we would like to estimate the causal impact [16] of receiving damage products on customer's spend patterns and factor it into our optimization algorithm.

References

1. Bachu, Y.: Packaging demand forecasting in logistics using deep neural networks. Master's thesis, Blekinge Institute of Technology (2019)
2. Barlow, R., Bartholomew, D.J., Bremner, J., Brunk, H.: Statistical Inference Under Order Restrictions; the Theory and Application of Isotonic Regression. Wiley, New York (1972)
3. Cortes, C., Vapnik, V.: Support-vector networks. Mach. Learn. **20**(3), 273–297 (1995). https://doi.org/10.1007/BF00994018
4. Corliss, G.: Which root does the bisection algorithm find? SIAM Rev. **19**(2), 325–327 (1977)
5. Diamond, S., Boyd, S.: CVXPY: a Python-embedded modeling language for convex optimization. J. Mach. Learn. Res. **17**(83), 1–5 (2016)
6. Ivanov, V.: The Theory of Approximate Methods and their Application to the Numerical Solution of Singular Integral Equations. Springer, New York (1976)
7. King, G., Zeng, L.: Logistic regression in rare events data. Polit. Anal. **9**, 137–163 (2001)
8. Knoll, D., Neumeier, D., Prüglmeier, M., Reinhart, G.: An automated packaging planning approach using machine learning. In: 52nd CIRP Conference on Manufacturing Systems, pp. 576–581 (2019)
9. Li, L., Lin, H.T.: Ordinal regression by extended binary classification. In: 19th International Conference on Neural Information Processing Systems, pp. 865–872. MIT Press (2006)
10. Machuca, M.M., Sainz, M., Costa, C.M.: A review of forecasting models for new products. Intangible Capital **10**(1), 1–25 (2014)
11. Mao, F., et al.: Small boxes big data: a deep learning approach to optimize variable sized bin packing. CoRR abs/1702.04415 (2017)
12. Martínez, M.T.A.: Models and algorithms for solving packing problems in logistics. Ph.D. thesis, Universidad de Castilla - La Mancha (2015)
13. Oneto, L., Ridella, S., Anguita, D.: Tikhonov, Ivanov and Morozov regularization for support vector machine learning. Mach. Learn. **103**(1), 103–136 (2015). https://doi.org/10.1007/s10994-015-5540-x
14. Papadimitriou, C.H., Steiglitz, K.: Combinatorial Optimization: Algorithms and Complexity. Dover, Mineola, New York (1998)
15. Platt, J.: Probabilistic outputs for support vector machines and comparisons to regularized likelihood methods. Adv. Large Margin Classifiers **10**(3), 61–74 (1999)
16. Rubin, D.: Causal inference using potential outcomes. J. Amer. Statist. Assoc. **100**(469), 322–331 (2005)
17. Tikhonov, A., Arsenin, V., John, F.: Solutions of ill-posed problems. Winston, Washington, D.C. (1977)
18. Wilson, R.C.: A packaging problem. Manage. Sci. **12**(4), B135–B145 (1965)

Topics in Financial Filings and Bankruptcy Prediction with Distributed Representations of Textual Data

Ba-Hung Nguyen[1(✉)], Shirai Kiyoaki[2], and Van-Nam Huynh[1]

[1] School of Knowledge Science, Japan Advanced Institute of Science and Technology, Nomi, Japan
{hungba,huynh}@jaist.ac.jp
[2] School of Information Science, Japan Advanced Institute of Science and Technology, Nomi, Japan
kshirai@jaist.ac.jp

Abstract. We uncover latent topics embedded in the management discussion and analysis (MD&A) of financial reports from the listed companies in the US, and we examine the evolution of topics found by a dynamic topic modelling method - Dynamic Embedding Topic Model. Using more than 203k reports with 40M sentences ranging from 1997 to 2017, we find 30 interpretable topics. The evolution of topics follows economics cycles and major industrial events. We validate the significance of these latent topics by the state-of-the-art performance of a simple bankruptcy ensemble classifier trained on both novel features - topical distributed representation of the MD&A, and accounting features.

Keywords: Financial filings · Bankruptcy prediction · Topic modelling

1 Introduction

Using EDGAR[1] system, the US Securities and Exchanges Commission (SEC) requires listed US enterprises to file their financial reports with the 10-Ks and 10-Qs forms for annual and quarterly reports, respectively. In these forms, along with the detail financial accounting statements, enterprises need also to include their Management Discussion & Analysis[2] (MD&A), which is a forward-looking statement. In that section, the top managers explain their business performance, address the compliance & risks, and express their views on the company future

[1] Electronic Data Gathering, Analysis, and Retrieval system - SEC.
[2] https://www.sec.gov/corpfin/cf-manual/topic-9.

Electronic supplementary material The online version of this chapter (https://doi.org/10.1007/978-3-030-67670-4_19) contains supplementary material, which is available to authorized users.

Y. Dong et al. (Eds.): ECML PKDD 2020, LNAI 12461, pp. 306–322, 2021.
https://doi.org/10.1007/978-3-030-67670-4_19

goals and projects. Together with examining and monitoring the company performance through traditional financial statements, these forms provide a rich-feature dataset that could be effectively exploited for understanding the evolution of embedded aspects in manager sentiment and further assist on other predictive tasks. In addition, common predictive models based on accounting data are limited due to the window dressing problem in reported earnings [9,11].

There are many benefits of building supportive systems based on the alternative data to complement the traditional predicting or scoring systems. Most current work in the textual analysis in the financial industry was quantifying the textual data to (i) form the predictors for future company financial performance by explaining the manager sentiment and stock returns [21,22]; (ii) understand the role of investment analyst report [13]; or (iii) improve the manager sentiment tone understanding [22]. However, little work has been paid to understanding what these predictors actually represent for and how they evolve through economic cycles and crisis. Our study contributes to the current literature of textual analysis for the financial reports by first uncovering the latent topics from the MD&A of the filings, investigating their evolution, and then further examining the predictive power of those textual topics and investigating how they could improve the traditional bankruptcy prediction methods. Specifically, by using Latent Dirichlet Allocation [5] and Dynamic Embedded Topic Models [7], we find the hidden, yet interpretable topics in the MD&A section and show that they are significant and deliver comparable predictive performance for models only using the MD&A textual data. More importantly, the topics follow major economic events and show effective complement role to the traditional accounting-based bankruptcy prediction model.

In what follows, we first present the relevant literature on the topic modelling and the textual analysis in bankruptcy modelling in Sect. 2. Section 3 devotes to the process of mining MD&As, estimating the latent topics, a summary of the data, and experimental settings. The results are presented and discussed in Sect. 4, and we conclude our work in Sect. 5.

2 Literature Reviews

In uncovering the latent topics embedded in textual data, the pioneering work of Blei et al. [5] introduced latent Dirichlet allocation (LDA) which uses the variational Bayesian inference to infer the latent topics from a large corpus. Then, based on an intuition that the document collections should reflect evolving content, Blei and Lafferty [4] proposed the dynamic topic models (DTM), which is a dynamic version of LDA to examine the evolution of topics over time, and showed the superior performance in term of the likelihood for the hold-out dataset, compared with the traditional LDA. However, as existing topic models fail to learn interpretable topics when working with large and heavy-tailed vocabularies, Embedding Topic Models [8] (ETM) bridges this gap by utilising the word embeddings [19]. ETM incorporates the embeddings into the inference procedure of the traditional LDA, specifically, it combines traditional topic models with word embeddings and models each word with a categorical distribution whose natural parameter is the inner product between a word embedding and the embedding of its assigned topic. ETM discovers interpretable topics even with large, imbalanced vocabularies that include rare words and stop words. And to model topic evolution in embedding spaces, Dieng et al. [7] further introduced Dynamic Embedding Topic Models (DETM) which combines the ETM with DTM by modelling each word with a categorical distribution parameterised by the inner product between the word embedding and a per-time-step embedding representation of its assigned topic. DETM is fitted using structured amortized variational inference with a recurrent neural network and it can learn smooth topic trajectories by defining a random walk prior over the embedding representations of the topics.

In applications of topic modelling for financial data, one of the most initial work in summarising the textual financial data is from Bao and Datta [3], where they employed the modified LDA to fit the sentence level analysis with the assumption of one topic for each sentence. Their empirical analysis focused on the risk-disclosure in the filings and found 30 risk types (topics). Among them, there are new and significant risk types to predict the risk perceptions of investors. Significantly, they discovered five more important topics than the topics found by the large-scaled supervised learning [14]. In the application of topic modelling in stock market analysis, Nguyen et al. [21] utilised the sentiment analysis, specifically, they analysed the financial social media data using a combination of topic modelling and sentiment classifier in predicting stock price movements. They showed that their ensemble model achieves better performance in predicting stock price movements compared with the traditional time series and human sentiment methods. Recently, some scholars examined the relationship between homogeneous and heterogeneous sources of financial texts and indicated that higher manager sentiment followed by lower earnings disruption and higher investment growth [15,22].

There are urgent needs not only on summarising the MD&A hidden topics but also to further leveraging these representations to enhance other causal or predictive models. To our knowledge, there are two work in examining the bankruptcy of firms by utilising the textual data, which are Gandhi et al. [9] and Mai et al. [18]. The first one employed word-level sentiment analysis based on financial wordlists [17] to examine the relationships of sentiment words with financial distress of the US banks, they suggested that more negative words in the reports are related to a higher likelihood of distressed delisting. The latter based on deep learning models which are a convolutional neural network model and embedding approaches to examine the gain in predictive power. Their results showed that simple averaging embedding is more effective than convolutional neural networks. However, neither the relationships between those sentiment words and the likelihood of bankruptcy have been investigated for non-banking sector data nor how the topics in textual reports relate to firm bankruptcy. In addition, since deep learning models are black-box models, few works have been devoted to the performance of classification models trained on more intuitive and interpretable textual features. Hence, we aim to bridge these gaps by employing topic models as an explorer for hidden topics in MD&A of the financial filings, and as a feature engineering method for enhancing the traditional bankruptcy prediction model through leveraging the textual data.

3 Data and Methods

3.1 Data

Accounting Data. In this study, we collect the accounting data from the Wharton Research Data Services (WRDS) for all listed firms in the US from 1997 to 2016. The detail statistics of their financial report elements are presented in Table 1 and Table 2 of the supplemental document. We exclude firms in the financial and regulated utilities sectors with SIC from 6000 to 6999 and from 4900 to 4949, respectively. As for the bankruptcy flags, a firm is marked as bankruptcy if it filed for bankruptcy under Chap. 7 or Chap. 11 bankruptcy filings[3].

10-K Filings. The raw filings of listed firms in the US could be retrieved from EDGAR. After removing for Tables, Figures, attached PDFs, and other redundant elements, we extract the MD&A section in each filing. We cover 10-K and 10-KSB (SB is for small business) filings in this study as other types of filing either notice a delay in document filings (10-K405) or a transition of the accounting period (10-KT and 10-K405T). We present the descriptive statistics of MD&A data extracted from SEC filings in Table 1. At the final stage, we merge the financial data and SEC filings by matching the CIK and the fiscal year-end of financial reports. We shift the bankruptcy indicator by one year and further remove the continuing operating reports for repayment plan if the business files for Chap. 11 bankruptcy. The final data consist of 50,786 firm-year observations, of which 174 firms are bankruptcy (approximately 0.34%).

3.2 Latent Dirichlet Allocation

We present LDA's terminologies in Table 2. Latent Dirichlet Allocation [5] (LDA) is considered as a generative probabilistic model, it assumes that each document in the corpus is represented as random mixtures over latent topics. Each topic is characterised by a distribution over terms in a vocabulary. Algorithm 1 shows the LDA generative process:

Blei et al., 2006 [4] introduced the dynamic topic model (DTM) which analyses the evolution of topics in large document collections over time. Essentially, DTM is an extension to LDA to adapt with sequential documents. LDA assumes that the word order within a document and the document order within the corpus are processed as in the same priority. In DTM, words are still assumed to be exchangeable, but the document order holds a vital role. Particularly, the

[3] https://www.sec.gov/reportspubs/investor-publications/investorpubsbankrupthtm.html.

Table 1. MD&As extracted from filings

Year	#MD& As	#NF	#unc	#omitted	#Sent.	Mean	Median	#tokens
1997	8711	422	7	561	740208	85	64	47328
1998	8931	454	4	563	862791	96.6	73	51372
1999	8934	391	2	611	1044325	116.9	89	54665
2000	9508	458	8	580	1065612	112.1	79	57295
2001	9447	424	1	510	1131682	119.8	83	59454
2002	10179	434	2	806	1488084	146.2	89	64979
2003	11878	419	6	1236	2107816	177.5	117	74079
2004	12124	390	4	1496	2274859	187.6	115	76837
2005	12475	635	2	2016	2501571	200.5	120	81210
2006	12251	678	5	1863	2472559	201.8	135	81163
2007	12087	485	5	1840	2526728	209	145	83147
2008	11432	443	6	1470	2519519	220.4	156	83350
2009	9919	366	4	769	2525077	254.6	189	79620
2010	9165	190	3	676	2405854	262.5	199	78967
2011	8840	162	2	659	2290261	259.1	193	78147
2012	8393	175	1	693	2214756	263.9	195	75322
2013	8105	186	1	677	2183919	269.5	203	74772
2014	8084	184	1	751	2193408	271.3	202	76434
2015	7985	182	1	912	2181273	273.2	204	76066
2016	7589	158	2	1081	2077774	273.8	201	74669
2017	7248	184	1	1113	1931412	266.5	192	71713
Total	203,285	7,420	68	20,833	40,739,488			

#MD&As is the total number of MD&As; #NF is the number of filings that do not have MD&A; #unc is the number of uncommon MD&As that we are unable to trace the sections they begin or end with; #omitted is the number of filings that have the MD&A section omitted; #Sent. is the total number of sentences of all MD&As; mean and median are the mean and median of the number of sentences in MD&As; #tokens is the total number of unique words in all MD&As.

Algorithm 1. Generative Process of LDA [5]

1: **for** each topic $k \in [1, K]$ **do**
2: sample mixture proportion $\phi \sim Dir(\beta)$
3: **end for**
4: **for** each document $m \in [1, M]$ **do**
5: Sample mixture proportion $\phi \sim Dir(\alpha)$
6: Sample document length $N_m \sim Poiss(.)$
7: **for** each word $n \in [1, N_m]$ **do**
8: Sample topic index $z_{m,n} \sim Mult()$
9: Sample a term for word $w_{m,n} \sim Mult(\phi_{z_{m,n}})$
10: **end for**
11: **end for**

Table 2. Notations for LDA model

M	# documents in the corpus (constant scalar)
K	# topics (constant scalar)
V	# terms in vocabulary (constant scalar)
N_d	length of the document d
α	hyper-parameter for topic proportions (vector in $\mathbb{R}^K$ space or a scalar if symmetric)
β	hyper-parameter for term proportion (vector in $\mathbb{R}^V$ or a scalar if symmetric)
θ	topic mixture proportion for document m One proportion for each topic in the document
ϕ	term mixture proportion for topic k One proportion for each term in the vocabulary
$w_{d,n}$	the term indicator for n^{th} word in document d
$z_{d,n}$	topic assignment of nth word in the document d

documents are divided into groups by time slice (e.g. month, quarter, year) and DTM assumes that the documents in each group come from a set of latent topics that evolved from the ones in the previous time slice. In this paper, we employ the online variational inference [12] of LDA for the textual data to examine the optimal number of topics, we then use the dynamic embedding topic model to capture the evolution of financial topics over time. In the next section, we briefly review the main idea of dynamic topic model in embedding spaces.

3.3 Topic Modelling in Embedding Spaces

Existing topic models fail to learn interpretable topics when working with large and heavy-tailed vocabularies. Embedding Topic Models [8] (ETM) bridge this gap by utilising the word embeddings (CBOW [19]):

$$w_{dn} \sim \text{softmax}(\rho^\top \alpha_{dn}). \tag{1}$$

The embedding matrix ρ is a $L \times V$ matrix whose columns contain the embedding representations of the vocabulary, $\rho_v \in \mathbb{R}^L$. The vector α_{dn} is the *context embedding*. The context embedding is the sum of the context embedding vectors (α_v for each word v) of the words surrounding w_{dn}. And to effectively model the dynamic of topics in embedding spaces taken into consideration of the imbalanced word distribution and topic evolution, Dieng et al. [7] introduced Dynamic Embedding Topic Models (DETM) which inherits the strengths of the ETM and DTM by modelling each word as a categorical distribution parameterised by the inner product of the word embedding and a per-time-slice topic embedding.

Denote $\alpha_k^{(t)}$ as *topic embedding* [8] of the kth topic at time slice t. In DETM, the probability of a word belongs to a topic is given by the (normalised) exponentiated inner product between the word and the topic's embedding at the corresponding time slice,

$$p(w_{dn} = v | z_{dn} = k, \alpha_k^{(t_d)}) \propto \exp\{\rho_v^\top \alpha_k^{(t_d)}\}. \quad (2)$$

The topic embeddings evolve under Gaussian noise with variance γ^2,

$$p(\alpha_k^{(t)} | \alpha_k^{(t-1)}) = \mathcal{N}(\alpha_k^{(t-1)}, \gamma^2 I). \quad (3)$$

The prior over θ_d depends on a latent variable η_{t_d}, where t_d is the time slice of document d:

$$p(\theta_d | \eta_{t_d}) = \mathcal{LN}(\eta_{t_d}, a^2 I) \quad (4)$$

where $p(\eta_t | \eta_{t-1}) = \mathcal{N}(\eta_{t-1}, \delta^2 I)$ and $\mathcal{LN}$ denotes a logistic-normal distribution. And the generative process of DETM is as follows:

Algorithm 2. Generative process of DETM [7]

1: Draw initial topic embedding $\alpha_k^{(0)} \sim \mathcal{N}(0, I)$
2: Draw initial topic proportion mean $\eta_0 \sim \mathcal{N}(0, I)$
3: **for** time step $t = 1, \ldots, T$ **do**
4: Draw topic embeddings $\alpha_k^{(t)} \sim \mathcal{N}(\alpha_k^{(t-1)}, \gamma^2 I)$ for $k = 1, \ldots, K$
5: Draw topic proportion means $\eta_t \sim \mathcal{N}(\eta_{t-1}, \delta^2 I)$
6: **end for**
7: **for** each document $d \in \mathcal{D}$ **do**
8: Draw topic proportions $\theta_d \sim \mathcal{LN}(\eta_{t_d}, a^2 I)$.
9: **for** each word n in the document **do**
10: Draw topic assignment $z_{dn} \sim \text{Cat}(\theta_d)$.
11: Draw word $w_{dn} \sim \text{Cat}(\text{softmax}(\rho^\top \alpha_{z_{dn}}^{(t_d)}))$.
12: **end for**
13: **end for**

The inference procedure in DETM is also made possible by optimising the Kullback-Leibler divergence of the approximation to the true posterior distribution $p(\theta, \eta, \alpha | \mathcal{D})$, in addition, the authors speed up algorithm via amortisation inference, a black box VI with the distribution over the topic proportions $q(\theta_d | \eta_{t_d}, \mathbf{w}_d)$ parameterised by a neural networks (either recurrent neural network or long short-term memory). In this paper, we employ DETM to create the hidden-topic vector representation of the textual data and use them as input features for predicting corporate bankruptcy.

3.4 Number of Topics Assessments

Identifying the optimal number of topics is of the most important tasks in topic modelling. We compute the perplexity of a held-out test set to evaluate the model with different setting of number of topics [5]. The perplexity is monotonically decreasing in the likelihood of the test data, and it is algebraically equivalent to the inverse of the geometric mean per-word likelihood. A lower perplexity score indicates better generalisation performance. The perplexity of test set D_{test} of M documents is defined as follows [5]:

$$perplexity(D_{test}) = 2^{\left\{-\frac{\sum_{d=1}^{M} log[p(\mathbf{w}_d)]}{\sum_{d=1}^{M} N_d}\right\}} \tag{5}$$

Since we cannot directly compute $log[p(\mathbf{w}_d)]$, we compute the lowerbound of perplexity [12]:

$$perplexity(n^{test}, \lambda, \alpha) \leq 2^{\{-bound\}} \tag{6}$$

where

$$bound = \frac{\sum_i (\mathbb{E}_q[log_p(n_i^{test}, \theta_i, z_i|\alpha, \beta)] - \mathbb{E}_q[log_q(\theta_i, z_i)])}{\sum_{i,w} n_{iw}^{test}} \tag{7}$$

and n_i^{test} denotes the vector of word count of ith document in a test set of M documents. The per-word perplexity in Eq. 7 is obtained by computing the probability of each word in the second half of a test document, conditioned on the first half.

3.5 Bankruptcy Prediction Feature Sets

One of our experiment in this study is to examine to which extend, the hidden topics in MD&A, once inferred, could help on leveraging the real-world practice in bankruptcy prediction. Thus, along with our topic distributed representations, we compare the distributed representation of MD&A with the industrial standard feature set in predicting corporate bankruptcy which is z-score, and we further compare it with other baselines including (i) dictionary-based count vectorisation based on a financial dictionary, and (ii) traditional word embedding. Particularly, we construct the following feature sets:

Table 3. Altman's factors and formulas

Panel I	z-score 5-factor [1]
Z1	Working Capital/Total Assets
Z2	Retained Earnings/Total Assets
Z3	EBIT/Total Assets
Z4	Market Value of Equity/Total Liabilities
Z5	Sales/Total Assets
Panel II	SME 5-factor [2]
A1	Cash Flow from Operating Activities/Current Liabilities
A2	Short Term Debt/Equity Book Value
A3	Cash/Total Assets
A4	EBIT/Interest Expenses
A5	Account Receivable/Liabilities

Table 4. Loughran and McDonald Wordlists

Notation	Description	Sample words
POS	Positive	Enthusiastically, assures, improve, empower, complimented
NEG	Negative	Investigating, complaining, confusion, severe, exculpation
UNC	Uncertainty	Risking, anticipated, hidden, unobservable, imprecision
LIT	Litigious	Prosecute, referenda, presumptively, licensable, sequestrator
SUF	Superfluous	Propagation, expeditiously, cessions, cognizable, assimilate
MS	Modal strong	Highest, strongly, will, unequivocally, lowest
MM	Modal moderate	Probably, rarely, regularly, should, tends
MW	Modal weak	Might, nearly, occasionally, perhaps, possible

- S_1 (z-score): Altman z-score with 5 factors [1] for the general corporate and 5 factors for the SMEs [2] bankruptcy prediction
- S_2 (DICT): Dictionary-based count (relative to the length of MD&A) vectorisation of the sentiment wordlists in Loughran and McDonald, 2011 [17]
- S_3 (D2V): Distributed representation of MD&A using *doc2vec* [16]
- S_4 (LDA): Topic representation for the MD&A using LDA model
- S_5 (DETM): Topic representation for the MD&A using DETM model

We present the feature sets S_1 and S_2 in Table 3 and 4, respectively:

3.6 Experimental Setup

We form the following features of S2: Count of negative, positive, uncertainty, litigious, and modal strong words namely *NEG*, *POS*, *UNC*, *LIT*, and *MS*, respectively[4]. We train the Doc2Vec [16] with distributed memory using 300 dimension. We use online learning [12] to estimate LDA models. As for the DETM, we follow [7] to train Word2Vec [19] with Skip-gram, 300 dimensions; batch size of 200; Relu activation; the feed-forward inference network consists of two hidden layers with 800 nodes each and drop-out of 0.1. We run DETM for 100 epochs.

We filter words with document frequency above 70%, as well as standard stop words from a list. Additionally, we remove low-frequency words that appear in less than 10 documents.

[4] We did not use the modal weak words because the uncertainty wordlist includes all words in that list, and the modal moderate list is also excluded since it does not have words with strong sentiment modification.

To determine the number of topics, first we run 10 epochs of the traditional LDA model with online learning [12] for all MD&As, the number of topics ranging from 10 to 100, and we compute the perplexity on a hold-out dataset to determine the optimal number of topics. We then run the LDA and DETM using full dataset of 22 years from 1997 to 2017 to examine the topics and their evolution. Finally, we set the predictive scenario as out-of-sample and out-of-time prediction commonly used in credit risk. Specifically, a moving temporal window of 10 years of data is used to train the classifier, which is logistic regression with l_2 regularisation, and the next year as the test data. For example, the data from 1997 to 2006 will be used to train and then predict the bankruptcy in 2007 data. After that, the training and testing window will shift ahead one year and we repeat the whole process.

At the final stage, we employ logistic regression with l_2 regularisation and inverse strength λ ranging from 0.001 to 0.01 (the smaller the λ the stronger the penalty to less influential features). Despite being simple, logistic regression is an industrial standard in bankruptcy prediction and is proved to give comparable performance with other advanced classifiers [1,20], and more importantly it is straightforward to explain its predictors. The final performance is reported for each year and averaged for all years.

4 Experimental Results and Discussions

4.1 Number of Topics in MD&A

As stated in Sect. 3, we find the optimal number of topics based on the bound of the perplexity of the second half of the test set conditioned on its first half. Figure 1 presents the bounds for different settings of number of topics ranging from 5 to 100, each LDA model [12] is trained using asymmetric α learned from corpus with 10 epochs, and maximum number of iterations through the corpus when inferring the topic distribution of a corpus is 100. There is a sharp dip

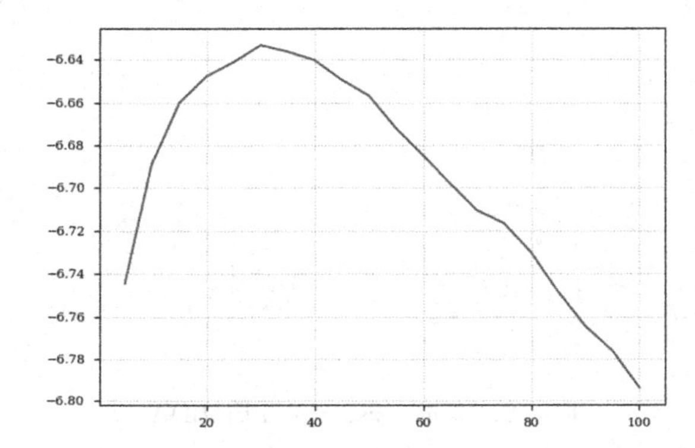

Fig. 1. Bound of perplexity on the test set

in $k = 30$ which shows that the perplexity is deteriorating, and it is encouraging that the possible number of topics could be 30. In addition, according to FASB, the MD&A should include "mission, activities, program and financial performance, systems, controls, legal compliance, financial position, and financial condition"[5] where, for example, the "program and financial performances" and "systems" should be divided further to smaller topics including liquidity, capital resources, results of operations, material changes, and so forth. Hence, we qualitatively consider 30 as a reasonable number of topics and based on this we estimate the full topic models for topic exploration task and moving-window based topic models for bankruptcy prediction task.

4.2 Topics in MD&A and Their Evolution

MD&A, as discussed in Sect. 1, is of important source of corporate information which reflects not only its past and current financial strength but emphasises the significant of the new products, services, collaboration projects, strategic partners, potential M&A deals, and so forth. However, our study, to the best of

Fig. 2. Topics discovered by LDA

[5] In Statement of Recommended Accounting Standards No. 15, Financial Accounting Standards Board (FASB).

our knowledge, is the first to utilise this large-scale dataset to provide an meaningful representation overtime of hidden topics using unsupervised methods. We present some significant topics inferred through LDA using wordcloud in Fig. 2, where the size of each word corresponding to its probability in a topic:

The wordclouds show some related, interesting words to name topics such as partnerships, research&development, energy price, investment loss, sale&store, tax¤cy rate. This shows that our approach using topic modelling is effective to uncover the hidden topics in the management discussion and analysis of the corporate filings. For the full list of wordcloud figures, please refer to Fig. 2 and Fig. 3 of the supplemental document.

One of the essential follow-up concerns that might attract the policymakers and macroeconomists is how these topics evolve through the market cycles. We provide further explanation to this question by using DETM which can leverage the word embedding with a smooth transition of topics over time. This is made possible by plotting the tensor $\beta = \text{softmax}(\rho^\top \alpha)$ of words in a topic over all time slices of our data. The following Fig. 3 presents the evolution of words in their associated topic.

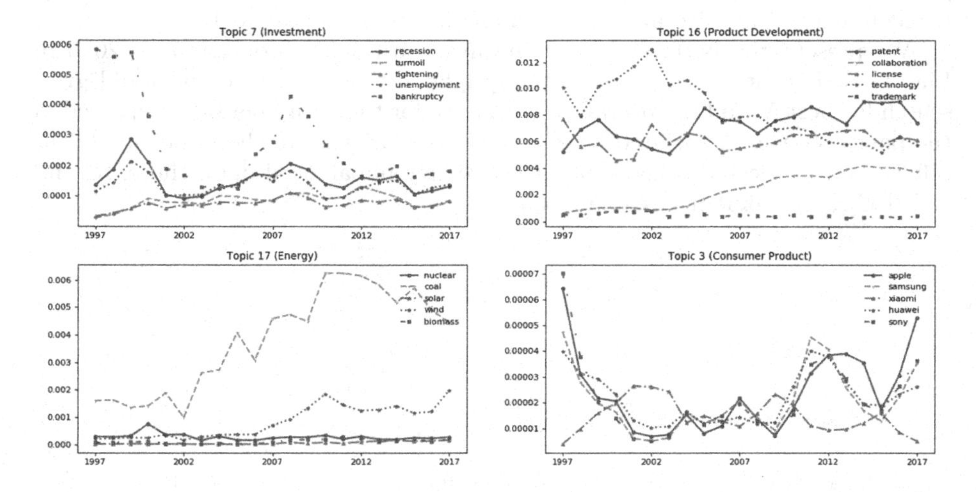

Fig. 3. The evolution of topics

Specifically, in *investment loss* topic, words associated with financial crisis such as 'recession', 'turmoil', 'tightening', and 'unemployment' show strong movement correlation and peak at the crisis period accordingly; the word 'collaboration' is progressing in topic *product development* whereas 'patent' and 'license' share their top weights; along with the decreasing of 'coal' following the sharp decline in coal production in 2015[6], the weights of renewable sources

[6] "U.S. coal production dropped by more than 10% in 2015 to 897 million short tons, the lowest production level since 1986", US Energy Information Administration, https://www.eia.gov/todayinenergy/detail.php?id=28732, Retrieved 3rd Mar, 2020.

such as 'wind' or 'solar' are increasing in *energy* topic; and the *consumer product* topic reveals the competition of several main electronic device producers over the last two decades, partially reflecting their up and down as well as their current positions in the US market. We then proceed to examine to which extend the topic distributed representation of the MD&A section in 10-K filings could help on bankruptcy prediction task in the section below.

4.3 Predictive Performance

In this prediction task, we compare the topic representation of MD&A using both LDA and DETM with three baselines: z-score of Altman [1], dictionary-based count of sentiment words based on the financial dictionary of Loughran and McDonald [17], and *doc2vec* [16]. We use three performance metrics which are area under the ROC curve (AUC), area under the precision-recall curve (AUPRC), and Brier score (BS). The first metric is commonly employed in assessing predictive performance, however, in the classification task with extremely imbalanced data, this might not present the general view on classifier performance because of the high prevalence of the majority class, and AUPRC is recommended by several research in credit risk modelling [6,10].

We present the performance of the five feature sets from 2007 to 2016 in Fig. 4a to Fig. 4c. Each point in the graphs is the mean value of 5-fold grid-search for best λ, and the overall performance for the entire period is reported in the parentheses of associated labels. While the Brier score shows no significant difference as expected because of our extremely imbalanced data, the AUC and AUPRC reveal some noticeable results.

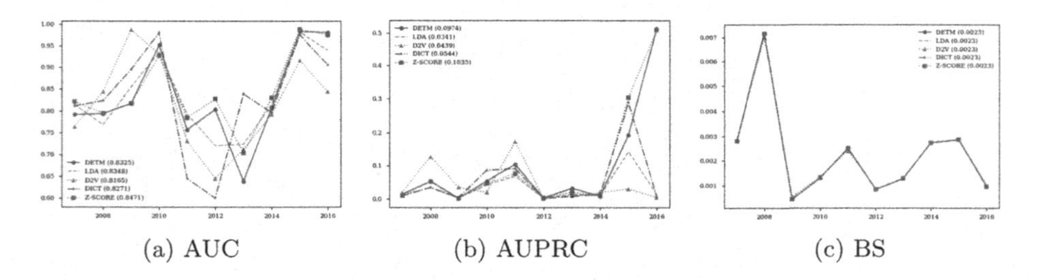

(a) AUC (b) AUPRC (c) BS

Fig. 4. Performance of classifier

First, in terms of AUC, the textual features provide comparable performance with z-score although they do not use any numerical or financial accounting data (AUC of the textual features ranging from 81.7% of D2V to 83.3% of DETM compared with 84.7% of z-score features). And in AUPRC, the DETM model produces the features as good as z-score in predicting bankruptcy (0.097 and 0.103). These observations provide important evidence on the significant role of the MD&A textual data in bankruptcy prediction. Second, for both AUC and AUPRC, despite trained on the same text corpus, the topic modelling approaches present higher overall performance compared with dict-based count vectorisation or distributed representation using skip-gram. One possible explanation for this is these two approaches do not have the topical structure for the input text data, hence the lower predictive performance. This demonstrates the significance of building topical structure of financial text. For the detail performance of five feature sets, the reader could refer to Table 3 in the supplemental document.

The seemingly reverse performance trend of LDA topical feature set and z-score suggests that their combination could benefit the classifier. We then examine the improvement in utilising the topical features along with traditional accounting features by simple concatenation of both feature sets. The predictive power of the combined feature set is compared with z-score features using the same performance measures and is presented in the following Table 5:

Table 5. Comparison of z-score and combined feature sets

	z-score			LDA + z-score			DETM + z-score		
	AUC	AUPRC	BS	AUC	AUPRC	BS	AUC	AUPRC	BS
2007	0.8217	0.0110	0.0028	0.8231	0.0250	0.1501	0.8765	0.0274	0.0028
2008	0.7946	0.0525	0.0072	0.8744	0.0936	0.1644	0.8651	0.0785	0.0071
2009	0.8179	0.0023	0.0005	0.8412	0.0027	0.1666	0.7241	0.0016	0.0005
2010	0.9268	0.0459	0.0013	0.9710	0.1298	0.1363	0.9715	0.1485	0.0013
2011	0.7850	0.0756	0.0025	0.8165	0.0947	0.1315	0.7835	0.0650	0.0025
2012	0.8271	0.0036	0.0009	0.9242	0.0145	0.1544	0.9483	0.0459	0.0008
2013	0.7039	0.0203	0.0013	0.8075	0.0061	0.1510	0.7033	0.0047	0.0013
2014	0.8299	0.0098	0.0027	0.8609	0.0283	0.1544	0.8322	0.0102	0.0027
2015	0.9887	0.3044	0.0029	0.9599	0.1429	0.1630	0.9586	0.2456	0.0028
2016	0.9756	0.5098	0.0010	0.9717	0.1088	0.1464	0.9730	0.0382	0.0010
Mean	0.8471	**0.1035**	0.0023	**0.8850**	0.0646	**0.1518**	0.8636	0.0666	0.0023

Despite a small average increment in AUC of 1.55 points for using DETM, we observe better enhancement in AUC of 3.89 points by using LDA to build MD&A representation, however, there are declines in AUPRC for both combinations. The proposed combinations are actually worse than the traditional z-score in 2015 and 2016 performance wise when z-score achieved almost perfect AUC of 98.87% and 97.56%, respectively.

We compare our simple model performance with state-of-the-art of Mai et al. [18] in Table 6. Using the similar MD&A data to predict corporate bankruptcy, our proposed features not only have an edge on the feature interpretability but also leverage simple logistic regression to gain 0.8–2.9 points in AUC over deep learning-based classifiers. This provides further evidence on the essential of constructing the interpretable features in financial risk applications.

Table 6. Comparison with state-of-the-art

	LDA	DETM	Mai et. at., 2019 [18]
Feature	Topic + z-score	Topic + z-score	Deep learning + ratios
AUC	88.5%	86.4%	85.6% and 84.2%

Deep learning models are trained using MD&A from 1994 to 2014 filings with 31 accounting ratios. 85.6% and 84.2% are for 80-20 and pre-2007/post-2008 train-val splittings for predicting one-year ahead out-of-sample data, respectively.

5 Conclusions

Understanding management discussion and analysis will help the investors and policy markers to response better to corporate business changes and prospects, especially when the financial statements are limited in plain numerical values and when there are problems of manipulating financial statements and reports [9]. By utilising the topic modelling approaches on more than 203 thousand 10-K filings, our first contribution is uncovering 30 topics embedded in the MD&A which reflect important business aspects such as 'energy', 'partnership', 'research and development', 'loan and interest rate', and so forth. In addition, the evolution of words in topics are in line with crucial economics events such as the financial crisis, the big reduction in coal production in 2015, or the cycles of competition of electronic devices producers.

Crucially, when the problem of window dressing to make up the credit quality becomes popular [9,11], there are increasing needs for employing alternative data to complement with traditional scoring methods, especially the data that reliably represent the forward-looking or future prospect of the businesses. We made this possible by our second contribution in out-of-time and out-of-sample bankruptcy prediction by examining the predictivity of topical features inferred from MD&A. And we showed that the topical features alone could provide comparable performance with industrial standard using z-score. More importantly, by simply concatenating the topical features and z-score features, we demonstrate the state-of-the-art performance in corporate bankruptcy prediction.

There are several potential extensions based on our current approach. The first is to employ higher frequency textual data to reflect more timely bankruptcy events, i.e using quarterly filings 10-Qs. Or to investigate the effects of balancing treatments such as oversampling and undersampling to tackle the severe imbalance classes and improve the distributed representation of the filings. Other research could devote to investigate a better combination of textual features with traditional features or to examine different training coverage to optimise the prediction performance.

Acknowledgments. We appreciate the fruitful discussions with Professor Jonathan Crook, Professor Galina Andreeva, Professor Raffaella Calabrese and other researchers at Business School, University of Edinburgh when Hung Ba was supported by JAIST

Research Grant and DRF Grant No. 238003 to work as a visiting scholar. Other errors retain our own.

References

1. Altman, E.I., Iwanicz-Drozdowska, M., Laitinen, E.K., Suvas, A.: Financial distress prediction in an international context: a review and empirical analysis of Altman's Z- score model. J. Int. Financ. Manag. Acc. **28**(2), 131–171 (2017)
2. Altman, E.I., Sabato, G.: Modelling credit risk for SMEs: evidence from the U.S. market. Abacus **43**(3), 332–357 (2007)
3. Bao, Y., Datta, A.: Simultaneously discovering and quantifying risk types from textual risk disclosures. Manag. Sci. **60**(6), 1371–1391 (2014)
4. Blei, D.M., Lafferty, J.D.: Dynamic topic models. In: Proceedings of the 23rd International Conference on Machine Learning - ICML 2006, pp. 113–120. ACM Press, Pittsburgh (2006)
5. Blei, D.M., Ng, A.Y., Jordan, M.I.: Latent Dirichlet allocation. J. Mach. Learn. Res. **3**, 993–1022 (2003)
6. Davis, J., Goadrich, M.: The relationship between Precision-Recall and ROC curves. In: Proceedings of the 23rd International Conference on Machine Learning, pp. 233–240. ACM (2006)
7. Dieng, A.B., Ruiz, F.J.R., Blei, D.M.: The dynamic embedded topic model. arXiv:1907.05545 [cs, stat] (2019)
8. Dieng, A.B., Ruiz, F.J., Blei, D.M.: Topic modeling in embedding spaces. arXiv preprint arXiv:1907.04907 (2019)
9. Gandhi, P., Loughran, T., McDonald, B.: Using annual report sentiment as a proxy for financial distress in U.S. banks. J. Behav. Finance **20**(4), 424–436 (2019)
10. García, V., Marqués, A.I., Sánchez, J.S.: Exploring the synergetic effects of sample types on the performance of ensembles for credit risk and corporate bankruptcy prediction. Inf. Fusion **47**, 88–101 (2019)
11. Guan, L., He, S.D., McEldowney, J.: Window dressing in reported earnings. Com. Lending Rev. **23**, 26 (2008)
12. Hoffman, M., Bach, F.R., Blei, D.M.: Online learning for latent Dirichlet allocation. In: Lafferty, J.D., Williams, C.K.I., Shawe-Taylor, J., Zemel, R.S., Culotta, A. (eds.) Advances in Neural Information Processing Systems 23, pp. 856–864. Curran Associates, Inc. (2010)
13. Huang, A.H., Lehavy, R., Zang, A.Y., Zheng, R.: Analyst information discovery and interpretation roles: a topic modeling approach. Manag. Sci. **64**(6), 2833–2855 (2018)
14. Huang, K.W., Li, Z.: A multilabel text classification algorithm for labeling risk factors in SEC form 10-K. ACM Trans. Manag. Inf. Syst. **2**(3), 1–19 (2011)
15. Jiang, F., Lee, J., Martin, X., Zhou, G.: Manager sentiment and stock returns. J. Financ. Econ. **132**(1), 126–149 (2019)
16. Le, Q., Mikolov, T.: Distributed representations of sentences and documents. In: International Conference on Machine Learning (2014)
17. Loughran, T., Mcdonald, B.: When is a liability not a liability? Textual analysis, dictionaries, and 10-Ks. J. Finance **66**(1), 35–65 (2011)
18. Mai, F., Tian, S., Lee, C., Ma, L.: Deep learning models for bankruptcy prediction using textual disclosures. Eur. J. Oper. Res. **274**(2), 743–758 (2019)

19. Mikolov, T., Sutskever, I., Chen, K., Corrado, G.S., Dean, J.: Distributed representations of words and phrases and their compositionality. In: Burges, C.J.C., Bottou, L., Welling, M., Ghahramani, Z., Weinberger, K.Q. (eds.) Advances in Neural Information Processing Systems 26, pp. 3111–3119. Curran Associates, Inc. (2013)
20. Nguyen, H.B., Huynh, V.N.: On sampling techniques for corporate credit scoring. J. Adv. Comput. Intell. Intell. Inform. **24**(1), 48–57 (2020)
21. Nguyen, T.H., Shirai, K., Velcin, J.: Sentiment analysis on social media for stock movement prediction. Expert Syst. Appl. **42**(24), 9603–9611 (2015)
22. Zhou, G.: Measuring investor sentiment. Ann. Rev. Financ. Econ. **10**, 239–259 (2018)

Why Did My Consumer Shop? Learning an Efficient Distance Metric for Retailer Transaction Data

Yorick Spenrath[1](✉), Marwan Hassani[1], Boudewijn van Dongen[1], and Haseeb Tariq[2]

[1] Eindhoven University of Technology, Eindhoven, Netherlands
{y.spenrath,m.hassani,b.f.v.dongen}@tue.nl
[2] BrandLoyalty, 's-Hertogenbosch, Netherlands
haseeb.tariq@brandloyalty-int.com

Abstract. Transaction analysis is an important part in studies aiming to understand consumer behaviour. The first step is defining a proper measure of similarity, or more specifically a distance metric, between transactions. Existing distance metrics on transactional data are built on retailer specific information, such as extensive product hierarchies or a large product catalogue. In this paper we propose a new distance metric that is retailer independent by design, allowing cross-retailer and cross-country analysis. The metric comes with a novel method of finding the importance of categories of products, alternating between unsupervised learning techniques and importance calibration. We test our methodology on a real-world dataset and show how we can identify clusters of consumer behaviour.

Keywords: Distance metric · Transaction categorization · Clustering · Optimization

1 Introduction

The analysis of consumer transaction data is an important task in creating value out of shopper interactions with online and physical stores. In recent years, the literature has extended rapidly on analyzing transactions to provide retailers with recommendations on a variety of topics [1,7,15,25,26]. One important part of such an analysis is to be able to tell how transactions of a single consumer and between several consumers relate. This paper is part of a larger project on consumer journey analysis and optimization. We intend to mine consumer behaviour from transactional data provided by retailers, to improve effectiveness of promotional programs. The starting point of this project is classifying each individual visit of a consumer to the store; by analyzing the corresponding receipt, and determining the *purpose* of their visit.

While receipts can be hand-labelled using domain knowledge, the vast quantity of data makes this infeasible. For the same reason, supervised learning strategies will be subject to overfitting small quantities of labelled data. To overcome

Y. Dong et al. (Eds.): ECML PKDD 2020, LNAI 12461, pp. 323–338, 2021.
https://doi.org/10.1007/978-3-030-67670-4_20

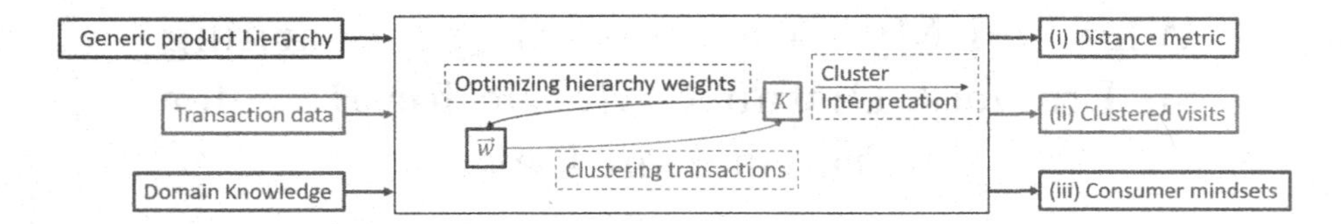

Fig. 1. Overview of the contributions in this paper

this, we resort to unsupervised learning strategies. Our starting point is a dataset of receipts; indicating which product categories (see Fig. 3) are bought during a trip to the retailer. For meaningful comparisons we require a distance metric between two receipts. First, this allows us to present the domain experts with representative examples in the dataset, and as such reduce the overfitting in supervised learning methods. Second, this allows us to use neighbourhood-based machine learning models such as k-medoids [17,21], spectral clustering [16] or hierarchical clustering [20], all of which can be used in an unsupervised setting as well. In this paper we present a novel framework for learning a distance metric that measures the similarity between consumer receipts. Our introduced framework learns the weights of the different components of the distance metric by optimizing the clusters of receipts. The overview of our contribution is presented schematically in Fig. 1.

In this paper, we make the following contributions: (i) we consider relations between products on a generic hierarchy which is fine-tuned by assigning different weights to different parts of the hierarchy. To find the best weights, we use clustering, but for this clustering, the weights serve as input again, i.e. (ii) we provide a fixed-point algorithm to find both, the weights and the clustering, together. Finally, (iii) we provide extensive experimental evaluation of our work using domain expert knowledge in a setting where huge quantities of unlabelled data is available. Part of this domain expert knowledge are the mindsets of consumer behaviour which we explain next.

Table 1. The 5 shopper mindsets identified by domain experts.

BULK	Buying large quantities of specific products to optimize discount
PLANNED	Plan ahead and buy everything needed for the week in one go
FRESH	Smaller trips with the key motivation of being flexible in when and what to eat, in particular for fresher fruits, vegetables and meats
QUALITY	Going to specific stores for specific products, like going to the butcher for meat
ON THE GO	Also smaller trips, but with an on-the-go character

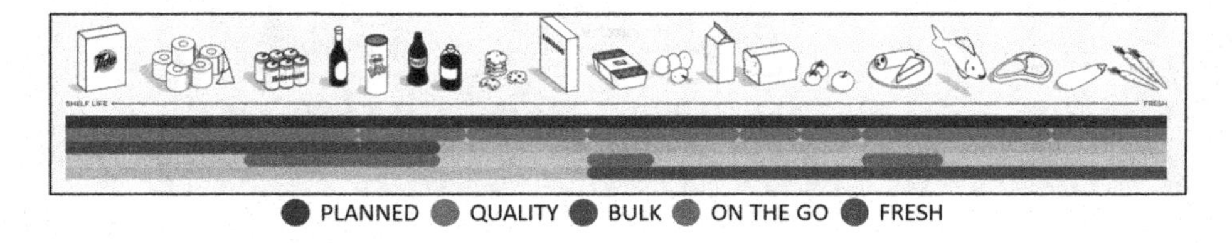

Fig. 2. Examples of products bought in each mindset.

One important challenge of most clustering algorithms is determining the number of clusters k one expects. Setting the value of k needs careful consideration; finding the best number of clusters is a difficult task in many unsupervised learning projects [10,13,19]. Considering that our end goal is to be able to tell for each receipt (and accompanying visit) *why* a shopper visits the retailer; it makes sense to base k on the number of visit types we expect. Earlier research[1] by domain experts has indicated five main reasons why shoppers visit retailers. Each of these 'mindsets' are an explanation of what moves consumers to a store, and where decisions are made. We have summarized these mindsets in Table 1, and a visual example of the products in each mindset is represented in Fig. 2.

The rest of this paper is organised as follows. We first discuss our framework in Sect. 2. Following this; we discuss experiments on a real-world dataset in Sect. 3. We then discuss how our solution relates to existing literature on transaction distance metrics in Sect. 4. We conclude the paper in Sect. 5.

2 A Framework for Learning the Distance Metric

In this section we discuss our framework for learning the distance metric. We first present the idea and formalization of the metric in Sects. 2.1 and 2.2. We present a novel idea for fine-tuning the weights of the distance metric in Sect. 2.3 and formalize it in Sect. 2.4.

2.1 Weighing a Product Hierarchy

The concept of a hierarchy is fundamental in our approach. This hierarchy relates categories of products to each other. An example of a hierarchy is presented in Fig. 3. The hierarchy consists of super-categories (3 in this example), each of which consists of categories (2 in each super-category in the example). By design, the hierarchy is retailer independent and can as such be applied to all consumer transactional data from supermarkets.

We make an abstraction from the actual receipt: instead of using all information on the products, their quantities and their price; we only consider which

[1] The full article on this study is found at https://medium.com/icemobile/mindset-moments-a-new-way-to-pinpoint-purchasing-decisions-18c58a574c02.

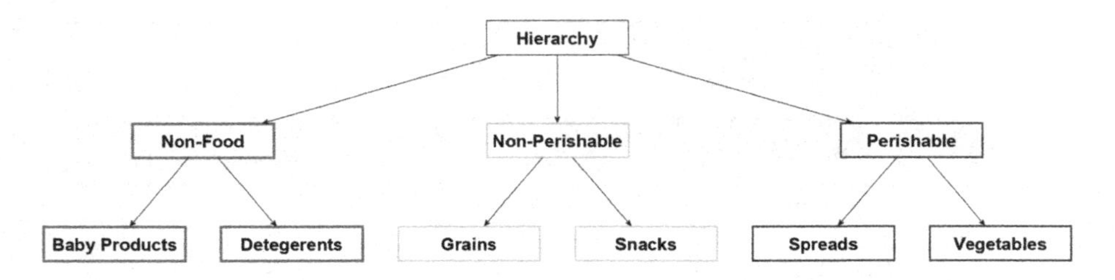

Fig. 3. Example of a retailer hierarchy. This hierarchy consists of three super-categories, each of which consists of two categories.

(super-)categories are included in the receipt. This abstraction allows for numerous improvements in computation time while, as we show in Sect. 3, it is still expressive enough to make a sensible distance metric and discover clusters of consumer behaviours. For our explanation we consider a small example of two receipts. Receipt a has two Non-Food categories, Baby Products and Detergents, and one Non-Perishable category, Grains. Receipt b has the categories Baby Products, Snacks and Spreads. The receipts, along with the distance metric computations are shown in Fig. 4.

We make a total of four comparisons between these two receipts. Each comparison is a Jaccard[2] distance on a specific subset of the hierarchy. The first three comparisons are within each super-category. For the super-category *Non-Food* we have one common category (Baby Products) and two total categories (Baby Products and Detergents) over the two receipts. The Jaccard distance for *Non-Food* is then $1 - \frac{1}{2} = \frac{1}{2}$. We can similarly analyze *Non-Perishable* and *Perishable*, they both have a Jaccard distance of 1. Despite having the same Jaccard distance, we do not capture the fact that both receipts do contain a *Non-Perishable* product, whereas this does not hold for *Perishable*. To overcome this we make a fourth comparison on the super-category level. The receipts have two super-categories in common, and three in total. As such, the *inter-super-category* distance is $1 - \frac{2}{3} = \frac{1}{3}$. We finally combine the four values using a weighted average. The weight of each distance allows domain expert knowledge

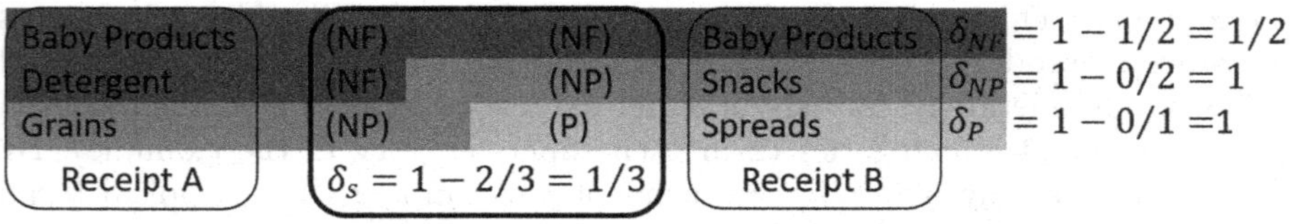

Fig. 4. Example receipts, along with computations for the distance metric.

[2] Given two sets $\mathcal{S}_1$ and $\mathcal{S}_2$, the Jaccard distance [23] between them is defined as $1 - \frac{|\mathcal{S}_1 \cap \mathcal{S}_2|}{|\mathcal{S}_1 \cup \mathcal{S}_2|}$.

to put emphasis on certain super-categories or can be fine-tuned to available transaction data.

2.2 Formalization

We now generalize the given example in a formal manner, all notations are summarized in Table 2. Let $C = C_1 \cup C_2 \cup \ldots \cup C_h$ be a set of *categories*. Each C_i is a *super-category* containing categories. We further have that $C_i \cap C_j = \emptyset$ for all $i \neq j$, i.e. each category is in exactly one super-category, and $C_i \neq \emptyset$ for all i, i.e. each super-category is non-empty. The set $\{C_1, C_2, \ldots C_h\}$ is called a *hierarchy.* Note that the hierarchy is a partition over C. The details of the used hierarchy are presented in Sect. 3.1. Without loss of generality, our target metric in this work considers a product hierarchy of height 2.

Let $\mathcal{D} \subseteq \{0,1\}^{|C|}$ be a set of receipts. A *receipt* is a description of the categories purchased in a *visit.* More specifically, for a receipt $\vec{a} \in \mathcal{D}$ we have $a_c = 1$ if a product of category c was bought, and $a_c = 0$ otherwise. For a receipt $\vec{a}$ we also need to know which super-categories were bought. For this purpose we define $s(\vec{a}, i)$ which tells us if any category in C_i was included in $\vec{a}$, formally:

$$s(\vec{a}, i) = \begin{cases} 0 & \forall c \in C_i : a_c = 0 \\ 1 & otherwise. \end{cases} \tag{1}$$

This function extends to two receipts, for $\vec{a}, \vec{b} \in \mathcal{D}$, we define $s(\vec{a}, \vec{b}, i)$ which tells us if any category in C_i was included in $\vec{a}$ or $\vec{b}$, formally:

$$s(\vec{a}, \vec{b}, i) = \begin{cases} 0 & \forall c \in C_i : a_c = b_c = 0 \\ 1 & otherwise. \end{cases} \tag{2}$$

Returning to the example of Sect. 2.1, the hierarchy consists of:

- C_1 = *Non-Food* = {*Baby Products, Detergents*}
- C_2 = *Non-Perishable* = {*Grains, Snacks*}
- C_3 = *Perishable* = {*Spread, Vegetables*}.

Furthermore[3], $\vec{a} = (1,1,1,0,0,0)$ and $\vec{b} = (1,0,0,1,1,0)$. Some examples of s include $s(\vec{a}, 3) = 0$, $s(\vec{b}, 2) = 1$, and $s(\vec{a}, \vec{b}, i) = 1$ for all i.

Given receipts $\vec{a}$ and $\vec{b}$, we define a total of $h+1$ distances between them. The first h each apply to a single super-category, these are *intra-super-category* distances as they make a comparison within a super-category. For a super-category C_i, the distance between $\vec{a}$ and $\vec{b}$ is defined as

$$\delta_i(\vec{a}, \vec{b}) = \begin{cases} 1 - \frac{\sum_{c \in C_i} \vec{a}_c \cdot \vec{b}_c}{\sum_{c \in C_i} \vec{a}_c + \vec{b}_c - \vec{a}_c \cdot \vec{b}_c} & \text{if } s(\vec{a}, \vec{b}, i) = 1 \\ 0 & \text{if } s(\vec{a}, \vec{b}, i) = 0 \end{cases} \tag{3}$$

[3] We order (super-)categories alphabetically in $\vec{a}$, $\vec{b}$ and s.

Put differently, the distance is the Jaccard distance between the set of categories in C_i included in the receipts. Because the Jaccard distance between empty sets is invalid, we assign the value 0 if both receipts are missing super-category C_i. We explain the choice of this value below. We next define the *inter-super-category* distance $\delta_s(\vec{a}, \vec{b})$. It is a Jaccard distance on the super-categories: comparing which super-categories are included in $\vec{a}$ and $\vec{b}$. For this we can make use of s as defined in Eqs. 1 and 2.

$$\delta_s(\vec{a}, \vec{b}) = 1 - \frac{\sum_{i=1...h} s(\vec{a}, i) \cdot s(\vec{a}, i)}{\sum_{i=1...h} s(\vec{a}, \vec{b}, i)}. \tag{4}$$

We define the distance between $\vec{a}$ and $\vec{b}$ as the weighted average of δ_s and all δ_i for which either $\vec{a}$ or $\vec{b}$ has products from C_i. Put differently, we ignore the super-categories for which neither receipt includes categories. This is because there is no concept of (dis)similarity for that super-category. This decision is partly inspired by [8]. Let w_s be the weight for δ_s in Eq. 4 and let w_i be the weight for each δ_i in Eq. 3. Let $\vec{w} = (w_s, w_1, w_2, \ldots w_h)$. Our assignment of 0 for invalid intra-super-category distances allows us to write:

$$\delta_{\vec{w}}(\vec{a}, \vec{b}) = \frac{w_s \cdot \delta_s(\vec{a}, \vec{b}) + \sum_{i=1...h} w_i \cdot \delta_i(\vec{a}, \vec{b})}{w_s + \sum_{i=1...h} w_i \cdot s(\vec{a}, \vec{b}, i)}. \tag{5}$$

Note that any scalar multiple of $\vec{w}$ results in the same distance metric. When considering values for $\vec{w}$ we add the restriction that the weights need to sum up to 1, though Eq. 5 can still be computed without this restriction.

Table 2. Symbols and parameters

Symbol	Description
C	Categories
C_i	Super-category
h	Number of super-categories
$\mathcal{D}$	Set of receipts
$\vec{a}$	(a_c) for $c \in C$. We have $a_c = 1 \Leftrightarrow$ category c was bought in receipt $\vec{a}$
$s(\vec{a}, i)$	Whether super-category i was bought in receipt $\vec{a}$ (1) or not (0)
$s(\vec{a}, \vec{b}, i)$	Whether super-category i was bought in either $\vec{a}$ or $\vec{b}$ (1) or not (0)
δ_x	Distance of super-category C_i (δ_i) or the inter-super-category (δ_s)
$\vec{w}$	$(w_s, w_1, w_2, \ldots w_h)$, weights of each distance
K	$\{K_1, K_2, \ldots K_k\}$ Clustering (partition) over $\mathcal{D}$
$\Phi(K, \vec{w})$	An internal evaluation measure, a quantitative value of the quality of a clustering K given weights $\vec{w}$

2.3 Finding Optimal Weights

To find $\vec{w}$ we use a clustering over $\mathcal{D}$ and an internal evaluation measure [11]. Given a clustering, we want to find the weights that minimize[4] the internal evaluation measure on the clustering.

Let $K = \{K_1 \cup K_2 \cup \ldots \cup K_k\}$ be a partition over $\mathcal{D}$. K is a clustering of the receipts. Using a clustering and a value of $\vec{w}$ we can define an internal evaluation measure $\Phi(\vec{w}, K)$ which describes the quality of clustering K with respect to weights $\vec{w}$. Following the concept of internal evaluation measures, Φ decreases if receipts within a cluster are closer together (lower value for Eq. 5) and if points in different clusters are further apart. For a given K, we want to find $\vec{w}$ that minimizes $\Phi(\vec{w}, K)$. While our approach can deal with any internal evaluation measure that considers compactness and separation [11], in this paper we use the one defined in [7], which was used there for a similar studies:

$$\Phi(\vec{w}, K) = \log\left\{\sum_{K_i \in K} \frac{1}{2|K_i|} \sum_{\vec{a},\vec{b} \in K_i} \delta_{\vec{w}}(\vec{a}, \vec{b})\right\} - \log\left\{\sum_{\vec{a},\vec{b} \in \mathcal{D}} \delta_{\vec{w}}(\vec{a}, \vec{b})\right\}. \tag{6}$$

A large part of the computation of Eq. 6 consists of summing over the distance between all points. This can be costly, especially because we need to do this for every $\vec{w}$ we wish to evaluate. We reduce the time complexity by precomputing parts of Eq. 6, the details are found in the repository discussed in Sect. 3.

The problem however, is that we *do not* have such a clustering K; as finding it requires a distance metric on $\mathcal{D}$ or a large set of labelled data. This means we need to find the solutions to two dependent problems: finding a clustering using the distance metric with $\vec{w}$; and finding the optimal $\vec{w}$ using a clustering. The solution to this is presented in the next section.

2.4 Finding Fixed Points

The target is to find *fixed points* that solve the two problems simultaneously. In other words, we need to find a combination $\Lambda := (\vec{w}, K)$ such that: (1) a clustering algorithm applied to $\mathcal{D}$ with a distance metric as defined by $\vec{w}$ and Eq. 5 finds clustering K, and (2) Eq. 6 with K is minimized by $\vec{w}$. A dataset may have multiple *fixed points* that satisfy these conditions. In this section, we explain first how we can find any of those fixed points, and then explain which of them is *the* fixed point of a dataset $\mathcal{D}$.

Finding *a* Fixed Point. We alternate between solving (1) the clustering given a set of weights and (2) the optimization given a clustering, until the solutions do not change. The starting point for this is a random set of weights $\vec{w}^{(0)}$. The details of finding $\vec{w}$ and K simultaneously are presented in Algorithm 1. All input parameters are summarized in Table 3. The algorithm requires the dataset, an

[4] Without loss of generality, we discuss minimization of the internal evaluation measure, the problem and solution is similar for internal evaluation measures that need to be maximized.

initial set of weights, and k as the number of clusters to find. Furthermore, we need two parameters for the stopping criteria: ϵ to check if weights vary significantly between two consecutive iterations, and n_{max} to set a maximum number of iterations. Next, we need two separate functions. `cluster` takes a dataset, a distance metric and an initial set of medoids to find the clusters and their medoids with the lowest internal distance. `minimize` takes a function and provides a numerical minimization. We finally need a parameter r to break cyclicity as explained below.

We initiate the first set of k medoids (as needed by `cluster`) in Line 1, using the **kmeans++** algorithm [2]. We use this to find an initial clustering in Line 2. In Line 4 through Line 10 we find the optimal weights and update the clustering. The weights are updated in Line 6. If the difference between weights of consecutive iterations is smaller than ϵ or if we reached the maximum number of iterations, we terminate the algorithm by saving the final weight as $\vec{w}$ and the final clustering as K. If neither stopping criterion is met we move on to the clustering step. During initial experiments, it turned out that sometimes the algorithm gets stuck in a cycle of several iterations (returning to a combination of weights and clusters which was found at an earlier iteration). To solve this, we increase all weights randomly by up to $r \leq \epsilon$. We then use these adapted weights for the clustering computation.

Selecting *the* Fixed Point. We apply the above procedure n_{rep} times, seeding the algorithm with a different, random set of weights $\vec{w}^{(0)}$ each repetition. Each time this results in one fixed point of the dataset. A fixed point can be found multiple times, formally let the *support* of a fixed point Λ^X be the fraction of the n_{rep} runs that ended with Λ^X. We then define Λ, *the* fixed point of $\mathcal{D}$, as the fixed point with the highest support. This entire procedure is schematically presented in Fig. 5.

Algorithm 1: Procedure to find $\vec{w}^*$, given an initial $\vec{w}^{(0)}$

---

input : A dataset $\mathcal{D}$, an initial $\vec{w}^{(0)}$, $\epsilon \geq 0$, $r \geq 0$, n_{max}, k, `cluster`, `minimize`
output: The optimal weights $\vec{w}^*$ and medoids m

1 $m^{(-1)} = \texttt{kmeans++}\left(\mathcal{D}, \delta_{\vec{w}^{(0)}}, k\right)$
2 $K^{(0)}, m^{(0)} = \texttt{cluster}\left(\mathcal{D}, \delta_{\vec{w}^{(0)}}, m^{(-1)}\right)$
3 $n = 0$
4 **while** *True* **do**
5 $\quad n \leftarrow n + 1$
6 $\quad \vec{w}^{(n)} \leftarrow \texttt{normalize}\left(\texttt{minimize}\left(\Phi(K^{(n-1)}, \vec{w})\right)\right)$
7 $\quad$ **if** $\forall i = 0 \dots h : \left|\vec{w}_i^{(n)} - \vec{w}_i^{(n-1)}\right| < \epsilon \vee n = n_{max}$ **then**
8 $\quad\quad$ **Return** $\vec{w}^{(n)}$ and $K^{(n)}$
9 $\quad \vec{w}^{(n')} \leftarrow \texttt{normalize}\left(\vec{w}^{(n)} + U(0, r)\right)$
10 $\quad K^{(n)}, m^{(n)} \leftarrow \texttt{cluster}\left(\mathcal{D}, \delta_{\vec{w}^{(n')}}, m^{(n-1)}\right)$

Table 3. Parameters for Algorithm 1

Symbol	Description
$\vec{w}^{(0)}$	Initial weight vector
ϵ	Maximum change of $\vec{w}$ for termination
r	Random change applied to $\vec{w}$ before clustering to avoid cyclicity
n_{max}	Maximum number of iterations
k	The number of clusters to find in the clustering step

3 Experimental Evaluation

In this section we discuss an experimental evaluation of our method. We describe the general settings for all experiments in Sect. 3.1, and discuss the experiments and their the results in Sects. 3.2 through 3.4. The entire code, experimental evaluation, and the used datasets are provided at https://github.com/YorickSpenrath/PKDD2020.

3.1 Experimental Settings

Our hierarchy consists of 29 categories, divided over 7 super-categories, shown in Table 4. We mapped all products of the retailer onto this generic product hierarchy with the help of domain expert knowledge by aligning the lowest level product groups from the original retailer to one of the 29 categories. We set ϵ to 0.01, which is less than one-tenth of a weight if all weights would be distributed equally ($\frac{1}{8}$). We set r to one-tenth of ϵ at 0.001, to not have it interfere with stability but at the same time still have effect on breaking cyclicity. We set n_{max} to 200 in order to prematurely end the algorithm before it consumes too much computation time.

For k we base our choice on the number of mindsets as discussed in Sect. 1. By design; our methodology does not account for the quantity and price of products; and it is therefore not possible to distinguish the mindset `BULK` (Table 1) from the other mindsets. As such we are explicitly searching for $k = 4$ clusters.

For `cluster` we use the Voronoi [17] algorithm, which we implemented using Numba [14]; favouring lower time complexity over a better clustering, different to for instance [21]. For `minimize` we use the differential evolution method proposed by [22] and implemented in SciPy [24]. We finally execute Algorithm 1 a total number of $n_{rep} = 1000$ times.

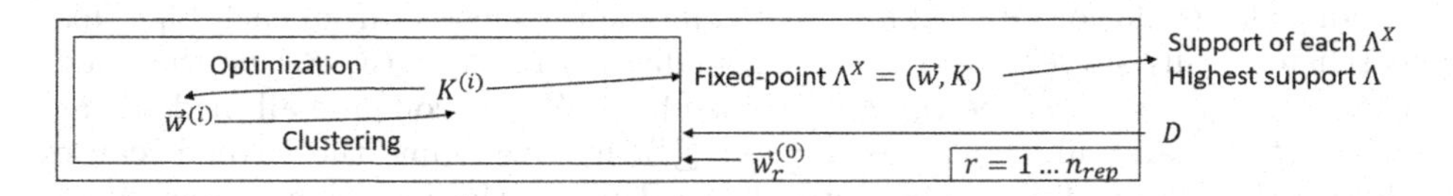

Fig. 5. Overview of the proposed algorithm.

Table 4. Hierarchy used in the experiments. (N)P stands for (Non-)Perishable

Super-category	Categories
Non Food	Pet articles, Personal care, Non food, Baby products, Washing and cleaning
Perishable A	Bread, Desserts, Milk, Fruits and vegetables P, Eggs, Meat and fish P, Cheese
Perishable B	Prepared meals P, Spread P, Carbs P, Soft drinks P, Condiments P
Non Perishable A	Condiments NP, Grains, Carbs NP, Meat and fish NP, Fruits and vegetables NP, Prepared meals NP
Non Perishable B	Snacks, Sweets
Non Perishable C	Spread NP, Hot drinks
Non Perishable D	Alcohols and tobacco, Soft drinks NP

3.2 Weights Learning

Apart from finding the optimal weights, we also want to verify that it is independent of the used dataset. As such, we apply the method of Sect. 2.3 to 100 datasets each having $|\mathcal{D}_i| = 1000$ visits. This computes the support of each of the fixed points of $\mathcal{D}_i$. Although these fixed points will have different medoids for each dataset, we can still compare the weights found for each $\mathcal{D}_i$. We compute the average and confidence interval of the support of each set of weights; the 5 weights with the highest average support are reported in Table 5.

We can see that the most supported set of weights is a clear winner: it is found almost two and a half times as often as the second most supported set of weights. This set of weights gives most importance on the Non Perishable B super-category; the one consisting of Sweets and Snacks. We further learn that Perishable B and the inter-super-category distance are also important for make clusters in our data.

3.3 Qualitative Cluster Analysis

The previous section focussed on the final weights; which are in fact only one part of our fixed points. In this section we analyze the contents of the discovered clusters using domain expert knowledge. The goal is to provide descriptive statistics on the contents of the clusters: which (super-)categories are characteristic for which clusters, and which (super-)categories are mostly missing.

In order to do so, we need a way to combine the clusters from each $\mathcal{D}_i$. This is schematically presented in Fig. 6. Formally, (1) let $\Lambda_i = (\vec{w}_i, K_i)$ be the fixed point for $\mathcal{D}_i$ and let m_i be the k medoids of K_i. We (2) combine all medoids as $M = \bigcup_{i=1}^{100} m_i$ and all weights as $\vec{w} = \frac{1}{100}\sum_{i=1}^{100} \vec{w}_i$. We define the latter because the weights of the fixed points are identical up to the precision presented in Sect. 3.2. We then (3) use `cluster` to create a clustering K over M using the

Table 5. The most frequently found weights, showing the percentage of repetitions that found this combination of weights, averaged over the 100 datasets, with their 95%-confidence intervals.

Support	43.5 ± 1.2%	16.2 ± 1.1%	7.5 ± 0.6%	4.0 ± 0.5%	3.2 ± 0.5%
Non Food	0.00	0.00	0.11	0.18	0.28
Non Perishable A	0.05	0.10	0.03	0.03	0.08
Non Perishable B	0.44	0.20	0.25	0.25	0.09
Non Perishable C	0.04	0.07	0.04	0.12	0.12
Non Perishable D	0.01	0.33	0.25	0.14	0.01
Perishable A	0.06	0.00	0.01	0.01	0.00
Perishable B	0.17	0.13	0.07	0.03	0.30
w_S	0.22	0.16	0.24	0.24	0.11

distance metric with weights $\vec{w}$. (4) All receipts in the same cluster in K are given the same label, and subsequently (5) for each $\mathcal{D}_i$, each cluster of K_i is given the label as their respective medoid. For each of the 400 clusters we compute how often a (super-)category is included in a receipt, and aggregate these values over clusters with the same labels. The results of this are presented in Table 6.

We analyzed the table together with the same domain experts that initially researched the mindsets of Fig. 2 and created the hierarchy of Table 4. Despite the abstraction of receipts to their categories (rather than quantities and prices), we can still distinguish properties of the four mindsets in the clusters. K_1 consists primarily of super-category Perishable A: fresh fruits, vegetables, meat and fish. This is typical for the `FRESH` mindset. Similarly; it has a relatively low inclusion of the Non Perishable categories. Compared to the other clusters, K_2 has a rather low inclusion of Perishable A, while having a high inclusion of Non Perishable D: alcohols, tobacco and non-perishable soft-drinks. One explanation for these two values is a `QUALITY` mindset, where people prefer specific stores like a butcher or greengrocer for the products from Perishable A, while preferring the retailer for their alcohols and tobacco. The value for Non Perishable A also supports this; as such non-perishable products from these categories are typically not found at butchers and groceries. However; within the `QUALITY` mindset; visits that focus specifically on meats or vegetables (but not both) do also exist. It appears this type of `QUALITY` mindset was not found. The difference between K_3 and K_4 is unfortunately less obvious, as both have characteristics regarding the mindsets

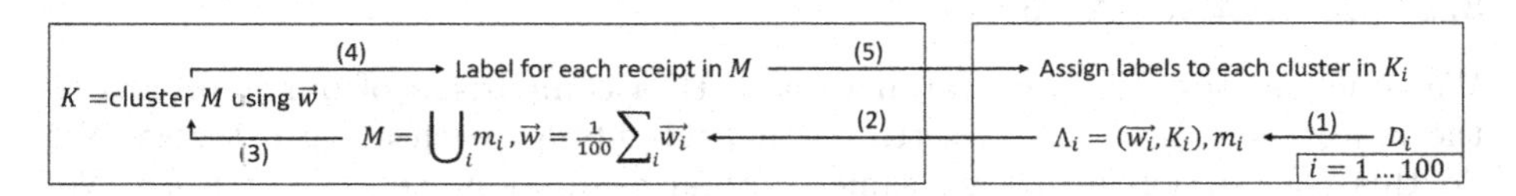

Fig. 6. Overview of the cluster analysis learning experiment.

Table 6. Descriptive Statistics of each cluster. The abbreviations are CONNP: Condiments NP, FnVP: Fruits and vegetables P, MIL: Milk, MnFP: Meat and fish P, PRENP: Prepared meals NP, SNA: Snacks, SOFNP: Soft drinks NP, SWE: Sweets.

Cluster size	K_1 259.6 ± 8.3	K_2 188.1 ± 14.9	K_3 270.3 ± 9.0	K_4 233.0 ± 4.0	K_{All} 1000.0±0.0
Non Food	22.5 ± 1.4%	23.9 ± 3.4%	44.1 ± 1.1%	31.7 ± 0.8%	31.7 ± 0.3%
Non Perishable A	32.8 ± 2.0%	15.7 ± 1.5%	64.0 ± 1.6%	42.2 ± 0.8%	43.6 ± 0.3%
Non Perishable B	0.1 ± 0.1%	0.2 ± 0.2%	74.5 ± 5.5%	99.6 ± 0.5%	45.0 ± 0.3%
Non Perishable C	6.5 ± 0.4%	5.8 ± 0.8%	15.0 ± 0.7%	10.0 ± 0.4%	9.8 ± 0.2%
Non Perishable D	35.7 ± 2.6%	66.4 ± 9.7%	53.5 ± 1.2%	40.2 ± 0.9%	44.6 ± 0.3%
Perishable A	70.8 ± 2.8%	28.7 ± 2.6%	71.5 ± 1.8%	68.7 ± 0.7%	67.0 ± 0.3%
Perishable B	39.1 ± 5.2%	41.1 ± 10.1%	52.5 ± 3.1%	41.5 ± 0.8%	43.2 ± 0.3%
Category 1	FnVP	SOFNP	SNA	SWE	FnVP
	39.2 ± 1.3%	38.1 ± 5.1%	72.9 ± 5.4%	99.3 ± 0.5%	41.3 ± 0.3%
Category 2	MnFP	AnT	FnVP	FnVP	SWE
	27.9 ± 1.0%	34.0 ± 5.5%	50.0 ± 1.7%	42.9 ± 0.7%	33.2 ± 0.3%
Category 3	SOFNP	PREP	MnFP	MnFP	MnFP
	21.6 ± 1.5%	29.5 ± 7.9%	41.9 ± 1.6%	33.7 ± 0.7%	32.0 ± 0.3%

of ON THE GO and PLANNED. In contrast to the other two clusters; K_3 and K_4 have a relative high inclusion on most super-categories. It further makes sense that K_3 and K_4 are separated from the other two because of their high inclusion for Non-Perishable B. They get separated from each-other because of the focus on the underlying category (snacks for K_3 and sweets for K_4, which together make up the Non Perishable B super-category). Although not shown due to space restrictions, $K_3's$ sweets at 34.4% and $K_4's$ snacks at 7.2% are much lower, making up the seventh and twenty-first largest category respectively. One way to explain this difference is that K_3 is more of an ON THE GO mindset; where the consumer gets some snacks for the evening. This is further supported by the high value of Non Perishable D; the drinks that go with these snacks. At the same time, $K_3's$ inclusion is higher on almost every super-category, which would be more fitting for a PLANNED mindset. Future research should focus on including quantities and prices of products, which is expected to help in distinguishing between K_3 and K_4.

3.4 Competitor Analysis

We conclude our experimental analysis with a comparison of our approach to the more naive approach of setting equal weights for each super-category. We do this comparison using the same internal evaluation measure of Eq. 6. We apply several clustering algorithms using our distance metric, but setting all weights equal to $\frac{1}{8}$ (i.e. without using our two-problem framework). We use the

Table 7. Comparison of our approach to using various clustering algorithms and our distance metric with uniform weight.

Index	Φ
Our framework	-7.917 ± 0.002
Voronoi [21]	-7.720 ± 0.003
Spectral [16]	-7.735 ± 0.003
HAC - complete [20]	-7.698 ± 0.004
HAC - average [20]	-7.698 ± 0.005
HAC - single [20]	-7.605 ± 0.000

algorithms as implemented in Scikit-Learn [18], by setting the number of clusters to 4 and by using default values for the other parameters. We compute Φ for each $\mathcal{D}_i$ as used in the previous experiments; and report the mean and the 95% confidence interval for each approach (Table 7).

The internal evaluation measure (which needs to be minimized) clearly indicates that our approach with learned weights performs better than the alternatives that do not learn weights; there is also no overlap in the confidence intervals. This shows the added benefit of learning the weights in improving the found clustering in terms of the internal evaluation measure.

4 Related Work

In this section we present some related work and how our approach fits into the literature.

Receipt distance metrics have been considered in different publications. Most metrics are based on either specific products of a retailer, or a large product hierarchy. Both these artefacts make the distance metrics more specific towards a single retailer.

In [15], the authors discuss several metric types on transaction data. They first consider affinity items; where two products that can for example substitute each-other (like two coke cans of different brands) are considered affinity products. Receipt distance is roughly based on the number of affinity products between them. In a sense; our method uses a similar concept taken to the extreme: we consider products in the same category to be equivalent. The authors then consider product distance, where the distance between two receipts is based on the closest or furthest two products, amongst others. A disadvantage is that it requires a distance between products, which is not suitable to our global target of developing retailer independent analytics.

The ROCK clustering algorithm for categorical attributes [9] uses transaction data as main use case. The clustering does not directly depend on the distance between datapoints, but on the number of common neighbour points. Contrary to our (super)-categories, ROCK does not assume relations between products, but uses the frequency of individual products. The disadvantage of the latter is that this makes inter-retailer comparisons infeasible. The clustering of categorical features proposed in [25] uses a similar approach. Their target is to have transactions that contain certain products frequently bought together to be in the same cluster. This approach does not focus on finding a distance metric but more on finding the clusters. Other work that computes clusters through common items can be found in [1].

The product hierarchy we use in our approach is balanced: all leafs of the tree are at the same depth. In [26], the authors do not restrict this and develop an approach for dealing with unbalanced product trees. The advantage of this is that it allows retailers focussed towards one category of products to have a deeper hierarchy on that category, while having a more shallow hierarchy for other categories. Our approach implicitly solves this by considering a generic hierarchy, which would effectively transform an unbalanced product tree in to the hierarchy such as presented in Table 4.

The work presented in [7], using earlier work in [4–6], has a similar approach to ours; in the sense that a set of weights over a hierarchy is optimized while clustering the transactional data. Fundamentally different however, is that the approach uses the full product hierarchy of a retailer, making cross-retailer and cross-country analysis more difficult.

5 Conclusion and Future Work

In this paper we presented a novel method for determining the distance between consumer transactions, guided by consumer behaviour using k-medoid clustering. The framework as-is is able to self-optimize and explain consumer behaviour, despite being based on a high level of abstraction of consumer transaction data. We showed that our approach not only finds sensible and separable clusters of consumer behaviour, but also provide reasonable stability in doing so.

Nonetheless, extensions to the algorithm are considered for future work. There is potential to improve on the final clusters; but also on more accurately matching consumer behaviour types. First and foremost is using quantities instead of inclusion; this gives a more descriptive distance metric at the cost of increased computation time. We have explicitly set $k = 4$. While this makes sense based on the original mindset study, it is interesting to see what happens for higher values of k, seeing if new or 'sub-mindsets' are found. Another way to approach this by applying density based clustering methods [3,12] instead. In the clustering step, we traded quality for efficiency by deploying Voronoi-clustering over PAM clustering, another optimization focus of our approach lies there. In the inter-super-category distance of Eq. 4 we treated every super-category with the same weight; i.e. both receipts having *Non-Food* products is treated the same as both receipts having *Non-Perishable* products. Adding weights within this distance, with or without using quantities instead of inclusion, is another next step in our research. Finally, it will be interesting to see if we can find the same quality of results regardless of the retailer and respective product hierarchy.

Acknowledgements. The authors kindly thank BrandLoyalty's domain experts, in particular Anna Witteman, Hanneke van Keep, Lenneke van der Meijden, and Steven van den Boomen, for their contributions to the research presented in this paper.

References

1. Aggarwal, C., Procopiuc, C., Yu, P.: Finding localized associations in market basket data. IEEE Trans. KDE **14**(1), 51–62 (2002)
2. Arthur, D., Vassilvitskii, S.: K-Means++: the advantages of careful seeding. In: Proceedings of the Eighteenth Annual ACM-SIAM Symposium on Discrete Algorithms. SODA 2007, pp. 1027–1035. Society for Industrial and Applied Mathematics, USA (2007)
3. Campello, R.J.G.B., Moulavi, D., Sander, J.: Density-based clustering based on hierarchical density estimates. In: Pei, J., Tseng, V.S., Cao, L., Motoda, H., Xu, G. (eds.) PAKDD 2013. LNCS (LNAI), vol. 7819, pp. 160–172. Springer, Heidelberg (2013). https://doi.org/10.1007/978-3-642-37456-2_14
4. Chen, X., Fang, Y., Yang, M., Nie, F., Zhao, Z., Huang, J.Z.: PurTreeClust: a clustering algorithm for customer segmentation from massive customer transaction data. IEEE Trans. Knowl. Data Eng. **30**(3), 559–572 (2018)
5. Chen, X., Huang, J.Z., Luo, J.: PurTreeClust: a purchase tree clustering algorithm for large-scale customer transaction data. In: 2016 IEEE 32nd International Conference on Data Engineering (ICDE), pp. 661–672, May 2016
6. Chen, X., Peng, S., Huang, J.Z., Nie, F., Ming, Y.: Local PurTree spectral clustering for massive customer transaction data. IEEE Intell. Syst. **32**(2), 37–44 (2017)
7. Chen, X., Sun, W., Wang, B., Li, Z., Wang, X., Ye, Y.: Spectral clustering of customer transaction data with a two-level subspace weighting method. IEEE Trans. Cybern. **49**(9), 3230–3241 (2019)
8. Gower, J.C.: A general coefficient of similarity and some of its properties. Biometrics **27**(4), 857–871 (1971)
9. Guha, S., Rastogi, R., Shim, K.: Rock: a robust clustering algorithm for categorical attributes. Inf. Syst. **25**, 345–366 (2000)
10. Hamerly, G., Elkan, C.: Learning the k in K-means. In: Proceedings of the 16th International Conference on Neural Information Processing Systems. NIPS 2003, pp. 281–288. MIT Press, Cambridge (2003)
11. Hassani, M., Seidl, T.: Using internal evaluation measures to validate the quality of diverse stream clustering algorithms. Vietnam J. Comput. Sci. **4**(3), 171–183 (2016). https://doi.org/10.1007/s40595-016-0086-9
12. Hassani, M., Spaus, P., Seidl, T.: Adaptive multiple-resolution stream clustering. In: Perner, P. (ed.) MLDM 2014. LNCS (LNAI), vol. 8556, pp. 134–148. Springer, Cham (2014). https://doi.org/10.1007/978-3-319-08979-9_11
13. Jain, A.K.: Data clustering: 50 years beyond K-means. Pattern Recogn. Lett. **31**(8), 651–666 (2010)
14. Lam, S.K., Pitrou, A., Seibert, S.: Numba. In: Proceedings of the Second Workshop on the LLVM Compiler Infrastructure in HPC - LLVM 2015, pp. 1–6. ACM Press, New York (2015)
15. Lu, K., Furukawa, T.: Similarity of transactions for customer segmentation. In: Quirchmayr, G., Basl, J., You, I., Xu, L., Weippl, E. (eds.) CD-ARES 2012. LNCS, vol. 7465, pp. 347–359. Springer, Heidelberg (2012). https://doi.org/10.1007/978-3-642-32498-7_26

16. Ng, A.Y., Jordan, M.I., Weiss, Y.: On spectral clustering: analysis and an algorithm. In: Proceedings of the 14th International Conference on Neural Information Processing Systems: Natural and Synthetic. NIPS 2001, pp. 849–856. MIT Press, Cambridge (2001)
17. Park, H.S., Jun, C.H.: A simple and fast algorithm for K-medoids clustering. Expert Syst. Appl. **36**(2 PART 2), 3336–3341 (2009)
18. Pedregosa, F., et al.: Scikit-learn: machine learning in Python. J. Mach. Learn. Res. **12**, 2825–2830 (2011)
19. Pelleg, D., Moore, A.W.: X-means: extending k-means with efficient estimation of the number of clusters. In: Proceedings of the Seventeenth International Conference on Machine Learning. ICML 2000, pp. 727–734. Morgan Kaufmann Publishers Inc., San Francisco (2000)
20. Rokach, L., Maimon, O.: Clustering methods. In: Maimon, O., Rokach, L. (eds.) Data Mining and Knowledge Discovery Handbook, pp. 321–352. Springer, Boston (2005). https://doi.org/10.1007/0-387-25465-X_15
21. Schubert, E., Rousseeuw, P.J.: Faster *k*-medoids clustering: improving the PAM, CLARA, and CLARANS algorithms. In: Amato, G., Gennaro, C., Oria, V., Radovanović, M. (eds.) SISAP 2019. LNCS, vol. 11807, pp. 171–187. Springer, Cham (2019). https://doi.org/10.1007/978-3-030-32047-8_16
22. Storn, R., Price, K.: Differential evolution - a simple and efficient heuristic for global optimization over continuous spaces. J. Global Optim. **11**(4), 341–359 (1997). https://doi.org/10.1023/A:1008202821328
23. Tan, P.N., Steinbach, M., Kumar, V.: Introduction to Data Mining, 1st edn. Addison-Wesley Longman Publishing Co., Inc., Boston (2005)
24. Virtanen, P., et al.: SciPy 1.0: fundamental algorithms for scientific computing in Python. Nat. Methods **17**, 261–272 (2020)
25. Wang, K., Xu, C., Liu, B.: Clustering transactions using large items. In: Proceedings of the Eighth International Conference on Information and Knowledge Management. CIKM 1999, pp. 483–490. ACM, New York (1999)
26. Wang, M.T., Hsu, P.Y., Lin, K.C., Chen, S.S.: Clustering transactions with an unbalanced hierarchical product structure. In: Song, I.Y., Eder, J., Nguyen, T.M. (eds.) DaWaK 2007. LNCS, vol. 4654, pp. 251–261. Springer, Heidelberg (2007). https://doi.org/10.1007/978-3-540-74553-2_23

Fashion Outfit Generation for E-Commerce

Elaine M. Bettaney(✉), Stephen R. Hardwick, Odysseas Zisimopoulos, and Benjamin Paul Chamberlain

ASOS.com, London, UK
elaine.bettaney@asos.com

Abstract. The task of combining complimentary pieces of clothing into an outfit is familiar to most people, but has thus far proved difficult to automate. We present a model that uses multimodal embeddings of pieces of clothing based on images and textual descriptions. The embeddings and a shared style space are trained end to end in a novel deep neural network architecture. The network is trained on the largest and richest labelled outfit dataset made available to date, which we open source. This is the first public expert created, labelled dataset and contains 586,320 labelled outfits. We evaluate the performance of our model using an AB test and compare it to a template based model that selects items from the correct classes, but ignores style. Our experiments show that our model outperforms by 21% and 34% for womenswear and menswear respectively.

Keywords: Representation learning · Fashion · Multi-modal deep learning

1 Introduction

User needs based around outfits include answering questions such as "What trousers will go with this shirt?", "What can I wear to a party?" or "Which items should I add to my wardrobe for summer?". The key to answering these questions requires an understanding of *style*. Style encompasses a broad range of properties including but not limited to, colour, shape, pattern and fabric. It may also incorporate current fashion trends, user's style preferences and an awareness of the context in which the outfits will be worn. In the growing world of fashion e-commerce it is becoming increasingly important to be able to fulfill these needs in a way that is scalable, automated and ultimately personalised.

This paper describes a system for Generating Outfit Recommendations from Deep Networks (GORDN) under development at ASOS. ASOS is a global e-commerce company focusing on fashion and beauty. With approximately 87,000 products on site at any one time, it is difficult for customers to perform an exhaustive search to find products that can be worn together. Each fashion product

Supported by ASOS.com.

Y. Dong et al. (Eds.): ECML PKDD 2020, LNAI 12461, pp. 339–354, 2021.
https://doi.org/10.1007/978-3-030-67670-4_21

added to our catalogue is photographed on a model as part of an individually curated outfit of compatible products chosen by our stylists to create images for its Product Description Page (PDP). The products comprising the outfit are then displayed to the customer in a *Buy the Look* (BTL) carousel. This offering however is not scalable as it requires manual input for every outfit. We aim to learn from the information encoded in these outfits to automatically generate an unlimited number of outfits.

A common way to compose outfits is to first pick a seed item, such as a patterned shirt, and then find other compatible items. We focus on this task: completing an outfit based on a seed item. This is useful in an e-commerce setting as outfit suggestions can be seeded with a particular product page or a user's past purchases. Our dataset comprises a set of outfits originating from BTL carousels on PDPs. These contain a seed, or 'hero product', which can be bought from the PDP, and compatible items we refer to as 'styling products'.

There is an asymmetry between hero and styling products. Whilst all items are used as hero products (in an e-commerce setting), styling products are selected as the best matches for the hero product and this matching is directional. For example when the hero product is a pair of Wellington boots it may create an engaging outfit to style them with a dress. However if the hero product is a dress then it is unlikely a pair of Wellington boots would be the best choice of styling product to recommend. Hence in general styling products tend to be more conservative than hero products. Our approach takes this difference into account by explicitly including this information as a feature.

We formulate our training task as binary classification, where GORDN learns to tell the difference between BTL and randomly generated negative outfits. We consider an outfit to be a set of fashion items and train a model that projects items into a single *style space*. Compatible items will appear close in style space enabling good outfits to be constructed from nearby items. GORDN is a neural network which combines embeddings of multi-modal features for all items in an outfit and outputs a single score. When generating outfits, GORDN is used as a scorer to assess the validity of different combinations of items.

In summary, our contributions are:

1. A novel model that uses multi-modal data to generate outfits that can be trained on images in the wild i.e. dressed people rather than individual item flat shots. Outfits generated by our model outperform a challenging baseline by 21% for womenswear and 34% for menswear.
2. A new research dataset consisting of 586,320 fashion outfits (images and textual descriptions) composed by professional stylists[1]. This is the world's largest annotated outfit dataset and is the first to contain menswear items.

[1] Dataset available at http://osf.io/cg6eq/?view_only=f446b0f51bca409da4175a4ce8d9000a.

Table 1. Statistics of the outfits dataset.

	Num. outfits	Num. items	Outfits of size 2	Outfits of size 3	Outfits of size 4	Outfits of size 5
WW	314,200	321,672	155,083	109,308	42,028	7,781
MW	272,120	270,053	100,395	102,666	58,544	10,515

2 Related Work

Our work follows an emerging body of related work on learning clothing style [11,24], clothing compatibility [19,21,24] and outfit composition [4,7,10,23]. Successful outfit composition encompasses an understanding of both style and compatibility.

A popular approach is to embed items in a latent *style* or *compatibility* space often using multi-modal features [10,21,22,24]. A challenge with this approach is how to use item embeddings to measure the overall outfit compatibility. This challenge is increased when considering outfits of multiple sizes. Song et al. [21] only consider outfits of size 2 made of top-bottom pairs. Veit et al. [24] use a Siamese CNN, a technique which allows only consideration of pairwise compatibilities. Li et al. [10] combine text and image embeddings to create multi-modal item embeddings which are then combined using pooling to create an overall outfit representation. Pooling allows them to consider outfits of variable size. Tangseng et al. [22] create item embeddings solely from images. They are able to use outfits of variable size by padding their set of item images to a fixed length with a 'mean image'. Our method is similar to these as we combine multi-modal item embeddings, however we aim not to lose information by pooling or padding.

Vasileva et al. [23] extend this concept by noting that compatibility is dependent on context - in this case the pair of clothing types being matched. They create learned type-aware projections from their style space to calculate compatibility between different types of clothing.

3 Outfit Datasets

The outfits dataset consists of 586,320 outfits, each containing between 2 and 5 items (see Table 1). In total these outfits contain 591,725 unique items representing 18 different womenswear (WW) product types and 22 different menswear (MW) product types. As all of our outfits have been created by professional stylists, they are representative of a particular fashion style.

Most previous outfit generators have used either co-purchase data from Amazon [13,24] or user created outfits taken from Polyvore [4,5,10,15,21–23], both of which represent a diverse range of styles and tastes. Co-purchase is not a strong signal of compatibility as co-purchased items are typically not bought with the intention of being worn together. Instead it is more likely to reflect a user's style preference. Data collected from Polyvore gives a stronger signal of compatibility and furthermore provide complete outfits.

The largest previously available outfits dataset was collected from Polyvore and contained 68,306 outfits and 365,054 items entirely from WW [23]. Our dataset is the first to contain MW as well. Our WW dataset contains an order of magnitude more outfits than the Polyvore set, but has slightly fewer fashion items. This is a consequence of stylists choosing styling products from a subset of items held in our studios leading to styling products appearing in many outfits.

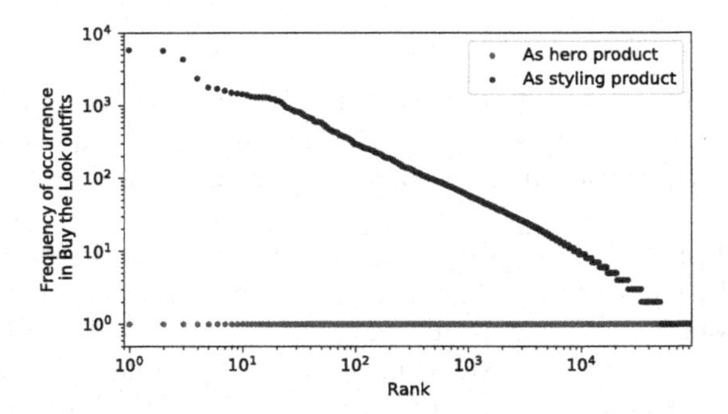

Fig. 1. Frequency of occurrence of each WW item in our outfits dataset. Items are ranked by how frequently they occur as styling products. Each item appears once at most as a hero product (red), while there is a heavily skewed distribution in the frequency with which items appear as styling products (blue). (Color figure online)

For each item we have four images, a text title and description, a high-level product type and a product category. We process both the images and the text title and description to obtain lower-dimensional embeddings, which are included in this dataset alongside the raw images and text to allow full reproducibility of our work. The methods used to extract these embeddings are described in Sect. 4.3 and 4.4, respectively. Although we have four images for each item, in these experiments we only use the first image as it consistently shows the entire item, from the front, within the context of an outfit, whilst the other images can focus on close ups or different angles, and do not follow consistent rules between product types.

4 Methodology

Our approach uses a deep neural network. We acknowledge some recent approaches that use LSTM neural networks [4,15]. We have not adopted this approach because fundamentally an outfit is a set of fashion items and treating it as a sequence is an artificial construct. Furthermore LSTMs are designed to progressively forget past items when moving through a sequence which in this context would mean that compatibility is not enforced between all outfit items.

We consider an outfit to be a set of fashion items of arbitrary length which match stylistically and can be worn together. In order for the outfit to work, each item must be compatible with all other items. Our aim is to model this by embedding each item into a latent space such that for two items (I_i, I_j) the dot product of their embeddings $(\mathbf{z}_i, \mathbf{z}_j)$ reflects their compatibility. We aim for the embeddings of compatible items to have large dot products and the embeddings of items which are incompatible to have small dot products. We map input data for each item I_i to its embedding $\boldsymbol{z}_i$ via a multi-layer neural network. As we are treating hero products and styling products differently, we learn two embeddings in the same space for each item; one for when the item is the hero product, $\boldsymbol{z}_i^{(h)}$ and one for when the item is a styling product, $\boldsymbol{z}_i^{(s)}$; which is reminiscent of the context specific representations in language modelling [14,16].

4.1 Network Architecture

For each item, the inputs to our network are a textual title and description embedding (1024 dimensions), a visual embedding (512 dimensions), a pre-trained GloVe embedding [16] for each product category (50 dimensions) and a binary flag indicating the hero product. First, each of the three input feature vectors is passed through their own fully connected ReLU layer. The outputs from these layers, as well as the hero product flag, are then concatenated and passed through two further fully connected ReLU layers to produce an item embedding with 256 dimensions (Fig. 2). We use batch normalization after each fully connected layer and a dropout rate of 0.5 during training.

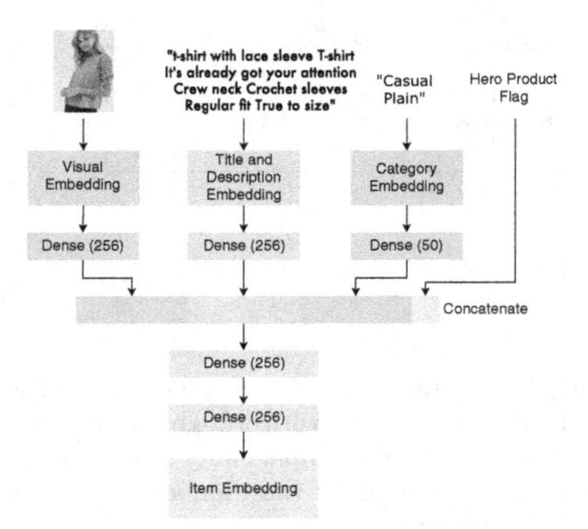

Fig. 2. Network architecture of GORDN's item embedder. For each item the embedder takes visual features, a textual embedding of the item's title and description, a pre-trained GloVe embedding of the item's product category and a binary flag indicating if the item is the outfit's hero product. Each set of features is passed through a dense layer and the outputs of these are concatenated along with the hero product flag before being passed through two further dense layers. The output is an embedding for the item in our *style* space. We train separate item embedders for WW and MW items.

4.2 Outfit Scoring

We use the dot product of item embeddings to quantify pairwise compatibility. Outfit compatibility is then calculated as the sum over pairwise dot products for all pairs of items in the outfit (Fig. 3).

For an outfit $\mathcal{S} = \{I_1, I_2, ..., I_N\}$ consisting of N items, the overall outfit score is defined by

$$y(\mathcal{S}) = \sigma\left(\frac{1}{N(N-1)} \sum_{\substack{i,j=1 \\ i<j}}^{N} \mathbf{z}_i \cdot \mathbf{z}_j\right), \tag{1}$$

where σ is the sigmoid function. The normalisation factor of $N(N-1)$, proportional to the number of pairs of items in the outfit is required to deal with outfits containing varying numbers of items. The sigmoid function is used to ensure the output is in the range [0,1].

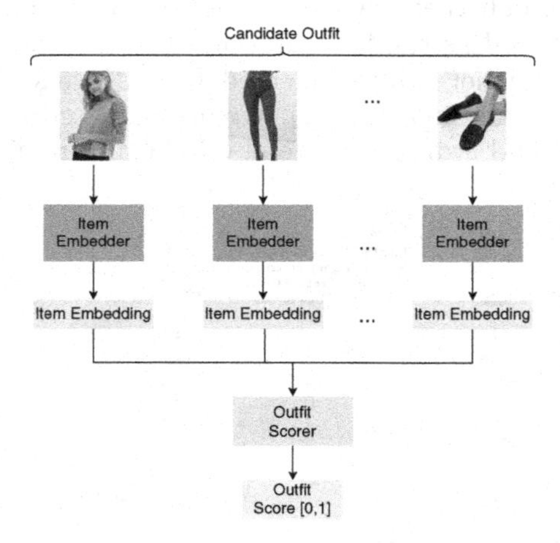

Fig. 3. GORDN's outfit scorer takes the learnt embeddings of each item in an outfit and produces a score in the range [0,1], with a high value representing a compatible outfit. The scorer takes the sum of the compatibility scores (dot product) for each pair of items within the outfit, normalised by the number of pairs of items. This is then passed through a sigmoid function.

4.3 Visual Feature Extraction

As described in Sect. 1, items are photographed as part of an outfit and therefore our item images frequently contain the other items from the BTL outfit. Feeding the whole image to the network would result in features capturing information for the entire input, that is both the hero and the styling products, leaking information to GORDN. It was therefore necessary to localise the target item within the image. To extract visual features from the images in our dataset we use VGG [20] and to specifically extract features focused on the most relevant areas of the image, we adopt an approach based on Class Activation Mapping (CAM) [25]. Weakly-supervised object localisation is performed by calculating a heatmap (CAM) from the feature maps of the last convolutional layer of a CNN, which highlights the discriminative regions in the input used for image classification. The CAM is calculated as a linear combination of the feature maps weighted by the corresponding class weights.

Before using the CAM model to extract image features, we fine-tune it on our dataset. Similar to [25] our model architecture combines VGG with a Global Average Pooling (GAP) layer and an output classification layer. We initialize VGG with weights pre-trained on ImageNet and fine-tune it towards product type classification (e.g. Jeans, Dresses, etc.). After training we pass each image to the VGG and obtain the feature maps.

To produce localised image embeddings, we use the CAM to spatially re-weight the feature maps. Similar to Jimenez et al. [8], we perform the re-weighting by a simple spatial element-wise multiplication of the feature maps with the CAM. Our pipeline is shown in Fig. 4. This re-weighting can be seen as a form of attention mechanism on the area of interest in the image. The final image embedding is a 512-dimensional vector. The same figure illustrates the effect of the re-weighting mechanism on the feature maps.

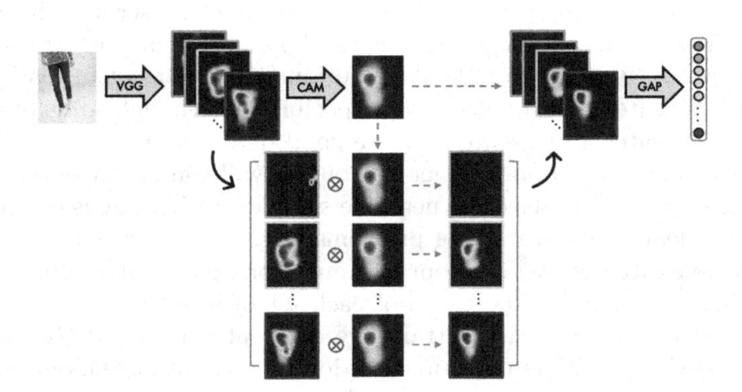

Fig. 4. Pipeline for extracting image embeddings. An image is passed to VGG and the final convolutional feature maps are used to calculate the class activation map (CAM). The CAM is then used to spatially re-weight the feature maps by element-wise multiplication. Finally, a global average pooling (GAP) layer averages each re-weighted feature map to calculate a single value (shown in same colours) and outputs a 512-dimensional image embedding. During training, re-weighting is ignored and the output of the model is passed into a fully-connected layer for product type classification. We can see the effect of feature map re-weighting in the brackets for the case of an item with trousers as the hero product and top and shoes as styling products. (Color figure online)

4.4 Title and Description Embeddings

Product titles typically contain important information, such as the brand and colour. Similarly, our text descriptions contain details such as the item's fit, design and material. We use pre-trained text embeddings of our item's title and description. These embeddings are learned as part of an existing system that predicts product attributes [3]. Vector representations for each word are passed through a simple 1D convolutional layer, followed by a max-over-time pooling layer and finally a dense layer, resulting in 1024 dimensional embeddings.

4.5 Training

We train GORDN in a supervised manner using a binary cross-entropy loss. Our training data consists of positive outfit samples taken from the outfits dataset and randomly generated negative outfit samples. We generate negative samples for our training and test sets by randomly replacing the styling products in each outfit with another item of the same type. For example, for an outfit with a top as the hero product and jeans and shoes as styling products, we would create a negative sample by replacing the jeans and shoes with randomly sampled jeans and shoes. We ensure that styling products appear with the same frequency in the positive and negative samples by sampling styling products from their distribution in the positive samples. This is important as the frequency distribution of styling products is heavily skewed (Fig. 1) and without preserving this GORDN could memorise frequently occurring items and predict outfit compatibility based on their presence. By matching the distribution GORDN must instead learn the characteristics of items which lead to compatibility. Although some of the negative outfits generated in this way may be good quality outfits, we assume that the majority of these randomly generated outfits will contain incompatible item combinations. Randomly selecting negative samples in this way is common practice in metric learning and ranking problems (e.g. [6,17]). In both training and testing, we generate one negative outfit sample for each positive outfit sample.

To assess the relative importance of each set of input features, we conduct an ablation study. We separately train five different versions of GORDN using only the textual title and description embeddings as input (text), only the visual embeddings (vis), both text and visual embeddings (text + vis), text, visual and category embeddings (text + vis + cat), and finally the full set of inputs (text + vis + cat + hero). For each of these configurations we trained 20 models from scratch using Adam [9] for 30 epochs.

4.6 Outfit Generation Method

Once trained, GORDN can generate novel outfits of any length by sequentially adding items and re-scoring the new outfit. Each outfit starts with a hero product from our catalogue. We then define an outfit template $\mathcal{P} = \{T^{(h)}, T_1, ..., T_{N-1}\}$ as a set of product types including the hero product type $T^{(h)}$ and $N-1$ other

Table 2. The number of outfits and items in our training and test partitions after applying the Louvain community detection method to the full outfits dataset. There are no items which appear in both the training and test set.

Department	Dataset	Number of outfits	Number of items
Womenswear	Train	237,478	239,818
	Test	76,722	81,854
Menswear	Train	201,844	198,947
	Test	70,276	71,106

compatible styling product types. Our aim is to find the set of items of the appropriate product types that maximises the outfit score y.

An exhaustive search over every possible combination of styling products cannot be computed within a reasonable e-commerce latency budget. Instead, we map the maximum inner product search in Eq. 1 to a Euclidean nearest neighbour problem that is solved approximately and combine this with a beam search (illustrated in Fig. 5). The approximate nearest neighbours algorithm uses a PCA-tree that has been adapted for recommendations problems [1]. We use a beam width of three because it returned the optimal outfit 77.5% of the time. The beam search algorithm is repeated for all $(N-1)!$ permutations of the styling product types in the template as different outfits may be generated depending on the order in which product types are added. The outfit returned is the one that has the maximal score across all template permutations. For each step of the beam search, we calculate the resultant vector of the partial outfit and find the w approximate nearest neighbours from the product catalogue (where w is the beam width). With each step searching through 2000–5000 products we achieved a ten times speed up in outfit generation whilst still maintaining a precision@5 of over 80%.

The choice of template $\mathcal{P}$ depends on the use case. Templates for each hero product type can be found from our outfits dataset. The distribution of templates can be used to introduce variety into the generated outfits. For the purposes of our AB test, we picked the most frequently occurring template for each hero product type.

5 Evaluation

We evaluate the performance of GORDN on two tasks. The first task is binary classification of genuine and randomly generated outfits, using a held out test set. The second task is user evaluation of outfits generated by GORDN in comparison to randomly generated outfits from a simple baseline model.

5.1 Train/test Split

We split the outfits dataset first into WW and MW and each of these into a training and test set ensuring that no items appeared in both sets. To achieve

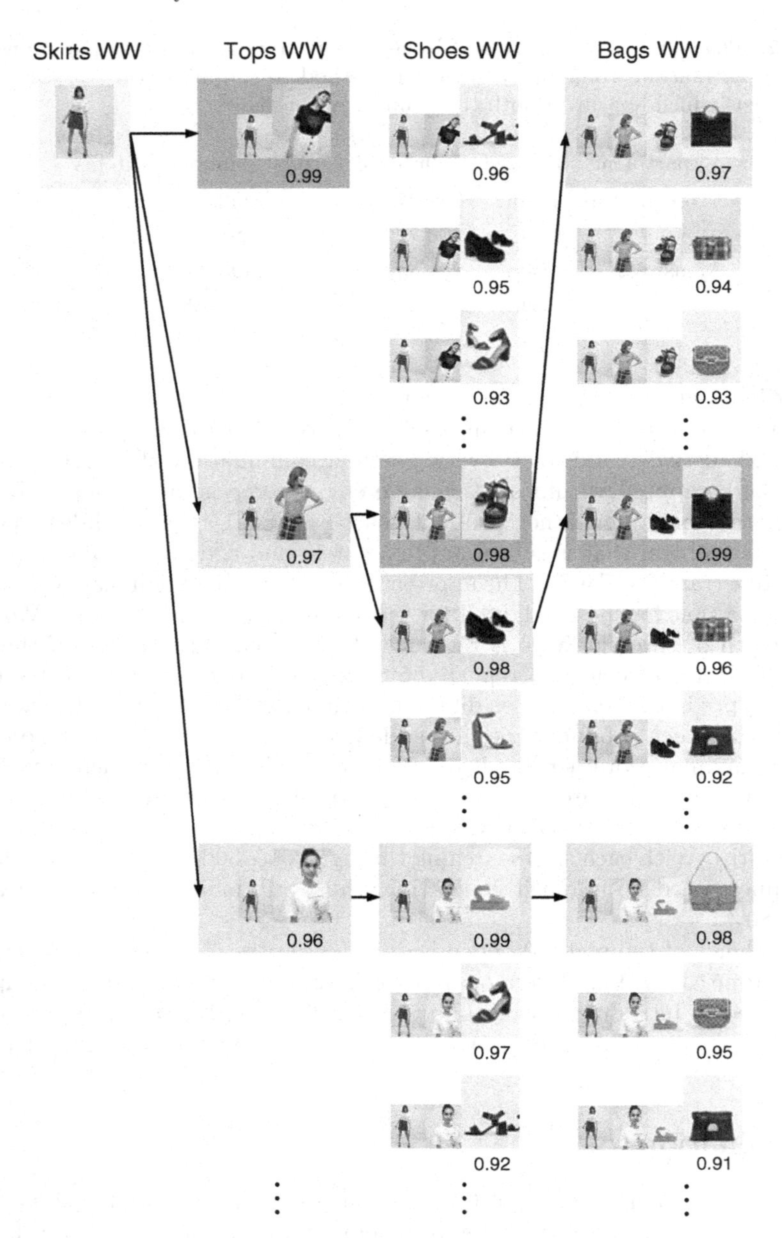

Fig. 5. Beam search in the context of outfit generation. Starting with an outfit template of product types and a hero product (highlighted in yellow) each product type in the template is filled sequentially by finding the products from the catalogue which when added to the outfit give the highest outfit score. After each step the number of outfits retained is reduced to the beam width (set to 3). The retained outfits after each step are highlighted in green with the highest scoring outfit in dark green. (Color figure online)

Table 3. Comparison of GORDN when using different features on the binary classification task. Scores are the mean over 20 runs of the test set AUC after 30 epochs.

Features	AUC	
	WW	MW
vis	0.66	0.55
text	0.80	0.66
text + vis	0.82	0.66
Text + vis + cat	0.82	0.67
text + vis + cat + hero	**0.83**	**0.67**

this we first represented the outfits dataset as a graph where the nodes are items and edge weights are defined by the number of outfits pairs of items are found together in. We then used the Louvain community detection method [2] to split the graph into communities which maximise the modularity. This resulted in many small communities which could then be combined together to create the train and test sets. When re-combining communities care was taken firstly to respect the desired train-test split ratio as far as possible and secondly to ensure items from each season are proportionally split between the train and test sets. This resulted in 76:24 and 74:26 train-test splits in terms of outfits for WW and MW respectively (see Table 2). The use of disjoint train and test sets provides a sterner test for GORDN as it is unable to simply memorise which items frequently co-occur in outfits in the training set. Instead, the embeddings GORDN learns must represent product attributes that contribute to fashion compatibility.

5.2 Outfit Classification Results

The test set contains BTL outfits and an equal number of negative samples. We use GORDN to predict compatibility scores for the test set outfits and then calculate the AUC of the ROC curve. We found that training separate versions of GORDN for WW and MW produced better results and so we report the performance of these here.

Table 3 shows the AUC scores achieved for different combinations of features. As we add features to GORDN we increase its performance, with the best performing model including text, visual, category and hero item features. The majority of the performance benefit came from the text embeddings with visual embeddings adding a small improvement. We expected our visual embeddings to be of poorer quality than those for Polyvore datasets as our images show whole outfits on people as opposed to a photograph of the fashion item in isolation. In contrast the success of our text embeddings could be due to the attribution task on which they were trained [3]. A total of 34 attributes were predicted, including many attributes that are directly applicable for outfit composition e.g. 'pattern', 'neckline', 'dress type' and 'shirt style'.

Table 4. Relative differences between the test and control group user scores. All results are significant at the 1% level.

	Ctrl score	Test score	Rel. diff. (%)	p-val
WW				
all	0.49	0.60	21.28	< 0.01
Dress | Shoes	0.54	0.78	46.12	< 0.01
Tops | Jeans | Shoes	0.61	0.64	4.53	< 0.01
Skirts | Tops | Shoes	0.33	0.36	10.77	< 0.01
MW				
all	0.49	0.66	34.16	< 0.01
T-Shirts | Jeans | Shoes, Boots & Trainers	0.63	0.76	19.07	< 0.01
Shirts | Trousers & Chinos | Shoes, Boots & Trainers	0.42	0.60	44.35	< 0.01
Shorts | T-Shirts | Shoes, Boots & Trainers	0.43	0.63	47.24	< 0.01

For all feature combinations the WW model greatly outperforms the MW one. This could be due to fashion items being more interchangeable in MW than in WW hence having more similar embeddings making the training task harder. For example the mean correlations between the text embeddings for the most prevalent product type in the WW and MW training sets are 0.041 (dresses) and 0.077 (T-shirts) respectively. More simply, there are many combinations of MW T-shirts and jeans that make equally acceptable outfits whereas there are far fewer for WW dresses and shoes.

Using GORDN to predict compatibility scores for our test set is equivalent to the *outfit compatibility* task used by [4] and [23]. As noted by Vasileva et al., Han's negative samples contain outfits that are incompatible due to multiple occurrences of product types e.g. multiple pairs of shoes in the same outfit. Since our negative samples were generated using templates respecting product type our data does not have this characteristic and hence we compare only to results in [23]. Our WW model achieves an AUC score just slightly less than Vasileva et al.'s compatibility AUC on their disjoint Polyvore outfits dataset.

5.3 Generated Outfit Evaluation

We perform an AB test to evaluate the quality of outfits generated by GORDN. We select six popular outfit templates to test, three each for WW and MW (shown in Table 4), and generate 100 WW and 100 MW outfits split evenly across the templates. We use a large pool of in stock products from which we randomly select hero products of the required product types. The remaining items in the outfits were generated using the beam search method described in Sect. 4.6 and illustrated in Fig. 5. These outfits constitute our test group. For a control group we take the same hero products and templates and generate outfits by randomly selecting items of the correct type from the same pool of products. By using outfit templates we ensure that none of the outfits contain incompatible product type combinations, such as by pairing a dress and a skirt, or by placing two pairs of shoes in one outfit. Instead, the quality of the outfits depends solely on style compatibility between items.

To run the AB test we developed a web app. The app displayed an outfit to a user asking them to decide if the items in the outfit work stylistically. The outfits were shown one at a time to each user with the order randomised. WW and MW outfits were only shown to female and male users respectively and each user rated 200 outfits from their corresponding gender.

The data collected from the app comprised a binary score for each user-outfit pair. The data exhibit two way correlation—all scores from the same user are correlated due to the inherent user preferences and all scores on the same outfit are also correlated. We therefore used a two-way random effects model as described in [18] to calculate the variance of the sample mean. We could then use a t-test for the difference between means to calculate if the difference between the test and control groups was significant.

The results are shown in Table 4. We analyse the results for WW and MW separately as the WW and MW models were trained separately. We collected 1,200 observations per group for WW and 900 for MW. We found the relative difference between the test and control groups to be 21.28% and 34.16% for WW and MW respectively. Testing at the 1% level these differences were sig-

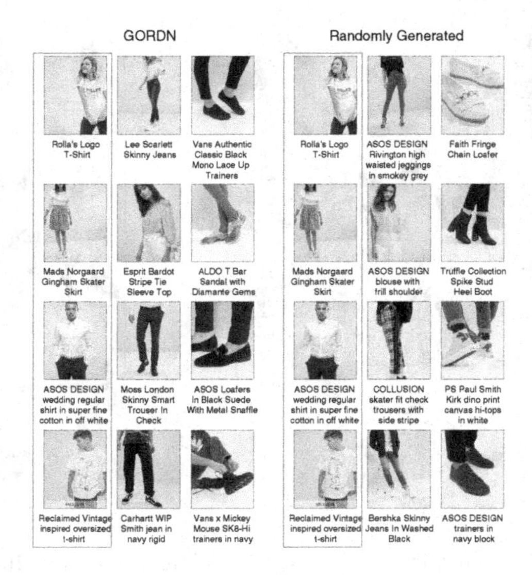

Fig. 6. Example outfits generated by GORDN (left column) and by random selection of items (right column) that were used in our AB test. In each row the same hero product is used (red box) and each model is given the same template of product types. (Color figure online)

nificant. We were able to further break down our results to find that GORDN outperformed the control significantly for all templates.

Examples of outfits generated for our AB test are shown in Fig. 6. For each hero product we show the outfit produced by GORDN alongside the randomly generated outfit. Many of the random examples appear to be reasonable outfits. Although the random model is simple, the use of outfit templates, combined with selecting only products that were in stock in the catalogue on the same day makes this a challenging baseline.

5.4 Style Space

We visualise our style space using a t-Distributed Stochastic Neighbour Embedding (t-SNE) [12] plot in two dimensions (Fig. 7). While similar items have similar embeddings, we can also see that compatible items of different product types

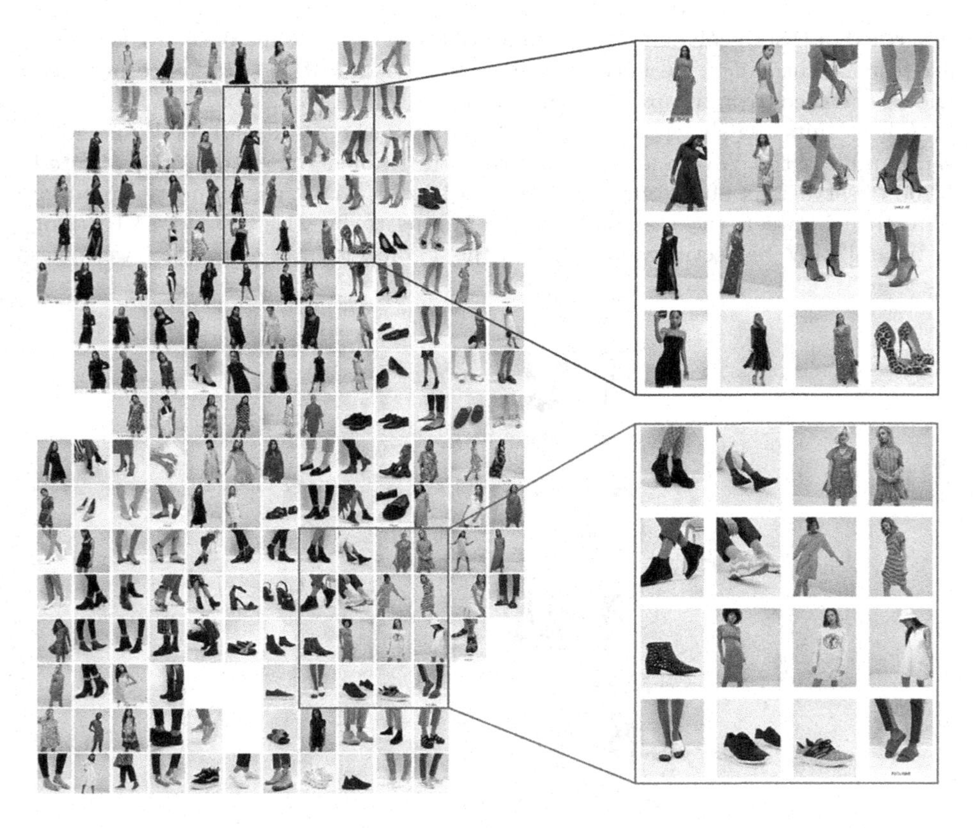

Fig. 7. A section of a t-Distributed Stochastic Neighbour Embedding (t-SNE) visualisation of the embeddings learnt by GORDN for WW dresses and shoes. Similar items have similar embeddings, but so do compatible items of different types. The two highlighted areas illustrate that casual dresses are embedded close to casual shoes (red), while occasion dresses are embedded close to occasion shoes (blue). (Color figure online)

have similar embeddings. Rather than dresses and shoes being completely separate in style space, these product types overlap, with casual dresses having similar embeddings to casual shoes and occasion dresses having similar embeddings to occasion shoes.

6 Conclusion

We have described GORDN, a multi-modal neural network for generating outfits of fashion items. GORDN learns to represent items in a latent style space, such that compatible items of different types have similar embeddings. GORDN is trained on a new outfits dataset, a resource that we provide for the research community, which contains over 500,000 outfits curated by professional stylists. The results of an AB test show that users approve of outfits generated by GORDN 21% and 34% more frequently than those generated by a templated baseline model for WW and MW, respectively.

References

1. Bachrach, Y., et al.: Speeding up the Xbox recommender system using a euclidean transformation for inner-product spaces. In: Proceedings of the 8th ACM Conference on Recommender Systems, pp. 257–264. ACM (2014)
2. Blondel, V.D., Guillaume, J.L., Lambiotte, R., Lefebvre, E.: Fast unfolding of communities in large networks. J. Stat. Mech. **2008**, 1–12 (2008)
3. Cardoso, Â., Daolio, F., Vargas, S.: Product characterisation towards personalisation. In: Proceedings of the 24th ACM SIGKDD International Conference on Knowledge Discovery & Data Mining - KDD '18, pp. 80–89 (2018)
4. Han, X., Wu, Z., Jiang, Y.G., Davis, L.S.: Learning fashion compatibility with bidirectional LSTMs. In: Proceedings of the 2017 ACM on Multimedia Conference - MM '17, pp. 1078–1086 (2017)
5. He, T., Hu, Y.: FashionNet: Personalized Outfit Recommendation with Deep Neural Network (2018). http://arxiv.org/abs/1810.02443
6. Hoffer, E., Ailon, N.: Deep metric learning using triplet network. In: Feragen, A., Pelillo, M., Loog, M. (eds.) SIMBAD 2015. LNCS, vol. 9370, pp. 84–92. Springer, Cham (2015). https://doi.org/10.1007/978-3-319-24261-3_7
7. Hu, Y., Yi, X., Davis, L.S.: Collaborative fashion recommendation: a functional tensor factorization approach. In: Proceedings of the 23rd ACM International Conference on Multimedia, pp. 129–138. Brisbane, Australia (2015)
8. Jimenez, A., Alvarez, J.M., Giro-i Nieto, X.: Class-weighted convolutional features for visual instance search. In: 28th British Machine Vision Conference (BMVC) (Sept 2017)
9. Kingma, D.P., Ba, J.: Adam: a method for stochastic optimization. In: International Conference for Learning Representations, San Diego (2015)
10. Li, Y., Cao, L., Zhu, J., Luo, J.: Mining fashion outfit composition using an end-to-end deep learning approach on set data. IEEE Trans. Multimedia **19**(8), 1946–1955 (2017). https://doi.org/10.1109/TMM.2017.2690144
11. Ma, Y., Jia, J., Zhou, S., Fu, J., Liu, Y., Tong, Z.: Towards better understanding the clothing fashion styles: a multimodal deep learning approach. In: AAAI Conference on Artificial Intelligence, pp. 38–44 (2017)

12. van der Maaten, L., Hinton, G.: Visualizing data using t-SNE. J. Mach. Learn. Res. **9**, 2579–2605 (2008)
13. McAuley, J., Targett, C., Shi, Q., van den Hengel, A.: Image-based recommendations on styles and substitutes. In: SIGIR Conference on Research and Development in Information Retrieval, pp. 43–52 (2015)
14. Mikolov, T., Chen, K., Corrado, G., Dean, J.: Distributed representations of words and phrases and their compositionality. In: Neural Information Processing Systems, pp. 3111–3119 (2013). https://doi.org/10.1162/jmlr.2003.3.4-5.951
15. Nakamura, T., Goto, R.: Outfit generation and style extraction via bidirectional LSTM and autoencoder. In: The Third International Workshop on Fashion and KDD (2018). http://arxiv.org/abs/1807.03133
16. Pennington, J., Socher, R., Manning, C.: Glove: global vectors for word representation. In: Proceedings of the 2014 Conference on Empirical Methods in Natural Language Processing (EMNLP) (2014). https://doi.org/10.3115/v1/D14-1162
17. Rendle, S., Freudenthaler, C., Gantner, Z., Schmidt-Thieme, L.: BPR: Bayesian personalized ranking from implicit feedback. In: Conference on Uncertainty in Artificial Intelligence, vol. 1120, pp. 452–461 (2009)
18. Ribeiro, F., Florêncio, D., Zhang, C., Seltzer, M.: CROWDMOS: an approach for crowdsourcing mean opinion score studies. In: ICASSP, IEEE International Conference on Acoustics, Speech and Signal Processing - Proceedings, pp. 2416–2419 (2011). https://doi.org/10.1109/ICASSP.2011.5946971
19. Shih, Y.S., Chang, K.Y., Lin, H.T., Sun, M.: Compatibility family learning for item recommendation and generation. In: AAAI Conference on Artificial Intelligence, pp. 2403–2410 (2018). https://arxiv.org/pdf/1712.01262.pdf
20. Simonyan, K., Zisserman, A.: Very deep convolutional networks for large-scale image recognition. CoRR abs/1409.1556 (2014)
21. Song, X., Feng, F., Han, X., Yang, X., Liu, W., Nie, L.: Neural compatibility modeling with attentive knowledge distillation. In: SIGIR Conference on Research & Development in Information Retrieval, pp. 5–14 (2018)
22. Tangseng, P., Yamaguchi, K., Okatani, T.: Recommending outfits from personal closet. In: Proceedings of the IEEE Conference on Computer Vision and Pattern Recognition, pp. 269–277 (2018)
23. Vasileva, M.I., Plummer, B.A., Dusad, K., Rajpal, S., Kumar, R., Forsyth, D.: Learning type-aware embeddings for fashion compatibility. In: Ferrari, V., Hebert, M., Sminchisescu, C., Weiss, Y. (eds.) ECCV 2018, Part XVI. LNCS, vol. 11220, pp. 405–421. Springer, Cham (2018). https://doi.org/10.1007/978-3-030-01270-0_24
24. Veit, A., Kovacs, B., Bell, S., Mcauley, J., Bala, K., Belongie, S.: Learning visual clothing style with heterogeneous dyadic co-occurrences. In: IEEE International Conference on Computer Vision, pp. 4642–4650 (2015)
25. Zhou, B., Khosla, A., Lapedriza, A., Oliva, A., Torralba, A.: Learning deep features for discriminative localization. In: Computer Vision and Pattern Recognition (2016)

Improved Identification of Imbalanced Multiple Annotation Intent Labels with a Hybrid BLSTM and CNN Model and Hybrid Loss Function

Supawit Vatathanavaro[1], Kitsuchart Pasupa[1(✉)], Sorratat Sirirattanajakarin[2], and Boontawee Suntisrivaraporn[2]

[1] Faculty of Information Technology King Mongkut's Institute of Technology Ladkrabang, Bangkok 10520, Thailand
62606061@kmitl.ac.th, kitsuchart@it.kmitl.ac.th
[2] Data Analytics, Chief Data Office Siam Commercial Bank, Bangkok 10900, Thailand
{sorratat.sirirattanajakarin,boontawee.suntisrivaraporn}@scb.co.th

Abstract. Payment or fund transfer transactions can be annotated by users when they are made through a mobile banking app, for example, *SCB Easy app*—a mobile banking app by Siam Commercial Bank—allows users to annotate transactions with 40 character texts. The AI^2 framework was used to identify user intentions with the transactions, so that the bank can offer the right product to the right customer at the right time. The framework employed Long Short-Term Memory (LSTM). Commonly, one annotated sample can be interpreted as representing multiple intents, thus we had a multiple label classification problem. However, the original model did not consider the class imbalance, that caused the model to bias toward the majority class. We introduced a new hybrid Bidirectional LSTM and Convolutional Neural Network model in conjunction with a new hybrid loss function to tackle the imbalance. Our model with hybrid loss function performed better than the AI^2 framework with a 4.5% improvement in F_1-score. Moreover, our hybrid loss function enabled the model to classify minority classes better, when the imbalance ratio became higher, compared with a conventional cross-entropy loss function. In other words, our hybrid loss function made the model to be more efficient in real-world multiple label imbalance problem.

Keywords: Deep learning · Hybrid model · Multi-label classification · Text classification · Natural Language Processing · Imbalanced data

1 Introduction

A fast-growing mobile banking app in Thailand is *SCB Easy app* with over 10 million users [36]. Nowadays, people make many financial transactions through

Y. Dong et al. (Eds.): ECML PKDD 2020, LNAI 12461, pp. 355–368, 2021.
https://doi.org/10.1007/978-3-030-67670-4_22

mobile banking, so it is important to annotate each transaction to remind a user what they have spent on. The Siam Commercial Bank's *SCB Easy app* allows users to annotate a transaction up to 40 characters in English or Thai through the "note-taking" feature [34]. From the record, more than 10% of transactions had been annotated, mostly in Thai. These annotations can refer to customer intention and can be used to understand customer behavior, allowing the bank to offer relevant products and services to the customer to increase their sales. The large volume of these texts make it unfeasible to usefully process them manually, therefore, Natural Language Processing (NLP) was used to automatically identify intents from these annotations.

In the past decade, NLP has a played significant role in text classification problems [13,16]: many people now use social media to communicate with each other, resulting in a large volumes of text data. Many researchers have applied NLP to tackle social media classification problems, *e.g.* Medication-related text classification from Twitter [32], Personality classification based on Twitter text [30], Twitter trending topic classification [21], *etc.*. Furthermore, social media data provides useful consumer information, that can enhance market research [9]. Hence, building an accurate model, for text classification to obtain the right information from text data, is important to launch a marketing campaign at the right time to the right group.

Most NLP research is based on English, resulting in a significant resources in English. On the other hand, limited work on Thai language NLP means that it remains a low resource language. Thai differs from English in word separation—Thai does not use spaces to segment words—and adjectives and adverbs follow the expressed word [28]. In the past, little research has attempted sentimental analysis in social media, due to the highly ambiguous syntax of Thai [3]. However, recently, many have attempted to analyse sentiments in Thai [28,33,35]. The Annotation Intent Identification framework–AI^2 [34] has been used to identify customer intention in Thai: it tackles a multi label classification problem and aims to predict user intentions in order to offer relevant products and services to the right customers at the right time. It has been used in the Siam Commercial Bank (SCB), Thailand, to classify annotation text from *SCB Easy app* [34]. Using this model improved the campaign response rates up to 3 to 5 times higher than the conventional model. The AI^2 model used Long Short-Term Memory (LSTM) and a Multi-Layer Perceptron (MLP) as the basic classifier. However, this model did not concern the class imbalance problem, which made the model easily biased towards majority classes, resulting in a poor prediction [10].

Here, we describe a new model, for use in the AI^2 framework, using a hybrid Bidirectional-LSTM (BLSTM) and a Convolutional Neural Network (CNN). Additionally, we defined a hybrid loss function, combining the conventional cross-entropy loss and similarity coefficient loss, which is suitable for class imbalance in multi-label classification. Our new model and loss function were evaluated in real-world banking data.

2 Related Work

Recently, many researchers have been applied deep learning to tackle text classification problems. Early, Recurrent Neural Networks (RNN) [6] were applied in NLP. However, the RNN model cannot capture all the information in a long sentence, because of the gradient vanishing problem [22]. Hence, LSTM [11] was designed to tackle this problem. LSTM was successfully applied to real-world data, *e.g.* text classification in a social media application [38], multi-label document classification [41]. Later, BLSTM was used to detect information from left to right and right to left. Xu *et al.* [40] applied BLSTM in order to tackle the short text comments in sentiment analysis. Zhou *et al.* [44] combined CNN and LSTM for text classification. They applied one-dimensional (1D) convolution for N-gram feature extraction and then fed a sequence of higher-level features to LSTM.

Zhou *et al.* [45] integrated BLSTM with 2D convolution, which was able to better capture the meaningful information than 1D convolution. This model performed better than conventional models on six text classification tasks, including sentiment analysis, question classification, subjectivity classification and newsgroup classification. Nonetheless, the class imbalance data in social media text remained a challenge in NLP [32]. In general, the class imbalance problem in text classification has been insufficiently studied [23]: class imbalance in real-world data is still found, since reported by Hospedales *et al.* [12].

Class imbalance can cause the model to perform well only on majority classes and fail on minority classes [10]. Commonly, methods of handling poorly balanced problems can be divided into two main levels—data-level and algorithm-level [42].

Data-level algorithms try to balance the class distributions, which can be distributed into three methods: over-sampling, under-sampling and hybrid methods [8]. Over-sampling balances the class distribution by randomly duplicating minority class samples. Under-sampling randomly discards majority class samples to balance the class distribution. Hybrid methods combine over-sampling and under-sampling methods.

An algorithm-level approach intends to emphasise the minority class, *e.g.* cost-sensitive learning that assigns higher misclassification costs to the minority class [42]. However, assigning an appropriate cost for each class is difficult. Another method is to develop a new loss function or modify an existing loss function. Lin *et al.* [24] devised a 'focal loss' function to handle class imbalance in dense object detection, where a foreground-background class is severely imbalanced: they achieved better performance than a faster Region-CNN (RCNN) model [31]. Pasupa *et al.* [29] modified the focal loss function to handle imbalance in multiple classes in an image classification task to classify canine red blood cells and showed that the modified focal loss function performed better than a conventional cross-entropy loss function, in combination with over sampling with lower computational complexity. Milletari *et al.* [25] used a 'Dice loss' function based on the Dice coefficient [5], which tackled a strong imbalance between the number of foreground and background voxels in volumetric medical image segmentation and achieved good performances and with less processing time than a fully automatic segmentation model [37].

3 Methodology

3.1 Hybrid BLSTM-2DCNN Model

We designed a BLSTM-2DCNN model, modified from Zhou *et al.* [45]. BLSTM, in combination with the 2D-CNN model, is shown in Fig. 1. Our model has four main sections:

(i) Word Embedding: this part transforms words into vectors, in which related words will have a similar representation.
(ii) BLSTM: we use BLSTM to extract information from left to right and right to left. This makes the model find features from both directions.
(iii) 2DCNN-Max pooling: we use 2D-Convolution layers follow by a rectifier linear unit (ReLU) and 1D-Max pooling for sentence classification, inspired by Kim [17]. Kim used a 2D-CNN layer, in which the filter height is the number of words that produce a new feature map and the width is equal to the number of dimensions in word vectors.
(iv) Fully connected layer: important features from the previous part are concatenated and fed into a fully connected layer or directly connected to an output layer.

3.2 Hybrid Loss Function

The cross-entropy loss function is commonly used in multiple label classification problems. Nam *et al.* [26] demonstrated that a cross-entropy loss function performed better than BP-MLL ranking loss function [43] in multiple label classification. A cross-entropy loss function can be defined as:

$$CE(p, y) = CE(p_i) = -\sum_{i}^{n}(y_i \log(p_i) + (1 - y_i)\log(1 - p_i)) \tag{1}$$

where p_i is the prediction probability, y_i is the target for label i and n is the number of target labels.

Later, Milletari *et al.* [25] used an objective function based on the Dice coefficient. They aimed to handle a problem where the region of interest was very small in a volumetric medical image segmentation, resulting in imbalance, between foreground and background voxels, by maximising the objective function, which ranges between 0 and 1. In image segmentation problem, when one sample contains multiple voxels, this becomes a multiple label classification problem, because one sample may belong in more than one category. Here, we used the Dice coefficient [5] to handle the class imbalance in multiple label classification. Since the Dice coefficient is a similarity coefficient, we used a Jaccard Tanimoto coefficient [14], which always shows superior performance than the Dice

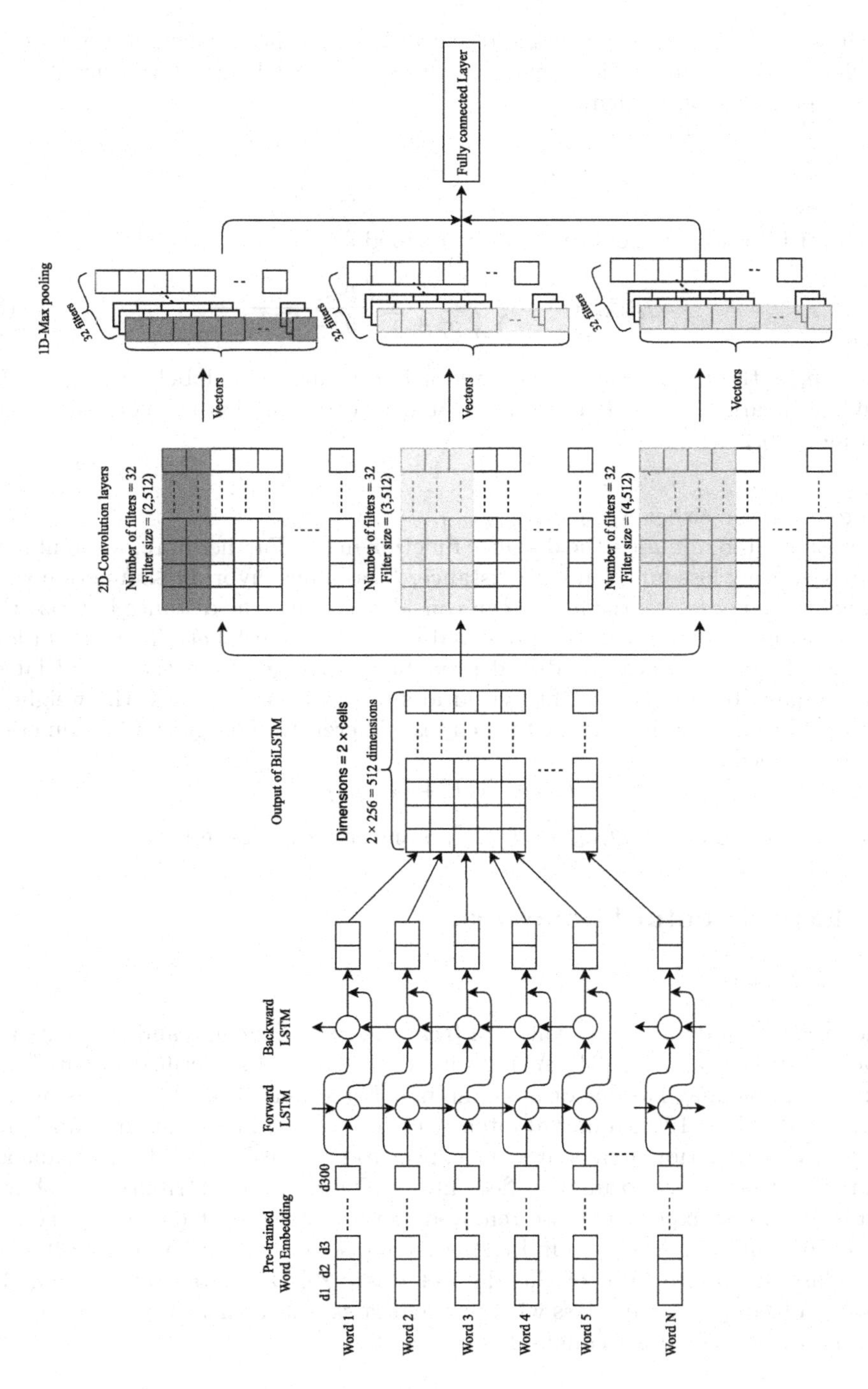

Fig. 1. Hybrid BLSTM-2DCNN architecture

coefficient [20]. Dice coefficients and Jaccard coefficients are commonly used as overlap measures in medical image analysis [4] and Chemoinformatics [27]. A Dice coefficient can be defined as:

$$S_{Dice}(p) = \frac{2\sum_i^n p_i y_i}{\sum_i^n p_i^2 + \sum_i^n y_i^2} \tag{2}$$

A Jaccard Tanimoto coefficient can be defined as:

$$S_{Jaccard}(p) = \frac{\sum_i^n p_i y_i}{\sum_i^n p_i^2 + \sum_i^n y_i^2 - \sum_i^n p_i y_i} \tag{3}$$

where p_i is the prediction probability, y_i is the target for label i and n is the number of target labels. It is known that similarity function is a complement of distance [20,27]:

$$D = 1 - S \tag{4}$$

where S can be either S_{Dice} or $S_{Jaccard}$.

We aimed to minimise the distance function in (4). Furthermore, we combined a cross-entropy loss function with distance, D, as a new hybrid loss function with a weight, w. Therefore, the model can consider not only the distance but also the cross-entropy loss between the predicted and the actual labels. The hybrid loss function increases as the predicted probability diverges from the actual label. Additionally, to get the best-fit hybrid loss function for this task, the weight w is applied to give an important to each loss differently. The hybrid loss function can be defined as:

$$\mathcal{L} = w \cdot CE + (1 - w)D \tag{5}$$

where D can be either D_{Dice} or $D_{Jaccard}$ and w ranges from 0 to 1.

4 Experimental Framework

4.1 Dataset

The annotation dataset was collected from the *SCB Easy app* and the texts are labeled by three experts [34]. A majority vote was used as verified intent. Each textual sample may has one or more intents. There were 80,000 sentences, which were divided into 12 meaningful intents, including eat, invest, learn, loan, rent, shop, travel, donation, home, insurance, peer-to-peer, utility and 1 miscellaneous (`'misc'`), that refers to unidentified intent. The dataset distribution is shown in Fig. 2. In our experiment, we removed `'misc'`, because it did not carry any useful information. As shown in Fig. 2, this dataset contained 5 minority classes. The class imbalance ratio of this dataset was ~29, calculated by dividing the number of samples in the class with the highest sample count—peer-to-peer—by the class with the lowest count—Insurance [2].

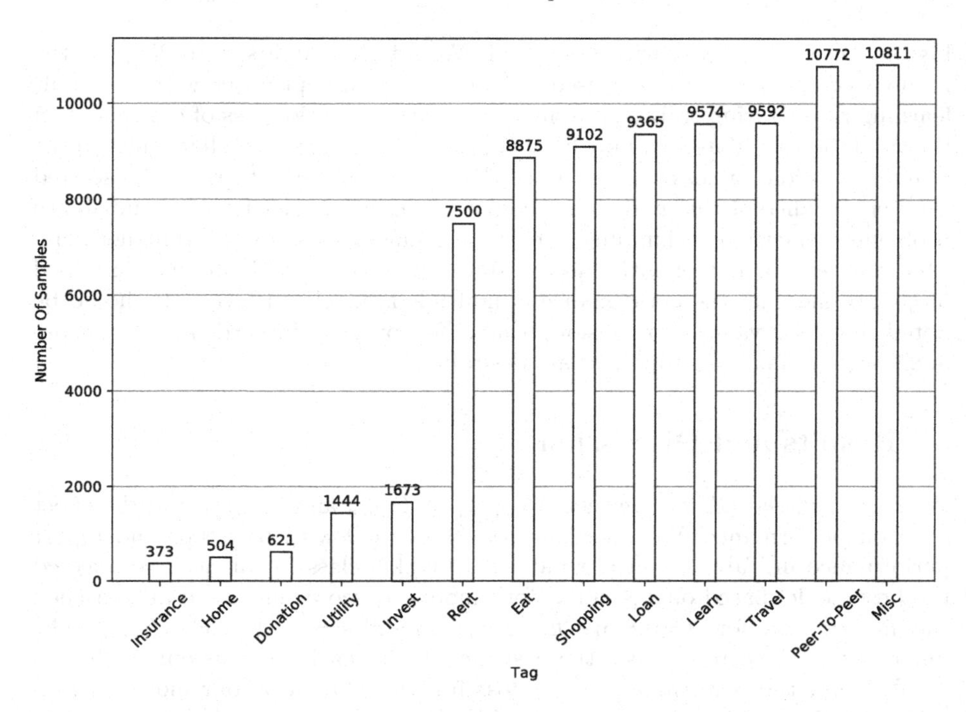

Fig. 2. Annotation of intent distribution

4.2 Experiment Setting

Following the described model in Sect. 3, samples were fed into the Word Embedding Layer with 23 words—the maximum number of Thai words that were found in 40 character annotations in this dataset—and then we used the dropout layer with a 0.2 drop rate. For Thai word vectors in our word embedding, we used pre-trained 300 dimensions word vectors trained by Grave *et al.* [7], which were trained on Common Crawl and Wikipedia [39] corpus using fastText [1]. This was followed by BLSTM—1 layer and 256 cells, which extracted information in both directions. Therefore, the output had was 512 dimensions. Next, in 2D-CNN, we were interested in bigrams, trigrams and 4-grams, so we set three filters with widths equal to the number of dimensions and heights (2, 3 or 4). Then 1D-Max pooling was applied to extract the most important features. The number of filters in each filter size of 2D-CNN was 32. In the last part, we concatenated all features, then applied a dropout layer with a 0.2 drop rate and passed to a 12-node output layer, matching the number of intents.

For the training step, we first tokenised each Thai sentence, using DeepCut—a Thai word tokenisation library [19]. Then we randomly split our dataset into 70% for training and 30% for testing. Further, we used 20% of the training set as a validation set. For the hybrid loss function, D was either D_{Dice} or $D_{Jaccard}$. We tested both similarity coefficients distance in combination with cross-entropy loss and used $w \in (0, 1)$ with a step 0.1. We set the batch size to 256 and the number of epochs to 30. We used Adam [18] as our optimiser with a default learning rate. During training, we monitored the validation loss of our model on the validation set. After training finished, we selected the model that achieved the minimum validation loss and evaluated that model on the test set. We illustrated the performance of the model on a commonly used metrics for class imbalance problems—macro-precision, macro-recall and macro-F_1-score [15]. Furthermore, we evaluated our model with Jaccard similarity score and Hamming loss based on [34] to measure the performance in multiple label classification. It should be noted that high values of precision, recall, F_1-score and Jaccard similarity score and low values of Hamming loss are desirable.

5 Results and Discussion

We combined the CE with either $D_{Jaccard}$ or D_{Dice} and investigated the effect of w on performance. We varied w from 0.1 to 0.9 with 0.1 steps and report performance in Table 1. Since we aimed to tackle class imbalance, we selected the best model based on F_1-score. Both model F_1-scores increased and reached the highest score for $w = 0.8$ in $CE + D_{Jaccard}$ and $w = 0.4$ in $CE + D_{Dice}$. The difference of F_1-score between the best model with each loss was only 0.3%.

We show the performance of the loss function, based on our model including the individual loss functions *e.g.* CE, $D_{Jaccard}$, D_{Dice} and the hybrid loss functions *e.g.* $CE + D_{Jaccard}(w = 0.8)$, $CE + D_{Dice}(w = 0.4)$ in Table 2. It can be seen that both $D_{Jaccard}$ and D_{Dice} performed better than the model with CE. The model recall was improved by 2.75% with $D_{Jaccard}$ and by 2.55% with D_{Dice} resulting in an F_1-score improvement of 1.24% with $D_{Jaccard}$ and by 1.16% with D_{Dice} compared with CE. We also investigated the Jaccard similarity score and the Hamming loss of both D. $D_{Jaccard}$ enabled the model to achieve the highest Jaccard similarity score *vs* CE and D_{Dice}. On the other hand, CE achieved the lowest Hamming loss and F_1-score compared to either $D_{Jaccard}$ or D_{Dice}. This confirmed that the model with CE was biased towards majority classes. Furthermore, we compared each individual loss function with a hybrid loss function and showed that the model with the hybrid loss function showed better improvement than the model with either $D_{Jaccard}$ or D_{Dice}. The model with $CE + D_{Jaccard}$ improved the model F_1-score by 1.68% and $CE + D_{Dice}$ improved by 1.98% compared with CE. We further compared the performance of our model with hybrid loss function with a conventional AI^2 model [34]. Our $CE + D_{Dice}(w = 0.4)$ model achieved F_1-score = 93.10, while the conventional model achieved F_1-score = 89.08, *i.e.* our model showed a 4.5% relative improvement. Additionally, our method improved Jaccard similarity scores by 3.69% and lowered Hamming loss by 0.0042.

As shown in Table 3, we examined the classification performance of the majority classes and the minority classes, when using CE and $CE+D_{Dice}$ with $w = 0.4$, which achieved highest F_1-score. The model with $CE + D_{Dice}$ achieved higher recall by 0.73% and F_1-score by 0.35%, when classifying majority classes and

Table 1. Effect of w in the hybrid loss function on precision, recall, F_1-score, Jaccard similarity score and Hamming loss. Bold values flag the best results.

Loss function	w	Macro precision	Macro recall	Macro F_1-score	Jaccard similarity	Hamming loss
$CE + D_{Jaccard}$	0.1	93.54	90.92	92.13	96.08	0.0062
	0.2	92.88	**92.03**	92.42	96.07	0.0065
	0.3	**94.83**	90.49	92.44	96.06	0.0060
	0.4	93.71	91.85	92.71	**96.24**	0.0060
	0.5	94.68	90.97	92.64	96.13	**0.0059**
	0.6	94.71	90.49	92.40	96.06	0.0061
	0.7	94.03	91.69	92.77	96.14	**0.0059**
	0.8	94.22	91.56	**92.80**	96.10	0.0060
	0.9	94.15	91.21	92.55	96.04	0.0060
$CE + D_{Dice}$	0.1	93.47	91.84	92.59	96.22	0.0062
	0.2	94.28	91.97	93.05	96.29	**0.0058**
	0.3	93.69	**92.31**	92.94	**96.33**	0.0060
	0.4	**94.72**	91.73	**93.10**	96.31	**0.0058**
	0.5	93.91	91.97	92.85	96.29	0.0060
	0.6	94.14	91.65	92.79	96.12	0.0060
	0.7	93.62	91.81	92.66	96.12	0.0060
	0.8	94.31	89.32	91.54	95.73	0.0064
	0.9	94.15	91.63	92.80	96.07	0.0062

Table 2. Precision, Recall, F_1-score, Jaccard Similarity Score and Hamming Loss achieved by the models with each loss function. The best performances are expressed in bold text.

Loss function	Macro precision	Macro recall	Macro F_1-score	Jaccard similarity	Hamming loss
CE(AI^2 [34])	91.08	88.27	89.08	92.62	0.0100
CE	94.55	88.54	91.12	95.56	0.0066
$D_{Jaccard}$	93.69	91.29	92.36	95.91	0.0067
D_{Dice}	93.63	91.09	92.28	95.72	0.0069
$CE + D_{Jaccard}(w = 0.8)$	94.22	91.56	92.80	96.10	0.0060
$CE + D_{Dice}(w = 0.4)$	**94.72**	**91.73**	**93.10**	**96.31**	**0.0058**

higher precision by 0.4%, recall by 6.63% and F_1-score by 4.24% in classifying minority classes.

Moreover, we plotted the F_1-score relative improvement of the $CE + D_{Dice}$ model, $w = 0.4$, *vs* the CE model in minority classes classification—see Fig. 3. The relative improvement of the $CE + D_{Dice}$ model increased when the imbalance

Table 3. Precision, Recall and F_1-score for CE and $CE + D_{Dice}$ with $w = 0.4$

Class size	Tag	CE			$CE + D_{Dice}$ with $w = 0.4$		
		Macro precision	Macro recall	Macro F_1-score	Macro precision	Macro recall	Macro F_1-score
Majority	Eat	98.06	97.01	97.53	98.15	97.66	97.90
	Learn	97.82	98.16	97.99	98.26	98.33	98.30
	Loan	96.89	96.79	96.84	98.19	96.51	97.34
	Rent	94.42	94.80	94.61	92.34	97.35	94.78
	Shop	97.06	96.66	96.86	96.62	97.64	97.12
	Travel	98.46	97.67	98.06	98.40	98.33	98.37
	Peer-To-Peer	96.02	95.14	95.58	96.75	95.51	96.12
	Average	96.96	96.60	96.78	96.96	97.33	97.13
Minority	Invest	88.34	90.59	89.45	90.06	90.59	90.32
	Home	85.71	59.26	70.07	87.31	72.22	79.05
	Utility	90.66	85.61	88.07	90.31	88.63	89.46
	Donation	95.30	69.95	80.68	93.06	79.31	85.64
	Insurance	95.88	80.87	87.74	97.14	88.70	92.73
	Average	91.18	77.26	83.20	91.58	83.89	87.44

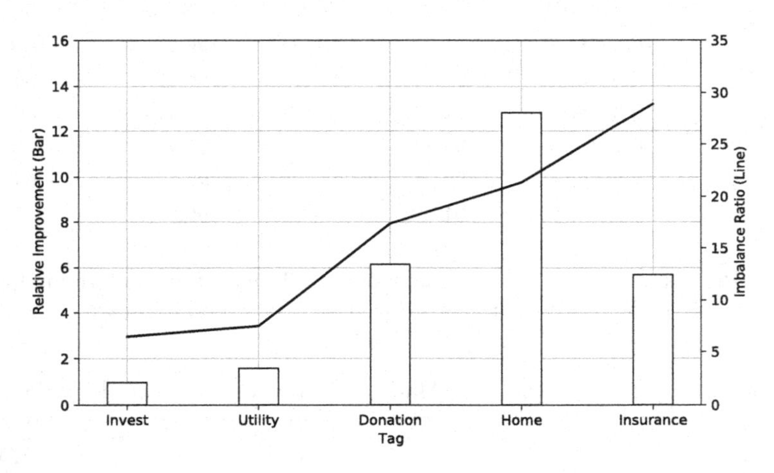

Fig. 3. Relative improvement of F_1-score of the model with $CE + D_{Dice}$ with $w = 0.4$ *vs* the model with CE in minority classes classification

ratio increased. To sum up, the $CE + D_{Dice}$ model was able to classify each minority class more accurately than CE, resulting in a more generalised model.

6 Conclusion

We described a model that analysed text messages, which were annotations for financial transactions, to discover the user intents: this enabled the bank to offer a relevant products to the right customer at the right time. Our hybrid model,

BLSTM-2DCNN, was combined with a hybrid loss function to solve the real-world multiple label imbalance problem found in the data we examined. Our model with either similarity coefficient loss functions or hybrid loss functions enabled the model to achieve higher F_1-scores, than the model with a cross-entropy loss function. The relative improvement in F_1-score of our model with hybrid loss function against the conventional model is 4.5%. Our model performance was a dramatic improvement on the conventional AI^2 model.

In future work, we will apply cost-sensitive learning, in combination with our losses to make the model more focused on the minority class. To find an appropriate values of the input parameters, including weights of each loss and the cost of each class, where it is hard to assign a suitable cost for each class, we will use an evolutionary algorithm to search for the best parameters.

References

1. Bojanowski, P., Grave, E., Joulin, A., Mikolov, T.: Enriching word vectors with subword information. Trans. Assoc. Comput. Linguist. **5**, 135–146 (2017). https://doi.org/10.1162/tacl_a_00051
2. Buda, M., Maki, A., Mazurowski, M.A.: A systematic study of the class imbalance problem in convolutional neural networks. Neural Netw. **106**, 249–259 (2018). https://doi.org/10.1016/j.neunet.2018.07.011
3. Chumwatana, T.: Using sentiment analysis technique for analyzing Thai customer satisfaction from social media. In: Proceedings of the 5th International Conference on Computing and Informatics (ICOCI 2015), 11–13 August 2015, Istanbul, Turkey (2015)
4. Crum, W.R., Camara, O., Hill, D.L.: Generalized overlap measures for evaluation and validation in medical image analysis. IEEE Trans. Med. Imaging **25**(11), 1451–1461 (2006). https://doi.org/10.1109/TMI.2006.880587
5. Dice, L.R.: Measures of the amount of ecologic association between species. Ecology **26**(3), 297–302 (1945)
6. Elman, J.L.: Finding structure in time. Cognitive Sci. **14**(2), 179–211 (1990). https://doi.org/10.1207/s15516709cog1402_1
7. Grave, E., Bojanowski, P., Gupta, P., Joulin, A., Mikolov, T.: Learning word vectors for 157 languages. In: Proceedings of the International Conference on Language Resources and Evaluation (LREC 2018), Miyazaki, Japan (2018)
8. Haixiang, G., Yijing, L., Shang, J., Mingyun, G., Yuanyue, H., Bing, G.: Learning from class-imbalanced data: review of methods and applications. Expert Syst. Appl. **73**, 220–239 (2017). https://doi.org/10.1016/j.eswa.2016.12.035
9. Hartmann, J., Huppertz, J., Schamp, C., Heitmann, M.: Comparing automated text classification methods. Int. J. Res. Mark. **36**(1), 20–38 (2019). https://doi.org/10.1016/j.ijresmar.2018.09.009
10. He, H., Garcia, E.A.: Learning from imbalanced data. IEEE Trans. Knowl. Data Eng. **21**(9), 1263–1284 (2009). https://doi.org/10.1109/TKDE.2008.239
11. Hochreiter, S., Schmidhuber, J.: Long short-term memory. Neural Comput. **9**(8), 1735–1780 (1997). https://doi.org/10.1162/neco.1997.9.8.1735
12. Hospedales, T.M., Gong, S., Xiang, T.: Finding rare classes: active learning with generative and discriminative models. IEEE Trans. Knowl. Data Eng. **25**(2), 374–386 (2011). https://doi.org/10.1109/TKDE.2011.231

13. Howard, J., Ruder, S.: Universal language model fine-tuning for text classification. arXiv preprint arXiv:1801.06146 (2018)
14. Jaccard, P.: Distribution de la flore alpine dans le bassin des dranses et dans quelques régions voisines. Bulletin de la Société Vaudoise des Sciences Naturelles **37**, 241–272 (1901)
15. Johnson, J.M., Khoshgoftaar, T.M.: Survey on deep learning with class imbalance. J. Big Data **6**(1), 1–54 (2019). https://doi.org/10.1186/s40537-019-0192-5
16. Joulin, A., Grave, E., Bojanowski, P., Douze, M., Jégou, H., Mikolov, T.: Fast-text. zip: Compressing text classification models. arXiv preprint arXiv:1612.03651 (2016)
17. Kim, Y.: Convolutional neural networks for sentence classification. arXiv preprint arXiv:1408.5882 (2014)
18. Kingma, D.P., Ba, J.: Adam: a method for stochastic optimization. arXiv preprint arXiv:1412.6980 (2014)
19. Kittinaradorn, R., et al.: Deepcut: a Thai word tokenization library using deep neural network (2019). https://doi.org/10.5281/zenodo.3457707
20. Kudisthalert, W., Pasupa, K.: A coefficient comparison of weighted similarity extreme learning machine for drug screening. In: Proceedings of 2016 8th International Conference on Knowledge and Smart Technology (KST), Chiangmai, Thailand, pp. 43–48 (2016). https://doi.org/10.1109/KST.2016.7440525
21. Lee, K., Palsetia, D., Narayanan, R., Patwary, M.M.A., Agrawal, A., Choudhary, A.: Twitter trending topic classification. In: Proceedings of the 11th International Conference on Data Mining Workshops (ICDM 2011), Vancouver, BC, Canada, pp. 251–258 (2011). https://doi.org/10.1109/ICDMW.2011.171
22. Li, D., Qian, J.: Text sentiment analysis based on long short-term memory. In: Proceedings of the International Conference on Computer Communication and the Internet (ICCCI 2016), Wuhan, China, pp. 471–475 (2016). https://doi.org/10.1109/CCI.2016.7778967
23. Li, Y., Guo, H., Zhang, Q., Gu, M., Yang, J.: Imbalanced text sentiment classification using universal and domain-specific knowledge. Knowl. Based Syst. **160**, 1–15 (2018). https://doi.org/10.1016/j.knosys.2018.06.019
24. Lin, T., Goyal, P., Girshick, R., He, K., Dollár, P.: Focal loss for dense object detection. IEEE Trans. Pattern Anal. Mach. Intell. **42**(2), 318–327 (2020)
25. Milletari, F., Navab, N., Ahmadi, S.A.: V-net: fully convolutional neural networks for volumetric medical image segmentation. In: Proceedings of the 4th International Conference on 3D Vision (3DV 2016), Stanford, CA, USA, pp. 565–571 (2016). https://doi.org/10.1109/3DV.2016.79
26. Nam, J., Kim, J., Mencía, E.L., Gurevych, I., Fürnkranz, J.: Large-scale multi-label text classification-revisiting neural networks. In: Proceedings of Joint European Conference on Machine Learning and Knowledge Discovery in Databases (ECML-PKDD 2014), Berlin, Heidelberg, pp. 437–452 (2014). https://doi.org/10.1007/978-3-662-44851-9_28
27. Pasupa, K., Kudisthalert, W.: Virtual screening by a new clustering-based weighted similarity extreme learning machine approach. PLoS ONE **13**(4), e0195478 (2018). https://doi.org/10.1371/journal.pone.0195478
28. Pasupa, K., Ayutthaya, T.S.N.: Thai sentiment analysis with deep learning techniques: a comparative study based on word embedding, POS-tag, and sentic features. Sustain. Cities Soc. **50**, 101615 (2019). https://doi.org/10.1016/j.scs.2019.101615

29. Pasupa, K., Vatathanavaro, S., Tungjitnob, S.: Convolutional neural networks based focal loss for class imbalance problem: a case study of canine redblood cells morphology classification. J. Ambient Intell. Humanized Comput. (2020). https://doi.org/10.1007/s12652-020-01773-x
30. Pratama, B.Y., Sarno, R.: Personality classification based on Twitter text using Naive Bayes, KNN and SVM. In: Proceedings of 2015 International Conference on Data and Software Engineering (ICoDSE 2015), Yogyakarta, Indonesia, pp. 170–174 (2015). https://doi.org/10.1109/ICODSE.2015.7436992
31. Ren, S., He, K., Girshick, R., Sun, J.: Faster R-CNN: towards real-time object detection with region proposal networks. In: IEEE Transactions on Pattern Analysis and Machine Intelligence, vol. 39 (06 2015). https://doi.org/10.1109/TPAMI.2016.2577031
32. Sarker, A., et al.: Data and systems for medication-related text classification and concept normalization from Twitter: insights from the social media mining for health (smm4h)-2017 shared task. J. Am. Med. Inform. Assoc. **25**(10), 1274–1283 (2018). https://doi.org/10.1093/jamia/ocy114
33. Ayutthaya, T.S.N., Pasupa, K.: Thai sentiment analysis via bidirectional LSTM-CNN model with embedding vectors and sentic features. In: Proceedings of 2018 International Joint Symposium on Artificial Intelligence and Natural Language Processing (iSAI-NLP), Pattaya, Thailand, pp. 1–6 (2018). https://doi.org/10.1109/iSAI-NLP.2018.8692836
34. Sirirattanajakarin, S., Suntisrivaraporn, B.: Annotation intent identification toward enhancement of marketing campaign performance. In: Proceedings of the 11th International Conference on Knowledge and Systems Engineering (KSE 2019), Da Nang, Vietnam, pp. 1–5 (2019). https://doi.org/10.1109/KSE.2019.8919386
35. Srinilta, C., Sunhem, W., Tungjitnob, S., Thasanthiah, S., Vatathanavaro, S., et al.: Lyric-based sentiment polarity classification of Thai songs. In: Proceedings of the International Multi Conference of Engineers and Computer Scientists (IMECS 2017), Hong Kong (2017)
36. The Siam Commercial Bank: SCB EASY Changes Digital-platform Scenes Once Again, Makes Tomorrow's Services Happen Today, Highlights Moment Banking to Fulfil the Needs in Each of Your Moments. https://www.scb.co.th/en/about-us/news/oct-2018/nws-scbeasy-moment-banking.html (2018). Accessed Feb 2020
37. Vincent, G., Guillard, G., Bowes, M.: Fully automatic segmentation of the prostate using active appearance models. MICCAI Grand Challenge Prostate MR Image Segmentation **2012**, 2 (2012)
38. Wang, J.H., Liu, T.W., Luo, X., Wang, L.: An LSTM approach to short text sentiment classification with word embeddings. In: Proceedings of the 30th Conference on Computational Linguistics and Speech Processing (ROCLING 2018), Hsinchu, Taiwan, pp. 214–223 (2018)
39. Wikipedia Developers: Wikipedia. https://www.wikipedia.org/
40. Xu, G., Meng, Y., Qiu, X., Yu, Z., Wu, X.: Sentiment analysis of comment texts based on BiLSTM. IEEE Access **7**, 51522–51532 (2019). https://doi.org/10.1109/ACCESS.2019.2909919
41. Yan, Y., Wang, Y., Gao, W.-C., Zhang, B.-W., Yang, C., Yin, X.-C.: LSTM^2: multi-label ranking for document classification. Neural Proc. Lett. **47**(1), 117–138 (2017). https://doi.org/10.1007/s11063-017-9636-0
42. Zhang, C., Tan, K.C., Li, H., Hong, G.S.: A cost-sensitive deep belief network for imbalanced classification. IEEE Trans. Neural Netw. Learn. Syst. **30**(1), 109–122 (2018). https://doi.org/10.1109/TNNLS.2018.2832648

43. Zhang, M.L., Zhou, Z.H.: Multilabel neural networks with applications to functional genomics and text categorization. IEEE Trans. Knowl. Data Eng. **18**(10), 1338–1351 (2006). https://doi.org/10.1109/TKDE.2006.162
44. Zhou, C., Sun, C., Liu, Z., Lau, F.: A c-lstm neural network for text classification. arXiv preprint arXiv:1511.08630 (2015)
45. Zhou, P., Qi, Z., Zheng, S., Xu, J., Bao, H., Xu, B.: Text classification improved by integrating bidirectional lstm with two-dimensional max pooling. arXiv preprint arXiv:1611.06639 (2016)

Measuring Immigrants Adoption of Natives Shopping Consumption with Machine Learning

Riccardo Guidotti[1(✉)], Mirco Nanni[2], Fosca Giannotti[2], Dino Pedreschi[1], Simone Bertoli[3], Biagio Speciale[4], and Hillel Rapoport[4]

[1] University of Pisa, Pisa, Italy
{riccardo.guidotti,dino.pedreschi}@unipi.it
[2] ISTI-CNR, Pisa, Italy
{mirco.nanni,fosca.giannotti}@isti.cnr.it
[3] Université Clermont Auvergne, CNRS, CERDI, Clermont-Ferrand, France
simone.bertoli@uca.fr
[4] Paris School of Economics, Paris, France
{biagio.speciale,hillel.rapoport}@univ-paris1.fr

Abstract. "Tell me what you eat and I will tell you what you are". Jean Anthelme Brillat-Savarin was among the firsts to recognize the relationship between identity and food consumption. Food adoption choices are much less exposed to external judgment and social pressure than other individual behaviours, and can be observed over a long period. That makes them an interesting basis for, among other applications, studying the integration of immigrants from a food consumption viewpoint. Indeed, in this work we analyze immigrants' food consumption from shopping retail data for understanding if and how it converges towards those of natives. As core contribution of our proposal, we define a score of adoption of natives' consumption habits by an individual as the probability of being recognized as a native from a machine learning classifier, thus adopting a completely data-driven approach. We measure the immigrant's adoption of natives' consumption behavior over a long time, and we identify different trends. A case study on real data of a large nation-wide supermarket chain reveals that we can distinguish five main different groups of immigrants depending on their trends of native consumption adoption.

Keywords: Immigrants shopping consumption · Human migration analysis · Machine-learning-based measure · Adoption trends · Integration

1 Introduction

Moving across borders exposes people to the norms that are adopted by the natives in their countries of destination. As time passes by, immigrants might

Y. Dong et al. (Eds.): ECML PKDD 2020, LNAI 12461, pp. 369–385, 2021.
https://doi.org/10.1007/978-3-030-67670-4_23

progressively adopt these norms [18,37]. Adoption is a dynamic process that requires time. A crucial choice that all immigrants must make is whether to adopt the habits of their host society [11]. Examples of this choice are decisions concerning ethnic-sounding vs. native-sounding names [1,20], whether to change the surname [6], the language spoken [4], and whether to marry with people of the same ethnic group [34]. All these measures reflect choices that are easily observed by one's peers and, thus, potentially exposed to social sanctions and not fully reflecting one's own preferences. In addition, these measures are usually observed at one point in time, while integration is an inherently dynamic phenomenon.

Jean Anthelme Brillat-Savarin, a lawyer, politician, and famous gastronome was among the firsts to recognize the relationship between identity and food consumption. In his book [12], he wrote his well-known aphorism: "Tell me what you eat and I will tell you what you are". Following this intuition, in this paper we rely on immigrants' detailed food consumption choices from shopping retail data, over a long time after immigration. We highlight that food adoption choices are not readily observed or inferred (unless the retail stores under study are very biased, e.g. very cheap or specialized in ethnic food, which is not the case of the data used in our study), and hence less exposed to peer pressure [3]. Moreover, a measure of adoption based on food shopping consumption has a dynamic nature and can be observed several times in the host country.

We use information from retail data and from the country of birth of the customers to develop a measure evaluating whether immigrants adopt consumption habits similar to those of natives. We name this measure Native Consumption Adoption (NCA). Given a temporal granularity, we model the shopping behavior of a customer as a vector containing information about shopping habits with respect to different food categories. Then, we exploit the knowledge of the country of birth, and we build a machine learning classifier able to distinguish between natives and immigrants. In this work, instead of measuring the compliance of customers to a pre-defined model of *native behaviour*, we propose to compute the NCA as the probability of being a native that the classifier learnt from training data assigns to each customer. We adopted classification methods instead of simpler solutions based on distances between vector representations of customers' purchases because preliminary experiments showed that the latter are unable to separate natives and immigrants in reasonable ways. Finally, we estimate the NCA for the immigrants over time, and we identify different trends of adoptions of the native' shopping consumption. We name TINCA the proposed methodology aimed at discovering Trends of Immigrants' Native Consumption Adoption.

We investigate whether immigrants' consumption in terms of NCA converges towards those of natives on a case study over real data describing the purchases of the customers (immigrants and natives) of a large Italian supermarket chain between 2008 and 2015. Experimental results reveal five different groups of immigrants depending on their trends of NCA, namely increasing and stable native-refusers, strong and weak native-adopters, and native-like customers. The proposed methodology and the results that can be derived from its application can be of interest to multiple subjects. For instance, social scientists analyzing the process of integration in the host society, or collaborating with institutions and organizations responsible for the integration of immigrants. In addition, the customer management of retail sales companies can benefit from the proposed methodology and the analysis that can be done on top of the outcomes returned.

The rest of the paper is structured as follows. Section 2 summarizes related work on migration. In Sect. 3, we formalize the problem, while in Sect. 4 we define the methodology proposed to solve it. Section 5 presents experiments in the form of a case study in which we employ the proposed methodology. Finally, Sect. 6 concludes the paper and illustrates future research directions.

2 Related Work

Human migration has been a constant of human history, from the earliest ages until now. The study of migration spans various research fields, including anthropology, sociology, economics, statistics, and, more recently, physics and computer science. In this paper, we focus on studying the *"stay"* phase of migration [36], i.e., immigrant integration, and changes in life and habits.

The complexity of the immigrants' choice, whether to assimilate or not in terms of food consumption, has been widely analyzed in economics. In [35] it is studied how Bengali Indian households in the US converged towards the host country's norm for their breakfast eaten at home. Economists recently started to consider the analysis of immigrants' consumption as a subject ripe for empirical investigation [10] and recognized the importance of understanding immigrants' consumption behavior to assess the labor market consequences of immigration [19]. The differences in food preferences across social groups are analyzed in [7], and it is shown that internal migrants in India bring their origin-state food preferences with them during migration. In [31], the authors find that immigrants' expenditure shares for different types of food in the nineteenth-century are predicted by past relative prices in their countries of origin. Similarly, [13] finds that the current purchases of consumers who migrate across states in the US depend on both where they live currently, and where they lived in the past. Our work is also related to the economics literature on the heterogeneity in cultural traits. In [5] it is shown that in the last decades economic convergence across European countries was not accompanied by cultural convergence. The authors of [9] rely on machine learning to measure cultural distance in terms of how predictable group membership is from media consumption, consumer behavior, time use and social attitudes. They show that cultural distances in the US have remained broadly constant over time, with few exceptions.

As previously discussed, migrant integration is generally measured through indicators related to the labor market and economic status. These statistics are available with low resolution and not for all countries. A new direction is that of observing integration through big data analysis. For instance, the analysis of online social networks can allow evaluating the level of adoption of a culture. There is a vast literature regarding immigration and social networks, especially Twitter. Most of the studies start from the language in which a post is written [30]. In [29] it is proposed an approach to detect English linguistic variation and quantify its significance among geographic regions, while in [33] geolocated tweets are exploited to analyze the language diversity over different countries. Also, the topological graph-composition of social networks is analyzed for studying migration. In [27], community-centric metrics are used to study cultural assimilation as a function of the number of social ties between migrant communities and natives using the set of friendship links extracted from Facebook. In [30], a bipartite graph structure, connecting tweet languages and cities, is used for studying again cultural assimilation but from the spatial segregation point of view. Finally, call data records (CDR) from the D4R challenge[1] are exploited in [8] to observe that integration seems to increase in time for refugees, and also that the presence of refugees influences the house market in Turkey, decreasing housing prices. The discussion above is not intended to be a complete review of data-driven analytical methods for studying immigrants integration. For a more comprehensive review, refer to [14,36].

In addition, to complement the perspectives of works in economics, sociology and computer science, our paper is, to the best of our knowledge, the first data-driven analytical method not using twitter or CDR for estimating immigrants integration in terms of the adoption of native consumption trends. Furthermore, it is also the first work not using retail data with the final purpose of customer profiling [15,21,24], customer segmentation [22,23,39], or pattern discovery [2,26]. We stress that, like in [17,32], the analytical process proposed in this paper takes into account the evolution of a customer and the changes in her behavior. The work in [17] exploits behavioral and demographic variables, and a transaction database for designing measures of similarity and unexpectedness for mining change patterns to analyze the degree of resemblance at different time periods. In [32] it is proposed a customer segmentation model that allows to track the evolution of a customer, including the splitting and merging of customer groups and allowing to observe how groups evolve and how individuals shift across groups. Similar aspects are addressed in the proposed approach.

3 Problem Formulation

In this section we define the context and the problem we want to solve.

Let $P = \{p_1 \dots p_q\}$ be a set of q products, we define a basket b (or transaction) as a subset of products such that $\emptyset \subset b \subseteq P$. Given a customer i, we name $B_i^{(t)} = \langle b_1 \dots b_n \rangle$ the set of temporally ordered n baskets purchased by i in the time interval t. t can represent a week, a month, etc., and therefore the baskets in $B_i^{(t)}$ correspond to those purchased in a certain week, month, etc., respectively. With $D(b)$, we refer to the day in which the basket b is purchased. We highlight that $\forall b \in B_i^{(t)}$, $D(b)$ lies within the time interval t. Given a product p, the function $E(p)$ returns the expenditure[2] required to purchase p.

[1] Data for refugees of Turkey http://d4r.turktelekom.com.tr/.

[2] We assume that the expenditure function E also accounts for the quantity.

Given a finite set of countries $\mathcal{C} = \{China, France, Italy, \dots\}$, we define C as the function that returns the country of birth of a customer i, i.e., $C(i) \in \mathcal{C}$. Given a reference country $r \in \mathcal{C}$, and a customer i, we can distinguish whether i is a *native* customer, i.e., $C(i) = r$, or i is a *foreign* customer, i.e., $C(i) \neq r$.

Let $\mathcal{B}^{(t)} = \{B_1^{(t)}, \dots, B_N^{(t)}\}$ be the set of baskets purchased by N different customers at time interval t. Given the set of sets of baskets $\{\mathcal{B}^{(t_1)}, \dots, \mathcal{B}^{(t_k)}\}$ purchased between time intervals t_1 and t_k, and a reference country r, our goal is to measure the degree with which foreign customers adopt a shopping behavior similar to that one of native customers along time.

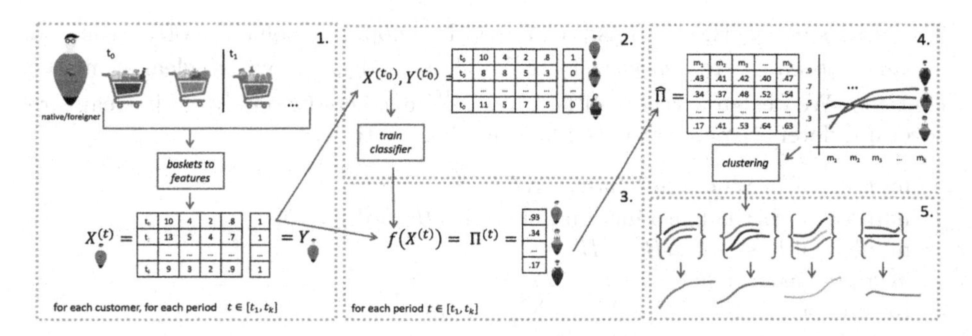

Fig. 1. TINCA analytic framework workflow. *1.* The baskets of each customer are turned into features describing the customer shopping behavior in various time periods. *2.* The feature matrix X describing the customers and their country of birth Y are used to train a machine learning classifier f. *3.* The classifier f is used to estimate the level of NCA for the various time periods $\Pi^{(t)} = f(X^{(t)}) \forall t \in [t_1, t_k]$. *4.* The NCA values of foreign customers $\hat{\Pi}$ is arranged in a matrix with respect to the time at which the customer started to purchase and a clustering algorithm is used to extract groups of customers with similar trends of NCA. *5.* From each group is extracted a representative trend of adoption/rejection of the shopping behavior of natives.

4 Trends of Immigrants Native Consumption Adoption

In this section, we describe our proposal for measuring immigrants' level of adoption of native shopping consumption over time. We name it TINCA as its aim is to discover Trends of Immigrants' Native Consumption Adoption. We indicate the Native Consumption Adoption with Π. Given the set of sets $\{\mathcal{B}^{(t_1)}, \dots, \mathcal{B}^{(t_k)}\}$, a reference country r, a prediction function f, TINCA consists of the following 5 steps, also illustrated in Fig. 1:

1. **Models** in the matrix $X^{(t)}$ the customers' shopping behavior for each time interval $t \in \{t_1, \dots, t_k\}$, such that every row $X_i^{(t)}$ is a vector of features describing the purchases of customer i;
2. **Learns** the machine learning classifier prediction function f on the training dataset $\langle X^{(t_0)}, Y^{(t_0)} \rangle$ where $Y^{(t_0)}$ indicates if the customer i is native or not.

3. **Measures** the NCA as $\Pi^{(t)} = f(X^{(t)})$, i.e., as the probability of being classified *native* by the classifier f for each time interval t between t_1 and t_k;
4. **Groups** immigrants with respect to the fluctuations of their NCA in Π over time periods, obtaining clusters of non natives with a similar NCA evolution;
5. **Extracts** from each group a representative trend of adoption/rejection.

In the following we provide the details of each step and explain the notation adopted in the above pseudo-code.

4.1 Modeling Customer Shopping Behavior

The first step of TINCA aims to model the shopping behavior of customer i at time interval t, represented by $B_i^{(t)}$ (shortened as B, where clear from the context). We design a vector of features $X_i^{(t)}$ extracted from $B_i^{(t)}$, that capture several different aspects of individual purchase activity:

- number of products purchased: $NP = \sum_{b \in B} |b|$
- number of distinct products purchased: $DP = \left|\bigcup_{b \in B} b\right|$
- number of baskets: $NB = |B|$
- average basket length[3]: $AL = \frac{NP}{NB}$
- total expenditure: $TE = \sum_{b \in B} \sum_{p \in b} E(p)$
- average expenditure: $AE = \frac{TE}{NB}$
- average period between purchases: $AP = \frac{D(b_{NB}) - D(b_1)}{NB - 1}$
- number of purchases of product p: $NB_p = \sum_{b \in B} \mathbb{1}_{p \in b}$
- total expenditure of product p: $TE_p = \sum_{b \in B} \mathbb{1}_{p \in b} E(p)$
- average period of product p: $AP_p = \frac{\max_{b \in B|_p} D(b) - \min_{b \in B|_p} D(b)}{(NB_p - 1)}$

where the operator $\mathbb{1}_{cond}$ returns one if the condition *cond* is verified, zero otherwise; and the operator $|_p$ applied to B returns the subset of baskets containing product p, i.e., $B|_p = \{b \in B \mid p \in b\}$. The features adopted are commonly used in various work in the literature analyzing transactional data for recommendation, classification or analysis purposes [16,17]. The first seven features capture *general* shopping behavior. Besides the total and average indicators of quantities NB, DP, NP, TE, AL and AE, we highlight how AP captures the average *frequency* of the period within which a customer makes a purchase. This is an important temporal indicator that can vary a lot even for customers having similar shopping habits in terms of items purchased. The other features, namely NB_p, TE_p, AP_p, capture the *specific* shopping behavior for each one of the various products $p \in P$. More precisely, since working with single products might lead to very sparse data, we suggest to group them into product categories collecting items of similar type, e.g., bread, pasta, tomatoes, milk, etc.

[3] We consider also features derived from others, like AL and AE, since they might capture different aspects of the customer shopping behavior. Where needed, redundant features can be removed at the preprocessing stage preceding the training phase of the machine learning classifier.

This abstraction has also the effect of reducing possible effects of the market dynamics, where a product might be easily replaced by others in a short time. Details on product categories adopted and data dimensionality for our case study are provided in Sect. 5. An important aspect to remark is that all the features we are adopting have a clear meaning and are easily interpretable, therefore they can be exploited for further analyses.

Given $\mathcal{B}^{(t)}$, we name $X^{(t)}$ the feature matrix modeling in each row $X_i^{(t)}$ the shopping behavior of a customer among those in $\mathcal{B}^{(t)}$, where $X_i^{(t)} = \langle NP, DP, NB, AL, TE, AE, AP, NB_1, TE_1, AP_1, \ldots, NB_n, TE_n, AP_n \rangle$ for $n = |P|$. The features in $X_i^{(t)}$ describe the food consumption of customer i at time t.

4.2 Learning and Measuring Native Consumption Adoption (NCA)

The second and third steps of TINCA consist in training a machine learning classifier f for estimating the degree of adoption of natives' shopping habits by foreign customers in a specific time interval t. To this aim, our definition of the Native Consumption Adoption (NCA) score starts from the observation that customers with shopping behaviours very different from natives' will be recognized very easily as non-natives by a classifier, with a high confidence; on the opposite, the more similar to natives is the purchase behaviour, the lower will be the confidence, till the point where the customer will be recognized as native, again with confidence dependent on the closeness to native behaviours. Based on these observations, in this work we compute NCA as the probability of being recognized as a native returned by the classifier [38]. This methodology advances state-of-the-art since, to the best of our knowledge, it is the first attempt to adopt machine learning classification to implicitly build a completely data-driven model of what it means to purchase like a native, thus not depending on preconceived hypotheses or handcrafted rules.

The simplest way of implementing the principles introduced above would consist in comparing the purchase features of a customer with those of natives, for example by applying a basic k-NN classification with an Euclidean distance, working either on the raw training set or on prototypes extracted beforehand. However, preliminary experiments showed that there is not a clear distinction between the overall features distribution of natives and foreign customers (see Sect. 5.1) and that solutions based on simple combinations of features fail (e.g. linear regression, shown in Sect. 5.2), thus calling for more complex analyses of the features and requiring more sophisticated machine learning classifiers.

More in detail, let $f : X \rightarrow [0, 1]$ be the prediction function of a machine learning classifier, that takes as input the features modeling the customer shopping behavior, and returns the probability that this customer is recognized as a native. Thus, given the model of shopping behavior X_i of customer i, we indicate with $\Pi_i = f(X_i)$ her NCA, and i is recognized as a foreign when $\Pi_i \approx 0$, while i is recognized as a native when $\Pi_i \approx 1$. We obtain the prediction function f from the learning function $l : X \times Y \rightarrow f$ of a machine learning classifier. The learning function l takes as input a set of N models of shopping behavior, i.e., the feature matrix X describing the customers' shopping behaviors, and a vector Y specifying if each customer is native or not with respect to a reference country r, i.e., $Y = \langle \mathbb{1}_{C(i)=r} \rangle_{i=1,\ldots,N}$; and returns the prediction function f. Coherently with the objective of our study, the prediction function is learnt from a dataset $\langle X^{(t_0)}, Y^{(t_0)} \rangle$ containing, on one hand, vectors of features $X_i^{(t_0)}$ modeling the initial purchases of immigrants (i.e. features related to their early purchase history, which is less likely to be affected by their possible integration); on the other hand, vectors of features $X_i^{(t_0)}$ modeling the shopping behavior of natives[4].

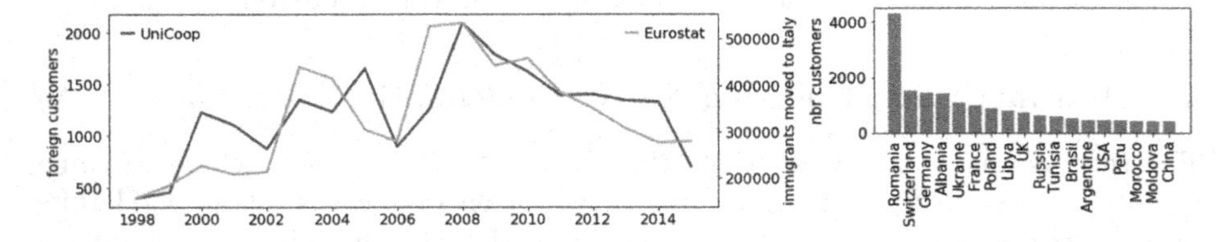

Fig. 2. *(Left)*: Eurostat trend of immigrants moved to Italy vs trend of foreign customers of UniCoop. *(Right)*: 18 most represented countries in the dataset.

Table 1. Number of baskets, total expenditure, average basket length, average expenditure, and average frequency means and standard deviations for natives and immigrants.

	Nbr baskets	Tot exp	Avg Basket len	Avg Exp	Avg freq
natives	3.95 ± 0.13	180.63 ± 13.67	8.21 ± 0.30	40.55 ± 2.42	3.29 ± 0.09
immigrants	4.39 ± 0.16	175.73 ± 12.59	7.26 ± 0.22	33.95 ± 1.93	3.28 ± 0.13

4.3 Grouping and Monitoring NCA Trends of Foreign Customers

The objective of steps four and five is to monitor how the NCA values of foreign customers (who are the focus of our study) evolve in time. We implement these steps by exploiting a centroid-based clustering [38] that simultaneously *groups* immigrants with respect to their NCA trends and computes *representative trends* as centroids. Each customer i might start her purchase history at a time $S(i)$ that differs from other customers. In particular, while some of them start with our observation period (i.e., $S(i) = t_1$), others might start later (i.e., $t_1 < S(i) \leq t_k$). In order to take that into consideration, and to perform unbiased comparisons of the trends, before clustering the NCA trends we align them with respect to $S(i)$. More formally, let $\Pi^{(t_1)}, \ldots, \Pi^{(t_k)}$ be the NCA for all the customers for

[4] Notice that we are implicitly assuming that the food consumption habits of natives do not change over time. While not true in general, we empirically observed that it holds for the vast majority of customers in our data. Studying natives' evolution in time is part of our future works.

each time interval t between t_1 and t_k. We create a matrix $\hat{\Pi}$ where each row $\hat{\Pi}_i$ corresponds to a foreign customer i, i.e., $C(i) \neq r$, and each column j corresponds to the j-th time period of the customer starting from her $S(i)$, i.e., $t = S(i)+j-1$. The trend analysis focuses on the first m time periods of $\hat{\Pi}$, where m is a parameter. The missing values in the NCA sequences (either because the customer has less than m time periods, or because of gaps) are replaced with the most recent available NCA value. Thus, through the matrix $\hat{\Pi}$ we can compare the NCA trends of foreign customers. Indeed, $\hat{\Pi}$ is used as input for the clustering algorithm and for extracting the representative trends for the various clusters. In particular, the representative trends are cleaned and studied exploiting methods from time series analysis [28].

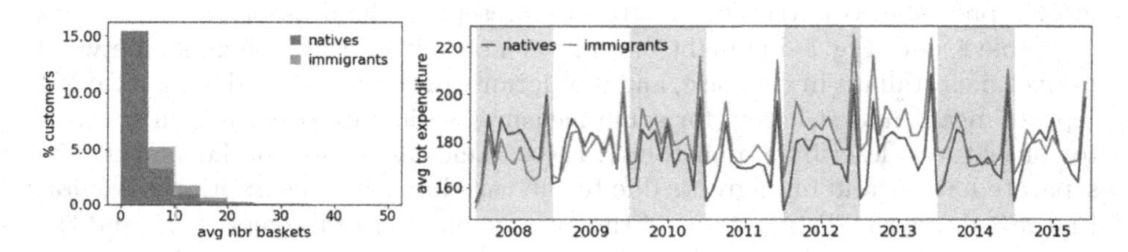

Fig. 3. *(Left)*: distributions of the average number of baskets for natives and immigrants. *(Right)*: trends of average total expenditure for natives and immigrants.

5 Experiments

In order to use TINCA[5] to study if and how immigrants' consumption converges towards those of natives, we exploited the purchases of food products made by a high number of documented immigrants (and natives) in *UniCoop Tirreno*[6], a large Italian supermarket chain. Thus, we set *Italy* as reference country r. Customers are provided with a loyalty card which allows linking different shopping sessions. We consider only customers labeled as *resident* according to [22], i.e., the customers who stably perform a minimum number of purchases over the years. In particular, we analyzed about 30 millions of transactions made by about 160k customers in 128 different shops, over the years 2008–2015. In our experiments we cover a month with each time interval t, while we use a set of products P with $|P| = 100$ distinct products where each product refer to a certain group of semantically similar items, i.e., "milk" identifies all the different types of milk[7]. Thus, we have a dimensionality of 307 for the feature matrix X

[5] The source code of TINCA is available here: https://github.com/riccotti/TINCA.
[6] https://www.unicooptirreno.it/, data: https://sobigdata.d4science.org/.
[7] The 100 product groups are available in the shared repository. The grouping was performed manually to respect the implicit semantic meaning. Each product models on average 1.9 ± 2.0 categories of items of the UniCoop dataset. The largest product groups are those modeling "bread", "fish", and "vegetables".

describing the customers shopping behavior. This choice with time intervals of one month, and with the selected machine learning classifier avoids any issue related with data sparsity for the classification task and the other steps.

5.1 Data Analysis

Before discussing the results obtained with TINCA, we briefly present the *UniCoop* dataset and the reasons for using it to study immigrants' adoption of native consumption. Figure 2 *(left)* shows the trend of immigrants moved to Italy according to Eurostat[8], and the trend of memberships with UniCoop Tirreno of foreign-born customers. This high correlation[9] confirms the suitability of the *UniCoop* dataset for this kind of study. We report in Fig. 2 *(right)* the 18 most represented countries out of the 158 present in the dataset.

Table 1 and Fig. 3 highlight how it is not trivial to distinguish between natives, i.e., Italians in this case, and immigrants in the *UniCoop* dataset. Table 1 reports means and std. dev. for some measures which are contained in the vector of features describing a customer: these values are quite similar and do not separate natives and immigrants due to the variability in consumption behavior. Figure 3 shows the distributions of the average number of baskets *(left)*, and the trends of average total expenditure *(right)*. In both cases we observe that natives and immigrants have rather similar distributions, without a strong separation.

	F1-score	Recall	Precision
RF	.37 ± .02	.58 ± .10	.29 ± .05
DT	.35 ± .01	.55 ± .09	.26 ± .01
LR	.10 ± .02	.05 ± .01	.70 ± .03

Fig. 4. Average performance and its variability on the twelve training datasets for the class not native.

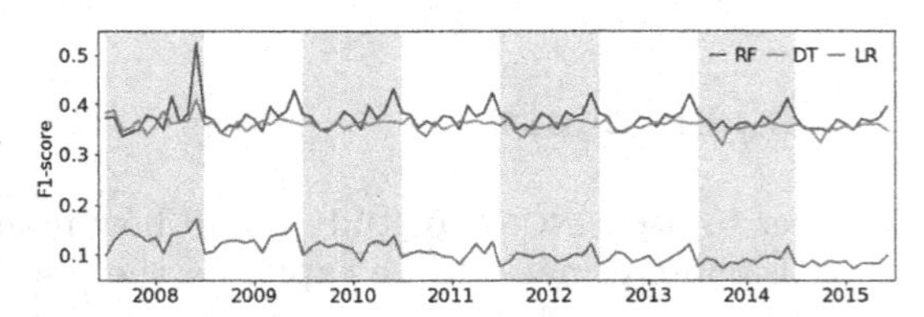

Fig. 5. F1-score along months when the classifiers are applied to estimate the values of NCA.

5.2 Machine Learning and Classification Performance Analysis

In this section, we provide details for the training of the machine learning classifiers, and we analyze their performance. We remark that our final objective is to obtain a model with sufficient discrimination power to identify and monitor in time native-looking vs. immigrant-looking behaviours. While, obviously, the

[8] Eurostat data: https://ec.europa.eu/eurostat/statistics-explained/index.php/Migration_and_migrant_population_statistics.

[9] Pearson of 0.75 and Spearman of 0.78 in both cases with p-value < 0.0005.

more accurate are the models, the better, our emphasis is on capturing the non-native class, which is fundamental for the analysis. As it will be shown, this is a difficult classification problem.

We account for seasonality [23] by training a machine learning classifier f for each month of the first year, i.e., $f^{(1)}, f^{(2)}, \ldots, f^{(12)}$ for January, February, etc., respectively. As the overall dataset is highly imbalanced, with only 3% of immigrants, we adopt an undersampling strategy to reach a better equilibrium and mitigate the well-known issues of classification with small minority classes. In particular, as learning datasets $X^{(1)}, X^{(2)}, \ldots, X^{(12)}$ we consider vectors of features selected as follows: *(i)* for all immigrants, consider the vectors modeling the first year of purchases; and *(ii)* take a random selection of vectors for natives. The result of the random undersampling is a set of training datasets (one per month) that contain 20% of immigrants and 80% of natives. Experiments are performed using a 10-fold cross validation approach for evaluation purposes and for parameters estimation over each month. Then, for each month the best classification models $f^{(1)}, f^{(2)}, \ldots, f^{(12)}$ are trained on the the entire learning datasets $X^{(1)}, X^{(2)}, \ldots, X^{(12)}$ of the first year. Finally, they are tested and employed to estimate the NCA on all the subsequent years.

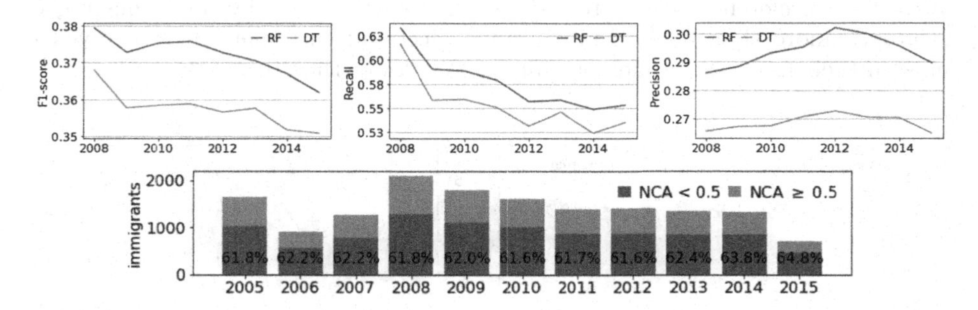

Fig. 6. *Top*: Average trends of performance for DTs and RFs along years. *Bottom*: Number of immigrants members from 2005 to 2015 divided by NCA. The percentages specify the portion of immigrants classified as *not natives*, i.e., with NCA < 0.5.

Among existing machine learning classifiers [38], we test Decision Trees (DT) and Random Forests (RF), because of their (partial) interpretability, since understanding the reasons behind the classification is one of our side goals [25]. We optimize the parameters selection search for maximizing the F1-score [38] with respect to the not native class $C(r) \neq r$. Also, we adopt Linear Regressors (LR) as a baseline. In Fig. 4, we report the average performance over the twelve months on the training data with respect to the class not native, i.e., immigrant. We observe that the baseline (LR) yields a relatively high precision, but a very poor recall, basically showing its inability in most cases to capture our class of interest, and also leading to a very low F1-score. DT and RF both largely outperform the baseline in terms of F1-score, with RF showing slightly better values than DT on all the measures. Hence, RF results to provide the best trade-off in performances, with a high recall (it captures 60% of non-native customers) and a lower yet acceptable precision. Further improving the performances is not the focus of this study, and will be the subject of future work.

Figure 5 confirms these intuitions showing the trends of F1-score across the various months when the classifiers $f^{(1)}, f^{(2)}, \ldots, f^{(12)}$ are applied to estimate the values of NCA. From Fig. 5 it is also clear that RFs, like in [9], overcome other classifiers and better fit our purposes. Moreover, we observe how, in certain months, it seems easier to distinguish natives from immigrants. Finally, Fig. 6 *(top)* shows the average F1-score, precision, and recall of DTs and RFs over the years. Besides remarking that RFs perform better than DTs, this supports the correctness of our intuition of monitoring changes in purchase habits. Indeed, we notice a drop in the F1-score due to a drop in the recall, while precision remains stable, with just a slight improvement. The decrease of the recall from 0.64 to 0.55 indicates that, as time passes, RFs decrease their power in recognizing immigrants based on the patterns learnt at the beginning of the observation period. The most natural explanation of this effect is that immigrants changed their shopping behavior, and started adopting consumptions closer to those of natives. We can discard the possibility that immigrants arriving in more recent years have preferences closer to natives than earlier ones, by observing Fig. 6 *(bottom)*: more than 60% of immigrants becoming customers in recent years have in their first year of membership an NCA lower than 0.5.

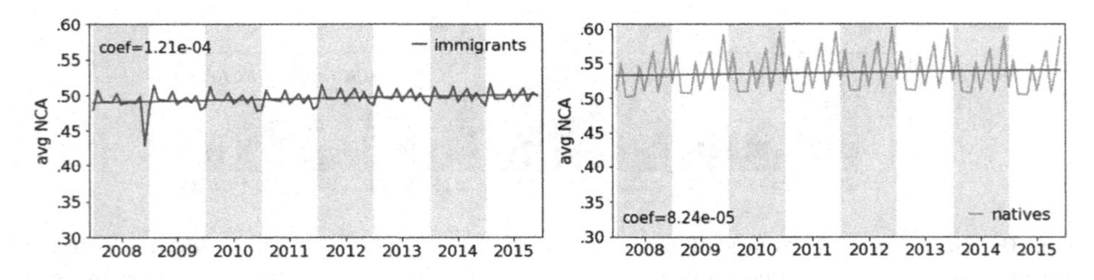

Fig. 7. Average NCA for immigrants *(left)* and natives *(right)*. In red are reported the linear regression trends and their coefficients.

5.3 Trends of Native Consumption Adoption Analysis

In this section, we analyze the trends of NCA observing their evolution across time. Before presenting the results obtained with TINCA relative to groups of immigrants having similar trends of NCA, we show that indeed NCA is different for immigrants and natives. In Fig. 7 we report the average NCA for immigrants (left) and natives (right). As expected, the NCA for immigrants is lower than 0.5, and for natives is higher than 0.5. We highlight that the limited difference between the NCA for immigrants and natives is due to the not easy task resolved by the machine learning classifier. Figure 7 also highlights in red the linear regressions (and their coefficients) showing that the average NCA of immigrants tends to 0.5, making immigrants more and more indistinguishable from natives.

We implement the centroid-based clustering of TINCA by exploiting the K-Means algorithm [38] to simultaneously group immigrants having similar NCA and to extract a representative trend of adoption/rejection. As already mentioned, we align trends of NCA of different customers with respect to the time $S(i)$ at which customer i started to perform purchases, that is, the number of *months since membership subscription*. We fill gaps in trends of NCA due to months without purchases by linear interpolation. As distance function, we rely on the Euclidean distance[10]. We apply K-Means only for immigrants, and we run K-means ranging the number of clusters from 2 to 150. We select 18 as the best number by observing the knee in the Sum of Squared Error (SSE) curve reported in Fig. 8 *(bottom right)*. Finally, in order to remove seasonality and

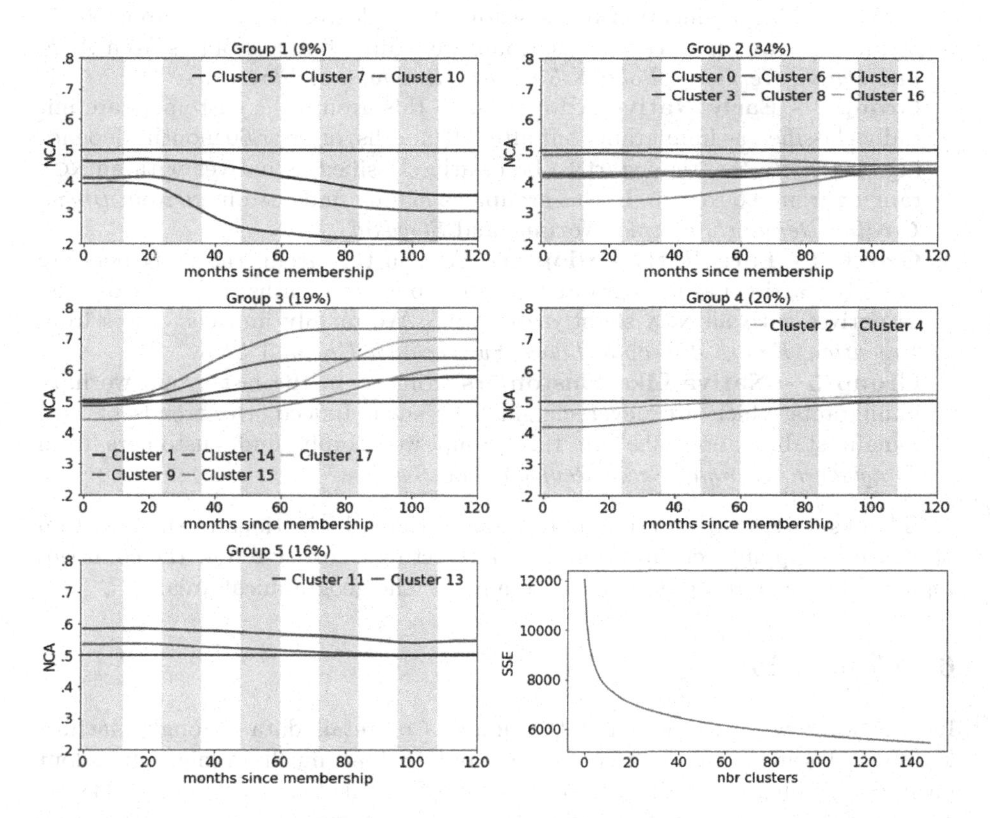

Fig. 8. Trends of NCA grouped according to the coefficient of the trend and its distance from the NCA classification threshold, i.e., 0.5. The size of each group is reported as percentage in the titles. The bottom right plot reports the Sum of Squared Error curve.

[10] We leave to future works the study of the effect of other specific functions for time series clustering like dynamic time warping.

noise, the representative NCA trend of each group is computed through time series decomposition [28] applied to the centroid of every cluster.

In the following, we analyze the 18 trends of NCA (Fig. 8) aggregated into five groups according to the trend's coefficient and its distance from the classification threshold, i.e., 0.5. We report each country within the group that captures the largest share of its customers, focusing on most representative ones[11].

- **Group 1 - Increasingly Native-Refuser.** In the first group, we find immigrants, classified as immigrants that maintain their "status" of immigrants and also increase their discrepancies against natives after 20/40 months. It is the smallest group detected with only 9% of all the immigrants. This group captures large portions of customers from *Iraq* and *Uzbekistan.*
- **Group 2 - Stable Native-Refuser.** In the second group, the largest one with 32% of immigrants, we observe customers classified as immigrants that maintain this classification but assessing their degree of NCA at about 0.45, i.e., not far from natives. In this group, we mainly find customers from *Albania*, *China*, *Germany*, *Poland*, *Romania*, *Russia*, and *Ukraine.*
- **Group 3 - Early Native-Adopters.** In this group, the customers are initially classified as immigrants but after 10 months, or after 60 months depending on the cluster, they start to be clearly classified as natives with an NCA ranging from 0.6 to 0.75. In this group, we mainly find customers from *Brazil*, *Croatia*, *Denmark*, *Eritrea*, *Norway*, and *Slovakia.*
- **Group 4 - Late Native-Adopters.** Also in this group the customers are initially classified as immigrants but after about 80 months they are classified as natives with an NCA slightly above 0.5. We mainly find customers from *Argentine*, *France*, *Ethiopia*, *Libya*, *Switzerland*, *UK*, and *USA.*
- **Group 5 - Native-Like Customers.** Finally, in the last group we have immigrants which are never classified as such. Indeed, the trends of NCA remain stably above 0.5. In this group we mainly find customers from *Bangladesh*, *Georgia*, *Czech Republic*, and *Sweden.*

Thanks to TINCA, we are able to retrieve these distinctions in the adoption of native's shopping consumption. It is interesting to observe how the countries reported for each group do not match any specific geographical area.

6 Conclusion

In this work we investigated if the analysis of retail data through machine learning classifiers can help in understanding how much immigrants adopt natives' shopping consumption. We accomplished this task by designing TINCA, a methodology aimed at discovering Trends of Immigrants Native Consumption

[11] For each group we emphasize the countries having the largest relative number of customers in that group normalized on the total number of customers from that specific country. Focusing on countries with larger absolute presence would be less interesting, as a few countries with overall very large presence (e.g. Romania, Switzerland and Germany) would simply overwhelm the others in all groups.

Adoption measuring the native adoption level by means of machine learning classifiers. Experiments on a real dataset revealed that foreign-born customers stably resident in Italy, i.e., immigrants, can be distinguished into five different groups depending on their trends of native consumption adoptions.

By design, the methodology and the results obtained are clearly of interest for the research community on social studies, and for the public bodies managing the integration of immigrants in the territory. However, we also expect that the information extracted can be useful for the retail sale companies themselves to improve their customer management, since they provide insights on the needs and behaviour of a significant portion of customers that are typically difficult to grasp, as their purchase habits derive from a time-evolving mix of culture, traditions, taste and financial resources.

This work is preliminary and it is intended to be extended in various directions besides testing it on other available datasets with similar characteristics and with different reference countries. First, we would like to validate the results with null models and extensive clustering evaluation. Second, we could add useful features taking into account spatio-temporal aspects like the typical hour and day of purchase, the number of visits at small-size and large-size shops, and a distinction between high-prices vs. low-prices products. Third, since the convergence between immigrants and natives might be explained by immigrants switching over time from low-quality to higher quality products, we could incorporate in the customer models aspects characterizing the variety of different products purchased, like the relative price within the product category. In a way this could also mean to consider aspects related to the economic status of the customers. Fourth, we could improve the performance of the machine learning classifier with a finer parameter tuning, by adopting other classifiers like support vector machines or deep neural networks, or by modeling customers with products in the word2vec fashion [40]. Fifth, we would like to explain the reasons for the classification by interpreting the decisions made by the classifiers [25] and therefore describe the retrieved groups also with respect to food consumption and products purchases. Finally, we would like to deepen the study with respect to specific countries by either developing classifiers aimed at recognizing specific nationalities, or by clustering the trends for each country separately.

Acknowledgment. This work is partially supported by the European Community H2020 programme under the funding schemes: H2020-INFRAIA-2019-1: Res. Infr. G.A. 871042 *SoBigData++*, G.A. 825619 *AI4EU*, G.A. 761758 *Humane AI*, and G.A. 780754 *Track &Know*. We thank UniCoop Tirreno for providing the data, and Roberto Zicaro for preliminary studies on the proposed methodology and analysis.

References

1. Abramitzky, R., et al.: Cultural assimilation during the age of mass migration. Technical report, National Bureau of Economic Research (2016)
2. Agrawal, R., et al.: Fast algorithms for mining association rules. In Proceedings of 20th International Conference Very Large Data Bases, VLDB, vol. 1215, pp. 487–499 (1994)

3. Akerlof, G.A., et al.: Identity economics. Econ. Voice **7**(2), 1–3 (2010)
4. Alba, R., et al.: Only english by the third generation? Demography **39**(3), 467 (2002)
5. Alesina, A., Tabellini, G., Trebbi, F.: Is Europe an optimal political area?. Technical report, National Bureau of Economic Research (2017)
6. Arai, M., et al.: Renouncing personal names: an empirical examination of surname change and earnings. J. Labor Econ. **27**(1), 127–147 (2009)
7. Atkin, D.: The caloric costs of culture: Evidence from Indian migrants. Am. Econ. Rev. **106**(4), 1144–1181 (2016)
8. Bertoli, S., et al.: Integration of Syrian refugees: insights from D4R, media events and housing market data. In: Salah, A.A., Pentland, A., Lepri, B., Letouzé, E. (eds.) Guide to Mobile Data Analytics in Refugee Scenarios, pp. 179–199. Springer, Cham (2019). https://doi.org/10.1007/978-3-030-12554-7_10
9. Bertrand, M., Kamenica, E.: Coming apart? cultural distances in the united states over time. Technical report, National Bureau of Economic Research (2018)
10. Borjas, G.J.: The analytics of the wage effect of immigration. IZA J. Migr. **2**(1), 1–25 (2013). https://doi.org/10.1186/2193-9039-2-22
11. Borjas, G.J.: Unraveling the immigration narrative. N&C (2016)
12. Brillat-Savarin, J.A.: Physiologie du goût. Charpentier (1841)
13. Bronnenberg, B.J., et al.: The evolution of brand preferences: evidence from consumer migration. Am. Econ. Rev. **102**(6), 2472–2508 (2012)
14. Bucheli, J.R., Fontenla, M., Waddell, B.J.: Return migration and violence. World Dev. **116**, 113–124 (2019)
15. Chaffey, D., Ellis-Chadwick, F., Mayer, R., Johnston, K.: Internet Marketing: Strategy, Implementation and Practice. Pearson Education, London (2009)
16. Chamberlain, B.P., et al.: Customer lifetime value prediction using embeddings. In: ACM SIGKDD, pp. 1753–1762 (2017)
17. Chen, M.-C., Chiu, A.-L., Chang, H.-H.: Mining changes in customer behavior in retail marketing. Expert Syst. Appl. **28**(4), 773–781 (2005)
18. Docquier, F., et al.: Emigration and democracy. The World Bank (2011)
19. Dustmann, C., et al.: Labor supply shocks, native wages, and the adjustment of local employment. Q. J. Econ. **132**(1), 435–483 (2017)
20. Fryer Jr., R.G., Levitt, S.D.: The causes and consequences of distinctively black names. Q. J. Econ. **119**(3), 767–805 (2004)
21. Guidotti, R., Coscia, M., Pedreschi, D., Pennacchioli, D.: Behavioral entropy and profitability in retail. In: 2015 IEEE DSAA, pp. 1–10. IEEE (2015)
22. Guidotti, R., Gabrielli, L.: Recognizing residents and tourists with retail data using shopping profiles. In: Guidi, B., Ricci, L., Calafate, C., Gaggi, O., Marquez-Barja, J. (eds.) GOODTECHS 2017. LNICST, vol. 233, pp. 353–363. Springer, Cham (2018). https://doi.org/10.1007/978-3-319-76111-4_35
23. Guidotti, R., Gabrielli, L., Monreale, A., et al.: Discovering temporal regularities in retail customers' shopping behavior. EPJ Data Sci. **7**(1), 1–26 (2018)
24. Guidotti, R., Monreale, A., Nanni, M.: Clustering individual transactional data for masses of users. In: KDD, pp. 195–204. ACM (2017)
25. Guidotti, R., Monreale, A., Ruggieri, S., et al.: A survey of methods for explaining black box models. ACM Comput. Surv. (CSUR) **51**(5), 1–42 (2018)
26. Guidotti, R., Rossetti, G., et al.: Personalized market basket prediction with temporal annotated recurring sequences. IEEE TKDE **31**(11), 2151–2163 (2018)
27. Herdağdelen, A., State, B., Adamic, L., Mason, W.: The social ties of immigrant communities in the united states. In: ACM WEBSCI, pp. 78–84 (2016)

28. Hyndman, R.J., et al.: Forecasting: Principles and Practice. OTexts, Melbourne (2018)
29. Kulkarni, V., et al.: Freshman or fresher? quantifying the geographic variation of language in online social media. In: AAAI ICWSM, pp. 615–618 (2016)
30. Lamanna, F., Lenormand, M., et al.: Immigrant community integration in world cities. PloS one **13**(3), e0191612 (2018)
31. Logan, T.D., Rhode, P.W.: Moveable feasts: A new approach to endogenizing tastes. manuscript (The Ohio State University) (2010)
32. L. Luo, et al. Tracking the evolution of customer purchase behavior segmentation via a fragmentation-coagulation process. In: IJCAI, pp. 2414–2420 (2017)
33. Magdy, A., Ghanem, T.M., Musleh, M., Mokbel, M.F.: Exploiting geo-tagged tweets to understand localized language diversity. In: GeoRich, pp. 1–6 (2014)
34. Qian, Z., et al.: Social boundaries and marital assimilation: Interpreting trends in racial and ethnic intermarriage. Am. Sociol. Rev. **72**(1), 68–94 (2007)
35. Ray, K.: The Migrants Table: Meals And Memories In. Temple University Press, Philadelphia (2004)
36. Sîrbu, A., et al.: Human migration: the big data perspective. Int. J. Data Sci. Anal. 1–20 (2020). https://doi.org/10.1007/s41060-020-00213-5
37. Spilimbergo, A.: Democracy and foreign education. AER **99**(1), 528–43 (2009)
38. Tan, P.-N., et al.: Introduction to Data Mining. Pearson Education India, Noida (2016)
39. Wedel, M., Kamakura, W.A.: Market segmentation: Conceptual and Methodological Foundations, vol. 8. Springer, New York (2012)
40. Yoshua, B., Réjean, D., Pascal, V., Christian, J.: A neural probabilistic language model. J. Mach. Learn. Res. **3**, 1137–1155 (2003)

Applied Data Science: Computational Social Science

Model Bridging: Connection Between Simulation Model and Neural Network

Keiichi Kisamori[1,2(✉)], Keisuke Yamazaki[2], Yuto Komori[2], and Hiroshi Tokieda[2]

[1] NEC Corporation, Kanagawa, Japan
k-kisamori@nec.com

[2] National Institute of National Institute of Advanced Industrial Science and Technology, Tokyo, Japan

Abstract. The interpretability of machine learning, particularly for deep neural networks, is crucial for decision making in real-world applications. One approach is replacing the un-interpretable machine learning model with *a surrogate model*, which has a simple structure for interpretation. Another approach is understanding the target system by using a simulation modeled by human knowledge with interpretable simulation parameters. Recently, simulator calibration has been developed based on kernel mean embedding to estimate the simulation parameters as posterior distributions. Our idea is to use a simulation model as an interpretable surrogate model. However, the computational cost of simulator calibration is high owing to the complexity of the simulation model. Thus, we propose a "model-bridging" framework to bridge machine learning models with simulation models by a series of kernel mean embeddings to address these difficulties. The proposed framework enables us to obtain predictions and interpretable simulation parameters simultaneously without the computationally expensive calculations of the simulations. In this study, we apply the proposed framework to essential simulations in the manufacturing industry, such as production simulation and fluid dynamics simulation.

Keywords: Interpretability · Simulation model · Kernel mean embedding · Data assimilation

1 Introduction

The interpretability of machine learning, especially for deep neural networks, is crucial for decision making in real-world applications. In recent years, many studies have addressed the interpretability of neural networks [4,6,15]. One of the approaches is replacing the un-interpretable machine learning model with *a surrogate model*, which has a simple structure for interpretation. This approach is a type of model compression. For instance, the "distillation" of a neural network model [7] is one of the representative methods for model compression for

Y. Dong et al. (Eds.): ECML PKDD 2020, LNAI 12461, pp. 389–405, 2021.
https://doi.org/10.1007/978-3-030-67670-4_24

Table 1. Comparison between machine learning models and simulation models.

	Machine learning model	Simulation model
Interpretability of parameter	Un-interpretable	Interpretable
Computational cost of the model	Not expensive	Expensive

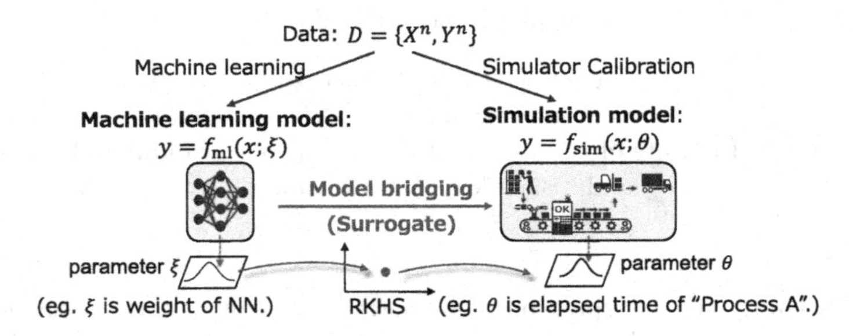

Fig. 1. Basic idea of the model-bridging framework.

replacing a complex model with a simplified model; however, there is no interpretability for a small surrogate neural network model. There are some methods to obtain an interpretable model, such as LIME [20], SHAP [14], and a method combined with a rule-based model [22]. These methods do not provide a clear pathway toward obtaining the interpretability of a neural network, as there are limitations to obtain local interpretability regarding the decision boundary of the prediction result [4,6].

Another approach for understanding the target system is by employing a simulation that might be outside the scope of conventional machine learning. In some application fields, simulations such as multi-agent simulation, traffic simulation, production simulation, or simulation of the dynamics of the physical system have already been used to understand the target system and to predict future behavior. Simulation modeling is implemented to describe the fundamental law of the objective system, using human knowledge with interpretable simulation parameters. The recently developed "simulator calibration" [3,9,10] is a method in which the simulation parameters are estimated as posterior distributions in the context of machine learning. Simulator calibration can provide a predictive result with interpretable simulation parameters. Our idea is to use a simulation model as an interpretable surrogate model. However, the difficulty of simulator calibration is attributed to a substantial computational cost; it typically takes more than one hour owing to the complexity of the simulation model (Table 1). In real-world applications, a predictive result and its reason often should be required to obtain within a minute.

We propose a "model-bridging (MB)" framework to predict using a machine learning model as well as obtain interpretable simulation parameters simultaneously without expensive calculation of a simulation model. The idea of this framework is to map the un-interpretable parameters of the machine learning model and the interpretable parameters of the simulation model (Fig. 1). The algorithm has to learn the relation of the posterior distribution estimated from each dataset between the machine learning model and the simulation model in advance; this framework can be considered as a meta-learning for each dataset.

Let us consider the example of production simulation for predicting the efficiency of manufacturing production, implementing a series of processes for production (example in Fig. 4). The production simulation aims to obtain a production efficiency and the reason for it simultaneously within a minute to improve the production efficiency. We formulate this problem setting. Assume that we obtain the dataset $\{X^n, Y^n\} = \{X_1, ..., X_n, Y_1, ..., Y_n\}$, where input $X_i \in \mathbb{R}^{d_x}$ is the number of products to be manufactured in unit time and output $Y_i \in \mathbb{R}^{d_y}$ is the efficiency of production. The simulation parameter $\theta \in \mathbb{R}^{d_\theta}$ is the elapsed time for each process, which undergoes a probabilistic behavior. The parameter θ is interpretable and helpful in understanding the system and decision making. Thus, we need to obtain the prediction $\hat{Y}_{n+1}$ for new data X_{n+1} as well as obtain the interpretable simulation parameters θ representing the elapsed time of each process, which provides information regarding the occurrence of "bottleneck processes." Here, the observed data and its generation process are considered to drift gradually, for example, the daily production of the factory due to the load of labors and machine environment factors such as temperature. The detailed assumption is described in a later section.

Note that this study considered a different problem setting from the conventional methods with simplified surrogate models, such as LIME, SHAP, and rule-based model; the interpretable model of the proposed method, i.e., simulation model, is complex and computationally expensive. There is no existing method for solving this new problem setting, where it is difficult to show the baseline for the evaluation. Experimentally, we confirm that the estimation of model bridging is reasonable in comparison with simulator calibration as a baseline with a significantly fast process owing to no execution of the simulation.

The main contribution of this paper is to propose a novel framework for bridging machine learning and simulation, which has never been discussed before from the context of machine learning and to demonstrate its effectiveness in real-world applications. The technical contribution is to expend the distribution-to-distribution regression on reproducing kernel Hilbert space (RKHS), as a suitable method for bridging function. The rest of this paper is organized as follows. We briefly review a series of applications of kernel mean embedding as the building blocks for the proposed framework. Subsequently, we propose the model-bridging framework. Finally, we confirm the accuracy of the proposed method for three cases of simulation.

2 Related Works

We briefly introduce simulator calibration and distribution regression based on kernel mean embedding as a building block of the proposed framework.

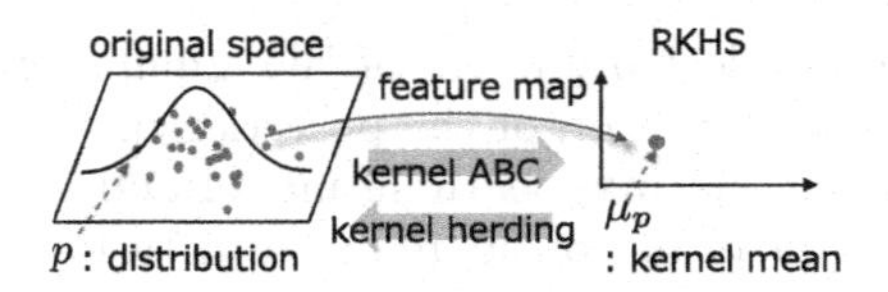

Fig. 2. Schematics of kernel mean embedding as a tool of model bridging.

2.1 Simulator Calibration

"Simulator calibration" [10] is a method for estimating the simulation parameter as the posterior distribution to reproduce real data. Simulator calibration is an example of data assimilation. The simulation model is treated as a regression function $f_{\rm sim}(x;\theta)$ by combining a series of kernel mean embedding methods. The conventional statistical methods of parameter estimation are not applicable to simulator calibration owing to the properties of the likelihood function: intractable or nondifferentiable. When Gaussian noise is employed with regression function $f_{\rm sim}(x;\theta)$, the likelihood is expressed as

$$p(y|x,\theta) = \frac{1}{\sqrt{2\pi\sigma_0^2}} \exp\left\{-\frac{1}{2\sigma_0^2}\|y - f_{\rm sim}(x;\theta)\|^2\right\},$$

where $\sigma_0 > 0$ is a constant of observation noise. This likelihood function is nondifferentiable owing to the simulation model $f_{\rm sim}(x;\theta)$. The posterior mean to be obtained is formulated as $p(\theta|X^n,Y^n) = p(Y^n|X^n,\theta)\pi(\theta)/Z(X^n,Y^n)$, where $\pi(\theta)$ is the prior distribution and $Z(X^n,Y^n)$ is the regularization constant. In this application, simulator calibration estimated the simulation parameter θ as a kernel mean of the posterior distribution by using kernel approximated Bayesian computation (kernel ABC) [5,17]. After obtaining the kernel mean of the posterior distribution, a posterior sample is obtained using kernel herding [1].

2.2 Application of Kernel Mean Embedding

As an application of kernel mean embedding [16], we briefly review the kernel ABC and kernel herding. The kernel mean embedding is a framework for mapping distributions into a RKHS $\mathcal{H}$ as a feature space. Kernel herding is a sampling method from the embedded distribution in RKHS that has the opposite operation of kernel mean embedding. Figure 2 shows a schematic of the relation of kernel ABC and kernel herding.

Kernel ABC: Kernel ABC [5,17] is a method for computing the kernel mean of the posterior distribution from a sample of parameter θ, generated by the prior distribution $\pi(\theta)$. The assumption is that the explicit form of the likelihood function is unavailable, while the sample from the likelihood function is available. The kernel ABC allows us to calculate the kernel mean of the posterior distribution as follows. First, the sample $\{\theta_1, ..., \theta_m\}$ is generated from prior distribution $\pi(\theta)$ and pseudo-data $\{\bar{Y}_1^n, ..., \bar{Y}_m^n\}$, as a sample from $p(y|x, \theta_j)$ for $j = 1, ..., m$. Next, the empirical kernel mean of the posterior distribution

$$\hat{\mu}_{\theta|YX} = \sum_{j=1}^{m} w_j k_\theta(\cdot, \theta_j) \tag{1}$$

is calculated, where k_θ is a kernel of θ. Weight w_j is calculated by

$$\begin{aligned} (w_1, ..., w_m)^T &= (G_y + m\delta I)^{-1} \mathbf{k}_y(Y^n) \ \in \mathbb{R}^m \\ G_y &= \{k_y(\bar{Y}_j^n, \bar{Y}_{j'}^n)\}_{j,j'=1}^m \ \in \mathbb{R}^{m \times m} \\ \mathbf{k}_y(Y^n) &= (k_y(\bar{Y}_1^n, Y^n), ..., k_y(\bar{Y}_m^n, Y^n)) \ \in \mathbb{R}^m. \end{aligned} \tag{2}$$

The $\delta \geq 0$ is a regularization constant, I is an identity matrix, and k_y is a kernel of y. The kernel $k_y(\bar{Y}_j^n, Y^n)$ indicates the "similarity" between pseudo-data $\bar{Y}_j^n$ and real data Y^n. The calculations of the kernel mean corresponds to the estimation of the posterior distribution as an element in $\mathcal{H}$.

Kernel Herding: Kernel herding [1] is a method used for sampling data from the kernel mean representation of a distribution, which is an element of the RKHS. Kernel herding can be considered as an opposite operation to that of kernel ABC. Kernel herding greedily obtains samples of θ by updating Eqs. (1) and (2) as given in Chen et al. [1].

2.3 Distribution Regression

Distribution regression is a regression for d_x-dimensional "distributions" represented by samples. In contrast, normal regression is regression for d_x-dimensional "point." There are several studies of distribution regression, including distribution-to-distribution regression [19] and distribution-to-point regression [12,21]. Oliva et al. [19] employed the idea of approximating a density function by kernel density estimation, rather than using RKHS. Szabó et al. [21] proposed the distribution-to-point with the kernel ridge regression method on RKHS; however, no methods are available for distribution-to-distribution regression on RKHS.

3 Proposed Framework: Model Bridging

We propose a novel framework to bridge the un-interpretable machine learning model and the interpretable simulation model. In this study, we assume a machine learning model, such as a Bayesian neural network (BNN) [18], as a parametric model and a Gaussian process as a non-parametric model. This proposed framework is applicable to any model. In this section, first, we confirm the problem setting and framework of model bridging. Second, we propose the algorithm of distribution-to-distribution regression, which is suitable for the proposed framework. Thereafter, we propose the formulation of the input of distribution-to-distribution regression for the parametric model, assuming BNN, and the non-parametric model, assuming the Gaussian process. Figure 3 and Algorithm 1 shows an overview of the framework.

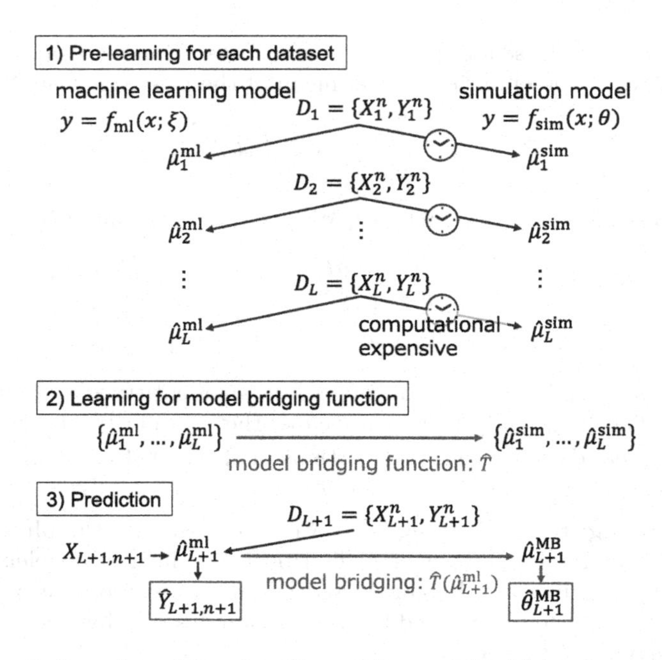

Fig. 3. Overview of the algorithm of the model-bridging framework.

3.1 Problem Setting, Assumption, and Usage of Model Bridging

We define the problem setting of the model-bridging framework. Let L be dataset $\{X_l^n, Y_l^n\}_{l=1}^L$ ($X_l^n \in \mathbb{R}^{n \times d_x}, Y_l^n \in \mathbb{R}^{n \times d_y}$), given in the pre-learning phase. For simplicity of explanation, we use the unique number of the data n and sample size m for all datasets. However, it can be different numbers generally, such as n_l and m_l. The purpose is to predict $\hat{Y}_{L+1,n+1}$ and simultaneously obtain interpretable simulation parameter $\hat{\theta}_{L+1}^{\mathrm{MB}}$ to reproduce $Y_{L+1,n+1} = f_{\mathrm{sim}}(X_{L+1,n+1}; \hat{\theta}_{L+1}^{\mathrm{MB}})$ without the expensive calculation of simulation model $f_{\mathrm{sim}}(x;\theta)$, when we obtain new dataset $\{X_{L+1}^n, Y_{L+1}^n\}$. The assumptions of the problem setting are as follows. These assumptions are prevalent for many applications of a simulation.

- The existing simulation model $f_{\mathrm{sim}}(x;\theta)$ with interpretable simulation parameter $\theta \in \mathbb{R}^{d_\theta}$ and a machine learning model $f_{\mathrm{ml}}(x;\xi)$ that is sufficiently accurate to predict a typical regression problem while having un-interpretable parameter $\xi \in \mathbb{R}^{d_\xi}$.
- The cost of simulator calibration is much higher than that of learning for the machine learning model. For instance, it takes more than one hour for simulator calibration of one dataset $\{X_l^n, Y_l^n\}$, while learning of BNN takes less than a minute.
- Dataset $\{X_l^n, Y_l^n\}$ has dependency of parameter θ_l for each $l = 1, \ldots, L$. Let us assume the following situation. $\{X_l^n, Y_l^n\}$ is obtained in one day with the same conditions, described as parameter θ_l, while the conditions are changed for the following day, described as θ_{l+1}.

Algorithm 1: Model bridging

1) Pre-learning for each dataset :
Input: Dataset $\{X_l^n, Y_l^n\}_{l=1}^L$,
machine learning model $f_{\text{ml}}(x, \xi)$
and simulation model $f_{\text{sim}}(x, \theta)$
Output: $\{\hat{\mu}_1^{\text{ml}}, ..., \hat{\mu}_L^{\text{ml}}\}$ and $\{\hat{\mu}_1^{\text{sim}}, ..., \hat{\mu}_L^{\text{sim}}\}$
for $l = 1$ to L **do**
Estimation for $\hat{\mu}_l^{\text{ml}}$ by Eq. (7)
Estimation for $\hat{\mu}_l^{\text{sim}}$ by Eq. (1)
end for
2) Learning for model-bridging function $\hat{T}$:
Input: $\{\hat{\mu}_1^{\text{ml}}, ..., \hat{\mu}_L^{\text{ml}}\}$ and $\{\hat{\mu}_1^{\text{sim}}, ..., \hat{\mu}_L^{\text{sim}}\}$
Output: Model-bridging function $\hat{T}$
Learning for $\hat{T}$ by Eq. (3)
3) Prediction:
Input: Dataset $\{X_{L+1}^n, Y_{L+1}^n\}$ and $X_{L+1,n+1}$
Output: $\hat{Y}_{L+1,n+1}$ and $\hat{\theta}_{L+1}$
Estimation for ξ_{L+1} and $\hat{\mu}_{L+1}^{\text{ml}}$
Prediction for $\hat{Y}_{L+1,n+1}$ by $f_{\text{ml}}(x; \xi_{L+1})$
Estimation for $\hat{\mu}_{L+1}^{\text{MB}} = \hat{T}(\hat{\mu}_{L+1}^{\text{ml}})$ by Eq. (5)
Sampling for $\hat{\theta}_{L+1}^{\text{MB}}$ by Eq. (6)

- The time for offline calculation of simulator calibration is sufficient, while the time for prediction is restricted.

Once we obtain the model-bridging function as a mapping from the machine learning model to the simulation model, we can obtain an accurate prediction for $\hat{Y}_{L+1,n+1}$ by both the machine learning model and interpretable $\hat{\theta}_{L+1}^{\text{MB}}$ by the simulation model for new dataset $\{X_{L+1}^n, Y_{L+1}^n\}$ without an expensive calculation from the simulation model.

3.2 Distribution-to-Distribution Regression

We present the regression algorithm between the conditional kernel mean of the machine learning model $\mu^{\text{ml}} \in \mathcal{H}$ and that of the simulation model $\mu^{\text{sim}} \in \mathcal{H}$, as a model-bridging function $\mu^{\text{sim}} = T(\mu^{\text{ml}})$. We develop the algorithm based on kernel ridge regression, which is suitable for kernel mean input and output on RKHS. This is the extension of the distribution-to-point regression method proposed by Szabó et al. [21] for the distribution output.

Kernel Ridge Regression for Kernel Mean. The formulation to be solved as an analogy of normal kernel ridge regression is as follows:

$$\hat{T} = \arg\max_{T \in \mathcal{F}} \frac{1}{L} \sum_{l=1}^{L} \|\hat{\mu}_l^{\text{sim}} - T(\hat{\mu}_l^{\text{ml}})\|_{\mathcal{F}}^2 + \lambda \|T\|_{\mathcal{F}}^2, \tag{3}$$

where $\lambda \geq 0$ is a regularization constant. $\mathcal{F}$ is a function space of kernel mean embeddings following Christmann et al. [2] and $\|\cdot\|_{\mathcal{F}}$ is its norm. The difference from ordinary kernel ridge regression is that the inputs and outputs are kernel means. Therefore, we define kernel $\kappa \in \mathcal{F}$, as a function of kernel mean $\mu \in \mathcal{H}$. We employ a Gaussian-like kernel as

$$\kappa(\mu, \mu') = \exp\left\{-\frac{1}{2\sigma_\mu^2} \|\mu - \mu'\|_{\mathcal{H}}^2\right\} \in \mathcal{F}, \tag{4}$$

where constant $\sigma_\mu > 0$ is the width of kernel κ and $\|\cdot\|_{\mathcal{H}}$ is RKHS norm. The kernel κ is also a positive definite kernel [2].

Following the representor theorem of kernel ridge regression [?], the estimated model-bridging function $\hat{T}$ for new $\hat{\mu}_{L+1}^{\mathrm{ml}}$ is described as

$$\hat{\mu}_{L+1}^{\mathrm{MB}} = \hat{T}(\hat{\mu}_{L+1}^{\mathrm{ml}}) = \sum_{l=1}^{L} v_l \hat{\mu}_l^{\mathrm{sim}} \quad \in \mathcal{F}, \tag{5}$$

where $\mathbf{v} = (v_1, ..., v_L)^T = (G_\mu + \lambda LI)^{-1} \mathbf{k}_\mu(\hat{\mu}_{L+1}^{\mathrm{ml}}) \in \mathbb{R}^L$. Gram matrix G_μ and the vector $\mathbf{k}_\mu(\hat{\mu}_{L+1}^{\mathrm{ml}})$ are described as follows:

$$G_\mu = \left\{\kappa(\hat{\mu}_l^{\mathrm{ml}}, \hat{\mu}_{l'}^{\mathrm{ml}})\right\}_{l,l'=1}^{L} \in \mathbb{R}^{L \times L}$$
$$\mathbf{k}_\mu(\hat{\mu}_{L+1}^{\mathrm{ml}}) = \left(\kappa(\hat{\mu}_1^{\mathrm{ml}}, \hat{\mu}_{L+1}^{\mathrm{ml}}), ..., \kappa(\hat{\mu}_L^{\mathrm{ml}}, \hat{\mu}_{L+1}^{\mathrm{ml}})\right)^T \in \mathbb{R}^L.$$

Kernel Herding from Kernel Mean $\hat{\mu}_l^{\mathrm{MB}}$. After obtaining the kernel mean of $\hat{\mu}_{L+1}^{\mathrm{MB}}$, kernel herding can be applied to sample $\hat{\theta}_{L+1}^{\mathrm{MB}} = \{\hat{\theta}_{L+1,1}, ..., \hat{\theta}_{L+1,m}\}$, where $\hat{\theta}_{L+1,j} \in \mathbb{R}^{d_\theta}$. The explicit form of the update equation for sample $j = 1, ..., m$ iteration of kernel herding with kernel mean $\hat{\mu}_{L+1}^{\mathrm{MB}}$ is as follows:

$$\hat{\theta}_{L+1,j} = \arg\max_\theta \sum_{l=1}^{L} \sum_{j'=1}^{m} v_l w_{l,j'} k_\theta(\theta, \theta_{l,j'}) - \frac{1}{j} \sum_{j'=1}^{j-1} k_\theta(\theta, \theta_{j'}) \in \mathbb{R}^{d_\theta}, \tag{6}$$

for $j = 2, ..., m$. For initial state $j = 1$, the update equation constitutes only the first term of Eq. (6). The weight of $w_{l,j}$ is calculated by kernel ABC for dataset $\{X_l^n, Y_l^n\}$ in Eq. (2).

3.3 Input of Distribution-to-Distribution Regression

We present the explicit formulation for calculating the kernel means of the machine learning model $\hat{\mu}_l^{\mathrm{ml}}$, as an input of the distribution-to-distribution regression. First, we present the formulation of BNN, as BNN is a useful model for many applications as a parametric Bayesian model. Second, we present the formulation for Gaussian process regression as a non-parametric Bayesian model. We consider the Gaussian process regression as a non-parametric alternative to BNN. The equivalence between the Gaussian process and BNN with one hidden layer with infinite nodes is well known [18]. Furthermore, a recent study reveals the kernel formulation that is equivalent to multi-layered BNN, as an extension of the Gaussian process [13]. We directly obtain empirical kernel mean without calculation of kernel from parameters in the parametric model for using Gaussian process regression.

Parametric Model: Bayesian Neural Network. We assume the BNN model $f_{\mathrm{ml}}(x;\xi)$ with a few hidden layers, where ξ is a parameter, such as weights for each node and bias terms of each layer. We can obtain the posterior distribution of ξ_l for $l = 1, ..., L$ by the Markov Chain Monte Carlo method or variational approximation. Then, the empirical kernel mean of the posterior distribution is represented as $\hat{\mu}_l^{\mathrm{ml}} = \sum_{j=1}^{m} k_\xi(\cdot, \xi_{l,j}) \in \mathcal{H}$ for $l = 1, ..., L$ dataset, where k_ξ is kernel of ξ.

We employ Gaussian-like kernel κ as an function of $\hat{\mu}_l^{\mathrm{ml}} \in \mathcal{H}$ as

$$\kappa(\hat{\mu}_l^{\mathrm{ml}}, \hat{\mu}_{l'}^{\mathrm{ml}}) = \exp\left\{-\frac{1}{2\sigma_\mu^2}\left\|\hat{\mu}_l^{\mathrm{ml}} - \hat{\mu}_{l'}^{\mathrm{ml}}\right\|_{\mathcal{H}}^2\right\} \in \mathcal{F}$$
$$= \exp\left\{-\frac{1}{\sigma_\mu^2}\left(1 - \sum_{j=1}^{m}\sum_{j'=1}^{m'} k_\xi(\xi_{l,j}, \xi_{l',j'})\right)\right\}.$$

The relation $\langle\hat{\mu}_l^{\mathrm{ml}}, \hat{\mu}_{l'}^{\mathrm{ml}}\rangle = \sum_{j=1}^{m}\sum_{j'=1}^{m'} k_\xi(\xi_{l,j}, \xi_{l',j'})$ is used, where $\langle\cdot,\cdot\rangle$ represents the inner product.

Non-Parametric Model: Gaussian Process Regression. We use the Gaussian process regression as a non-parametric model. In this case, we can directly obtain empirical kernel mean without the calculation of kernel from parameters, such as ξ in the parametric model. We present that the prediction with Gaussian process regression can be considered as a conditional kernel mean. As a result of Gaussian process regression, we can express the mean of the predictive distribution in general for l-th dataset $\{X^n, Y^n\}$ as

$$y = \hat{\mu}_{Y|X,l}(x) = \sum_{i=1}^{n} u_{l,i}(x) k_y(\cdot, Y_{l,i}), \tag{7}$$

where $u_{l,i}(x) = \{(G_x + n\lambda' I)^{-1}\mathbf{k}_x(x)\}_i$. The G_x is the Gramm matrix, $\lambda' \geq 0$ is regularization constant, and k_x is kernel of x. This formulation is clear if we remember the equivalence between Gaussian process regression and kernel ridge regression [8]. As a predictor of $\hat{Y}_{l,n+1}$ for new $X_{l,n+1}$, we can calculate $\hat{Y}_{l,n+1} = \hat{\mu}_{Y|X,l}(X_{l,n+1})$. Note that this $\hat{\mu}_{Y|X,l}$ is interpreted as the kernel mean of the Y_l^n conditioned by X_l^n. Thus, we can use $\hat{\mu}_{Y|X,l}$ as the input of the distribution-to-distribution regression, represented as $\hat{\mu}_l^{\mathrm{ml}}$. We employ Gaussian-like kernel κ as an function of $\hat{\mu}_l^{\mathrm{ml}}$ as

$$\kappa(\hat{\mu}_l^{\mathrm{ml}}, \hat{\mu}_{l'}^{\mathrm{ml}}) = \exp\left\{-\frac{1}{\sigma_\mu^2}\left(1 - \sum_{i=1}^{n}\sum_{i'=1}^{n'} u_{l,i} u_{l',i'} k_y(Y_{l,i}, Y_{l',i'})\right)\right\}.$$

There is a difference between the proposed non-parametric method and parametric method. In the parametric method, the distribution of parameter ξ_l is the input of the distribution-to-distribution regression, while in the non-parametric method, the distribution of data Y_l^n conditioned by X_l^n is the input.

4 Experiment

This framework is widely applicable to various domains for industries that include multi-agent simulation, traffic simulation, and simulation of dynamics of physical systems, such as thermomechanics, structural mechanics, and electromagnetic mechanics. We present the applications of the model-bridging framework for three simulations: 1) a simple production simulation to show and explain the framework effectiveness in detail, 2) a realistic production simulation to show the capability for realistic application, and 3) a simulation of fluid dynamics to show of the applicability to a wide variety of simulation fields. The detailed information on the three experiments is provided in the supplemental material owing to page limitations.

Through the three experiments, we confirm that model bridging enables us to predict $\hat{Y}_{L+1,n+1}$ for new $X_{L+1,n+1}$, using a machine learning model, and obtain the interpretable simulation parameter $\hat{\theta}^{\mathrm{MB}}_{L+1}$ without an expensive calculation of the simulation when we obtain the $L+1$-th dataset $\{X^n_{L+1}, Y^n_{L+1}\}$. We also investigate the accuracy of the estimation of the parameter $\hat{\theta}^{\mathrm{MB}}_{L+1}$ and compared the execution time with simulator calibration as a baseline. Note that these experiments cannot be compared with other state-of-the-art surrogate approaches, such as LIME, SHAP, and other methods with model compression, because no other methods exist that simultaneously obtain the prediction result and interpretable simulation parameters.

4.1 Common Setting of Experiments

In practice, the effective hyperparameter for model-bridging function to be tuned is the regularization constant λ for distribution-to-distribution regression. The hyperparameter λ can stabilize the calculation of the inverse Gram matrix. This hyperparameter should be determined by cross-validation. Further, as a common hyperparameter of the kernel method, the width of the kernel must be selected to measure the similarity between the data. We employed a Gaussian kernel for k_y, k_x, k_θ, and k_ξ for all experiments. The typical setting of the width of the kernel, in practice, is the median of Euclid distance of the input data of a kernel. In all the experiments performed in this study, we apply this setting and confirmed that all kernels perform adequately. We used a PC equipped with a 3.4-GHz Intel core i7 quad-core processor and 16GB memory. The main computational cost is for a simulation in the pre-learning phase of this framework.

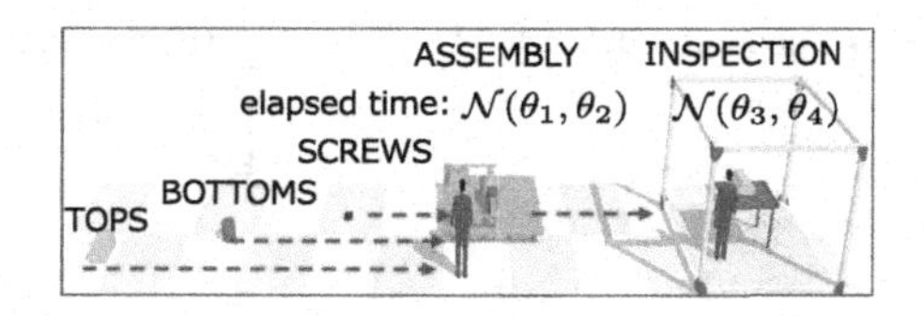

Fig. 4. Production simulation model for the experiment.

4.2 Experiment with Simple Production Simulator

Setting. Production simulators are widely used simulation software for discrete and interconnection systems to model various processes, such as production, logistics, transportation, and office works. We used a *WITNESS*, a popular software package of production simulation[1]. We examined the regression problem using a simulation model that has a simple four-dimensional simulation parameter $\theta \in \mathbb{R}^4$ (Fig. 4). We defined the simulation input $x = X_i \in \mathbb{R}$ as the number of products to be manufactured, output $Y_i = f_{\text{sim}}(X_i, \theta) \in \mathbb{R}$ as the total time to manufacture all the X_i-th products, and parameter θ as the time required for each procedure on the production line. Moreover, the time required for "ASSEMBLY" is $\mathcal{N}(\theta_1, \theta_2)$, and that for "INSPECTION" is $\mathcal{N}(\theta_3, \theta_4)$, where $\mathcal{N}(\mu_{\text{ND}}, \sigma_{\text{ND}})$ is the normal distribution with mean μ_{ND} and standard deviation σ_{ND}. We assumed that the elapsed time of each process would increase considerably, owing to an increasing load, if the number of products to be manufactured also increases. To create this situation artificially, we set different true parameters between the observed data region $\theta^{(0)}$ and the predictive region $\theta^{(1)}$. We set $\theta^{(0)} = (2, 0.5, 5, 1)^T$ if $x \leq 110$ and $\theta^{(1)} = (3.5, 0.5, 7, 1)^T$ if $x > 110$. The shift in parameters θ_1 and θ_3 between $\theta^{(0)}$ and $\theta^{(1)}$ is the sigmoid function. For each l-th dataset, the observed data of size $n = 50$ and sample size $m = 100$ are generated by $q_l(x) = \mathcal{N}(\chi_l, 5)$, where χ_l is uniform distribution in $[70, 130]$ for $l = 1, ..., L$. The number of training datasets, L, is 100. We defined the prior distribution as the uniform distribution over $[0, 5] \times [0, 2] \times [0, 10] \times [0, 2]$. We used a BNN having two fully connected hidden layers with three nodes and bias nodes for each layer, as a machine learning model. The activation function is ReLU. The regularization constant is $\lambda = 1.0 \times 10^{-6}$ for this experiment.

Result. The execution time of model bridging is 9.6 [s] in the presented computational environment, while simulator calibration requires about 3.1 [h] for $L + 1$-th dataset. Simulator calibration requires $m \times n$ execution of simulation and each simulation takes 2 [s] in this case. As representatives of all test datasets, the results of the model-bridging framework for two different datasets (l-th and l'-th dataset), which are randomly selected, are shown in Fig. 5. Figure 5 (A) shows the observed data for l-th dataset as red squares and the l'-th dataset as red triangles. The solid line and dashed line are the fitted results by BNN with

[1] https://www.lanner.com/en-us/technology/witness-simulation-software.html.

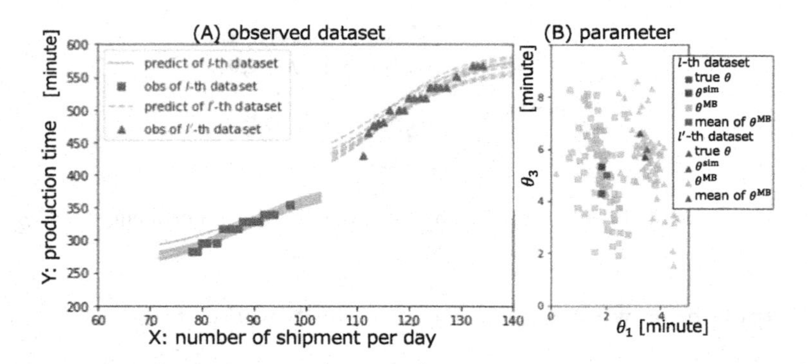

Fig. 5. As representatives for all dataset, two test datasets are shown: l-th dataset as square markers and l'-th dataset as triangle markers. (A) Observed data and fitted result by BNN. (B) Estimated distribution of simulation parameters by model bridging. (Color figure online)

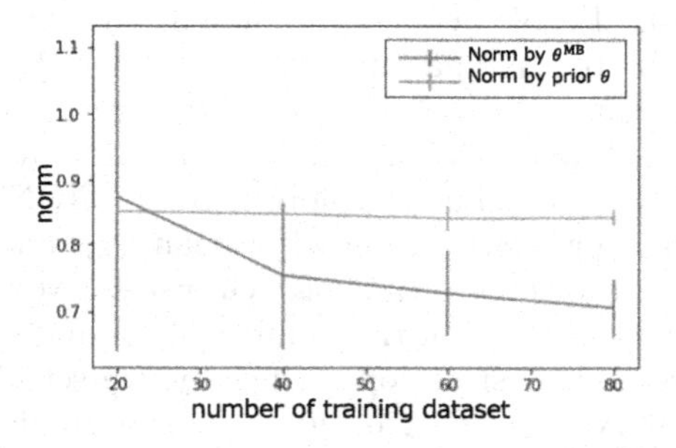

Fig. 6. Estimation of norm for $\hat{\mu}$ by the model-bridging function for number of training datasets. The blue line represents the estimated result of the norm, and the orange line represents the norm with prior information only. (Color figure online)

variational approximation. Figure 5 (B) shows the estimated posterior distributions of simulation parameters by model bridging $\hat{\theta}_l^{\mathrm{MB}}$ each dataset. The red markers show the true parameter, green markers show the estimated result of simulator calibration, and blue markers show the mean of the distribution. Each square denotes the l-th dataset and each triangle denotes the l'-th dataset. We can see a reasonably accurate estimation of θ_l^{MB} and $\theta_{l'}^{\mathrm{MB}}$ model-bridging framework in comparison with simulator calibration for the two different datasets with two different true parameters θ.

Note that from the perspective of interpretability, we can clearly see the practical effectiveness of simultaneously obtaining the prediction result with interpretable parameters, such as "elapsed time of a process." From these interpretable parameters, we can understand that the production efficiency is decreased (l'-th dataset in Fig. 5) mainly because of the increased elapsed time of "INSPECTION" ($= \theta_1$).

Table 2. Summary of true and estimated parameters in the experiment for realistic production simulation. T_{BF} represents the mean time between failures, and T_R represents the mode time of repair for each process. The estimated parameters are mean and standard deviation (in parentheses) of posterior mean for one-leave-out cross-validation.

Process	Saw		Coat		Inspection		Harden		Grind		Clean	
Param.	T_{BF} θ_1	T_{R} θ_2	T_{BF} θ_3	T_{R} θ_4	T_{BF} θ_5	T_{R} θ_6	T_{BF} θ_7	T_{R} θ_8	T_{BF} θ_9	T_{R} θ_{10}	T_{BF} θ_{11}	T_{R} θ_{12}
$\theta^{(0)}$	100	25	150	5	**100**	20	150	5	75	15	120	20
$\theta^{(1)}$	100	25	150	5	**80**	20	150	5	75	15	120	20
$\hat{\theta}^{\mathrm{sim}}$	100.5	25.1	153.2	5.0	**104.2**	17.3	146.5	5.1	73.7	14.6	96.8	20.2
$(x \leq 30)$	(9.5)	(2.2)	(14.2)	(0.4)	(10.1)	(1.9)	(14.2)	(0.3)	(6.7)	(1.2)	(15.7)	(2.3)
$\hat{\theta}^{\mathrm{sim}}$	100.5	24.7	153.6	5.0	**84.3**	22.8	148.2	5.0	73.8	15.2	115.4	20.4
$(x > 30)$	(8.3)	(1.4)	(10.6)	(0.3)	(7.4)	(0.9)	(8.9)	(0.3)	(4.6)	(1.0)	(9.8)	(1.8)
$\hat{\theta}^{\mathrm{MB}}$	102.5	21.2	165.1	6.0	**98.6**	17.0	179.0	5.5	74.0	17.8	94.6	12.7
$(x \leq 30)$	(12.7)	(1.8)	(12.6)	(0.4)	(9.9)	(1.8)	(12.1)	(0.3)	(6.8)	(1.6)	(14.5)	(2.3)
$\hat{\theta}^{\mathrm{MB}}$	104.1	21.1	165.3	6.1	**86.7**	19.1	180.8	5.4	74.0	18.1	99.3	13.2
$(x > 30)$	(10.1)	(2.3)	(14.2)	(0.5)	(11.8)	(2.1)	(8.1)	(0.5)	(4.5)	(1.2)	(21.2)	(1.6)

We also investigated $\|\hat{\mu}^{\mathrm{MB}}_{L+1} - \hat{\mu}^{\mathrm{sim}}_{L+1}\|^2_{\mathcal{H}}$ to confirm the convergence of the proposed distribution-to-distribution regression in the model-bridging framework. The detailed formulation for the numerical calculation is presented in the supplementary material. Figure 6 shows the mean and standard deviation of one-leave-out cross-validation of the test dataset. The horizontal axis shows the number of training datasets. We can see the convergence for bias that originates from simulator calibration.

4.3 Experiment with Realistic Production Simulator

Setting. We used a model to reproduce a real metal-processing factory that manufactures valves from metal pipes, with six primary processes: "saw," "coat," "inspection," "harden," "grind," and "clean," in the order shown in the supplementary material. Each process is composed of complex procedures, such as the preparation rule, waiting, and machine repair during trouble. The purpose of this production simulation is also to predict the total production time $Y_i \in \mathbb{R}$ when the number of units $X_i \in \mathbb{R}^3$ for three types of products to be manufactured is set. Each of the six processes contains two parameters of machine downtime owing to failure: mean time between failures (T_{BF}) and mode time required for repair (T_{R}). We defined these parameters as twelve-dimensional parameter $\theta \in \mathbb{R}^{12}$ (see Table 2). The distribution of the mean time between failures is represented as a negative exponential distribution. The distribution of the time required for repair is represented as an Erlang distribution with the mode time and shape parameter set at three.

Similar to the simple experiment discussed in the previous section, we set the true parameter as $\theta^{(0)}$ if $x \leq 30$ and $\theta^{(1)}$ if $x > 30$. The summary of the true

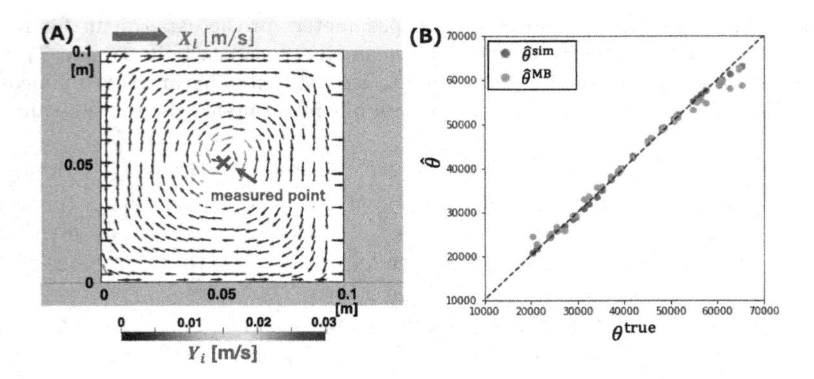

Fig. 7. (A) The experiment of "cavity," which is a two-dimensional square space surrounded by walls (gray) on three sides while moving material (light blue) is located on top of the space. Input X_i is the velocity of the material on top of the cavity, while output Y_i is the velocity at the point depicted by the x-mark at a specific time. (B) The estimated result of Reynolds number by simulator calibration ($\hat{\theta}^{\text{sim}}$) and model bridging ($\hat{\theta}^{\text{MB}}$) as a function of true θ (θ^{true}). (Color figure online)

parameter is shown in Table 2. The shift in parameter θ_5 between $\theta^{(0)}$ and $\theta^{(1)}$ is a sigmoid function. The observed data of size is $n = 30$ by $\mathcal{N}(\chi_l, 3)$ where χ_l is generated by uniform distribution in $[20, 40]$. The number of parameter samples is $m = 50$ and the number of datasets is $L = 40$. We defined the prior distribution as the uniform distribution over $[60, 140] \times [15, 35] \times [100, 200] \times [3, 10] \times [60, 140] \times [15, 35] \times [100, 200] \times [3, 10] \times [50, 100] \times [10, 20] \times [100, 200] \times [15, 35]$. We use Gaussian process regression as a machine learning model. The hyperparameter of the regularization constant λ is 0.1.

Result. The execution time of model bridging is 1.1 [s] in the presented computational environment, while simulator calibration requires about 1.3 [h] for $L+1$-th dataset. The simulator calibration requires $m \times n$ execution of simulation, and each simulation takes 3 [s] in this case. The results of the mean and standard deviation of the estimated parameters for one-leave-out cross-validation are shown in the bottom rows in Table 2. All parameters of estimation by the model-bridging framework are accurate within the standard deviation for $\theta^{(0)}$ and $\theta^{(1)}$, respectively. We can see the effectiveness of a high-dimensional parameter space with a realistic experiment. From the estimated result of simulation parameters, we can understand that the difference in θ_5 results in different predictions for each situation, while other parameters are constant. This insight obtained from the interpretable parameters leads to improvements in the production process.

4.4 Experiment with Simulator for Fluid Dynamics

Setting. Through computer-aided engineering (CAE) simulations, we confirmed that our model-bridging algorithm is applicable to the simulation of fluid-dynamics systems. We employed the typical benchmark in this field, called "cavity flow experiment," with OpenFOAM®[2,3] (Fig. 7 (A)). We considered a two-dimensional squared space called "cavity" fulfilled with a fluid having an unknown Reynolds number. The Reynolds number is used to help predict flow patterns and velocities in fluid dynamics. Turbulent flow is somewhat challenging to predict, even though it is ubiquitous in real-world situations. In this experiment, input $X_i \in \mathbb{R}$ is the velocity of the material on top of the cavity; the output $Y_i \in \mathbb{R}$ is the velocity at the particular point (see Fig. 7 (A)); and parameter $\theta \in \mathbb{R}$ is the Reynolds number (see supplementary material for details). The number of data $n = 50$, the number of samples $m = 41$, and the number of dataset $L = 41$ are generated by different true $\theta_l (= \theta_l^{\mathrm{true}})$. The prior distribution is defined as the uniform distribution over $[20000, 65000]$. We used Gaussian process regression as a machine learning model. The hyperparameter of regularization is $\lambda = 1.0^{-5}$.

Result. The execution time of model bridging is 2.6 [s] in the presented computational environment, while the simulator calibration requires about 9 [h]. Each simulation takes about 17 [s] in this case. Figure 7 (B) shows the estimated result of $\hat{\theta}^{\mathrm{sim}}$ by simulator calibration and $\hat{\theta}^{\mathrm{MB}}$ by model bridging as a function of true θ for $L = 41$ dataset with one-leave-out cross-validation. The dashed line in Fig. 7 (B) shows $\theta^{\mathrm{true}} = \hat{\theta}^{\mathrm{MB}} (= \hat{\theta}^{\mathrm{sim}})$ to ensure that the estimation is accurate if the result is on the dashed line. We can see a reasonable estimation of $\hat{\theta}^{\mathrm{MB}}$. The result of the velocity prediction of velocity Y_i is also reasonably accurate (see the supplementary material). Human experts can understand why the Reynolds number causes such flow of fluid.

5 Discussion

There are many possible options to be discussed in the proposed framework for the individual-use case. In this study, we assume the given observed dataset as the problem setting. Further, there are two other possible ways for problem setting with the assumption of the data generation process: 1) generate data from $f_{\mathrm{ml}}(x; \xi)$ and 2) generate data from $f_{\mathrm{sim}}(x; \theta)$. Considering another case with these assumptions of data generation might be meaningful, e.g., when the real observed data are limited or when the simulation has high confidence. Another option to be discussed is the parametric or non-parametric regression model for the model-bridging function $\hat{T}$. In this study, we present the practical effectiveness of the model-bridging framework, while a theoretical analysis of the asymptotic behavior of this framework is still desired.

[2] https://www.openfoam.com/.

[3] https://www.openfoam.com/documentation/tutorial-guide.

6 Conclusion

We propose a novel framework named "model bridging" to bridge from the uninterpretable machine learning model to the simulation model with interpretable parameters. The model-bridging framework enables us to obtain precise predictions from the machine learning model as well as obtain the interpretable simulation parameter simultaneously without the expensive calculations of a simulation. We confirmed the effectiveness of the model-bridging framework and accuracy of the estimated simulation parameter using production simulation and simulation of fluid dynamics, which are widely used in the real-world manufacturing industry.

References

1. Chen, Y., Welling, M., Smola, A.: Super-samples from kernel herding. In: Proceedings of the Twenty-Sixth Conference Annual Conference on Uncertainty in Artificial Intelligence, pp. 109–116 (2010)
2. Christmann, A., Steinwart, I.: Universal Kernels on Non-Standard Input Spaces. In: Advances in Neural Information Processing Systems (2010)
3. Cleary, E., Garbuno-Inigo, A., Lan, S., Schneider, T., Stuart, A.M.: Calibrate, Emulate, Sample, pp. 1–27 (2020) http://arxiv.org/abs/2001.03689
4. Doshi-Velez, F., Kim, B.: Towards a rigorous science of interpretable. Mach. Learn. (2017). https://doi.org/10.1016/j.intell.2013.05.008
5. Fukumizu, K., Song, L., Gretton, A.: Kernel Bayes' rule: Bayesian inference with positive definite kernels. J. Mach. Learn. Res. **14**, 3753–3783 (2013)
6. Guidotti, R., Monreale, A., Ruggieri, S., Turini, F., Pedreschi, D., Giannotti, F.: A survey of methods for explaining black box models (2018). https://doi.org/10.1145/3236009
7. Hinton, G., Vinyals, O., Dean, J.: Distilling the Knowledge in a Neural Network. arXiv:1503.02531v1 (2015)
8. Kanagawa, M., Hennig, P., Sejdinovic, D., Sriperumbudur, B.K.: Gaussian Processes and Kernel Methods: A Review on Connections and Equivalences. arXiv:1807.02582v1 (2018)
9. Kennedy, M.C., O'Hagan, A.: Bayesian calibration of computer models. J. Royal Stat. Soc. Ser. B (Statistical Methodology) **63**(3), 425–464 (2001)
10. Kisamori, K., Kanagawa, M., Yamazaki, K.: Simulator calibration under covariate shift with kernels. In: Proceedings of the 23rd International Conference on Artificial Intelligence and Statistics (2020)
11. Kung, S.Y.: Kernel Methods and Machine Learning. Cambridge University Press, Cambridge (2014). https://doi.org/10.1017/CBO9781139176224
12. Law, H.C.L., Sutherland, D.J., Sejdinovic, D., Flaxman, S.: Bayesian Approaches to Distribution Regression. In: Proceedings of the 21st International Conference on Artificial Intelligence and Statistics (2017)
13. Lee, J., Bahri, Y., Novak, R., Schoenholz, S.S., Pennington, J., Sohl-Dickstein, J.: Deep Neural Networks as Gaussian Process. In: Proceedings of The International Conference on Learning Representations (2018)
14. Lundberg, S.M., Lee, S.I.: A unified approach to interpreting model predictions. In: Advances in Neural Information Processing Systems (2017)

15. Molnar, C.: Interpretable Machine Learning. Christoph Molnar, Babson Park (2019)
16. Muandet, K., Fukumizu, K., Sriperumbudur, B., Scholkopf, B.: Kernel Mean Embedding of Distributions: A Review and Beyonds. arXiv:1605.09522 p. 133 (2016)
17. Nakagome, S., Fukumizu, K., Mano, S.: Kernel approximate Bayesian computation in population genetic inferences. Stat. Appl. Genet. Mol. Biol. **12**(6), 667–678 (2013). https://doi.org/10.1515/sagmb-2012-0050
18. Neal, R.M.: Bayesian Learning for Neural Networks. Springer, New York (1996). https://doi.org/10.1007/978-1-4612-0745-0
19. Oliva, J.B., Schneider, J.: Distribution to Distribution Regression. In: Proceedings of The 30th International Conference on Machine Learning (2013)
20. Ribeiro, M.T., Singh, S., Guestrin, C.: Why Should I Trust You? explaining the predictions of any classifier. In: Proceedings of the 22nd ACM SIGKDD International Conference on Knowledge Discovery and Data Mining (2016)
21. Szabo, Z., Sriperumbudur, B., Poczos, B., Gretton, A.: Learning theory for distribution regression. J. Mach. Learn. Res. **17**, 1–40 (2016)
22. Wang, T.: Gaining free or low-cost transparency with interpretable partial substitute. In: Proceedings of the 36th International Conference on Machine Learning (2019)

Semi-supervised Multi-aspect Detection of Misinformation Using Hierarchical Joint Decomposition

Sara Abdali[1](✉), Neil Shah[2](✉), and Evangelos E. Papalexakis[1]

[1] Department of Computer Science and Engineering, University of California Riverside, 900 University Avenue, Riverside, CA, USA
sabda005@ucr.edu, epapalex@cs.ucr.edu

[2] Snap Inc., CA, USA
nshah@snap.com

Abstract. Distinguishing between misinformation and real information is one of the most challenging problems in today's interconnected world. The vast majority of the state-of-the-art in detecting misinformation is fully supervised, requiring a large number of high-quality human annotations. However, the availability of such annotations cannot be taken for granted, since it is very costly, time-consuming, and challenging to do so in a way that keeps up with the proliferation of misinformation. In this work, we are interested in exploring scenarios where the number of annotations is limited. In such scenarios, we investigate how to tap into a diverse number of resources that characterize a news article, henceforth referred to as "aspects" can compensate for the lack of labels. In particular, our contributions in this paper are twofold: 1) We propose the use of three different aspects: article content, context of social sharing behaviors, and host website/domain features, and 2) We introduce a principled tensor based embedding framework that combines all those aspects effectively. We propose `HiJoD` a 2-level decomposition pipeline which not only outperforms state-of-the-art methods with F1-scores of 74% and 81% on `Twitter` and `Politifact` datasets respectively, but also is an order of magnitude faster than similar ensemble approaches.

Keywords: Misinformation detection · Hierarchical tensor decomposition · Multi-aspect modeling · Ensemble learning · Semi-supervised classification

1 Introduction

In recent years, we have experienced the proliferation of websites and outlets that publish and perpetuate misinformation. With the aid of social media platforms, such misinformation propagates wildly and reaches a large number of the population, and can, in fact, have real-world consequences. Thus, understanding

Y. Dong et al. (Eds.): ECML PKDD 2020, LNAI 12461, pp. 406–422, 2021.
https://doi.org/10.1007/978-3-030-67670-4_25

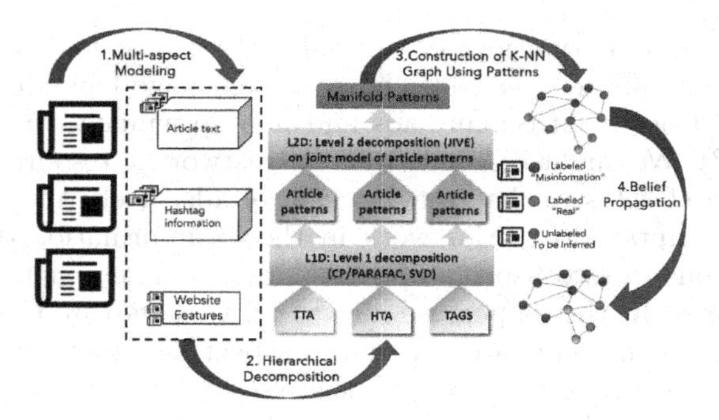

Fig. 1. Overview: We propose using three aspects1: a) content, b) social context (in the form of hashtags), and c) features of the website serving the content. We, further, propose `HiJoD` hierarchical approach for finding latent patterns derived from those aspects, generate a graphical representation of all articles in the embedding space, and conduct semi-supervised label inference of unknown articles.

and flagging misinformation on the web is an extremely important and timely problem, which is here to stay.

There have been significant advances in detecting misinformation from the article content, which can be largely divided into knowledge-based fact-checking and style-based approaches; [18,28] survey the landscape. Regardless of the particular approach followed, the vast majority of the state-of-the-art is fully supervised, requiring a large number of high-quality human annotations in order to learn the association between content and whether an article is misinformative. For instance, in [3], a decision-tree based algorithm is used to assess the credibility of a tweet, based on Twitter features. In another work, Rubin et al. [26] leverage linguistic features and a SVM-based classifier to find misleading information. Similarly, Horne and Adali [10] apply SVM classification on content-based features. There are several other works [6,12,13] for assessing credibility of news articles, all of which employ propagation models in a supervised manner. Collecting human annotation for misinformation detection is a complicated and time consuming task, since it is challenging and costly to identify human experts who can label news articles devoid of their own subjective views and biases, and possess all required pieces of information to be a suitable "oracle." However, there exist crowd-sourced schemes such as the browser extension "BS Detector"[1] which provide coarse labels by allowing users to flag certain articles as different types of misinformation, and subsequently flagging the entire source/domain as the majority label. Thus, we are interested in investigating methods that can compensate for the lack of large amounts of labels with leveraging different signals or aspects that pertain to an article. A motivating consideration is that we as humans empirically consider different aspects of a particular article in order to distinguish between misinformation and real information. For example, when we review a news article on a web site, and we have no prior knowledge about

[1] http://bsdetector.tech/.

the legitimacy of its information, we not only take a close look at the content of the article but also we may consider how the web page looks (e.g. does it look "professional"? Does it have many ads and pop-up windows that make it look untrustworthy?). We might conclude that untrustworthy resources tend to have more "messy" web sites in that they are full of ads, irrelevant images, pop-up windows, and scripts. Most prior work in the misinformation detection space considers only one or two aspects of information, namely content, headline or linguistic features. In this paper, we aim to fill that gap by proposing a comprehensive method that emulates this multi-aspect human approach for finding latent patterns corresponding to different classes, while at the same time leverages scarce supervision in order to turn those insights into actionable classifiers. In particular, our proposed method combines multiple aspects of article, including (a) article content (b) social sharing context and (c) source webpage context each of which is modeled as a tensor/matrix. The rationale behind using tensor based model rather than state-of-the-art approaches like deep learning methods is that, such approaches are mostly supervised methods and require considerable amount of labeled data for training. On the contrary, we can leverage tensor based approaches to find meaningful patterns using less labels. Later on, we will compare the performance of tensor-based modeling against deep learning methods when there is scarcity of labels. We summarize the contributions as follows:

- **Leveraging different aspects of misinformation**: In this work, we not only consider content-based information but also we propose to leverage multiple aspects for discriminating misinformative articles. In fact, we propose to create multiple models, each of which describes a distinct aspect of articles.
- **A novel hierarchical tensor based ensemble model**: We leverage a hierarchical tensor based ensemble method i.e., `HiJoD` to find manifold patterns that comprise multi-aspect information of the data.
- **Evaluation on real data**: We extensively evaluate `HiJoD` on two real world datasets i.e., `Twitter` and `Politifact`. `HiJoD` not only outperforms state-of-the-art alternatives with F1 score of 74% and 81% on above datasets respectively, but also is significantly faster than similar ensemble methods. We make our implementation publicly available[2] to promote the reproducibility.

2 Problem Formulation

Considering the following formulation of misinformation detection problem:

> **Given** N articles with associated 1) article text, social sharing context (hashtags that are used when sharing the article) 2) HTML source of the article's publisher webpage, and, 3) binary (misinformative/real) labels for p% of articles, **Classify** the remaining articles into the two classes.

[2] https://github.com/Saraabdali/HiJoD-ECMLPKDD.

At a high level, we aim to demonstrate the predictive power of incorporating multiple aspects of article content and context on the misinformation detection task, and consider doing so in a low-label setting due to practical challenges in data labeling for this task. Our proposed method especially focuses on (a) multi-aspect data modeling and representation choices for downstream tasks, (b) appropriate triage across multiple aspects, and (c) utility of a semi-supervised approach which is ideal in sparse label settings. In the next section, we detail our intuition and choices for each of these components.

We aim to develop a manifold approach which can discriminate misinformative and real news articles by leveraging content-based information in addition to other article aspects i.e. social context and publisher webpage information. To this end, we propose using tensors and matrices to analyze these aspects jointly. We develop a three-stage approach: (a) *multi-aspect article representation*: we first introduce three feature context representations (models) which describe articles from different points of view (aspects), (b) *Manifold patterns finding using a hierarchical approach*: In proposed `HiJoD`, we first decompose each model separately to find the latent patterns of the articles with respect to the corresponding aspect, then we use a strategy to find shared components of the individual patterns (c) *semi-supervised article inference*: finally, we focus on the inference task by construing a K-NN graph over the resulted manifold patterns and propagating a limited set of labels.

3 Proposed Methodology

3.1 Multi-aspect Article Representation

We first model articles with respect to different aspects. We suggest following tensors/matrix to model content, social context and source aspects respectively:

- **(Term $\times$ Term $\times$ Article) Tensor (`TTA`)**: The most straightforward way that comes into mind for differentiating between a fake news and a real one is to analyze the content of the articles. Different classes of news articles, i.e., fake and real classes tend to have some common words that co-occur within the text. Thus, we use a tensor proposed by [11] to model co-occurrence of these common words. This model not only represents content-based information but also considers the relations between the words and is stronger than widely used bag of words and tf/idf models. To this end, we first create a dictionary of all articles words and then slide a window across the text of each article and capture the co-occurred words. As a result, we will have a co-occurrence matrix for each article. By stacking all these matrices, we create a three mode tensor where the first two dimensions correspond to the indices of the words in the dictionary and the third mode indicates the article's ID. We may assign binary values or frequency of co-occurrence to entries of the tensors. However, as shown in [2] binary tensor is able to capture more nuance patterns. So, in this work we also use binary values.

- **(Hashtag × Term × Article) Tensor (HTA)**: Hashtags often show some trending across social media. Since, hashtags assigned to an article usually, convey social context information which is related to content of news article, we propose to construct a hashtag-content tensor to model such patterns. In this tensor, we want to capture co-occurrence of words within the articles and the hashtags assigned to them. For example, an article tagged with a hashtag `#USElection2016` probably consists of terms like "Donald Trump" or "Hillary Clinton". These kind of co-occurrences are meaningful and convey some patterns which may be shared between different categories of articles. The first two modes of this tensor correspond to hashtag and word indices respectively and the third mode is article mode.
- **(Article × HTML features) Matrix (TAGS)**: Another source of information is the trustworthiness of serving webpage. In contrast to reliable web resources which usually have a standard form, misleading web pages may often be messy and full of different advertisement, pop-ups and multimedia features such as images and videos. Therefore, we suggest to create another model to capture the look and the feel of the web page serving the content of news article. We can approximate look of a web page by counting HTML features and tags and then represent it by a (article, HTML feature) matrix. The rows and columns of this matrix indicate the article and hashtag IDs respectively. We fill out the entries by frequency of HTML tags in the web source of each article domain. Figure 2 demonstrates aforementioned aspects.

3.2 Hierarchical Decomposition

Now, the goal is to find manifold patterns with respect to all introduced aspects. To this end, we look for shared components of the latent patterns.

Level-1 Decomposition: Finding Article Patterns with Respect to Each Aspect. As mentioned above, first, we find the latent patterns of the articles with respect to each aspect. To do so, we decompose first two tensor models using Canonical Polyadic or CP/PARAFAC decomposition [8] into summation of rank one tensors, representing latent patterns within the tensors as follows:

$$\mathcal{X} \approx \Sigma_{r=1}^{R} \mathbf{a}_r \circ \mathbf{b}_r \circ \mathbf{c}_r \tag{1}$$

where $\mathbf{a}_r \in \mathbb{R}^I$, $\mathbf{b}_r \in \mathbb{R}^J$, $\mathbf{c}_r \in \mathbb{R}^K$ [24] and R is the rank of decomposition. The factor matrices are defined as $\mathbf{A} = [\mathbf{a}_1\ \mathbf{a}_2 \ldots \mathbf{a}_R]$, $\mathbf{B} = [\mathbf{b}_1\ \mathbf{b}_2 \ldots \mathbf{b}_R]$, and $\mathbf{C} = [\mathbf{c}_1\ \mathbf{c}_2 \ldots \mathbf{c}_R]$ where for TTA and HTA, $\mathbf{C}$ corresponds to the article mode and comprises the latent patterns of the articles with respect to content and social context respectively. The optimization problem for finding factor matrices is:

$$\min_{\mathbf{A},\mathbf{B},\mathbf{C}} = \left\| \mathcal{X} - \Sigma_{r=1}^{R} \mathbf{a}_r \circ \mathbf{b}_r \circ \mathbf{c}_r \right\|^2 \tag{2}$$

To solve the above optimization problem, we use Alternating Least Squares (ALS) approach which solves for any of the factor matrices by fixing the others

[24]. For the tag matrix, to keep the consistency of the model we suggest to use SVD decomposition ($\mathbf{X} \approx \mathbf{U\Sigma V}^T$) because CP/PARAFAC is one extension of SVD for higher mode arrays. In this case, $\mathbf{U}$ comprises the latent patterns of articles with respect to overall look of the serving webpage.

Level-2 Decomposition: Finding Manifold Patterns. Now, we want to put together article mode factor matrices resulted from individual decompositions to find manifold patterns with respect to all aspects. Let's $\mathbf{C_{TTA}} \in \mathbb{R}^{N \times r_1}$ and $\mathbf{C_{HTA}} \in \mathbb{R}^{N \times r_2}$ be the third factor matrices resulted from CP decomposition of TTA and HTA respectively and let's $\mathbf{C_{TAGS}} \in \mathbb{R}^{N \times r_3}$ be $\mathbf{U}$ matrix resulted from rank r_3 SVD of TAGS where N is the number of articles. Since all these three matrices comprise latent patterns of the articles, we aim at finding patterns which are shared between all. To do so, we concatenate the three article embedding matrices ($\mathbf{C}$) and decompose the joint matrix of size $(r_1 + r_2 + r_3) \times N$ to find shared components. Like level-1 decomposition, we can simply take the SVD of joint matrix. SVD seeks an accurate representation in least-square setting but to find a meaningful representation we need to consider additional information i.e., the relations between the components. Independent components analysis (ICA) tries to find projections that are statistically independent as follows:

$$[\mathbf{C_{TTA}}; \mathbf{C_{HTA}}; \mathbf{C_{TAGS}}] = \mathbf{AS} \tag{3}$$

where $\mathbf{A}$ is the corresponding shared factor of size $(r_1+r_2+r_3) \times N$ and $\mathbf{S}$ is the mixed signal of size $(r_1 + r_2 + r_3) \times R$. To achieve the statistical independence, it looks for projections that leads to projected data to be as far from Gaussian distribution as possible. The problem is that, SVD (PCA) and ICA may result in identical subspaces. In fact, this happens when the direction of greatest variation and the independent components span the same subspace [29].

In order to consider relations between components and find a meaningful representation, we can consider shared and unshared components for $\mathbf{C}$ matrices such that unshared components are orthogonal to shared ones. To this end, we propose to use the Joint and Individual Variation Explained (JIVE) for level-2 decomposition [22]. More precisely, let's consider $\mathbf{A_i}$ and $\mathbf{J_i}$ to be the matrices representing the individual structure and submatrix of the joint pattern for $i \in \{\mathtt{HTA}, \mathtt{TTA}, \mathtt{TAGS}\}$ respectively such that they satisfy the orthogonality constraint. Using JIVE method, we decompose each article mode factor matrix $\mathbf{C}$ as follows:

$$\mathbf{C_i} = \mathbf{J_i} + \mathbf{A_i} + \epsilon_i \tag{4}$$

To find $\mathbf{A}$ and $\mathbf{J}$ matrices, we use the approach presented in [22]. In other words, we fix $\mathbf{A}$ and find $\mathbf{J}$ that minimizes the following residual matrix:

$$\|\mathbf{R}\|^2 = \|\epsilon_{TTA}; \epsilon_{HTA}; \epsilon_{Tags}\|^2 \tag{5}$$

The joint structure $\mathbf{J}$ which minimizes $\|\mathbf{R}\|^2$ is equal to the rank r SVD of joint matrix when we remove the individual structure and in the same way individual structure for each $\mathbf{C}$ matrix is the rank r_i SVD of $\mathbf{C}$ matrix when we remove the joint structure [22]. Figure 2 and Algorithm 1 demonstrate the details.

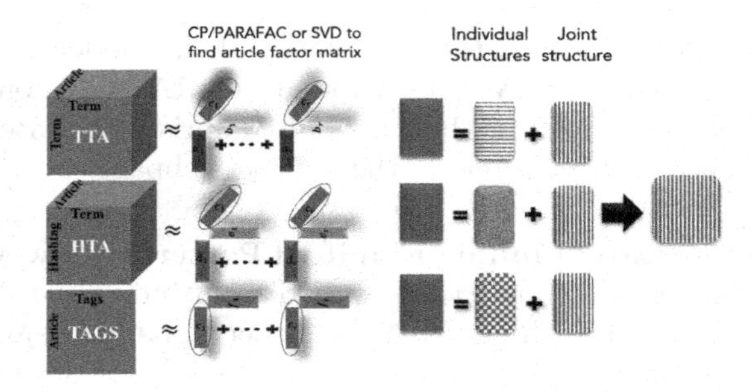

Fig. 2. `HiJoD` finds manifold patterns of the articles that can be used for classification.

Algorithm 1: `HiJoD` Hierarchical decomposition

```
1  Input:TTA, HTA, TAGS
2  Output: J_joint
3  First level of Decomposition
4  C_TTA =CP-ALS (TTA,r_1);
5  C_HTA =CP-ALS (HTA,r_2);
6  C_Tags =SVD (TAGS,r_3);
7  C^Joint = [C_TTA; C_HTA; C_Tags]
8  Second level of Decomposition
9  while ||R||^2 < ε do
10     J = UΣV^T
11     [J_TTA; J_HTA; J_Tags] =SVD (C_Joint, r_joint)//Calculate r_joint using
       Algorithm 2
12     for i ∈ { HTA,TTA,TAGS } do
13         A_i = C_i − J_i
14         A_i =SVD (A_i * (I − VV^T), r_i) //To satisfy the orthogonality constraint
15     end
16     C_i^new = C_i^joint − A_i
17     ε_i = |C_i^joint − C_i^new|
18     C^joint = C^new
19 end
```

3.3 Semi-supervised Article Inference

Previous step, provides us with a $N \times r$ matrix which comprises manifold patterns of N articles. In this step, we leverage this matrix to address the semi-supervised problem of classifying misinformation. Row i of this matrix represents article i in r dimensional space, we suggest to construct a K-NN graph using this matrix such that each node represents an article and edges are Euclidean distances between articles (rows of manifold patterns matrix) to model the similarity of articles. We utilize the Fast Belief Propagation (FaBP) algorithm as is described in [2,17] to propagate labels of known articles (fake or real) throughout K-NN graph. Belief propagation is a message passing-based algorithm. Let's $m_{j \hookrightarrow i}(x_i)$

Algorithm 2: Calculating the rank for joint and individual matrices

1 **Input:**$\alpha \in (0, 1)$, n_perm
2 **Output:** r
3 Let's λ_j be the j'th singular value of $\mathbf{X}, i = 1, \ldots, rank(\mathbf{X})$.
4 **while** $n \leq n\_perm$ **do**
5 Permute the columns within each X_i, and calculate the singular values of the resulting $\mathbf{C_{joint}}$
6 **end**
7 $\lambda_i^{perm} = 100(1 - \alpha)$ percentile of j'th singular values
8 Choose largest r such that $\forall j \leq r, \lambda_j \geq \lambda_j^{perm}$

indicates the message passes from node j to node i and N_i denotes all the neighboring nodes of node i. $m_{j \hookrightarrow i}(x_i)$ conveys the opinion of node j about the belief (label) of node i. Each node of a given graph G leverages messages received from neighboring nodes and calculates its belief iteratively as follows:.

$$b_i(x_i) \propto \prod_{j \in (N_i)} m_{j \hookrightarrow i}(x_i) \tag{6}$$

4 Experimental Evaluation

In this section, we discuss the datasets, baselines and experimental results.

4.1 Dataset Description

`Twitter` **dataset** To evaluate `HiJoD`, we created a new dataset by crawling Twitter posts contained links to articles and shared between June and August 2017. This dataset comprises 174 k articles from more than 652 domains. For labeling the articles, we used BS-Detector, which is a *crowd-sourced* toolbox in form of a browser extension, as ground truth. BS-Detector categorizes domains into different categories such as bias, clickbait, conspiracy, fake, hate, junk science, rumor, satire, and unreliable. We consider above categories as "misinformative" class. A key caveat behind BS-Detector, albeit being the most scalable and publicly accessible means of labeling articles at-large, is that labels actually pertains to the *domain* rather than the article itself. At the face of it, this sounds like the labels obtained are for an entirely different task, however, Helmstetter et al. [9] show that training for this "weakly labeled" task (using labels for the domains), and subsequently testing on labels pertaining to the articles, yields minimal loss in accuracy and labels still hold valuable information. Thus, we choose BS-Detector for ground truth. However, in order to make our experiments as fair as possible, in light of the above fact regarding the ground truth, we do as follows:

- We restrict the number of articles per domain we sample into our pool of articles. So, we experimented with randomly selecting a single article per

domain and iterating over 100 such sets. Since we have 652 different domains in `Twitter` dataset, in each iteration we chose 652 non-overlapping articles so, totally we examine different approaches for 65.2 K different articles.
- In order to observe the effect of using more articles per domain, we repeated the experiments for different number of articles per domain. We observed that the embedding that uses HTML tags receives a disproportionate boost in its performance. We attribute this phenomenon to the fact that all instances coming from the same domain have exactly the same HTML features, thus classifying correctly one of them implies correct classification for the rest, which is proportional to the number of articles per domain.
- We balance the dataset so that we have 50% fake and 50% real articles at any given run per method. We do so to have a fair evaluation setting and prevent the situation in which there is a class bias. To show the insensitivity of `HiJoD` to class imbalance, we also experiment on an imbalanced dataset.

`Politifact` dataset: For second dataset, we leverage FakeNewsNet dataset that the authors of [14] used for their experiments[3] [15,28]. This dataset consists of 1056 news articles from the `Politifact` fact checking website, 60% being real and 40% being fake. Using this imbalanced dataset, we can experiment how working on an imbalanced dataset may affect the proposed approach. Since not all of the signals we used from `Twitter` dataset exist in `Politifact` dataset, For this dataset we created the following embeddings:

- `TTA` tensor: To keep the consistency, we created `TTA` from articles text.
- User-News Interaction Embedding: We create a matrix which represent the users who tweets a specific news article, as proposed in [14].
- Publisher-News Interaction Embedding: We create another matrix to show which publisher published a specific news article, as proposed in [14].

Using the aforementioned signals, we can also test the efficacy of `HiJoD` when we leverage aspects other than those we proposed in this work.

4.2 Baselines for Comparison

As mentioned earlier, the two major contributions of `HiJoD` are: 1) the introduction of different aspects of an article and how they influence our ability to identify misinformation more accurately, and 2) how we leverage different aspects to find manifold patterns which belongs to different classes of articles. Thus, we conduct experiments with two categories of baseline to test each contribution separately:

Content-based Approaches to Test the Effect of Additional Aspects. We compare with state-of-the-art content-based approaches to measure the effect of introducing additional aspects (hashtags and HTML features) into the mix, and whether the classification performance improves. We compare against:

[3] https://github.com/KaiDMML/FakeNewsNet.

- `TTA` In [2], Bastidas et al. effectively use the co-occurrence tensor in a semi-supervised setting. They demonstrate how this tensor embedding outperforms other purely content-based state-of-the-art methods such as SVM on content-based features and Logistic regression on linguistic features [7,10]. Therefore, we select this method as the first baseline, henceforth referenced as "`TTA`". The differences between our results and the results reported by Bastidas et al. [2] is due to using different datasets. However, since we used the publicly available code by Bastidas et al.[4] [2], if we were to use the same data in [2] the results would be exactly the same.
- `tf_idf/SVM` is the well-known term frequency–inverse document frequency method widely used in text mining and information retrieval and illustrates how important a word is to a document. We create a `tf_idf` model out of articles text and apply SVM classifier on the resulted model.
- `Doc2Vec/SVM` is an NLP toolbox proposed by Le et al. [20] from Google. This model is a shallow, two-layer neural network that is trained to reconstruct linguistic contexts of document. This algorithm is an extension to word2vec which can generate vectors for words. Since the SVM classifier is commonly used on this model, we also leverage SVM for document classification.[5]
- `fastText` is an NLP library by Facebook Research that can be used to learn word representations to efficiently classify document. It has been shown that `fastText` results are on par with deep learning models in terms of accuracy but an order of magnitude faster in terms of performance.[6]
- `GloVe/LSTM` GloVe is an algorithm for obtaining vector representations of the words. Using an aggregated global word-word co-occurrence, this method results in a linear substructures of the word vector space. We use the method proposed in [19,21] and we create a dictionary of unique words and leverage Glove to map indices of words into a pre-trained word embedding. Finally, as suggested, we use LSTM to classify articles.[7]

Ensemble Approaches to Test the Efficacy of Our Fused Method. Another way to jointly derive the article patterns, specially when the embeddings are tensors and matrices, is to couple matrices and tensors on shared mode(s) i.e., article mode in this work. In this technique which called coupled Matrix and Tensor Factorization (`CMTF`), the goal is to find optimized factor matrices by considering all different optimization problems we have for individual embeddings. Using our proposed embeddings, we the optimization problem is:

$$\min_{\mathbf{A},\mathbf{B},\mathbf{C},\mathbf{D},\mathbf{E},\mathbf{F}} \|\texttt{TTA} - [\mathbf{A},\mathbf{B},\mathbf{C}]\|^2 + \|\texttt{HTA} - [\mathbf{D},\mathbf{E},\mathbf{C}]\|^{\mathbf{2}} + \|\texttt{TAGS} - \mathbf{C}\mathbf{F}^{\mathbf{T}}\|^{\mathbf{2}} \quad (7)$$

where $[\mathbf{A},\mathbf{B},\mathbf{C}]$ denotes $\Sigma_{r=1}^{R}\mathbf{a}_r \circ \mathbf{b}_r \circ \mathbf{c}_r$, and $\mathbf{C}$ is the shared article mode as shown in Fig. 3. In order to solve the optimization problem above, we use the

[4] https://github.com/Saraabdali/Fake-News-Detection-_ASONAM-2018.
[5] https://github.com/seyedsaeidmasoumzadeh/Binary-Text-Classification-Doc2vec-SVM.
[6] https://github.com/facebookresearch/fastText.
[7] https://github.com/prakashpandey9/Text-Classification-Pytorch.

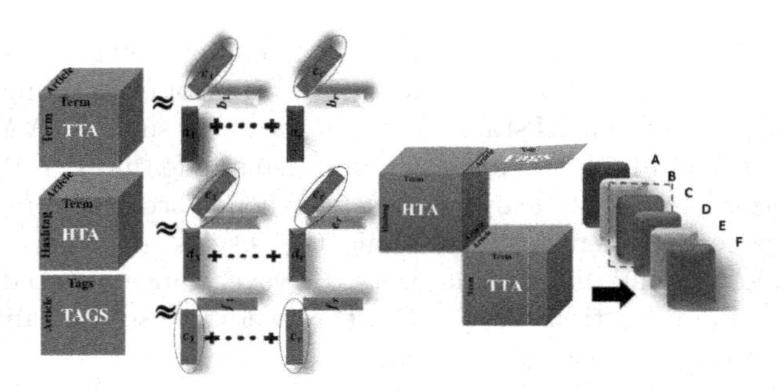

Fig. 3. Couple tensor matrix factorization for finding shared patterns of the articles.

approach introduced in [5], which proposes *all-at-once-optimization* by computing the gradient of every variable of the problem, stacking the gradients into a long vector, and applying gradient descent. There is an advanced version of coupling called `ACMTF` that uses weights for rank-one components to consider both shared and unshared ones [1]. We applied `ACMTF` as well and it led to similar results, however, slower than standard `CMTF`. Thus, we just report the result of `CMTF`.

Moreover, as mentioned earlier, to derive the manifold pattern which is shared between the aspects, we can leverage different mathematical approaches such as:

- `SVD` (Singular Value Decomposition)
- `JICA` (Joint Independent Component Analysis)

on $\mathbf{C_{joint}}$. To measure the performance of `JIVE` method for finding manifold patterns, we will also examine the above approaches for level-2 decomposition.

4.3 Comparing with Baselines

Implementation. We implemented `HiJoD` and all other approaches in MATLAB using Tensor Toolbox version 2.6. Moreover, to implement `JICA` approach, we used FastICA[8] , a fast implementation of ICA. For the baseline approach `CMTF` we used the Toolbox[9] in [4,5]. For the belief propagation step, we used implementation introduced in [17]. Based on the experiments reported in [2], we employed a sliding window of size 5 for capturing the co-occurring words. Using AutoTen [25], which finds the best rank for tensors, we found out the best rank for `TTA` and `HTA` is 10 and 40 respectively. For the `TAGS` embedding, we took the full SVD to capture the significant singular values and set the `TAGS` SVD rank to 20. Since for `CMTF` approach we have to choose the same rank, as a heuristic, we used the range of ranks for individual embeddings (10 for `HTA` and 40 for `TTA`) and again searched for the ensemble rank in this range. Based on

[8] https://github.com/aludnam/MATLAB/tree/master/FastICA_25.
[9] http://www.models.life.ku.dk/joda/CMTF_Toolbox.

Table 1. F1 score of HiJoD outperforms all content-based methods on Twitter dataset.

%Labels	tf_idf/SVM	Doc2Vec/SVM	fastText	GloVe/LSTM	TTA	HiJoD
10	0.500 ± 0.032	0.571 ± 0.092	0.562 ± 0.031	–	0.582 ± 0.018	**0.693 ± 0.009**
20	0.461 ± 0.013	0.558 ± 0.067	0.573 ± 0.027	–	0.598 ± 0.018	**0.717 ± 0.010**
30	0.464 ± 0.015	0.548 ± 0.048	0.586 ± 0.024	0.502 ± 0.087	0.609 ± 0.019	**0.732 ± 0.011**
40	0.475 ± 0.023	0.547 ± 0.034	0.592 ± 0.028	0.503 ± 0.060	0.614 ± 0.022	**0.740 ± 0.010**

Table 2. HiJoD outperforms coupling approaches in terms of F1 score in both datasets.

%Labels	Twitter			Politifact	
	CMTF	CMTF++	HiJoD	CMTF	HiJoD
10	0.657 ± 0.009	0.657 ± 0.010	**0.693 ± 0.009**	0.733 ± 0.007	**0.766 ± 0.007**
20	0.681 ± 0.010	0.681 ± 0.009	**0.717 ± 0.010**	0.752 ± 0.012	**0.791 ± 0.007**
30	0.691 ± 0.010	0.692 ± 0.009	**0.732 ± 0.011**	0.774 ± 0.006	**0.802 ± 0.007**
40	0.699 ± 0.010	0.699 ± 0.009	**0.740 ± 0.010**	0.776 ± 0.004	**0.810 ± 0.008**

our experiments, rank 30 leads to the best results in terms of F1-score. Moreover, grid search over 1–30 nearest neighbors yielded choice of 5 neighbors for CMTF and 15 for HiJoD. For the rank of joint model as well as individual and joint structures i.e., $\mathbf{A_i}$ and $\mathbf{J_i}$, we used the strategy proposed in [22] and for reproducibility purposes is demonstrated in Algorithm 2. The intuition behind this approach is to find the rank of joint structure i by comparing the singular values of the original matrix with the singular values of n_{perm} randomly permuted matrices. If the jth singular value in the original matrix is $\geq 100(1-\alpha)$ percentile of the jth singular value of n_{perm} matrices, we keep it as a significant one. Number of these significant singular values shows the rank. n_{perm} and α are usually set to 100 and 0.05 respectively. For timing experiment, we used the following configuration:

- CPU: Intel(R) Core(TM) i5-8600 K CPU @3.60 GHz
- OS: CentOS Linux 7 (Core)
- RAM: 40 GB

Testing the Effect of Different Aspects. This experiment refers to the first category of baselines i.e., state-of the-art content-based approaches introduced earlier. Table 1 demonstrates the comparison; label% shows the percentage of known data used for propagation/training of the models. As reported, HiJoD outperforms all content-based approaches significantly. For example, using only 10% of known labels the F1 score of HiJoD is around 12% and 13% more than Doc2Vec/SVM and fastText respectively. In case of GloVe/LSTM, due to small size of training set i.e., 10 and 20%, the model overfits easily which shows the strength of HiJoD against deep models when there is scarcity of labeled data for training. Moreover, our ensemble model that leverages TTA as one of its embeddings beats the individual decomposition of TTA which illustrates the effectiveness of adding other aspects of the data for modeling the news articles.

Table 3. Comparing execution time (Secs.) of CMTF against HiJoD on Twitter dataset shows that HiJoD is an order of magnitude faster than CMTF approach.

	L1D		L2D+Rank finding		Total time (Secs.)	
L1D Rank	CMTF	CP/SVD	CMTF	JIVE	CMTF	HiJoD
5	352.96	7.63	–	13.69	352.96	21.32
10	1086.70	16.35	–	49.69	1086.70	66.04
20	11283.51	44.25	–	133.70	11283.51	177.95
30	13326.80	84.76	–	278.43	13326.80	363.19

Testing the Efficacy of Fused Method in HiJoD. For the second category of baselines, we compare HiJoD against the recent work for joint decomposition of tensors/matrices i.e., CMTF. As discussed earlier, we couple TTA, HTA and TAGS on shared article mode. Since the"term" mode is also shared between TTA and HTA, we also examine the CMTF by coupling on both article and term modes, henceforth referenced as CMTF++ in the experimental results. For the Politifact dataset, there is only one shared mode i.e., article mode. therefore, we only compare against CMTF approach. The experimental results for theses approaches are shown in Table 2. As illustrated, HiJoD leads to higher F1 score which confirms the effectiveness of HiJoD for jointly classification of articles. One major drawback of CMTF approach is that, we have to use a unique decomposition rank for the joint model which may not fit all embeddings and may lead to losing some informative components or adding useless noisy components due to inappropriate rank of decomposition. Another drawback of this technique is that, as we add more embeddings, the optimization problem becomes more and more complicated which may cause the problem become unsolvable and infeasible in terms of time and resources. The time efficiency of HiJoD against CMTF approach is reported in Table 3; we refer to level-1 and level-2 decompositions as L1D and L2D respectively. As shown, for all ranks, HiJoD is an order of magnitude faster than CMTF due to the simplicity of optimization problem in comparison to Eq. 7 which means HiJoD is more applicable for real world problems.

Testing the Efficacy of Using JIVE for Level-2 Decomposition. In this experiment we want to test the efficacy of JIVE against other approaches for deriving the joint structure of $\mathbf{C}$. The evaluation results for this experiment are reported in Table 4. As shown, SVD and JICA resulted in same classification performance which as explained earlier indicates that the directions of greatest variation and the independent components span the same subspace [29]. However, the F1 score of HiJoD is higher than two other methods on both datasets which practically justifies that considering orthogonal shared and unshared parts and minimizing the residual can improve the performance of naive SVD.

HiJoD vs. Single Aspect Modeling. Finally, we want to investigate how adding other aspects of the articles affect the classification performance. To this end, we decompose our proposed embeddings i.e., TTA, HTA and TAGS extracted

Table 4. Comparing the efficacy of different approaches for level-2 decomposition illustrates that using JIVE approach for HiJoD ouperforms other methods.

%Labels	Twitter			Politifact		
	JICA	SVD	JIVE	JICA	SVD	JIVE
10	0.684 ± 0.009	0.685 ± 0.017	**0.693 ± 0.009**	0.749 ± 0.006	0.756 ± 0.009	**0.766 ± 0.007**
20	0.710 ± 0.009	0.712 ± 0.014	**0.717 ± 0.010**	0.775 ± 0.006	0.783 ± 0.009	**0.791 ± 0.007**
30	0.724 ± 0.009	0.724 ± 0.018	**0.732 ± 0.011**	0.786 ± 0.006	0.796 ± 0.012	**0.802 ± 0.007**
40	0.734 ± 0.009	0.735 ± 0.012	**0.740 ± 0.010**	0.797 ± 0.006	0.807 ± 0.008	**0.810 ± 0.008**

from Twitter dataset individually and leverage the **C** matrices for classification. The result of this experiment is demonstrated in Fig. 4. As mentioned before, in contrast to CMTF, in HiJoD we can use different ranks for different aspects as we did so previously. However, in order to conduct a fair comparison between individual embeddings and HiJoD, we merge the embeddings of the same rank. It is worth mentioning that, performance of HiJoD is higher than what is shown in Fig. 4 due to concatenation of **C** matrices of the best rank. So, this result is just for showing the effect of merging different aspects by fixing other parameters i.e., rank and k. As shown, the HiJoD even when we join embeddings of the same rank and do not use the best of which outperforms individual decompositions.

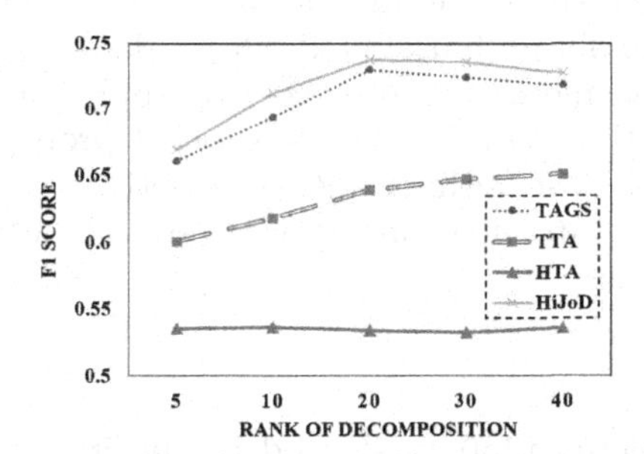

Fig. 4. F1-score of using individual embeddings vs. HiJoD. Even when the best rank of each embedding is not used, HiJoD outperforms individual decompositions

5 Related Work

Ensemble Modeling for Misinformation Detection. A large number of misinformation detection approaches focus on a single aspect of the data, such as article content [2,30], user features [31], and temporal properties [18]. There exist, however, recent approaches that integrate various aspects of an article in the same model. For example, in [14] the authors propose an ensemble model for finding fake news. In this approach, a bag of words embedding is used to model content-based information, while in this work, we leverage a tensor model i.e.,

TTA which not only enables us to model textual information, but also is able to capture nuanced relations between the words. The different sources of information used in [14] (user-user, user-article and publisher-article interactions) do not overlap with the aspects introduced here (hashtags and HTML features), however, in our experiments we show that HiJoD effectively combines both introduced aspects as well as the ones in [14]. In another work [16], news contents and user comments are exploited jointly to detect fake news. Although user comment is a promising aspect, still the main focus is on the words of comments. However, we use HTML tags and hashtags in addition to the textual content.

Semi-supervised Learning/Label Propagation Models. The majority of mono-aspect modeling proposed so far leverage a supervised classifier. For instance, in [7] a logistic regression classifier is used which employs linguistic and semantic features for classification. In [10], authors apply a SVM classifier for content based features. Moreover, some works have been done using recurrent neural network (RNN) and Dynamic Series-Time Structure (DSTS) models [23,27]. In contrary to the aforementioned works, we use a model which achieves very precise classification when leverage very small amount of ground truth. There are some proposed methods that mainly rely on propagation models. For example, in [12] the authors proposed a hierarchical propagation model on a suggested three-layer credibility network. In this work, a hierarchical structure is constructed using event, sub-event and message layers, even though a supervised classifier is required to obtain initial credibility values. In [13], a PageRank-like credibility propagation method is proposed to apply on a network of events, tweets and users. In this work, we leverage belief propagation to address the semi-supervised problem of misinformation detection. We show that proposed approach outperforms state-of-the-art approaches in label scarcity settings.

6 Conclusion

In this paper, we propose HiJoD, a 2-level decomposition pipeline that integrate different aspects of an article towards more precise discovery of misinformation on the web. Our contribution is two-fold: we introduce novel aspects of articles which we demonstrate to be very effective in classifying misinformative vs. real articles, and we propose a principled way of fusing those aspects leveraging tensor methods. We show that HiJoD is not only able to detect misinformation in a semi-supervised setting even when we use only 10% of the labels but also an order of magnitude faster than similar ensemble approaches in terms of execution time. Experimental results illustrates that HiJoD achieves F1 score of roughly 74% and 81% on Twitter and Politifact datasets respectively which outperforms state-of-the-art content-based and neural network based approaches.

Acknowledgments. The authors would like to thank Gisel Bastidas for her invaluable help with data collection. Research was supported by a UCR Regents Faculty Fellowship, a gift from Snap Inc., the Department of the Navy, Naval Engineering Education Consortium under award no. N00174-17-1-0005, and the National Science

Foundation Grant no. 1901379. Any opinions, findings, and conclusions or recommendations expressed in this material are those of the author(s) and do not necessarily reflect the views of the funding parties.

References

1. Acar, E., Levin-Schwartz, Y., Calhoun, V., Adali, T.: ACMTF for fusion of multimodal neuroimaging data and identification of biomarkers. In: 2017 25th European Signal Processing Conference (EUSIPCO) (2017)
2. Bastidas, G.G., Abdali, S., Shah, N., Papalexakis, E.E.: Semi-supervised content-based detection of misinformation via tensor embeddings. In: ASONAM (2018)
3. Castillo, C., Mendoza, M., Poblete, B.: Information credibility on twitter. In: International Conference on World Wide Web, pp. 675–684. WWW 2011 (2011)
4. Acar, E., Lawaetz, A.J., Rasmussen, M.A., Bro, R.: Structure-revealing data fusion model with applications in metabolomics. In: IEE EMBS (2013)
5. Acar, E., Kolda, T., Dunlavy, D.: All-at-once optimization for coupled matrix and tensor factorizations. In: KDD Workshop on Mining and Learning with Graphs (2011)
6. Gupta, M., Zhao, P., Han, J.: Evaluating event credibility on twitter. In: SIAM, pp. 153–164 (2012)
7. Hardalov, M., Koychev, I., Nakov, P.: In search of credible news. In: Dichev, C., Agre, G. (eds.) AIMSA 2016. LNCS (LNAI), vol. 9883, pp. 172–180. Springer, Cham (2016). https://doi.org/10.1007/978-3-319-44748-3_17
8. Harshman, R.A.: Foundations of the PARAFAC procedure: models and conditions for an "explanatory" multi-modal factor analysis. UCLA Working Papers in Phonetics, 16(1), 84 (1970)
9. Helmstetter, S., Paulheim, H.: Weakly supervised learning for fake news detection on twitter. In: (ASONAM), pp. 274–277 (2018)
10. Horne, B.D., Adali, S.: This just. In: Fake news packs a lot in title, uses simpler, repetitive content in text body, more similar to satire than real news. CoRR (2017)
11. Hosseinimotlagh, S., Papalexakis, E.E.: Unsupervised content-based identification of fake news articles with tensor decomposition ensembles (2017)
12. Jin, Z., Cao, J., Jiang, Y.G., Zhang, Y.: News credibility evaluation on microblog with a hierarchical propagation model, pp. 230–239. (ICDM) (2014)
13. Jin, Z., Cao, J., Zhang, Y., Luo, J.: News verification by exploiting conflicting social viewpoints in microblogs (2016)
14. Shu, K., Sliva, A., Wang, S., Liu, H.: Beyond news contents: the role of social context for fake news detection. In: WSDM, pp. 312–320 (2019)
15. Shu, K., Cui, L., Wang, S., Liu, D., Liu, H.: Exploiting tri-relationship for fake news detection. arXiv preprint arXiv:1712.07709 (2017)
16. Shu, K., Cui, L., Wang, S., Lee., D., Liu, H.: Defend: explainable fake news detection. In: Proceedings of 25th ACM SIGKDD International Conference on Knowledge Discovery and Data Mining (2019)
17. Koutra, D., Ke, T.-Y., Kang, U., Chau, D.H.P., Pao, H.-K.K., Faloutsos, C.: Unifying guilt-by-association approaches: theorems and fast algorithms. In: Gunopulos, D., Hofmann, T., Malerba, D., Vazirgiannis, M. (eds.) ECML PKDD 2011. LNCS (LNAI), vol. 6912, pp. 245–260. Springer, Heidelberg (2011). https://doi.org/10.1007/978-3-642-23783-6_16
18. Kumar, S., Shah, N.: False information on web and social media: A survey. arXiv preprint arXiv:1804.08559 (2018)

19. Lai, S., Xu, L., Liu, K., Zhao, J.: Recurrent convolutional neural networks for text classification. In: Bonet, B., Koenig, S. (eds.) AAAI, vol. 333, pp. 2267–2273 (2015)
20. Le, Q., Mikolov, T.: Distributed representations of sentences and documents. In: ICML (2014)
21. Lin, Z., et al.: A structured self-attentive sentence embedding (2017)
22. Lock, E., Hoadley, K., Marron, J., Nobel, A.: Joint and individual variation explained (JIVE) for integrated analysis of multiple data types. Ann. Appl. Stat. **7**, 523–542 (2013)
23. Ma, J., Gao, W., Wei, Z., Lu, Y., Wong, K.F.: Detect rumors using time series of social context information on microblogging websites. In: CIKM (2015)
24. Papalexakis, E.E., Faloutsos, C., Sidiropoulos, N.D.: Tensors for data mining and data fusion: Models, applications, and scalable algorithms. ACM Trans. Intell. Syst. Technol. **8**, 16:1–16:44 (2016)
25. Papalexakis, E.E: Automatic unsupervised tensor mining with quality assessment. In: SIAM SDM (2016)
26. Rubin, V.L., Conroy, N.J., Chen, Y., Cornwell, S.: Fake news or truth? using satirical cues to detect potentially misleading news (2016)
27. Ruchansky, N., Seo, S., Liu, Y.: CSI: A hybrid deep model for fake news. CoRR (2017)
28. Shu, K., Sliva, A., Wang, S., Tang, J., Liu, H.: Fake news detection on social media: a data mining perspective. SIGKDD Explor. Newsl. **19**(1), 22–36 (2017)
29. Vasilescu, M.A.O.: A Multilinear (Tensor) Algebraic Framework for Computer Graphics, Computer Vision and Machine Learning. Ph.D. thesis, Citeseer (2012)
30. Wu, L., Li, J., Hu, X., Liu, H.: Gleaning wisdom from the past: early detection of emerging rumors in social media. In: SIAM SDM, pp. 99–107 (2017)
31. Wu, L., Liu, H.: Tracing fake-news footprints: characterizing social media messages by how they propagate. In: WSDM, pp. 637–645 (2018)

A Deep Dive into Multilingual Hate Speech Classification

Sai Saketh Aluru, Binny Mathew(✉), Punyajoy Saha, and Animesh Mukherjee

Indian Institute of Technology Kharagpur, Kharagpur, India
{saisakethaluru,binnymathew,punyajoys}@iitkgp.ac.in,
animeshm@cse.iitkgp.ac.in

Abstract. Hate speech is a serious issue that is currently plaguing the society and has been responsible for severe incidents such as the genocide of the Rohingya community in Myanmar. Social media has allowed people to spread such hateful content even faster. This is especially concerning for countries which lack hate speech detection systems. In this paper, using hate speech dataset in 9 languages from 16 different sources, we perform the first extensive evaluation of multilingual hate speech detection. We analyze the performance of different deep learning models in various scenarios. We observe that in low resource scenario LASER embedding with Logistic regression perform the best, whereas in high resource scenario, BERT based models perform much better. We also observe that simple techniques such as translating to English and using BERT, achieves competitive results in several languages. For cross-lingual classification, we observe that data from other languages seem to improve the performance, especially in the low resource settings. Further, in case of zero-shot classification, evaluation on Italian and Portuguese dataset achieve good results. Our proposed framework could be used as an efficient solution for low-resource languages. These models could also act as good baselines for future multilingual hate speech detection tasks. Our code (Code: https://github.com/punyajoy/DE-LIMIT) and models (Models: https://huggingface.co/Hate-speech-CNERG) are available online.

Warning: contains material that many will find offensive or hateful.

Keywords: Hate speech · Multilingual · Classification · BERT · Embeddings

1 Introduction

Online social media has allowed dissemination of information at a faster rate than ever [22,23]. This has allowed bad actors to use this for their nefarious purposes such as propaganda spreading, fake news, and *hate speech*. Hate speech is defined as a "direct and serious attack on any protected category of people based on their

S. S. Aluru and B. Mathew—Equal Contribution.

Y. Dong et al. (Eds.): ECML PKDD 2020, LNAI 12461, pp. 423–439, 2021.
https://doi.org/10.1007/978-3-030-67670-4_26

Table 1. Examples of hate speech.

Text	Hate speech?
I f**king hate ni**ers!	Yes
Jews are the worst people on earth and we should get rid of them	Yes
"6 million was not enough. next time ovens will be the least of your concerns # sixmillionmore"	Yes
Mexicans are f**king great people!	No

race, ethnicity, national origin, religion, sex, gender, sexual orientation, disability or disease" [11]. Representative examples of hate speech are provided in Table 1.

Hate speech is increasingly becoming a concerning issue in several countries. Crimes related to hate speech have been increasing in the recent times with some of them leading to severe incidents such as the genocide of the Rohingya community in Myanmar, the anti-Muslim mob violence in Sri Lanka, and the Pittsburg shooting. Frequent and repetitive exposure to hate speech has been shown to desensitize the individual to this form of speech and subsequently to lower evaluations of the victims and greater distancing, thus increasing outgroup prejudice [35]. The public expressions of hate speech has also been shown to affect the devaluation of minority members [18], the exclusion of minorities from the society [26], and the discriminatory distribution of public resources [12].

While the research in hate speech detection has been growing rapidly, one of the current issues is that majority of the datasets are available in English language only. Thus, hate speech in other languages are not detected properly and this could be detrimental. While there are few datasets [3,27] in other language available, as we observe, they are relatively small in size.

In this paper, we perform the first large scale analysis of multilingual hate speech by analyzing the performance of deep learning models on 16 datasets from 9 different languages. We consider two different scenarios and discuss the classifier performance. In the first scenario (monolingual setting), we only consider the training and testing from the same language. We observe that in low resource scenario models using LASER embedding with Logistic regression perform the best, whereas in high resource scenario, BERT based models perform much better. We also observe that simple techniques such as translating to English and using BERT, achieves competitive results in several languages. In the second scenario (multilingual setting), we consider training data from all the other languages and test on one target language. Here, we observe that including data from other languages is quite effective especially when there is almost no training data available for the target language (aka zero shot). Finally, from the summary of the results that we obtain, we construct a catalogue indicating which model is effective for a particular language depending on the extent of the data available. We believe that this catalogue is one of the most important contributions of our work which can be readily referred to by future researchers working to advance the state-of-the-art in multilingual hate speech detection.

The rest of the paper is structured as follows. Section 2 presents the related literature for hate speech classification. In Sect. 3, we present the datasets used for the analysis. Section 4 provides details about the models and experimental settings. In Sect. 5, we note the key results of our experiments. In Sect. 6 we discuss the results and provide error analysis.

2 Related Works

Hate speech lies in a *complex nexus with freedom of expression, individual, group and minority rights, as well as concepts of dignity, liberty and equality* [16]. Computational approaches to tackle hate speech has recently gained a lot of interest. The earlier efforts to build hate speech classifiers used simple methods such as dictionary look up [19], bag-of-words [6]. Fortuna *et al.* [13] conducted a comprehensive survey on this subject.

With the availability of larger datasets, researchers started using complex models to improve the classifier performance. These include deep learning [37] and graph embedding techniques [30] to detect hate speech in social media posts. Zhang *et al.* [37] used deep neural network, combining convolutional and gated recurrent networks to improve the results on 6 out of 7 datasets used. In this paper, we have used the same CNN-GRU model for one of our experimental settings (monolingual scenario).

Research into the multilingual aspect of hate speech is relatively new. Datasets for languages such as Arabic and French [27], Indonesian [21], Italian [33], Polish [29], Portuguese [14], and Spanish [3] have been made available for research. To the best of our knowledge, very few works have tried to utilize these datasets to build multilingual classifiers. Huang *et al.* [20] used Twitter hate speech corpus from five languages and annotated them with demographic information. Using this new dataset they study the demographic bias in hate speech classification. Corazza *et al.* [8] used three datasets from three languages (English, Italian, and German) to study the multilingual hate speech. The authors used models such as SVM, and Bi-LSTM to build hate speech detection models. Our work is different from these existing works as we perform the experiment on a much larger set of languages (9) using more datasets (16). Our work tries to utilize the existing hate speech resources to develop models that could be generalized for hate speech detection in other languages.

3 Dataset Description

We looked into the datasets available for hate speech and found 16 publicly[1] available sources in 9 different languages[2]. One of the immediate issues, we observed was the mixing of several types of categories (offensive, profanity, abusive, insult

[1] Note that although Table 2 contains 19 entries, there are three occurrences of Ousidhoum *et al.* [27] and two occurrences of Basile *et al.* [3] for different languages.

[2] We relied on http://hatespeechdata.com for most of the datasets.

etc.). Although these categories are related to hate speech, they should not be considered as the same [9]. For this reason, we only use two labels: *hate speech* and *normal*, and discard other labels. Next, we explain the datasets in different languages. The overall dataset statistics are noted in Table 2.

Arabic: We found two arabic datasets that were built for hate speech detection.

- Mulki *et al.* [25] : A Twitter dataset[3] for hate speech and abusive language. For our task, we ignored the abusive class and only considered the hate and normal class.
- Ousidhoum *et al.* [27]: A Twitter dataset[4] with multi-label annotations. We have only considered those datapoints which have either hate speech or normal in the annotation label.

Table 2. Dataset details

Language	Dataset	Source	Hate	Non-hate	Total
Arabic	Mulki *et al.* [25]	Twitter	468	3,652	4,120
	Ousidhoum *et al.* [27]	Twitter	755	915	1,670
English	Davidson *et al.* [9]	Twitter	1,430	4,163	5,593
	Gibert *et al.* [17]	Stormfront	1,196	9,748	10,944
	Waseem *et al.* [36]	Twitter	759	5,545	6,304
	Basile *et al.* [3]	Twitter	5,390	7,415	12,805
	Ousidhoum *et al.* [27]	Twitter	1,278	661	1,939
	Founta *et al.* [15]	Twitter	4,948	53,790	58,738
German	Ross *et al.* [32]	Twitter	54	315	369
	Bretschneider *et al.* [5]	Facebook	625	5,161	5,786
Indonesian	Ibrohim *et al.* [21]	Twitter	5,561	7,608	13,169
	Alfina *et al.* [1]	Twitter	260	453	713
Italian	Sanguinetti *et al.* [33]	Twitter	231	1,329	1,560
	Bosco *et al.* [4]	Facebook & Twitter	3,355	4,645	8,000
Polish	Ptaszynski *et al.* [29]	Twitter	598	9,190	9,788
Portuguese	Fortuna *et al.* [14]	Twitter	1,788	3,882	5,670
Spanish	Basile *et al.* [3]	Twitter	2,228	3,137	5,365
	Pereira *et al.* [28]	Twitter	1,567	4,433	6,000
French	Ousidhoum *et al.* [27]	Twitter	399	821	1,220
Total			32,890	126,863	159,753

[3] https://github.com/Hala-Mulki/L-HSAB-First-Arabic-Levantine-HateSpeech-Dataset.

[4] https://github.com/HKUST-KnowComp/MLMA_hate_speech.

English: Majority of the hate speech datasets are available in English language. We select six such publicly available datasets.

- Davidson *et al.* [9] provided a three class Twitter dataset[5], the classes being hate speech, abusive speech, and normal. We have only considered the hate speech and normal class for our task.
- Gibert *et al.* [17] provided a hate speech dataset[6] consisting sentences from Stormfront[7], a white supremacist forum. Each sentence is tagged as either hate or normal.
- Waseem *et al.* [36] provided a Twitter dataset[8] annotated into classes: sexism, racism, and neither. We considered the tweets tagged as sexism or racism as hate speech and neither class as normal.
- Basile *et al.* [3] provided multilingual Twitter dataset[9] for hate speech against immigrants and women. Each post is tagged as either hate speech or normal.
- Ousidhoum *et al.* [27] provided Twitter dataset (See Footnote 6) with multi-label annotations. We have only considered those datapoints which have either hate speech or normal in the annotation label.
- Founta *et al.* [15] provided a large dataset[10] of 100K annotations divided in four classes: hate speech, abusive, spam, and normal. For our task, we have only considered the datapoints marked as either hate or normal, and ignored the other classes.

German: We select two datasets available in German language.

- Ross *et al.* [32] provided a German hate speech dataset[11] for the refugee crisis. Each tweet is tagged as hate speech or normal.
- Bretschneider *et al.* [5] provided a Facebook hate speech dataset[12] against foreigners and refugees.

Indonesian. We found two datasets for the Indonesian language.

- Ibrohim *et al.* [21] provided an Indonesian multi-label hate speech and abusive dataset[13]. We only consider the hate speech label for our task and other labels are ignored.
- Alfina *et al.* [1] provided an Indonesian hate speech dataset[14]. Each post is tagged as hateful or normal.

[5] https://github.com/t-davidson/hate-speech-and-offensive-language.
[6] https://github.com/aitor-garcia-p/hate-speech-dataset.
[7] www.stormfront.org.
[8] https://github.com/zeerakw/hatespeech.
[9] https://github.com/msang/hateval.
[10] https://github.com/ENCASEH2020/hatespeech-twitter.
[11] https://github.com/UCSM-DUE/IWG_hatespeech_public.
[12] http://www.ub-web.de/research/.
[13] https://github.com/okkyibrohim/id-multi-label-hate-speech-and-abusive-language-detection.
[14] https://github.com/ialfina/id-hatespeech-detection.

Italian. We found two datasets for the Italian language.

- Sanguinetti *et al.* [33] provided an Italian hate speech dataset[15] against the minorities in Italy.
- Bosco *et al.* [4] provided hate speech dataset[16] collected from Twitter and Facebook.

Polish. We found only one dataset for the Polish language

- Ptaszynski *et al.* [29] provided a cyberbullying dataset[17] for the Polish language. We have only considered hate speech and normal class for our task.

Portuguese. We found one dataset for the Portuguese language

- Fortuna *et al.* [14] developed a hierarchical hate speech dataset[18] for the Portuguese language. For our task, we have used the binary class of hate speech or normal.

Spanish. We found two dataset for the Spanish language.

- Basile *et al.* [3] provided multilingual hate speech dataset (See Footnote 11) against immigrants and women.
- Pereira *et al.* [28] provided hate speech dataset[19] for the Spanish language.

French

- Ousidhoum *et al.* [27] provided Twitter dataset (See Footnote 6) with multi-label annotations. We have only considered those data points which have either hate speech or normal in the annotation label.

4 Experiments

For each language, we combine all the datasets and perform stratified train/ validation/ test split in the ratio 70%/10%/20%. For all the experiments, we use the same splits of train/val/test. Thus, the results are comparable across different models and settings. We report macro F1-score to measure the classifier performance. In case we select a subset of the dataset for the experiment, we repeated the subset selection with 5 different random sets and report the average performance. This would help to reduce the performance variation across different sets. In our experiments, the subsets are stratified samples of size $16, 32, 64, 128, 256$.

[15] https://github.com/msang/hate-speech-corpus.
[16] https://github.com/msang/haspeede2018.
[17] http://poleval.pl/tasks/task6.
[18] https://github.com/paulafortuna/Portuguese-Hate-Speech-Dataset.
[19] https://zenodo.org/record/2592149.

4.1 Embeddings

In order to train models in multilingual setting, we need multilingual word/sentence embeddings. For sentences, LASER embeddings were used and for words MUSE embeddings were used.

Laser embeddings: LASER[20] denotes Language-Agnostic SEntence Representations [2]. Given an input sentence, LASER provides sentence embeddindgs which are obtained by applying max-pooling operation over the output of a BiLSTM encoder. The system uses a single BiLSTM encoder with a shared BPE vocabulary for all languages.

Muse embeddings: MUSE[21] denotes Multilingual Unsupervised and Supervised Embeddings. Given an input word, MUSE gives as output the corresponding word embedding [7]. MUSE builds a bilingual dictionary between two languages without using any parallel corpora, by aligning monolingual word embedding spaces in an unsupervised way.

4.2 Models

CNN-GRU (Zhang *et al.* [37]): This model initially maps each of the word in a sentence into a 300 dimensional vector using the pretrained Google News Corpus embeddings [24]. It also pads/clips the sentences to a maximum of 100 words. Then this 300×100 vector is passed through drop layer and finally to a 1-D convolution layer with 100 filters. Further, a maxpool layer reduces the dimension to 25×100 feature matrix. Now this is passed through a GRU layer and it outputs a 100×100 dimension matrix which is globally max-pooled to provide a 1×100 vector. This is further passed through a softmax layer to give us the final prediction.

BERT: BERT [10] stands for Bidirectional Encoder Representations from Transformers pretrained on data from english language. It is a stack of transformer encoder layers with multiple "heads", i.e. fully connected neural networks augmented with a self attention mechanism. For every input token in a sequence, each head computes key value and query vectors which are further used to create a weighted representation. The outputs of each head in the same layer are combined and run through a fully connected layer. Each layer is wrapped with a skip connection and a layer normalization is applied after it. In our model we set the token length to 128 for faster processing of the query[22].

mBERT: Multilingual BERT (mBERT[23]) is a version of BERT that was trained on Wikipedia in 104 languages. Languages with a lot of data were sub-sampled and others were super sampled and the model was pretrained using the same

[20] https://github.com/facebookresearch/LASER.

[21] https://github.com/facebookresearch/MUSE.

[22] In the total data 0.17% datapoints have more than 128 tokens when tokenized, thus justifying our choice.

[23] https://tinyurl.com/yxh57v3a.

method as BERT. mBERT generalizes across some scripts and can retrieve parallel sentences. mBERT is simply trained on a multilingual corpus with no language IDs, but it encodes language identities. We used mBERT to train hate speech detection model in different languages once again limiting to a maximum of 128 tokens for sentence representation.

Translation: One simple way to utilize datasets in different languages is to rely on translation. Simple techniques of translation has shown to give good results in tasks such as sentiment analysis [34]. We use Google Translate[24] to convert all the datasets in different languages to English since translation to English from other languages typically have less errors in comparison to the other way round.

For our experiments we use the following four models:

1. **MUSE + CNN-GRU:** For the given input sentence, we first obtain the corresponding MUSE embeddings which are then passed as input to the CNN-GRU model.
2. **Translation + BERT:** The input sentence is first translated to the English language which are then provided as input to the BERT model.
3. **LASER + LR:** For the given input sentence, we first obtain the corresponding LASER embeddings which are then passed as input to a Logistic Regression (LR) model.
4. **mBert:** The input sentence is directly fed to the mBert model.

4.3 Hyperparameter Optimization

We use the validation set performance to select the best set of hyperparameters for the test set. The hyperparameters used in our experiments are as follows: batch size: 16, learning rate: $2e^{-5}, 3e^{-5}, 5e^{-5}$ and epochs: $1, 2, 3, 4, 5$.

5 Results

5.1 Monolingual Scenario

In this setting, we use the data from the same language for training, validation and testing. This scenario commonly occurs in the real world where monolingual dataset is used to build classifiers for a specific language.

Observations: Table 3 reports the results of the monolingual scenario. As expected, we observe that with increasing training data, the classifier performance increases as well. However, the relative performance seem to vary depending on the language and the model. We make several observations. First, **LASER + LR** performs the best in low-resource settings (16, 32, 64, 128, 256) for all the languages. Second, we observe that **MUSE + CNN-GRU** performs the worst in almost all the cases. Third, **Translation + BERT** seems to achieve

[24] https://github.com/sergei4e/gtrans.

competitive performance for some of the languages such as German, Polish, Portuguese, and Spanish. Overall we observe that there is no 'one single recipe' for all languages; however, **Translation + BERT** seems to be an excellent compromise. We believe that improved translations in some languages can further improve the performance of this model.

Although **LASER + LR** seems to be doing good in low resource setting, if enough data is available, we observe that BERT based models: **Translation + BERT** (English, German, Polish, and French) and **mBERT** (Arabic, Indonesian, Italian, and Spanish) are doing much better. However, what is more interesting is that although BERT based models are known to be successful when a larger number of datapoints are available, even with 256 datapoints some of these models seem to come very close to **LASER + LR**; for instance, **Translation + BERT** (Spanish, French) and **mBERT** (Arabic, Indonesian, Italian).

5.2 Multilingual Scenario

In this setting, we will use the dataset from all the languages expect one $(N-1)$, and use the validation and test set of the remaining language. This scenario represents when one wishes to employ the existing hate speech dataset to build a classifier for a new language. We have considered **LASER + LR** and **mBERT** that are most relevant for this analysis. In the **LASER + LR** model, we take the LASER embeddings from the $(N-1)$ languages and add to this the target language data points in incremental steps of $16, 32, 64, 128$ and 256. The logistic regression model is trained on the combined data, and we test it on the held out test set of the target language.

For using the multilingual setting in **mBERT** we adopt a two-step fine-tuning method. For a language L, we use the dataset for $N-1$ languages (except the Lth language) to train the **mBERT** model. On this trained **mBERT** model, we perform a second stage of fine-tuning using the training data of the target language in incremental steps of $16, 32, 64, 128, 256$. The model was then evaluated on the test set of the Lth language.

We also test the models for zero shot performance. In this case, the model is not provided any data of the target language. So, the model is trained on the $(N-1)$ languages and directly tested on the Nth language test set. This would be the case in which we would like to directly deploy a hate speech classifier for a language which does not have any training data.

Observations: Table 4 reports the results of the multilingual scenario. Similar to the monolingual scenario, we observe that with increasing training data, the classifier performance increases in general.

This is especially true in low resource settings of the target languages such as English, Indonesian, Italian, Polish, Portuguese.

In case of zero shot evaluation, we observe that **mBERT** performs better than **LASER + LR** in three languages (Arabic, German, and French). **LASER + LR** perform better on the remaining six languages with the results in Italian

Table 3. Monolingual scenario: the training, validation and testing data is used from the same language. Here, Full D represents the full training data. The **bold** figures represent the best scores and underline represents the second best.

Language	Model	Training size					
		16	32	64	128	256	Full D
Arabic	MUSE + CNN-GRU	0.4412	0.4438	0.4486	0.4664	0.5818	0.7368
	Translation + BERT	0.4555	0.4495	0.5551	0.5448	0.7017	0.8115
	LASER + LR	**0.5533**	**0.6755**	**0.7304**	**0.7488**	**0.7698**	0.7920
	mBert	0.4588	0.4533	0.4408	0.6486	0.7295	**0.8320**
English	MUSE + CNN-GRU	0.4580	0.4594	0.4653	0.4646	0.4813	0.6441
	BERT	0.4071	0.3925	0.4260	0.4720	0.4578	**0.7143**
	LASER + LR	**0.4617**	**0.4899**	**0.5376**	**0.5624**	**0.5885**	0.6526
	mBert	0.1773	0.3251	0.4488	0.4578	0.4578	0.7101
German	MUSE + CNN-GRU	0.4708	0.4708	0.4708	0.4708	0.4762	0.5756
	Translation + BERT	0.4812	0.4758	0.4719	0.4729	0.4724	**0.7662**
	LASER + LR	0.4974	**0.5201**	**0.5465**	**0.5925**	**0.6488**	0.6873
	mBert	**0.5037**	0.4750	0.4708	0.4717	0.5022	0.6517
Indonesian	MUSE + CNN-GRU	0.4250	0.4823	0.5263	0.5354	0.5890	0.7110
	Translation + BERT	0.4957	0.5003	0.5179	0.5682	0.6341	0.7670
	LASER + LR	**0.5226**	**0.5376**	**0.5882**	**0.6259**	**0.6890**	0.7872
	mBert	0.5106	0.5219	0.5414	0.6016	0.6530	**0.8119**
Italian	MUSE + CNN-GRU	0.4055	0.4476	0.4461	0.5206	0.5965	0.7349
	Translation + BERT	0.5006	0.5943	0.6215	0.6678	0.6919	0.7922
	LASER + LR	0.5688	**0.6210**	**0.6843**	**0.7175**	**0.7347**	0.7996
	mBert	**0.5774**	0.4567	0.5834	0.6664	0.7026	**0.8260**
Polish	MUSE + CNN-GRU	0.4842	0.4842	0.4841	0.4842	0.5180	0.6337
	Translation + BERT	0.4842	0.4853	0.4842	0.4842	0.5066	**0.7161**
	LASER + LR	**0.4889**	**0.4879**	**0.5360**	**0.5739**	**0.6172**	0.6439
	mBert	0.4829	0.4847	0.4842	0.4842	0.4842	0.7069
Portuguese	MUSE + CNN-GRU	0.4480	0.3807	0.4184	0.4228	0.4562	0.6100
	Translation + BERT	0.4532	0.4893	0.4712	0.5102	0.5994	0.6935
	LASER + LR	**0.5194**	**0.5536**	**0.6070**	**0.6210**	**0.6412**	**0.6941**
	mBert	0.5154	0.4245	0.4148	0.5493	0.5745	0.6713
Spanish	MUSE + CNN-GRU	0.4382	0.3354	0.3558	0.4203	0.4995	0.6364
	Translation + BERT	0.4598	0.4722	0.5080	0.4576	0.6035	0.7237
	LASER + LR	**0.5168**	**0.5434**	**0.5521**	**0.5938**	**0.6153**	0.6997
	mBert	0.4395	0.4285	0.4048	0.4861	0.5999	**0.7329**
French	MUSE + CNN-GRU	0.4878	0.4683	0.5008	0.5222	0.5250	0.5619
	Translation + BERT	0.4173	0.4260	0.4429	0.4749	0.6037	**0.6595**
	LASER + LR	**0.5058**	**0.5486**	**0.6136**	**0.6302**	**0.6085**	0.6172
	mBert	0.4818	0.4139	0.4053	0.4355	0.5701	0.6165

Table 4. Multilingual scenario: the training data is from all the languages except one and the validation and testing data is from the remaining language. The **bold** figures represent the best scores.

Testing language	Model	Training size						
		Zero shot	16	32	64	128	256	Full D
Arabic	LASER + LR	0.4645	**0.4651**	0.4664	0.4704	0.4784	0.4930	0.6751
	mBert	**0.6442**	0.4535	**0.4738**	**0.5302**	**0.7331**	**0.7707**	**0.8365**
English	LASER + LR	**0.6050**	**0.6051**	**0.6052**	**0.6053**	**0.6054**	0.6060	0.6808
	mBert	0.4971	0.4750	0.4670	0.5044	0.5242	**0.6091**	**0.7374**
German	LASER + LR	0.4695	0.4661	0.4727	0.4729	**0.4740**	0.4784	0.5622
	mBert	**0.5437**	**0.5146**	**0.4927**	**0.4733**	0.4718	**0.4786**	**0.6651**
Indonesian	LASER + LR	**0.6263**	**0.6251**	**0.6252**	**0.6241**	**0.6182**	0.6151	0.5977
	mBert	0.5113	0.5186	0.5049	0.4871	0.5864	**0.6318**	**0.8044**
Italian	LASER + LR	**0.6861**	**0.6857**	**0.6855**	**0.6855**	**0.6860**	0.6867	0.7071
	mBert	0.5335	0.5318	0.5444	0.6696	0.6704	**0.7189**	**0.8147**
Polish	LASER + LR	**0.5912**	**0.5926**	**0.5931**	**0.5935**	**0.5901**	**0.5829**	0.5672
	mBert	0.0725	0.4961	0.5049	0.4841	0.4842	0.4842	**0.6670**
Portuguese	LASER + LR	**0.6567**	**0.6565**	**0.6566**	**0.6563**	**0.6565**	**0.6573**	**0.6755**
	mBert	0.5995	0.5526	0.5694	0.5961	0.6148	0.6294	0.6660
Spanish	LASER + LR	**0.5408**	**0.5415**	**0.5417**	**0.5406**	0.5434	0.5437	0.5708
	mBert	0.2677	0.4464	0.4751	0.5126	**0.6080**	**0.6302**	**0.7383**
French	LASER + LR	0.4228	0.4180	0.4171	0.4180	0.4181	0.4198	0.4684
	mBert	**0.5487**	**0.5310**	**0.5138**	**0.5698**	**0.5849**	**0.5948**	**0.5968**

and Portuguese being pretty good. In case of Portuguese, zero shot **Laser + LR** (without any Portuguese training data) obtains an F-score of 0.6567, close to the best result of 0.6941 (using full Portuguese training data).

For the languages such as Arabic, German, and French, **mBERT** seems to be performing better than **LASER + LR** is almost all the cases (low resource and Full D). **LASER + LR**, on the other hand, is able to perform well for Portuguese language in all the cases. For the rest of the five languages, we observe that **LASER + LR** is performing better in low resource settings, but on using the full training data of the target language, **mBERT** performs better.

5.3 Possible Recipes Across Languages

As we have used the same test set for both the scenarios, we can easily compare the results to access which is better. Using the results from monolingual and multilingual scenario, we can decide the best kind of models to use based on the availability of the data. The possible recipes are presented as a catalogue in Table 5. Overall we observe that **LASER + LR** model works better for low resource settings while BERT based models work well for high resource settings. This possibly indicates that BERT based models, in general can work well when there is larger data available thus allowing for a more accurate fine-tuning. We believe that this catalogue is one of the most important contributions of our work which can be readily referred to by future researchers working to advance the state-of-the-art in multilingual hate speech detection.

Table 5. The table describes the best model to use in low and high resource scenario. In general, LASER + LR performs well in low resource setting and BERT based models are better in high resource settings

Language	Low resource	High resource
Arabic	Monolingual, LASER + LR	Multilingual, mBERT
English	Multilingual, LASER + LR	Multilingual, mBERT
German	Monolingual, LASER + LR	Translation + BERT
Indonesian	Multilingual, LASER + LR	Monolingual, mBERT
Italian	Multilingual, LASER + LR	Monolingual, mBERT
Polish	Multilingual, LASER + LR	Translation + BERT
Portuguese	Multilingual, LASER + LR	Monolingual, LASER+LR
Spanish	Monolingual, LASER + LR	Multilingual, mBERT
French	Monolingual, LASER + LR	Translation + BERT

6 Discussion and Error Analysis

6.1 Interpretability

In order to compare the interpretability of **mBERT** and **LASER + LR**, we use LIME [31] to calculate the average importance given to words by a particular model. We compute the top 5 most predictive words and their attention for each sentence in the test set. The total score for each word is calculated by summing up all the attentions for each of the sentences where the word occurs in the top 5 LIME features. The average predictive score for each word is calculated by dividing this total score by the occurrence count of each word. In Table 6 we note the top 5 words having the highest attention scores and compare them qualitatively across models.

While comparing the models' interpretability in Table 6, we see that **LASER + LR** focuses more on the hateful keywords compared to **mBERT**, i.e., words like 'pigs' etc. **mBERT** seems to search for some context of the hate keywords as shown in Table 7. Models dependent on the keywords can be useful when we are in a highly toxic environment such as GAB[25] since most of the derogatory keywords typically occur very close or at least simultaneously along with the hate target, for e.g., the first case in Table 1. In sites which are less toxic like Twitter, complex methods giving attention to the context like **mBERT** might be more helpful,for e.g., the third case in Table 1.

6.2 Error Analysis

In order to delve further into the models, we conduct an error analysis[26] on both the **mBERT** and **LASER + LR** models using a sample of posts where the

[25] https://en.wikipedia.org/wiki/Gab_(social_network).

[26] Note that we rely on translation for interpretations of the errors and the translation itself might also have some error.

output was wrongly classified from the test set.We analyze the common errors and categorize them into the following four types:

1. **Wrong classification due to annotation's dilemma (AD):** These error cases occur due to ambiguous instances where according to us the model predicts correctly but the annotators have labelled it wrong.
2. **Wrong classification due to confounding factors (CF):** These error cases are caused when the model predictions rely on some irrelevant features like normalized form of mentions (*@user*) and links (*URL*) in the text.
3. **Wrong classification due to hidden context (HC):** These error cases are caused when the model fails to capture the context of the post.
4. **Wrong classification due to abusive words (AW):** These error cases are caused by over-dependence of the model on the abusive words.

Table 8 shows the errors of the **mBERT** and **LASER + LR** models. For **mBERT**, the first example has no specific indication of being a hate speech and is considered an error on the part of annotators. In the second example the author of the post actually wants the reader to not use the abusive terms, i.e., sl*t

Table 6. Interpretations of the model outcomes.

German		**Indonesian**	
mBERT	LASER + LR	mBERT	LASER + LR
spendieren (spend)	fotzen (pu**ies)	loo (loo)	NAJIS (unclean)
drogen (drugs)	Trottel (fool)	rusak (broken)	bajingan (son of a bi**h)
schœn (beautiful)	abschaum (scum)	makhluk (creature)	MAMPUS (dead)
kastrieren (castrate)	WICHSER (w**ker)	pengkhianatan (betrayal)	Idiot (idiot)
einsetzen (deploy)	Scheissen (shit)	celeng (wild boar)	F**kYou (f**k you)
Italian		**Polish**	
mBERT	LASER + LR	mBERT	LASER + LR
innervosirmi (get nervous)	Schifo (schifo)	stanowisk (posts)	pieprzysz (f**k)
vomitata (vomited)	demoliscile (demoliscile)	pomysł (idea)	gówno (shit)
cascarci (fall for)	disonesti (dishonest)	powiedzieli (they said)	idiota (idiot)
italioti (italioti)	massacrale (massacrale)	cwelica (cwelica)	Idiotów (idiots)
annegano (drown)	schifoso (lousy)	obrazka (picture)	świry (suck)
Portuguese		**Spanish**	
mBERT	LASER + LR	mBERT	LASER + LR
fuder (f**k)	FOFURA (cuteness)	Hxrry_again (hxrry_again)	piratas (pirates)
heterofobicos (heterophobic)	tretas (fights)	majisimos (majestic)	MARICA (sissy)
vagabunda (slut)	porcaria (filth)	mate (mate)	perseguidos (persecuted)
cracuda (crunchy)	foda (f**k)	publicidad (advertising)	pegaso6038 (pegasus6038)
femimimismo (feminism)	heterofobicos (heterophobic)	sevilla (seville)	Putas (wh**es)
French		**English**	
mBERT	LASER + LR	mBERT	LASER + LR
mongol (mongolian)	jérusalem (jerusalem)	n**guh	Rapist
medelin (medelin)	ptdrrrrrrrrrr (ptdrrrrrrrrrr)	fa**otnwe	BlackB4Illegal
arabe (arab)	negrophobe (ne*rophobe)	n**let	Misttgg
barges (barges)	juifs (jews)	nig	sexualevak
marocains (moroccons)	bf (bf)	h*ed	prostitute
Arabic			
mBERT	LASER + LR		
تساعده (help him)	خنازير(pigs)		
ورد(flower)	رووووح(rooooh)		
يامطي (my run)	ياكلبة(eat it)		
خارجية(external)	خنزير(pig)		
لهجة (dialect)	قذره (dirty)		

Table 7. Examples showing word with the highest predictive word for both mBERT and *LASER + LR*.

Sentences with hate label
das *pack* muss tag und nacht gejagt werden,ehe sie es mit den deutschen machen !! (**Translated :-** the *pack* must be hunted day and night before they do it with the Germans !!)
absolument ! il faut l'arraisonner en mer par la marin nationale arrêter tous les occupants expulser les *migrant*... @url (**Translated :-** absolutely! it must be boarded at sea by the navy national arrest all occupants expel *migrants*... @url)

Table 8. Various types of errors **(E)** for the models **(M)** : mBERT and LASER + LR. The ground truth **(GT)** and prediction **(P)** consist of 0 (Non-Hate)/1 (Hate) label.

M	Sentences	GT	P	E
mBERT	Arabic Translation: He and his father, and Abu Alto and Abu Israel, are doomed to go to Israel to blind, insolent Syrian opponents, and to betray that I have not seen and my eyes have seen	1	0	AD
	"If you have tries to get w/a girl you are not allowed to call her demeaning names like "slut whore etc." sorry bout yall"	0	1	AW
	"Könnten wir Schmarotzer und Kriminelle loswerden würde die Asylanten-Schwemme auf beherrschbare Zahlen runtergehen" **Translation:** If we could get rid of parasites and criminals, the asylum seeker flood would drop to manageable numbers	1	0	HC
LASER + LR	"Die hat jede Art von Realität verloren und braucht dringend Hilfe am besten ne Einweisung in die Geschlossene für immer und Ewig und ihr Gefolge gleich mit" **Translation:** She has lost all kind of reality and urgently needs help, best a briefing in the closed forever and ever and her followers at the same time	1	0	AD
	"USER USER Gw mah tetep anti cina... gara gara gw ngga bisa sipit dan putih kayak mereka...wkwkwk" **Translation:** USER USER I am still anti-Chinese ... because I can't be narrow and white like them ... hahaha	0	1	CF
	"RT @mundodrogado: Antes o homossexualismo era proibido.Depois passou a ser tolerado.Hoje é normal. Eu vou embora antes que vire obrigatór" **Translation:** RT @mundodrogado: Before homosexuality was forbidden. Then it became tolerated. Today it's normal. I'm leaving before it becomes mandatory...	1	0	HC
	this movie is actually good cuz its so retarded	0	1	AW

and wh*re (*found using LIME*) but the model picks them as indicators of hate speech. The third example has mentioned the term "parasite" as a derogatory remark to refugees and the model did not understand it.

For the **LASER + LR** model, the first example is an error on the part of the annotators. In the second case the model captures the word "USER" (*found using LIME*), a confounding factor which affects the models' prediction. For the third case, the author says (s)he will leave before homosexuality gets normalized which shows his/her hatred toward the LGBT community but the model is unable to capture this. In the last case the model predicts hate speech based on the word "retarded" (*found using LIME)* which should not be the case.

7 Conclusion

In this paper, we perform the first large scale analysis of multilingual hate speech. Using 16 datasets from 9 languages, we use deep learning models to develop classifiers for multilingual hate speech classification. We perform many experiments under various conditions – low and high resource, monolingual and multilingual settings – for a variety of languages. Overall we see that for low resource, LASER + LR is more effective while for high resource BERT models are more effective. We finally suggest a catalogue which we believe will be beneficial for future research in multilingual hate speech detection. Our Code (Code: https://github.com/punyajoy/DE-LIMIT) and Models (Models: https://huggingface.co/Hate-speech-CNERG) are available online for other researchers to use.

References

1. Alfina, I., Mulia, R., Fanany, M.I., Ekanata, Y.: Hate speech detection in the Indonesian language: a dataset and preliminary study. In: 2017 International Conference on Advanced Computer Science and Information Systems (ICACSIS), pp. 233–238. IEEE (2017)
2. Artetxe, M., Schwenk, H.: Massively multilingual sentence embeddings for zero-shot cross-lingual transfer and beyond. Trans. Assoc. Comput. Linguist. **7**, 597–610 (2019)
3. Basile, V., et al.: Semeval-2019 task 5: multilingual detection of hate speech against immigrants and women in twitter. In: Proceedings of the 13th International Workshop on Semantic Evaluation, pp. 54–63 (2019)
4. Bosco, C., Felice, D., Poletto, F., Sanguinetti, M., Maurizio, T.: Overview of the evalita 2018 hate speech detection task. In: EVALITA 2018-Sixth Evaluation Campaign of Natural Language Processing and Speech Tools for Italian. vol. 2263, pp. 1–9. CEUR (2018)
5. Bretschneider, U., Peters, R.: Detecting offensive statements towards foreigners in social media. In: Proceedings of the 50th Hawaii International Conference on System Sciences (2017)
6. Burnap, P., Williams, M.L.: Us and them: identifying cyber hate on twitter across multiple protected characteristics. EPJ Data Sci. **5**(1), 11 (2016)
7. Conneau, A., Lample, G., Ranzato, M., Denoyer, L., Jégou, H.: Word translation without parallel data. arXiv preprint arXiv:1710.04087 (2017)

8. Corazza, M., Menini, S., Cabrio, E., Tonelli, S., Villata, S.: A multilingual evaluation for online hate speech detection. ACM Trans. Internet Technol. (TOIT) **20**(2), 1–22 (2020)
9. Davidson, T., Warmsley, D., Macy, M., Weber, I.: Automated hate speech detection and the problem of offensive language. In: Eleventh International AAAI Conference on Web and Social Media (2017)
10. Devlin, J., Chang, M.W., Lee, K., Toutanova, K.: Bert: pre-training of deep bidirectional transformers for language understanding (2018)
11. ElSherief, M., Kulkarni, V., Nguyen, D., Wang, W.Y., Belding, E.: Hate lingo: a target-based linguistic analysis of hate speech in social media. In: Twelfth International AAAI Conference on Web and Social Media (2018)
12. Fasoli, F., Maass, A., Carnaghi, A.: Labelling and discrimination: do homophobic epithets undermine fair distribution of resources? Br. J. Soc. Psychol. **54**(2), 383–393 (2015)
13. Fortuna, P., Nunes, S.: A survey on automatic detection of hate speech in text. ACM Comput. Surv. (CSUR) **51**(4), 85 (2018)
14. Fortuna, P., da Silva, J.R., Wanner, L., Nunes, S., et al.: A hierarchically-labeled Portuguese hate speech dataset. In: Proceedings of the Third Workshop on Abusive Language Online, pp. 94–104 (2019)
15. Founta, A.M., et al.: Large scale crowdsourcing and characterization of twitter abusive behavior. In: Twelfth International AAAI Conference on Web and Social Media (2018)
16. Gagliardone, I., Gal, D., Alves, T., Martinez, G.: Countering Online Hate Speech. Unesco Publishing (2015)
17. de Gibert, O., Perez, N., Pablos, A.G., Cuadros, M.: Hate speech dataset from a white supremacy forum. In: Proceedings of the 2nd Workshop on Abusive Language Online (ALW2), pp. 11–20 (2018)
18. Greenberg, J., Pyszczynski, T.: The effect of an overheard ethnic slur on evaluations of the target: how to spread a social disease. J. Exper. Soc. Psychol. **21**(1), 61–72 (1985)
19. Guermazi, R., Hammami, M., Hamadou, A.B.: Using a semi-automatic keyword dictionary for improving violent web site filtering. In: 2007 Third International IEEE Conference on Signal-Image Technologies and Internet-Based System, pp. 337–344. IEEE (2007)
20. Huang, X., Xing, L., Dernoncourt, F., Paul, M.J.: Multilingual twitter corpus and baselines for evaluating demographic bias in hate speech recognition. arXiv preprint arXiv:2002.10361 (2020)
21. Ibrohim, M.O., Budi, I.: Multi-label hate speech and abusive language detection in indonesian twitter. In: Proceedings of the Third Workshop on Abusive Language Online, pp. 46–57 (2019)
22. Mathew, B., Dutt, R., Goyal, P., Mukherjee, A.: Spread of hate speech in online social media. In: Proceedings of the 10th ACM Conference on Web Science, pp. 173–182 (2019)
23. Mathew, B., Illendula, A., Saha, P., Sarkar, S., Goyal, P., Mukherjee, A.: Temporal effects of unmoderated hate speech in gab. arXiv preprint arXiv:1909.10966 (2019)
24. Mikolov, T., Chen, K., Corrado, G., Dean, J.: Efficient estimation of word representations in vector space. arXiv preprint arXiv:1301.3781 (2013)
25. Mulki, H., Haddad, H., Ali, C.B., Alshabani, H.: L-hsab: A levantine twitter dataset for hate speech and abusive language. In: Proceedings of the Third Workshop on Abusive Language Online, pp. 111–118 (2019)

26. Mullen, B., Rice, D.R.: Ethnophaulisms and exclusion: the behavioral consequences of cognitive representation of ethnic immigrant groups. Personal. Soc. Psychol. Bull. **29**(8), 1056–1067 (2003)
27. Ousidhoum, N., Lin, Z., Zhang, H., Song, Y., Yeung, D.Y.: Multilingual and multi-aspect hate speech analysis. In: Proceedings of the 2019 Conference on Empirical Methods in Natural Language Processing and the 9th International Joint Conference on Natural Language Processing (EMNLP-IJCNLP), pp. 4667–4676 (2019)
28. Pereira-Kohatsu, J.C., Quijano-Sánchez, L., Liberatore, F., Camacho-Collados, M.: Detecting and monitoring hate speech in twitter. Sensors (Basel, Switzerland), **19**(21) (2019)
29. Ptaszynski, M., Pieciukiewicz, A., Dybała, P.: Results of the poleval 2019 shared task 6: first dataset and open shared task for automatic cyberbullying detection in polish twitter. In: Proceedings of the PolEval2019Workshop, p. 89 (2019)
30. Ribeiro, M.H., Calais, P.H., Santos, Y.A., Almeida, V.A., Meira Jr, W.: Characterizing and detecting hateful users on twitter. In: Twelfth International AAAI Conference on Web and Social Media (2018)
31. Ribeiro, M.T., Singh, S., Guestrin, C.: Why should i trust you? explaining the predictions of any classifier. In: Proceedings of the 22nd ACM SIGKDD International Conference on Knowledge Discovery and Data Mining, pp. 1135–1144 (2016)
32. Ross, B., Rist, M., Carbonell, G., Cabrera, B., Kurowsky, N., Wojatzki, M.: Measuring the reliability of hate speech annotations: The case of the European refugee crisis. arXiv preprint arXiv:1701.08118 (2017)
33. Sanguinetti, M., Poletto, F., Bosco, C., Patti, V., Stranisci, M.: An Italian twitter corpus of hate speech against immigrants. In: Proceedings of the Eleventh International Conference on Language Resources and Evaluation (LREC 2018) (2018)
34. Singhal, P., Bhattacharyya, P.: Borrow a little from your rich cousin: using embeddings and polarities of English words for multilingual sentiment classification. In: Proceedings of COLING 2016, the 26th International Conference on Computational Linguistics: Technical Papers, pp. 3053–3062 (2016)
35. Soral, W., Bilewicz, M., Winiewski, M.: Exposure to hate speech increases prejudice through desensitization. Aggressive Behav. **44**(2), 136–146 (2018)
36. Waseem, Z., Hovy, D.: Hateful symbols or hateful people? predictive features for hate speech detection on twitter. In: Proceedings of the NAACL Student Research Workshop, pp. 88–93 (2016)
37. Zhang, Z., Robinson, D., Tepper, J.: Detecting hate speech on twitter using a convolution-GRU based deep neural network. In: Gangemi, A., et al. (eds.) ESWC 2018. LNCS, vol. 10843, pp. 745–760. Springer, Cham (2018). https://doi.org/10.1007/978-3-319-93417-4_48

Spatial Community-Informed Evolving Graphs for Demand Prediction

Qianru Wang[1], Bin Guo[1(✉)], Yi Ouyang[1], Kai Shu[2], Zhiwen Yu[1], and Huan Liu[3]

[1] School of Computer Science and Engineering, Northwestern Polytechnical University, Xi'an, People's Republic of China
guob@nwpu.edu.cn

[2] Department of Computer Science, Illinois Institute of Technology, Chicago, USA

[3] College of Computer Science and Engineering, Arizona State University, Tempe, USA

Abstract. The rapidly increasing number of sharing bikes has facilitated people's daily commuting significantly. However, the number of available bikes in different stations may be imbalanced due to the free check-in and check-out of users. Therefore, predicting the bike demand in each station is an important task in a city to satisfy requests in different stations. Recent works mainly focus on demand prediction in settled stations, which ignore the realistic scenarios that bike stations may be deployed or removed. To predict station-level demands with evolving new stations, we face two main challenges: (1) How to effectively capture new interactions in time-evolving station networks; (2) How to learn spatial patterns for new stations due to the limited historical data. To tackle these challenges, we propose a novel Spatial Community-informed Evolving Graphs (SCEG) framework to predict station-level demands, which considers two different grained interactions. Specifically, we learn time-evolving representation from fine-grained interactions in evolving station networks using EvolveGCN. And we design a Bi-grained Graph Convolutional Network(B-GCN) to learn community-informed representation from coarse-grained interactions between communities of stations. Experimental results on real-world datasets demonstrate the effectiveness of SCEG on demand prediction for both new and settled stations. Our code is available at https://github.com/RoeyW/Bikes-SCEG

Keywords: Spatial-temporal analysis · Urban computing · Demand prediction · Graph neural network

1 Introduction

Sharing-bike is becoming an increasingly popular means for commuting due to its cheap cost, easy access and convenient usage. Bike riders can check-in to ride

This work was partially supported by the National Key R&D Program of China (2019YFB1703901), the National Natural Science Foundation of China (No. 61772428,61725205,61902320,61972319) and China Scholarship Council.

Y. Dong et al. (Eds.): ECML PKDD 2020, LNAI 12461, pp. 440–456, 2021.
https://doi.org/10.1007/978-3-030-67670-4_27

the bike from one station and check-out at any other stations. According to the reports of Citi Bike, the largest bike-sharing company in New York City, the number of total bikes grew up to 12,000 and over 800 stations were added in 2017. Also, it is estimated that bike riders take 10 million trips in one year[1]. Due to free check-in and check-out demands of bike rides, available bikes in some stations are shortage while available bikes in others are redundant. The bike company needs to balance bikes in stations manually according to demands of stations. Therefore, demand prediction is important to balance bikes for the stations in a city.

The majority of recent works on sharing-bike demand prediction focus on settled stations, which are all existing during training and test phases, using geographical distance and temporal correlations between stations. For example, Lin *et al.* [4] used a data-driven method to learn the relationship between stations with graph neural networks to predict future demands. Chai *et al.* [9] fused multiple graphs constructed from the perspectives of the spatial distance, trip records and check-in/check-out correlations between settled stations to predict bike's demands. They delete the new stations from the dataset, which only appear few days. In the real-world scenario, new stations may be added or removed by the bike-sharing company to better satisfy users' demands. For example, according to the report of Citi Bike, the bike company added over 100 stations in one year (See footnote 1). At the early stage of adding a new station, it has limited historical demands and interactions (trips to/from other stations). But existing works may not be directly applied to predict demands for new stations due to relying on long term historical data. Therefore, it is necessary to propose a more effective model to predict demands for stations when new stations are added.

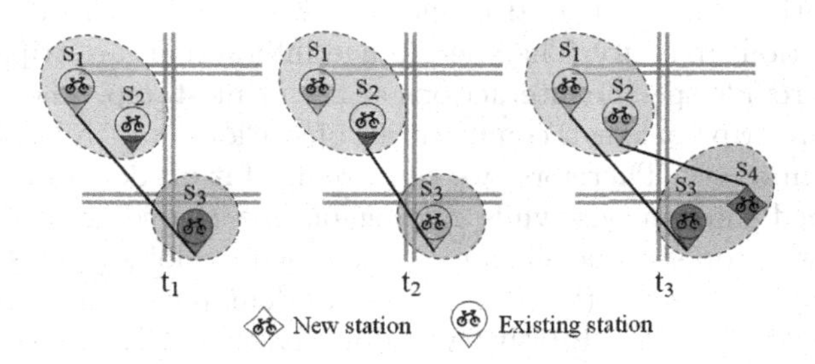

Fig. 1. The Illustration of time evolving station networks. The color of stations changes from green to red when the number of available bikes decrease. The lines between stations indicate that there are riding trips between them. The blue and red dash circles indicate communities of stations. (Color figure online)

However, it is a non-trivial task to perform prediction when new stations are added. We face two main challenges to deal with evolving new stations. From the **temporal** aspects, it is necessary to capture new interactions, which

[1] https://www.citibikenyc.com/about.

continuously appear especially when new stations are added. Besides changing of interactions between settled stations, new interactions between new stations and settled stations may appear over time, which affect demands of new stations and settled stations. As shown in Fig. 1, the demands of stations are different due to the different interactions at each timestamp. For example, users would like to ride from S_1 to S_3 at t_1. And some users chose S_2 to start the trips at t_2. Specifically, there would be some new interactions that haven't appeared before a new station was added. For example, S_4 was added at t_3. Then the interactions between S_2 and S_4 appeared at that time, which will affect the demands of S_2 and S_4. Therefore, to predict demands of stations, we need to consider dynamic interactions over time, especially for new interactions.

From the **spatial** aspects, spatial interaction patterns of new stations are difficult to be learned due to the limited historical interactions, which make it difficult to predict the demands of new stations. A straightforward way to provide information for the new station is using spatial interactions of the nearby stations directly. However, station-level interactions exist fluctuation and random [23], which cannot learn the interaction patterns of new station accurately. Compared to station-level interactions, stations within a spatial community have similar demand trends and more distinct interaction patterns as observed. Thus, when a new station is added into a spatial community, we can use the historical information of its community to supply demand trends and possible interactions for this new station. For example, stations in the blue community usually interact with stations in the red community as shown in Fig. 1. When the new station S_4 is added into the red community, S_4 may interact with stations in the blue community. Therefore, it is necessary to consider possible communities that new stations will interact with based on community-level interactions.

To tackle the aforementioned challenges, we consider two different-grained spatial interactions in a city. The fine-grained interactions are trips between stations, which provide specific interactions at each timestamp. The coarse-grained interactions are trips between communities of stations, which have more steady distributions in a city. Therefore, we can predict future demands by extracting the fine-grained interactions while relying on the coarse-grained interactions. To this end, we propose a novel framework which exploits Spatial Community-informed Evolving Graphs (SCEG) to predict demands with evolving new stations in station networks. Firstly, we adopt EvolveGCN [5] to represent the dynamic interactions in each timestamp to handle time-evolving station networks, which is called as *Time-evolving representation*. To deal with different-grained spatial graphs, we then design a Bi-grained Graph Convolution Network (B-GCN) to represent station-level representation based on community-level interactions, which is called as *Community-informed representation*. Finally, we use variational autoencoder [11] to fuse two kind of representations while considering some variances in the real world. In summary, the main contributions of our paper are as follows:

1. We consider a novel scenario for demand prediction that station networks would involve some new stations.

2. We propose a novel framework (SCEG), which exploits spatial community informed time-evolving graphs to predict demands for settled stations and new stations.
3. We design a Bi-grained Graph Convolutional Network (B-GCN), which assigns community-level interactions to stations.
4. We conduct experiments on the real-world datasets in New York City and Washington D.C.. Experimental results show that our model outperforms existing state-of-the-art baselines both on settled and new stations, especially when new stations are more than settled stations.

2 Problem Statement

To better represent interactions, we use graphs to model the relationship between communities or stations. To represent coarse-grained interactions between spatial communities and fine-grained interactions between stations, we use three kinds of graphs to define the problem: time-evolving station graphs, a spatial communities graph and bi-grained graphs.

Definition 1. *Time-evolving station graphs* $G^S = (G_1^S, ..., G_T^S)$.

At the tth timestamp, $G_t^S = (S_t, E_t^S)$, $S_t = (s_1, ..., s_i, ...)$ is the station set in a city. $e_t^{ij} \in E_t^S$ represents the number of interactions between station s_i and s_j. A_t^S denotes the binary adjacent matrix.

Definition 2. *A spatial community graph* $G^C = (C, E^{Com})$ *is constructed to indicate the interactions between spatial communities, which is a weighted graph.*

We denote C and E^{Com} as the set of communities and edges between communities. Each community consists of stations with similar attributes (such as spatial distance, demands on weekdays). A^C is a weighted adjacent matrix. $a_{ij}^C \in A^C$ represents the probability of riding from one community to another, which is calculated by the number of trips between c_i and c_j in the number of trips from/to c_i (i.e., $a_{ij}^C = \frac{\#trips(c_i,c_j)}{\#trips(c_i)}$).

Definition 3. *Bi-grained graphs* $G^B = (G_1^B, ..., G_T^B)$ *is used to link time-evolving station graph* G_t^S *and the spatial community graph* G^C.

At the tth timestamp, $G_t^B = (S_t, C, E_t^B)$, $E_t^B \in \mathbb{R}^{|S_t| \times |C|}$ represents edges between stations and communities. $a_{t,ij}^B \in A_t^B$ is equal to 1 if the ith station belongs to the jth community.

Additionally, we use $D_t = \{D_t^i, i \in \{in, out\}\}$ as check-in and check-out demands of stations. F^t denotes the external temporal features, e.g., temperature, wind speed, weather and weekday. These temporal features significantly affect the demand of stations. For example, demands on rainy days are less than sunny days. We give the formal problem definition as follows:

Given G^C, $G_{t-\Delta t}^B$ and a Δt period of historical data set $X^E = \{G_{t-\Delta t:t-1}^S, D_{t-\Delta t:t-1}, F_{t-\Delta t:t-1}\}$, we predict the check-in and check-out demand D_t for stations at t^{th} timestamp.

3 The Proposed Model

To predict station-level demands, we proposed SCEG to exploit time-evolving station networks informed by spatial community. As shown in Fig. 2, our model consists of two phases: *learning phase* and *prediction phase*. In the learning phase, we learn time-evolving representation and community-informed representation of stations from the dynamic interactions of station graphs and interactions between spatial communities. In the prediction phase, we predict future demands conditioned on these two latent representations.

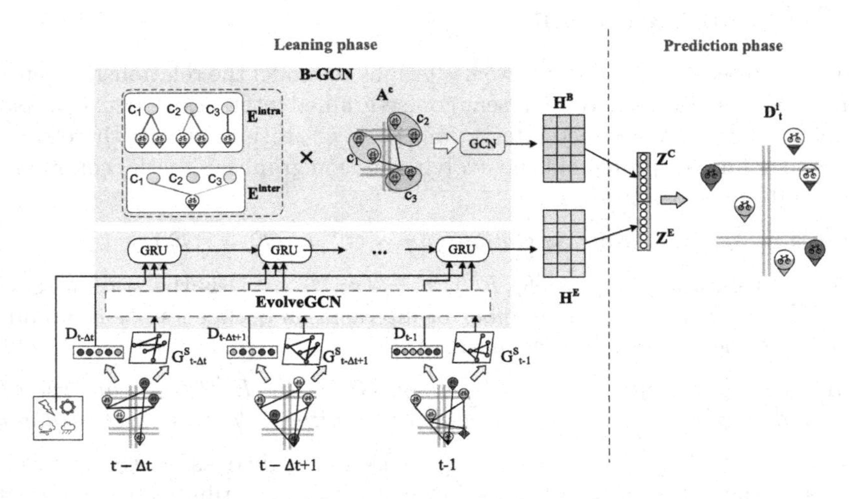

Fig. 2. The architecture of SCEG model. The learning phase is used to infer latent representations of spatial communities and time-evolving graphs. The prediction phase is used to predict future demands conditioned on these two latent representations

Learning phase is used to encode the spatial community graph and time-evolving information into two latent representations: *time-evolving representation* Z^E and *community-informed representation* Z^C separately. Since the true posterior distributions $p(Z^E|X^E)$ and $p(Z^C|G^C, G^B_{t-\Delta t})$ are intractable, the inference model $q_\phi(Z^E|X^E)$ and $q_\phi(Z^C|G^C, G^B_{t-\Delta t})$ are introduced to approximate them, where $q_\phi(Z^E|X^E) = \mathcal{N}(\mu^E, (\sigma^E)^2 I)$, $q_\phi(Z^C|G^C, G^B_{t-\Delta t}) = \mathcal{N}(\mu^C, (\sigma^C)^2 I)$. ϕ represents encoder parameters $\{\mu^E, \sigma^E, \mu^C, \sigma^C\}$.

Prediction phase is used to predict future demand based on the fusion of the two latent representations. The future check-in/out demand distribution $p(D^i_t)$ is written as:

$$\log p(D_t^i) = \log \iint p(D_t^i|Z^E, Z^C)p(Z^E)p(Z^C)d_{Z^E}d_{Z^C} \tag{1}$$

3.1 Learning Phase

Time-Evolving Graphs Representation. To capture evolving patterns for stations at each timestamp, existing works used a single GCN to encode station graphs in each timestamp, which shared parameters among the various timestamps. However, a single GCN for graphs at each timestamp is hardly to predict demands of new stations. Therefore, we need to learn interactions between settled stations and new stations in time-evolving station graphs. Besides time-evolving station graphs, there are some temporal features (e.g., temperature, wind speed, demands of stations), which also contribute to demands of stations. We need to combine these temporal features while encoding the time-evolving graphs.

Firstly, we use EvolveGCN to encode changes of nodes and edges for the time-evolving graphs. We use GRU (gated recurrent unit) [13] to update the hidden state on $t-1$ timestamp as shown in Eq. (2).

$$\begin{aligned} W_t^{(l)} &= GRU_1(W_{t-1}^{(l)}) \\ H_t^{(l)} &= ReLU(A_t^S \times H_t^{(l-1)} \times W_t^{(l)}) \end{aligned} \tag{2}$$

where $W_t^{(l)}$ and $H_t^{(l)}$ represent a weight matrix and a hidden state for the lth layer at timestamp t. $ReLU(\cdot)$ is a rectified linear unit. The symbol $\times$ means matrix multiplication.

Secondly, we combine the embedding of time-evolving graphs with temporal features. To obtain temporal embedding of the temporal information, we use another GRU to encode them as shown in Eq. (3).

$$\widetilde{H}_t = GRU_2(H_{t-1}^{(l)}, D_{t-1}, F_{t-1}) \tag{3}$$

Based on the embedding of temporal information(X^E), we can approximate the distribution $Z^E \sim q_\phi(Z^E|X^E)$ by estimating μ^E and σ^E as shown in Eq. (4). Due to some random between station-level interaction, we sample Z^E using a random normal distribution ϵ_1.

$$\begin{aligned} H^E &= \tanh(\widetilde{H}_t \times W^E + b^E) \\ \mu^E &= FC_1(H^E), \quad \log(\sigma^E)^2 = FC_2(H^E) \\ Z^E &= \mu^E + \sigma^E \cdot \epsilon_1, \quad where \quad \epsilon_1 \sim \mathcal{N}(0, I) \end{aligned} \tag{4}$$

where W^E, b^E are trainable variables. $FC(\cdot)$ denotes fully connected layers. The symbol $\cdot$ means element-wise multiplication.

Bi-Grained Graph Representation. Due to frequently changing of time-evolving graphs, we involve a steady and coarse-grained graph, also called as a spatial community graph, which also helps infer demand trends and possible interactions of new stations. Based on this idea, we need to assign the interactions of one cluster to the stations in it. As we all know, stations' demands are different even though they have similar demand trends in the same community. Therefore, we should consider the differences between stations in a community when assigning the community's information. On the other hand, each station also interacts with other communities. So it is necessary to explore how a station is affected by other communities.

To find spatial communities in a city, we group stations in a long period (i.e., 6 months) by K-Means, which uses Euclidean distance and their demands on different weekdays as features. After the period, we add a new station into a spatial community according to the distance and its few historical demands.

To address issues above, we firstly designed a Bi-grained Graph Convolution Network (B-GCN), which leverages a bi-grained graph to learn Community-informed representation. Specifically, we use a single GCN to represent interactions between spatial communities.

$$E^C = ReLU(A^C \times H^C \times W^C) \tag{5}$$

where H^C and W^C are trainable variables.

To transfer the embedding of community-level interactions to station-level, we use a bi-grained graph ($G^B_{t-\Delta t}$) to weight the affect of the community stations' belong to and other communities separately. Intra-weight represents how we assign the information of a community to stations in it as shown in Eq. (6). Equation (7) represents inter-weight, which calculates how a station interacts with other communities.

$$e^{intra}_{t,ij} = \begin{cases} \frac{w^{intra}_{ij} \cdot a^B_{t-\Delta t,ij}}{\sqrt{\sum_{i|s_i \in c_j} (w^{intra}_{ij} \cdot a^B_{t-\Delta t,ij})^2}}, & s_i \in c_j \\ 0, & s_i \notin c_j \end{cases} \tag{6}$$

$$e^{inter}_{t,ij} = \begin{cases} \frac{w^{inter}_{ij} \cdot (1-a^B_{t-\Delta t,ij})}{\sqrt{\sum_{j|s_i \notin c_j} (w^{inter}_{ij} \cdot (1-a^B_{t-\Delta t,ij}))^2}}, & s_i \notin c_j \\ 0, & s_i \in c_j \end{cases} \tag{7}$$

where $w^{intra}_{ij} \in W^{intra}$, $w^{inter}_{ij} \in W^{inter}$,$e^{intra}_{t,ij} \in E^{intra}_t$, $e^{inter}_{t,ij} \in E^{inter}_t$. And $W^{intra}, W^{inter}, E^{intra}_t, E^{inter}_t \in \mathbb{R}^{|S_{t-\Delta t}| \times |C|}$.

Using intra-weight and inter-weight, we calculate stations' representations H^B which is transferred from communities' representations.

$$H^B = (E^{intra}_t + E^{inter}_t) \times E^C \tag{8}$$

Secondly, we learn the latent community-informed representation. Though interactions between spatial communities seem regular than those between stations, they still have some randomness. Therefore, we need to consider some randomness based on $q_\phi(Z^C|G^C, G^B_{t-\Delta t})$ we learned. To obtain the latent variable $Z^C \sim q_\phi(Z^C|G^C, G^B_{t-\Delta t})$, we estimate μ^C and σ^C of $q_\phi(Z^C|G^C, G^B_{t-\Delta t})$ as follows:

$$\begin{aligned} \mu^C = FC_3(H^B), \quad \log(\sigma^C)^2 = FC_4(H^B) \\ Z^C = \mu^C + \sigma^C \cdot \epsilon_2 \quad where \quad \epsilon_2 \sim \mathcal{N}(0, I) \end{aligned} \tag{9}$$

where W^C is a weight matrix, Z^C is sampled from $q_\phi(Z^C|G^C, G^B_{t-\Delta t})$ using a random normal distribution ϵ_2.

3.2 Prediction Phase

Future demand is affected by the fusion of spatial community and time-evolving information. We need to consider specific information in time-evolving information while relying on coarse-grained distribution in a city. Therefore, conditioned on two independent latent variables Z^C and Z^E, we estimate the future demand distribution $p_\theta(D^i_t|Z^C, Z^E)$ as follows:

$$p_\theta(D^i_t|Z^C, Z^E) = FC_5(Z) \tag{10}$$

where Z is concatenation of $[Z^E, Z^C]$. θ is a set of trainable variables in $FC_5(\cdot)$.

To learn the encoder and decoder parameters (ϕ and θ), we need to maximize the lower bound of $p(D^i_t)$ as shown in Eq. (11), which is derived from Eq. (1):

$$\begin{aligned}
&\log p(D^i_t) \\
&\geq \iint q_\phi(Z^C, Z^E|G^C, G^B_{t-\Delta t}, X^E) \log \frac{p_\theta(D^i_t, Z^C, Z^E)}{q_\phi(Z^C, Z^E|G^C, G^B_{t-\Delta t}, X^E)} d_{Z^C} d_{Z^E} \\
&= \mathbb{E}_{q_\phi(Z^C, Z^E|G^C, G^B_{t-\Delta t}, X^E)} \left[p_\theta(D^i_t|Z^C, Z^E)\right] \\
&- D_{KL}\left(q_\phi(Z^C|G^C, G^B_{t-\Delta t}) \parallel p(Z^C)\right) \\
&- D_{KL}\left(q_\phi(Z^E|X^E) \parallel p(Z^E)\right) \triangleq \mathcal{L}(D^i_t; \phi, \theta)
\end{aligned} \tag{11}$$

where $p(Z^C) = p(Z^E) = \mathcal{N}(0, I)$.

Meanwhile, we also need to measure how accurately the model predicts future demand. We use a mean squared error as a loss function to measure the prediction error:

$$\mathcal{L}_{mse} = \frac{1}{n}(\widetilde{D}^i_t - D^i_t)^2 \tag{12}$$

Therefore, we need to optimize the loss function as follows to train the model:

$$\mathcal{L} \simeq \mathcal{L}_{mse} + \alpha \left[D_{KL}\left(q_\phi(Z^C|G^C, G^B_{t-\Delta t}) \parallel p(Z^C)\right) + D_{KL}\left(q_\phi(Z^E|X^E) \parallel p(Z^E)\right)\right] \tag{13}$$

4 Experiments

In this section, we conduct several experiments to evaluate our model on settled stations and new stations.

4.1 Datasets

We evaluate our model on two public real-world datasets as shown in Table 1. Bike-sharing datasets in New York City and Washington D.C. are collected from

Citi Bike website[2] and Capital Bike website[3], respectively. A station, which starts to be checked in/out from a certain timestamp, is defined as a new station. Compared to the dataset in New York City, new stations are more than settled stations in Washington D.C.. Besides the bike-sharing data, we use some external temporal data for two cities which are used to characterize urban dynamics, such as meteorology data (e.g., weather type, wind speed, temperature)[4, 5] and holiday data (e.g., workday, holiday)[6]. Temperature and wind speed are normalized as continuous variables which range from 0 to 1. Other temporal data is encoded to one-hot variables. We choose the data of last 14 d as test set in each city. 85% of the remained data are training set and 15% of the remained data are validation set.

Table 1. The statistics of the dataset.

Data	NYC	Washington D.C
Duration	16/10/1–17/10/27	11/01/01–12/12/9
Records	17,726,635	3,166,051
# stations	846	193
# new stations	259	105

4.2 Experiment Settings

Experimental Setup. We predict daily demand for each station. To predict the daily demand, we use a sliding window to get the historical data. For new stations, we mask their data before they appear. We use one GCN layer, one EvolveGCN layer and one GRU layer on both datasets. We adopt different hidden units of layers in different cities. For dataset in New York City, we cluster the stations into 20 communities. We set 64 hidden units for the GCN layer in B-GCN, 128 hidden units for the EvolveGCN layer and 128 hidden units for the GRU_2 layer. FC_1 and FC_2 are set as 128 hidden units. The learning rate is set as 5×10^{-4}. For dataset in Washington D.C., we cluster the stations into 20 communities. We separately use 32, 64, 128 hidden units for the GCN layer, EvolveGCN layer and GRU_2 layer. And FC_3 and FC_4 are 64 hidden units. The learning rate is also set as 5×10^{-4}.

Baselines. We compare our model (SCEG) with the following state-of-the-art works on demand prediction:

[2] https://www.citibikenyc.com/system-data.
[3] https://www.capitalbikeshare.com/system-data.
[4] https://www.kaggle.com/selfishgene/historical-hourly-weather-data.
[5] https://www.kaggle.com/marklvl/bike-sharing-dataset.
[6] https://www.opm.gov/policy-data-oversight/pay-leave/federal-holidays.

- **GRU** [13]: it only uses historical demand, meteorology data and holiday data as input.
- **T-GCN** [4,6]: They use a single GCN to encode the time-evolving station graphs into spatial embedding in each timestamp, then use GRU to encode the temporal features and spatial embedding over time.
- **E-GCN** [5]: It uses EvolveGCN to encode spatial information in each timestamp, then uses GRU to encode the temporal features and spatial embedding.
- **Multi-graph** [9]: It uses multiple graphs including distance graph, correlation graph, and interaction graph as input of one timestamp, then uses LSTM to encode information in each timestamp.

In addition to the above state-of-the-art methods, we provide another two variants related to our model:

- **CT-GCN**: Based on T-GCN, it uses another GCN to encode the spatial community graph. The embedding of the time-evolving information and spatial communities are concatenated to predict demands.
- **SCEG-w/oBI**: It directly flattens community embedding after GCN when calculating community-informed representation, which does not use B-GCN to assign community interactions to stations.

Evaluation Metrics. We use two evaluation metrics: mean absolute percentage error (MAPE) and root mean square percentage error (RMSPE) similar to [14].

$$MAPE^* = \frac{1}{t \times n_t^*} \sum_{t,n_t^*} \frac{\left|\widetilde{D}_t^{i*} - D_t^{i*}\right|}{D_t^{i*}} \quad RMSPE^* = \frac{1}{t} \sum_t \sqrt{\frac{1}{n_t^*} \sum_{n_t^*} \left(\frac{\widetilde{D}_t^{i*} - D_t^{i*}}{D_t^{i*}}\right)^2} \tag{14}$$

where $\widetilde{D}_t^i$ is prediction demand and D_t^i is the ground-truth demand. n_t^* is the number of stations at t^{th} timestamp.

We evaluate the results in three perspectives: performance on all stations in t^{th} timestamp ($MAPE^{all}$, $RMSPE^{all}$), performance on settled stations ($MAPE^{settled}$, $RMSPE^{settled}$) and performance on new stations ($MAPE^{new}$, $RMSPE^{new}$).

4.3 Prediction Results

We evaluate check-in and check-out demands prediction with 6 baselines as shown in Table 2. Values in bold represent the best performance. The results show that the overall performance of our model is better than other baselines. Our model performs much better than other baselines when the number of new stations is more than the number of settled stations.

We first predict demand using GRU on both datasets, which only uses historical demands and external temporal data. Compared to GRU, other methods perform better as a result of considering the interaction between stations. It infers that station-level demand prediction is significantly related to spatial interactions.

Table 2. Prediction errors of check-in and check-out demands.

City	Method	Check-in			
		$MAPE^{all}$	$RMSPE^{all}$	$MAPE^{settled}$	$RMSPE^{settled}$
N.Y.C.	GRU	0.611	2.144	0.407	0.848
	T-GCN	0.582	2.098	0.393	0.791
	CT-GCN	0.496	1.425	0.349	0.721
	E-GCN	0.446	1.323	<u>0.304</u>	<u>0.644</u>
	Multi-graph	0.460	1.100	0.366	0.772
	SCEG-w/oBI	<u>0.426</u>	<u>1.068</u>	0.345	0.726
	SCEG	**0.383**	**0.969**	**0.271**	**0.591**
W.D.C.	GRU	0.936	1.894	1.083	1.910
	T-GCN	0.699	1.411	0.679	1.225
	CT-GCN	0.602	1.104	0.650	1.408
	E-GCN	0.583	1.026	0.550	1.086
	Multi-graph	0.515	0.970	0.545	1.045
	SCEG-w/oBI	<u>0.508</u>	<u>0.949</u>	<u>0.445</u>	<u>0.795</u>
	SECG	**0.453**	**0.899**	**0.437**	**0.763**
City	Method	Check-out			
		$MAPE^{all}$	$RMSPE^{all}$	$MAPE^{settled}$	$RMSPE^{settled}$
N.Y.C.	GRU	0.888	3.381	0.704	2.114
	T-GCN	0.694	1.973	0.706	2.010
	CT-GCN	0.533	1.689	0.460	1.359
	E-GCN	0.494	1.435	<u>0.307</u>	<u>0.602</u>
	Multi-graph	0.511	1.310	0.355	0.773
	SCEG-w/oBI	<u>0.471</u>	<u>0.959</u>	0.437	0.850
	SCEG	**0.357**	**0.825**	**0.256**	**0.488**
W.D.C.	GRU	0.750	1.340	0.719	1.316
	T-GCN	0.649	1.118	0.678	1.234
	CT-GCN	0.646	1.148	0.629	1.103
	E-GCN	0.586	1.042	0.605	1.004
	Multi-graph	0.559	1.016	0.566	1.092
	SCEG-w/oBI	<u>0.516</u>	<u>0.998</u>	<u>0.561</u>	<u>1.088</u>
	SECG	**0.491**	**0.945**	**0.449**	**0.833**

- $MAPE^{all}$ and $RMSPE^{all}$ are metrics to evaluate performance on all stations. $MAPE^{settled}$ and $RMSPE^{settled}$ are metrics to evaluate performance on settled stations.
- Numbers in bold denote the best performance and numbers with underlines denote the second best performance.

Compare with T-GCN, CT-GCN performs a little better as it consider the spatial community. It infers that the steady and coarse-grained graph helps reduce some prediction errors. But it still performs worse than E-GCN, which doesn't use the embedding of spatial community. Due to using a single GCN, T-GCN and CT-GCN cannot characterize the time-evolving interactions between stations. Though multi-graph performs better than T-GCN and CT-GCN by characterizing the relationship between stations from different views, it still perform worse than E-GCN as it doesn't have a time-evolving structure. Therefore, capturing time-evolving interactions accurately plays an important role in prediction. Compared to E-GCN, SCEG-w/oBI performs better on W.D.C dataset, whose new stations are more than settled stations. It infers that E-GCN do well in encoding dynamic interactions between settled stations. But E-GCN ignores the spatial community, which doesn't help prediction for new stations with patterns of the whole city. And the latent representation combining spatial community and time-evolving graphs help infer possible interactions. Meanwhile, SCEG-w/oBI considers some random when models latent representations, which is more suitable for real-word situations. Compared to SCEG-w/oBI, SCEG uses B-GCN, which relates stations and communities reasonably, which improves the performance significantly.

In summary, the overall performance of our model is better than other baselines. And our model significantly improves overall performance when new stations are more than settled stations.

4.4 Capability of Predicting New Stations

To evaluate the performance on the scenario of new stations, we calculate $MAPE$ and $RMSPE$ of new stations. Meanwhile, we illustrate changes of prediction errors for a new station with different lengths of historical data.

Prediction Errors of New Stations. We show the prediction results of check-out demands for new stations in Table 3. The results show that our model performs better than other baselines on new stations. As expected, GRU and T-GCN have poor performance on handling new stations, which are not trained before. Though CT-GCN improves a little performance by involving spatial community, it cannot learn dynamic information from time-evolving graphs. Compared with CT-GCN, E-GCN performs better. It reveals that the spatial community without some time-evolving information will not predict demands of new stations well, especially in the situation that there are more new stations added. SCEG-w/oBI improve performance obviously as the latent spatial representation is helpful to predict future interactions and demands of new stations. And adding some random is useful for new stations whose spatial patterns is unknown. SCEG weights the affect of communities on each stations, which contribute to analyze the trend of new stations. Therefore, SCEG outperforms than other baselines by using time-evolving representation and community-informed representation. We decrease at least 23.2% on the dataset of New York City and 5.1% on the dataset of Washington D.C. in terms of $MAPE$. Compared to Multi-graph which is also a demand prediction work, we decrease 58.1% on dataset of N.Y.C. and 11.3% on dataset of W.D.C. in terms of $MAPE$.

Evaluation on Different Lengths of Historical Data. We select one new station to find the changes in prediction errors over the length of historical data. The station was deployed on 2012/11/29 in Washington D.C., which was not trained before. We use different lengths of historical data, which range from 1 day to 6 d. As shown in Fig. 3, the prediction error decreases obviously when we involve more historical data. However, the prediction error decreases slowly when the length of historical data reaches 6 d. The reason may be that distant data cannot provide a lot of relevant information. It infers that involving more than 6 d' data cannot help a lot that will cost more time to predict demands. So we select 6 d' historical data, considering a balance between performance and time cost.

4.5 Visualization of Intra-weights and Inter-weights

To analyze the affect of communities on stations, we demonstrate the visualization of intra-weights and inter-weights on 2012/11/29 as shown in Fig. 4. Figure 4(a) shows the intra-weights in Community #5, which consists of 10 stations. From the result, we learn that occupation rates of stations are different. Station #154, #90 and #80 are popular stations in this community. And Station #64 and #61 have fewer demands. Figure 4(b) shows the inter-weights of the settled stations and the new stations. The settled station belongs to Community #12 and the new station belongs to Community #10. The result shows that the settled station frequently interacts with Community #18. The new station rarely interacts with Community #19, Community #6 and Community #8. From the result, we learn that the new station has much fewer interactions than the settled stations. But it's still obvious that a station has different interactions with communities even though the new station has few interactions currently.

4.6 Parameter Analysis

In this part, we evaluate α in the loss function. Because of the difficulty of training VAE, we train our model by warming up. α is changed in each epoch to balance $\mathcal{L}_{mse}$ and KL divergences ($\alpha = epoch * \alpha_0$). To evaluate the effect of different α, we choose the different α_0 at the initial stage which ranges from 1×10^{-4} to 1×10^{-1} and train the model with the same number of epochs. The results shown in Fig. 5 are prediction errors of check-in demands in New York City. With increasing of α_0, the prediction errors of new stations are significantly affected. But prediction errors of all stations and settled stations are little affected. The prediction errors of new stations (yellow line) reach the lowest when α_0 is equal to 1×10^{-3}. When α_0 is equal to 1×10^{-3}, the model reaches better performance and consumes as much time as other parameter settings. The results illustrate

Table 3. Prediction errors for new stations

City	Method	$MAPE^{new}$	$RMSPE^{new}$
NYC	GRU	2.380	6.663
	T-GCN	1.193	2.760
	CT-GCN	1.189	2.648
	E-GCN	1.184	2.570
	Multi-graph	1.256	3.020
	SCEG-w/oBI	0.907	1.329
	SCEG	**0.675**	**1.275**
W.D.C.	GRU	0.976	1.582
	T-GCN	0.865	1.397
	CT-GCN	0.838	1.361
	E-GCN	0.636	1.145
	Multi-graph	0.643	1.150
	SCEG-w/oBI	0.581	1.110
	SCEG	**0.530**	**0.996**

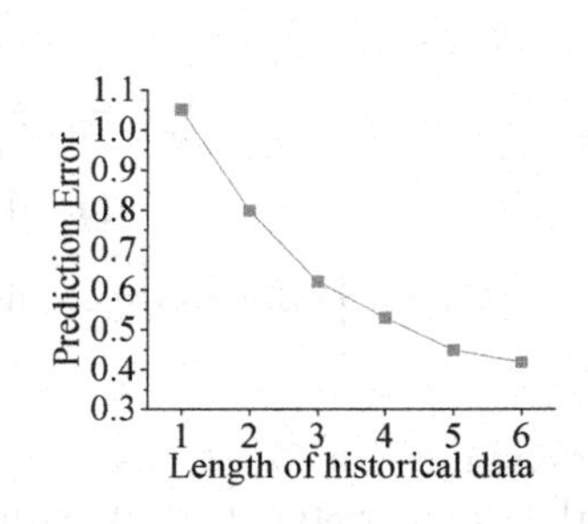

Fig. 3. Evaluation on different lengths of historical data

that the model relies more on $\mathcal{L}_{mse}$ at the beginning of training, which help learn the representations of spatial demands. Due to few historical data, demands of new stations will be predicted better after learning the representation of spatial demands well. However, it is difficult to learn the latent spatial representation directly, we should gradually adjust based on $\mathcal{L}_{mse}$.

5 Related Work

In this section, we briefly introduce related works about demand prediction and spatial-temporal computing.

5.1 Demand Prediction

Demand prediction is a popular topic in urban computing, which is helpful to balance resources. Some works divided the city into grids to predict grid-level

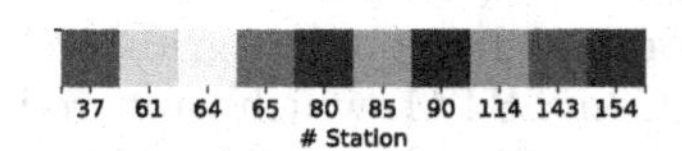

(a) Intra-weights in Community #5. The square which is more dark denotes more occupation rate.

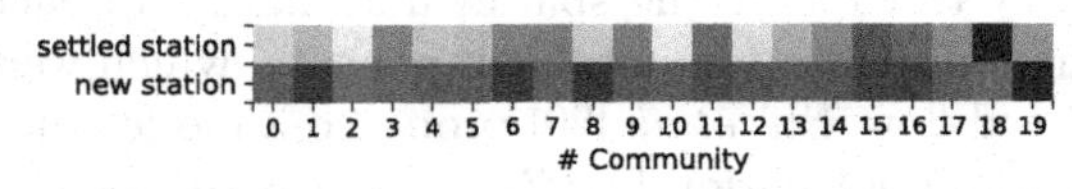

(b) Inter-weights for the settled station and the new station. Squares' colors change from dark blue to dark red mean the weights change from low to high.

Fig. 4. Visualization of intra-weights and inter-weights (Color figure online)

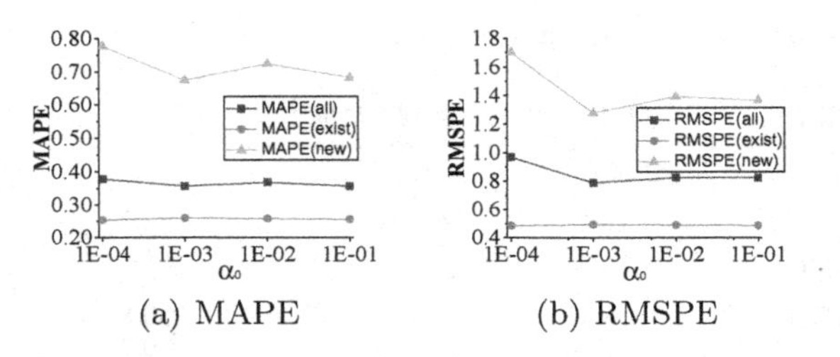

(a) MAPE (b) RMSPE

Fig. 5. Performance on different α_0 in N.Y.C. Dataset. (Color figure online)

demands [15,22]. Considering about the changing of a city, some works used dynamic clusters to mine some demand patterns in a city. Chen *et al.* [3] proposed a dynamic cluster method according to correlations between stations over time. Li *et al.* [14,21] used interactions and correlations of demands to find clusters and then considered inter-cluster transition. To predict the fine-grained demands, researchers focus on station-level demands, which is more challenging. Yoon *et al.* [8] extracted a temporal pattern of stations by using similarity and then built a temporal model based on ARIMA. Hulot *et al.* [1] extracted traffic behaviors and used four different machine learning methods to help make online balancing operations based on predicting demands. Lately, several works involved graphs to characterize the relationship between stations. Chai *et al.* [9] used multi-graph to extract spatial information from different views. Lin *et al.* [4] proposed two architectures based on GCN. The first architecture linearly combined the hidden states from multiple GCN layers. The second one used LSTM to encoder temporal information over the hidden states from single GCN layers.

When existing works predicted station-level demands, they only predicted for the stations that existed all the time. But we consider the real-world scenario that there will be some new stations built at some timestamps.

5.2 Spatial-Temporal Computing

With the increasing number of spatial-temporal data, spatial-temporal computing is necessary to be involved to analyze urban trends. Recently, deep learning methods are widely used to fit the complex problems and big data in spatial-temporal computing. The main method is using CNN and RNN to extract spatial and temporal information [19,20]. Zhang *et al.* [18] and Lin *et al.* [17] leveraged CNN to capture the spatial information of each timestamp and fed them into three different temporal components, which were combined to predict grid-level crowd flow. Shi *et al.* [20] proposed a model named ConvLSTM, which captured spatial information by CNN and learned temporal information by LSTM. After GCN achieved a great success, it is widely used in spatial computing. Chen *et al.* [16] conducted a graph of the traffic network and used GCN to extract spatial information. To consider the multi-view of a city, Sun *et al.* [10] fused views of different periods to predict the flows in irregular regions.

Although the works above achieved great success in spatial-temporal computing, they cannot be used directly for station-level demand prediction. The reason is that they didn't consider time-evolving networks with new nodes.

6 Conclusion

To predict station-level demands, we proposed a novel model named SCEG to exploit time-evolving station graphs informed by the spatial community graph. The spatial community graph, which has coarse-grained interactions in a city, provided some possible interactions and demand trends for new stations. Meanwhile, SCEG represented fine-grained station-level interactions at each timestamp using EvolveGCN. As experimental results shown, time-evolving station graphs and the spatial community graph both contributed to demand prediction for new stations. SCEG performed better than 6 baselines on both settled stations and new stations.

References

1. Hulot, P., Aloise, D., Jena, S.D.: Towards station-level demand prediction for effective rebalancing in bike-sharing systems. In: Proceedings of the 24th ACM SIGKDD International Conference on Knowledge Discovery & Data Mining, pp. 378–386. ACM (2018)
2. Borgnat, P., Abry, P., Flandrin, P., Robardet, C., Rouquier, J.B., Fleury, E.: Shared bicycles in a city: a signal processing and data analysis perspective. Adv. Complex Syst. **14**(3), 415–438 (2011)
3. Chen, L., et al.: Dynamic cluster-based over-demand prediction in bike sharing systems. In: Proceedings of the 2016 ACM International Joint Conference on Pervasive and Ubiquitous Computing, pp. 841–852. ACM(2016)
4. Lin, L., He, Z., Peeta, S.: Predicting station-level hourly demand in a large-scale bike-sharing network: a graph convolutional neural network approach. Transp. Res. Part C: Emerg. Technol. **97**, 258–276 (2018)
5. Pareja, A., et al.: Evolvegcn: evolving graph convolutional networks for dynamic graphs. In: Proceedings of the AAAI Conference on Artificial Intelligence, pp. 1–8 (2020)
6. Zhao, L., et al.: T-gcn: a temporal graph convolutional network for traffic prediction. IEEE Trans. Intell. Transp. Syst. **21**(9), 3848–3858 (2019)
7. Manessi, F., Rozza, A., Manzo, M.: Dynamic graph convolutional networks. Pattern Recogn. **97**, 1–18 (2020)
8. Yoon, J.W., Pinelli, F., Calabrese, F.: Cityride: a predictive bike sharing journey advisor. In: IEEE 13th International Conference on Mobile Data Management, pp. 306–311. IEEE (2012)
9. Chai, D., Wang, L., Yang, Q.: Bike flow prediction with multi-graph convolutional networks. In: Proceedings of the 26th ACM SIGSPATIAL International Conference on Advances in Geographic Information Systems, pp. 397–400. ACM (2018)
10. Sun, J., Zhang, J., Li, Q., Yi, X., Zheng, Y.: Predicting citywide crowd flows in irregular regions using multi-view graph convolutional networks. In: arXiv preprint arXiv:1903.07789 (2019)

11. Kingma, D.P., Welling, M.: Auto-encoding variational bayes. In: arXiv preprint arXiv:1312.6114 (2013)
12. Kipf, T.N., Welling, M.: Semi-supervised classification with graph convolutional networks. In: arXiv preprint arXiv:1609.02907 (2016)
13. Cho, K., et al.: Learning phrase representations using RNN encoder-decoder for statistical machine translation. In: arXiv preprint arXiv:1406.1078 (2014)
14. Li, Y., Zheng, Y.: Citywide bike usage prediction in a bike-sharing system. IEEE Trans. Knowl. Data Eng. **32**(6), 1079–1091 (2019)
15. Ye, J., Sun, L., Du, B., Fu, Y., Tong, X., Xiong, H.: Co-prediction of multiple transportation demands based on deep spatio-temporal neural network. In: Proceedings of the 25th ACM SIGKDD International Conference on Knowledge Discovery & Data Mining, pp. 305–313. ACM (2019)
16. Chen, C., et al.: Gated residual recurrent graph neural networks for traffic prediction. In: Proceedings of the AAAI Conference on Artificial Intelligence, pp. 485–492. ACM (2019)
17. Lin, Z., Feng, J., Lu, Z., Li, Y., Jin, D.: DeepSTN+: context-aware spatial-temporal neural network for crowd flow prediction in Metropolis. In: Proceedings of the AAAI Conference on Artificial Intelligence, pp. 1020–1027. ACM (2019)
18. Zhang, J., Zheng, Y., Qi, D.: Deep spatio-temporal residual networks for citywide crowd flows prediction. In: Proceedings of the AAAI Conference on Artificial Intelligence, pp. 1655–1661. ACM (2017)
19. Qin, D., Yu, J., Zou, G., Yong, R., Zhao, Q., Zhang, B.: A novel combined prediction scheme based on CNN and LSTM for urban PM 2.5 concentration. IEEE Access **7**, 20050–20059 (2019)
20. Xingjian, S.H.I., Chen, Z., Wang, H., Yeung, D.Y., Wong, W.K., Woo, W.C.: Convolutional LSTM network: a machine learning approach for precipitation nowcasting. In: NIPS 2015, pp. 802–810 (2015)
21. Li, Y., Zheng, Y., Zhang, H., Chen, L.: Traffic prediction in a bike-sharing system. In: Proceedings of the 23rd SIGSPATIAL International Conference on Advances in Geographic Information Systems, pp. 1–10. ACM (2015)
22. Wei, H., Wang, Y., Wo, T., Liu, Y., Xu, J.: Zest: a hybrid model on predicting passenger demand for chauffeured car service. In: Proceedings of the 25th ACM International on Conference on Information and Knowledge Management, pp. 2203–2208 (2016)
23. Singla, A., Marco, S., Gábor, B., Pratik, M., Moritz, M., Andreas, K.: Incentivizing users for balancing bike sharing systems. In: Proceedings of the AAAI Conference on Artificial Intelligence, pp. 1–7. ACM (2015)

Applied Data Science: Sports

SoccerMix: Representing Soccer Actions with Mixture Models

Tom Decroos(✉), Maaike Van Roy, and Jesse Davis

Department of Computer Science and Leuven.AI, KU Leuven, Leuven, Belgium
{tom.decroos,maaike.roy,jesse.davis}@cs.kuleuven.be

Abstract. Analyzing playing style is a recurring task within soccer analytics that plays a crucial role in club activities such as player scouting and match preparation. It involves identifying and summarizing prototypical behaviors of teams and players that reoccur both within and across matches. Current techniques for analyzing playing style are often hindered by the sparsity of event stream data (i.e., the same player rarely performs the same action in the same location more than once). This paper proposes SoccerMix, a soft clustering technique based on mixture models that enables a novel probabilistic representation for soccer actions. SoccerMix overcomes the sparsity of event stream data by probabilistically grouping together similar actions in a data-driven manner. We show empirically how SoccerMix can capture the playing style of both teams and players and present an alternative view of a team's style that focuses not on the team's own actions, but rather on how the team forces its opponents to deviate from their usual playing style.

1 Introduction

Style of play, which refers to the behavior on the field of the teams and players during a game, is an important concept in soccer. There is substantial value in gaining a better understanding of playing style as this can be leveraged in areas such as player scouting and match preparation. Because simple descriptive statistics such as pass completion percentage or shot count are usually insufficient to capture playing style, media and fans have traditionally assessed playing style via manual video analysis. However, the advent of novel data sources such as optical tracking and event stream data have motivated an explosion of interest in applying automated techniques to try to glean insights into both player and team behaviors [2,5,7–10,16,18].

Because it is much more widely accessible than optical tracking data, most techniques focus on analyzing event stream data which describes all on-the-ball actions performed by players during a match. Vendors such as WyScout, StatsBomb, and Opta collect this data using human annotators. While watching video feeds of soccer match, annotators record attributes such as the timestamp, location, type (e.g., pass, dribble, shot), involved player, etc. per on-the-ball action. Depending on the type of the action, the annotator also collects additional information such as the end location of a pass or the outcome of a tackle.

Y. Dong et al. (Eds.): ECML PKDD 2020, LNAI 12461, pp. 459–474, 2021.
https://doi.org/10.1007/978-3-030-67670-4_28

Analyzing the playing style of a team or player based on event stream data often involves constructing a so-called *fingerprint* of that team or player which summarizes their actions and captures distinguishing behaviors such as where on the field they tend to perform certain actions. This is often done by dividing actions into groups of similar actions and counting how often players or teams perform actions within each group. However, assessing similarity is difficult because actions are described by various attributes (e.g., type, location) which lay in different domains (e.g., discrete, continuous).

One approach is to lay a grid over the field and proclaim two actions to be similar when they are of the same type and fall in the same grid cell [5,7,16,17]. However, this approach has three downsides. First, the somewhat arbitrary and abrupt boundaries between grid cells can make certain spatially close actions appear dissimilar. Second, choosing the best resolution for the grid is non-trivial as a coarse grid ignores important differences between locations, while a more fine-grained grid will drastically increase the sparsity of the data as a smaller number of actions will fall in a single grid cell. Third, ideally we would like to group actions on additional attributes such as ball direction, but considering more attributes makes each action more unique, which increases the sparsity in the data. Hence, most approaches only include one or two attributes in their analysis [2,7,14]. Rarely do approaches consider three or more attributes [8,17].

In this paper, we make three contributions. Our first contribution is SoccerMix, a novel mixture-model approach for analyzing on-the-ball soccer actions that addresses the shortcomings of grid-based approaches. On the one hand, it alleviates the problem of sparsity by grouping actions in a data-driven manner. On the other hand, SoccerMix's probabilistic nature alleviates the issues of the arbitrary and abrupt boundaries imposed by grid cells. More uniquely, SoccerMix also considers the direction that actions tend to move the ball in, which is an important property for capturing style of play that has received little attention thus far. For example, it allows distinguishing among players or teams that play probing forward passes versus those that play safer lateral passes in a specific zone of the pitch. Intuitively, the action groups produced by SoccerMix can be thought of as describing *prototypical* actions of a certain type, location, and direction. Our second contribution is that we provide a number of use cases that illustrate how SoccerMix can aid in scouting and match analysis by capturing the playing styles of both teams and players. In contrast to existing approaches which solely focus on the offensive style of a team, SoccerMix can also yield insights into a team's defensive style. Specifically, we model how a team can force its opponent to deviate from its typical style of play. Our third contribution is that we provide a publicly available implementation of SoccerMix.[1]

2 Methodology

Our goal is to capture the playing style of either a player or a team. As in past papers [7,10,16,17], our intuition is that playing style is tied to where on the

[1] https://github.com/ML-KULeuven/soccermix.

pitch a player (or team) tends to carry out certain types of actions. Most playing style analysis techniques follow the same two-step approach:

Step 1: Partition all on-the-ball actions into groups of similar actions and represent each action by its membership to one or more of these groups.
Step 2: Transform the group membership counts of a player's or team's actions into a human-interpretable summary of playing style.

Traditionally, most research has focused on the second step [5,7,8,16]. However, picking sub-optimal groups in the first step can introduce significant problems such as sparsity down the line. In fact, many sophisticated data aggregation methods such as pattern mining [8] and matrix factorizaton [7] are often only used in step two to combat the problems introduced by the sub-optimal groups established in the first step.

In this paper, we attempt to tackle the first step in a more intelligent manner than before in order to greatly simplify the second step. More specifically, we aim to find groups of similar actions such that players' or teams' group membership counts are already human-interpretable and informative of playing style. This way, no additional sophisticated transformation is needed in step two. Finding these groups of similar actions involves answering four questions:

1. Which properties of actions are relevant for capturing playing style?
2. How can we group actions based on both discrete and continuous properties?
3. How can we prevent sparsity (many groups with little or no actions in them)?
4. How can we group actions based on properties with different notions of similarity (e.g., linear data vs. circular data)?

2.1 Describing Actions

Various companies provide event stream data and each one uses a different format, has varying definitions of events, and records different sets of events. Moreover, the data also contains extraneous information such as changes in weather that are not crucial for analysis. The SPADL representation [6] addresses these concerns by converting event streams to a uniform representation designed to facilitate analysis.[2] Hence, we first transform our data into this format.

Typically, playing style analysis focuses on action types and locations. One piece of data that is important for style of play that has received little attention is the direction of actions. For example, it is important to differentiate among players who tend to play probing forward passes versus those that tend to play safer, lateral passes. Unfortunately, the direction of the ball is only implicitly present in the SPADL representation through the start and end locations of actions. Therefore, in this paper, we post process SPADL's output and represent each action as a tuple (t, x, y, θ) where t is the type of the action (e.g., shot, tackle, pass, receival), $x \in [0, 105]$ (meters) and $y \in [0, 68]$ (meters) denote the location on the field where the action happened, and $\theta \in [-\pi, \pi]$ (radians) denotes the direction the ball travels in following the action (Fig. 1).

[2] https://github.com/ML-KULeuven/socceraction.

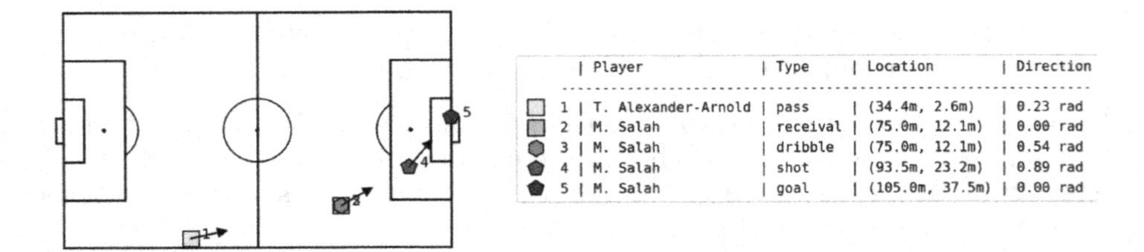

	Player	Type	Location	Direction
1	T. Alexander-Arnold	pass	(34.4m, 2.6m)	0.23 rad
2	M. Salah	receival	(75.0m, 12.1m)	0.00 rad
3	M. Salah	dribble	(75.0m, 12.1m)	0.54 rad
4	M. Salah	shot	(93.5m, 23.2m)	0.89 rad
5	M. Salah	goal	(105.0m, 37.5m)	0.00 rad

Fig. 1. This phase of Liverpool scoring a goal illustrates the event stream data used in this paper. Actions are described by their type t, location (x, y), and direction θ.

2.2 Grouping Actions with Mixture Models

Grouping actions on multiple attributes is non-trivial as it requires fusing together both discrete attributes (i.e., the action type) and continuous attributes (i.e., the location and direction). Past work has mostly ignored direction and focused on fusing action type and location. The most common approach is to lay a grid over the field and for each action type count the number of times it occurs in each zone [7,16,17]. However, this approach has two significant problems. First, this approach ignores the fact that some actions only ever occur in certain areas of the pitch (e.g., throw-ins only occur on the outer edges of the field, shots typically only occur on the attacking half of the field). Second, the boundaries between grid cells are arbitrary and abrupt, which can disrupt the spatial coherence. This can make some actions that occurred in nearby locations appear dissimilar because they fall in different location groups.

This paper takes a different approach and uses mixture models to cluster actions. Mixture models are probabilistic models that assume that all the data points are generated from a mixture of a finite number of distributions with unknown parameters [12]. Formally, a mixture model calculates the probability of generating observation x as:

$$p(x) = \sum_{j=1}^{k} \alpha_j \cdot F_j(x|\Theta_j) \tag{1}$$

where k is the number of components in the mixture model, α_j is the probability of the jth component, and F_j is a probability distribution or density parameterized by Θ_j for the jth component. Intuitively, mixture models can be thought of as a soft clustering variant of k-means clustering. Mixture models address all the drawbacks of the grid-based approach. First, they perform a more data-driven as opposed to hand-crafted partitioning of the pitch. This results in a more nuanced partitioning as the mixture model can learn a more fine-grained representation in zones where lots of actions take place and a more course-grained one in zones where actions are less frequent. Second, by performing a soft grouping each action has a probability of belonging to each cluster, which alleviates the arbitrariness of grid boundaries.

SoccerMix hierarchically groups actions with mixture models in two stages:

Stage 1. For each action type, fit a mixture model to the locations (x, y) of the actions of that type. This allows SoccerMix to model that certain action types usually occur in specific areas of the field (e.g., shots only occur close to the goal, see Fig. 2)

Stage 2. For each component of each mixture model in stage 1, fit a new mixture model to the directions θ of the actions in that component. This allows SoccerMix to model that the direction that a specific action tends to move the ball in, depends on the location where the action occurred (e.g., passes in central midfield are usually lateral or backwards, rarely forwards, see Fig. 3).

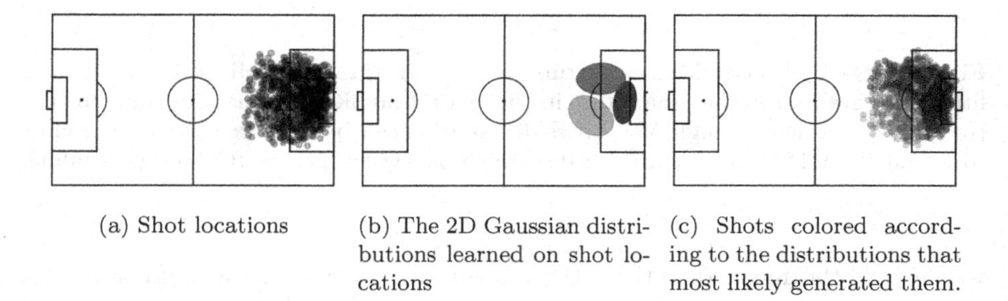

(a) Shot locations

(b) The 2D Gaussian distributions learned on shot locations

(c) Shots colored according to the distributions that most likely generated them.

Fig. 2. Stage 1 of SoccerMix: a mixture model with three 2D Gaussian distributions is fitted to shot locations.

2.3 Distributions of Locations and Directions

The next question to consider is which distributions to use as the components of the mixture models. Locations and directions require a different notion of similarity. In the spatial domain, nearby locations are similar, which we can naturally model using a 2D Gaussian distribution (Fig. 2) [15]:

$$pdf(\mathbf{x}) = \frac{1}{\sqrt{(2\pi)^2|\boldsymbol{\Sigma}|}} \exp\left(-\frac{1}{2}(\mathbf{x}-\boldsymbol{\mu})^{\mathrm{T}}\boldsymbol{\Sigma}^{-1}(\mathbf{x}-\boldsymbol{\mu})\right) \tag{2}$$

where $\boldsymbol{\mu}$ is the mean and $\boldsymbol{\Sigma}$ is the covariance matrix of the distribution.

When viewed as directions, $-\pi + \epsilon_1$ and $\pi - \epsilon_2$ are similar because directions can be seen as values on a circle rather than on a line. However, a Gaussian distribution would not consider these directions to be similar. Therefore, we model the directions using a Von Mises distribution which arises in the directional statistics literature [3,11]. Unlike a Gaussian, Von Mises distributions allow for the possibility that observations close to $-\pi$ and observations close to π can be generated by the same distribution (Fig. 3). The probability density function of a Von Mises distribution is:

$$pdf(\theta) = \frac{1}{2\pi I_0(\kappa)} \exp\left(\kappa \cos(\theta - \mu)\right) \tag{3}$$

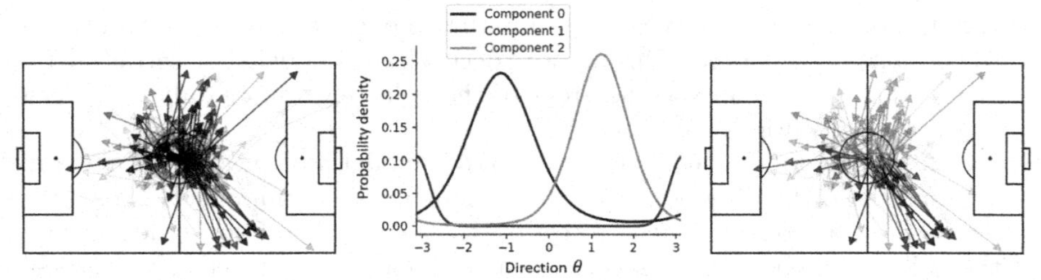

(a) Passes that start in the central midfield

(b) The Von Mises distributions learned on pass directions

(c) The passes colored according to the distribution that most likely generated them.

Fig. 3. Stage 2 of SoccerMix: a mixture model with three Von Mises distributions is fitted to a group of passes that start in the central midfield. In Fig. 3b, component 1 (red) illustrates how a single Von Mises distribution can be fitted to observations close to $-\pi$ and π and is thus essential for describing backwards passes. (Color figure online)

where μ is the mean direction (the distribution is centered around μ) and κ is a measure of concentration ($\kappa = 0$ means that the distribution is uniform over the circle while a high value for κ means that the distribution is strongly concentrated around the angle μ). Finally, $I_0(\kappa)$ is the modified Bessel function of order 0, whose exact definition lies beyond the scope of this paper [11].

2.4 Fitting a Mixture Model to the Data

Fitting the parameters of a mixture model to a data set is typically done using the Expectation Maximization algorithm [1]. Given n observations $\{x_1, \dots, x_n\}$, k distributions $\{F_1, \dots, F_k\}$, and nk latent variables r_{ij} which denote how likely it is that distribution F_j generated observation x_i, the algorithm iteratively performs the following two steps:

Expectation. For each observation x_i and distribution F_j, compute the responsibility r_{ij}, i.e., how likely it is that F_j generated x_i:

$$r_{ij} = \alpha_j \cdot F_j(x_i|\Theta_j).$$

Maximization. For each distribution F_j, compute its weight α_j and its parameter set Θ_j. α_j is the prior probability of selecting component j and can be computed as follows:

$$\alpha_j = \frac{\sum_{i=1}^{n} r_{ij}}{\sum_{j=1}^{k} \sum_{i=1}^{n} r_{ij}}.$$

Θ_j is the parameter set that maximizes the likelihood of distribution F_j having generated each observation x_i with probability r_{ij}. To update Θ_j, we employ the distribution-specific update rules detailed below.

It is straightforward to compute the maximum likelihood estimates for the Gaussian distribution's parameter set $\Theta_j = \{\boldsymbol{\mu}_j, \boldsymbol{\Sigma}_j\}$:

$$\boldsymbol{\mu}_j = \frac{1}{\sum_{i=1}^{n} r'_{ij}} \sum_{i=1}^{n} r'_{ij} \cdot x_i \tag{4}$$

$$\boldsymbol{\Sigma}_j = \frac{1}{\sum_{i=1}^{n} r'_{ij}} \sum_{i=1}^{n} r'_{ij} \cdot (x_i - \boldsymbol{\mu}_j)(\boldsymbol{\mu}_j - x_i)^T \tag{5}$$

where r'_{ij} is a normalized responsibility computed as:

$$r'_{ij} = \frac{r_{ij}}{\sum_{j=1}^{k} r_{ij}}.$$

Computing the maximum likelihood estimates for the Von Mises distributions is more challenging for two reasons. First, we use the output of the learned location mixture models as input for the direction mixture models. More specifically, each observation x_i has a respective weight $w_i = \alpha_{loc} \cdot F_{loc}(x_i|\Theta_{loc})$ (where F_{loc} is the location distribution we wish to further decompose) that represents the probability of observation x_i being part of the input set of observations for the direction mixture model. These weights w_i necessitate slightly altering how the responsibilities r_{ij} are normalized. Second, learning the parameters for a Von Mises distribution is inherently harder than for Gaussians. Directly estimating κ_j is impossible as its exact equations cannot be analytically solved. Luckily, an approximation using the mean result distance R_j exists that works remarkably well for many practical purposes (Eq. 7) [11]. We first construct normalized responsibilities r''_{ij} that pretend that each observation x_i in the data set was generated by the mixture model with a probability of w_i and then update the parameter set $\Theta_j = \{\mu_j, \kappa_j\}$ as follows:

$$\mu_j = \text{atan2}\left(\mu_j^{sin}, \mu_j^{cos}\right) \tag{6}$$

$$\kappa_j \approx \frac{R_j(2 - R_j^2)}{(1 - R_j^2)} \tag{7}$$

where

$$\mu_j^{sin} = \frac{1}{\sum_{i=1}^{n} r''_{ij}} \sum_{i=1}^{n} r''_{ij} \cdot \sin x_i \qquad \mu_j^{cos} = \frac{1}{\sum_{i=1}^{n} r''_{ij}} \sum_{i=1}^{n} r''_{ij} \cdot \cos x_i$$

$$R_j = \sqrt{(\mu_j^{sin})^2 + (\mu_j^{cos})^2} \qquad r''_{ij} = w_i \cdot \frac{r_{ij}}{\sum_{j=1}^{k} r_{ij}}.$$

One of the contributions in this paper is that we publicly release our implementation of mixture models at https://github.com/ML-KULeuven/soccermix. This implementation supports learning a mixture of any type of distribution from a weighted input set of observations.

2.5 Practical Challenges

When applying SoccerMix to real-world event stream data, three practical challenges arise. First, the locations in event stream data are approximations. For some actions, such as goal kicks, annotators use a set of predefined start locations instead of its actual location. Therefore we add random noise to the locations and directions of actions to ensure that we do not simply recover the annotation rules for some actions. Second, the mixture models are sensitive to outliers (e.g., actions with highly irregular locations). Therefore, we preprocess the event stream data to remove outliers using the Local Outlier Factor algorithm [4]. Third, we need to select the number of components used in each mixture model. The number of components needed depends on the action type. For example, passes need more components than corners; a team can perform passes anywhere on the field, but they can take corners from only two locations (the corner flags). We select the number of components in each mixture model by formulating an integer linear programming problem where the goal is to optimize the total Bayesian Information Criterion (BIC) of the entire set of mixture models.[3]

2.6 Capturing Playing Style with SoccerMix

Our goal is to construct a vector that describes a specific player's or team's style. Intuitively, SoccerMix discovers groups of similar actions, where each group describes a *prototypical* action of a certain type, location, and direction. Hence, we can use the learned mixture models to encode each action as a probability distribution over all prototypical actions and encode this in a weight vector. We can then build a style vector for a player (team) by summing the weight vectors of all actions performed by that player (team) in a specific time frame (e.g., a game or a season). In the style vector, the weight of an action group can be interpreted as how often a player (team) performed that prototypical action.

3 Experiments

In our experiments, we use event stream data provided by Statsbomb for the 2017/18 and 2018/19 seasons of the English Premier League (EPL). Using 400,000 actions sampled from the 2017/18 season, we fitted 2D Gaussian mixture models to the locations of the 23 action types to produce 147 location groups. Next, we fitted Von Mises mixture models to the directions of the actions in those groups to produce 247 groups that describe prototypical actions of a certain type, location, and direction (Fig. 4). Learning all mixture models took approx. 30 min on a computer with 32 GB RAM and an Intel i7-6700 CPU @ 3.40 GHz with 8 cores. We used these mixture models to produce weight vectors for ± 2,300,000 actions in 760 games and used those to construct style vectors for 676 players and 23 teams.

[3] More details on our approach to select the number of components used in each mixture model can be found in the public implementation.

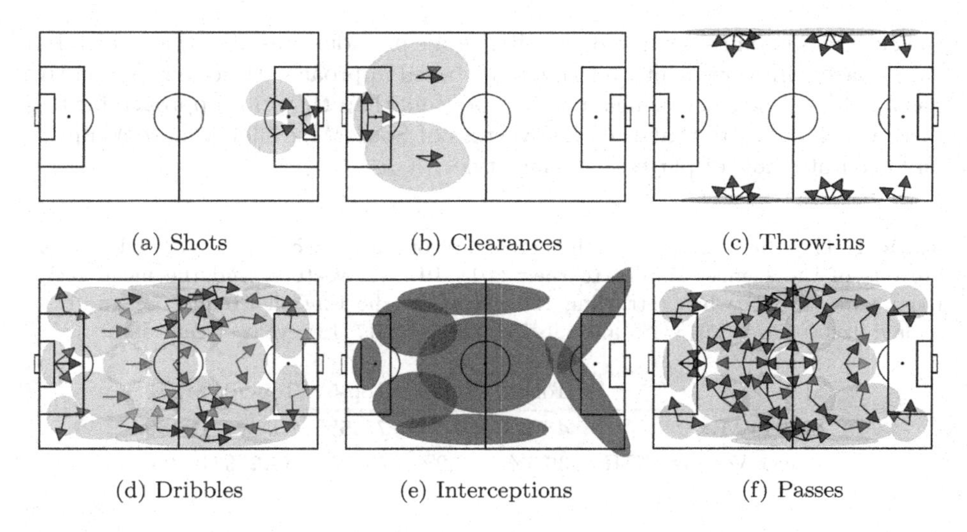

(a) Shots (b) Clearances (c) Throw-ins

(d) Dribbles (e) Interceptions (f) Passes

Fig. 4. Examples of the prototypical actions discovered by SoccerMix. Ellipses denote 2D Gaussian distributions that describe locations. Arrows denote the center of the Von Mises distributions that describe ball directions. Some action types do not directly move the ball and are thus only grouped on location (e.g., interceptions in Fig. 4e).

In this section, we first show how the style vectors produced by SoccerMix can be used to identify players based on their playing style. Next, we show how to compare the playing styles of teams and players, along with an approach for capturing the defensive style of teams. Finally, we use our style vectors to take a closer look at the game that cost Liverpool the title to Manchester City in the 2018/19 season and investigate what exactly went wrong.

3.1 De-anonymizing Players

No objective definition of playing style exists, which creates challenges. Intuitively, one would expect that in the short-term (i.e., across consecutive seasons) a player's style will not change substantially. Based on this insight, Decroos et al. [7] proposed the following evaluation setup: Given anonymized event stream data for a player, is it possible to identify the player based on his playing style in the previous season?

We perform the exact same player de-anonymization experiment as Decroos et al. and compare SoccerMix to their approach: player vectors based on non-negative matrix factorization (NMF). For both approaches, we used the actions of 193 players that played at least 900 min in both seasons. Then, for each player, we create a rank-ordered list of his most similar players by comparing that player's style vector constructed over the 2018/19 season to the style vectors of all players constructed over the 2017/18 season. Table 1 shows how SoccerMix is more successful than the NMF-based player vectors on nearly all ranking metrics. In 48.2% of the cases, SoccerMix correctly identifies a player's style for the current season as being most similar to his previous season's style, which is a 33% relative improvement over the NMF-based approach. Moreover, SoccerMix has a substantially better mean reciprocal rank than the prior approach for this task, which suggests that the style vectors of SoccerMix offer a more complete and accurate view of players' playing style.

Table 1. The top-k results (i.e., the percentage of players whose 2017/18 style vectors are one of the k most similar to their 2018/19 style vectors) and the mean reciprocal rank (MRR) when retrieving 193 players in the English Premier League from anonymized (season 2018/19) and labeled (season 2017/18) event stream data.

Method	Top-1	Top-3	Top-5	Top-10	MRR
SoccerMix	**48.2%**	**62.7%**	**71.5%**	80.8%	**0.589**
Player Vectors (NMF)	36.5%	53.2%	66.5%	**83.2%**	0.505

3.2 Comparing the Playing Style of Players

The style vectors produced by SoccerMix can be used to illustrate the differences in playing style between two players. As an illustrative use case, consider comparing the playing style of Manchester City forward Sergio Agüero and Liverpool forward Roberto Firmino who are both world-class center forwards playing for top teams. Figure 5 illustrates the differences in their style vector for shots, take-ons, interceptions, passes, dribbles, and receivals during the 2018/19 EPL season. Spatially, Agüero is more active in the penalty box as he performs more take-ons, dribbles, and ball receivals in that area than Firmino. In contrast, Firmino performs these actions more in the midfield. Finally, the interception map shows that while Agüero does not completely neglect his defensive duties, Firmino plays a more expansive role that sees him also intercept the ball on the flanks and near the penalty box. These insights correspond to Agüero's reputation of being an out-and-out striker who camps out near the opponent's penalty box whereas Firmino often drops deep to facilitate for his attacking partners Mohammed Salah and Sadio Mané. SoccerMix allows generating such figures for any two players which has the potential to aid clubs in player scouting as they can identify players whose style fits how they wish to play.

3.3 Comparing the Playing Style of Teams

SoccerMix's style vectors can also be used to compare the playing style of teams. To illustrate this use case, we compare the playing styles of Manchester City and Liverpool, who both completely dominated the 2018/19 English Premier League, finishing at the top of the table with 98 and 97 points respectively with a large 25-point gap to distant third contender Chelsea.

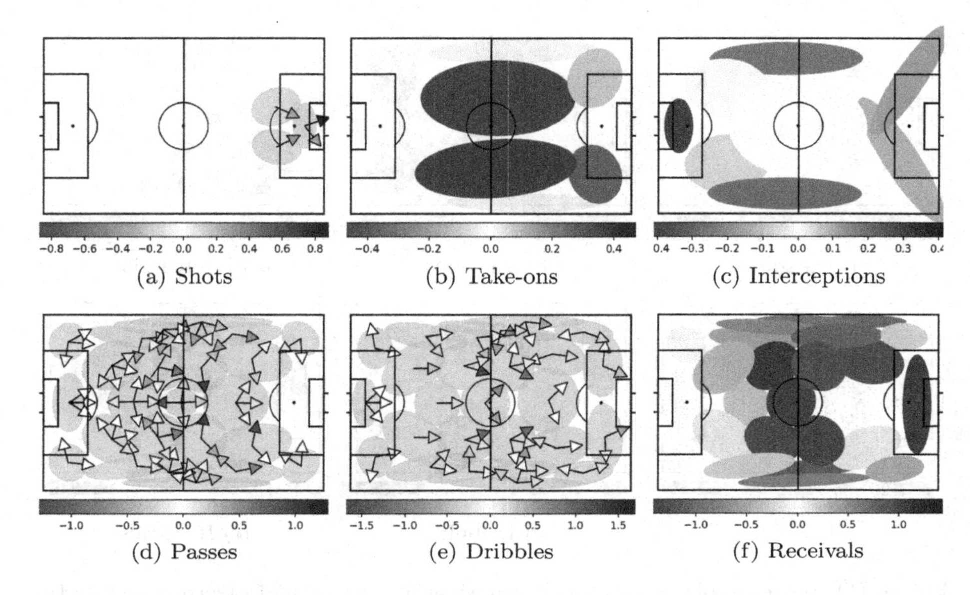

(a) Shots (b) Take-ons (c) Interceptions

(d) Passes (e) Dribbles (f) Receivals

Fig. 5. Differences in playing style between Manchester City forward Sergio Agüero and Liverpool forward Roberto Firmino during the 2018/19 EPL season. Blue (red) actions indicate that Agüero (Firmino) performed more of these actions than the other. Both players are shown as playing left to right (→). Agüero is more active in the penalty box, while Firmino's actions are more spread out over the midfield. (Color figure online)

Figure 6 shows how Manchester City performs noticeably more take-ons, passes, dribbles, and receivals in the heart of the opponent's half compared to Liverpool. This illustrates how the coaches of both teams have shaped their team's playing style to their own soccer philosophy. Under Jürgen Klopp, Liverpool have perfected the art of frequent counter-pressing and speedy counter-attacks. Under Pep Guardiola, Manchester City at times mimics the possession-based, tiki-taka style of its coach's ex-club (FC Barcelona), passing and moving the ball high up on the field.

Additionally, Liverpool seems to funnel the play towards their right side, performing noticeably more clearances, take-ons, and interceptions on their right flank. The most likely source of this uptick is Trent Alexander-Arnold, a right-back at Liverpool who is widely regarded as one of the best attacking full-backs in professional soccer and is a spearhead of Liverpool's transitional, counter-attacking style of play.[4]

[4] https://sport.optus.com.au/articles/os6422/trent-alexander-arnold-is-changing-the-full-back-position.

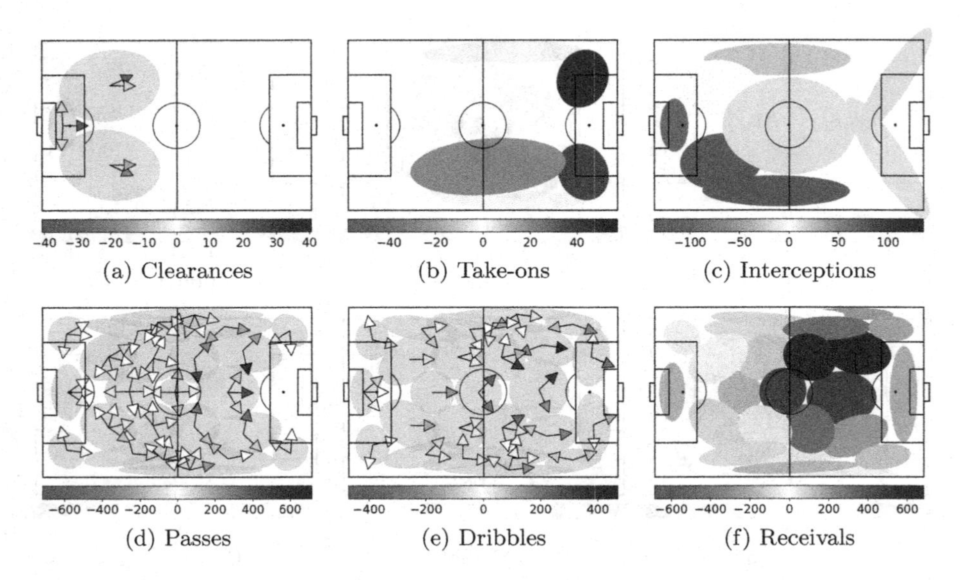

Fig. 6. Differences in playing style between Manchester City and Liverpool during the 2018/19 EPL season based on the prototypical action groups obtained with SoccerMix. Blue (red) actions indicate that Manchester City (Liverpool) performed more of these actions than the other team. Both teams are shown as playing left to right (→). Liverpool funnels play towards their right side, while Manchester City generally plays higher up the field. (Color figure online)

3.4 Capturing the Defensive Playing Style of Teams

Approaches that capture playing style usually focus on offensive playing style, i.e., what does a team do when in possession of the ball? Analyzing defensive style is much harder as it involves off-the-ball actions such as correct positioning and putting pressure on attackers, which are not recorded in event streams. Our insight is that these off-the-ball actions are often performed with the intention of *preventing certain actions from occurring*. This suggests that we can gain a partial understanding of defensive style by measuring the effects that a team's off-the-ball actions have on what on-the-ball actions their opponent performs. More precisely, we analyze how a team forces its opponents to deviate from their usual playing style.

To illustrate this, we measure the mean difference between teams' style vectors constructed using (1) only the matches against Liverpool and (2) all other matches (i.e., those not involving Liverpool). Figure 7 shows how Liverpool causes their opponents, playing left to right, to be flagged more for offside than is typical. This indicates a well-synchronized line of defense that employs a very effective offside trap. The crosses show that, although Liverpool limits the number of crosses its opponents perform, this restriction is not symmetric: they allow fewer crosses from the left of defense (the offense's right) than the right. Lastly, as a combination of both offensive and defensive playing style, Liverpool generally forces the other teams to play more on their own half than on Liverpool's half.

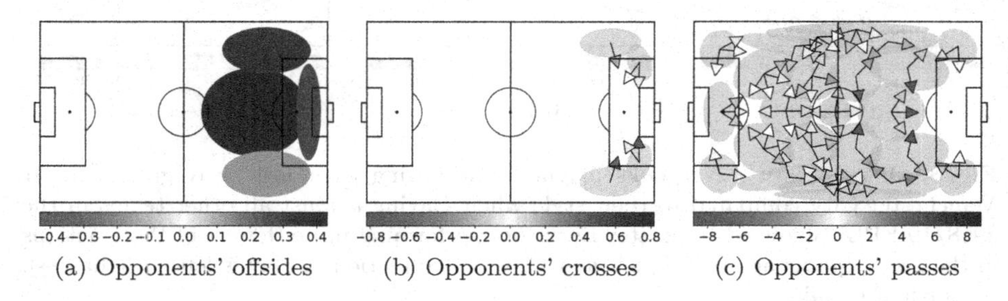

(a) Opponents' offsides (b) Opponents' crosses (c) Opponents' passes

Fig. 7. Illustrations of how Liverpool (a) employs a good offside trap, (b) has a weaker defense at their right flank when it comes to preventing their opponents from crossing the ball, and (c) forces other teams to play more on their own half. Blue (red) indicates that teams perform more (fewer) of these actions when playing against Liverpool. (Color figure online)

3.5 Case Study: How Liverpool Lost the Title to Manchester City in a Single Game

On January 3rd, 2019, Liverpool held a 6 point lead atop the EPL table when they traveled to play Manchester City in a highly anticipated match. Alas, in their only league loss of the season, Liverpool fell 2–1 and ended up missing out on the title to Manchester City by a single point. It is not a stretch to say that this was the game that cost them the title. Using the concept of style difference vectors from the previous section, Fig. 8 illustrates how Liverpool's playing style in this game drastically deviated from how they played against other teams. In short, Manchester City maintained their typical high defensive line and forced Liverpool to remain on their own side of the field. This is apparent in both the higher number of passes, dribbles, and receivals Liverpool had to perform deep in their own half as well as the fact that they performed significantly fewer actions than normal in Manchester City's half.

While interesting, it is not completely surprising that Liverpool's offensive output suffered against its only decent rival that season, Manchester City. To dig deeper, we adjust for the level of the opponent and compare Liverpool's playing style in their away game (loss) and home game (draw) against Manchester City in 2018/19 (Fig. 9). In its away game, Liverpool made noticeably less use of its left flank, performing fewer passes, dribbles, and receivals in that area. This suggests that Liverpool's left flank players were not functioning very well that game, which is further evidenced by midfielder James Milner and winger Sadio Mané on Liverpool's left flank being substituted out in the 57th and 77th minute of the game.

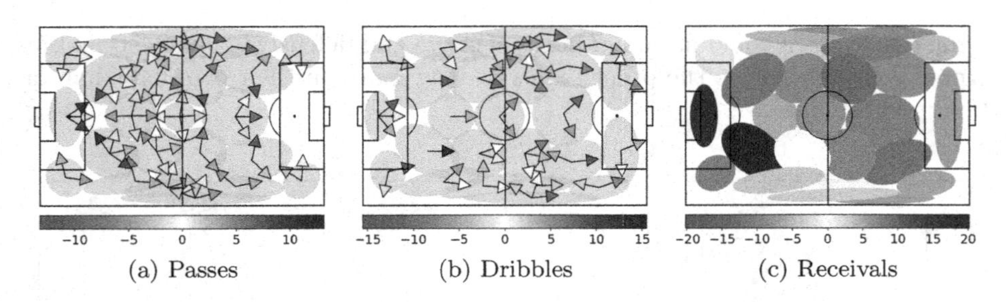

(a) Passes (b) Dribbles (c) Receivals

Fig. 8. Differences in Liverpool's playing style during their lost away game against Manchester City compared to their style when playing against all other teams in the 2018/19 EPL. Blue (red) indicates Liverpool performing more (fewer) of these actions in their away game against Manchester City. The direction of play is left to right (→). (Color figure online)

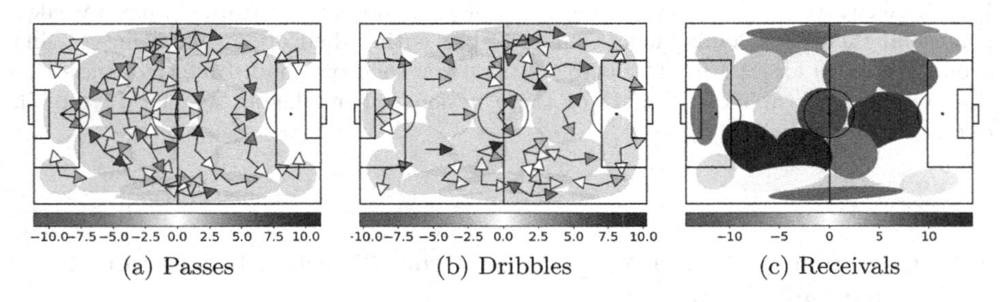

(a) Passes (b) Dribbles (c) Receivals

Fig. 9. Differences in Liverpool's playing style between their away game and their home game against Manchester City in the 2018/19 EPL. Red (blue) indicates fewer (more) of these actions in the away game than in the home game. Liverpool's left flank players were having a bad day in the away game, as evidenced by the fewer passes, dribbles, and receivals in that area than normal. (Color figure online)

4 Related Work

Many approaches group actions by overlaying a grid on the field [7,8,16]. How they differ is in how they combat the challenges associated with this grid. Decroos et al. [8] avoid the sparsity issues of a fine-grained grid by dividing the field into only four zones (left-flank, midfield, right-flank, and penalty box), as the performance of their pattern mining algorithm rapidly declined when using a more fine-grained grid. However, their patterns can then only describe ball movements between these four zones and are thus too broad and simple to be able to identify unique characteristics related to playing style. Van Haaren et al. [16] attempted to combine the advantages of both coarse and fine-grained grids by encoding action locations on multiple granularity levels. However, they found that this multi-level representation of actions blew up the search space of their inductive logic programming approach and led to heavy computational costs.

Decroos and Davis [7] apply a post-processing step to the counts of a fine-grained grid. More specifically, the count of each grid cell is replaced by a weighted mean of itself and its neighboring grid cells, which promotes spatial coherence between grid cells and combats issues such as sparsity and abrupt boundaries. However, there are two downsides to this approach. First, a new technique with its own parameters (that are non-trivial to tune) is added to the analysis pipeline. Second, this approach encourages dividing actions into a number of groups that is excessive for representing the characteristics of the data, which makes it difficult for automated systems to numerically process the new data representation and for humans to interpret the end results. For example, Decroos and Davis use 50×50 grid cells to represent shot behavior of players (of which most will be empty), while SoccerMix only needs 3 location groups to represent shot behavior.

5 Conclusion

Capturing the playing style of teams and players in soccer can be leveraged in areas such as player scouting and match preparation. In this paper we introduced SoccerMix: an approach to intelligently partition player actions into groups of similar actions. Intuitively, each group describes a prototypical action with a specific type, location, and direction. We have shown how SoccerMix can be used to capture the playing style of both teams and players. Additionally, we introduced a new way to capture the defensive playing style of a team by using deviations in the actions of that team's opponents. Finally, we have publicly released SoccerMix's implementation at https://github.com/ML-KULeuven/soccermix.

Acknowledgements. Tom Decroos is supported by the Research Foundation-Flanders (FWO-Vlaanderen). Maaike Van Roy is supported by the Research Foundation-Flanders under EOS No. 30992574. Jesse Davis is partially supported by KU Leuven Research Fund (C14/17/07), Research Foundation - Flanders (EOS No. 30992574, G0D8819N). Thanks to StatsBomb for providing the data used in this paper.

References

1. Bailey, T.L., Elkan, C., et al.: Fitting a mixture model by expectation maximization to discover motifs in bipolymers (1994)
2. Bekkers, J., Dabadghao, S.: Flow motifs in soccer: what can passing behaviortell us? J. Sports Anal. (Preprint), 1–13 (2017)
3. Best, D., Fisher, N.I.: Efficient simulation of the von mises distribution. J. Royal Stat. Soc. Ser. C (Applied Statistics) **28**(2), 152–157 (1979)
4. Breunig, M.M., Kriegel, H.P., Ng, R.T., Sander, J.: Lof: identifying density-based local outliers. In: Proceedings of the 2000 ACM SIGMOD International Conference on Management of data, pp. 93–104 (2000)
5. Cintia, P., Rinzivillo, S., Pappalardo, L.: A network-based approach to evaluate the performance of football teams. In: Machine Learning and Data Mining for Sports Analytics Workshop, Porto, Portugal (2015)

6. Decroos, T., Bransen, L., Van Haaren, J., Davis, J.: Actions speak louder than goals: Valuing player actions in soccer. In: Proceedings of the 25th ACM SIGKDD International Conference on Knowledge Discovery and Data Mining, KDD 2019, pp. 1851–1861. ACM, New York (2019). https://doi.org/10.1145/3292500.3330758
7. Decroos, T., Davis, J.: Player vectors: characterizing soccer players' playing style from match event streams. In: Brefeld, U., Fromont, E., Hotho, A., Knobbe, A., Maathuis, M., Robardet, C. (eds.) ECML PKDD 2019. LNCS (LNAI), vol. 11908, pp. 569–584. Springer, Cham (2020). https://doi.org/10.1007/978-3-030-46133-1_34
8. Decroos, T., Van Haaren, J., Davis, J.: Automatic discovery of tactics in spatio-temporal soccer match data. In: Proceedings of the 24th ACM SIGKDD International Conference on Knowledge Discovery & Data Mining, pp. 223–232 (2018)
9. Gyarmati, L., Hefeeda, M.: Analyzing in-game movements of soccer players at scale. arXiv preprint arXiv:1603.05583 (2016)
10. Gyarmati, L., Kwak, H., Rodriguez, P.: Searching for a unique style in soccer. arXiv preprint arXiv:1409.0308 (2014)
11. Mardia, K.V., Jupp, P.E.: Directional Statistics, vol. 494. Wiley, Chichester (2009)
12. McLachlan, G.J., Basford, K.E.: Mixture Models: Inference and Applications to Clustering, vol. 38. M. Dekker, New York (1988)
13. Pedregosa, F., et al.: Scikit-learn: machine learning in python. J. Mach. Learn. Res. **12**, 2825–2830 (2011)
14. Pena, J.L.: A Markovian model for association football possession and its outcomes. arXiv preprint arXiv:1403.7993 (2014)
15. Reynolds, D.A.: Gaussian mixture models. Encycl. Biometrics **741**, 659–663 (2009)
16. Van Haaren, J., Dzyuba, V., Hannosset, S., Davis, J.: Automatically discovering offensive patterns in soccer match data. In: Fromont, E., De Bie, T., van Leeuwen, M. (eds.) IDA 2015. LNCS, vol. 9385, pp. 286–297. Springer, Cham (2015). https://doi.org/10.1007/978-3-319-24465-5_25
17. Van Haaren, J., Hannosset, S., Davis, J.: Strategy discovery in professional soccer match data. In: Proceedings of the KDD-16 Workshop on Large-Scale Sports Analytics, pp. 1–4 (2016)
18. Wang, Q., Zhu, H., Hu, W., Shen, Z., Yao, Y.: Discerning tactical patterns for professional soccer teams: an enhanced topic model with applications. In: Proceedings of the 21th ACM SIGKDD International Conference on Knowledge Discovery and Data Mining, pp. 2197–2206 (2015)

Automatic Pass Annotation from Soccer Video Streams Based on Object Detection and LSTM

Danilo Sorano[1], Fabio Carrara[2], Paolo Cintia,[1] Fabrizio Falchi[2], and Luca Pappalardo[2(✉)]

[1] Department of Computer Science, University of Pisa, Pisa, Italy
[2] ISTI-CNR, Pisa, Italy
luca.pappalardo@isti.cnr.it

Abstract. Soccer analytics is attracting increasing interest in academia and industry, thanks to the availability of data that describe all the spatio-temporal events that occur in each match. These events (e.g., passes, shots, fouls) are collected by human operators manually, constituting a considerable cost for data providers in terms of time and economic resources. In this paper, we describe PassNet, a method to recognize the most frequent events in soccer, i.e., passes, from video streams. Our model combines a set of artificial neural networks that perform feature extraction from video streams, object detection to identify the positions of the ball and the players, and classification of frame sequences as passes or not passes. We test PassNet on different scenarios, depending on the similarity of conditions to the match used for training. Our results show good classification results and significant improvement in the accuracy of pass detection with respect to baseline classifiers, even when the match's video conditions of the test and training sets are considerably different. PassNet is the first step towards an automated event annotation system that may break the time and the costs for event annotation, enabling data collections for minor and non-professional divisions, youth leagues and, in general, competitions whose matches are not currently annotated by data providers.

Keywords: Sports analytics · Computer vision · Applied data science · Deep learning · Video semantics analysis

1 Introduction

Soccer analytics is developing nowadays in a rapid way, thanks to sensing technologies that provide high-fidelity data streams extracted from every match and training session [10,20,22]. In particular, the combination of video-tracking data and soccer-logs, which describe the movements of players and the spatio-temporal events that occur during a match, respectively, allows sophisticated technical-tactical analyses [3,5,6,18,19,27]. However, from a data provider's perspective, the collection of soccer-logs is expensive, time-consuming, and not free

Y. Dong et al. (Eds.): ECML PKDD 2020, LNAI 12461, pp. 475–490, 2021.
https://doi.org/10.1007/978-3-030-67670-4_29

from errors [16]. It is indeed still performed manually through proprietary software for the annotation of events (e.g., passes, shots, fouls) from video streams, a procedure that requires around three human operators and about two hours per match [20]. Given these costs and the enormous number of matches played every day around the world, data providers collect data regarding relevant professional competitions only, of which they sell data to clubs, companies, websites, and TV shows. For all these reasons, an automated event annotation system would provide many benefits to the sports industry. On the one hand, it would bring a reduction of errors, time, and costs of annotation for data providers: an automatic annotation system may substitute one of the human operators, or it may be used to check the reliability of the events collected manually. On the other hand, it would enable data collections for non-professional divisions, youth leagues and, in general, competitions whose matches data providers have no economic convenience to annotate.

Most of the works in the literature focus on video summarization, i.e., the detection from video streams of salient but infrequent episodes in matches, such as goals, replays, highlights, and play-breaks [1,12,17,23,28]. Another strand of research focuses on the identification of players through their jersey number [9] or on the detection of ball possession [14]. Nevertheless, there are no approaches that focus specifically on the recognition of *passes* from video streams, although passes correspond to around 50% of all the events in a soccer match [20]. It hence goes without saying that a system that wants to drastically reduce errors, time and economic resources required by event annotation must be able to accurately recognize passes from video streams.

In this paper, we propose PassNet, a computer vision system to detect passes from video streams of soccer matches. We define a *pass* as a game episode in which a player kicks the ball towards a teammate, and we define *automatic pass detection* as the problem of detecting all sequences of video frames in which a pass occurs. PassNet solves pass detection combining three models: ResNet18 for feature extraction, a Bidirectional-LSTM for sequence classification, and YOLOv3 for the detection of the position of the ball and the players. To train PassNet, we integrate video streams of four official matches with data describing when each pass begins and ends, collected manually through a pass annotation application we implement specifically to this purpose. We empirically prove that PassNet overtakes several baselines on different scenarios, depending on the similarity of conditions of the match's video stream used for training, and that it has good agreement with the sets of passes annotated by human operators of a leading data collection company. Given its flexibility and modularity, PassNet is a first step towards the construction of an automated event annotation tool for soccer.

2 Related Work

De Sousa et al. [24] group event detection methods for soccer in low-level, medium-level, and high-level analysis. The low-level analysis concerns the recognition of basic marks on the field, such as lines, arcs, and goalmouth. The middle-level analysis aims at detecting the behavior of the ball and the players. The high-level analysis regards the recognition of events and video summarization.

Video Summarization. Most of the works in the literature focus on the detection from video streams of salient actions such as goals, replays, highlights, and play-breaks. Bayat et al. [1] propose a heuristic method to detect goals from video streams based on the audio intensity, color intensity, and the presence of goalmouth. Zawbaa et al. [29] perform video summarization through shot-type classification, play-break classification, and replay detection, while Kapela et al. [13] can detect scores and near misses that do not result in a score. Jiang et al. [12] detect goals, shots, corner kicks, and cards through a combination of convolutional and recurrent neural networks that proceeds in three progressive steps: play-break segmentation, feature extraction, and event detection. Yu et al. [28] use deep learning to identify replays and associated events such as goals, corners, fouls, shots, free-kicks, offsides, and cards. Saraogi et al. [23] develop a method to recognize notable events combining generic event recognition, event's active region recognition, and shot classification. Liu et al. [17] use 3D convolutional networks to perform play-break segmentation and action detection from video streams segmented with shot boundary detection. Similarly, Fakhar et al. [8] address the problem of highlight detection in video streams in three steps: shot boundary detection, shot view classification, and replay detection.

Ball, Player and Motion Detection. Kahn et al. [14] use object detection methods based on deep learning to identify when a player is in possession of the ball. Gerke et al. [9] use a convolutional neural network to recognize the jerseys from labeled images of soccer players. Carrara et al. [4] use recurrent neural networks to annotate human motion streams, in which each step represents 31 joints of a skeleton, each described as a point in a 3D space.

Contribution of Our Work. An overview of the state of the art cannot avoid noticing that there are no approaches that focus on the recognition of *passes* from video streams. Nevertheless, automatic pass detection is essential, considering that passes correspond to around 50% of all the events in a soccer match [20]. In this paper, we fill this gap by providing a method, based on a combination of artificial neural networks, that can recognize passes from video streams.

3 Pass Detection Problem

An *event* is any relevant episode that occurs at some point in a soccer match, e.g., pass, shot, goal, foul, save. The type of events annotated from video streams is similar across different data collection companies, although there may be differences in the way annotated events are structurally organized [16,20]. From a video stream's perspective, an event is the sequence of n frames $\langle k_t, k_{t+1}, \ldots, k_{t+n} \rangle$ in which it takes place.

Nowadays, events are annotated from video streams through a manual procedure performed through a proprietary software (the tagger) by expert video analysts (the operators) [20]. For example, the company Wyscout[1] uses one

[1] https://wyscout.com/.

operator per team and one operator acting as a responsible supervisor of the output of the whole match [20]. Each operator annotates each relevant episode during the match, hence defining the event's type, sub-type, coordinates on the pitch, and additional attributes [20]. Finally, the operators perform quality control by checking the coherence between the events that involve both teams, and through manually scanning the annotated events. Manual event annotation is time-consuming and expensive: since the annotation of a single match requires about two hours and three operators, the effort and the costs needed to tag an entire match-day are considerable [20].

We define automatic event detection as the problem of annotating sequences of frames in a video stream with a label representing the corresponding event that occurs during that frame sequence. In particular, in this paper we focus on *automatic pass detection*: detecting all sequences of video frames in which a *pass* occurs. A pass is a game episode in which a player in possession of the ball tries to kick it towards a teammate.

4 PassNet

Our solution to the automatic pass detection problem is PassNet, the architecture of which is shown in Fig. 1.[2] It combines three tasks: *(i) feature extraction* reduces the dimensionality of the input using ResNet18 (Sect. 4.1); *(ii) object detection* detects the players and the ball in the video frames using YOLOv3 (Sect. 4.2); *(iii) sequence classification* classifies sequences of frames as containing a pass or not using a Bi-LSTM [4] (Sect. 4.3).

In PassNet, each frame has dimension $3 \times 352 \times 240$, where 3 indicates the RGB channel, and 352×240 is the size of an input frame. The sequence of frames that composes a video stream (Fig. 1a) is provided in input to *(i)* a feature extraction module (Fig. 1b), which outputs a sequence of vectors of 512 features, and *(ii)* to an object detection module (Fig. 1c), which outputs a sequence of vectors of 24 features describing the positions of the ball and the closest players to it. The two outputs are combined into a sequence of vectors of 536 features (Fig. 1d) and provided as input to a sequence classification module (Fig. 1e), which generates a pass vector (Fig. 1f) that indicates, for each frame of the original sequence, whether or not it is part of a pass sequence.

4.1 Feature Extraction

The sequence of frames is provided in input to the Feature Extraction module frame by frame, each of which is transformed into a feature vector by the image classification model ResNet18 [11]. In the end, the feature vectors are combined again into a sequence. ResNet18 consists of a sequence of convolution and pooling layers [11]. Convolution layers use convolution operations to produce a feature map by sliding a kernel of fixed size over the input tensor and computing the

[2] PassNet's code and data are available at https://github.com/jonpappalord/PassNet.

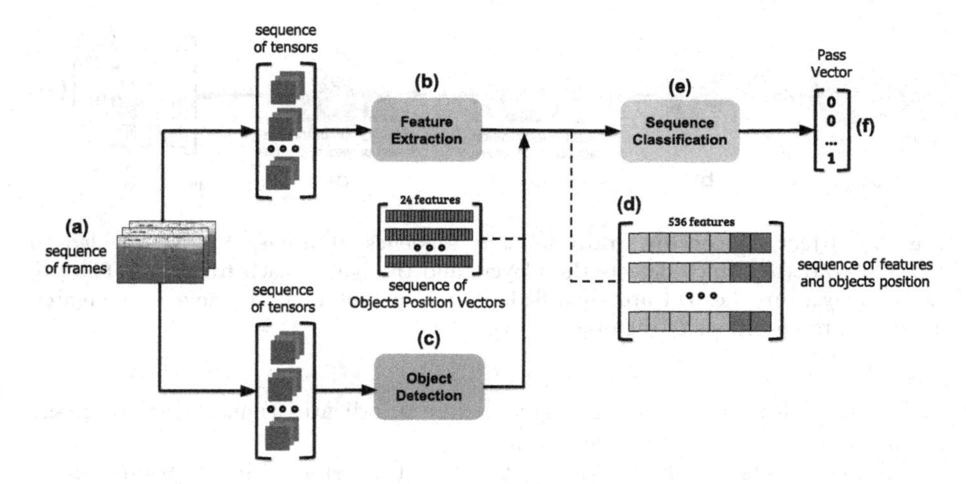

Fig. 1. Architecture of PassNet. The sequence of frames of the video stream (a) is provided to a Feature Extraction module (b) that outputs a sequence of vectors of 512 features, and to an Object Detection module (c) that outputs a sequence of vectors of 24 features describing the position of the ball and the closest players to it. The two outputs are combined into a sequence of vectors of 536 features (d) and provided to a Sequence Classification module (e), which outputs a pass vector (f) that indicates, for each frame of the original sequence, whether or not it is part of a pass sequence.

dot-product between the covered input and the kernel weights. Pooling layers reduce the dimensionality of each feature map while retaining the most important information. ResNet18 returns as output a feature vector of 1,000 elements, which represents all possible classes (objects) in the ImageNet-1K data set [7] used to train the model. For our purpose, we take the output of the last average pooling layer of ResNet18, which generates a vector of 512 features.

4.2 Object Detection

To identify the position of relevant objects in each frame, such as the players and the ball, we use YOLOv3 [21], a convolutional neural network for real-time object detection. In particular, we use a version of YOLOv3 pre-trained on the COCO dataset [15], which contains images labeled as balls or persons. YOLOv3 assigns to each object detected inside a frame a label and a bounding box, the center of which indicates the position of the object inside the frame. Figure 2 summarizes the process of extraction of the position of the objects using YOLOv3.

We provide to YOLOv3 a match's video stream frame by frame. Based on the objects identified by YOLOv3, we convert each frame into a vector of 24 elements, that we call *Objects Position Vector* (OPV). OPV combines six vectors of length four, describing the ball and the five closest players to the ball. In particular, each of the six vectors has the following structure:

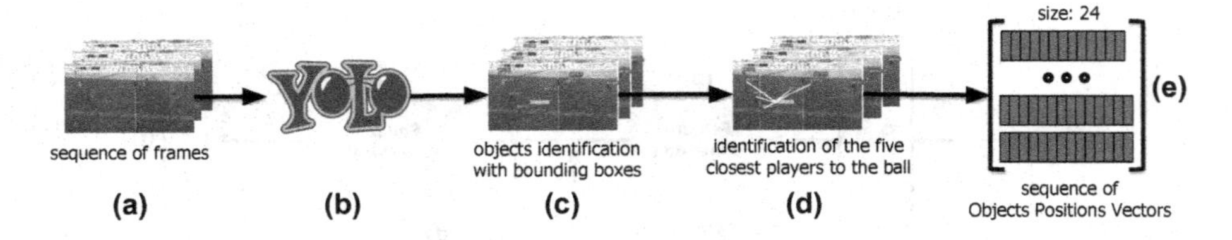

Fig. 2. Object Detection module. The sequence of frames (a) is provided to YOLOv3 [21] (b), which detects the players and the ball in each frame (c). The five closest players to the ball are identified (d) and a vector of 24 elements is created describing the positions of the objects (e).

- The first element is a binary value, where 0 indicates that the vector refers to the ball and 1 that it refers to a player;
- The second element has value 1 if the object is detected in the frame, and 0 that the vector is a flag vector (see below);
- The third and the fourth elements indicate the coordinates of the object's position, normalized in the range $[-1, +1]$, where the center of the frame has coordinates $(0, 0)$ (Fig. 3a).

For example, vector $[0, 1, 0.8, -0.1]$ indicates that YOLOv3 detects that ball in a frame at position $(0.8, -0.1)$, while $[1, 1, -0.1, 0.4]$ indicates that YOLOv3 detects a player at position $(-0.1, 0.4)$. Note that YOLOv3 may identify no objects in a frame, even if they are actually present [21].[3] When an object is not detected, we substitute the corresponding vector with a *flag vector*. Specifically, if in a frame the ball is not detected, we describe the ball using flag vector $[0, 0, 0, 0]$.

We detect the five closest players to the ball by computing the distance to it of all the players detected in a frame (Fig. 3b). If less than five players are detected, we describe a player using the flag vector $[1, 0, 2, 2]$. When no player or ball is detected, we use flag vectors for both the ball and the players. If at least one player is detected, but the ball is not detected, we assume that the ball is located at the center of the frame and identify the five closest players to it.

4.3 Sequence Classification

The outputs of the Feature Extraction module and the Object Detection module are combined into a sequence of vectors of 536 features and provided in input to the Sequence Classification module (Fig. 4). We use a sliding window δ to split the sequence of vectors into sub-sequences of length δ. Each sub-sequence (Fig. 4a) goes in input to a Bidirectional-LSTM (Fig. 4b), followed by two dense layers (Fig. 4c) that output a vector of δ values. Each element of this vector is transformed into 1 (`Pass`) or 0 (`No Pass`) according to a sigmoid activation function (Fig. 4d) and an activation threshold (Fig. 4e). The best hyper-parameter values of the Sequence Classification module (e.g., number of hidden units in the dense layers, value of activation threshold) are determined experimentally.

[3] In our experiments, we find that this situation happens for 0.66% of the frames.

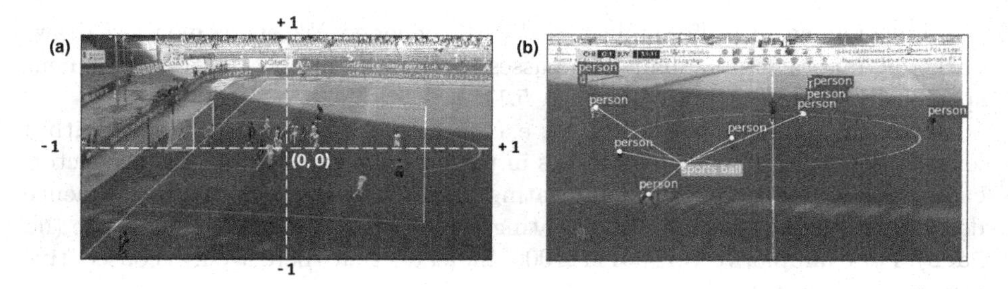

Fig. 3. Construction of the Object Position Vectors. (a) Normalization of the coordinates in the range $[-1, +1]$, where the center of the frame has coordinates $(0, 0)$. (b) The five players identified by YOLOv3 with the shortest distance from the ball.

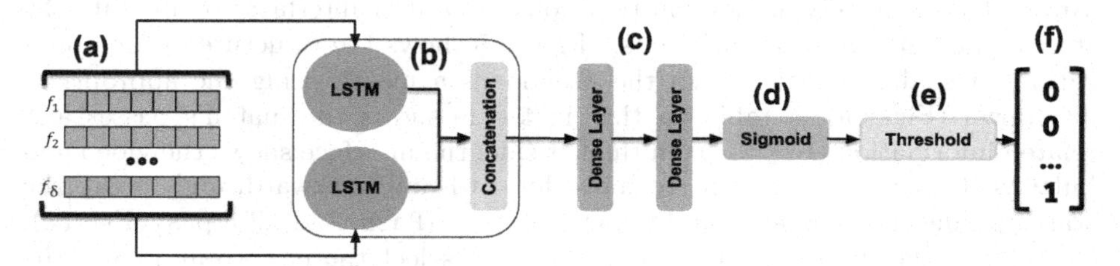

Fig. 4. Sequence Classification Module. Each sub-sequence of δ vectors (a) is provided to a Bi-LSTM (b), which processes the sub-sequence in two directions: from the first frame f_1 to the last frame f_δ and from f_δ to f_1. The output of the Bi-LSTM goes to two dense layers (c) with ReLu activation functions and dropout = 0.5. A sigmoid activation function (d) and an activation threshold (e) are used to output the pass binary vector, in which 1 indicates the presence of a pass in a frame.

5 Data Sets

We have video streams corresponding to four matches in the Italian first division: AS Roma vs. Juventus FC, US Sassuolo vs. FC Internazionale, AS Roma vs. SS Lazio from season 2016/2017, and AC Chievo Verona vs. Juventus FC from season 2017/2018. All of them are video broadcasts on TV and have resolution 1280×720 and 25 frames per second. We eliminate the initial part of each video, in which the teams' formations are presented and the referee tosses the coin, and we split the videos into the first and second half. For computational reasons, we reduce the resolution of the video to 352×240 and 5 frames per second.

We associate each video with an external data set containing all the spatio-temporal events that occur during the match, including passes. These events are collected by Wyscout through the manual annotation procedure described in Sect. 3. In particular, each pass event describes the player, the position on the field, and the time when the pass occurs [20].

We use the time of a pass to associate it with the corresponding frame in the video. Unfortunately, an event indicates the time when the pass starts, but not when it ends. Moreover, by comparing the video and the events, we note that the time of an event is often misaligned with the video. We overcome these drawbacks by annotating manually the passes through an application we implement specifically to this purpose (see Sect. 5.1).

After the manual annotation, for each match, we construct a vector with a length equal to the number of frames in the corresponding video. In this vector, each element can be either 1, indicating that the frame is part of a sequence describing a pass (`Pass`), or 0, indicating the absence of a pass in the frame (`No Pass`). For example, vector [0011111000] indicates that there are five consecutive frames in which there is a pass.

5.1 Pass Annotation Application

We implement a web application that contains a user interface to annotate the starting and ending times of a pass.[4] Figure 5 shows the structure of the application's visual interface. When the user loads a match using the appropriate dropdown (Fig. 5a), a table on the right side shows the match's passes and related information (Fig. 5b). On the left side, the interface shows the video and buttons to play and pause it, to move forward and backward, and to tag the starting time (`Pass Start`) and the ending time (`Pass End`) of a pass (Fig. 5c). When the user clicks on a row in the table to select the pass to annotate, the video automatically goes at the frame two seconds before the pass time. At this point, the user can use the `Pass Start` and `Pass End` buttons to annotate the starting and ending times, that will appear in the table expressed in seconds since the beginning of the video. In total, we annotate 3,206 passes, which are saved into a file that will be used to train PassNet and evaluate its performance.

6 Experiments

We compare PassNet with the following models:

- ResBi uses just the Feature Extraction module and the Sequence Classification module, i.e., it does not use the Object Detection module for the recognition of the position of the ball and the players;
- Random predicts the label randomly;
- MostFrequent predicts always the majority class, `No Pass` (71% of the frames);
- LeastFrequent predicts always the minority class, `Pass` (29% of the frames).

To classify a frame as `Pass` or `No Pass` we try several activation thresholds (Fig. 1g): 0.5, 0.9, and the threshold that maximizes the Youden Index (YI) [2], where YI $= Rec + TrueNegativeRate - 1$, and YI $\in [0, 1]$. YI represents in the model's ROC curve the farthest point from the random classifier's curve. We compare the models in terms of accuracy (Acc), F1-score ($F1$), precision on class `Pass` ($Prec$), precision on class `No Pass` ($PrecNo$), recall on class `Pass` (Rec), recall on class `No Pass` (RecNo) [25].

[4] The application is developed using python framework Flask, and is available at https://github.com/jonpappalord/PassNet.

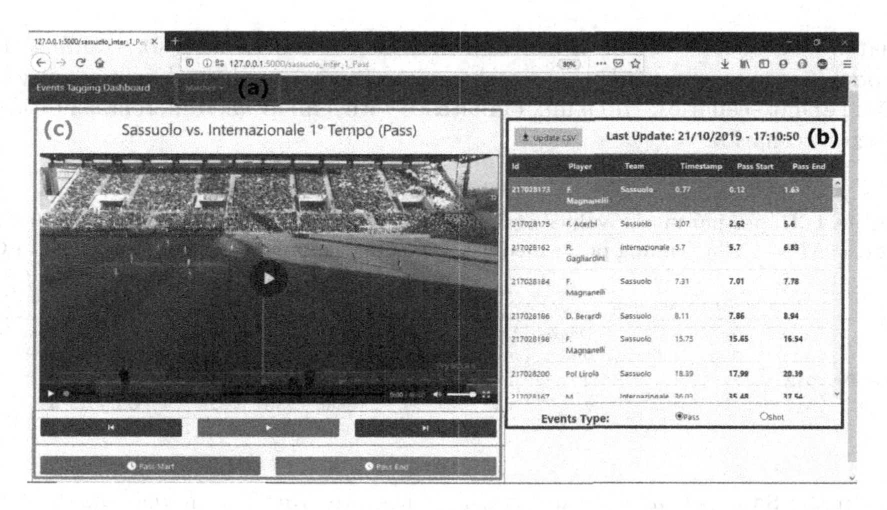

Fig. 5. Visual interface of the manual annotation application. The user can load a match using the appropriate dropdown (a). On the right side (b), a table shows all the pass events of the match and related information. On the left side (c), the interface shows the video and buttons to start and pause it, to move backward and forward, and to annotate the starting and ending times. When the user clicks on a row in the table, the video moves to two seconds before the event. A video that illustrates the functioning of the application is here: https://youtu.be/vO98f3XuTAU.

We perform the experiments on a computer with CPU Intel(R) Core(TM) i7-6800K CPU @ 3.40 GHz, 32 GB RAM, GPU GeForce GTX 1080. The time required to train the model (11 epochs) on a match half is around 11 h; the time required for the annotation of a match half is on average 30 min.

6.1 Results

We fix some hyper-parameter values and use the first half of AS Roma vs. Juventus FC to tune the learning rate, the sequence dimension δ, and the hidden dimension of the Sequence Classification module of PassNet and ResBi (Table 1). We use each hyper-parameter configuration to train PassNet and ResBi on a validation set extracted from the training set and test the configuration on the second half of AS Roma vs. Juventus FC. In particular, we try values 128, 256 and 512 for the hidden dimension, $\delta = 10, 25$, and values 0.01, 0.001 and 0.0001 for the learning rate. We evaluate the performance of each configuration in terms of average precision (AP), the weighted mean of precision at several thresholds. Formally, $AP = \sum_n (Rec_n - Rec_{n-1}) Prec_n$, where $Prec_n$ and Rec_n are precision and recall at the n-th threshold, respectively, and Rec_{n-1} is the recall at the $(n-1)$-th threshold [25]. Table 1 shows the hyper-parameter values corresponding to the best configuration of PassNet and ResBi.

We test PassNet, ResBi and the baselines on four scenarios, in which we use: *(i)* the same match used for training the model (`Same` scenario); *(ii)* a match that shares similar video conditions as the match used for training the model (Similar scenario); *(iii)* matches with different teams and light conditions (Different scenario); *(iv)* a mix of matches with similar and different conditions (Mixed scenario).

Table 1. Hyper-parameter values of the best configuration of PassNet (AP = 74%) and ResBi (AP = 75%). Tuning performed using the first half of AS Roma vs. Juventus FC.

	Fixed hyper-parameters							Tuned hyper-parameters			
	Input dim	Yolo dim	Layer dim	Dense dim	Drop out	Batch size	Opti-mizer	Hidden dim	Seq dim	Learn. rate	Best epoch
PassNet	512	24	1	2	0.5	1	Adam	**128**	**25**	**0.0001**	**6**
ResBi	512	-	1	2	0.5	1	Adam	**256**	**25**	**0.0001**	**4**

In the Same scenario we use two matches: we train the models on the first half of AS Roma vs. Juventus FC and test them on the second half of the same match; similarly we train the models on the first half of US Sassuolo vs. Internazionale FC and test them on the second half of the match. On AS Roma vs. Juventus FC, PassNet and ResBi have similar performance in terms of F1-score ($F1_{\text{PassNet}} = 70.21\%$, $F1_{\text{ResBi}} = 70.50\%$), and they both outperform the baseline classifiers with an improvement of 21.24% absolute and 43% relative with respect to the best baseline, LeastFrequent ($F1_{\text{Least}} = 49.26\%$, see Table 3). On US Sassuolo vs. FC Internazionale, PassNet has lower performance than the other match but still outperforms ResBi and the baselines ($F1_{\text{PassNet}} = 54.44\%$, $F1_{\text{ResBi}} = 53.72\%$), with an improvement of 15.75% absolute and 41% relative with respect to LeastFrequent ($F1_{\text{Least}} = 38.69\%$, Table 3).

We then test the models on the Similar scenario using the first half of AS Roma vs. Juventus FC as training set and the first half of match AS Roma vs. SS Lazio as test set. Note that this match is played by one of the teams that played the match used for training the model (AS Roma), and it is played in the same stadium (Stadio Olimpico, in Rome) and with the same light conditions (in the evening). PassNet outperforms ResBi and the baselines ($F1_{\text{PassNet}} = 61.74\%$, $F1_{\text{ResBi}} = 59.34\%$), with an improvement of 20.19% absolute and 48% relative w.r.t. LeastFrequent ($F1_{\text{Least}} = 41.55\%$, Table 4, right).

PassNet outperforms all models on the Mixed scenario, too. Here we train the models using the first halves of AS Roma vs. Juventus FC and US Sassuolo vs. FC Internazionale and test them on AC Chievo Verona vs. Juventus FC. We obtain $F1_{\text{PassNet}} = 63.73\%$ and $F1_{\text{ResBi}} = 58.17\%$, with an improvement of 11.96% absolute and 23% relative with respect to LeastFrequent ($F1_{\text{Least}} = 51.77\%$, see Table 4, left). Finally, we challenge the models on the Different scenario, in which we use match AS Roma vs. Juventus FC to train the models and we test them on matches AC Chievo vs. Juventus FC and US Sassuolo vs. FC Internazionale. PassNet and ResBi have similar performance in terms of F1-score (Table 5) and they both outperform LeastFrequent. Figure 6a compares the ROC curves of PassNet and ResBi on the four experimental scenarios.

Table 2. F1-score (in percentage) of PassNet at different thresholds for each match/scenario. The best value for each combination of metric and threshold is highlighted in **grey**. For YI, we specify the value of the threshold in parenthesis.

	Scenario	YI	TH = .5	TH = .9
AS Roma vs. Juventus FC 2H	Same	**70.21 (.19)**	69.64	65.20
US Sassuolo vs. FC Inter 2H	Same	**54.44 (.0005)**	40.60	28.26
US Chievo vs. Juventus FC 1H	Mixed	**63.73 (.003)**	42.55	27.22
AS Roma vs. SS Lazio 1H	Similar	**61.74 (.15)**	61.70	55.24
US Sassuolo vs. FC Inter 2H	Different	**53.92 (.004)**	47.00	34.62
US Chievo vs. Juventus FC 1H	Different	**59.51 (.0003)**	23.17	8.45

Table 3. Comparison of PassNet, ResBi and the baselines (Random, Most, Least), on matches AS Roma vs. Juventus FC (left) and US Sassuolo vs. FC Internazionale (right), of the Same scenario. The metrics are specified in percentage. YI = Youden Index.

	2H of AS Roma vs. Juventus FC					2H of US Sassuolo vs. FC Inter				
	PassNet	ResBi	Baselines			PassNet	ResBi	Baselines		
	YI = .19	YI = .05	Random	Most	Least	YI = .0005	YI = .01	Random	Most	Least
Acc	76.07	77.18	50.09	67.32	32.68	63.74	67.88	50.61	76.02	23.98
F1	**70.21**	**70.50**	**39.74**	**0.0**	**49.26**	**54.44**	**53.72**	**32.81**	**0.0**	**38.69**
Prec	59.17	61.03	32.82	0.0	32.68	38.96	41.04	24.35	0.0	23.98
Rec	86.32	83.45	50.36	0.0	100	90.34	77.73	50.29	0.0	100
PrecNo	91.45	90.22	67.46	67.32	0.0	94.78	90.21	76.38	76.02	0.0
RecNo	71.09	74.13	49.95	100	0.0	55.34	64.77	50.70	100	0.0

In summary, PassNet and ResBi significantly outperform the baselines on all four scenarios, indicating that our approach is able to learn from data to annotate pass events. We find that PassNet outperforms ResBi on all scenarios but the Different one, on which the two models have similar performance in terms of F1-score. In particular, PassNet achieves the best performance on the Same scenario ($AUC = 0.87$), followed by the Similar ($AUC = 0.84$), the Mixed ($AUC = 0.79$), and the Different ($AUC = 0.72$) scenarios (Fig. 6a).

In general, our results highlight that the detection of the ball and the closest players to it makes a significant contribution to pass detection. This is not surprising on the Same scenario, in which we use the second half of the match the first half of which is used for training. However, the fact that the performance on the Similar scenario is better than the performance on the Mixed and the Different scenarios suggests that, provided the availability of matches for an entire season, we may build different models for different teams, for different light conditions, or a combination of both. Moreover, the similar performance of PassNet and ResBi on the Different scenario suggests that the contribution of the object detection module is weaker for matches whose video conditions differ significantly from those of the match used for training the model. Note that the threshold with the maximum Youden Index provides also the best results in terms of F1-score (Table 2).

Table 4. Comparison of PassNet, ResBi and the baselines (Random, Most, Least), on matches AC Chievo vs. Juventus FC (left, Mixed scenario) and AS Roma vs. SS Lazio (right, Similar scenario). The metrics are specified in percentage. YI = Youden Index.

	1H of AC Chievo vs. Juventus FC					1H of AS Roma vs. SS Lazio				
	PassNet	ResBi	Baselines			PassNet	ResBi	Baselines		
	YI = .003	YI = .0064	Random	Most	Least	YI = .15	YI = .07	Random	Most	Least
Acc	70.00	63.37	49.99	65.08	34.92	73.13	72.32	50.30	73.78	26.22
F1	**63.73**	**58.17**	**39.59**	**0.0**	**51.77**	**61.74**	**59.34**	**34.90**	**0.0**	**41.55**
Prec	55.15	48.38	32.82	0.0	34.92	49.26	48.25	26.58	0.0	26.22
Rec	75.48	72.95	49.89	0.0	100	82.68	77.05	50.81	0.0	100
PrecNo	83.60	80.05	67.13	65.08	0.0	91.89	89.65	74.14	73.78	0.0
RecNo	67.06	58.23	50.04	100	0.0	69.73	70.63	50.12	100	0.0

Table 5. Comparison of PassNet, ResBi and the baselines (Random, Most, Least) on AC Chievo vs. Juventus FC (left) and US Sassuolo vs. FC Internazionale (right), of the Different scenario. The metrics are specified in percentage. YI = Youden Index.

	1H of AC Chievo vs. Juventus FC					2H of US Sassuolo vs. FC Inter				
	PassNet	ResBi	Baselines			PassNet	ResBi	Baselines		
	YI = .0003	.00001	Random	Most	Least	YI = .004	.0005	Random	Most	Least
Acc	63.33	59.82	49.99	65.08	34.92	66.25	67.13	50.61	76.02	23.98
F1	**59.51**	**53.42**	**39.59**	**0.0**	**51.77**	**53.91**	**54.08**	**32.81**	**0.0**	**38.69**
Prec	48.43	44.88	32.82	0.0	34.92	40.08	40.66	24.35	0.0	23.98
Rec	77.16	65.97	49.89	0.0	100	82.34	80.71	50.29	0.0	100
PrecNo	82.02	75.58	67.13	65.08	0.0	91.65	91.17	76.38	76.02	0.0
RecNo	55.91	56.52	50.04	100	0.0	61.17	62.85	50.70	100	0.0

Figure 6b shows how recall and precision change varying the threshold used to construct the passing vector, a useful tool for possible users of PassNet, such as data collection companies: if they want to optimize the precision of pass detection, the plot indicates that high thresholds must be used. In contrast, if they want to optimize recall, thresholds in the range $[0.0, 0.4]$ must be preferred.

To visualize the limits of PassNet, we create some videos that show the results of its predictions as the match goes by. Figure 7a shows the structure of these videos. On the left side, we show the match, in which a label "Pass" appears every time PassNet detects a pass. On the right side, we show two animated plots that compare the real label with the model's prediction. In these plots, value 0 indicates no pass, value 1 indicates that there is a pass. The observation of these videos reveals that PassNet sometimes classifies consecutive passes that come in a close interval of time as a single pass (AS Roma vs. Juventus FC). This is presumably because the YI threshold cannot detect the slightest changes that occur between two consecutive passes. Interestingly, the videos also reveal the presence of errors during the manual annotation. For example, in US Sassuolo vs. FC Internazionale, PassNet recognizes passes that actually take place but that were not annotated by the human operators. Another error, that may constitute

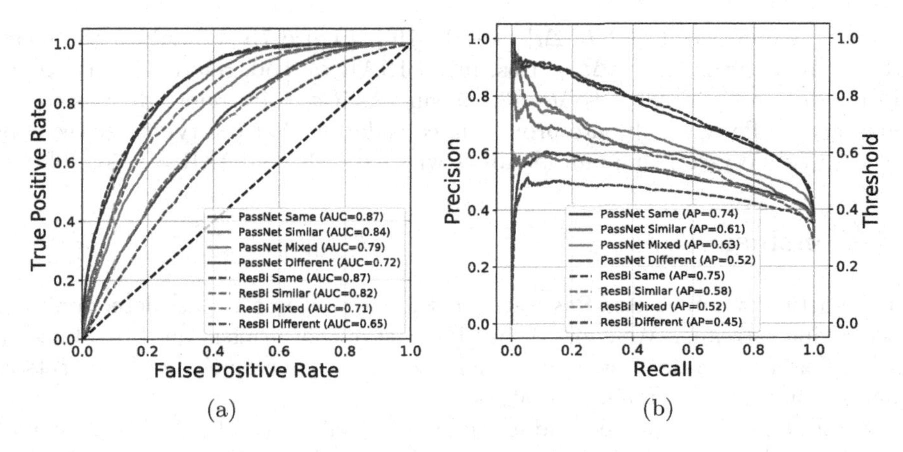

Fig. 6. Classification Performance. (a) ROC curves and (b) precision and recall vayring the threshold, for PassNet and ResBi on the `Same` (AS Roma vs. Juventus FC), `Similar`, `Mixed` and `Different` (AC Chievo vs. Juventus FC) scenarios.

room for future improvement, is that PassNet usually misclassifies as passes situations in which a player runs in possession of the ball.

As a further assessment of the reliability of the annotation made by our system, we evaluate the degree of agreement on matches in Tables 3, 4, and 5 between PassNet and Wyscout's human operators computing the Inter-Rater Agreement Rate, defined as IRAR $= 1 - \frac{1-p_o}{1-p_e} \in [0, 1]$, where p_o is the relative agreement among operators (ratio of passes detected by both) and p_e is the probability of chance agreement [16]. In order to compute IRAR, we first associate each pass annotated by PassNet at time t with the Wyscout pass (if any)

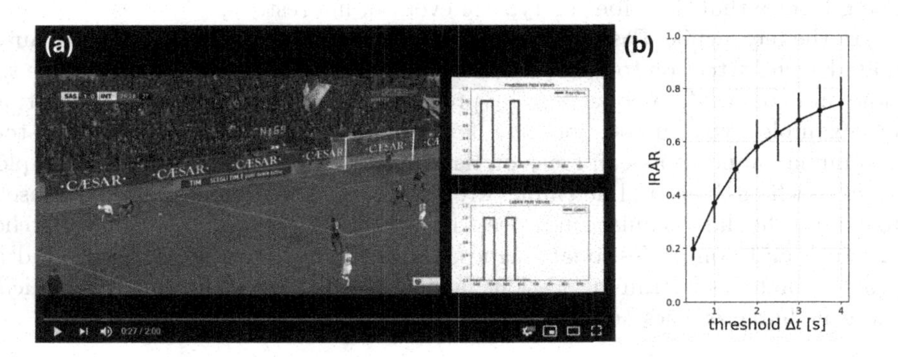

Fig. 7. PassNet **in action.** (a) Structure of the video showing how PassNet annotates US Sassuolo vs. FC Internazionale as the match goes by. The left side shows the match, a label "Pass" appears every time a pass is detected. The right side shows two animated plots comparing the real (red) and the predicted (blue) labels. (b) Average IRAR w.r.t. Wyscout operators varying the time threshold Δt. (Color figure online)

in the time interval $[t-\Delta t, t+\Delta t]$ (see [16,19]). Figure 7b shows how the mean IRAR varies with Δt: at $\Delta t = 1.5$s, mean IRAR ≈ 0.50, referred to as "moderate agreement" in [26], at $\Delta t = 3$s, mean IRAR ≈ 0.70, referred to as "good agreement". These results are promising considering that the typical agreement between two Wyscout human operators with $\Delta t = 1.5$ is IRAR $= 0.70$ [19].

7 Conclusion

In this article, we presented PassNet, a method for automatic pass detection from soccer video streams. We showed that PassNet outperforms several baselines on four different scenarios, and that it has a moderate agreement with the sets of passes annotated by human operators.

PassNet can be improved and extended in several ways. First, in this article, we use broadcast videos that contain camera view changes and play-breaks such as replays, checks at the VAR, and goal celebrations. These elements may introduce noise and affect the performance of the model. The usage of fixed camera views and play-break detection models may be used to clean the video streams, reduce noise, and further improve the performance of PassNet. Second, we use a pre-trained YOLOv3 that can recognize generic persons and balls. Although our results show that it provides a significant contribution to the predictions, we may build an object detection module to recognize specifically the soccer players and ball. Given the flexibility of YOLOv3's architecture, this may be achieved simply by training the model on a labeled data set of soccer players and balls. Moreover, we may integrate in PassNet existing methods to detect the identity of players, for example by recognizing their jersey number. Finally, PassNet can be easily adapted to annotate other crucial events, such as shots, fouls, saves, and tackles, or to discriminate between accurate and inaccurate passes. Given the modularity of our model, this may be achieved simply by training the model using frames that describe the type of event of interest.

In the meanwhile, PassNet is a first step towards the construction of an automated event detection tool for soccer. On the one hand, this tool may reduce the time and cost of data collection, by providing a support to manual annotation. For example, event annotation may be partially delegated to the automatic tool and human annotators can focus on data quality control, especially for complex events such as duels and defending events. On the other hand, it would consent to extend the data acquisition process to unexplored directions, such as matches in youth and non-professional leagues, to leagues far away in the past, and to practice matches in training sessions, allowing researchers to compare technical-tactical characteristics across divisions, times and phases of the season.

Acknowledgments. This work has been supported by project H2020 SoBigData++ #871042.

References

1. Bayat, F., Moin, M.S., Bayat, F.: Goal detection in soccer video: role-based events detection approach. Int. J. Electr. Comput. Eng. **4**(6), 2088–8708 (2014)
2. Berrar, D.: Performance measures for binary classification. In: Encyclopedia of Bioinformatics and Computational Biology, pp. 546–560 (2019)
3. Bornn, L., Fernandez, J.: Wide open spaces: a statistical technique for measuring space creation in professional soccer. In: MIT Sloan Sports Analytics Conference (2018)
4. Carrara, F., Elias, P., Sedmidubsky, J., Zezula, P.: LSTM-based real-time action detection and prediction in human motion streams. Multimedia Tools Appl. **78**(19), 27309–27331 (2019). https://doi.org/10.1007/s11042-019-07827-3
5. Cintia, P., Giannotti, F., Pappalardo, L., Pedreschi, D., Malvaldi, M.: The harsh rule of the goals: data-driven performance indicators for football teams. In: IEEE International Conference on Data Science and Advanced Analytics, pp. 1–10 (2015)
6. Decroos, T., Bransen, L., Van Haaren, J., Davis, J.: Actions speak louder than goals: valuing player actions in soccer. In: 25th ACM SIGKDD International Conference on Knowledge Discovery and Data Mining, pp. 1851–1861 (2019)
7. Deng, J., Dong, W., Socher, R., Li, L., Li, K., Fei-Fei, L.: Imagenet: a large-scale hierarchical image database. In: 2009 IEEE Conference on Computer Vision and Pattern Recognition, pp. 248–255 (2009)
8. Fakhar, B., Kanan, H.R., Behrad, A.: Event detection in soccer videos using unsupervised learning of spatio-temporal features based on pooled spatial pyramid model. Multimedia Tools Appl. **78**(12), 16995–17025 (2019)
9. Gerke, S., Muller, K., Schafer, R.: Soccer jersey number recognition using convolutional neural networks. In: Proceedings of the IEEE International Conference on Computer Vision Workshops, pp. 17–24 (2015)
10. Gudmundsson, J., Horton, M.: Spatio-temporal analysis of team sports. ACM Comput. Surv. **50**(2), 1–34 (2017)
11. He, K., Zhang, X., Ren, S., Sun, J.: Deep residual learning for image recognition. In: IEEE Conference on Computer Vision and Pattern Recognition, pp. 770–778 (2016)
12. Jiang, H., Lu, Y., Xue, J.: Automatic soccer video event detection based on a deep neural network combined cnn and rnn. In: 28th IEEE International Conference on Tools with Artificial Intelligence, pp. 490–494 (2016)
13. Kapela, R., McGuinness, K., Swietlicka, A., O'Connor, N.E.: Real-time event detection in field sport videos. In: Moeslund, T.B., Thomas, G., Hilton, A. (eds.) Computer Vision in Sports. ACVPR, pp. 293–316. Springer, Cham (2014). https://doi.org/10.1007/978-3-319-09396-3_14
14. Khan, A., Lazzerini, B., Calabrese, G., Serafini, L.: Soccer event detection. In: 4th International Conference on Image Processing and Pattern Recognition, pp. 119–129 (2018)
15. Lin, T.-Y., et al.: Microsoft COCO: common objects in context. In: Fleet, D., Pajdla, T., Schiele, B., Tuytelaars, T. (eds.) ECCV 2014. LNCS, vol. 8693, pp. 740–755. Springer, Cham (2014). https://doi.org/10.1007/978-3-319-10602-1_48
16. Liu, H., Hopkins, W., Gómez, A.M., Molinuevo, S.J.: Inter-operator reliability of live football match statistics from opta sportsdata. Int. J. Perform. Anal. Sport **13**(3), 803–821 (2013)
17. Liu, T., et al.: Soccer video event detection using 3d convolutional networks and shot boundary detection via deep feature distance. In: International Conference on Neural Information Processing, pp. 440–449 (2017)

18. Pappalardo, L., Cintia, P.: Quantifying the relation between performance and success in soccer. Adv. Complex Syst. **20**(4), 1750014 (2017)
19. Pappalardo, L., Cintia, P., Ferragina, P., Massucco, E., Pedreschi, D., Giannotti, F.: Playerank: data-driven performance evaluation and player ranking in soccer via a machine learning approach. ACM Trans. Intell. Syst. Technol. **10**(5), 1–27 (2019)
20. Pappalardo, L., et al.: A public data set of spatio-temporal match events in soccer competitions. Nat. Sci. Data **6**(236), 1–15 (2019)
21. Redmon, J., Farhadi, A.: Yolov3: An incremental improvement. arXiv preprint arXiv:1804.02767 (2018)
22. Rossi, A., Pappalardo, L., Cintia, P., Iaia, F.M., Fernàndez, J., Medina, D.: Effective injury forecasting in soccer with gps training data and machine learning. PLoS One **13**(7), 1–15 (2018)
23. Saraogi, H., Sharma, R.A., Kumar, V.: Event recognition in broadcast soccer videos. In: Proceedings of the Tenth Indian Conference on Computer Vision, Graphics and Image Processing, p. 14. ACM (2016)
24. de Sousa, S.F., Araújo, A.D.A., Menotti, D.: An overview of automatic event detection in soccer matches. In: IEEE Workshop on Applications of Computer Vision, pp. 31–38 (2011)
25. Tan, P.N., Steinbach, M., Kumar, V.: Introduction to Data Mining. Pearson Education India, Chennai (2016)
26. Viera, A.J., Garrett, J.M.: Understanding interobserver agreement: the kappa statistic. Fam. Med. **37**(5), 360–363 (2005)
27. Wei, X., Sha, L., Lucey, P., Morgan, S., Sridharan, S.: Large-scale analysis of formations in soccer. In: 2013 International Conference on Digital Image Computing: Techniques and Applications, pp. 1–8 (2013)
28. Yu, J., Lei, A., Hu, Y.: Soccer video event detection based on deep learning. In: Kompatsiaris, I., Huet, B., Mezaris, V., Gurrin, C., Cheng, W.-H., Vrochidis, S. (eds.) MMM 2019. LNCS, vol. 11296, pp. 377–389. Springer, Cham (2019). https://doi.org/10.1007/978-3-030-05716-9_31
29. Zawbaa, H.M., El-Bendary, N., Hassanien, A.E., Kim, T.H.: Event detection based approach for soccer video summarization using machine learning. Int. J. Multimedia Ubiquit. Eng. **7**(2), 63–80 (2012)

SoccerMap: A Deep Learning Architecture for Visually-Interpretable Analysis in Soccer

Javier Fernández[1,2(✉)] and Luke Bornn[3]

[1] Polytechnic University of Catalonia, Barcelona, Spain
javier.fernandez.de.la.rosa@upc.edu, javier.fernandezr@fcbarcelona.cat
[2] FC Barcelona, Barcelona, Spain
[3] Simon Fraser University, British Columbia, Canada
lbornn@sfu.ca

Abstract. We present a fully convolutional neural network architecture that is capable of estimating full probability surfaces of potential passes in soccer, derived from high-frequency spatiotemporal data. The network receives layers of low-level inputs and learns a feature hierarchy that produces predictions at different sampling levels, capturing both coarse and fine spatial details. By merging these predictions, we can produce visually-rich probability surfaces for any game situation that allows coaches to develop a fine-grained analysis of players' positioning and decision-making, an as-yet little-explored area in sports. We show the network can perform remarkably well in the estimation of pass success probability, and present how it can be adapted easily to approach two other challenging problems: the estimation of pass-selection likelihood and the prediction of the expected value of a pass. Our approach provides a novel solution for learning a full prediction surface when there is only a single-pixel correspondence between ground-truth outcomes and the predicted probability map. The flexibility of this architecture allows its adaptation to a great variety of practical problems in soccer. We also present a set of practical applications, including the evaluation of passing risk at a player level, the identification of the best potential passing options, and the differentiation of passing tendencies between teams.

Keywords: Soccer analytics · Spatio-temporal statistics · Representation learning · Fully convolutional neural networks · Deep learning · Interpretable machine learning

1 Introduction

Sports analytics is a fast-growing research field with a strong focus on data-driven performance analysis of professional athletes and teams. Soccer, and many other team-sports, have recently benefited from the availability of high-frequency tracking data of both player and ball locations, facilitating the development of

Y. Dong et al. (Eds.): ECML PKDD 2020, LNAI 12461, pp. 491–506, 2021.
https://doi.org/10.1007/978-3-030-67670-4_30

fine-grained spatiotemporal performance metrics [14]. One of the main goals of performance analysis is to answer specific questions from soccer coaches, but to do so we require models to be robust enough to capture the nuances of a complex sport, and be highly interpretable so findings can be communicated effectively. In other words, we need models to be both accurate and also translatable to soccer coaches in visual terms.

The majority of existing research in soccer analytics has focused on analyzing the impact of either on-ball events, such as goals, shots, and passes, or the effects of players' movements and match dynamics [5]. Most modeling approaches share one or more common issues, such as: heavy use of handcrafted features, little visual interpretability, and coarse representations that ignore meaningful spatial relationships. We still lack a comprehensive approach that can learn from lower-level input, exploit spatial relationships on any location, and provide accurate predictions of observed and unobserved events at any location on the field.

The main contributions of our work are the following:

- We present a novel application of deep convolutional neural networks that allows calculating full probability surfaces for developing fine-grained analysis of game situations in soccer. This approach offers a new way of providing coaches with rich information in a visual format that might be easier to be presented to players than the usual numerical statistics.
- We show how this architecture can ingest a flexible structure of layers of spatiotemporal data, and how it can be easily adapted to provide practical solutions for challenging problems such as the estimation of pass probability, pass selection likelihood and pass expected value surfaces.
- We present three novel practical applications derived from pass probability surfaces, such as the identification of optimal passing locations, the prediction of optimal positioning for improving pass probability, and the prediction of team-level passing tendencies.

The presented approach successfully addresses the challenging problem of estimating full probability surfaces from single-location labels, which corresponds to an extreme case of weakly-supervised learning.

2 Related Work

From an applied standpoint, our work is related to several other approaches aimed at estimating pass probabilities and other performance metrics derived from spatiotemporal data in soccer. Regarding the technical approach, we leverage recent findings on weakly-supervised learning problems and the application of fully convolutional neural networks for image segmentation.

Soccer Analytics. Pass probability estimation has been approached in several ways. A physics-based model of the time it takes each player to reach and control the ball has been used to derive pass probabilities on top of tracking data [15].

Other approaches include the use of dominant regions to determine which player is most likely to control the ball after a pass [5] or using a carefully selected set of handcrafted features to build linear prediction models [13]. The related problem of pass selection has been approached by applying convolutional neural networks that predict the likelihood of passing to a specific player on the attacking team [7]. The estimation of pass value has been approached either by the expert-guided development of algorithmic rules [1], the application of standard machine learning algorithms on a set of handcrafted features [13], or problem-specific deep learning models with dense layers and single output prediction [4]. While some of the related work has estimated probability surfaces by inference on a set of discrete pass destination locations [4,15], none has yet approached the learning of probability surfaces directly.

Fully Convolutional Networks and Weakly-Supervised Learning. Fully convolutional networks have been extensively applied to semantic image segmentation, specifically for the pixel-labeling problem to successfully detect broad pixel areas associated with objects in images. The approach most related to our work builds a hierarchy of features at different sampling levels that are merged to provide segmentation regions that preserve both fine and coarse details [8]. From a learning perspective, image segmentation has been approached as either supervised [8], weakly-supervised [12], and semi-supervised learning problems [11]. Commonly, available labels are associated with many other pixels in the original image. However, in our case, labels are only associated with a single location in the desired probability map, transforming our learning problem into an unusual case of weakly-supervised learning.

3 A Deep Model for Interpretable Analysis in Soccer

We build our architecture on top of tracking data extracted from videos of professional soccer matches, consisting of the 2D-location of players and the ball at 10 frames per second, along with manually tagged passes. At every frame we take a snapshot of the tracking data and create a representation of a game situation consisting of a $l \times h \times c$ matrix, where c channels of low-level information are mapped to a $l \times h$ coarse spatial representation of the locations on the field. We seek an architecture that can learn both finer features at locations close to a possible passing destination and features considering information on a greater spatial scale. For passes, local features might be associated with the likelihood of nearby team-mates and opponents reaching the destination location and information about local spatial pressure. On the other hand, higher scale features might consider player's density and interceptability of the ball in its path from the location of origin. Finally, we seek to estimate this passing probability to any other location on the $l \times h$ spatial extent of the field.

This game state representation is processed by the deep neural network architecture presented in Fig. 1. The network creates a feature hierarchy by learning convolutions at $1x$, $1/2x$, and $1/4x$ scales while preserving the receptive field of

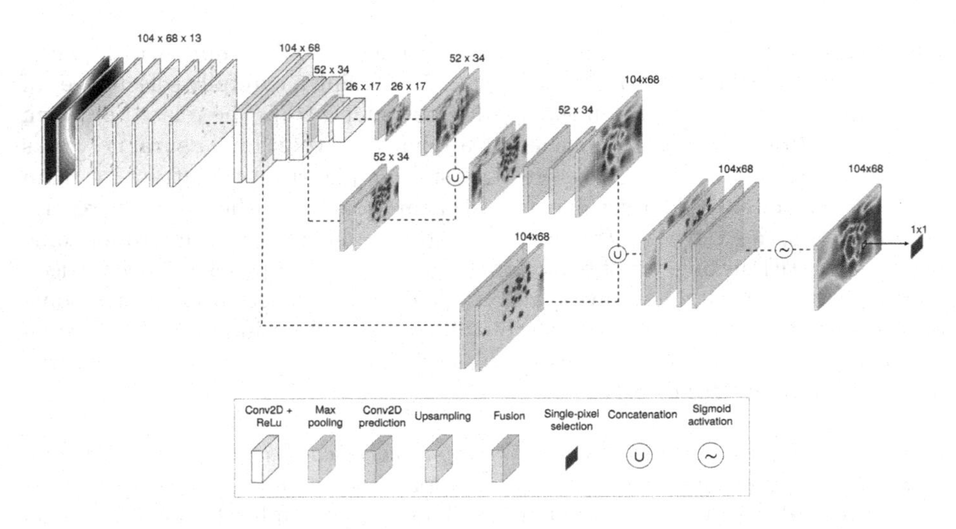

Fig. 1. SoccerMap architecture for a coarse soccer field representation of 104×68 and 13 input channels.

the filters. Predictions are produced at each of these scales, and then upsampled nonlinearly and merged through fusion layers. A sigmoid activation layer is applied to the latest prediction to produce pass probability estimations at every location, preserving the original input scale. During training, a single-location prediction, associated with the destination of a sample pass is selected to compute the log-loss that is backpropagated to adjust the network weights.

3.1 The Reasoning Behind the Choice of Layers

The network incorporates different types of layers: max-pooling, linear, rectified linear unit (ReLu) and sigmoid activation layers, and 2D-convolutional filters (conv2d) for feature extraction, prediction, upsampling and fusion. In this section we present a detailed explanation of the reasoning behind the choice of layers and the design of the architecture.

Convolutions for Feature Extraction. At each of the $1x$, $1/2x$, and $1/4x$ scales two layers of conv2d filters with a 5×5 receptive field and stride of 1 are applied, each one followed by a ReLu activation function layer, in order to extract spatial features at every scale. In order to keep the same dimensions after the convolutions we apply symmetric padding to the input matrix of the convolutional layer. We chose symmetric-padding to avoid border-image artifacts that can hinder the predicting ability and visual representation of the model.

Fully Convolutional Network. There are several conceptual and practical reasons for considering convolutional neural networks (convnets) for this problem. Convolutional filters are designed to recognize the relationships between nearby pixels, producing features that are spatially aware. Convnets have been proven successful in data sources with a Euclidean structure, such as images and videos, so a 2D-mapping of soccer field location-based information can be expected to be an ideal data structure for learning essential features. Also, these features are expected to be non-trivial and complex. Convnets have been proven to learn what are sometimes more powerful visual features than handcrafted ones, even given large receptive fields and weak label training [9]. Regarding the architecture design, we are interested in learning the full $l \times h$ mapping of passing probabilities covering the extent of a soccer field, for which fully convolutional layers are more appropriate than classical neural networks built for classification when changing dense prediction layers for 1 x 1 convolution layers.

Pooling and Upsampling. The network applies downsampling twice through max-pooling layers to obtain the $1/2x$ and $1/4x$ representations. Since activation field size is kept constant after every downsampling step, the network can learn filters of a wider spatial extent, leading to the detection of coarse details. We learn non-linear upsampling functions at every upsampling step by first applying a $2x$ nearest neighbor upsampling and then two layers of convolutional filters. The first convolutional layer consists of 32 filters with a 3×3 activation field and stride 1, followed by a ReLu activation layer. The second layer consists of 1 layer with a 3×3 activation field and stride 1, followed by a linear activation layer. This upsampling strategy has been shown to provide smoother outputs and to avoid artifacts that can be usually found in the application transposed convolutions [10].

Prediction and Fusion Layers. Prediction layers consist of a stack of two convolutional layers, the first with 32 1×1 convolutional filters followed by an ReLu activation layer, and the second consists of one 1×1 convolutional filter fo-llowed by a linear activation layer. Instead of reducing the output to a single prediction value, we keep the spatial dimensions at each step and use 1×1 convolutions to produce predictions at each location. The stack learns a non-linear prediction on top of the output of convolutional layers. To merge the outputs at different scales, we concatenate the pair of matrices and pass them through a convolutional layer of one 1×1 filter.

3.2 Learning from Single-Location Labels

We seek a model that can produce accurate predictions of the pass probability to every location on a $l \times h$ coarsed representation of a soccer field, given a $l \times h \times c$ representation of the game situation at the time a pass is made. In training, we only count with the manually labeled location of the pass destination and a binary label of the outcome of the pass.

Definition 1 (SoccerMap). *Let* $X = \{x | x \in \mathbb{R}^{l \times h \times c}\}$ *be the set of possible game state representations at any given time, where* $l, h \in \mathbb{N}_1$ *are the height and length of a coarse representation of soccer field, and* $c \in \mathbb{N}_1$ *the number of* *data channels, a SoccerMap is a function* $f(x;\theta), f : \mathbb{R}^{l\times h\times c} \rightarrow \mathbb{R}^{l\times h}_{[0,1]}$*, where* f *produces a pass probability map, and* θ *are the network parameters.*

Definition 2 (Target-Location Loss). *Given the sigmoid function* $\sigma(x) = \frac{e^x}{e^x+1}$ *and let* $y_k \in \{0,1\}$ *be the outcome of a pass at time* $t(x_k)$*, for a game state* x_k*,* d_k *the destination location of the pass* k*,* f *a SoccerMap with parameters* θ*, and logloss the log-loss function, we define the target-location loss as*

$$L(f(x_k;\theta), y_k, d_k) = logloss(f(x_k;\theta)_{d_k}, y_k)$$

We approach the training of the model as a weakly-supervised learning task, where the ground truth labels only correspond to a single location in the full mapping matrix that needs to be learned. The target-location loss presented in Definition 2 essentially shrinks the output of a SoccerMap f to a single prediction value by selecting the prediction value at the destination of the pass, and then computes the log-loss between this single prediction and the ground-truth outcome value.

3.3 Spatial and Contextual Channels from Tracking Data

Our architecture is designed to be built on top of two familiar sources of data for sports analytics: tracking data and event data. Tracking data consists of the location of the players and the ball at a high frequency-rate. Event-data corresponds to manually labeled observed events, such as passes, shots, and goals, including the location and time of each event. We normalize the players' location and the ball to ensure the team in possession of the ball attacks from left to right, thus standardizing the representation of the game situation. On top of players' and ball locations, we derive low-level input channels, including spatial (location and velocity) and contextual information (ball and goal distance). Channels are represented by matrices of $(104, 68)$ where each cell approximately represents 1m^2 in a typical soccer field.

Definition 3 (Tracking-Data Snapshot). *Let* $Z_p(t), Z_d(t), Z_b(t), Z_g(t) \in \{z | z \in \mathbb{R}^{l\times h}\}$ *be the locations of the attacking team players, the location of the defending team players, the location of the ball, and the location of the opponent goal, respectively, at time* t*, then a tracking-data snapshot at time* t *is defined as the 4-tuple* $Z(t) = (Z_p(t), Z_d(t), Z_b(t), Z_g(t))$*.*

In order to create a game state representation $X(t) \in X$ as described in Definition 1 we produce 13 different channels on top of each tracking-data snapshot Z where a pass has been observed, which constitute the game-state representation for the pass probability model. Each channel corresponds to either a sparse or dense matrix of size (h, l), according to the chosen dimensions for the coarse field representation. The game-state representation is composed of the following channels:

- Six sparse matrices with the location, and the two components of the velocity vector for the players in both the attacking team and the defending team, respectively.

- Two dense matrices where every location contains the distance to the ball and goal location.
- Two dense matrices containing the sine and cosine of the angle between every location to the goal and the ball location, and one dense matrix containing the angle in radians to the goal location.
- Two sparse matrices containing the sine and cosine of the angle between the velocity vector of the ball carrier and each of the teammates in the attacking team.

4 Experiments and Results

4.1 Dataset

We use tracking-data, and event-data from 740 English Premier League matches from the 2013/2014 and 2014/2015 season, provided by *STATS LLC*. Each match contains the (x, y) location for every player, and the ball sampled at 10 Hz. The event-data provides the location, time, player, team and outcome for 433,295 passes. From this data, we extract the channels described in Sect. 3.3 for a coarse $(104, 68)$ representation of a soccer field to obtain a dataset of size $433295 \times 104 \times 68 \times 13$. There are 344,957 successful passes and 88,338 missed passes.

4.2 Benchmark Models

We compare our results against a series of benchmark models of increasing levels of complexity. We define a baseline model *Naive* that for every pass outputs the known average pass completion in the full dataset (80%) following a similar definition in [13]. We build two additional models *Logistic Net* and *Dense2 Net* based on a set of handcrafted features built on top of tracking-data. Logistic Net is a network with a single sigmoid unit, and Dense2 Net is a neural network with two dense layers followed by ReLu activations and a sigmoid output unit.

Handcrafted Features. We build a set of spatial features on top of tracking-data based on location and motion information on players and the ball that is similar to most of the features calculated in previous work on pass probability estimation [5,13,15]. We define the following set of handcrafted features from each pass: origin and destination location, pass distance, attacking and defending team influence at both origin and destination, angle to goal at origin and destination, and the maximum value of opponent influence in a straight line between origin and destination. The team's spatial influence values are calculated following the model presented in [3].

4.3 Experimental Framework

In this section, we describe the experimental framework for testing the performance of the proposed architecture for the pass success probability estimation problem.

Training, Validation, and Test Set. We randomly selected matches from both available seasons and split them into a training, validation, and test set with a 60 : 20 : 20 distribution. We applied a stratified split, so the successful/missed pass class ratio remains the same across datasets. The validation set is used for model selection during a grid-search process. The test set is left as hold-out data, and results are reported on performance for this dataset. For the benchmark models, datasets are built by extracting the features described in Sect. 4.2, and an identical split is performed. Features are standardized column-wise by subtracting the mean value and dividing by the standard deviation.

Optimization. Both the SoccerMap network and the baseline models are trained using adaptive moment estimation (Adam). Model selection is achieved through grid-search on learning rates of 10^{-3}, 10^{-4} and 10^{-5}, and batch sizes of 1, 16 and 32, while β_1, β_2 are set to 0.9 and 0.999, respectively. We use early stopping with a minimum delta rate of 0.001. Optimization is computed on a single Tesla M60 GPU and using Tensorflow 1.5.0. During the optimization, the negative log-loss is minimized.

Metrics. Let N be the number of examples in the dataset, y the ground-truth labels for pass events and $\hat{y}$ the model predictions. We report the negative log-loss

$$\mathcal{L}(\hat{y}, y) = -\frac{1}{N}\sum_i y_i \cdot log(\hat{y}_i) + (1 - y_i) \cdot log(1 - \hat{y}_i).$$

In order to validate the model calibration we use a variation of the expected calibration error (ECE) presented in [6] which computes the expected difference between accuracy and confidence of the model on a finite set of samples split into K bins of size $1/K$, according to the predicted confidence or probability for every sample. Since our model is not designed for classification, we use the count of the number of examples of the positive class rather than accuracy for ECE.

Let B_k be a bin where $k \in [1, K]$ then

$$ECE = \sum_{k=1}^{K} \frac{|B_k|}{N} \left| \left(\frac{1}{|B_k|} \sum_{i \in B_k} 1(y_i = 1) \right) - \left(\frac{1}{|B_k|} \sum_{i \in B_k} \hat{y}_i \right) \right|.$$

A perfectly calibrated model will have a ECE value of 0. Additionally, we provide a calibration reliability plot [6] showing the mean confidence for every bin B_k.

4.4 Results

Table 1 presents the results for the benchmark models and SoccerMap for the pass probability dataset. We can observe that SoccerMap achieves remarkably lower error than the other models and produces a calibrated estimation of pass probability. Despite the considerably large number of parameters in SoccerMap, the inference time for a single sample is low enough to produce a real-time estimation for frame rates 200 Hz. Figure 2 presents a calibration reliability plot for each of the models. Both Logistic Net and SoccerMap produce well-calibrated estimations of pass probabilities, however, SoccerMap is proven to be considerably more precise as shown by the difference in log-loss between both.

Table 1. Results for the benchmark models and SoccerMap for the pass probability dataset.

Model	Log-loss	ECE	Inference time	Number of parameters
Naive	0.5451	–	–	0
Logistic Net	0.384	**0.0210**	0.00199 s	11
Dense2 Net	0.349	0.0640	0.00231 s	231
SoccerMap	**0.217**	**0.0225**	0.00457 s	401, 259

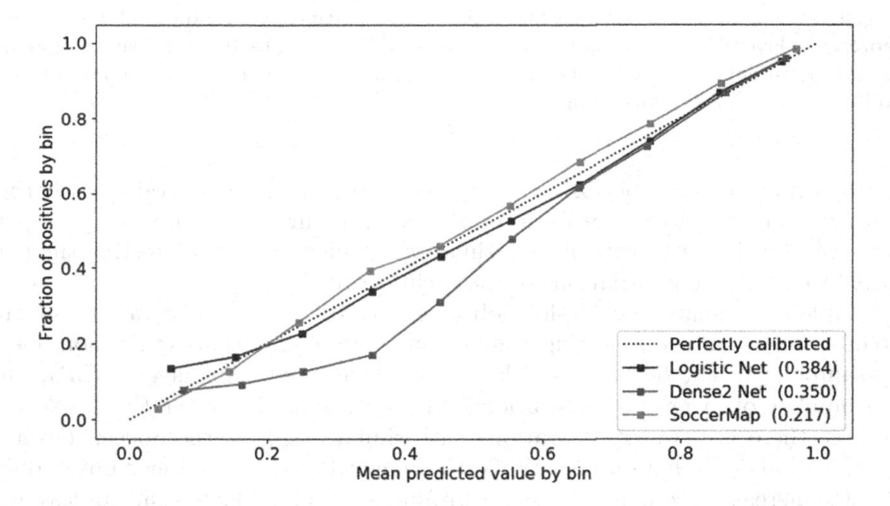

Fig. 2. A calibration reliability plot, where the X-axis presents the mean predicted value for samples in each of 10 bins, and the Y-axis the fraction of samples in each bin containing positive examples.

Figure 3 presents the predicted pass probability surface for a specific game situation during a professional soccer match. We observe that the model can capture both fine-grained information, such as the influence of defending and attacking players on nearby locations and coarse information such as the probability of reaching more extensive spatial areas depending on the distance to the ball and the proximity of players. We can also observe that the model considers the player's speed for predicting probabilities of passing to not-yet occupied spaces, a critical aspect of practical soccer analysis.

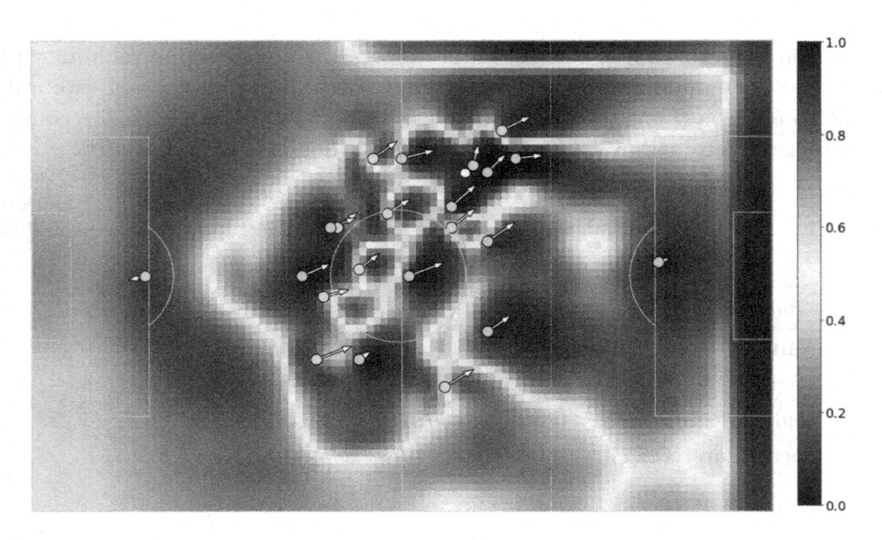

Fig. 3. Pass probability surface for a given game situation. Yellow and blue circles represent players' locations on the attacking and defending team, respectively, and the arrows represent the velocity vector for each player. The white circle represents the ball location. (Color figure online)

Ablation Study. We performed an ablation study in order to evaluate whether the different components of the proposed architecture allow improving its performance on the pass probability estimation problem or not, by testing the performance of different variations of the architecture.

Table 2 presents the log-loss obtained on different configurations of the architecture with the following components: skip-connections (SC), non-linear up-sampling (UP), fusion layer (FL), non-linear prediction layer (NLP), and the number of layers of convolutional filters by sampling layer (NF). We can observe there are two configurations with similar log-loss: the SoccerMap and SoccerMap-UP configurations. While the removal of the non-linear upsampling slightly increases the performance, it produces visual artifacts that are less eye-pleasing when inspecting the surfaces. Given that the surfaces are intended to be used by soccer coaches in practice, SoccerMap provides a better option for practical purposes.

5 Practical Applications

In this section, we present a series of novels practical applications that make use of the full probability surface for evaluating potential passing actions and assessing player's passing and positional skills.

Table 2. Ablation study for subsets of components of the SoccerMap architecture.

Architecture	SC	UP	FL	NLP	NF	Log-loss
SoccerMap	YES	YES	YES	YES	2	**0.217**
SoccerMap-NLP	YES	YES	YES	NO	2	0.245
SoccerMap-FL	YES	YES	NO	YES	2	0.221
SoccerMap-FL-NLP	YES	YES	NO	NO	2	0.292
SoccerMap-UP	YES	NO	YES	YES	2	**0.216**
SoccerMap-UP-FL	YES	NO	NO	YES	2	0.220
SoccerMap-UP-NLP	YES	NO	YES	NO	2	0.225
SoccerMap-FL-NLP	YES	NO	NO	NO	2	0.235
Single Layer CNN-D4	NO	YES	YES	YES	2	0.256
Single Layer CNN-D8	NO	YES	YES	YES	4	0.228

5.1 Adapting SoccerMap for the Estimation of Pass Selection Likelihood and Pass Value

One of the main advantages of this architecture is that is can be easily adapted to other challenging problems associated with the estimation of pass-related surfaces, such as the estimation of pass selection and pass value.

Pass Selection Model. An interesting and unsolved problem in soccer is the estimation of the likelihood of a pass being made towards every other location on the field, rather than to specific player locations. We achieve this by directly modifying the sigmoid activation layer of the original architecture by a softmax activation layer, which ensures that the sum of probabilities on the output surface adds up to 1. For this case, instead of pass success, we use a sparse matrix as a target output and set the destination location of the pass in that matrix to 1.

Pass Value Model. While a given pass might have a low probability of success, the expected value of that pass could be higher than a different passing option with higher probability, thus in some cases, the former could be preferable. We can directly adapt SoccerMap to estimate a pass value surface by modifying the target value and the loss function to be used. For this case, we use as an outcome the expected goals value [2] of the last event in possession of any given pass, which can be positive or negative depending on whether the attacking or defending team had the last action in that possession.

Figure 4 presents the surfaces for pass selection and pass value models derived from this architecture. With these surfaces, we can provide direct visual guidance to coaches to understand the value of the positioning of its team, the potential value gains of off-ball actions, and a team's likely passes in any given situation.

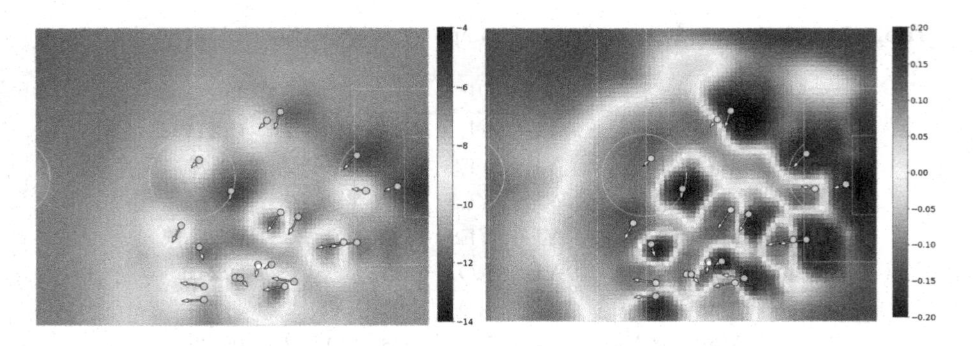

Fig. 4. On the left column, the pass selection surface for a give game-state, presented on a logarithmic scale. On the right column, a pass value surfaces for the same game-state, with the color scale constrained to a [−0.2,0.2] range. (Color figure online)

5.2 Assessing Optimal Passing and Location

While soccer analytics have long-focused on using pass probabilities to evaluate a player's passing skills based on observed pass accuracy [13,15], there are still two main challenging problems that remain unattended: the identification of optimal passing locations and optimal off-ball positioning for improving pass probability.

Visual Assessment of Optimal Passing. Given a game-state, where a player is in possession of the ball, we define the optimal and sub-optimal pass destinations as the locations near the teammates than provide a higher pass probability than the current location of the corresponding teammate. To obtain the optimal passing locations we first calculate the pass probability surface of a given game-state and then evaluate the probability of every possible location in a 5×5 grid around the expected teammate location in the next second, based on the current velocity. The location within that grid with the highest probability difference with the current player's location is set as the optimal passing location. Additionally, a set of sub-optimal passing locations are obtained by identifying locations with positive probability difference and that are at least 5 m away from previous sub-optimal locations. In the left column of Fig. 5 we present in red circles the set of best passing locations for each of the possession team players for a given game state. This kind of visualization provides a coach the ability to perform a direct visual inspection of passing options and allows her to provide direct feedback to players about specific game situations, improving the coach's effective communication options.

Visual Assessment of Optimal Positioning. Following a similar idea, we can leverage pass probabilities surfaces to detect the best possible location a player could occupy to increase the probability of receiving a pass directly. To obtain the optimal location for each player, we recalculate the pass probability surface of the same game situation but translating the location of the player (one player at a time) to any other possible location in the 5×5 grid. We analogously obtain the optimal locations, as described before. In the right column of Fig. 5 we observe in green circles the expected pass probability added if the player would have been placed in that location instead. Again, this tool can be handy for coaches to instruct players on how to improve their off-ball game.

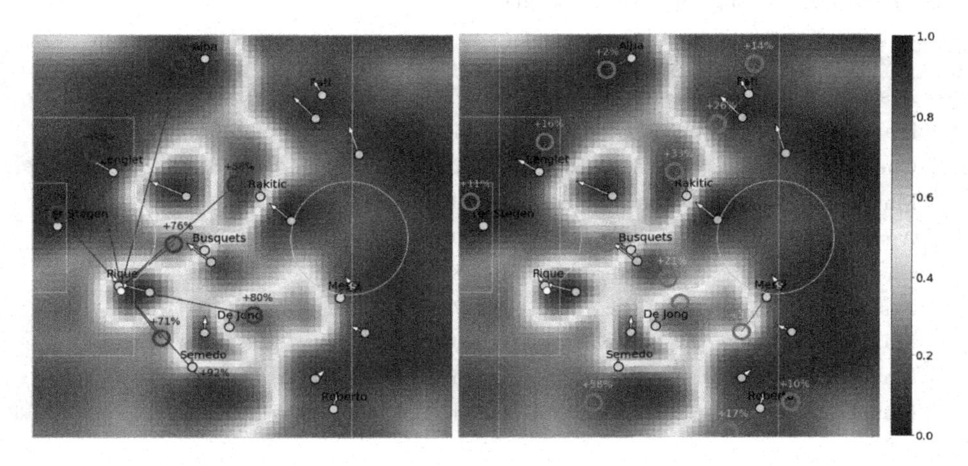

Fig. 5. In the left column, we present a game-state where red circles represent the optimal passing location for each teammate, and the expected pass probability. In the right column, the green circles represent the optimal positioning of players increasing the expected pass probability if the players were placed in those locations at that time. (Color figure online)

Assessing Passing Skill. We propose a new metric *pass completion added (PPA)* to evaluate the quality of a players' selection of the passing destination location. For each observed pass, we calculate the difference between the probability of the optimal pass and the probability of the selected pass. This metric is formally defined in Eq. 1, where S and M are the set of successful and missed passes, respectively, $\hat{y}$ is the optimal pass probability, and y is the selected pass probability. Intuitively a player reward is discounted if the selected pass was not optimal. In the case of the pass being unsuccessful, the player is only penalized in proportion to the probability difference with the optimal location, rewarding the player's pass selection.

$$PPA = \sum_{s=1}^{S}(1 - \hat{y}^s)(1 - (\hat{y}^s - y^s)) - \sum_{m=1}^{M}(\hat{y}^s)(\hat{y}^s - y^s) \tag{1}$$

In Table 3 we present the best ten players in pass completion added for the 2014–2015 season of the English Premier League, where the cumulative PPA of a player is normalized by 90 min played. The table includes the estimated player price in 2014, provided by www.transfermarkt.com. We can observe that the list contains a set of the best players in recent times in this league, including creative midfielders such as Oezil,Silva, Hazard and Fabregas, deep creative wingers such as Navas and Valencia, and Rosicky, a historical player.

Table 3. Ranking of the best ten players in pass completion added for the season 2014–2015 of the English Premier League.

Team	Player Name	PPA/90m	Age in 2014	Player price (2014)
Arsenal	Mesut Oezil	0.0578	24	€45M
Manchester City	David Silva	0.0549	28	€40M
Chelsea	Eden Hazard	0.0529	23	€48M
Manchester United	Antonio Valencia	0.0502	29	€13M
Arsenal	Tomas Rosicky	0.0500	33	€2M
Chelsea	Cesc Fabregas	0.0484	27	€40M
Arsenal	Santi Cazorla	0.0470	29	€30M
Manchester City	Jesus Navas	0.0469	28	€20M
Manchester City	Yaya Toure	0.0466	30	€30M
Manchester City	Samir Nasri	0.0447	26	€22M

5.3 Team-Based Passing Selection Tendencies

The pass selection adaptation of SoccerMap, presented in Sect. 5.1, provides a fine-grained evaluation of the passing likelihood in different situations. However, it is clear to observe that passing selection is likely to vary according to a team's player style and the specific game situation. While a league-wide model might be useful for grasping the expected behavior of a typical team in the league, a soccer coach will be more interested in understanding the fine-grained details that separate one team from the other. Once we train a SoccerMap network to obtain this league-wide model, we can fine-tune the network with passes from each team to grasp team-specific behavior. In this application example, we trained the pass selection model with passes from all the teams from English Premier League season 2014–2015. Afterward, we retrained the initial model with passes from two teams with different playing-styles: Liverpool and Burnley.

In Fig. 6 we compare the pass selection tendencies between Liverpool (left column) and Burnley (right column). On the top left corner of both columns, we show a 2D plot with the difference between the league mean passing selection heatmap, and each team's mean passing selection heatmap, when the ball is within the green circle area. We can observe that Liverpool tends to play short passes, while Burnley has a higher tendency of playing long balls to the forwards or opening on the sides. However, this kind of information would not escape from the soccer coach's intuition, so we require a more fine-grained analysis of each team's tendencies in specific situations. In the two plots of Fig. 6 we show over each players' location the percentage increase in passing likelihood compared with the league's mean value. In this situation, we can observe that when a left central defender has the ball during a buildup, Liverpool will tend to play short passes to the closest open player, while Burnley has a considerably higher tendency to play long balls to the forwards, especially if forwards are starting a run behind the defender's backs, such as in this case. Through a straightforward fine-tuning of the SoccerMap-based model, we can provide detailed information to the coach for analyzing specific game situations.

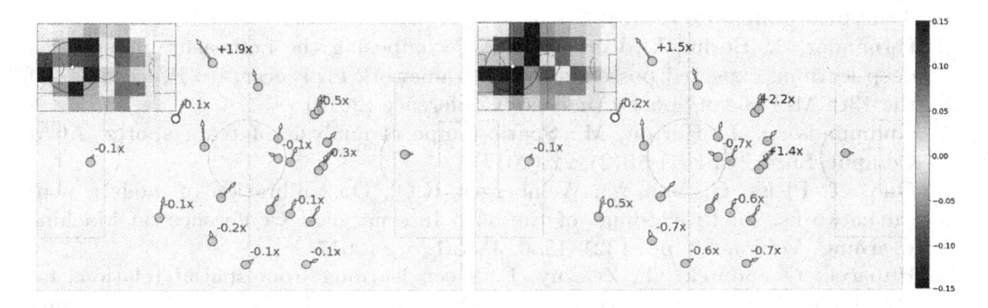

Fig. 6. A game-state representation of a real game situation in soccer. Above each player (circles) we present the added percentage difference of pass likelihood in that given situation in comparison with the league for two teams: Liverpool (left column) and Burnley (right column). The heatmaps in both top left corners of each column represent the mean difference in pass selection likelihood with the league, when the ball is located within the green circle. (Color figure online)

6 Discussion and Future Work

The estimation of full probability surfaces provides a new dimension for soccer analytics. The presented architecture allows generating visual tools to help coaches perform fine-tuned analysis of opponents and own-team performance, derived from low-level spatiotemporal soccer data. We show how this network can be easily adapted to many other challenging related problems in soccer, such as the estimation of pass selection likelihood and pass value, and that can perform remarkably well at estimating the probability of observed passes. By merging features extracted at different sampling levels, the network can extract both fine and coarse details, thereby managing to make sense of the complex spatial dynamics of soccer. We have also presented several novels practical applications on soccer analytics, such as evaluating optimal passing, evaluating optimal positioning, and identifying context-specific and team-level passing tendencies. This framework of analysis derived from spatiotemporal data could also be applied directly in many other team sports, where the visual representation of complex information can bring the coach and the data analyst closer.

References

1. Cakmak, A., Uzun, A., Delibas, E.: Computational modeling of pass effectiveness in soccer. Adv. Complex Syst. **21**(03n04), 1850010 (2018)
2. Eggels, H.: Expected goals in soccer: explaining match results using predictive analytics. In: The Machine Learning and Data Mining for Sports Analytics Workshop, p. 16 (2016)
3. Fernández, J., Bornn, L.: Wide open spaces: a statistical technique for measuring space creation in professional soccer. In: Proceedings of the 12th MIT Sloan Sports Analytics Conference (2018)
4. Fernández, J., Bornn, L., Cervone, D.: Decomposing the immeasurable sport: a deep learning expected possession value framework for soccer. In: Proceedings of the 13th MIT Sloan Sports Analytics Conference (2019)
5. Gudmundsson, J., Horton, M.: Spatio-temporal analysis of team sports. ACM Comput. Surv. (CSUR) **50**(2), 22 (2017)
6. Guo, C., Pleiss, G., Sun, Y., Weinberger, K.Q.: On calibration of modern neural networks. In: Proceedings of the 34th International Conference on Machine Learning- Volume 70, pp. 1321–1330. JMLR. org (2017)
7. Hubáček, O., Šourek, G., Železný, F.: Deep learning from spatial relations for soccer pass prediction. In: Brefeld, U., Davis, J., Van Haaren, J., Zimmermann, A. (eds.) MLSA 2018. LNCS (LNAI), vol. 11330, pp. 159–166. Springer, Cham (2019). https://doi.org/10.1007/978-3-030-17274-9_14
8. Long, J., Shelhamer, E., Darrell, T.: Fully convolutional networks for semantic segmentation. In: Proceedings of the IEEE Conference on Computer Vision and Pattern Recognition, pp. 3431–3440 (2015)
9. Long, J., Zhang, N., Darrell, T.: Do convnets learn correspondence? In: Advances in Neural Information Processing Systems, pp. 1601–1609 (2014)
10. Odena, A., Dumoulin, V., Olah, C.: Deconvolution and checkerboard artifacts. Distill **1**(10), e3 (2016)
11. Papandreou, G., Chen, L.C., Murphy, K.P., Yuille, A.L.: Weakly-and semi-supervised learning of a deep convolutional network for semantic image segmentation. In: Proceedings of the IEEE International Conference on Computer Vision, pp. 1742–1750 (2015)
12. Pathak, D., Krahenbuhl, P., Darrell, T.: Constrained convolutional neural networks for weakly supervised segmentation. In: Proceedings of the IEEE International Conference on Computer Vision, pp. 1796–1804 (2015)
13. Power, P., Ruiz, H., Wei, X., Lucey, P.: Not all passes are created equal: objectively measuring the risk and reward of passes in soccer from tracking data. In: Proceedings of the 23rd ACM SIGKDD International Conference on Knowledge Discovery and Data Mining, pp. 1605–1613. ACM (2017)
14. Rein, R., Memmert, D.: Big data and tactical analysis in elite soccer: future challenges and opportunities for sports science. SpringerPlus **5**(1), 1410 (2016)
15. Spearman, W., Basye, A., Dick, G., Hotovy, R., Pop, P.: Physics-based modeling of pass probabilities in soccer. In: Proceeding of the 11th MIT Sloan Sports Analytics Conference (2017)

Stop the Clock: Are Timeout Effects Real?

Niander Assis(✉), Renato Assunção, and Pedro O. S. Vaz-de-Melo

Departmento de Ciência da Computação, Universidade Federal de Minas Gerais, Belo Horizonte, Brazil
{niander,assuncao,olmo}@dcc.ufmg.br

Abstract. Timeout is a short interruption during games used to communicate a change in strategy, to give the players a rest or to stop a negative flow in the game. Whatever the reason, coaches expect an improvement in their team's performance after a timeout. But how effective are these timeouts in doing so? The simple average of the differences between the scores before and after the timeouts has been used as evidence that there is an effect and that it is substantial. We claim that these statistical averages are not proper evidence and a more sound approach is needed. We applied a formal causal framework using a large dataset of official NBA play-by-play tables and drew our assumptions about the data generation process in a causal graph. Using different matching techniques to estimate the causal effect of timeouts, we concluded that timeouts have no effect on teams' performances. Actually, since most timeouts are called when the opposing team is scoring more frequently, the moments that follow resemble an improvement in the team's performance but are just the natural game tendency to return to its average state. This is another example of what statisticians call the *regression to the mean* phenomenon.

Keywords: Causal inference · Sports analytics · Timeout effect · Momentum · Bayesian networks

1 Introduction

In sports, timeout is a short interruption in a play commonly used to stop a negative flow in the game, to discuss a strategy change, or to rest the players [3]. As this is the most direct way coaches can intervene during a game, their influence and strategic ability is best expressed during these events. A timeout is usually called when a team has a rather long streak of score losses [7,24]. Popular belief and research [5,11,16,18,21] have found a positive effect on teams' performances after the timeout. That is, on average, the team asking for the

Electronic supplementary material The online version of this chapter (https://doi.org/10.1007/978-3-030-67670-4_31) contains supplementary material, which is available to authorized users.

Y. Dong et al. (Eds.): ECML PKDD 2020, LNAI 12461, pp. 507–523, 2021.
https://doi.org/10.1007/978-3-030-67670-4_31

timeout recovers from the losses by scoring positively immediately after. This observed difference has been wrongly used as evidence that the timeout has a real and positive effect on teams' performance. In order to answer such causal question, a formal counterfactual analysis should be used, and that is what we propose in this work.

There is an intense interest on causal models to analyze non-experimental data since causal reasoning can answer questions that machine learning itself cannot [15]. Our approach is built on top of these causal inference approaches that are briefly introduced in Sect. 3. For each timeout event at time t_r in the database, we found a paired moment t_c in the same game when no timeout has been called that serves as a *control* moment for t_r, reflecting what would have happened the timeout had not been called. This control moment is chosen based on other variables about the current game instant, which were drawn in a causal graph that depict our assumptions about the generation of the data and, as we will further discuss, asses if the causal effect can be estimated without bias. In order to quantify how the game changed just after a given moment t, we proposed *Short-term Momentum Change* (STMC), which is discussed in Sect. 4.1.

After using a *matching* approach to construct our matched data (pairs of (t_r, t_c)), we found virtually no difference between the distribution of the STMC for real timeouts and control instants, i.e., the estimated *timeout effect* is very close to zero or non-existent. Hence, we conclude that the apparent positive effect of timeouts is another example of the well-known *regression to the mean fallacy* [1]. The dynamic match score fluctuates naturally and, after an intense increase, commonly returns towards a mild variation. Thus, because timeouts are usually called near the extreme moments, as we will show, the game seems to benefit to those loosing. In summary, the main contributions of this paper are the following:

- We proposed a metric called *Short-term Momentum Change* (STMC) to quantify how much the game momentum changes after a time moment t_r associated with an event, such as a regular ball possession or an interruption of the game;
- We collected and organized a large dataset covering all the play-by-play information for all National Basketball Association (NBA) games of the four regular seasons from 2015 to 2018. A single season has over 280 thousands *game instants*, as we define in Sect. 4, and over 17 thousand timeout events;
- After a detailed causal inference analysis to evaluate the timeout effect of the *Short-term Momentum Change*, we did not find evidence that the timeout effect exists or that its effect size is meaningful. Inspired by others, we did also consider two other settings in which the timeout effect could be different: (i) only the last five minutes of the games and (ii) everything but the last five minutes.

The next section describes the previous work carried out on the effect of timeouts and discusses how our work distance from them. Sect. 3 gives a background on causal inference and the statistical models adopted. We start Sect. 4

by summarizing timeout rules in the NBA and describing our dataset, the play-by-play tables. In Sect. 4.1 we introduce our outcome variable of interest, the *Short-term Momentum Change*, and in Sect. 4.2 our causal model. In Sect. 4.3, we describe our treatment and control groups and, in Sect. 4.4, we explain our matching approaches. All the results are presented in Sect. 5. We close the paper in Sect. 6 with our conclusions.

2 Related Work

Timeouts are used and implemented in team sports for several reasons, such as to rest or change players, to inspire morale, to discuss plays, or to change the game strategy [20]. However, timeouts are mostly used to stop a negative flow in the game [7,24], which is popularly referred as "the game momentum." In basketball, momentum arises when one team is scoring significantly more than the other [11,22].

Several earlier studies analyzed the effect timeouts have for decreasing the opponent's momentum in the game [5,11,18,21]. These studies analyze the effect of timeouts on teams' performance just before and after it was called. For instance, by using a small sample of seven televised games from the 1989 National Collegiate Athletic Association (NCAA) tournament, Mace *et al.* [11] recorded specific events of interest, which were classified as either *Reinforcers* (e.g. successful shots) or *Adversities* (e.g. turnovers) and verified that the rate of these events change significantly among teams in the 3 min before and after each timeout. They found that while the team that called the timeout improved its performance, the opponent team decreased it. Other works reached the same conclusions using similar methodologies and different data sets [5,18,21].

To the best of our knowledge, Permutt [16] was the first to acknowledge the *regression to the mean* phenomenon in such analysis. Permutt considered specific game moments—timeouts called for after a team suffered a loss of six consecutive points. Similar to others, the short-term scoring ratio was observed to be higher after timeouts. However, in contrast to others, the paper compares real timeouts with other similar game moments without a timeout. With such analysis, Permutt found that timeouts can be effective at enhancing performance, but at a small magnitude. The most significant result shows that the home-team with a "first-half restriction" presents a 0.21 increase in average ratio for the next ten points. Calling a timeout predicts that the home-team will score 5.47 out of the next ten points as opposed to 5.26 points when a timeout is not called. Thus, the conclusion is that timeouts do not have any significant effect in changing the momentum of a game, i.e., using 6-0 runs as an indicator of instances where momentum would be a factor, teams were successful at "reversing" momentum even without the timeout as a mediator.

Although the work of Permutt [16] innovates by considering counterfactuals, the analyses still leave room for reasonable doubts about the reality of the timeout effect. It fails to take into account the existence of other important factors that could also influence on the momentum change and confound the true timeout effect. As a result, spurious correlations could have caused the lack of effect observed in the data. In our work, we take into account other factors such as coaches' and team's abilities, stadium and match conditions, clock time, quarter and relative score between the teams. More important, different from all the studies described in this section, we adopt a formally defined causal model approach [14] with a counterfactual analysis with its constructed control group. We show in a compelling way that timeouts do not have an effect in teams' performances.

3 The Causality Framework

For illustration, consider Y our outcome variable and $A \in \{0, 1\}$ the *treatment variable*. Regardless of the actual value of A, we define $Y_{A=1}$ to be the value of Y had A been set to $A = 1$ and $Y_{A=0}$ to be the value of Y had A been set to $A = 0$. We say there is a causal effect of A on Y if $Y_{A=0} \neq Y_{A=1}$ and, conversely, there is no causal effect or the effect is null if $Y_{A=0} = Y_{A=1}$. These defined values are called *potential outcomes* [12] because just one potential outcome is factual, truly observed, while the others are counterfactuals. Therefore, we cannot generally identify the causal effects of a single individual. This problem is known as the *Fundamental Problem of Causal Inference (FPCI)* [8].

Nevertheless, in most causal inference settings, the real interest is in the population level effect, or the average causal effect defined by $\mathbb{E}(Y_{A=1} - Y_{A=0})$. $\mathbb{E}(Y \mid A = 1) - \mathbb{E}(Y \mid A = 0)$ gives a reliable estimate in randomized experimental studies, where treatment A is assigned randomly to each unit. However, in observational ones, we need to collect more information to control for and to make assumptions. One important assumption is the *conditional ignorability* [19]. This assumption is satisfied if, given a vector of covariates $\mathbf{X}$, the treatment variable A is conditionally independent of the potential outcomes ($Y_{A=0} \perp\!\!\!\perp A \mid \mathbf{X}$ and $Y_{A=1} \perp\!\!\!\perp A \mid \mathbf{X}$), and there is a positive probability of receiving treatment for all values of $\mathbf{X}$ ($0 < \mathrm{P}(A = 1 \mid \mathbf{X} = \mathbf{x}) < 1$ for all $\mathbf{x}$). The conditional ignorability assumption allows us to state that $\mathbb{E}(Y_{A=1} - Y_{A=0} \mid \mathbf{X}) = \mathbb{E}(Y \mid A = 1, \mathbf{X}) - \mathbb{E}(Y \mid A = 0, \mathbf{X})$ for every value of $\mathbf{X}$. This represents the basic rationale behind the *matching* technique.

The simplest matching technique is the *exact matching*. For each possible $\mathbf{X} = \mathbf{x}$, we form two subgroups: one composed by individuals that received the treatment and have $\mathbf{X} = \mathbf{x}$, and the other by individuals that did not receive the treatment and have $\mathbf{X} = \mathbf{x}$. Unfortunately, exact matching is not feasible when the number of covariates is large or some are continuous. As an alternative, examples are usually matched according to a distance metric $d_{ij} = d(\mathbf{x}_i, \mathbf{x}_j)$ between the covariate configurations of pairs (i, j) of observations. The *Mahalanobis distance* [23] is a common choice for such distance metric as it takes into account the correlation between the different features in the vector $\mathbf{X}$. Another option is to use *propensity score* [19] for estimation of causal effects, which is defined as probability of receiving treatment given the covariates, i.e., $s(\mathbf{X}) = \mathbb{P}(A = 1 \mid \mathbf{X})$. Rosenbaum and Rubin proved that it is enough to just match on a distance calculated using the scalar scores $s(\mathbf{x})$, rather than the entire vector $\mathbf{x}$ [19].

In general, each matching approach can be implemented using algorithms that are mainly classified as either *greedy* or *optimal.* The *greedy* ones, also known as *nearest-neighbor matching*, matches the i-th treated example with the available control example j that has the smallest distance d_{ij}. *optimal matching*, however, takes into account the whole reservoir of examples since the goal is to generate a matched sample that minimizes the total sum of distances between the pairs. Such optimal approach can be preferred in situations where there are great competition for controls. For a good review on different matching methods for causal inference, see [23].

Whatever matching technique used, its success can be partially judged by how balanced out are the covariates in the treatment and control groups. By pruning unmatched examples from the dataset, the control and treated groups of the remaining matched sample should have similar covariate distributions, when we say that matching achieves *balance* of the covariates distribution.

4 The Causal Effect of Timeout

According to the NBA 2016–2017 season official rules, in a professional NBA regular game, each team is entitled to six full-length timeouts and one 20-second timeout for each half. A full-length timeout can be of 60 s or 100 s, depending on when the timeout was requested. Also, every game has four regular periods plus the amount of overtime periods necessary on the occurrence of ties. There is a specific amount of timeouts expected in each period for commercial purposes. If neither team calls a timeout before a specific time, thus not fulfilling the next expected timeout, the official scorer stops the game and calls a timeout. The timeout is charged to the team that has not been charged before, starting with the *home* team. These timeouts are called *mandatory* or *official* timeouts.

In basketball games, possessions are new opportunities to score in the game. Each possession starts from the moment a team gets hold of the ball until one of his players scores, commits a fault, or loses the ball in defensive rebounds or turnovers. The total number of possessions are guaranteed to be approximately the same for both teams at the end of a match, so it provides a good standardization for the points scored by each team [9]. Indeed, most of basketball statistics are already given in a per possession manner.

Play-by-play tables capture the main play events such as goal attempts, rebounds, turnovers, faults, substitutions, timeouts and end of quarters (periods). For each play, we have the time in which it happened, the players and/or the team involved and any other relevant information, e.g., the score just after the play is recorded. Each play event is recorded as a new line in the table. While ball possessions are not clearly recorded in play-by-play tables, one can identify every change of possession from observing the game events.

In this work, we use play-by-play tables to identify the ball possessions and use them to observe how the teams' performance change when timeouts are called. We have identified every change of possession alongside the main interruptions—timeouts and end of quarters—in each game. Every change of possession and every main interruption is considered a new *game instant*. We model each basketball game as a series of discrete *game instants*. More formally, a *game instant* is either (1) a regular ball possession; or (2) a major game interruption, which can be a regular timeout, an official scorer timeout, or the end of a quarter. Player substitutions and fouls were not considered as a main interruption. In fact, substitutions can happen when the ball stops and not only during timeouts.

4.1 Short-Term Momentum Change

Here we describe our outcome variable associated with teams' performance, for which we aim to estimate how it is affected by the timeouts. Let $\{P_t\}$ be an univariate stochastic process associated to a single match and indexed by the discrete game instants t. At any timeout moment, the team calling the timeout is defined as the *target team*. The P_t random variable at the end of the t-th game instant is the score of the target team minus the opposing team's score and it is called the scoring margin. Hence, P_t is a positive quantity when the target team is winning the match at game instant t and negative otherwise. At the end of the t-th game instant, $P_t = P_{t-1}$ in two situations. First, if t is a regular ball possession instant whose attacking team did not score, i.e., the possession ended with a turnover or defensive rebound. Second, if t is a main interruption instant and, consequently, none of the teams had the opportunity to score.

In order to evaluate how "momentum" changes after a game instant, we use the *Short-term Momentum Change* (STMC), which is the amount by which the *scoring margin per possession rate* changes right after an game instant. For any game instant t and a positive integer $\lambda > 0$, we define the STMC, Y_t^λ, as the average rate of change from P_t to $P_{t+\lambda}$ ($\Delta P_t^{t+\lambda}$) minus the average rate of change from $P_{t-\lambda-1}$ to P_{t-1} ($\Delta P_{t-\lambda}^t$). Note that we do not take into account the possible change in scoring margin caused in game instant t (the change from P_{t-1} to P_t):

$$Y_t^\lambda = \frac{P_{t+\lambda} - P_t}{\lambda} - \frac{P_{t-1} - P_{t-\lambda-1}}{\lambda} = \Delta P_t^{t+\lambda} - \Delta P_{t-\lambda}^t \tag{1}$$

for $t - \lambda \geq 0$ and $t + \lambda \leq n$, where n is the total number of game instants in a given game.

The hyper-parameter λ controls the time window used to evaluated how the game scoring dynamics changes around t. To balance out the offensive and defensive ball possessions, λ must be an even integer. Also, the variable Y_t^λ should only be evaluated if the interval $[t-\lambda,\ t+\lambda]$ contains no game interruptions, with the possible exception of t. In a causal perspective, λ represents our assumption for how many game instants that the interference (calling a timeout or not) at game instant t can influence and is influenced by, in the short-term.

Let A_t be the binary indicator that a timeout has been called at time t. We will denote $A_t = 1$ if a team calls a timeout right before the game instant t and $A_t = 0$ if t is a regular ball possession. If we find the set of covariates $\mathbf{X}$

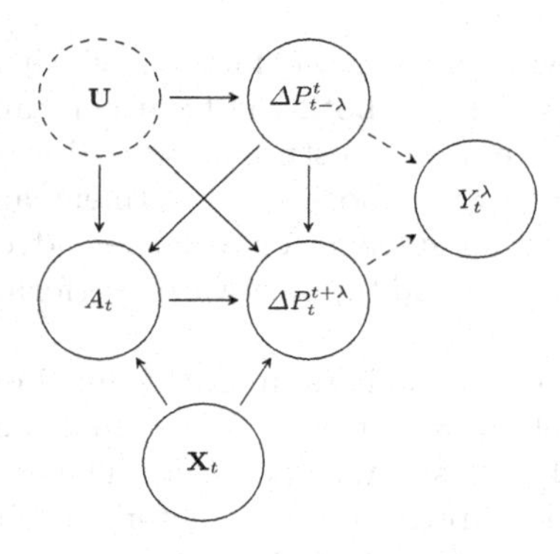

Fig. 1. A causal graph to model the timeout effect.

that satisfy the conditional ignorability assumption, we can apply a matching technique and our average causal effect of interest, $\mathbb{E}(Y^\lambda_{A_t=1} - Y^\lambda_{A_t=0})$, can be estimated taking the difference in means from the matched treatment and control groups. The estimated *timeout effect* TE is defined as:

$$TE = \mathbb{E}(Y^\lambda_t \mid A_t = 1) - \mathbb{E}(Y^\lambda_t \mid A_t = 0). \quad (2)$$

Every game is composed by two teams, the *home* and the *away* team. Because we want to estimate the causal effect of timeouts on the performance of the *team that actually asked for it*, we decided to estimate the average causal effect of timeouts called by the *home* teams (TE_h) and the *away* teams (TE_a), separately. We proceed now to present our causal model which encodes our assumptions.

4.2 The Causal Model

Pearl [14] suggests the use of directed acyclic graphs (DAGs) as a way of encoding causal model assumptions with nodes representing the random variables and the direct edges representing direct causal relationships. One can identify in such graph a set of variables (or nodes) that satisfies the so called *back-door criterion* [14]. These are variables that blocks all back-door paths from A (the treatment variable) to Y (the outcome variable) and does not include any descendants of A. Given that the graphical model includes all important confounding variables, it can be shown that conditioning on them suffices to remove all non-causal dependencies between A and Y. In other words, it leaves only causal dependence that corresponds to the causal effect.

There are many factors that can potentially influence the short-term performance (STMC) of the team that called a timeout after a given game instant. These can be *intra-game* factors, which vary along the game, such as the scoring margin, the quarter and the time since the start of the quarter, or *inter-game* factors, which vary from game to game, such as the venue conditions, the attendance at the venue, the specific adversary team, the players available and the teams' momentum in the season.

It is very intuitive why *intra-game* factors, which are specific to a game instant, are considered a cause of both the treatment and outcome, thus being considered a confounder. In a not so straightforward way, some *inter-game* factors are also very likely to affect both the treatment and outcome. For example, a team playing against a stronger or a weaker adversary would differently request the available timeouts and the afterwards performance may be differently affected.

Figure 1 shows our causal model graph. Each game instant t can either receive the treatment assignment $A_t = 1$ or $A_t = 0$, meaning that the game instant is a timeout or a regular ball possession, respectively. The variables $\mathbf{X}_t$ represent the observed covariates, which are *intra-game* factors specific to the game instant t: (i) the current quarter (period) (Q_t), (ii) the current scoring margin (P_t) and (iii) the current time in seconds since the start of the period (S_t). The variables represented by the node $\mathbf{U}$ are the *inter-game* factors, or the covariates related to a *specific game* that influence both the treatment assignment and the game outcome as exemplified in the last paragraph. Most of these variables are not directly observed or very difficult to measure—players and coach strategies, teams' relative skill difference and venue conditions. Hence, we include them in our graph as a dashed circle. The average rates of scoring margin change before ($\Delta P_{t-\lambda}^{t}$) and after ($\Delta P_{t}^{t+\lambda}$) the game instant t are also in the graph, as well as the outcome Y_t^{λ} that is connected by dashed edges since it is a deterministic node—a logical function of the other two stochastic nodes.

We are interested in the causal effect of A_t on Y_t^{λ}. Since $\Delta P_{t-\lambda}^{t}$ is a direct cause of A_t and not the reverse—for obvious chronological reasons—, we actually want to measure the causal effect of A_t on $\Delta P_{t}^{t+\lambda}$. According to the *back-door* criterion [14], if we adjust for $\mathbf{U}$, $\mathbf{X}_t$ and $\Delta P_{t-\lambda}^{t}$ we block any non-causal influence of A_t on $\Delta P_{t}^{t+\lambda}$.

4.3 Data

Because we want to estimate TE as defined in Eq. (2), our treated and control groups are formed by game instants' STMC, Y_t^{λ}. As discussed in Sect. 4.1, depending on which value we choose for λ, Y_t^{λ} is not valid—if the short-term window induced by λ includes another major interruption besides the possible t or is longer than the start or end of the game. Therefore, our inclusion criteria for both groups is that STMC exists and can be calculated. For a given λ, the treated group, $\{Y_t^{\lambda} | A_t = 1\}$, is formed by the valid real timeouts' STMC. The control group, $\{Y_t^{\lambda} | A_t = 0\}$, is formed by any valid game instant t's STMC that is not a timeout or any other kind of major interruption.

Since we want to estimate TE_h and TE_a, we have two treatment groups, one for timeouts called by *home* teams and one for timeouts called by *away* teams. On the other hand, it does not make sense to classify the control group as either *home* or *away*, thus we have just one control group. We will limit ourselves in the future to just mention theses treatment groups as either the *home* treatment group or *away* treatment group.

Our data consist of play-by-play information for every game from the 2014–2015, 2015–2016, 2016–2017 and 2017–2018 National Basketball Association NBA regular seasons. We crawled the data from the Basketball-Reference website[1]. Most of our analysis will consist only of games from the NBA 2016–2017 season because using more than a single season would lead to very big samples that are impractical to apply our matching approaches. Also, while we did perform the same analysis using only other seasons, achieving very similar results, the choice for the 2016–2017 season is arbitrary, mainly due to be the first season for which we collected the data.

The 2016–2017 season had a total of 30 teams and 1,309 games (1,230 for the regular season and 79 in the playoffs). Considering all games, we computed 281,373 game instants, including the 17,765 identified timeouts (7,754 were called by *home* teams and 8,011 by *away* teams), and the 2,000 *mandatory* timeouts. Our datasets, code and further instructions on how to reproduce our results can be found at our GitHub repository[2].

4.4 Matching

The variables $\mathbf{U}$, $\mathbf{X}_t$ and $\Delta P_{t-\lambda}^{t}$ should be controlled for. In other words, they should be considered as possible confounders. Consequently, all of these variables are included in our matching for a valid causal inference. While we consider $\mathbf{U}$, the *inter-game factors*, unobserved covariates, we can still control them by pairing timeout examples with non-timeout examples *taken from the same game*. Furthermore, the variable $\Delta P_{t-\lambda}^{t}$ is likely the most important confounder covariate in our model. Indeed, coaches tend to call a timeout when their teams are suffering from a bad "momentum", evidencing great influence on the treatment assignment A_t. Also, $\Delta P_{t}^{t+\lambda}$, the average rate of scoring margin change after a game instant t, should be highly causal dependent on $\Delta P_{t-\lambda}^{t}$. Therefore, in whatever matching approach we use, timeouts and control examples *taken from the same game* and with equal $\Delta P_{t-\lambda}^{t}$ are going to be matched, hopefully, achieving balance for $\mathbf{X}_t$. We also restrict our matches to be constructed with *non-overlapping ball possessions*. This restriction arises from our assumption that λ defines a range of game instants that are dependent and influence A_t as discussed in Sect. 4.1.

We applied three matching procedures: (1) *no-balance matching*; (2) *Mahalanobis matching*, and (3) *propensity score matching*. In the no-balance matching, each treatment example is paired with a valid control example that has the same $\Delta P_{t-\lambda}^{t}$ and is taken from the same game. We did not considered $\mathbf{X}_t$ in this matching. For the Mahalanobis matching, we applied the Mahalanobis distance using all covariates in $\mathbf{X}_t$, i.e., current quarter (Q_t), current scoring margin (P_t), and current clock time in seconds (S_t) since the start of the quarter. Finally, for the propensity score matching technique, we applied a simple euclidean distance match on the estimated scalar propensity score.

[1] http://www.basketball-reference.com.
[2] https://github.com/pkdd-paper/paper667.

The true propensity score $s(\mathbf{X}) = \mathbb{P}(A = 1 \mid \mathbf{X})$ is unknown and must be estimated. Since estimating $\mathbb{P}(A = 1 \mid \mathbf{X})$ can be seen as a classification task, any a supervised classification model could be used. While logistic regression is the most common estimation procedure for propensity score, Lee *et al.* [10] showed that, in a non-linear dependence scenario, the use of machine learning models such as boosting regression trees to estimate the propensity score achieves better covariate balance in the matched sample. Indeed, our treatment assignment present a great non-linear dependence on its covariates. Take the clock time S_t, for example. As explained in Sect. 4.3, the timeout rules of NBA stimulate coaches to call a timeout just before a mandatory timeout would have been called by the official scorer. We use the boosting regression tree algorithm implemented in the *gbm* R package [6] to estimate the propensity score using $\mathbf{X}_t$.

Because we restrict timeout and non-timeout pairs to be taken from the same game, we have a very sparse matching problem. The *rcbalance* R package [17] implementation of *optimal matching* exploits such sparsity of treatment-control links to reduce computational time for larger problems. We use the optimal algorithm implemented in this package for all the aforementioned matching approaches. In addition, before applying any matching technique, we retained in the control subpopulation only those non-timeout game instants t' ($A_{t'} = 0$) for which the value $\Delta P^{t'}_{t'-\lambda}$ is exactly equal to at least one $\Delta P^{t}_{t-\lambda}$ calculated to a real timeout instant t ($A_t = 1$) in the same game. This improved the running performance even more.

5 Experimental Results

In order to find out whether our data shows the generally accepted positive correlation between timeouts and improvements in the "momentum", we calculated Y_t^λ for every game instant associated with a timeout t in every single game using $\lambda = 2, 4, 6$. Figure 2 shows the estimated density distribution of the STMC Y_t^λ for all timeouts, including those called by both *home* and *away* teams, but removing the *official* timeouts. The sample means and number of valid timeout examples in each sample are 0.629 and 14,031 for $\lambda = 2$, 0.421 and 12,225 for $\lambda = 4$, 0.302 and 10,296 for $\lambda = 6$, respectively.

These results shows that, when a timeout is called by a team, its momentum improves by a small positive amount afterwards. For instance, with $\lambda = 4$, the average value of STMC is 0.421. That is, on average, there is an increase of 0.421 points for a team's scoring margin per possession after it called the timeout. These results are consistent with previous works mentioned in Sect. 2 [5,11,16, 18,21]. We applied the non-parametric ones-sample Wilcoxon statistical test and a bootstrap based test for the mean with the null hypothesis being that the mean is equal to zero. Both tests for the three different values of λ yielded p-values numerically equal to zero.

While these results could be used as evidence on why there is such common and widespread belief that timeouts improves teams' performance, or*breaks* the momentum, it is not an evidence of the causal effect of timeouts. We move on to consider the analysis under our causal framework discussed in Sect. 4.

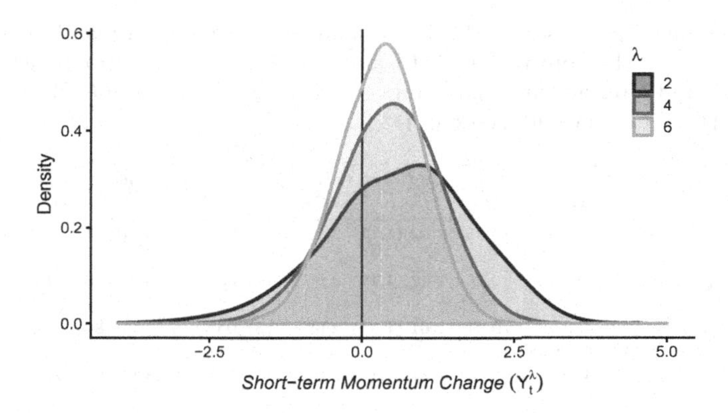

Fig. 2. The STMC (Y_t^λ) distribution for home and away timeouts, considering three different ball possession windows $\lambda = 2, 4, 6$.

5.1 Matching Results

Each of the three matching methods was applied twice: one time using the treatment group with *away* timeouts and the control group, and the other using the treatment group with *home* timeouts and the control group. Some examples did not find a valid match and, therefore, were not included in the matched samples. Also, it should be noted that all matches were performed without replacement.

To evaluate for proper covariate balance between the treatment groups, a common numerical discrepancy measurement is the difference in means divided by the pooled standard deviation of each covariate, known as the *standardized mean difference* (SMD) [4]. Unlike t-tests, SMD is not influenced by sample sizes and allows comparison between variables of different measured units. There is no general consensus on which value of SMD should denote an accepted imbalance level. Some researches, although, have proposed a threshold of 0.1 [13]. Table 1 summarizes the covariate distribution with its mean and standard deviation values and the SMD of our matched samples considering all different approaches.

For simplicity, we are only including here the results from the *home* timeouts matched samples. Indeed, the *away* samples showed very similar results and it can also be accessible from our GitHub repository[3]. Also, we do not show balance for $\Delta P_{t-\lambda}^t$ as it is perfectly balanced due to our perfect match on this covariate. The unmatched sample sizes for control groups are 172, 785, 101, 093 and 49, 403; and, for treatment groups, 6, 912, 6, 048 and 5, 127 with $\lambda = 2, 4, 6$, respectively. For the matched samples, because of our 1:1 matching approach, the sample sizes are equal in both treatment and control groups, even across the different matching techniques—for $\lambda = 2, 4, 6$, they were 6, 895, 5, 477 and 3, 832, respectively.

From an initial look we can see that there are very similar results in terms of covariate balance for all three different matching approaches. Also, with no

[3] https://github.com/pkdd-paper/paper667.

Table 1. Summary statistics and SMD for balance assessment for matching using *home* treatment group. The control ($A_t = 0$) and timeout ($A_t = 1$) groups are presented before (BM) and after all three matchings approaches: No-Balance (NB), Mahalanobis distance (M), and Propensity score (P)

λ	Method	S_t (mean (sd))			Q_t (mean (sd))			P_t (mean (sd))		
		$A_t = 0$	$A_t = 1$	SMD	$A_t = 0$	$A_t = 1$	SMD	$A_t = 0$	$A_t = 1$	SMD
2	BM	363.42 (198.77)	410.03 (168.47)	0.253	2.42 (1.12)	2.68 (1.15)	0.222	1.73 (10.81)	−0.00 (10.78)	0.161
	NB	363.11 (201.26)	410.20 (168.52)	0.254	2.47 (1.12)	2.68 (1.15)	0.178	0.56 (10.63)	0.00 (10.78)	0.052
	M	397.73 (168.83)	410.27 (168.50)	0.074	2.61 (1.11)	2.68 (1.15)	0.062	0.53 (10.86)	0.00 (10.78)	0.049
	P	403.85 (163.08)	410.14 (168.50)	0.038	2.67 (1.13)	2.68 (1.15)	0.009	0.23 (11.01)	0.00 (10.78)	0.021
4	BM	351.23 (187.75)	388.45 (153.17)	0.217	2.33 (1.11)	2.58 (1.13)	0.222	1.75 (10.69)	0.22 (11.04)	0.141
	NB	351.59 (191.15)	394.10 (151.23)	0.247	2.33 (1.11)	2.58 (1.14)	0.219	0.44 (10.44)	0.33 (11.14)	0.011
	M	381.35 (167.92)	393.79 (151.12)	0.078	2.43 (1.04)	2.57 (1.14)	0.135	0.50 (11.06)	0.32 (11.12)	0.016
	P	385.03 (162.81)	393.70 (151.37)	0.055	2.56 (1.11)	2.58 (1.14)	0.017	0.38 (11.33)	0.32 (11.12)	0.005
6	BM	334.96 (170.51)	380.80 (143.08)	0.291	2.19 (1.08)	2.49 (1.12)	0.270	1.67 (10.34)	0.33 (11.09)	0.124
	NB	332.33 (173.87)	389.79 (139.28)	0.365	2.20 (1.08)	2.48 (1.12)	0.253	0.71 (10.40)	0.79 (11.13)	0.007
	M	351.13 (162.90)	389.67 (138.95)	0.255	2.27 (1.02)	2.48 (1.12)	0.200	0.77 (10.80)	0.80 (11.17)	0.002
	P	356.99 (160.61)	389.72 (139.63)	0.218	2.35 (1.08)	2.48 (1.12)	0.122	0.72 (10.91)	0.81 (11.16)	0.008

surprise, the matched treatment and control samples have equal sizes and the $\Delta P_{t-\lambda}^{t}$ covariate is equally distributed. Comparing the *no-balance* matching with the balance in *before matching*, it is clear that this simple matching procedure reduces substantially the SMD for all covariates. However, this is not enough to make the SMD negligible for all cases and hence whatever conclusions based on the comparison of these no-balance matched samples will be rightly subjected to doubt. Considering the *Mahalanobis matching*, it achieved better covariate balance for all covariates and λ values in comparison with *before matching* and *no-balance matching*. For $\lambda = 2$ and $\lambda = 4$, all SMDs are bellow 0.1, with the exception of the Q_t covariate in $\lambda = 4$, in both analysis. The $\lambda = 6$ configuration, on the other hand, presented the worse SMDs—P_t is the only covariate with SMD bellow the 0.1 mark. The *Mahalanobis matching* was not as good as the *Propensity score matching*. All covariates have SMDs smaller than 0.1 for $\lambda = 2$ and $\lambda = 4$ in both analysis. For $\lambda = 6$, while Q_t and S_t still had SMD above the 0.1 mark, it is still smaller than the ones obtained in the previous matching for the same configuration.

Plotting the covariate distributions from both treatment and control groups is a qualitative alternative of checking proper balance. Figure 3 shows the covariate distributions for S_t, Q_t and P_t after all three matchings with $\lambda = 4$ on the analysis using timeouts called by the *home* team. We can see that *No-balance* and *Mahalanobis* matching presented some imbalance for S_t and Q_t. On the other hand, *propensity score matching* shows rather similar distributions suggesting a much better balance. The fact that timeouts and control examples are matched only if taken from the same game makes it more complicated to find matches with a more balanced $\mathbf{X}_t$. *Mahalanobis matching*, while trying to match samples as close as possible, encounters great difficulties. Propensity score, however, translates all interactions and non-linearities presented in the joint distribution of all covariates with the propensity score. This is why its matching was better.

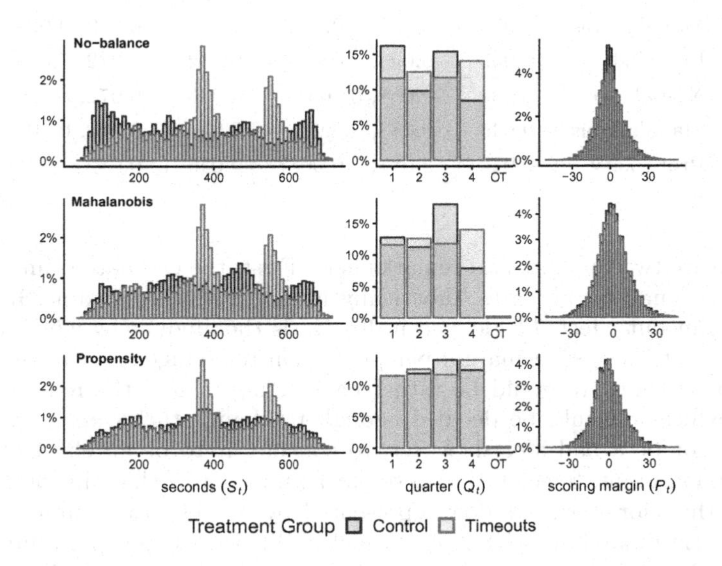

Fig. 3. The control (magenta) and timeout (turquoise) *seconds*, *quarter* and *scoring margin* covariate ($\mathbf{X}_t$) distributions after all matchings using the *home* timeouts treatment group and $\lambda = 4$.

5.2 Timeout Effect

We analyzed our matched data using a Monte Carlo permutation test [2, Vol II, Chap. 10] and difference of means between the two groups as our test statistic. When the null hypothesis is true, the timeout effect TE defined in (2) is equal to zero. The treatment label was permuted ten thousand times. It should be noted that we have large samples—the number of treatment-control pairs in each test varies from $3,766$ to $7,082$. Hence, we believe our analysis would mostly benefit if effect sizes, alongside statistical significance, are taken into account for our conclusions. Table 2 shows the estimated average timeout effects for the *home* (TE_h) and the *away* (TE_a) teams. In addition, we have also included in the table a 99% level confidence interval generated from the p-values obtained with the Monte Carlo permutation test.

Table 2. The estimated timeout effect for both *away* (TE_a) and *home* (TE_h) timeouts under different matching procedures and λ values. The respective confidence intervals for each timeout effect for 99% level is also shown. While some values are statistically significant, all timeout effects are negligible due to the small effect sizes.

λ	Method	TE_a	99% CI	TE_h	99% CI
2	No-balance	−0.028	(−0.075, 0.020)	−0.022	(−0.072, 0.028)
	Mahalanobis	−0.032	(−0.079, 0.015)	−0.017	(−0.067, 0.032)
	Propensity	−0.044	(−0.092, 0.004)	−0.021	(−0.071, 0.028)
4	No-balance	−0.043	(−0.082, −0.004)	−0.023	(−0.063, 0.017)
	Mahalanobis	−0.053	(−0.092, −0.013)	−0.013	(−0.053, 0.028)
	Propensity	−0.059	(−0.098, −0.020)	−0.032	(−0.072, 0.008)
6	No-balance	−0.031	(−0.068, 0.005)	−0.036	(−0.072, 0.000)
	Mahalanobis	−0.046	(−0.083, −0.010)	−0.036	(−0.072, 0.000)
	Propensity	−0.044	(−0.081, −0.008)	−0.046	(−0.082, −0.009)

There are two fundamental remarks here. First, the estimated timeout effect *TE* is very small for all cases, practically irrelevant during a game. Remember that the timeout effect *TE* defined in Eq. (2) is the amount by which the team should expect its scoring margin per possession to change if a timeout were to be called by them. It would be rather challenging to find the minimum effect size for which it would be deemed enough to change the teams performance. It would not be a bad idea, however, to consider it to be an effect of at least *one* marginal point. Specially because our assumption is that timeouts have an effect in the short-term window represented by λ. Yet, from our results, take the propensity matching with $\lambda = 4$, which yielded the largest absolute timeout effect equal to -0.059, for instance. This means that if a team calls a timeout, it should expect its scoring margin to decrease 0.059 points per possession in the next 4 possessions, which can be considered negligible in a basketball game. Second, by analyzing the confidence intervals, the number of significant tests is very small given the very large number of examples in each of them. From the 18 tests, only 6 were statistically significant in the $\alpha = 0.001$ level. In fact, these confidence intervals barely include the 0 effect value. Nevertheless, while these tests were statistically significant, they are still negligible effects. Also, we want to point it out that we did not perform any adjustment for multiplicity of tests. Indeed, such adjustment would yield a smaller number of statistically significant tests.

We went further and investigated the *timeout effect TE* for two particular cases: (i) when the last five minutes are excluded and (ii) when only the final five minutes of the games are taken into account. For both cases, we rerun the matching approaches on the new subsets of the data. Before executing the matching approaches, we filtered out from treatment and control groups examples that happened within the last five minutes of the last quarter (the 4th) of each game

Table 3. Additional analysis for excluding or considering only the last 5 min. For each analysis, the same matching approaches were applied using the new subsets of data. The results are very similar to the original analysis considering the whole game.

		Minus Last 5 Min				Only Last 5 Min			
λ	Method	TE_a	99% CI	TE_h	99% CI	TE_a	99% CI	TE_h	99% CI
2	No-balance	−0.029	(−0.082, 0.024)	−0.010	(−0.065, 0.045)	−0.083	(−0.115, −0.052)	−0.033	(−0.067, 0.001)
	Mahalanobis	−0.021	(−0.074, 0.032)	−0.010	(−0.065, 0.045)	−0.091	(−0.122, −0.060)	−0.040	(−0.074, −0.006)
	Propensity	−0.044	(−0.098, 0.009)	−0.012	(−0.067, 0.043)	−0.094	(−0.125, −0.063)	−0.038	(−0.072, −0.004)
4	No-balance	−0.036	(−0.078, 0.006)	−0.030	(−0.073, 0.013)	−0.006	(−0.068, 0.055)	0.038	(−0.029, 0.105)
	Mahalanobis	−0.055	(−0.097, −0.013)	−0.022	(−0.065, 0.020)	−0.019	(−0.081, 0.044)	0.026	(−0.041, 0.093)
	Propensity	−0.057	(−0.099, −0.015)	−0.038	(−0.081, 0.004)	−0.019	(−0.082, 0.043)	0.007	(−0.060, 0.074)
6	No-balance	−0.044	(−0.082, −0.005)	−0.041	(−0.079, −0.003)	−0.046	(−0.315, 0.223)	−0.032	(−0.250, 0.186)
	Mahalanobis	−0.051	(−0.089, −0.012)	−0.040	(−0.078, −0.002)	−0.028	(−0.299, 0.244)	−0.048	(−0.267, 0.172)
	Propensity	−0.046	(−0.085, −0.008)	−0.055	(−0.093, −0.017)	−0.046	(−0.317, 0.224)	−0.048	(−0.265, 0.170)

for (i). In a similar fashion, we filtered out from treatment and control groups examples that happened outside of the last five minutes of each game for (ii). However, because we ended up with fewer sample units available for matching, we included examples extracted from the other NBA seasons for the (ii) case, i.e., the 2014–2015, 2015–2016 and 2017–2018 seasons.

The results from both of these new analysis are in Table 3. For the case of excluding the last five minutes, we can see that we have slightly more statistical significant tests, 8 out of 18. Still, all of them have small effect sizes, making them not practically significant. The largest estimated effect is −0.057, which is found under the same configuration that we found the largest in the original analysis: TE_a for the propensity score matching with $\lambda = 4$. For the case of considering only the last five minutes, the only 5 statistically significant tests were all found with $\lambda = 2$, but again, with very small effect sizes. The largest absolute effect is −0.094 for propensity score matching.

6 Conclusion

In this work we proposed a causality framework to quantify the effect of timeouts on basketball games. For the best of our knowledge, we were the first to resort on the theory of causality to solve this problem. While all previous studies pointed to a positive timeout effect, by applying our causality model on a large dataset of official NBA play-by-play data, we concluded that timeouts have no effect on teams' performance. This is another example of what statisticians call the *regression to the mean* phenomenon. Since most timeouts are called when the opponent team is scoring more frequently, the moments that follow resemble an improvement in the team's performance, but are just the natural game tendency to return to its average state. We have also stratified our analysis by either including only the last five minutes or everything but the last five minutes of all games, but the results pointed to the same conclusion: timeouts have virtually no effect on team's performance.

Acknowledgments. This work is supported by the authors' individual grants from FAPEMIG, CAPES and CNPq.

References

1. Barnett, A.G., Van Der Pols, J.C., Dobson, A.J.: Regression to the mean: what it is and how to deal with it. Int. J. Epidemiol. **34**(1), 215–220 (2004)
2. Bickel, P.J., Doksum, K.A.: Mathematical Statistics: Basic Ideas and Selected Topics. Chapman and Hall/CRC, Volumes I-II (2015)
3. Coffino, M.J.: Odds-On Basketball Coaching: Crafting High-Percentage Strategies for Game Situations. Rowman & Littlefield Publishers, Lanham (2017)
4. Flury, B.K., Riedwyl, H.: Standard distance in univariate and multivariate analysis. Am. Stat. **40**(3), 249–251 (1986)
5. Gómez, M.A., Jiménez, S., Navarro, R., Lago-Penas, C., Sampaio, J.: Effects of coaches' timeouts on basketball teams' offensive and defensive performances according to momentary differences in score and game period. Eur. J. Sport Sci. **11**(5), 303–308 (2011)
6. Greenwell, B., Boehmke, B., Cunningham, J., Developers, G.: GBM: generalized Boosted Regression Models (2019). https://CRAN.R-project.org/package=gbm, r package version 2.1.5
7. Halldorsson, V.: Coaches use of team timeouts in handball: a mixed method analysis. Open Sports Sci. J. **9**(1), 143–152 (2016)
8. Holland, P.W.: Statistics and causal inference. J. Am. Stat. Assoc. **81**(396), 945–960 (1986)
9. Kubatko, J., Oliver, D., Pelton, K., Rosenbaum, D.T.: A starting point for analyzing basketball statistics. J. Quanti. Anal. Sports **3**(3), 1–24 (2007). https://econpapers.repec.org/article/bpjjqsprt/v_3a3_3ay_3a2007_3ai_3a3_3an_3a1.htm
10. Lee, B.K., Lessler, J., Stuart, E.A.: Improving propensity score weighting using machine learning. Stat. Med. **29**(3), 337–346 (2010)
11. Mace, F.C., Lalli, J.S., Shea, M.C., Nevin, J.A.: Behavioral momentum in college basketball. J. Appl. Behav. Anal. **25**(3), 657–663 (1992)
12. Neyman, J.: edited and translated by dorota m. dabrowska and terrence p. speed (1990). On the application of probability theory to agricultural experiments. essay on principles. section 9. Stat. Sci. **5**(4), 465–472 (1923)
13. Normand, S.L.T., et al.: Validating recommendations for coronary angiography following acute myocardial infarction in the elderly: a matched analysis using propensity scores. J. Clin. Epidemiol. **54**(4), 387–398 (2001)
14. Pearl, J.: Causality. Causality: Models, Reasoning, and Inference. Cambridge University Press, Cambridge (2009)
15. Pearl, J., Mackenzie, D.: The Book of Why: The New Science of Cause and Effect. Basic Books, New York (2018)
16. Permutt, S.: The Efficacy of Momentum-Stopping Timeouts on Short-Term Performance in the National Basketball Association. Ph.D. thesis, Haverford College. Department of Economics (2011)
17. Pimentel, S.D.: rcbalance: large, Sparse Optimal Matching with Refined Covariate Balance (2017). https://CRAN.R-project.org/package=rcbalance, r package version 1.8.5
18. Roane, H.S., Kelley, M.E., Trosclair, N.M., Hauer, L.S.: Behavioral momentum in sports: a partial replication with women's basketball. J. Appl. Behav. Anal. **37**(3), 385–390 (2004)
19. Rosenbaum, P.R., Rubin, D.B.: The central role of the propensity score in observational studies for causal effects. Biometrika **70**(1), 41–55 (1983)

20. Saavedra, S., Mukherjee, S., Bagrow, J.P.: Is coaching experience associated with effective use of timeouts in basketball? Sci. Rep. **2**, 676 (2012)
21. Sampaio, J., Lago-Peñas, C., Gómez, M.A.: Brief exploration of short and mid-term timeout effects on basketball scoring according to situational variables. Eur. J. Sport Sci. **13**(1), 25–30 (2013)
22. Siva, J.M., Cornelius, A.E., Finch, L.M.: Psychological momentum and skill performance: a laboratory study. J. Sport Exercise Psychol. **14**(2), 119–133 (1992)
23. Stuart, E.A.: Matching methods for causal inference: a review and a look forward. Stat. Sci. **25**(1), 1 (2010)
24. Zetou, E., Kourtesis, T., Giazitzi, K., Michalopoulou, M.: Management and content analysis of timeout during volleyball games. Int. J. Perform. Anal. Sport **8**(1), 44–55 (2008)

Demo Track

Deep Reinforcement Learning (DRL) for Portfolio Allocation

Eric Benhamou(✉), David Saltiel, Jean Jacques Ohana, Jamal Atif, and Rida Laraki

AI for Alpha, HOMA Capital, LAMSADE Dauphine, Neuilly-sur-Seine, France
eric.benhamou@aiforalpha.com

Abstract. Deep reinforcement learning (DRL) has reached an unprecedent level on complex tasks like game solving (Go [6], StarCraft II [7]), and autonomous driving. However, applications to real financial assets are still largely unexplored and it remains an open question whether DRL can reach super human level. In this demo, we showcase state-of-the-art DRL methods for selecting portfolios according to financial environment, with a final network concatenating three individual networks using layers of convolutions to reduce network's complexity. The multi entries of our network enables capturing dependencies from common financial indicators features like risk aversion, citigroup index surprise, portfolio specific features and previous portfolio allocations. Results on test set show this approach can overperform traditional portfolio optimization methods with results available at our demo website.

Keywords: Deep reinforcement learning · Portfolio selection · Convolutional networks · Index surprise · Risk aversion

1 Introduction

Markovitz Approach: Portfolio optimization and its implied diversification have been instrumental in the development of the asset management industry. The seminal work of [5] paved the way for a rational approach to allocate funds among the possible investment choices, quantifying returns and risk statistically. However, a straight out of the box Markovitz portfolio optimization tends to be unreliable in practice as risk estimations are very sensitive to input selections and leads to very unstable conclusions. In particular, it is well-known that naive equally weighted portfolios often outperform mean-variance optimized portfolios [2], that the latter can produce extreme or non-intuitive weights [1]. These well documented problems do not mean a quantitative approach is necessarily flawed but rather that it has to adapt to its environment and motivates for DRL.

Hence, we cast the portfolio optimization problem as a continuous control program with delayed rewards, using state of the art deep reinforcement learning methods. These methods do not aim at making predictions (like in supervised learning) but rather at learning the optimal policy for portfolio weights within

Y. Dong et al. (Eds.): ECML PKDD 2020, LNAI 12461, pp. 527–531, 2021.
https://doi.org/10.1007/978-3-030-67670-4_32

a dynamic and continuously changing market. This approach has the major advantage to adapt to changing market conditions.

Related Work: The idea of applying DRL to portfolio allocation has recently taking off in the machine learning community with some recent works on crypto currencies [4,8]. Compared to traditional approaches on financial time series, both [8] and [4] found that novel architectures such as Convolutional Neural Network (CNN) tend to perform better for crypto currencies and Chinese stock markets. In this work, we extend their work by integrating common financial states to incorporate in our deep network some common characteristics of financial markets. We also show that it increases the network efficiency to use previous allocations. Although [4] claimed that adding noise to the learning problem should ease model training we found the opposite conclusion. This may be explained by the larger number of features used in our work that should better capture the dynamic nature of financial markets. In this demo, we showcase how state-of-the-art DRL is able to pick the best portfolio allocation out of sample using financial features used by asset managers: risk aversion index, correlation between equities and bonds, Citi economic surprise index. We provide performance out of sample and test various configurations, that are summarized in a demo website.

Contributions: Our contributions are to provide critical insights for the design of this Deep RL problem validated empirically. In particular, we found that

- reward function is criticial. Sharpe ratio reward leads to worse results.
- CNN performs better than LSTM and captures implicit features.
- dependency to previous allocations enhances the model.
- adding noise does not improve the model.

2 Method Used

Network Architecture: Our network (as described in Fig. 1) uses three types of inputs: portfolio returns observed over the last two weeks, one month and one year (network 1); common asset features like correlation between equities and bonds, Citigroup economic surprise and risk aversion indexes (network 2) observed over the last week, as well as the previous portfolio allocation (network 3). We concatenate these 3 networks into a final one using two dense layers and a final softmax one to infer the portfolio weights. Our reward is either the sharpe ratio or the net value of the final portfolio. We also use either convolution layers for the network 1 and 2 (convolution 2D and 1D respectively) or LSTM units. We introduce noise in the training to challenge it but found that this does not improve contrarily to what [4] found. Train dataset is from 01-Jan-2010 to 31-Dec-2017, while test data set ranges from 01-Jan-2018 to 31-Dec-2019. Results of the various trained networks can be visualized in http://www.aiforalpha.com/deeprl/models.html.

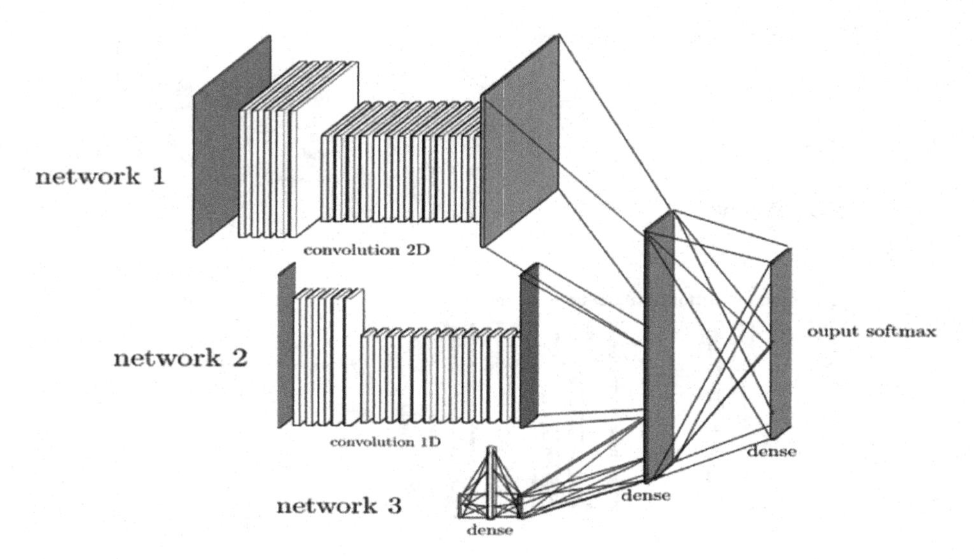

Fig. 1. Best DRL network architecture

DRL Algorithm: To find the optimal action $\pi^*(S_t)$ (the portfolio allocation) according to states S_t (financial information) given a reward R_t (the best net portfolio final performance), we represent the policy as our 3 entries network and use deep policy gradient method with non linear activation (Relu). We use buffer replay to memorize all marginal rewards, so that we can start batch gradient descent once we reached the final time step. We use traditional Adam optimization so that we have the benefit of adaptive gradient descent with root mean square propagation [3].

Results. Performance results are given below in Table 1. Overall, DRL is able to substantially overperform not only traditional methods like static Markovitz or even dynamic Markovitz but also the best portfolio (Naive winner) in terms of net performance and Sharpe ratio, with a final annual net return of 9.49% compared to 4.40% and 5.27% for static and dynamic Markovitz methods. Dynamic Markovitz method consists in computing the Markovitz optimal allocation every 3 months while the static Markovitz method simply uses the Markovitz optimal portfolio computed on train data set. The Naive winner method consists in just selecting the best portfolio over the train data set. It turns out this is also the best portfolio on the test data set (Fig. 2).

3 Target Users and Future Extension

Our system aims at asset managers to present this new approach. We differentiate from current offerings in portfolio optimization as they mostly rely on Markovitz optimization.

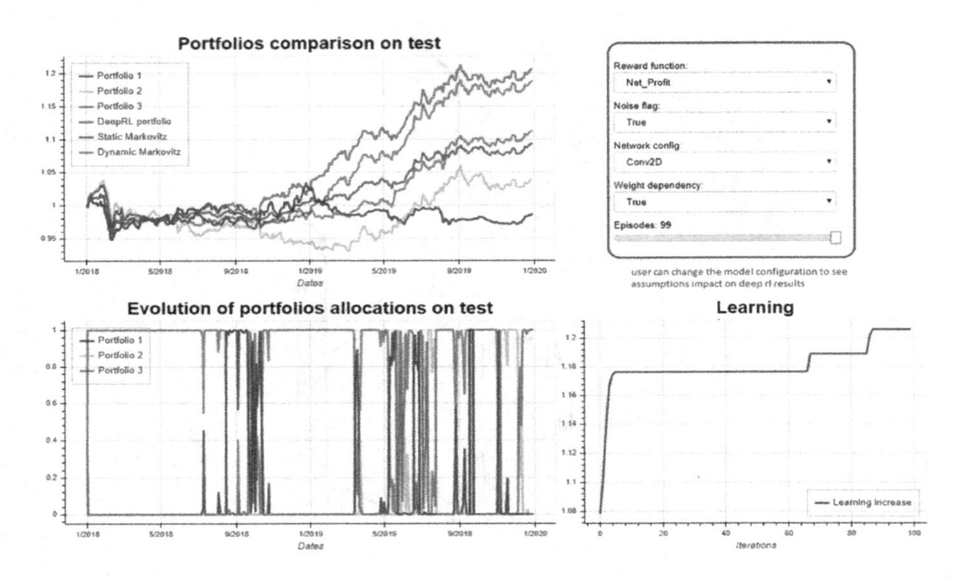

Fig. 2. Deep RL Portfolio Optimisation website

Table 1. Performance results

	Portfolio 1	Portfolio 2	Portfolio 3	Static Markovitz	Dynamic Markovitz	Deep RL	Naive winner
Performance	−0.55%	1.94%	8.47%	4.40%	5.27%	**9.49%**	8.47%
Std dev	4.57%	5.25%	4.91%	**3.55%**	4.22%	4.78%	4.91%
Sharpe ratio	Na	0.37	1.72	1.24	1.25	**1.99**	1.72

In conclusion, we found that DRL overperforms very substantially not only the static but also the dynamic Markovitz method suggesting that DRL captures dependency between the environment states and the reward. Future extensions to this work are to analyze more features and check whether we can enrich our RL problem with more predictive features, like news and market sentiments.

References

1. Black, F., Litterman, R.: Global portfolio optimization. Financ. Anal. **48**(5), 28–43 (1992)
2. DeMiguel, V., et al.: Optimal versus naive diversification: How inefficient is the 1/n portfolio strategy? Rev. Finan. Stud. **22**, 1915–1953 (2009)
3. Kingma, D., Ba, J.: Adam: A method for stochastic optimization (2014)
4. Liang, Z., et al.: Adversarial deep reinforcement learning in portfolio management (2018)
5. Markowitz, H.: Portfolio selection. J. Finance **7**, 77–91 (1952)
6. Silver, D., et al.: Mastering the game of go without human knowledge. Nature **550**(7676), 354–359 (2017)

7. Vinyals, O., et al.: Grandmaster level in starcraft ii using multi-agent reinforcement learning. Nature **575**(7782), 350–354 (2019)
8. Zhengyao, J., et al.: Reinforcement learning framework for the financial portfolio management problem. arXiv (2017)

Automated Quality Assurance for Hand-Held Tools via Embedded Classification and AutoML

Christoffer Löffler[1,2(✉)], Christian Nickel[2], Christopher Sobel[2], Daniel Dzibela[2], Jonathan Braat[2], Benjamin Gruhler[2], Philipp Woller[2], Nicolas Witt[2], and Christopher Mutschler[2,3]

[1] Friedrich-Alexander University Erlangen-Nuremberg (FAU), Erlangen, Germany
christoffer.loeffler@iis.fraunhofer.de
[2] Precise Positioning and Analytics Department, Fraunhofer IIS, Nuremberg, Germany
[3] Department of Statistics, Ludwig-Maximilians-University (LMU), Munich, Germany

Abstract. Despite the ongoing automation of modern production processes manual labor continues to be necessary due to its flexibility and ease of deployment. Automated processes assure quality and traceability, yet manual labor introduces gaps into the quality assurance process. This is not only undesirable but even intolerable in many cases.

We introduce a process monitoring system that uses inertial, magnetic field and audio sensors that we attach as add-ons to hand-held tools. The sensor data is analyzed via embedded classification algorithms and our system directly provides feedback to workers during the execution of work processes. We outline the special requirements caused by vastly different tools and show how to automatically train and deploy new ML models.

1 Introduction

In modern industrial assembly processes all vitally important tasks are already highly automated and hence provide quality assurance by design. To satisfy strict quality requirements, expensive automation for high quantity production steps is standard. However, for many other tasks (such as fixing faceplates or other embellishments) this expensive automation is not worth its cost. Here, highly flexible human labor is much more efficient. Human workers can quickly adapt to new tasks, using different tools, and only need an initial instruction.

This human versatility to new tasks and tools is both the strong point of manual labor but also its deficit. Humans rarely function perfectly according to a task definition, and thus, the quality of the produced items cannot be guaranteed to a certain standard as it can be for an automated assembly. To overcome this quality issue, there are expensive solutions using a four-eye principle of human error detection, i.e., acceptance tests, or even automated camera-based approaches. Such mechanisms are expensive, error-prone, and add testing delays.

Y. Dong et al. (Eds.): ECML PKDD 2020, LNAI 12461, pp. 532–535, 2021.
https://doi.org/10.1007/978-3-030-67670-4_33

In the ideal case we track manual processes that use hand-held (power) tools, e.g., electric screwdrivers, and directly detect whether a task was performed according to its specification. While this allows an automated quality assessment of human labor existing smart tools are expensive and limited in their variety.

For this to work, the assessment of manual work requires a reliable classification of the performed tasks. However, these processes are often not known in advance. Workers may use wildly different tools (e.g. hammers, electric screwdrivers, etc.) and even configurations of such tools may differ. The tasks can be as diverse as fixing parts on a stationary workbench or of a car's underbody in an assembly line. Finally, the working environment can differ, e.g., a metal hut environment influences the sensors differently from a saw mill.

Attaching external sensors to hand-held tools and using machine learning for classification enables the detection of high-quality process parameters [1]. Our proposed Tool Tracking system (https://www.iis.fraunhofer.de/tooltracking) uses an ML-based approach to fill the gap in quality assurance. It can be adapted to different tools and work processes, and can hence be used in combination with existing tools. We use inertial sensors, i.e., accelerometers and gyroscopes, a magnetic field sensor, and a microphone to assess the process. Our AutoML pipeline supports all steps starting from data collection for new tasks, over automatic training up to the deployment on the target sensor module.

2 System Architecture

The sensor module is attached to the current hand-held tool via an easily removable fixture that can be used on many tools. If required by a task an optional (optical or radio-based) positioning system can be added to the system. The sensor module and the user interface are wirelessly connected to a back-end server, hosting a service architecture. The worker interacts with our system via web browser, e.g., using a tablet computer.

Inferencing. Every sensor module analyzes the raw data directly, using efficient C/C++ implementations of the feature extraction and models, and only transmits relevant results. It is unaware of any work process logic, and solely detects

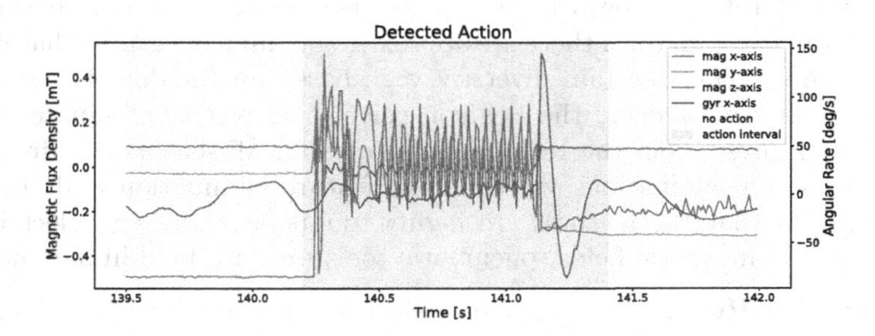

Fig. 1. The electric screwdriver's magnetic field indicates activity, and the gyroscope's *x-axis* spike that the torque is reached (*tightening* action highlighted).

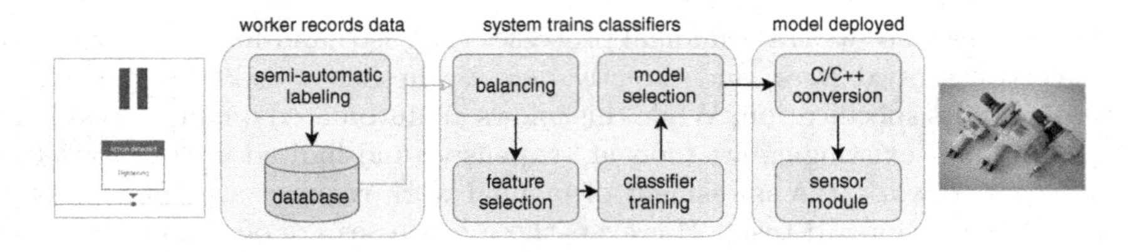

Fig. 2. The AutoML pipeline for training new classifier models.

and classifies actions in a more power-efficient two-step classification approach. A `stage-1` model continuously analyzes the sensor stream and detects the presence of *actions* with very low error rate and power consumption. A `stage-2` model is triggered by `stage-1` and classifies action intervals into distinct *classes*, such as *tightening* or *riveting*. An example of raw data with action detection is given in Fig. 1. The back-end then uses a state machine to verify the work process logic, e.g., whether the actions were performed in the correct order, within the required pitch or roll angle interval, and given a positioning system, even at the correct position at a work piece.

Teaching. The users can define new work processes, e.g., sequences of *tightening*, that rely on previously trained models by recording demonstrations of successive steps, optionally enriching the state machine with position constraints. Moreover, they can teach new tools to the system that, in contrast to new processes, require the automatic training of new classifiers via our AutoML pipeline, see Fig. 2. First, an existing `stage-1` model labels new data with user-defined class labels. Next, we either extract features (similar to `tsfuse` [3]) and deploy the trained `scikit-learn` pipelines [4] to the embedded module (inspired by `sklearn-porter`[1]), or we optimize deep learning models (e.g. fully convolutional networks (FCN), temporal convolutional networks (TCN), LSTMs, etc.) with Optuna [5]. Our C++ library receives the configuration and instantiates it.

Features. The AutoML pipeline includes a set of 32 statistical but also advanced features (for instance, we adapt `libxtract` [2]). We additionally create features encoding the temporal occurrences of the features within an action interval, e.g., a spike in rotation towards the end of the interval indicates reaching the torque of *tightening*. In total there are approx. 1,300 unique features that we then select by sensor type to retain diversity, e.g., based on Random Forest feature importance, and filter using the Pearson correlation test to eliminate strongly correlated features. From the remaining features for all sensors, the framework selects the best combinations via recursive feature elimination and trains the classifier. Note that the features are highly tool-dependent, e.g., electric tools generate strong magnetic fields, pneumatic tools produce loud hissing noises.

Experimental Results. Actions from electric screwdrivers, such as e.g., *un-/tightening*, are detected with 99% accuracy (trained on 20 min of data) with

[1] https://github.com/nok/sklearn-porter.

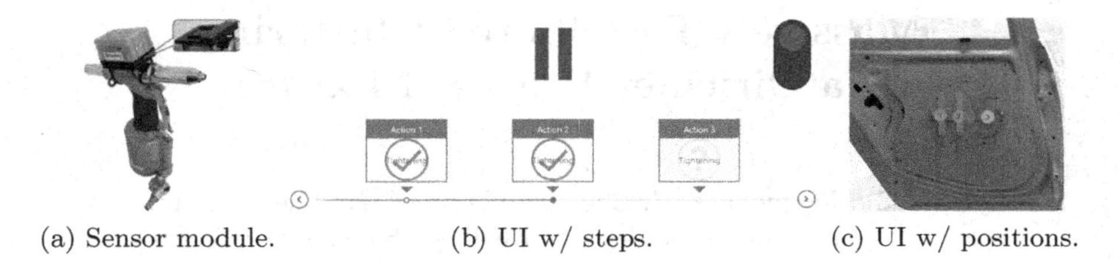

(a) Sensor module. (b) UI w/ steps. (c) UI w/ positions.

Fig. 3. Hardware used for the demonstration.

F1-scores of 93% for decision trees, and 96%, 98% for 94% for `FCN`, `TCN` or `LSTM`. We also detect short pneumatic rivet gun actions with error rates of $2-4\%$ using just 850 samples of 200 ms. However, pneumatic screwdrivers are more difficult to classify: on 2,130 samples `stage-1` achieves $91 - 94\%$ accuracy and `stage-2` only $73 - 80\%$ F1-score.

3 About the Demonstration

We demonstrate our system with two different hand-held tools, i.e., electric screwdriver and pneumatic rivet gun, see Fig. 3a. We record and execute new work processes, but rely on previously trained models, and detect successive un-/tightening or riveting actions on a work piece (Fig. 3c). The data is processed on the back-end server and visualized on a tablet computer (Fig. 3b). The `stage-1` classifier operates on 200 ms windows with 100 Hz sampling for inertial sensors, 150 Hz for the magnetic field sensor, and 8 kHz for the microphone. `Stage-2` operates on the whole action interval, and uses a more complex model. A demo video is available at https://youtu.be/MCpC2i-ZbDs.

Acknowledgements. This work was supported by the Bavarian Ministry of Economic Affairs, Infrastructure, Energy and Technology as part of the Bavarian project Leistungszentrum Elektroniksysteme (LZE) and through the Center for Analytics-Data-Applications (ADA-Center) within the framework of "BAYERN DIGITAL II".

References

1. Dörr, M., Ries, M., Gwosch, T., Matthiesen, S.: Recognizing product application based on integrated consumer grade sensors: a case study with handheld power tools. Proc. CIRP **84**, 798–803 (2019)
2. Bullock, J.: Libxtract: a lightweight library for audio feature extraction. In: International Computer Music Conf, Santiago, Chile (2007)
3. De Brabandere, A., Robberechts, P., De Beéck, T., Davis, J.: Automating feature construction for multi-view time series data. In: ECML-PKDD Workshop on Automating Data Science, pp. 1–16, Würzburg, Germany (2019)
4. Pedregosa, F., et al.: Scikit-learn: machine learning in Python. J. Mach. Learn. Res. **12**, 2825–2830 (2011)
5. Akiba, T., Sano, S., Yanase, T., Ohta, T., Koyama, M.: Optuna: a next-generation hyperparameter optimization framework. In: Proceedings of KDD (2019)

Massively Distributed Clustering via Dirichlet Process Mixture

Khadidja Meguelati[1(✉)], Benedicte Fontez[2], Nadine Hilgert[2], Florent Masseglia[1], and Isabelle Sanchez[2]

[1] Inria, University of Montpellier, CNRS, LIRMM, Montpellier, France
{khadidja.meguelati,florent.masseglia}@inria.fr
[2] MISTEA, University Montpellier, Montpellier SupAgro, INRAE, Montpellier, France
benedicte.fontez@supagro.fr, {nadine.hilgert,isabelle.sanchez}@inrae.fr

Abstract. Dirichlet Process Mixture (DPM) is a model used for multivariate clustering with the advantage of discovering the number of clusters automatically and offering favorable characteristics, but with prohibitive response times, which makes centralized DPM approaches inefficient. We propose a demonstration of two parallel clustering solutions : i) DC-DPM that gracefully scales to millions of data points while remaining DPM compliant, which is the challenge of distributing this process, ii) HD4C that addresses the curse of dimensionality by performing a distributed DPM clustering of high dimensional data such as time series or hyperspectral data.

Keywords: Gaussian random process · Dirichlet process mixture model · Clustering · Parallelism · Reproducing kernel hilbert space

1 Introduction

Clustering is the task of grouping similar data into the same cluster and separating dissimilar data in different clusters. It is a data mining technique intensively used for data analytics, with many applications. Clustering may be used for identification in the new challenge of digital agriculture or high throughput plant phenotyping, as illustrated in the demonstration. One of the main difficulties, for clustering, is the fact that we don't know, in advance, the number of clusters to be discovered. The Dirichlet Process Mixture (DPM) approach allows estimating that number and assigning observations to clusters, in the same process [1].

However, DPM is highly time consuming. Consequently, several attempts have been done to make it distributed but the resulting approaches usually suffer from convergence issues (imbalanced data distribution on computing nodes) [4] or do not fully benefit from DPM properties [3]. Furthermore, making DPM parallel is not straightforward since it must compare each record to the set of existing clusters, a highly repeated number of times. That impairs the global performances of the approach in parallel, since comparing all the records to

Y. Dong et al. (Eds.): ECML PKDD 2020, LNAI 12461, pp. 536–540, 2021.
https://doi.org/10.1007/978-3-030-67670-4_34

all the clusters would call for a high number of communications and make the process impracticable.

In this demonstration, we show how our solutions combine the advantages of DPM (clustering quality) and of distributed computing (fast response time on large datasets). The challenge is to build a consistent view of the clusters, despite a necessary split discovery mechanism. Obviously, such a mechanism need to guarantee low amounts of communications. This is one of the keys in distributed data science systems supporting algorithms often originally designed for centralized environments. Our solutions rely on sufficient statistics, by means of synchronization between machines in terms of proposed clusters at each step [1,2]. We provide an interactive demonstration of these previous results. This demonstration allows the user to check the results on various datasets and to evaluate the impact of parameter choices.

2 Distributed Clustering via Dirichlet Process Mixture

DC-DPM [1] presents a parallel clustering solution that takes advantage of the computing power of distributed systems by using parallel frameworks such as Spark.

The main novelty of DC-DPM is to propose a model and its estimation at the master level by exploiting the sufficient statistics from the workers, in a DPM compliant approach. The challenge of using sufficient statistics, in a distributed environment, is to remain in the DPM approach at all steps, including the synchronization between the worker and master nodes. Our approach approximates the DPM model even at the master level when local data is replaced by sufficient statistics between iterations.

In our work, the distributed DPM allows each node to have a view on the local results of all the other nodes, while avoiding exhaustive data exchanges. This is made by adding weights which represent proportions of observations from all clusters evaluated on the whole dataset. These weights are updated at the master level.

The workflow of DC-DPM is illustrated in Fig. 1. It consists in identifying local clusters on the workers and synchronizing these clusters on the master. These clusters are then communicated as a basis among workers for local clustering consistency. By iterating this process we seek global consistency of DPM in a distributed environment and obtain DPM compliant results as shown in this demonstration and detailed in Sect. 5.

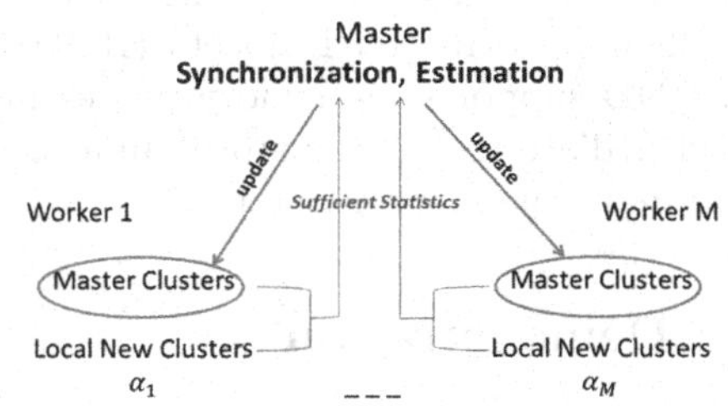

Fig. 1. Workflow of the DC-DPM approach

3 High Dimensional Data Distributed Dirichlet Clustering

HD4C [2] presents a parallel clustering approach adapted for high dimensional data. Actually, DC-DPM is a solution proposed to this issue when data is multivariate. This solution is based on a distributed DPM. In the case of high dimensional data or signals (infinite dimension), matrix computation is no more feasible (e.g., no inverse matrix, no matrix product). We need to replace a matrix product by an inner product in an adequate space of functions and to find the adequate measure. This inner product is mandatory to compute the likelihood and the posterior. To do that, HD4C [2] uses the properties of the Reproducible Kernel Hilbert Spaces (RKHS). We assume that the random variable takes its values in a space of infinite dimension. Therefore, high dimensional data will be seen as trajectories of a random process. Our work focuses on Gaussian random process because of its ability to avoid simple parametric assumptions and integrate a lot of structure. For implementation, data are defined as an autocorrelated Gaussian process called Ornstein-Uhlenbeck.

4 DC-DPM and HD4C in Spark

We implemented both DC-DPM and HD4C in Spark and made the code available for download and test at https://github.com/khadidjaM. To the best of our knowledge, this is the first time that a complete parallel solution for high dimensional data clustering is demonstrated, showing significant performance improvement over the state of the art and existing centralized solutions. Figure 2 illustrates the basic architecture of our solution, which is written in Scala. The main components of DC-DPM and HD4C are developed within Spark and deployed on top of Hadoop Distributed File System (HDFS) in order to efficiently read input data, as well as to store final results, and thus to overcome the bottleneck of centralized data storing.

The main feature of Spark is its distributed memory abstraction, called Resilient Distributed Datasets (RDD) and parallel operations used to handle it. RDD supports in-memory processing computation. As shown in Fig. 2, the intermediate results are stored in a distributed memory instead of disk storage and make the system faster.

5 Demonstration

Parameters and Datasets. In this demonstration, we can choose the type of data according to the dimension: multivariate or high dimensional data. For both, the following parameters can be configured by the user: the dataset, the display level (individuals or cluster centers) and the number of displayed points.

Two examples from digital agriculture were used in this demonstration: *i)* the herd monitoring where animal activity is monitored using a collar-mounted

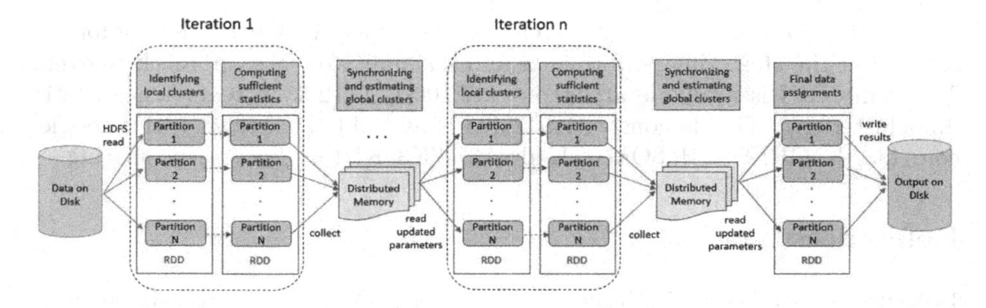

Fig. 2. Architecture of DC-DPM and HD4C in Spark

accelerometer (high dimensional dataset), *ii)* clustering of image pixels to define plant labels as pre-treatment for plant phenotyping (multivariate dataset).

Synthetic data were generated using a two-steps principle. First, we generated cluster centers according to a multivariate normal distribution (multivariate data) or according to some polynomials (high dimensional data). Then, for each center, we generated data by using a multivariate normal distribution or a gaussian process.

The last dataset corresponds to more than 4K spectrum of 680 dimensions representing a protein rate measured on 10 different products.

Scenarios. The demonstration GUI is divided into three tabs. First, the "Method" tab page explains the clustering approaches proposed for each type of data: DC-DPM for multivariate data and HD4C for high dimensional data. Next, in the other tabs, the GUI enables the user to use drop-downs to set the parameters (see Fig. 3 and 4). On the right, a plot of the clustering assignments for both the ground-truth and the chosen method: randomly

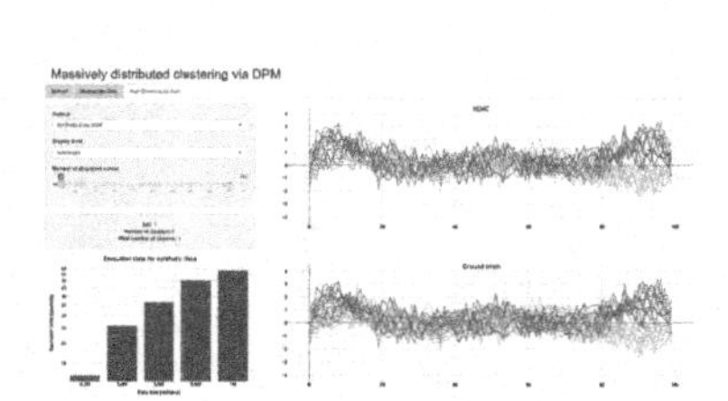

Fig. 3. Users can vary the input datasets, display level and number of displayed data. The demo will report back information about the clustering performances and plot the clustering assignments for both the ground-truth and the chosen method.

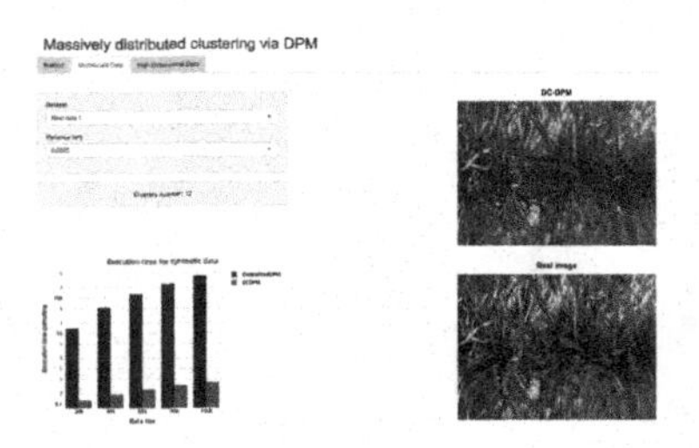

Fig. 4. The specific case of image pixel clustering. Users can vary the "variance" parameter σ^2 to control the number of clusters. Better view on: http://147.100.179.112:3838/team/kmeguelati/dpmclustering/.

selected individuals or all cluster centers are displayed. On the left, information about the clustering performance and scalability of each approach is given. The demo GUI is available at: http://147.100.179.112:3838/team/kmeguelati/dpmclustering/. The demonstration video is available at: https://drive.google.com/file/d/1GHLF5csHk8Oa7PZK4dwA3RTK35KNIbne/view?usp=sharing

References

1. Meguelati, K., Fontez, B., Hilgert, N., Masseglia, F.: Dirichlet process mixture models made scalable and effective by means of massive distribution. In: Proceedings of the 34th ACM/SIGAPP Symposium on Applied Computing, Limassol, Cyprus, pp. 502–509 (2019). https://doi.org/10.1145/3297280.3297327
2. Meguelati, K., Fontez, B., Hilgert, N., Masseglia, F.: High dimensional data clustering by means of distributed dirichlet process mixture models. In: IEEE International Conference on Big Data (IEEE BigData). Los-Angeles, United States, pp. 890–899. IEEE (2019)
3. Wang, R., Lin, D.: Scalable estimation of dirichlet process mixture models on distributed data. In: Proceedings of the 26th International Joint Conference on Artificial Intelligence. IJCAI 2017, AAAI Press, pp. 4632–4639 (2017)
4. Williamson, S., Dubey, A., Xing, E.: Parallel markov chain monte carlo for nonparametric mixture models. In: International Conference on Machine Learning, pp. 98–106 (2013)

AutoRec: A Comprehensive Platform for Building Effective and Explainable Recommender Models

Qing Cui(✉), Qitao Shi, Hao Qian, Caizhi Tang, Xixi Li, Yiming Zhao, Tao Jiang, Longfei Li, and Jun Zhou(✉)

Ant Financial Services Group, Z Space, No. 556 Xixi Road, Hangzhou, China
{cuiqing.cq,qianhao.qh}@alibaba-inc.com
{qitao.sqt,caizhi.tcz,muxi.lxx,desert.zym,lvshan.jt,longyao.llf, jun.zhoujun}@antfin.com

Abstract. This paper presents a comprehensive platform named AutoRec, which can help developers build effective and explainable recommender models all in one platform. It implements several well-known and state-of-art deep learning models in item recommendation scenarios, a AutoML framework with a package of search algorithms for automatically tuning of hyperparameters, and several instance-level interpretation methods to enable the explainable recommendation. The main advantage of AutoRec is the integration of AutoML and explainable AI abilities into the deep learning based recommender algorithms platform.

Keywords: Recommender system · AutoML · Explainable AI

1 Introduction

For recommender system, deep learning models have many advantages over the traditional models especially for the ranking task, including the high expression capacity, avoiding of extensive feature engineering and flexibility to handle sequential behavior features [6,19]. As more and more deep learning models are emerging, we have implemented several widely used deep learning models along with the core component layers as a modular and extendable package. Users can conveniently use the built-in models, build custom models with the built-in layers, or implement custom layers and models by inheriting the corresponding base class.

Despite the effectiveness, deep learning models often require an intensive hyperparameter tuning to approach a good local optimum. We integrate the hyperparameter tuning function into our platform to solve this problem. Except the straightforward method of grid search and random search, we have implemented several hyperparameter tuning methods, including population-based methods and model-based methods. We can also leverage these methods to automatically select the best model or layer to further reduce the labour cost. With

Y. Dong et al. (Eds.): ECML PKDD 2020, LNAI 12461, pp. 541–545, 2021.
https://doi.org/10.1007/978-3-030-67670-4_35

these AutoML abilities, the productivity can be extremely increased, so that the small recommender senario can also have a good enough model with much less effort.

Another shortcoming of deep learning models is the essence of black-box model, which is difficult to understand even for domain expert. With more applications of artificial intelligence and people's doubts about the fairness of artificial intelligence, there are already some regulations requiring that decisions based on artificial intelligence must be explained [1]. Considering the increasing demand for interpretability, we have integrated a class of attribution methods into our platform, so that the deep learning models trained through this platform have decision-level interpretability.

To sum up, we develop a comprehensive system which can build a state-of-art deep learning model for recommendation with little effort, tune the hyperparameters and network architecture automatically, and produce an decision level explanation while predicting. All of these can be done with a web-based GUI, which is easy to use for developers with less experience. We also provided a shell interface with an configuration to provide more flexibility.

Although there are already deep learning based model package [2], auto tuning and search toolkit [3], and gradient-based attribution method framework [4] respectively, these systems are seperated and it is difficult to leverage all these abilities in one platform. Microsoft Recommender [8] provides a recommender module along with tuning utilities, but the implemented algorithms mainly focus on collaborative filtering methods, and it does not provide the instance-level explanation. As far as we know, we are the first to integrate both AutoML (automatic machine learning) and XAI (explainable artificial intelligence) ability into a deep learning recommender algorithm library, which can together build an effective, efficient and trustful recommender system. Our platform is useful for those who want to automatically build a deep learning recommender model with instance-level explanation.

2 Overview of the System

Our system is composed of three parts, recommender algorithms, AutoML utilities, XAI utilities. The whole system is developed with Python and uses Tensorflow as its back-end computation engine.

Recommender Algorithms. We have implemented several wide-used recommender algorithms in recent years, and the full list is in Table 1. We leverage a three-level design, Layer, Model and Estimator which are all the standard interfaces of Tensorflow, to guarantee the extendibility and flexibility. Users can directly use the built-in models through Estimator interface, or build custom models by combining the existing core layers. The combination can also be done through a web-based GUI by dragging and dropping the layer components.

AutoML Utilities. We have also developed an AutoML framework, and integrated it into our system. The AutoML framework is composed of RESTful

Table 1. The implemented recommender models and layers

Models	Layers
Wide and Deep [7]	DNNLayer
DeepFM [9]	FMLayer, DNNLayer
Deep and Cross Network [17]	CrossLayer, DNNLayer
xDeepFM [12]	CINLayer, DNNLayer
Deep Interest Network [19]	FuseLayer
AutoInt [15]	AutoIntLayer
Neural FM [10]	FMLayer, DNNLayer
Bert4Seq [16]	MultiHeadAttentionLayer, SelfAttentionLayer, LayerNorm PositionWiseFeedForwardLayer, TransformerLayer
MultiCNN [11]	MultiCNNLayer, SimpleAttnLayer

service, StudyRunner, SearchAlgorithm and EarlyStopping components. The RESTful service is responsible for creating study, collecting evaluation metrics and managing status information. StudyRunner calls SearchAlgorithm component to get the candidate trials, then executes the trials, and get the evaluation metric as feedback. The SearchAlgorithm component updates the search space with regarding to the feedback, and recommend the next trial to StudyRunner. The EarlyStopping component is used to stop a trial early when certain criteria is reached, like the minimal accuracy. The algorithmic core of the AutoML framework is the SearchAlgorithm which determines the efficiency and effectiveness of the search procedure. We have implemented several kinds of search algorithms in this framework, including Grid Search, Random Search, Genetic Search, Differential Evolution, Bayesian Optimization and RACOS [18].

The AutoML framework acts like a wapper outside the training and evaluating processes of the recommender model. The hyperparameters of neural network as well as the type of layer can all be optimized by this framework.

XAI Utilities. We implemented an interface Interpetable which has a method *get_interpretation*, and a base class BaseModel that inherits the Interpretable interface. The interpretation methods are implemented as Interpreters that inherit the BaseModel class, and the gradient is calculated by registering a custom gradient operator in Tensorflow. As long as the recommender model inherits the Interpreter class, the model has the ability to produce the importance score along with the predicted output in the inference stage. These importance scores estimate how significantly the predicted out would change when the features change. Using these importance scores, we can further transform the scores into some understandable descriptions with some pre-defined rules on original features. Currently, we implemented three gradient-based instance-level interpretation methods, including Gradient * Input [14], layer-wise Relevance Propagation (LRP) [5] and DeepLIFT [13].

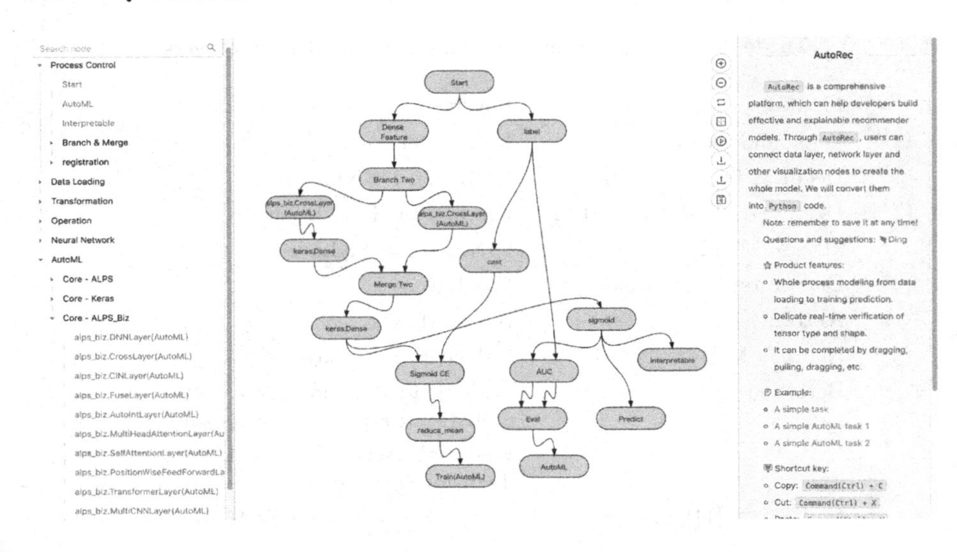

Fig. 1. The web-based GUI for AutoRec system

3 System Demonstration

The user can start a task through the command line providing a json configuration file. Checking the task status and trial status can also be done by the command line, and it will return a link of the dashboard which presents the progress of each trials and the links to more detailed logs. Beyond this, we also developed a web-based GUI to ease the effort to learn the system (see demo https://youtu.be/EF6suiajcU0). As shown in Fig. 1, we can choose the hyperparameter tuning algorithm and set the configurations of each component on the right panel in the page. The configuration panel will change according to the selected component in the main canvas. The recommender model can be constructed by dragging from the left panel and dropping the components in the main canvas, including the built-in layers, AutoML components and interpreter components. According to the configuration user selected in the GUI, the system will generate a complete runable model definition code and configuration, which will extremely increase the developing productivity and avoid the obvious mistakes.

References

1. Art. 22 gdpr automated individual decision-making, including profiling (2018). https://gdpr.eu/article-22-automated-individual-decision-making/
2. Deepctr (2019). https://deepctr-doc.readthedocs.io/en/latest/index.html
3. Neural network intelligence (2019). https://nni.readthedocs.io/en/latest/
4. Ancona, M., et al.: Towards better understanding of gradient-based attribution methods for deep neural networks. In: ICML (2018). https://github.com/marcoancona/DeepExplain

5. Bach, S., Binder, A., Montavon, G., Klauschen, F., Müller, K.R., Samek, W.: On pixel-wise explanations for non-linear classifier decisions by layer-wise relevance propagation. PloS One **10**(7), e0130140 (2015)
6. Chen, C., Zhao, P., Li, L., Zhou, J., Li, X., Qiu, M.: Locally connected deep learning framework for industrial-scale recommender systems. In: Proceedings of the 26th International Conference on World Wide Web Companion, WWW 2017 Companion, pp. 769–770 (2017). https://doi.org/10.1145/3041021.3054227
7. Cheng, H.T., et al.: Wide & deep learning for recommender systems. In: Proceedings of the 1st Workshop on Deep Learning for Recommender Systems, pp. 7–10. ACM (2016)
8. Graham, S., Min, J.K., Wu, T.: Microsoft recommenders: tools to accelerate developing recommender systems. In: RecSys 2019, pp. 542–543 (2019)
9. Guo, H., Tang, R., Ye, Y., Li, Z., He, X.: Deepfm: a factorization-machine based neural network for ctr prediction. arXiv preprint arXiv:1703.04247 (2017)
10. He, X., Chua, T.S.: Neural factorization machines for sparse predictive analytics. In: Proceedings of the 40th International ACM SIGIR Conference on Research and Development in Information Retrieval, pp. 355–364. ACM (2017)
11. Kim, Y.: Convolutional neural networks for sentence classification. In: Proceedings of the 2014 Conference on Empirical Methods in Natural Language Processing (EMNLP), pp. 1746–1751. Doha, Qatar, October 2014. https://doi.org/10.3115/v1/D14-1181
12. Lian, J., Zhou, X., Zhang, F., Chen, Z., Xie, X., Sun, G.: xdeepfm: combining explicit and implicit feature interactions for recommender systems. In: Proceedings of the 24th ACM SIGKDD International Conference on Knowledge Discovery & Data Mining, pp. 1754–1763. ACM (2018)
13. Shrikumar, A., Greenside, P., Kundaje, A.: Learning important features through propagating activation differences. In: Proceedings of the 34th International Conference on Machine Learning, ICML 2017, vol. 70, pp. 3145–3153. JMLR.org (2017)
14. Shrikumar, A., Greenside, P., Shcherbina, A., Kundaje, A.: Not just a black box: Learning important features through propagating activation differences. CoRR (2016). http://arxiv.org/abs/1605.01713
15. Song, W., et al.: Autoint: automatic feature interaction learning via self-attentive neural networks. In: Proceedings of the 28th ACM International Conference on Information and Knowledge Management, CIKM 2019, pp. 1161–1170. New York, NY, USA (2019). https://doi.org/10.1145/3357384.3357925
16. Sun, F., et al.: Bert4rec: sequential recommendation with bidirectional encoder representations from transformer. In: Proceedings of the 28th ACM International Conference on Information and Knowledge Management, CIKM 2019, pp. 1441–1450, New York, NY, USA (2019). https://doi.org/10.1145/3357384.3357895
17. Wang, R., Fu, B., Fu, G., Wang, M.: Deep & cross network for ad click predictions. In: Proceedings of the ADKDD 2017, p. 12. ACM (2017)
18. Yu, Y., Qian, H., Hu, Y.Q.: Derivative-free optimization via classification. In: AAAI 2016 (2016)
19. Zhou, G., et al.: Deep interest network for click-through rate prediction. In: Proceedings of the 24th ACM SIGKDD International Conference on Knowledge Discovery & Data Mining, pp. 1059–1068. ACM (2018)

multi-imbalance: Open Source Python Toolbox for Multi-class Imbalanced Classification

Jacek Grycza, Damian Horna, Hanna Klimczak, Mateusz Lango(✉), Kamil Pluciński, and Jerzy Stefanowski

Poznan University of Technology, Institute of Computer Science, Poznań, Poland
jacekgrycza.jg@gmail.com, horna.damian@gmail.com, hanna.klimczak@gmail.com, mlango@cs.put.edu.pl, kamil.plucinski97@gmail.com, jstefanowski@cs.put.poznan.pl

Abstract. This paper presents multi-imbalance, an open-source Python library, which equips the constantly growing Python community with appropriate tools to deal with multi-class imbalanced problems. It follows the code conventions of sklearn package. It provides implementations of state-of-the-art binary decomposition techniques, ensembles, as well as both novel and classic re-sampling approaches for multi-class imbalanced classification. For demonstration and documentation, consult the project web page: www.cs.put.poznan.pl/mlango/multiimbalance.php.

Keywords: Imbalanced data · Multi-class ensembles · Re-sampling

1 Introduction

Learning classifiers from imbalanced data concerns problems where it is necessary to accurately recognize instances of the minority class. While many methods for improving classifiers learned from imbalanced data have already been proposed (see [2] for review), there are still relevant challenges that need to be addressed. One of them is the ability to deal with multiple minority classes.

Learning multiple imbalanced classes is much more challenging than handling the binary case, as it is necessary to deal with more complex decision boundaries and class distributions [4]. Currently, proposed methods include either the decomposition of multi-class problems to special binary subtasks later aggregated within the ensemble framework or modifications of re-sampling in the pre-processing stage.

Another practical restriction to work with these methods is the lack of readily available software implementations that could support the comparative study of many algorithms or applying them to solve various multi-class problems. The most popular and representative packages for imbalanced data, such as the subpart of KEEL library or WEKA in Java, imbalanced-learn in Python and some

Y. Dong et al. (Eds.): ECML PKDD 2020, LNAI 12461, pp. 546–549, 2021.
https://doi.org/10.1007/978-3-030-67670-4_36

R libraries as IRIC, focus on binary imbalanced problems only. Although the author's implementations of single algorithms might sometimes be found on the Internet, there is only one recently proposed Matlab toolbox which includes several tree and ensemble algorithms for multi-class imbalanced learning [7].

In our opinion, there is still a need for an open-source framework implemented in a more popular programming language, covering other pre-processing and specialized ensemble methods. We have decided to choose Python due to its growing popularity in academic and industrial audiences. Various Python libraries, such as pandas, sklearn, numpy, or scipy, are intensively exploited in many machine learning sub-fields in recent years. Following the design patterns and programming interface of the popular sklearn project, we developed a novel `multi-imbalance` package, which is briefly presented in the next sections. The library is easily extendable and can be quickly integrated into projects relying on previously enumerated popular Python libraries, hopefully creating a good platform for more reproducible research in the imbalanced data community.

2 Overview of the Package

The package implements state-of-the-art approaches for multi-class imbalanced problems, which are divided into three general categories: (1) binary decomposition approaches, (2) specialized pre-processing, and (3) ensembles.

The approaches in the first category include decompositions with such ensembles as one-vs-one (OVO), one-vs-all (OVA), and Error-Correcting Output Codes (ECOC) with various encoding strategies (dense, sparse and complete [5]). Inside each of those frameworks, one can employ additional re-sampling strategies. The user can apply re-sampling methods provided by our package or other Python libraries that implement sklearn interface (including imbalanced-learn). A user's implementation of their own re-sampling method can also be applied.

The package also contains five re-sampling methods, in particular often applied approaches, i.e., Global-CS method that re-samples the data randomly, Mahalanobis Distance Oversampling (MDO), and Static-SMOTE, which create new artificial examples during pre-processing. Moreover, two recently proposed, more advanced pre-processing methods, namely SPIDER3 and SOUP (Similarity-based Oversampling and Undersampling Preprocessing) [4], are also included. SPIDER3 combines instances relabeling and local re-sampling, capturing interrelations between minority classes with misclassification costs [6]. A different approach to class interrelations is incorporated in SOUP method. It models them with class similarities and example difficulty levels [4].

Furthermore, the package contains a top-performing specialized ensemble model for multi-class imbalanced data: Multi-class Roughly Balanced Bagging (MRBB), which is our multi-class version of RBB proposed by Hido and Kashima for binary data [3], and an integration of SOUP re-sampling into bagging.

The detailed description of nearly all of these methods can be found in Chap. 8 of the recent book [2]. We also refer the reader to the web page of our `multi-imbalance` package, where we provide more detailed descriptions, demo movie, and more references.

3 Project Design

The package is based on common Python libraries for scientific computing, i.e. numpy, scipy, scikit-learn and imbalanced-learn. We followed best practices during development of our package when it comes to both architectural principles which were inspired by scikit-learn package and PEP8 coding conventions. The package consists of two main subpackages - ensemble and resampling. Each of the resampling methods we implemented can be used by calling fit_transform method. Furthermore, each of the ensembles can be used by calling fit to train and predict to perform the actual prediction on multi-class imbalanced data.

The project is developed collaboratively with the use of GitHub. Individual developers can contribute to our package by issuing a pull request. Each commit is automatically checked by the Travis CI tool with unit tests, which covers 97% of the codebase. The documentation is maintained with the use of sphinx library, hosted on https://multi-imbalance.readthedocs.io/en/latest/ webpage and continuously updated with each new release of our software package. The package is distributed under the permissive MIT licence and can be easily installed with Python package installer (pip).

4 An Example of Use

We will show a brief example[1] of the use of multi-imbalance for a selected multi-class imbalanced dataset. We choose glass dataset from UCI repository which is often used in experimental studies regarding multi-class imbalance [2–4] since it poses interesting challenges for standard learning classifiers (e.g. class overlapping). It is a real-life dataset containing 214 examples with 9 numeric attributes assigned to six classes with high cardinality differences ($IR > 7$).

```
1  from sklearn.tree import DecisionTreeClassifier
2  from multi_imbalance.resampling.mdo import MDO
3  from multi_imbalance.utils.data import load_arff_dataset
4
5  # Load dataset from e.g. *.arff file
6  X, y = load_arff_dataset(f'glass.arff')
7
8  # Perform MDO resampling
9  mdo = MDO(maj_int_min={ #MDO parameters:
10           'maj': [0, 1],  # indices of majority classes
11           'min': [2, 3, 4, 5]  # indices of minority classes
12      }) # (other parameters with default values)
13 X_train_res, y_train_res = mdo.fit_transform(X, y)
14 # Train a standard sklearn classifier
15 clf = DecisionTreeClassifier()
16 clf.fit(X_train_res, y_train_res)
```

[1] An extended version of this example is available at multi-imbalance Github repository https://github.com/damian-horna/multi-imbalance/blob/master/examples/use_case.ipynb.

The glass dataset is provided in an *.arff file, so our library automatically performs the required preprocessing for sklearn compatibility (line 6). As standard DecisionTreeClassifier achieves on glass G-mean of 0.22, we run MDO sampling [1] for improving this result. We instantiate the MDO object providing all required parameters of the method (lines 9–12) and later apply its `fit_transform` method producing a new resampled dataset (line 13). This dataset can be easily used for training any of sklearn's classifiers (e.g. decision tree, lines 16–17). The resulting tree classifier achieves 0.70 G-mean value showing the advantage of using the MDO resampling for this dataset.

5 Conclusions

This paper presents multi-imbalance package: a novel Python library dedicated to multi-class imbalanced data. It provides the first Python implementations of several state-of-the-art imbalanced learning methods, making them available to the growing community of machine learning practitioners using this programming language. The package is easy to install, extend, and use as it is fully integrated into the Python ecosystem. It allows straightforward comparisons of methods' performances, making an essential step towards more reproducible research.

Acknowledgements. This research was supported by PUT Statutory Funds.

References

1. Abdi, L., Hashemi, S.: To combat multi-class imbalanced problems by means of over-sampling techniques. IEEE Trans. Knowl. Data Eng. **28**, 238–251 (2016)
2. Fernández, A., García, S., Galar, M., Prati, R.C., Krawczyk, B., Herrera, F.: Learning from Imbalanced Data Sets. Springer, Cham (2018). https://doi.org/10.1007/978-3-319-98074-4
3. Hido, S., Kashima, H.: Roughly balanced bagging for imbalance data. In: Proceedings of the SIAM International Conference on Data Mining, pp. 143–152 (2008)
4. Janicka, M., Lango, M., Stefanowski, J.: Using information on class interrelations to improve classification of multiclass imbalanced data: a new resampling algorithm. Int. J. Appl. Math. Comput. Sci. **29**(4), 769–781 (2019)
5. Kuncheva, L.: Combining Pattern Classifiers: Methods and Algorithms. Wiley, Hoboken (2004)
6. Wojciechowski, S., Wilk, S., Stefanowski, J.: An algorithm for selective preprocessing of multi-class imbalanced data. In: Proceedings of International Conference on Computer Recognition Systems, pp. 238–247 (2017)
7. Zhang, C., et al.: Multi-imbalance: an open-source software for multi-class imbalance learning. Knowl. Based Syst **174**, 137–143 (2019)

VisualSynth: Democratizing Data Science in Spreadsheets

Clément Gautrais[1(✉)], Yann Dauxais[1], Samuel Kolb[1], Archit Jain[1], Mohit Kumar[1], Stefano Teso[2], Elia Van Wolputte[1], Gust Verbruggen[1], and Luc De Raedt[1]

[1] Department of Computer Science, KU Leuven, Leuven, Belgium
{clement.gautrais,yann.dauxais,samuel.kolb,arcchit.jain,mohit.kumar, elia.wolputte,gust.verbruggen,luc.raedt}@cs.kuleuven.be
[2] University of Trento, Trento, Italy
stefano.teso@unitn.it

Abstract. We introduce VisualSynth, a framework that wants to democratize data science by enabling naive end-users to specify the data science tasks that match their needs. In VisualSynth, the user and the spreadsheet application interact by highlighting parts of the data using *colors*. The colors define a partial specification of a data science task (such as data wrangling or clustering), which is then completed and solved automatically using artificial intelligence techniques. The user can interactively refine the specification until she is satisfied with the result.

1 Introduction

Most data science techniques are beyond the reach of typical end-users. For instance, spreadsheets – a key tool in data management – are analyzed using formulas, which most users can neither write nor understand [2]. Hence, they are often processed manually, preventing scalability and increasing the chance of mistakes. The question we ask is then: *is it possible to enable non-technical end-users to specify and solve data science tasks that match their needs*? Our key observation is that the data science task of interest can often be identified by looking at the data and with a little help from the user. This observation enables us to design an interactive framework, VisualSynth[1], in which the machine and the user collaborate towards designing and solving a data science task. VisualSynth combines two components: a minimal interaction protocol that allows non-technical users to (partially) specify a data science task using color highlighting, and a smart framework for solving the partially specified data science task automatically, based on inductive models.

[1] Demonstration video: https://youtu.be/df6JgHl28Vw.

This work has received funding from the European Research Council (ERC) under the European Union's Horizon 2020 research and innovation programme (grant agreement No [694980] SYNTH: Synthesising Inductive Data Models).

Y. Dong et al. (Eds.): ECML PKDD 2020, LNAI 12461, pp. 550–554, 2021.
https://doi.org/10.1007/978-3-030-67670-4_37

The intent of this setup is to combine the respective strengths of end-users, namely their domain knowledge, and of computers, namely their ability to quickly carry out large computations. In contrast to other automation approaches, like AutoML [3,9], in VISUALSYNTH the task to be performed (e.g., classification) is not assumed to be fixed or known. While interacting with spreadsheet users has been studied on a simple machine learning task [8], VISUALSYNTH considers a wide range of data science tasks using a unified interaction system.

2 Interacting Using Colors

Given a spreadsheet, we define a *sketch* to be a set of colors applied to one or more rows, columns, or cells in the spreadsheet (see Fig. 1, left). VISUALSYNTH employs colors to define the type, inputs, outputs and other parameters of a data science task. Taken together, the sketch and the spreadsheet can be mapped onto a concrete *data science task* (e.g., a wrangling task), which can then be solved and whose results (e.g., a set of new tables) can be placed into the original spreadsheet, yielding an extended spreadsheet. This idea is captured in the following schema: $\left.\begin{array}{c}\text{spreadsheet}\\+\\\text{sketch}\end{array}\right\} \rightarrow \begin{array}{c}\text{data science}\\\text{task}\end{array} \rightarrow \left.\begin{array}{c}\text{spreadsheet}\\+\\\text{model}\end{array}\right\} \rightarrow \begin{array}{c}\text{new}\\\text{spreadsheet}\end{array}$ Let us illustrate this schema for wrangling, the task of transforming data in the right format for downstream data science tasks. The user indicates which cells belong to the same row by coloring them using the same color [10]. A *wrangling sketch* is therefore a set of colored cells, where each color defines a partial example of the expected wrangling result and imposes a constraint on the output, i.e., that the partial example should be mapped onto a single row into the target spreadsheet. Figure 1 left presents an example of a wrangling sketch. A commonly used paradigm for data wrangling is programming by example (PBE) [5], in which a language of transformations $\mathcal{L}$ is defined and the wrangler searches for a program $P \in \mathcal{L}$ that maps the input examples to the corresponding outputs. In the context of VISUALSYNTH, the task is: **Given** a wrangling sketch and a spreadsheet, **Find** a program that transforms the spreadsheet so that cells with the same color end up in the same row, and no row can contain cells with multiple colors. In Fig. 1 left, the task is: **Given** the green and orange colorings, a spreadsheet, a language $\mathcal{L}$, **Find** a wrangling program $P \in \mathcal{L}$ such that green cells end up in a single row, and orange cells in another single row. Figure 1 middle shows the output of the wrangling program P. We refer the reader to [4] for a more in-depth presentation of sketches and their uses for different data science tasks.

3 Use Case: Ice Cream Sales Auto-completion

Let us illustrate VISUALSYNTH using an auto-completion task. Imagine that you are a sales manager at an ice cream factory. You have data about past sales and about your shops, as shown in Fig. 1. A first difficulty is that the sales is not nicely formatted. A first task is therefore to wrangle it into a format that is

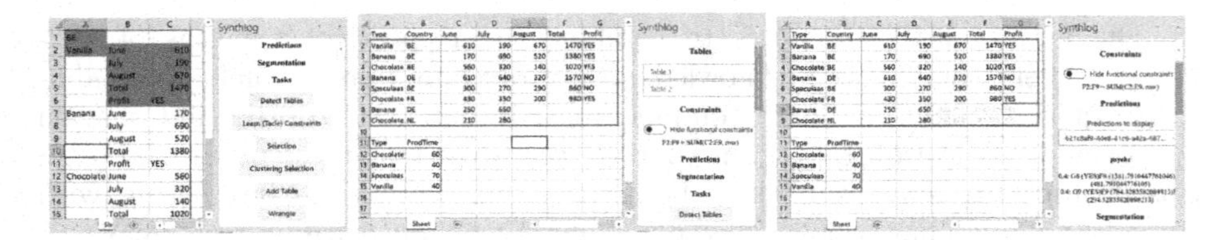

Fig. 1. Left: view of the VisualSynth add-in before performing the wrangling. User hints are represented as colors, that VisualSynth uses to infer a wrangling task (Sect. 2). The right panel displays a button for each suggested task. Middle: VisualSynth add-in after the wrangling execution. Constraints have also been learned by clicking on the associated button. The right panel displays the learned constraints. Right: view of the VisualSynth add-in after autocompletion has been performed. The add-in displays learned constraints, as well as the predictions and their confidence.

more amenable to data analysis. A sketch for wrangling consist in coloring cells that should belong to a single row with a single color. As soon as the user starts putting colors on cells, VisualSynth automatically suggest a wrangling task to the user in the add-in panel. The user can then perform the suggested task by clicking on the associated button in our add-in. Through this interaction, the data wrangling component can produce the table presented in Fig. 1.

As this table contains segments of consecutive numerical values, VisualSynth suggests to learn Excel constraints. By clicking on the associated button, Excel constraints will be detected using TaCLe [6], a spreadsheet constraint learner. The learned constraints are then displayed to the user who can inspect them. In this case the system learned that $Total = June + July + August$.

However, some past sales data are still missing. To determine which shops made a profit you need to obtain an estimate of the missing values. As the table contains missing values, VisualSynth suggests the Auto-completion task. A key feature of Auto-completion is that missing values are inferred using one or more predictive models (often classifiers or regressors [1]) while ensuring that predictions are compatible with the formulas and the constraints detected in the spreadsheet [7]. For example, in Fig. 1 right, the values inferred respect the constraint $Total = June + July + August$. If you click on the associated button, the machine starts suggesting values, which you can then either accept as is or correct them. Corrections trigger a new auto-completion loop, with additional constraints expressing that the user corrected some values.

This use case featured only a few of VisualSynth capabilities but it can handle a number of key data science tasks, namely data wrangling, concept learning, data selection, prediction, clustering, constraint learning and auto-completion. This use case also shows how VisualSynth uses sketches to perform non trivial data science tasks in Excel, a software that is familiar to spreadsheet users.

4 Architecture

VISUALSYNTH is implemented as a Microsoft Excel© add-in (see Fig. 2). The add-in is implemented with React (https://reactjs.org), the server with Node.js© (https://nodejs.org), and the Synth library with Python. Tasks can be implemented in any language, as long as their output correspond to state components.

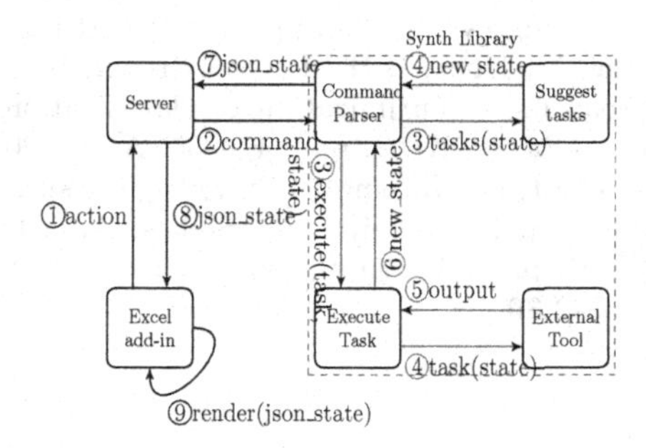

Fig. 2. Architecture of VISUALSYNTH. A state contains the spreadsheet data, the learned models and the interactions with the user: state = data+models+interactions. When an action is performed in Excel, VISUALSYNTH calls the Synth library through a server to perform a function. There are 2 main types of functions in the Synth library: 1) Getting task recommendations given the current state (in purple), and 2) Executing a task (in green). In both cases, the Synth library returns a new state, which is passed to the Excel add-in and rendered by mapping state components (tables, models, predictions...) to a visual element (border, font or fill formatting of a cell). For example, in Fig. 1, a table is mapped to a range border in Excel, the confidence of a cell value prediction is mapped to the cell font color. Task are recommended using the state [4]. (Color figure online)

References

1. Bishop, C.M.: Pattern Recognition and Machine Learning. Springer, Heidelberg (2006)
2. Chambers, C., Scaffidi, C.: Struggling to excel: a field study of challenges faced by spreadsheet users. In: 2010 IEEE Symposium on Visual Languages and Human-Centric Computing, pp. 187–194. IEEE (2010)
3. Feurer, M., Klein, A., Eggensperger, K., Springenberg, J., Blum, M., Hutter, F.: Efficient and robust automated machine learning. In: Advances in Neural Information Processing Systems, pp. 2962–2970 (2015)
4. Gautrais, C., Dauxais, Y., Teso, S., Kolb, S., Verbruggen, G., De Raedt, L.: Human-machine collaboration for democratizing data science. arXiv preprint arXiv:2004.11113 (2020)
5. Halbert, D.C.: Programming by example. Ph.D. thesis, University of California, Berkeley (1984)

6. Kolb, S., Paramonov, S., Guns, T., De Raedt, L.: Learning constraints in spreadsheets and tabular data. Mach. Learn. **106**(9), 1441–1468 (2017). https://doi.org/10.1007/s10994-017-5640-x
7. Kolb, S., Teso, S., Dries, A., De Raedt, L.: Predictive spreadsheet autocompletion with constraints. Mach. Learn. **109**(2), 307–325 (2019). https://doi.org/10.1007/s10994-019-05841-y
8. Sarkar, A., Jamnik, M., Blackwell, A.F., Spott, M.: Interactive visual machine learning in spreadsheets. In: 2015 IEEE Symposium on Visual Languages and Human-Centric Computing (VL/HCC), pp. 159–163. IEEE (2015)
9. Thornton, C., Hutter, F., Hoos, H.H., Leyton-Brown, K.: Auto-weka: combined selection and hyperparameter optimization of classification algorithms. In: Proceedings of the 19th ACM SIGKDD, pp. 847–855. ACM (2013)
10. Verbruggen, G., De Raedt, L.: Automatically wrangling spreadsheets into machine learning data formats. In: Duivesteijn, W., Siebes, A., Ukkonen, A. (eds.) IDA 2018. LNCS, vol. 11191, pp. 367–379. Springer, Cham (2018). https://doi.org/10.1007/978-3-030-01768-2_30

FireAnt: Claim-Based Medical Misinformation Detection and Monitoring

Branislav Pecher(✉), Ivan Srba, Robert Moro, Matus Tomlein, and Maria Bielikova

Slovak University of Technology in Bratislava, Faculty of Informatics and Information Technologies, Ilkovicova 2, 842 16 Bratislava, Slovakia
{branislav.pecher,ivan.srba,robert.moro,matus.tomlein, maria.bielikova}@stuba.sk

Abstract. Massive spreading of medical misinformation on the Web has a significant impact on individuals and on society as a whole. The majority of existing tools and approaches for detection of false information rely on features describing content characteristics without verifying its truthfulness against knowledge bases. In addition, such approaches lack explanatory power and are prone to mistakes that result from domain shifts. We argue that involvement of human experts is necessary for successful misinformation debunking. To this end, we introduce an end-to-end system that uses a claim-based approach (claims being manually fact-checked by human experts), which utilizes information retrieval (IR) and machine learning (ML) techniques to detect medical misinformation. As a part of a web portal called FireAnt, the results are presented to users with easy to understand explanations, enhanced by an innovative use of chatbot interaction and involvement of experts in a feedback loop.

Keywords: Medical misinformation · Claim-based detection · FireAnt

1 Introduction

Medical misinformation presents a serious problem as it can cause patients to reject scientifically-proven treatments or to take unproven or even dangerous ones. It may be harmful not only to individuals, but to society as well, for example by endangering the herd immunity when people refuse to receive vaccines for preventable diseases. Most recently, we witnessed attacks on 5G infrastructure due to a hoax blaming this technology to spread the new coronavirus[1].

Several systems for misinformation detection have already been deployed, such as ClaimBuster [2], which can automatically fact-check rumours in political discourses, or dEFEND [1], which focuses primarily on the explainability aspect of fake news detection. The existing tools are mostly focused on political false information and work mostly with posts from social networking sites

[1] https://www.bbc.com/news/52168096.

Y. Dong et al. (Eds.): ECML PKDD 2020, LNAI 12461, pp. 555–559, 2021.
https://doi.org/10.1007/978-3-030-67670-4_38

(Twitter being used the most), while news articles published on news sites are used only rarely. Their inner mechanisms are typically based on IR and ML approaches that consider content characteristics, such as writing style or shallow semantics (e.g., various textual representations), exposing them to the problem of weak explainability and domain shift (when the characteristics change). Especially in the case of medical domain, we emphasize that a robust and reliable misinformation detection cannot rely solely on the text style, but it must assess actual content veracity against prior knowledge (e.g., medical research and clinical trials). Such automatic detection may be very challenging and thus the current solutions must involve human experts as well as IR and ML techniques.

In this paper, we demonstrate an end-to-end system for detecting medical misinformation shared on the Web that utilizes medical claims manually evaluated by human experts. Our main contributions can be summarized as follows:

- We focus on misinformation detection in news articles and blogs (being the main origin of misinformation) instead of posts from social networking sites.
- Our veracity detection approach is based on expert evaluated fact-checks.
- By using the claims, we provide users with meaningful and easy to understand explanations of why an article was marked as false. The explanations are further enhanced by an innovative presentation using chatbots.
- We involve human experts in a feedback loop - users have an opportunity to ask experts in case of doubts about reliability of results.

2 System Overview

Our proposed end-to-end system consists of three main components and several substeps as depicted in Fig. 1. They are implemented within our platform MonAnt [3] that provides necessary tools for scraping web pages, "AI core" module for deploying and interconnecting ML-based methods and a web portal called FireAnt (**Fi**ghting and **re**ducing **Ant**isocial behaviour), which serves as a user interface to present the results to experts and general public.

2.1 Data Acquisition

Using the MonAnt platform, we collected more than 200 000 health-related English articles from various news sites and blogs and periodical extractions

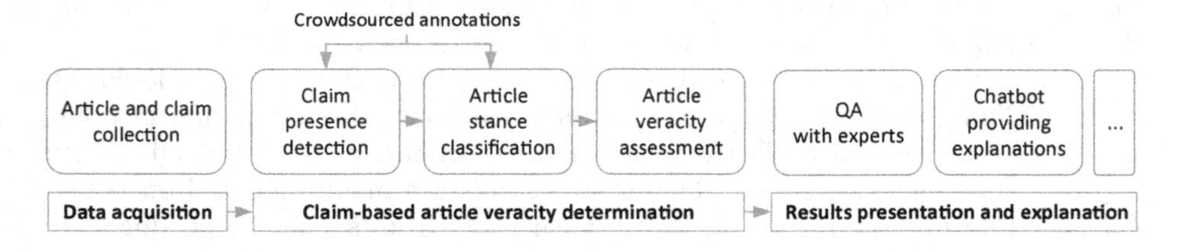

Fig. 1. An overview of components in the proposed system

continuously obtain new articles. We collect articles' content (e.g., title, body, images) as well as metadata (e.g., authors, publication date). We also collected more than 1000 medical claims (e.g., "vaccines cause autism"), the veracity of which was determined by human experts, from various fact-checking sites like Snopes, Metafact or Fullfact. We unified various veracity ratings used by these sites into the following values: *false*, *mostly false*, *mixed*, *mostly true*, *true* and *unknown*.

2.2 Claim-Based Article Veracity Determination

To perform claim-based veracity determination, we follow a similar methodology as the one used in [4]. First, we **represent articles and claims** using sentence embeddings calculated by universal sentence encoder.

Claim presence detection is performed using a novel unsupervised approach. As the number of possible article-claim pairings is high, with most of them being irrelevant, we first apply a pre-filtering step. An average score from cosine similarity of embedding representations between claim and article title; and between claim and K most similar sentences is calculated. Those pairings that surpass a defined similarity threshold are further analyzed using our IR approach. We extract 1-, 2- and 3-grams from the claim statement and for each n-gram, we calculate a score by multiplying: a) the similarity of sentence embeddings between the claim and the most similar sentence containing the given n-gram, and 2) TF-IDF score of the given n-gram in the whole corpus. The final claim-presence score is calculated as an average between all n-grams and compared with another threshold. Value of K and both thresholds are tuned using a small number of labelled pairings obtained by crowdsourcing.

In the next step, we **classify the stance of an article** to a claim into three classes: the article *agrees*, *disagrees* with the claim; or just *discusses* it without adopting any stance. We pass each pair to a bidirectional LSTM neural network. This network takes as an input the sentence embedding representation of the claim, followed by representation of all the sentences in the article, starting with the title. As not all of the sentences are equally relevant in deciding the stance, we introduce an attention mechanism into the network just before the fully-connected layers. Doing this, we guarantee that the neural network learns to prefer the article sentences that most resemble the claim representation.

Finally, we **assess article veracity** by combining veracity of the claim present in the article and the article stance to the claim. We consider three cases: 1) article agrees with the claim, so we directly use the claim veracity; 2) article disagrees, so we use a flipped veracity instead; 3) article discusses the claim, so we mark the pairing as *unknown*. When there are multiple claims detected in an article, its final veracity is determined by the veracity of its lowest (false being the lowest) rated pairing.

2.3 Results Presentation and Explanation

The collected, as well as predicted, data are continuously updated and presented to users by means of web portal FireAnt[2]. Currently out of ∼200 000 articles, 35.04% are mapped to at least one claim and 22.25% are assigned with a final validity. The most important features, which will also be the subject of demonstration, include: a) **checker**, where a user can see all collected, as well as predicted, data about a specific article, claim, or source; b) **chatbot**, which provides users with additional information about articles and claims, along with the reasoning and claim-based explanation of the determined veracity; c) **search**, which supports exploration of collected and predicted data by providing a multitude of search criteria; or d) **notification alerts**, which are sent when new articles/claims/sources satisfying search criteria appear. Moreover, we provide users with the possibility to **ask medical expert users** for more information and their opinion on the presented data. Even though we can achieve a high accuracy in our AI methods, they are not 100% correct all the time and thus further incorporation of an expert into the feedback loop is essential.

FireAnt provides tools and features for **the general public**, as well as **the experts**. From supported use cases, we depict two illustrative examples, one for each of our target users. Consider a user from the general public, who encountered an article with dubious veracity. When given the URL of this article, the system will display its determined veracity, expertly evaluated veracity of the source and evolution of the article's popularity on social network Facebook. Chatbot will guide such user through the presented results. An expert user may be interested in exploring emerging or widely shared misinformative claims or articles. Notifications can be used to get early warnings for new popular misinformation or to get a daily update about the number of newly created (mis)informative articles regarding some disease (e.g., COVID-19).

Acknowledgments. This work was partially supported by the Slovak Research and Development Agency under the contracts No. APVV-17-0267, APVV SK-IL-RD-18-0004; by the Scientific Grant Agency of the Slovak Republic, under the contracts No. VG 1/0725/19 and VG 1/0667/18. The authors wish to thank students, who contributed to the design and implementation of FireAnt portal.

References

1. Cui, L., Shu, K., Wang, S., Lee, D., Liu, H.: defend: a system for explainable fake news detection. In: Proceedings of the 28th ACM International Conference Information and Knowledge Management, pp. 2961–2964 (2019)
2. Hassan, N., Zhang, G., Arslan, F., Caraballo, J., Jimenez, D., et al.: ClaimBuster: the first-ever end-to-end fact-checking system. In: Proc. VLDB Endowment **10**(12), 1945–1948 (2017)

[2] Video presentation of the system as well as the system itself is available at: https://fireant.monant.fiit.stuba.sk/about.

3. Srba, I., Moro, R., Simko, J., et al.: Monant: universal and extensible platform for monitoring, detection and mitigation of antisocial behaviour. In: Proceedings of WS on Reducing Online Misinformation Exposure - ROME '19, pp. 1–7 (2019)
4. Wang, X., Yu, C., Baumgartner, S., Korn, F.: Relevant document discovery for fact-checking articles. In: Proceedings of the International Conference Computer on World Wide Web - WWW '18, pp. 525–533. ACM Press (2018)

GAMA: A General Automated Machine Learning Assistant

Pieter Gijsbers(✉) and Joaquin Vanschoren

Eindhoven University of Technology, Eindhoven, Netherlands
p.gijsbers@tue.nl

Abstract. The General Automated Machine learning Assistant (GAMA) is a modular AutoML system developed to empower users to track and control how AutoML algorithms search for optimal machine learning pipelines, and facilitate AutoML research itself. In contrast to current, often black-box systems, GAMA allows users to plug in different AutoML and post-processing techniques, logs and visualizes the search process, and supports easy benchmarking. It currently features three AutoML search algorithms, two model post-processing steps, and is designed to allow for more components to be added.

1 Introduction

Automated Machine Learning (AutoML) aims to automate the process of building machine learning models, for instance, by automating the selection and tuning of preprocessing and learning algorithms in machine learning pipelines. In recent years, many AutoML systems have been developed, such as Auto-WEKA [10], auto-sklearn [4], TPOT [9] and ML-Plan [8]. They vary in the types of pipelines they build (e.g. fixed or variable length), how they optimize them (e.g. using evolutionary or Bayesian optimization), and whether or how they employ meta-learning (e.g. warm-starting) or post-processing (e.g. ensembling).

We demonstrate[1] a new open-source AutoML system, GAMA[2] [6], which distinguishes itself by it modularity (allowing users to compose AutoML systems from sub-components), extensibility (allowing new components to be added), transparency (tracking and visualizing the search process to better understand what the AutoML system is doing), and support for research, such as integration with the AutoML benchmark [5]. The main difference to our earlier publication [6] is the redesign to allow for a modular AutoML pipeline and the addition of a graphical user interface.

As such, it caters to a wide range of users, from people without a deep machine learning background who want an easy-to-use AutoML tool, to those who want better control and understanding of the AutoML process, and especially researchers who want to perform systematic AutoML research.

[1] A video demonstration can be found at https://youtu.be/angsGMvEd1w.
[2] Code and documentation can be found at https://github.com/PGijsbers/gama/.

Y. Dong et al. (Eds.): ECML PKDD 2020, LNAI 12461, pp. 560–564, 2021.
https://doi.org/10.1007/978-3-030-67670-4_39

Currently, three different search algorithms and two post-processing techniques are available, but we welcome and plan to include more techniques in the future. For novice users, GAMA offers a default configuration shown to perform well in our benchmarks.

2 System Overview

Modular AutoML Pipeline

Rather than prescribing a specific combination of AutoML techniques, GAMA allows users to combine different search and post-processing algorithms into a flexible AutoML 'pipeline' that can be tuned to the problem at hand.

There are three optimization algorithms currently implemented in GAMA to search for optimal machine learning pipelines: random search [1], an asynchronous successive halving algorithm (ASHA) [7] which uses low-fidelity estimates to filter out bad pipelines early, and an asynchronous multi-objective evolutionary algorithm.

After the pipeline search has completed, a post-processing technique will be executed to construct the final model. It is currently possible to either train the single best pipeline or create an ensemble out of pipelines evaluated during search, as described in [2]. In subsequent work, we plan to expand the number of search and post-processing techniques available out-of-the-box.

Listing 1.1 shows how to configure GAMA with non-default search and post-processing methods and use it as a drop-in replacement for scikit-learn estimators.[3]

New AutoML algorithms or variations to existing ones can be included and tested with relative ease. For instance, each of the search algorithms described above has been implemented and integrated in GAMA with less than 170 lines of code, and they can all make use of shared functions for logging, parallel pipeline evaluation and adhering to runtime constraints. It also allows users to research other questions, such as how to choose the search algorithm for AutoML.

Interface

GAMA comes with a graphical web interface which allows novice users to start and configure GAMA. Moreover, it visualizes the AutoML process to enable researchers to easily monitor and analyse the behavior of specific AutoML configurations.

GAMA logs the creation and evaluation of each pipeline, including meta-data such as creation time and evaluation duration. For pipelines created through evolution, it also records the parent pipelines and how they differ. One can also compare multiple logs at once, creating figures such as Fig. 1 that shows the convergence rate of five different GAMA runs over time on the airline dataset[4].

[3] An always up-to-date version of this listing can be found at https://pgijsbers.github.io/gama/master/citing.html.

[4] https://www.openml.org/d/1169.

Listing 1.1. Configuring an AutoML pipeline with GAMA

```
from gama import GamaClassifier
from gama.search_methods import AsynchronousSuccessiveHalving
from gama.postprocessing import EnsemblePostProcessing

automl = GamaClassifier(
    search=AsynchronousSuccessiveHalving(),
    post_processing=EnsemblePostProcessing()
)
automl.fit(X, y)
automl.predict(X_test)
automl.fit(X_test, y_test)
```

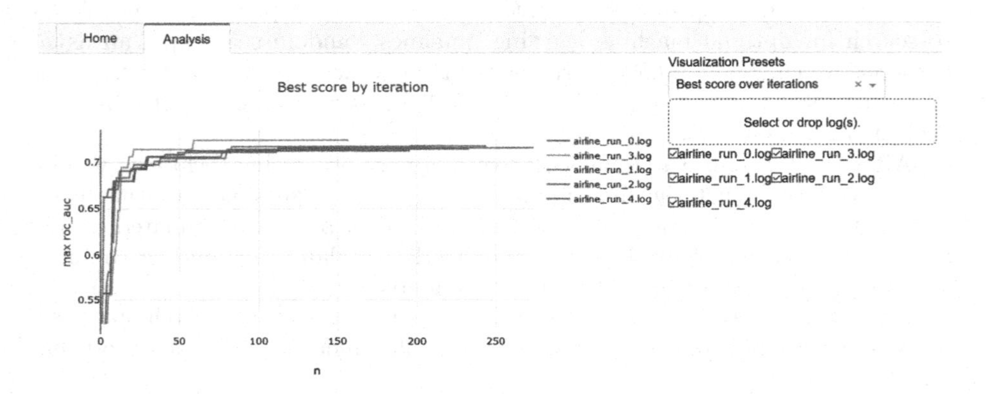

Fig. 1. Visualization of logs

Benchmarking

GAMA in integrated with the open-source AutoML Benchmark introduced in [5]. Figure 2 shows the results of running GAMA with its default settings some of the biggest and most challenging datasets for which each other framework had results in the original work.[5] The full and latest results will be made available in the GAMA documentation.

3 Related Work

GAMA compares most closely to auto-sklearn and TPOT as they also optimize scikit-learn pipelines. Auto-sklearn and GAMA both implement the same ensembling technique [2]. GAMA and TPOT both feature evolutionary search with NSGA2 selection [3], although GAMA's implementation uses asynchronous evolution, which is often faster. While TPOT and auto-sklearn have a fixed AutoML

[5] Although we could not run these experiments on the same (AWS) hardware, we took care to use the same computational constraints.

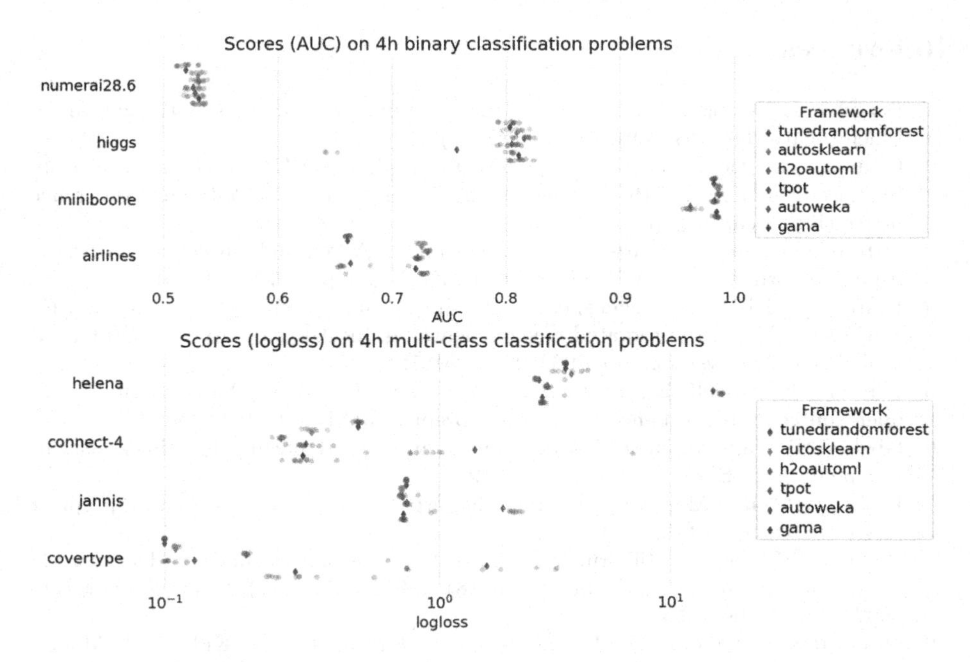

Fig. 2. Performance benchmark results

pipeline, they do allow modifications to their search space. To the best of our knowledge, GAMA is the only AutoML framework that offers a modular and extensible composition of AutoML systems, and extensive support for AutoML research.

4 Conclusion

In this proposal we presented GAMA, an open-source AutoML tool which facilitates AutoML research and skillful use through its modular design and built-in logging and visualization. Novice users can make use of the graphical interface to start GAMA, or simply use the default configuration which is shown to generate models of similar performance to other AutoML frameworks. Researchers can leverage GAMA's modularity to integrate and test new AutoML search procedures in combination with other readily available building blocks, and then log, visualize, and analyze their behavior, or run extensive benchmarks. In future work, we aim to integrate additional search techniques as well as extend the AutoML pipeline with additional steps, such as warm-starting the pipeline search with meta-data.

Acknowledgements. This software was developed with support from the Data Driven Discovery of Models (D3M) program run by DARPA and the Air Force Research Laboratory.

References

1. Bergstra, J., Bengio, Y.: Random search for hyper-parameter optimization. J. Mach. Learn. Res. **13**(Feb), 281–305 (2012)
2. Caruana, R., Niculescu-Mizil, A., Crew, G., Ksikes, A.: Ensemble selection from libraries of models. In: Proceedings of the Twenty-First International Conference on Machine Learning, p. 18 (2004)
3. Deb, K., Pratap, A., Agarwal, S., Meyarivan, T.: A fast and elitist multiobjective genetic algorithm: NSGA-II. IEEE Trans. Evol. Comput. **6**(2), 182–197 (2002)
4. Feurer, M., Klein, A., Eggensperger, K., Springenberg, J., Blum, M., Hutter, F.: Efficient and robust automated machine learning. In: Advances in Neural Information Processing Systems, pp. 2962–2970 (2015)
5. Gijsbers, P., LeDell, E., Thomas, J., Poirier, S., Bischl, B., Vanschoren, J.: An open source AutoML benchmark. arXiv preprint arXiv:1907.00909 (2019)
6. Gijsbers, P., Vanschoren, J.: GAMA: genetic automated machine learning assistant. J. Open Source Softw. **4**(33), 1132 (2019)
7. Li, L., et al.: Massively parallel hyperparameter tuning. arXiv preprint arXiv:1810.05934 (2018)
8. Mohr, F., Wever, M., Hüllermeier, E.: ML-Plan: automated machine learning via hierarchical planning. Mach. Learn. **107**(8), 1495–1515 (2018). https://doi.org/10.1007/s10994-018-5735-z
9. Olson, R.S., Urbanowicz, R.J., Andrews, P.C., Lavender, N.A., Kidd, L.C., Moore, J.H.: Automating biomedical data science through tree-based pipeline optimization. In: Squillero, G., Burelli, P. (eds.) EvoApplications 2016. LNCS, vol. 9597, pp. 123–137. Springer, Cham (2016). https://doi.org/10.1007/978-3-319-31204-0_9
10. Thornton, C., Hutter, F., Hoos, H.H., Leyton-Brown, K.: Auto-WEKA: combined selection and hyperparameter optimization of classification algorithms. In: Proceedings of KDD-2013, pp. 847–855 (2013)

Instructional Video Summarization Using Attentive Knowledge Grounding

Kyungho Kim, Kyungjae Lee, and Seung-won Hwang(✉)

Yonsei University, Seoul, South Korea
{ggdg12,lkj0509,seungwonh}@yonsei.ac.kr

Abstract. This demonstration considers the scenario of summarizing an instructional video, for query such as "how to cook galbi", to efficiently obtain a skillset. Specifically, we use the query to retrieve both the relevant video and the external procedural knowledge, such as a recipe document, and show summarization is more effective with such augmentation.

Keywords: Multimodality · Video summarization · Attentive knowledge grounding

1 Introduction

Instructional videos have become an effective means to learn new skills including cooking, gardening, and sports. Though more people rely on videos for searching instructional information, video search scenarios focus on returning the entire video matching user queries [3,5].

In this paper, we study a summarization scenario of effectively learning skill from video summaries, by providing the extracted key segments from the full video. Existing video summarization [6,7] extracts highlight clips, identifying whether each video frame is the keyframe (1) or not (0). For this purpose, most recently, multi-modality of both video clip and transcript has been used [10], for such classification.

Our key claim is that deciding keyframe should not be localized to the given clip (and its transcript), but globally matched with respect to the entire procedure. To validate, we propose to augment external procedural knowledge (e.g., cooking recipe) and leverage its relevance to the given clip. Figure 1 illustrates our scenario of summarizing an instructional video for user query Q, such as '*how to make Galbi*'. We use Q, not only to retrieve the video, but also to retrieve its recipe document, to decide a keyframe, using transcript, video clip, and recipe.

Inspired by conversational Question Answering (ConvQA) [11], retrieving a relevant document as knowledge to enrich the representation of utterances, we propose to use the recipe as a procedural knowledge to enrich that of video

K. Kim and K. Lee—Equally contributed to this work.

Y. Dong et al. (Eds.): ECML PKDD 2020, LNAI 12461, pp. 565–569, 2021.
https://doi.org/10.1007/978-3-030-67670-4_40

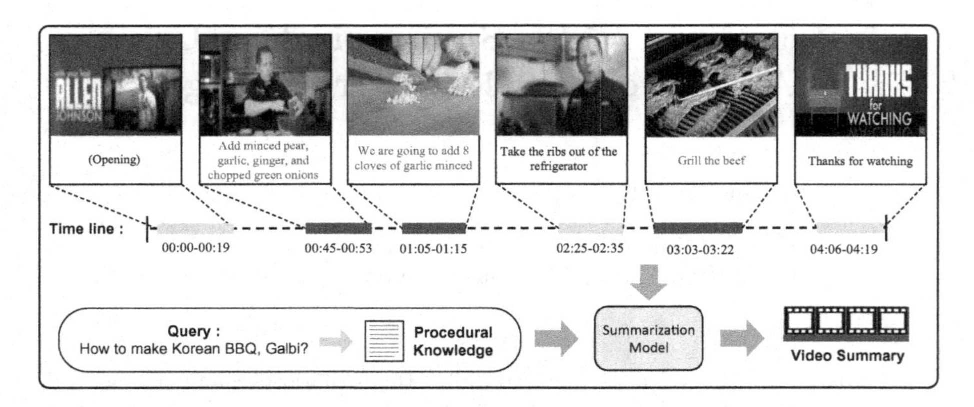

Fig. 1. An overview of our approach. Returned segments are marked in red, for users to either skip to specific instruction or watch all segments as a summary without inessential steps, such as opening and greeting. (Color figure online)

utterance, which is video clip and its transcript pair in our context. Specifically, clip is semantically matched to relevant steps in the recipe, without any relevance supervision unlike in [11], for which we propose attentive knowledge matching inspired by video QA model [1,4]. Details will be discussed more in Sect. 3 & 4. We release our demo at http://dilab.yonsei.ac.kr/IVSKS.php, built on instructional cooking video set.

2 Problem Formulation

We abstract our task as predicting multiple video segments given a query. More formally, let a given video be V containing image frames I and transcript T, where l image frames ($I = \{I_1, I_2, ..., I_l\}$) are sampled from V (1 frame per 1.5 s) and each I_k corresponds to a sentence T_k in the transcript. That is, a video V can be defined as a sequence of tuples, $V = \{(I_1, T_1), (I_2, T_2), ..., (I_l, T_l)\}$. Let a query be $\boldsymbol{Q} = \{w_1, w_2, ..., w_n\}$, where w means n words of the query. We expand the given query Q into textual descriptions answering the query using external procedural knowledge (*e.g.*, an encyclopedia of how-to queries). The knowledge often consists of multiple sentences, m sentences, corresponding to multiple steps required for Q, denoted as $\boldsymbol{K} = \{S_1, S_2, ..., S_m\}$, where S indicates each sentence in the descriptions. Our empirical finding (Sect. 4) suggests that K is most effective in the goal of generating an instructional summary.

With this formulation, our goal is to predict whether tuple (I_k, T_k) should be included in answer segment A or not, with respect to the given query Q_c and its finer counterpart K. For evaluation, the predicted answer segment $\boldsymbol{A}$ is compared with the ground-truth answer segment $\boldsymbol{A}^*$ with F1 score in Sect. 4.

3 Proposed Model

Our model consists of (a) **Video Encoding**, turning a sequence of (I, T) into multimodal representation, (b) **Knowledge-Video (KV) fusion**, aligning knowledge K with video segments. We are the first to deal with the expansion of query, with procedural knowledge grounding, and predict multiple segments as an answer.

Video Encoding: For multimodal encoding of a video, we use a pre-trained model, LXMERT [9], receiving an image and paired sentence as input. Since there is no knowledge paired with videos, thus we delay the knowledge representation to the next step of video encoding. The multimodal encoding using LXMERT with transcript and image is done as follows:

$$\boldsymbol{v_k} = LXMERT(I_k, T_k), \qquad (k \in [1, l]) \tag{1}$$

where $\boldsymbol{v_k}$ indicates multimodal representation in k-th image frame and transcript. We compute the multimodal embedding for each frame at the point of LXMERT operating in image units. After encoding each frame (I_k, T_k) to $\boldsymbol{v_k}$, we contextualize the video sequence using (1) self-attention for global context, and (2) bi-directional LSTM for considering temporal information.

KV Fusion: The first step of KV fusion is to expand Q into $K = \{S_1, ..., S_m\}$, augmented by external knowledge. We then encode sentence S_k using CLS representation from BERT model [2], as $\boldsymbol{s_k} = BERT(S_k)$. However, this BERT encoding is unaware of corresponding video frame representation. The goal of KV fusion is (a) to align S_i to corresponding video representation (if exists), such that (b) importance of S_i can be estimated by whether it can be aligned with video frame, and vice versa. Specifically, we can estimate importance of S_i from frame, modeled as **Frame-to-Knowledge (F2K)** attention, or that of frame from knowledge **Knowledge-to-Frame (K2F)** attention, inspired by attention flow layer of BiDAF [8]. Formally, we describe **F2K** attention as below:

$$\alpha_i^k = \underset{i}{\text{softmax}}(\boldsymbol{v_k} W_1 \boldsymbol{s_i}), \qquad \hat{s_k} = \sum_i \alpha_i^k \cdot \boldsymbol{s_i} \tag{2}$$

where W_1 indicates a trainable matrix and α_i^k is the attention weights. That is, given k-th image, $\hat{s_k}$ represents the weighted sum of descriptions s_i, which can attend specific description corresponding to the given image.

Likewise, **K2F** attention signifies which frame has relevant and critical information, with respect to alignment with frames, as we formally describe below:

$$\beta^k = \underset{k}{\text{softmax}}(\boldsymbol{v_k} W_2 \boldsymbol{s}), \qquad \hat{v_k} = \beta^k \cdot \hat{\boldsymbol{s_k}} \tag{3}$$

where W_2 indicates a trainable matrix, β^k is the attention weights and s means the embedding of procedural knowledge. In our experiment, we decide to use s as output of the convolution network of attentive descriptions. We use concatenation of $\hat{v_k}$ and v_k for final fusion features.

During training, we use cross-entropy loss with labels L_k where the ground-truth in from beginning to end position is equal to 1, otherwise 0. Finally, our loss function is described as below:

$$Loss = -\sum_k \{L_k \cdot log(FC([v_k|\hat{v_k}])) + (1 - L_k) \cdot (1 - log(FC([v_k|\hat{v_k}])))\} \quad (4)$$

where FC is fully-connected network with a sigmoid function, and | indicates concatenation. At inference time, we extract frames over threshold, and aggregate them as the final answer summary.

Table 1. Empirical validation

	Random	Summ (no K)	Ours (KV fusion)
F1 Score	40.0%	67.6%	**69.6%**

4 Data Collection and Demonstration

Dataset: We build demo upon YouCook2 dataset[1], containing 2000 query-video pairs. As annotators segment videos covering important steps, we can compute F1 score with respect to the chosen segments as ground-truth summary answer $\boldsymbol{A}^*$. For overall scores, we average each score over all of the instances. Our procedural knowledge reference documents were crawled from a collection of recipes[2], and we retrieve top 1 recipe by searching queries then use them as knolwedge K.

Our demo URL shows example queries and their results, and Table 1 reports our quantitative evaluation: **Random** is a naive baseline selecting a random start/end frame for each video. **Summ** is a summarization baseline without considering knowledge K, by removing $\hat{v_k}$ in Eq. (4). The rest is **Ours**. Ours with query expansion, augmented from external knowledge, and KV fusion using **F2K** and **K2F** improves the F1 score (∼70%), leading to 2% gain, compared to **Summ**.

Acknowledgements. Microsoft Research Asia and Artificial Intelligence Graduate School Program (2020-0-01361).

References

1. Colas, A., Kim, S., Dernoncourt, F., Gupte, S., Wang, D.Z., Kim, D.S.: TutorialVQA: Question answering dataset for tutorial videos. arXiv:1912.01046 (2019)
2. Devlin, J., Chang, M.W., Lee, K., Toutanova, K.: BERT: Pre-training of deep bidirectional transformers for language understanding. arXiv:1810.04805 (2018)

[1] http://youcook2.eecs.umich.edu/.
[2] https://www.allrecipes.com/.

3. Dong, J., et al.: Dual encoding for zero-example video retrieval. In: CVPR (2019)
4. Lee, K., Duan, N., Ji, L., Li, J., Hwang, S.W.: Segment-then-rank: non-factoid question answering on instructional videos. In: Proceedings of the AAAI Conference on Artificial Intelligence, vol. 34, pp. 8147–8154 (2020)
5. Liu, Y., Albanie, S., Nagrani, A., Zisserman, A.: Use what you have: Video retrieval using representations from collaborative experts. arXiv:1907.13487 (2019)
6. Otani, M., Nakashima, Y., Rahtu, E., Heikkila, J.: Rethinking the evaluation of video summaries. In: CVPR (2019)
7. Rochan, M., Ye, L., Wang, Y.: Video summarization using fully convolutional sequence networks. In: Ferrari, V., Hebert, M., Sminchisescu, C., Weiss, Y. (eds.) ECCV 2018, Part XII. LNCS, vol. 11216, pp. 358–374. Springer, Cham (2018). https://doi.org/10.1007/978-3-030-01258-8_22
8. Seo, M., Kembhavi, A., Farhadi, A., Hajishirzi, H.: Bidirectional attention flow for machine comprehension. arXiv abs/1611.01603 (2017)
9. Tan, H., Bansal, M.: LXMERT: Learning cross-modality encoder representations from transformers. arXiv:1908.07490 (2019)
10. Xu, F.F., et al.: A benchmark for structured procedural knowledge extraction from cooking videos. arXiv preprint arXiv:2005.00706 (2020)
11. Yang, L., et al.: Response ranking with deep matching networks and external knowledge in information-seeking conversation systems. In: SIGIR (2018)

PrePeP: A Light-Weight, Extensible Tool for Predicting Frequent Hitters

Christophe Couronne, Maksim Koptelov, and Albrecht Zimmermann(✉)

Normandie Univ, UNICAEN, ENSICAEN, CNRS – UMR GREYC, Caen, France
{christophe.couronne,maksim.koptelov,albrecht.zimmermann}@unicaen.fr

Abstract. We present PrePeP, a light-weight tool for predicting whether molecules are frequent hitters, and visually inspecting the subgraphs supporting this decision. PrePeP is contains three modules: a mining component, an encoding/predicting component, and a graphical interface, all of which are easily extensible.

1 Introduction

For more than a century, systematic drug development has led to a wide range of drugs for many illnesses and diseases. As a side-effect, much low-hanging fruit – easily identifiable components – has already been plucked. In addition, new pathogens arise, and known ones can develop resistances against existing drugs.

Modern drug development therefore involves *high-throughput screening* (HTS) [2], in which thousands or even millions of compounds are tested for activity against a given target. Since this remains a time-consuming process, HTS has been augmented by *virtual screening* [3], in which physical tests are replaced by predictions based on computational models.

In either process, the biggest hurdle are *false positives*, compounds that look like promising candidates during screening but turn into wasted time and money later on. This is annoying in the best of times but can have a much worse impact when time is of the essence, such as during the current Covid-19 pandemic, when large-scale efforts at underway to develop effective antiviral drugs.[1]

Among such false positives are so-called *frequent hitters* (FH), compounds that show activity in many assays, e.g. because they are non-specific *pan-assay interference compounds* (PAINS) [1]. While we addressed this problem in [4], the tool we presented there was an early prototype, and we have since improved it.

In particular, we have turned the tool more modular, breaking it down into:

- *The mining component*, which allows anyone with an sdf file containing frequent and non-frequent hitters to derive discriminative subgraphs.

[1] https://lejournal.cnrs.fr/articles/covid-19-15-milliard-de-molecules-passees-au-criblage-virtuel.

The tool can be downloaded at http://scientific-data-mining.org, “Software”.

Y. Dong et al. (Eds.): ECML PKDD 2020, LNAI 12461, pp. 570–573, 2021.
https://doi.org/10.1007/978-3-030-67670-4_41

- *The predictive component*, which, based on a training set and sets of subgraphs, predicts for molecules contained in an sdf file if they are FH or not.
- *The graphical interface*, which allows for each molecule predicted FH to visualize the subgraphs that support the prediction.

There exist other tools for FH prediction [5,6] but only in the form of web services. Running predictions for more than a few tens of molecules can take several hours during which the user gets no feedback on the process. Our tool, on the other hand, can be run locally, on as many machines as the user has available. It is easily extensible by users, is light-weight (2.9 MB in archived form), and depends only on a few widely available Python libraries.

2 Mining Discriminative Subgraphs

To explain the usage of the mining component, we quickly repeat the basics of the method proposed in [4]. A vital aspect of it is the mining of subgraphs discriminating between frequent and non-frequent hitters. Since frequent hitters are very much in the minority in most data sets, non-frequent hitters are subsampled to create balanced molecular data sets. This sampling is repeated to capture all information contained in the non-frequent hitters.

Structural information stored in sdf files is translated into gsp files, from which discriminating subgraphs are mined using a supervised *gSpan* implementation developed by Siegfried Nijssen. These subgraphs, finally, are used to encode the underlying data in terms of their presence and absence.

The code is written in Python and requires the networkx library. Launched on the command line, it takes as parameters the original sdf file and the number of subgraphs to be mined per sample. Its output is a folder containing the subgraph files as well as the encoded training data. For the classifier reported on in [4], based on ~150k molecules, 200 samples, and 100 subgraphs per sample, the zipped folder amounts to 2.8 MB, easily transferable. This is of particular interest because neither the definition of frequent hitters nor the data sets to build models from are clear. Researchers can therefore easily experiment with different options.

3 Predicting Frequent Hitters

Prediction is done by learning a decision tree on each sample and taking the majority vote of those trees for a molecule to be predicted. Given the results reported in [4], we refrain from using molecular descriptors and base predictions solely on the mined discriminative subgraphs.

Also written in Python, and making use of openbabel 3.0.0, the module includes by default the data and subgraphs used in our earlier publication. A folder containing different data and subgraphs can be passed as a parameter on the command line. The name of the sdf file containing the molecules to be predicted is passed as a mandatory parameter. The module learns the appropriate number of decision trees, encodes the molecules to be predicted using the available subgraphs, and predicts for each molecule whether it is a frequent hitter or not.

Separating this component from the graphical interface described in the next section is a conscious decision. If the two were tightly integrated, it would be impossible to launch predictions remotely on a high-performance server to leave them running unattended for a while.

4 Visualizing Graphs Supporting the Decision

While the predictive component would be the go-to option when working with a large data set, the biggest problem with PAINS is that the biochemical mechanism is in many cases not well understood. To support understanding the predictions, particularly in the case of molecules where the expert is not convinced by the prediction, the graphical interface *wraps* the predictive component.

The interface supports the operations of the predictive component: loading data and subgraphs, loading molecules to be predicted, encoding those molecules (prediction is done automatically). Additionally, however, the user can select individual molecules (listed on the left) and if they have been predicted as being frequent hitters, the subgraphs supporting this decision are shown in the right.

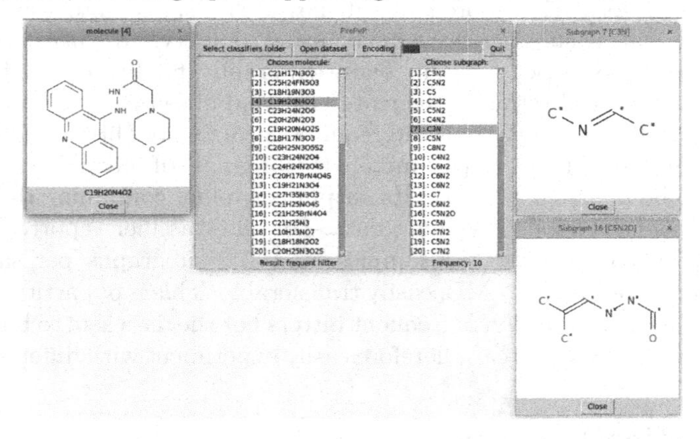

As described above, prediction is performed by majority vote of a number of decision trees and we consider subgraphs to *support the decision* if they

- occur in inner nodes of decision trees predicting the molecule as FH, and
- the test whether they are present in the molecule is *true*.

Clicking on the molecule opens a separate window showing a visualization. Clicking on subgraphs, in turn, also open separate windows visualizing them, which allows to see where in the molecule the subgraph occurs, and (given the necessary expertise) whether basing prediction on the subgraph makes sense. Subgraphs are ordered by how often they were involved in the decision, with that information given at the bottom of the window for the selected subgraph.

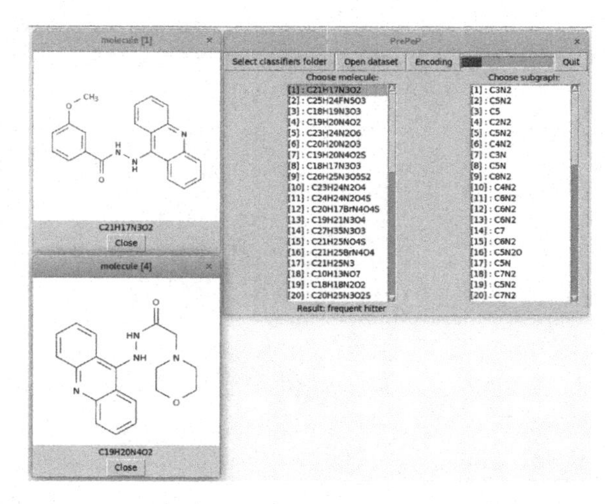

In the same manner as for subgraphs, the user can select several molecules to compare visually. For the sake of presentation, we have grouped windows beside each other but all windows can be freely moved around, the help grouping graphs that one wants to consider together, for example.

5 Conclusion

We consider PrePeP a useful tool for chemoinformaticians who want to do quick-shot frequent hitter prediction. It can be easily deployed, multiple instances can be run on high-performance servers, and it can be extended with new/additional data and subgraphs.

References

1. Aldrich, C., et al.: The ecstasy and agony of assay interference compounds (2017)
2. An, W.F., Tolliday, N.: Cell-based assays for high-throughput screening. Mol. Biotechnol **45**(2), 180–186 (2010)
3. Bajorath, J.: Integration of virtual and high-throughput screening. Nat. Rev. Drug Discov. **1**(11), 882–894 (2002)
4. Koptelov, M., Zimmermann, A., Bonnet, P., Bureau, R., Crémilleux, B.: Prepep: a tool for the identification and characterization of pan assay interference compounds. In: Guo, Y., Farooq, F. (eds.) KDD, pp. 462–471. ACM (2018)
5. Matlock, M.K., Hughes, T.B., Dahlin, J.L., Swamidass, S.J.: Modeling small-molecule reactivity identifies promiscuous bioactive compounds. J. Chem. Inf. Model. **58**(8), 1483–1500 (2018)
6. Stork, C., Chen, Y., Šícho, M., Kirchmair, J.: Hit dexter 2.0: machine-learning models for the prediction of frequent hitters. J. Chem. Inf. Model. **59**(3), 1030–1043 (2019)

Author Index

Printed in the United States
By Bookmasters